Antimicrobial Strategies in the Food System: Updates, Opportunities, Challenges

Dan Li
Editor

Antimicrobial Strategies in the Food System: Updates, Opportunities, Challenges

Springer

Editor
Dan Li
Department of Food Science & Technology
National University of Singapore
Singapore, Singapore

ISBN 978-3-031-95055-1 ISBN 978-3-031-95056-8 (eBook)
https://doi.org/10.1007/978-3-031-95056-8

This Springer imprint is published by the registered company Springer Nature Switzerland AG
The registered company address is: Gewerbestrasse 11, 6330 Cham, Switzerland

Preface

The battle against microorganisms in food has been an ongoing challenge throughout human history, driven by the need for both food safety and preservation. Today, global food systems face mounting pressures from climate change, population growth, and ecosystem degradation, necessitating a transition toward sustainability. However, the adoption of sustainable food production systems may introduce new hazards or create novel transmission pathways for existing risks within the food supply chain. At the same time, advancements in science and technology along with growing consumer awareness and demand for healthier dietary options have led to the development of numerous alternative antimicrobial strategies in recent years. This book provides a timely update on the latest progress in antimicrobial strategies within food systems. Looking ahead, new challenges will inevitably arise, and with them, new innovations and practices will emerge to address these evolving threats. When that time comes, we hope our work will remain a steadfast foundation in the ongoing advancement of this field.

Singapore, Singapore Dan Li

Contents

Part I
Antimicrobial Strategies: Reducing Hazards and Spoilage for a More Sustainable Food System

Emerging and Reemerging Food Microbial Hazards

Mieke Uyttendaele, Thomas De Bock, and Xingchen Zhao

1 Definition of a Human Pathogen

The epidemiological triad suggests that every microorganism can be a human pathogen if there is a susceptible human host, and the context or environmental factors enable the susceptible host to come into contact with the microbial agent (van Seventer & Hochberg, 2017). Often, further classification of the pathogen is inherently linked to the susceptibility of the host for infection (Casadevall & Pirofski, 2002, 2018). As such, human pathogens, although debatable, are often classified into **established human pathogens** and **opportunistic human pathogens** (Methot & Alizon, 2014), the latter being an infectious pathogen that is normally commensal in the body but can cause disease when the host's resistance is altered, e.g., immunocompromised individuals due to medical conditions or treatments, or due to age or pregnancy (Berg et al., 2014).

Furthermore, human illness due to specific pathogens can be attributed to specific foods (of animal or plant-based origin) but also to other sources like environmental exposure (e.g. waterborne), direct contact with animals and human-to-human contact (EFSA, 2008). The definition of a human pathogen as a **foodborne pathogen** is based on the (repeated) reporting of a microbial agent in a foodborne outbreak or in

M. Uyttendaele (✉) · T. De Bock
Food Microbiology and Food Preservation Research Unit, Department of Food Technology, Safety and Health, Faculty of Bioscience Engineering (FMFP-UGent), Ghent University, Ghent, Belgium
e-mail: Mieke.Uyttendaele@UGent.be

X. Zhao
Food Microbiology and Food Preservation Research Unit, Department of Food Technology, Safety and Health, Faculty of Bioscience Engineering (FMFP-UGent), Ghent University, Ghent, Belgium

Research Group for Food Microbiology and Hygiene, National Food Institute, Technical University of Denmark, Kgs. Lyngby, Denmark

D. Li (ed.), *Antimicrobial Strategies in the Food System: Updates, Opportunities, Challenges*, https://doi.org/10.1007/978-3-031-95056-8_1

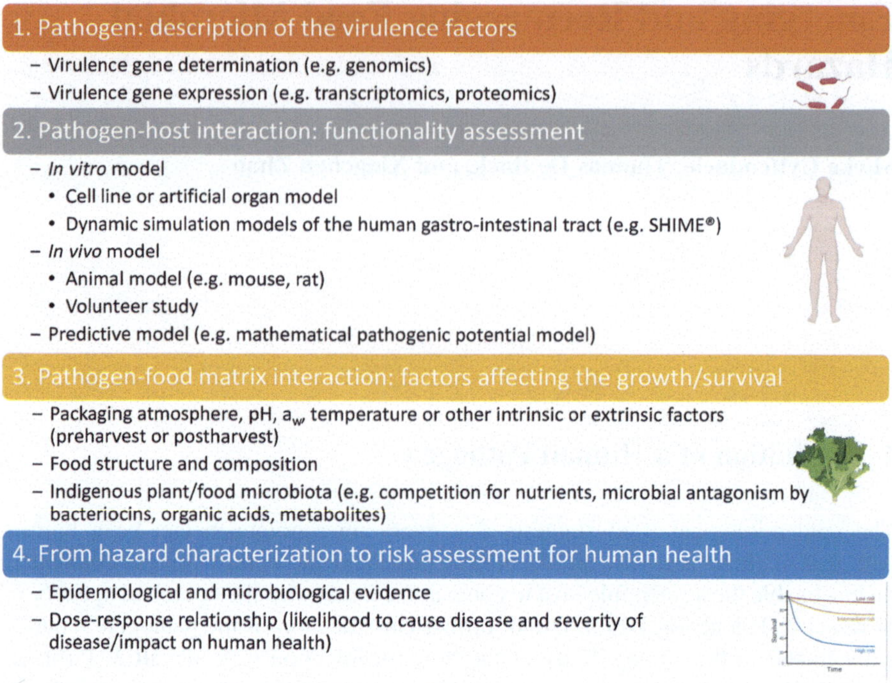

Fig. 1 Different factors to evaluate the weight of evidence for adverse health effects in humans, when undertaking hazard characterization for foodborne pathogens (Zhao et al., 2025)

individual reported cases. **Strong epidemiological** or **microbiological evidence** supports linking the human pathogen with the type of food involved, the transmission route, or the source of pathogen contamination during food processing or primary production (EFSA, 2013a).

Many elements should be considered in the characterization of the bacterial hazard that affects the ability of the bacterium to cause disease in humans via food transmission. When undertaking hazard characterization for foodborne pathogens, a range of factors need to be considered to evaluate the weight of evidence for adverse health effects in humans, needing a multi-method approach (Fig. 1). Overall, thus all these pieces of information, ranging from genomic data to data from infection models in cellular microbiology and real-life data from foodborne outbreaks, are used to describe the likelihood of infection and disease in human hosts, thus moving from the hazard to the risk (FAO and WHO, 2021a).

Pathogenicity and virulence of bacterial hazards have often been addressed and evaluated. At the most basic level, genetic content describes whether virulence factors are present or not, as well as their predicted function and how they may impact the host. To determine whether the virulence factors are likely to be active in a particular environmental niche, or a susceptible host, their gene expression first needs to be determined, coupled with production of the gene product (e.g. protein).

Table 1 General risk factors that cause increased susceptibility to infections

Risk factor	Reason for increased susceptibility
Age (Y and O)	In newborns, the immune system is not yet fully developed Gradual decline in the immune system with aging
Pregnancy (P)	Altered immunity during pregnancy
Primary immune deficiency (I)	The immune system is weakened by a congenital, usually genetic defect in one or more components of the immune system
Secundary immune deficiency (I)	The immune system is weakened by illness or medical intervention. Examples: Neutropenia due to chemotherapy Disruption of the gut microbiota due to long-term antibiotic use Weakened immune system due to HIV infection

This determines the activity of a particular virulence factor. To test the function of virulence factors and understand the human host-pathogen interaction, model systems of increasing complexity are used (Fig. 1) (Lopez-Jimenez & Mostowy, 2021). Many in vitro models are used, from layers of a single cell type right up to reconstructed artificial organs (Feaugas & Sauvonnet, 2021), from static to dynamic simulation models of the gastrointestinal tract with or without higher complexity of competing microbiota present (e.g. SHIME®) (Roussel et al., 2020). Specific measurements of pathogenicity in in vivo models including animal (Palmer & Slauch, 2017) and human volunteer studies (Mawer, 1988) can also be performed, but have drawbacks. Moreover, due to ethical reasons and improving animal welfare, animal tests are currently reduced and volunteer studies are quite costly and difficult to perform.

Casadevall (2017) introduced the concept and mathematical calculation of pathogenic potential, taking into account the contributions of both the (human) host and the microbe (i.e. the bacterial strain) to microbial pathogenesis. The increase (or decrease) of foodborne disease due or (re)emergence of pathogens thus depends not only on the characteristics of the pathogen (virulence and virulence factors) and factors that impact its (renewed) introduction or evolving occurrence in the food supply chain but is also influenced by the changes in the presence of susceptible hosts in the society. In general, a number of risk factors of a host can be defined, which make a human host more susceptible to the occurrence and in particular the severity of foodborne infections (Table 1). Although for established foodborne pathogens any individual within a population is at risk for contracting foodborne disease, the subgroups most vulnerable are also often referred to as YOPIs (Young, Old, Pregnant, Immunocompromised).

It is expected that if due to demographic changes in the society or other factors that impact on the overall health status of humans thus also fluctuations in the occurrence of foodborne diseases will occur. Information on (re)emerging microbial hazards can best be generated from epidemiological and microbiological data available from reported cases and (strong evidence) foodborne outbreaks.

2 Global Burden of Foodborne Diseases, Reported Cases and Outbreak Data

WHO first ever **estimates** of the global burden of foodborne diseases show almost 1 in 10 people fall ill every year from eating contaminated food and 420,000 die as a result (WHO, 2015). While the burden of foodborne diseases is a public health concern globally, the WHO African and South-East Asia Regions have the highest incidence. Children under 5 years of age are at particularly high risk. Diarrhea is often caused by eating raw or undercooked meat, eggs, fresh produce and dairy products contaminated by norovirus, *Campylobacter*, non-typhoidal *Salmonella* and pathogenic *E. coli*. Other major contributors to the global burden of foodborne diseases are typhoid fever, hepatitis A, *Taenia solium* (a tapeworm), and aflatoxin (produced by mould on grain that is stored inappropriately and as a mycotoxin overall classified as a chemical hazard, although produced by fungi). Certain diseases, such as those caused by non-typhoidal *Salmonella*, are a public health concern across all regions of the world, in high- and low-income countries alike. Other diseases, such as typhoid fever, foodborne cholera, and those caused by pathogenic *E. coli*, are much more common to low-income countries, while *Campylobacter* is an important pathogen in high-income countries. Parasites are often overlooked as foodborne pathogens due to a combination of factors such as their relatively lower incidence, delayed symptoms, diagnostic challenges, and the focus of food safety efforts on bacteria and viruses. Despite this, parasites can cause severe illness, particularly in vulnerable populations, and their inclusion in food safety monitoring is crucial for comprehensive public health protection. More information on the WHO estimate of global burden of foodborne disease are on-line available at the WHO website (http://www.who.int/foodsafety/areas_work/food borne-diseases/ferg/en/)

Several countries publish on a regular basis an overview of **reported individual cases** of human foodborne diseases as well of collective foodborne outbreaks. From the annual report issued by the European Food Safety Authority (EFSA) and the European Centre for Disease Prevention and Control (ECDC) the status of foodborne disease in the EU can be summarized. For example, in 2022, the two most commonly reported zoonoses in EU were the gastrointestinal diseases campylobacteriosis and salmonellosis, with a total of 137,107 and 65,208 human cases, respectively (EFSA/ECDC, 2023). The rates of confirmed cases as expressed per 100,000 inhabitants remained the same as those registered the year before i.e. 43.1 and 15.3 for campylobacteriosis and salmonellosis, respectively. The third most commonly reported zoonosis in humans in 2022 was yersiniosis, a disease caused by *Y. enterocolitica* (or *Y. pseudotuberculosis*), which is generally associated with the consumption of undercooked pork or vegetables. Next on the list in terms of the number of reported cases were Shiga toxin-producing *Escherichia coli* (STEC) infections and listeriosis caused by *L. monocytogenes*.

Furthermore, during 2022, a total of 5763 **foodborne outbreaks**, 48,605 human cases, 2783 hospitalisations and 64 deaths were reported by 27 EU Member States and the United Kingdom (Northern Ireland) although overall, the total number of outbreaks reported by each EU Member State varied considerably among countries (EFSA/ECDC, 2023). Multi-country outbreaks were caused by *Salmonella, Shigella* and *Cryptosporidium*. Being able to link outbreak cases occurring in multiple countries to the same cross-border outbreak depends on the capacity of each country to either investigate the food exposure history of the cases in detail, to trace-back and trace-forward distribution chains of contaminated food, or to perform highly discriminatory comparative analyses based on a pheno-genotypic characterisation of the causative agents in order to assess the level of relatedness among the agents isolated from human cases and/or food. Moreover, rapid exchange information systems such as the Early Warning Response System (EWRS), the EpiPulse and the Rapid Alert System for Food and Feed (RASFF) that is active within EU allow real-time dissemination of the alert among European countries, which greatly supports the investigation of multi-country outbreaks.

In 2022, 487 **strong-evidence** foodborne outbreaks were identified in the EU (EFSA/ECDC, 2023). These strong evidence can be analysed in more detail as presented in Table 2 for the relative **pathogen/food vehicle pair** ranking causing the highest number of strong-evidence outbreaks in 2022, as well as for the period 2018–2021 for trend watching purposes. Overall, *Salmonella* foodborne outbreaks ranked most often among the top positions in this top-10 classification with a variety of foods.

When looking at foodborne outbreak data from a different viewpoint and performing ranking in relation to **human health impact**, it showed that in 2022 *Clostridium perfringens* toxins was the causative agent involving the highest number of human **cases**, often associated with red meat or unspecified meat and meat products consumption (EFSA/ECDC, 2023). *Salmonella* caused the highest number of **hospitalisations** (related to egg and egg products and other foods mentioned in Table 2 but also related to some outbreaks of *Salmonella* in tap water including well water). *Listeria monocytogenes* was the agent causing the highest number of reported **deaths** caused by foodborne outbreaks in EU (N = 28; 43.8% of total deaths) paired with a wide variety of food vehicles. *Listeria* can be responsible for severe infections in humans (i.e. meningitis, sepsis, abortion, etc.), especially among vulnerable and fragile population groups (Goulet et al., 2012).

Further data and investigations of foodborne outbreaks can be found from other countries. e.g. in US at the website of the Center for Disease control (https://www.cdc.gov/foodborne-outbreaks/index.html). Monitoring of foodborne outbreak data and analysis of trends in individual reported cases of food infections or food intoxications is useful and needed for detection of (re)emerging pathogens i.e. pathogens whose incidence are increasing or who predominate (again) due to their physiological versatility and genetic adaptability or (renewed) introduction because of changing consumer preferences such as sustainable food consumption promoting a plant-based dietary shift and increased interest for unprocessed or fresh-like minimal processed foods with prolonged shelf life under refrigeration or (spontaneously) fermented foods, etc.

Table 2 Top-10 pathogen/food vehicle pairs causing the highest number of strong-evidence outbreaks (data derived from reporting by EU Member States in 2022 (EFSA/ECDC, 2023))

2022				2018–2021		Evaluation[a] (2022 versus 2018–2021)
Rank[b]	Causative agent	Food vehicle	Outbreaks (N)	Rank[b]	Outbreaks (N/year)	
1	*Salmonella*	Eggs and egg products	86	1	82.3 (37–135)	–
2	Histamine and scombrotoxin	Fish and fish products[c]	24	5	18.0 (14–23)	↑
2	*Salmonella*	Mixed food[e]	24	4	19.8 (3–27)	–
3	*Salmonella*	Pig meat and products thereof[d]	18	6	17.0 (11–26)	–
4	Norovirus (and other calicivirus)	Mixed food[e]	16	7	9.5 (5–13)	↑↑
5	*Salmonella*	Sweets and chocolate[f]	12	18	4.5 (2–8)	↑↑
5	*Staphylococcus aureus* toxins	Mixed food[e]	12	13	6.0 (1–12)	↑↑
6	*Salmonella*	Bakery products[g]	11	3	21.3 (9–32)	↓
7	Norovirus (and other calicivirus)	Crustaceans, shellfish, (live bivalve) molluscs and products thereof	10	2	58.5 (19–144)	↓↓
8	*Campylobacter*	Broiler meat (*Gallus gallus*) and products thereof	9	11	7.8 (4–10)	–

[a]A single arrow indicates variations of 25% and 50% in the number of outbreaks; double arrows indicate variations >50%; a 'stable' value indicates variations of between −25% and + 25%
[b]Ranking of the food vehicle based on the number of strong-evidence foodborne outbreaks in which the combination (causative agent/food vehicle) was implicated (rank 1 is the highest rank meaning the most commonly implicated). Strong-evidence outbreaks with unknown causative agents are not included
[c]Fish and fish products including fishery products from fish species associated with a high amount of histidine
[d]Pig meat and products thereof including fresh raw (minced) meat and meat specialities such as cured seasoned pork sausages
[e]Mixed food including ready-to-eat salads, sandwiches and other processed food products and prepared dishes such as pasta, finger food, pizza-like dishes, rice based dishes, meat-based or fish and seafood dishes etc.
[f]Sweets and chocolate including confectionery products and soft candy
[g]Bakery products include cakes, desserts sometimes containing raw eggs or raw cream

3 Foodborne Bacterial Hazards Causing Food Infections

A **food infection** is caused by the uptake of low numbers of micro-organisms which colonize the intestine after ingestion of the contaminated food. The normal flora of the intestine will be disrupted by the presence and growth of the pathogenic micro-organisms. The consequences of food infectants (e.g. *Salmonella* spp., *Campylobacter* spp., pathogenic *E. coli*) are the occurrence of **gastrointestinal symptoms** (stomach ache and diarrhea) after 8–16 h till a few days after consumption of the contaminated food. Typical for a food infection is the development of fever. The symptoms usually disappear after a few days or a week (with or without an antibiotic treatment and/or adequate rehydration). Sometimes a food infection can lead to hospitalization, especially for sensitive consumers indicated as the YOPI's (Young, Old, Pregnant, Immunodeficient). Some food infectants as *Listeria monocytogenes* are invasive, are able to overcome the intestinal barrier, enter the blood and partly escape the immune system by multiplying intracellularly in phagocytes (white blood cells). This gives the opportunity to further spreading of the germ in the body making the severity of the infection higher. The mortality of listeriosis has a magnitude of 15–30%. *Salmonella* spp., pathogenic shigatoxin-producing *E. coli, Campylobacter jejuni/coli, Listeria monocytogenes, Yersinia enterocolitica* and pathogenic *Vibrio* species are the most important bacterial infectious foodborne agents worldwide, although their relative importance may differ dependent upon the region in the world. An overview of their growth characteristics and predominant associated food products is shown in Table 3.

These pathogens will also be shortly discussed further highlighting some particularities in relationship to their versatility, and their status as established or (re)-emerging pathogen.

3.1 Salmonella

The genus *Salmonella* includes Gram-negative, non-spore forming, facultative anaerobic bacteria. They are usually motile, rod shaped (0.7–1.5 × 2.0–5.0 μm) and able to ferment glucose, thus producing acid and gas. According to the Kauffmann-White-LeMinor scheme, which is based on the immune reactivation with two surface antigens (flagellar (H) and somatic (O) antigens), there are more than 2600 known *Salmonella* serotypes. The genus *Salmonella* has two species: *Salmonella enterica* and *Salmonella bongori*. *S. bongori* is generally isolated from non-mammalian hosts. *S. enterica* itself is divided into six subspecies: *enterica, salamae, arizonae, diarizonae, houtenae* and *indica*. The majority of the serotypes belongs to *S. enterica* subspecies *enterica*; they are mainly associated with warm-blooded vertebrates and are usually transmitted by ingestion of food or water contaminated by infected faeces (Stevens et al., 2009).

Table 3 Overview of growth characteristics[a] of most important bacterial food infectants

Micro-organism	Opt T (°C)	T range (°C)	pH range	Min. a_w	O_2 need	Heat resistance[b]
Listeria monocytogenes	30–37	0–45	4.4–9.4	0.92	Facultative anaerobic	D_{60} 2.5–4.0 min D_{65} 0.75 min D_{70} 0.1–0.3 min
Salmonella spp	37	8–45	4.4–9.0	0.95	Facultative anaerobic	D_{60} 0.4–0.6 min
Campylobacter jejuni/coli	41–42	32–45	4.9–9.0	0.99	Micro aerophilic[c]	D_{55} 0.6–2.3 min
STEC (e.g. *E. coli* O157)	37	8–45	4.4–9.0	0.95	Facultative anaerobic	D_{58} 1.2 min D_{60} 0.46 min
Yersinia enterocolitica	30–37	(−2) - 44	4.5–9.6	0.96 (5% NaCl)	Facultative anaerobic	D_{58} 1.4–1.8 min
Pathogenic *Vibrio* (e.g. *Vibrio parahaemolyticus*)	37	5–43	4.8–11	0.94 (10% NaCl)[d]	Facultative anaerobic	D_{55} 2.5 min D_{65} < 1 min

[a]The growth characteristics are mainly indicative to the limits of the growth range and are subject to variation dependent of the strain. Moreover, they are only applicable if all other determining growth factors are optimal. With combination technology (hurdle technology) growth inhibition will occur much faster. It should be noted that outside the growth range or below the minimum value for growth, survival of the germ can occur

[b]The heat resistance is expressed as a D-value (the time needed at a specific temperature for 1 log reduction (90% inactivation of the population)). The D-value mentioned is only indicative. It is known that this value can vary dependent on different factors, such as the strain, and the type of food as well the a_w and pH of the food. Higher heat resistance is often observed with fatty food (with low water activity) whereby the bacterial strain finds protection from the matrix. In general it is assumed that those non-spore forming bacteria are sensitive for pasteurization and that a heating step of 2 minutes at 70 °C is sufficient for a 6 $\log_{10}$ reduction

[c]*Campylobacter* requires a microaerophilic atmosphere for growth existing of 5% O2, 10% CO_2, 85% N_2

[d]*Vibrio parahaemolyticus* is halotolerant with growth optimum at 3% NaCl (a_w 0.98)

Salmonella spp. is widely distributed in nature, its primary habitat is the intestinal tract of animal (both livestock and wild life) hosts. Being mainly intestinal, they are excreted in faeces (of animals but also occasional by humans that can be asymptomatic carriers) and may be found in faecal polluted water and thus also be introduced in the environment and survive for several weeks or months in water or soil or soil amendments.

Salmonella spp. is due to its large variety in serotypes and widespread prevalence a relevant pathogen in a wide range of products of animal origin: from eggs to milk and dairy products and bakery products and from fresh meat and poultry, fish and shellfish to all kinds of meat or fishery products. EU coordinated control programmes for *Salmonella* in products of animal origin have been a major success story in the EU. While before 2004, more than 200,000 human cases were reported each year in the 15 EU countries, it dropped to less than 70,000 cases in 2022 in 27 EU countries. Still *Salmonella* is the main cause of foodborne outbreaks (mainly due to *Salmonella* Enteritidis in eggs, Table 2) (EFSA/ECDC, 2023). Over the past two decades however, *Salmonella* outbreaks could also be linked world-wide with foods of non-animal origin: fruits (such as melons or papayas), fruit vegetables (such as bell peppers or tomatoes), (packaged) leafy greens, fresh or dried herbs, nuts and seeds (Hanning et al., 2009; Zweifel & Stephan, 2012). In these cases the likely cause of contamination was the use of contaminated irrigation water and/or import from countries where it is harder to fully respect good agricultural practices and hygiene. It has also been noticed that *Salmonella* spp. can survive well in products with a low water activity and outbreaks of milk powder, chocolate, peanut butter, etc. have been described (Morasi et al., 2022). With these products it is clear that the pathogen cannot grow but the presence and survival of *Salmonella* in these high-fat products can potentially cause infection.

Overall, it is to be noted that *Salmonella* is quite a versatile and adaptable pathogen that, even after more than 140 years of its first discovery in 1885 (Brands, 2006), it is world-wide still the leading cause of foodborne bacterial infections. Different serotypes or genotypes associated with other food types or ecological niches are emerging steadily. *Salmonella* is thus a pathogen needing continuous attention.

3.2 Human Pathogenic Escherichia coli

For the most part, *E. coli* is a group of harmless bacteria that are most often used as indicator organisms for faecal contamination and inadequate hygiene. However, several *E. coli* clones have acquired virulence factors that have allowed them to adapt to new niches and become pathogenic (Box 1.1). Shiga toxin producing *E. coli* (STEC) are the major group involved in foodborne disease and they can be classified as truly new emerging pathogens as they emerged de novo by transfer of mobile virulence factors (bacteriophages encoding Shiga-toxins) (Boerlin, 1999). The first recognized outbreak of STEC occurred in 1982 during an investigation of haemorrhagic colitis when several people were sickened by eating beef hamburger patties from a fast food chain that were found to be contaminated with *E. coli* serotype O157. Since then, *E. coli* O157 has been associated with various outbreaks, with ground beef remaining the most common vehicle among foodborne cases. However other major STEC serotypes or serogroups of concern are *E. coli* O26, O103, O145, O111, and O91 (EFSA, 2009). These serogroups are a small number in the entire family of STEC, which comprises over 400 serotypes that differs greatly in both their physiological characteristics and their pathogenic potential to humans. In 2022, there were 7117 confirmed cases of human STEC infections in the EU corresponding to a notification rate of 2.1 cases per 100,000 population (EFSA/ECDC, 2023).

Box 1.1: *Escherichia coli*: A Paradigm as a Versatile Bacterial Species

Escherichia coli exhibits considerable versatility and comprises a variety of non-pathogenic variants, as well as pathogenic variants that have been identified to cause intestinal or extraintestinal diseases in humans and many animal hosts. *E. coli* clearly exhibits fitness in multiple niches and can become naturalised in the environment. Most *E. coli* strains are not harmful and are an important part of the normal microbiota of the human gut, where they can be beneficial to humans. The better studied pathogenic *E. coli* is divided into two large groups: the intestinal pathogenic *E. coli* and the extraintestinal pathogenic *E. coli*, depending on the site of infection. Both are further divided into different pathotypes that cause distinct diseases. In food safety, intestinal pathogenic *E. coli* causing foodborne outbreaks are the most important. The most prominent intestinal pathogenic *E. coli* pathotype is Shiga toxin-producing *E. coli* (STEC), but also enteroaggregative *E. coli* (EAEC), enteroinvasive *E. coli* (EIEC), enteropathogenic *E. coli* (EPEC), enterotoxigenic *E. coli* (ETEC), diffusely adherent *E. coli* (DAEC) are important.

STEC may cause a range of symptoms varying from non-complicated diarrhea, bloody diarrhea and haemologic uremic syndrome (HUS). In some cases, the infection can lead to death. STEC always possess Shiga toxin genes (*stx*) which encode for Shiga toxins. There are two major types of Shiga toxins, namely Stx1 and Stx 2, and the *stx* genes have several subtypes, with four for *stx1* and at least 12 for *stx2*.

It is also worth mentioning that due to the development of phylogenetic studies, the relationship between *Shigella* species and *E. coli* (e.g. EIEC) needs to be reconsidered. *Shigella*, bacteria responsible for bacillary dysentery, was recognized in the late nineteenth century. Early phenotyping experiments revealed that it was closely related to *E. coli*, and DNA–DNA reassociation studies in the 1970s showed that *Shigella* species could not be distinguished from *E. coli* strains. Recent molecular studies have further validated the genetic relatedness of the *Shigella* species and EIEC There has been a consensus that the *Shigella* species do indeed fall within the definition of the *E. coli* species, but still, the nomenclature of the genus *Shigella* and species included has been retained for historical and medical reasons.

Transmission pathways for the pathogenic STEC human infection include faecal-oral, food-borne, environmental, and person-to-person (Caprioli et al., 2005). The food-borne route often involves consumption of undercooked meat, mainly bovine meat and products thereof represent a primary reservoir of STEC but also milk and dairy products made from raw milk (e.g. ice cream) or fresh produce (e.g. fresh spinach possibly contaminated by wild life around the fields or the use of pelleted animal manure as fertilizer) were implicated and also tap water including well water can be a source of STEC infection (Sharapov et al., 2016; EFSA, 2020a). Following the large and deadly outbreak of STEC O104:H4 food poisoning in Germany, related to the consumption of contaminated sprouted fennel seeds studies were undertaken demonstrating the unexpected long survival (up to 2 years) of *E. coli* O157 (and also *Salmonella*) on seeds during storage (Van der Linden et al., 2013). The study also showed that the pathogens maintain their ability to resuscitate and proliferate on the seedlings. Seeds or seedlings should thus be considered as a potential contamination source not only for sprouted seeds but for examples also for the cultivation of leafy vegetables, in particular also when hydroponics cultivation might become trending.

It is to be noted that in contrast to *Salmonella* where all *Salmonella enterica* strains are recognized as established pathogens, pathogenic *E. coli*, consist of different subgroups for which it remains a challenge to differentiate pathogenic STEC strains that are forming an effective risk for food safety and public health from the non-pathogenic "generic" *E. coli* strains (which are commensals).

3.3 *Campylobacter*

Campylobacter (mainly *C. jejuni* and *C. coli*) are the leading cause of bacterial foodborne infections in both developed and developing countries. In many countries *Campylobacter* still remains underreported, also due to the challenges in detection of this susceptible micro-organism amidst the competing microbiota both in clinical samples and foods (Harrison et al., 2022).

Campylobacter is a zoonosis, and an inherent likely to be part of typical intestinal bacteria of chickens (and birds in the broad sense) although a seasonal effect is noted worldwide. Annual increase in *Campylobacter* contamination in broilers and in parallel human infections begins in late spring and reaches a maximum in summer. A rise in ambient temperature and overall increase of wild life activity is corresponding to this periodic seasonality. Examination of risk factors that might explain this seasonal increase also identifies flies as a vector for transmitting *Campylobacter* in poultry flocks and between other agricultural animals (Nichols, 2005). Most human campylobacteriosis cases (50–90% depending upon the region) could be attributed to a poultry source, while 10–15% could be attributed to a cattle source (e.g. by drinking unpasteurized milk) (Boysen et al., 2014; Lake et al., 2021; EFSA, 2010). However, *Campylobacter* is also known as a waterborne pathogen (drinking contaminated well water or infections caused by swimming in recreational water). Poultry consumption per se is not necessarily the risk factor but rather poultry meat handling preparation and consumption. Several cross-contamination routes in the kitchen are confirmed, e.g. putting cooked poultry meat at the same plate of the raw poultry meat (11.2%), not washing hands after handling raw poultry meat (11%), using the same knife (6.7%) or cutting board (8.2%) for the vegetables which was prior used for the poultry meat; but also 16% would eat not well-cooked poultry meat if served at a barbecue (Sampers et al., 2010).

C. jejuni and *C. coli* are thermotolerant and thus cannot multiply under refrigeration and even not in case of temperature abuse (in contrast to *Salmonella*). It can only grow at specific microaerophilic conditions and is quite sensitive to heat. As soon as the temperature exceeds 50 °C, inactivation occurs. Other stressors (e.g. pH drop, drying, salting ...) also give rise to rapid inactivation and reduction of *Campylobacter* spp. *Campylobacter jejuni/coli* is also sensitive to freezing, an intervention strategy implemented in some countries (Iceland, Norway) for contaminated poultry to reduce exposure to the pathogen (Sampers et al., 2010). Risk analysis shows that particularly the higher numbers of *Campylobacter* in poultry meat are related to a risk to public health. In 2017 EU-wide a process hygiene criterium was introduced for *Campylobacter* on broiler carcasses: a quantitative target of 1000 *Campylobacter* spp. per gram (pooled) neck skin (after chilling) rather than the use of an approach of absence per 25 grams as is common for *Salmonella*. Information on contamination rates of broiler carcasses after chilling shows that, in comparison to *Salmonella* (3–12% present in broiler neck skin samples in EU in 2022) *Campylobacter* is commonly found and also present in relatively high numbers (10 to 1000 cfu/g) (ca. 17–20% of samples exceeded the process hygiene limit of 1000 CFU/g) (EFSA/ECDC, 2023).

Campylobacter spp. is widely perceived as a sensitive microorganism but can survive especially well in a cold, moist environment. They are genetically diverse and highly variable genotypes of *Campylobacter jejuni/coli* circulate in the food chain. *Campylobacter* has been since 1970s an emerging pathogen but still is reason for concern given the high number of reported cases, and the related burden of disease also taking into account that in a limited number of cases campylobacteriosis may lead to serious complications and long-term effects. *Campylobacter* infection may lead to reactive arthritis (a joint disorder that involves inflammation of one or more joints) (1–5%) and in a limited number of cases (ca. 0.1%) to the Guillain Barré (GB) syndrome (Pope et al., 2007). The GB syndrome is a disorder in which the body's immune system attacks part of the peripheral nervous system which potentially results in paralysis. Control of *Campylobacter* in broilers from farm to slaughterhouse up to the consumer is still a challenge and not an easy task (EFSA, 2020b). Not only because *Campylobacter* is ubiquitous in the environment and in much raw chicken and thus *Campylobacter*-free broilers and poultry meat are not feasible to achieve but also because human risk-behaviours that increase exposure to the pathogen are difficult to change by educational intervention efforts (Rushton et al., 2019). Moreover, given the seasonality already observed in campylobacteriosis, it is also expected that *Campylobacter* infections will further increase due to climate change both by the food but also the non-food transmission route (Kuhn et al., 2020).

3.4 Listeria monocytogenes

L. monocytogenes is a Gram-positive bacterium, non-spore forming, facultative anaerobic and has the important characteristic to be psychrotolerant and thus able to grow at refrigeration temperature. Overall, *Listeria* is particularly resistant to environmental stress and it is able to survive or even multiply under a wide range of unfavourable conditions of acidity (pH minimum 4.4) and water activity (a_w minimum 0.92). *L. monocytogenes* is an environmental bacterium known as the most heat resistant vegetative pathogen, but which is still sensitive to pasteurization (Table 3).

L. monocytogenes was initially described in 1926 as an animal pathogen (in rabbit) and named *Bacterium monocytogenes* (because infecting monocytes). In 1940 the genus was renamed as *Listeria* in honour of British pioneer of sterile surgery Joseph Lister. *L. monocytogenes* has been considered for long time an animal pathogen. It is only in 1980–90 that *L. monocytogenes* has been associated to food and listeriosis in humans. Although still considered a zoonotic pathogen, *L. monocytogenes* is widely distributed in nature and food processing environments. It has been isolated from soil, vegetation, sewage, water, animal feed, fresh and frozen meat, and in the faeces of healthy animals including humans (McLauchlin et al., 2004). Therefore, it is not surprising that *L. monocytogenes* was isolated from and associated with outbreaks or recalls from a variety of foodstuffs (e.g. cooked meat products, smoked fish, cheese, ice cream, pre-cut lettuce, sandwiches, deli-salads etc.) (Uyttendaele et al., 2009; EFSA, 2013b; Zhang et al., 2021; Conrad et al., 2023; Van Paepeghem et al., 2024). Ready-to-eat foods (RTE) with prolonged shelf-life are considered to be the major vehicle for listeriosis and to a lesser extent, fresh unprocessed foods (fresh meat or fish) with a short shelf life (up to 5 days). Increasingly minimal processed foods are kept 5–10 days or longer under refrigeration (e.g. multi-ingredient salad bowls, bagged leafy greens) and *L. monocytogenes* is the pathogen to take into consideration to ensure food safety when setting shelf life (Ceuppens et al., 2016; Van Paepeghem et al., 2024).

Although the number of recorded clinical cases is rather limited, an increase of listeriosis is observed the last years in the EU. In 2022, 2738 confirmed invasive human cases of *L. monocytogenes* were reported (EFSA/ECDC, 2023). This corresponded to 0.62 cases per 100,000 population, which was an increase of 15.9% compared with 2021 (0.53 cases per 100,000 population) and the highest rate and number of cases reported since 2007. The cause of this increase is at the moment not clear yet, but this leads to a constant vigilance towards this (re)emerging pathogen. It is a fact that there is an increase of populations in our society who are more susceptible to the risk of listeriosis. This is partly attributable to the aging of the population (21% of the European population is over 65). Moreover, in the modern, fast-paced lifestyle, the population is consuming more RTE and ready-to-cook food products that entail minimal processing. Also the demand for products with less salt has been increasing, which is attributed to an increasing awareness of consumers considering nutrition and health and consumers often prefer clean label foods (without additives). However, salt and preservatives are important factors in food preservation which puts pressure on product development ensuring growth inhibition of *L. monocytogenes*. Applying hurdle technology and increasing knowledge regarding the factors that impact the behaviour of *L. monocytogenes* during shelf life (intrinsic product characteristics, modified atmosphere packaging, microbial interaction, etc.) are important. As new food types are introduced, the processing steps, product formulation or storage conditions need to be validated in relationship to potential contamination routes and growth potential of *L. monocytogenes*. For example, in 2022 in France an outbreak was reported with a vegan cheese substitute, the product did not undergo a hygiene step such as a pasteurization step that could control contamination at the level of raw materials or control (post)contamination from processing environment (Leclercq et al., 2024).

In EU listeriosis cases resulted in 1330 hospitalisations and 286 deaths in 2022. Within fatal cases of listeriosis, 59.8% were in the age group 65–84 years, while 24.3% were in the age group >84 years. The overall EU case fatality rate among cases with known outcome was 18.1% indicating the severity of the disease (EFSA/ECDC, 2023). In South Africa a total of 1060 listeriosis cases were reported in an outbreak in 2017–2018 associated with prepacked 'polony', a cooked meat sausage made of pork and beef, which led to 216 deaths. It was the world's largest listeriosis outbreak globally (Smith et al., 2019). The public health risk associated with 'ready-to-eat' (RTE) foods depends mainly on the effectiveness of control measures of Good Manufacturing Practices (GMP) and Hazard Analysis and Critical Control Point (HACCP) programmes at manufacturing and distribution. Further investments in training and education, as well as the implementation of an appropriate food safety culture in food companies producing or assembling these types of pre-packed, plant-based RTE foods, are recommended (Spagnoli et al., 2023). However, it is also important to continue raising awareness of listeriosis and its risks associated with certain consumption habits and types of food in risk groups. For instance, in 2019, the consumption of hospital-provided pre-packed RTE sandwiches with cooked, diced chicken led to nine cases of illness and seven deaths in the UK (Public Health England, 2020). Targeted risk communication to susceptible groups (and health care institution employees) on microbial risks and the needs of strict adherence to the cold chain and serving and consumption of foods less prone to *L. monocytogenes* contamination might be recommended (De Bock et al., 2023). The adoption of this series of control measures will be needed to reverse the trend of increase of listeriosis cases and stop the (re)emergence of *L. monocytogenes* as a foodborne pathogen.

3.5 Pathogenic **Vibrio** *Species*

Pathogenic *Vibrio* species are Gram-negative pathogens that are native to marine coastal areas and brackish waters and can be found in contaminated fish and shellfish. Although strains of *Vibrio cholerae* O1 or O139 are responsible for cholera, a severe diarrheal illness that can rapidly progress to dehydration and death without timely intervention, there are three main non-cholera pathogenic Vibrio species. These include strains of V. cholerae non-O1/non-O139, as well as *Vibrio parahaemolyticus*. But not all *V. parahaemolyticus* strains can cause human disease (gastro-enteritis). A pathogenic strain carries the gene for at least one of the two hemolysins: TDH (thermostable direct hemolysin) or TRH (TDH-related hemolysin). Another significant pathogen is *Vibrio vulnificus*. Although the specific major pathogenic factors of *V. vulnificus* have not been definitively identified, all strains are considered pathogenic (Brauge et al., 2024).

Overall, from available data collected, *V. parahaemolyticus* is expected to be present in approximately 20% of seafood, with one in five positive samples containing pathogenic strains. Pathogenic *V. vulnificus* can be detected in about 6% of seafood samples, being most prevalent in bivalve molluscs. Non-choleragenic *V. cholerae* can be detected in about 4% of tested seafood samples (EFSA, 2024). The temperature of the water plays an important role in the occurrence of vibrios. They are seldom isolated when the temperature of the water is <15 °C. The spreading of these vibrios is monitored in the context of climate change and possible warming of the sea. Prevalence of these pathogens in seafood is expected to increase globally due to climate change (FAO and WHO, 2021b; EFSA, 2024).

3.6 *Pathogenic* Yersinia *Species*

Yersinia are a Gram-negative rod-shaped bacteria, belonging to the order Enterobacterales. Regarding foodborne illnesses, *Y. enterocolitica* and *Y. pseudotuberculosis* are the most important species. For *Y. enterocolitica* 6 biotypes are distinguished (1A, 1B, 2, 3, 4, 5) with over 70 different serotypes (Wauters et al., 1987), but few are pathogenic for humans. Biotype 1A strains are regularly isolated from the environment and non-virulent since they do not carry the plasmid Yersinia Virulence (pYV plasmid). *Y. enterocolitica* biotype 1B strains are highly pathogenic, while biotypes 2–5 are considered low or moderately pathogenic (ANSES, 2008). The latter are most commonly isolated in Europe. *Y. pseudotuberculosis* is subdivided in 4 biotypes and 15 serotypes. No clear differences in pathogenicity among these biotypes are described and, consequently, all isolates are considered potentially pathogenic for humans (EFSA, 2007). *Y. enterocolitica* bioserotype 4/O:3 (83%) and, to a lesser extent, *Y. enterocolitica* bioserotype 2/O:9 (15%) are reported to be the main causative agents of foodborne human yersiniosis infections in the EU. *Y. pseudotuberculosis* was only responsible for 1–2% of the confirmed cases (EFSA/ECDC, 2023).

Y. *enterocolitica* has already been isolated from different food products like water, milk, poultry, beef, pork and lamb. Of these products, pork and (raw) milk products are the main sources and it is especially in these food products that *Y. enterocolitica* is targeted and analysed. Distinguishing pathogenic from non-pathogenic *Yersinia* strains on selective media is reported to be difficult, due to their similar colony morphology (Fredriksson-Ahomaa & Korkeala, 2003). Pigs (in particular their tonsils) are assumed to be the main reservoir of for human pathogenic serotypes (Van Damme et al., 2015).

Yersiniosis was the third most commonly reported foodborne gastrointestinal infection in humans in the European Union. In 2022, there were 7919 confirmed cases of human yersiniosis (2.2 cases per 100,000 population) (EFSA/ECDC, 2023). Hence, even though yersiniosis has been less associated with large foodborne outbreaks (rather with sporadic infections) and its mortality rates were reported to be lower, in comparison with salmonellosis or listeriosis, yersiniosis should not be neglected in terms of foodborne bacterial illnesses. *Y. enterocolitica* is a psychrotrophic (cold tolerant) bacterium, so refrigeration of foods gives inadequate protection against propagation. This might indicate a higher risk for public health associated with the consumption of raw, fresh pork or other meat (Van Damme et al., 2017). Although yersiniosis has usually been associated with consumption of undercooked pork products or unpasteurised milk, it is important to sample more RTE foods, such as RTE vegetables, which are consumed without any treatment, to reduce or eliminate potential *Yersinia* contamination. There is the growing trend of RTE pre-packed (fresh-cut) vegetables in industrialised countries which have recently also been linked to sporadic outbreaks of *Y. enterocolitica* (Karlsson et al., 2021).

4 Foodborne Bacterial Hazards Causing Food Intoxications or Toxi-Infections

A **food intoxication** is a food poisoning caused by the consumption of a food product which contains a microbial toxin previously produced during outgrowth of a toxigenic micro-organism to high numbers (usually more than 10^5 CFU/g) in this food product. Implicated strains are known to produce a number of exotoxic factors that are absent in avirulent strains. During a food intoxication the presence of the toxin is crucial and the toxin molecule directly influences the biological reactions in the host's body and is the cause of the disease symptoms. The presence of the pathogenic micro-organism in the food product at the time of consumption is not necessary for food intoxication. Example of foodborne pathogens with this type of pathogenesis are *Staphylococcus aureus* and emetic *Bacillus cereus* strains. After outgrowth and toxin production the micro-organisms are often killed by a heating step, while the produced toxins—heat stable toxins—remain in the food product and cause symptoms, i.e. vomiting (emesis), in the case of *S. aureus* and some *B. cereus* strains, shortly (within 2-6 h) after consumption of the food product.

In case of a **food toxi-infection**, growth of the micro-organism in the food product to higher numbers is still required (possibly accompanied with toxin production), but in this case the toxins are unstable (sensitive to heat inactivation and/or proteolytic and pH degradation by gastric enzymes and gastric environment) and thus *de novo* toxin production is required in the gastrointestinal tract. This means that the consumption of food products with high numbers (usually more than 10^5 CFU/g) of viable toxin producing micro-organisms—in this case *Clostridium perfringens, Clostridium botulinum* and some *Bacillus cereus* strains—is required. A food toxi-infection is classified as a food intoxication because the toxin molecule is the sole cause of the disease symptoms. The reference to infection is made in the term toxi-infection because the consumption of viable cells is required, but always in large quantities. The presence of low numbers of micro-organisms (100–1000 up to max. 10^5 CFU/g) in food can be tolerated, since this holds limited risk for food safety. In contrast, micro-organisms causing food infections can also cause disease when present in low numbers (1 per 25 g, or dependent upon the infectious agent <10 to max. 100 CFU/g). The consumption of high numbers of micro-organisms during a food toxi-infection is probably necessary to resist gastric passage and reach the intestines in sufficient numbers to establish themselves within the intestinal microbial community and produce toxins during (brief) outgrowth and/or sporulation. This process of consumption of micro-organisms in food, gastrointestinal passage and toxin production results in slightly longer incubation times than for food intoxications, but the symptoms will still appear quickly (within 6-12 h). The symptoms of *Clostridium perfringens* toxi-infection are abdominal pain and diarrhea, as is the case for *Bacillus cereus* toxi-infection. In contrast, the symptoms of *Cl. botulinum* toxi-infection are more severe with symptoms of paralysis (dry mouth, drooping eyelids, muscle weakness) and possibly even resulting in death (caused by the paralysis of the heart muscle) when anti-toxin is not administered in time.

Based on growth characteristics (Table 4) or variations in the toxin structure, a sub-division can be made into certain "types" in some microbial species such as *Clostridium perfringens, Clostridium botulinum, Bacillus cereus* and *Staphylococcus aureus*. The strains which produce toxins against human are often clustered and mainly situated in a limited number of these "types". For example, toxi-infections by *Clostridium perfringens* are nearly exclusively caused by *Cl. perfringens* type A. Sometimes bacterial strains (isolates) of other "types" are associated with intoxications of animals but not humans (e.g. Cl. botulinum type C and D comprise strains which cause botulism in ducks, fish and cows but not in humans).

4.1 Staphylococcus aureus

Humans and animals are usual carrier of *S. aureus* (either on their skin, hair, nose or throat). Since this bacteria cannot grow at temperatures below 7 °C and that multiplication of *S. aureus* in raw products is inhibited or slowed down by competing accompanying microbiota on foods, elevated numbers and toxin production are normally not reached under the conditions of maintained cold chain and good hygienic practices. Nevertheless, Staphylococcal food poisoning is still a common cause of foodborne illness and often associated with foods prepared by commercial caterers sold in restaurants, pubs, street vendors, takeaway, etc. or in canteens at workplace, school, health care or residential facilities.

The known epidemiology of staphylococcal food poisoning suggests a food contaminated by an infected food handler which was subsequently subject to temperature abuse (Hennekinne et al., 2012; Pillsbury et al., 2013). Therefore, good personal hygiene is essential but also respect of appropriate time/temperature combinations during heat treatment and keeping hot or cool-down rapidly and refrigerate appropriately before or at distribution or serving. Nevertheless,

Table 4 Growth conditions of bacterial food intoxicants

Pathogen		Growth values			Oxygen requirement	D value
		T°	pH min	a_w min		
Bacillus cereus		5/15–35/50[a]	4.3	0.91-0.93	Facultative anaerobic[b]	D_{100} 1.2–8 (spores)
Clostridium botulinum	Type A, and proteolytic strains of type B and F	10–52	4.6	0.935	Anaerobic. Can tolerate traces of oxygen for growth Toxin production only under strict anaerobic conditions	D_{121} 0.2 D_{100} 25 (spores)
	Type E, and nonproteolytic strains of type B and F	3.3–45	5.0	0.97		D_{90} 3.1 D_{100} < 0.1 (spores)
Clostridium perfringens		12–50	5.5	0.93	Anaerobic, no growth in presence of O_2 Eh range for growth −125 to +350 mV	D_{90} 106–193 (spores) D_{55} 2-7 (vegetative cells)
Staphylococcus aureus		7–50 (10–48 for toxin production)	4	0.83 for growth (0.85 for toxin production)	Facultative anaerobe	D_{60} 0.5–1.3

[a]5/35 °C limits for psychrotrophic strains; 15/50 °C for mesophilic strains. Spore germination may occur up to about 60 °C
[b]Oxygen is required for the production of the emetic toxin. Certain controversy still exists in comparison of aerobic and anaerobic conditions in total amounts of enterotoxins produced (possible function of the strain used, nutritional environment, method used for determination, growth that precedes toxin production…)

S. aureus is also able to nest itself on the surfaces of process equipment due to inadequate and insufficient cleaning and disinfection and thus provoke (post)-contamination of (cooked) foods. This implies that in addition to the personal hygiene and respect of the cold chain also good cleaning and disinfection of equipment, utensils or other food contact surfaces is a key element in the control of *S. aureus*.

Staphylococcus aureus is an established pathogen. Still, in the rapid turnover of workforce involved in food preparation or distribution of meals in institutional catering or restaurants there is a need for knowledge of and adherence to good hygienic practices and the WHO five keys to safer foods (WHO, 2006) to prevent this type of food poisoning.

4.2 Bacillus cereus

The *B. cereus* group, also known as *B. cereus* sensu lato (s.l.), is a subdivision of the *Bacillus* genus and comprises numerous closely related species. All members are Gram-positive, rod-shaped, low GC%, spore forming aerobic or facultatively anaerobic bacteria that are widespread in natural environments. According to the current LPSN list (https://lpsn.dsmz.de/, accessed 09 July, 2024), the *B. cereus* group consists of 25 validated species including *B. anthracis, B. cereus, B. thuringiensis, B. mycoides, B. pseudomycoides, B. weihenstephanensis, B. cytotoxicus, B. toyonensis, B. wiedmannii, B. albus, B. luti, B. mobilis, B. nitratireducens, B. pacificus, B. paramycoides, B. paranthracis, B. proteolyticus, B. tropicus, B. fungorum, B. gaemokensis, B. manliponensis, B. hominis, B. sanguinis, B. paramobilis, B. rhizoplanae* and 2 not validated species called "*B. bingmayongensis*" and "*B. clarus*". The former five species were identified during the twentieth century, whereas the remaining species were classified in the twenty-first century.

B. anthracis, B. cereus and *B. thuringiensis* are the most frequently studied species of the *B. cereus* group, mainly due to their capacity of causing human diseases and for their use in biological control applications. *B. anthracis* is the etiological agent of anthrax, an acute lethal disease in humans and animals. It is a category A bio-threat agent and has since long been considered as a potential agent for biological warfare or bioterrorism. *B. cereus* sensu stricto is an established foodborne pathogen. It causes 2 syndromes: a diarrheal form (toxico-infection) and an emetic form (food intoxication). For the diarrheal form ingested *B. cereus* spores and/or some survived vegetative cells first adhere to the intestinal mucosa, followed by germination (in the case of spores), vegetative outgrowth and production of enterotoxins, which damage the nearby epithelium of the host. Emetic food poisoning is caused by the toxin cereulide produced during the growth of emetic *B. cereus* strains in food (Ceuppens et al., 2013). It has been confirmed that most cases of foodborne outbreaks caused by the *B. cereus* group have been associated with concentrations above 10^5 CFU/g, thus this level of contamination is in many countries recognized as a food safety action limit although individual cases of both emetic and diarrhoeal illness have ever been reported involving between 10^3 and 10^5 CFU/g of *B. cereus* (EFSA, 2016).

Characteristics of the diarrheal and emetic syndromes caused by *B. cereus* are summarized in Table 5.

B. cereus is thought to be underestimated as a foodborne pathogen. *Bacillus cereus* toxins ranked first for the number of reported foodborne outbreaks caused by bacterial toxins in the EU in 2022 and was responsible for two deaths (mainly severe impact in case of a foodborne outbreak in health care and residential facilities). A

Table 5 Characteristics of the two types of foodborne illness caused by *B. cereus*

Characteristics	Diarrhoeal syndrome (toxico-infection)	Emetic syndrome (intoxication)
Primary cause	Ingestion of vegetative cells and spores and toxin production in small intestine	Preformed toxin in food
Relevant number of bacteria	Infectious dose (concentration): $\geq 10^5$ CFU/g of food[a] The total number required is lower for spores compared to vegetative cells[b]	Toxin producing number: 10^5–10^8 cells/g of food Cereulide: 8 ~ 10 μg/kg body weight[c], 2 ~ 6 μg/g food[d]
Incubation time	8–16 h (occasionally >24 h)	0.5–5 h
Duration of illness	12–24 h (occasionally several days)	6–24 h
Symptoms	Abdominal pain, watery diarrhea and occasionally nausea	Nausea, vomiting and malaise and in some cases fatal liver failure
Foods most frequently implicated	Proteinaceous foods: Milk and dairy products, meat products, pudding, soups, sauces and vegetables	Starch-rich foods: Rice, pasta, potato, pastry and noodles
Toxin associated	Haemolysin BL (Hbl) Nonhaemolytic enterotoxin (Nhe) Cytotoxin K (CytK)	Cereulide (Ces)
Toxin properties	Heat labile proteins (inactivated by 56 °C/5 min) Inactivated by proteases and pH < 4.0	Heat stable cyclic peptide (no loss of activity at 121 °C for 90 min) Not inactivated by protease enzymes Stable at pH 2 to 11

Adapted from Granum and Lund (1997), Stenfors Arnesen et al. (2008), Kumari and Sarkar (2016), Webb et al. (2019)
[a]concentration unacceptable for ready-to-eat foods (Health Protection Agency, 2009; EFSA, 2016)
[b]Clavel et al. (2004), (Schoeni & Wong, 2005)
[c]quantity of cereulide necessary to cause illness to a 50 kg human assuming 8 μg cereulide per kg is the toxic dose for humans (Jaaskelainen et al., 2003)
[d]EFSA (2016)

considerable increase in the number of foodborne outbreaks caused by *Bacillus cereus* toxins was observed in 2022 (306 reported outbreaks including 3192 human cases, a relative increase of 251.7% compared with 2021!) (EFSA/ECDC, 2023). This increase might be attributable to the increased focus on investigating these most of the time rather mild toxi-infection/intoxication syndromes and thus better reporting of this pathogen implicated in outbreaks by some countries. *Bacillus cereus* toxins were identified as the causative agent of five very large foodborne outbreaks (i.e. ≥ 100 cases). While various factors may contribute to the contamination of food with bacterial toxins (including not only *B. cereus* toxins but also *Staphylococcus aureus* toxins and *Clostridium perfringens* toxins), mainly incorrect temperatures and storage times were reported (EFSA/ECDC, 2023).

Due to its ability to form spores *B. cereus* can easily survive for extended periods in the environment and is widely distributed in nature and our surroundings (air, soil, water). *Bacillus cereus* is often found in plant-based ingredients but is also a common contaminant of dairy products. It can contaminate milk at the time of milking. They can also often be found in cooked chilled foods with prolonged shelf life under refrigeration, most typical multi-ingredients foods including pasta, rice or potato puree, a cooked or minimal processed vegetable component and meat or another cooked protein component (Daelman et al., 2013). Sporeforming bacteria such as *Bacillus* spp. are heat-resistant and thus have the potential to survive also in processed foods that are subjected to a heat-treatment as part of the manufacturing process. Pasteurization treatment is not effective in reducing contamination. *Bacillus cereus* is also well known to produce biofilms, a serious concern for the food industry, with also sometimes spores that get resistant to cleaning and disinfection and are easily transferred. As the contamination of raw materials is not completely avoidable, it is clear that the correct application of hygienic procedures and keeping the cold chain to prevent further development of *B. cereus* to elevated numbers is extremely important in order to avoid and manage the circulation of *B. cereus* along the food supply chain (Tirloni et al., 2022).

Recently there has been more attention also to some psychrotolerant strains of *B. cereus*, which have been classified as *B. weihenstephanensis* because of their concern for food safety of foods with prolonged shelf life under refrigeration (Lechner et al., 1998). However, Liu et al. (2018) showed that psychrotolerant characteristic and signature sequences of 16S rRNA and *cspA* genes were incapable of distinguishing *B. weihenstephanensis* from some non-*B. weihenstephanensis* strains. Also *B. cytotoxicus* strains that have high cytotoxicity, are thermotolerant and can grow up to 50 °C have raised attention. However, it has been recently reported that the hazardous potential of *B. cytotoxicus* may be lower than initially assumed due to low or no toxicity to Vero cell cultures of *B. cytotoxicus* isolates from mashed potato powders (Burtscher et al., 2021). *B. wiedmannii* is a psychrotolerant and cytotoxic species initially identified in 2016 (Miller et al., 2016).

B. thuringiensis is an entomopathogenic bacterium to various insect pests, such as Lepidoptera, Coleoptera, and Diptera by producing insecticidal parasporal crystal proteins or δ-endotoxins that has recently raised attention as an emerging foodborne pathogen (De Bock et al., 2021). For this, it is important to know indeed that the **taxonomy and nomenclature of *Bacillus cereus* group members** is quite complex (Box 1.2). ***B. thuringiensis*** is a spore forming bacterium that is widely distributed in the environment (i.e. soil and plant), and some strains have been authorized and safely used as commercial biopesticides for decades. However, the few outbreaks that have been reported related to *B. thuringiensis* warrant attention and fuel the debate for its position as a foodborne pathogen. It has been reported that many *B. thuringiensis* isolates carry the diarrheal toxin and *cytK-2* genes, and have the ability to produce diarrheal toxins and express other virulence factors. Besides, the difficulty of differentiation within the *B. cereus* group between *B. cereus* and *B. thuringiensis* encountered in (food or clinical) diagnostic labs due to the high similarity in morphology and genome has contributed to this uncertainty about *B. thuringiensis* to be recognized as emerging pathogen.

Box 1.2: Taxonomy and Nomenclature of *Bacillus cereus* Group Members: A Complex Issue

For historical reasons, the taxonomy and the definition of these so-called classical species within the *B. cereus* group are predominantly based on phenotypes, related to their medical and economic significance, which are mainly associated with plasmid elements. The above reasons suggest that describing the taxonomy of the *B. cereus* group by phenotypically based species names alone is not adequate, but due to traditional and pragmatic reasons, the classical species names are still being used today. The classification and taxonomic separation of the *B. cereus* group needs to be supported by genetic evidence with modern molecular biology tools such as whole genome sequencing (WGS). However, numerous genomospecies thresholds (92.5–97% ANI i.e. average nucleotide identity) for *B. cereus* s. l. species classification have been used by researchers sometimes resulting in the same isolate or strain being assigned to different species using different taxonomic frameworks. It also worth to mention that not all *B. cereus* s. l. isolates are being subjected to WGS and many routine food laboratories only differentiate *B. cereus* group members phenotypically (e.g. FDA-BAM method) which is generally simple, rapid and cheap. Relying strictly on a phenotypic-based taxonomic classification at the species level can lead not only to misclassification of isolates but also to an inaccurate assessment of a given isolate's virulence potential because of the genetic heterogeneity and the horizontal gene transfer of plasmids that can occur among the *B. cereus* group. The bacterial genome includes both the chromosome and extrachromosomal elements like plasmids and phages. The phenotypic traits used to distinguish *B. anthracis*, *B. cereus* and *B. thuringiensis* are located on plasmids instead of chromosomes. Plasmids of *B. anthracis*, pXO1 and pXO2 carry the tripartite lethal toxin complex and the biosynthetic genes for the poly-γ-D-glutamic acid capsule, respectively. The emetic *B. cereus* strains contain several plasmids encoding cereulide toxin synthesis genes.

Insecticidal trait of *B. thuringiensis* is often based on the presence of plasmid-encoded large Cry protein inclusions of the δ-endotoxin. If these plasmids are lost, these strains will lose their species-specific virulence traits. Various molecular and biochemical methods have been used to explore the diversity and genetic relationship between species and strains composing the *B. cereus* group including MLST (multi-locus sequence typing) based on 7 housekeeping genes and *panC* typing which compares the *panC* gene sequence from new isolates with those of reference strains. (Zhao et al., 2023). Seven phylogenetic groups of the *B. cereus* group and their phenotypic properties are shown in Table 6.

(continued)

Box 1.2 (continued)

Table 6 Seven phylogenetic groups of 3193 isolates in the *B. cereus* group and their phenotypic properties (Guinebretiere et al., 2008, 2010; EFSA, 2016)

Phylogenetic group	Species	Temperature of growth (°C)	Thermo type
I	*B. mycoides, B. pseudomycoides, B. cereus, B. thuringiensis*	10–43	Mesophilic
II	*B. cereus, B. thuringiensis, B. anthracis, B. weihenstephanensis*	7–40	Psychrotolerant
III	*B. cereus, B. anthracis, B. thuringiensis, B. mycoides*	15–45	Mesophilic
IV	*B. thuringiensis, B. cereus, B. mycoides, B. weihenstephanensis*	10–45	Mesophilic
V	*B. cereus, B. thuringiensis, B. mycoides, B. toyonensis, B. anthracis*	8–40	Intermediate
VI	*B. thuringiensis, B. cereus, B. mycoides, B. weihenstephanensis*	5–37	Psychrotrophic
VII	*B. cytotoxicus*	20–50	Thermotolerant

Concerns on this food safety issue of biopesticides are raised in particular because during application pre-harvest deliberately quite high numbers of *B. thuringiensis* are deposited on the (edible) crop without the mandatory respect of a set pre-harvest interval (waiting time before harvest), which might result in contamination with elevated levels of this hazard situated at or above the current food safety action limit for (presumptive) *B. cereus* post-harvest when the plant becomes food. Moreover, *B. thuringiensis* is also naturally associated with soil and plants. No track record of association of fresh produce to foodborne outbreaks of *B. thuringiensis* due to its use as a biopesticide in agricultural edible crop production was available until a report of an alleged foodborne outbreak in one MS in the EU discussed by EFSA (2016). There are few studies investigating the prevalence of *B. thuringiensis* in fresh produce (lettuce, tomatoes, bell peppers, etc.) sampled at retail but the occurrence of *B. thuringiensis* in fresh produce higher than 10^5 CFU/g seemed to be rare (De Bock et al., 2021; Zhao et al., 2022, 2023). This is most probably because on fresh (raw) vegetables usually Gram-negative bacteria (in particular *Pseudomonas* spp.) dominate and act as competing microbiota leaving less opportunities for *B. cereus* group species (including *B. thuringiensis*) to multiply.

Overall, there is a long track record of safe use of these biopesticides. The first commercial *B. thuringiensis*-containing plant protection product was produced in 1938 in France. There is a need however to estimate the impact of agricultural practices and thus the pathogenicity of these biopesticide strains and their residual persistence or growth potential during post-harvest storage or processing of fresh produce. This will need collaboration of various scientific disciplines. Expertise is needed in plant protection, plant microbial ecology, food safety, pathogenicity, public health, etc. This case study of *B. thuringiensis* also shows the need to balance the interest of these various disciplines. A **debate** in society is on-going on the benefits and safety of sustainable agricultural production seeking to avoid the use of chemical pesticides and promoting biocontrol agents such as *B. thuringiensis* biopesticides but also other *Bacillus* spp. (or other microorganisms) as biocontrol agents or biostimulants (Etesami et al., 2023) while minimizing the risk of such plant beneficial micro-organisms to become emerging foodborne pathogens causing foodborne disease to susceptible human hosts when plant becomes food.

4.3 Clostridium perfringens

Cl. perfringens is an anaerobic spore forming Gram-positive food toxico-infectant. Enterotoxin producing strains of *Cl. perfringens* are present in the soil, water, dust and the intestinal tract of humans and animals. Because *Cl. perfringens* is associated with the intestinal tract of humans and animals, the bacterium is also used as an indicator of hygiene especially of the surface waters. Because of its spore forming nature *Cl. perfringens* survives probably longer extra-intestinal (in water for example) in comparison to *E. coli*, which is also often used as a hygiene indicator. It is mainly *Cl. pefringens* type A strains that produce enterotoxin (CPE- Clostridium perfringens enterotoxin) harmful to humans (McLane et al., 2012).

This pathogen is also frequently found in raw meat that can be contaminated with the feces from dirty animals during slaughter, slaughter/processing environment, equipment or kitchen staff during preparation. Foods involved in *Cl. perfringens* toxi-infection are often meat and meat products, thickened sauces (gravies, etc.) stews or soups with meat etc. Usual scenario involves preparation of the food in advance, inadequate cooling and storing until the next day and (insufficient) reheating for consumption. In the EU it showed that in 2022 *Clostridium perfringens* toxins was the causative agent involving the highest number of human cases in foodborne outbreaks, because many outbreaks related to mass catering (fairs or festivals) or occurred in schools or kindergarten (EFSA, 2013a, b).

Cl. perfringens is an established pathogen and similar as for *Staphylococcus aureus* (the second most frequently reported bacterial toxin-producing agent of foodborne outbreaks in the EU in 2022) needs attention for good hygienic practices and knowledge and adherence to the WHO five keys to safer foods (WHO, 2006) to prevent this type of food poisoning events.

4.4 Clostridium botulinum

In 1895 a sensational poisoning took place in Belgium. Some musicians of the village brass band died there after a joint funeral meal; others became seriously ill. From the food (raw cured smoked ham) and from the victims, the bacterium *Bacillus botulinus* (botulus, Latin for sausage) and later called *Clostridium botulinum* was isolated (Devriese, 2000). *Cl. botulinum* is an anaerobic spore forming Gram-positive food intoxicant. The bacterium is omnipresent in the nature and can be isolated from soil, water, bottom sediments of rivers, lakes and seawater. In addition, they are also found in the intestines of fish and mammals and the gills and viscera of crabs and other shellfish. *Cl. botulinum* can only grow under strict anaerobic conditions. This comprises both in-film packages and canned products, as well as large (pressed) product masses (e.g. ham, roasts) where the inner core of the products is anaerobic. The main reasons for the occurrence of growth of *Cl. botulinum* are insufficient heat treatment, leakage of the package, improperly adjusted intrinsic factors (mainly pH and nitrite concentration), and the long term storage above the minimal growth temperature. The incidence of the disease is extremely low, but it raises concerns due to high mortality rate.

Of the outbreaks that are reported annually most are related to preserved foods produced at home, e.g. "sterilized" at too low temperatures or during too short time. However, also commercial products are sometimes involved in an outbreak. Sausages, meat products, canned vegetables and seafood are the most common foods involved in human botulism (Uyttendaele, 2018). Especially fish packaged under modified atmosphere (and thus not kept on ice but stored at 4–7 °C) is perceived as a high risk product for non-proteolytic psychrotrophic *Cl. botulinum* type E. The type E prevalence in raw fish varied from 10% to 40%, depending on the fish species studied (Hyytia et al., 1998). It is therefore a usual practice, and advice to include small amount of oxygen in the package that would interrupt the anaerobic conditions and in that way inhibit growth and toxin production of *Cl. botulinum* (Uyttendaele, 2018). Nowadays more and more attention is being given to the vegetables-in-oil type of products, sometimes home-prepared and stored (Nummer et al., 2011) and some cheese products in which no fermentation has been used e.g. fresh Mascarpone cheese (an acid-set spreadable cheese) as possible risk carrier for botulism (Chaidoutis et al., 2022).

Some special attention should be given to the case of infant botulism. The only food item associated with infant botulism is honey. In honey samples associated with infant botulism cases, the quantity of *C. botulinum* spores may be as low as 5–70 spores/g, even though higher spore counts have also been reported. When spores of *Clostridium botulinum* are ingested by an infant less than 1 year of age, they may germinate in the intestinal lumen and produce botulinum neurotoxin The colonization is due to the immature infantile intestinal microflora although several other factors may also influence the susceptibility of an infant. Due to the risk for infant botulism, the health authorities in several countries have recommended that honey should not be given to infants less than 1 year of age (Nevas et al., 2005).

5 Foodborne Viruses

It is important to mention that next to bacteria, also viruses are taking up an important place in foodborne disease statistics. In 2009, they were responsible for 19% of all outbreaks in the EU causing over 1000 outbreaks and affecting more than 8700 citizens. The total number of outbreaks caused by viruses has been increasing since 2007 (EFSA, 2011). Foodborne viruses are characterized by a low infectious dose (10–100 virus particles). Viruses can't multiply on food, but are able to survive during the shelf life of the products. High amounts of virus particles ($>10^9$ particles/g feces) are excreted by infected persons, and this continues even after the disappearance of the symptoms emphasizing the need of good hygienic practices during food handling at all times, also by asymptomatic carriers.

The interest in the role of viruses as a cause of foodborne infections continues to rise mainly because (molecular based) detection methods have become available. At present, human norovirus was identified as the leading cause of viral foodborne illness, followed by hepatitis A and hepatitis E viruses. Whereas Norovirus is ranked highest concerning the frequency of illness, Hepatitis A virus and Hepatitis E virus ranks higher compared to norovirus in terms of clinical severity (FAO and WHO, 2024). Overall, as diagnostics improve, and these methods are increasingly applied in food monitoring and surveillance and foodborne outbreak investigation, more foodborne illnesses relating to enteric viruses are expected to be identified, especially in developing countries where there are little data today.

5.1 *Norovirus*

Noroviruses are genetically and antigenically diverse, being classified into 48 genetic clusters which fall within ten genogroups (Chhabra et al., 2020). Without envelopes, noroviruses are simply constructed of an exterior capsid protein and the interior viral genome, a positive-sense single-stranded RNA. The typical clinical manifestation of Norovirus infection include vomiting, non-bloody diarrhea, as well as low-grade fever, abdominal cramps, and headache after a 1 to 3 days incubation period. Norovirus infections are generally mild, transient, and self-limiting within 2–3 days. Many colonized individuals are even asymptomatic, although viral excretion may continue over days or weeks.

Norovirus is the leading cause of vomiting and diarrhea from acute gastroenteritis among people of all ages in the United States. Each year, there are about 2500 reported norovirus outbreaks and norovirus causes 58% of foodborne illnesses acquired in the United States occurring in a wide range of institutional settings, including schools, colleges, child care centers, cruise ships, etc. (CDC, 2024a). Over half of all norovirus outbreaks reported however occur in hospital or long-term health care facilities. Outbreaks in these settings can sometimes last months. Compared with healthy people, norovirus illnesses can be more severe—and occasionally even deadly—in patients in hospitals or long-term care facilities. Also in EU most viral outbreaks were associated with restaurants and catering services. In 2022 norovirus caused 332 outbreaks including 12 large outbreaks involving more than 100 cases in several EU Member States (EFSA/ECDC, 2023). Infected food workers are frequently the source of outbreaks in food-service settings, often by touching ready-to-eat foods (such as raw fruits and vegetables) before serving them. However, any food served raw or handled after being cooked can get contaminated with norovirus. It should be noted that repeated outbreaks may result from inappropriate clean-up and disinfection after vomiting or diarrhea. This is because norovirus can persist on surfaces for days or weeks and is resistant to many common disinfectants. This finding underlines the importance of strengthening hygiene standards and disseminating knowledge for food handlers (CDC, 2024a).

Norovirus outbreaks can also occur from food that is contaminated at the source or on the farm by human faecal material. This can include shellfish (e.g. oysters) harvested from contaminated water. Shellfish could bio-accumulate viral particles from a large quantity of water and are quite often consumed uncooked. While pre-harvest preventative interventions are preferable, post-harvest interventions such as depuration, relaying and thermal treatment have been used to mitigate the risk of norovirus infection associated with shellfish consumption (Gyawali et al., 2019). Vegetables (e.g. leafy greens) might be sprayed with contaminated water in the field or fruit (e.g. berries) manual harvested by Norovirus shedding food pickers or spraying contaminated pesticide-water (Jacxsens et al., 2017). Fresh produce is often globally sourced and generally consumed without effective treatment to get rid of the contaminated viruses. Accessible and understandable guidelines on *Good Agricultural Practices* and training programs to explain basic hygiene principles are to be promoted throughout the food supply chain (Nestec Ltd., 2016). Specific production-related intervention strategies should focus on water source, location, method and timing of application. Emerging treatments of water (e.g., ozone, photocatalysis, ultraviolet and ultrafiltration) demonstrate potential, but require infrastructure investment. Biochar filtration is a relatively inexpensive method that shows promise for treating reused water (FAO and WHO, 2024).

5.2 *Hepatitis A Virus*

HAV is classified as a Picornavirus, measures 27 nm and is non-enveloped. The virus contains four capsid proteins encompassing a positive polarity single-stranded RNA genome. There is only one HAV serotype and primates are the only natural host. Infection by Hepatitis A virus represents the most serious form of viral illness acquired from foods. After ingestion, uptake of Hepatitis A virus particles takes place in the gastrointestinal tract, and subsequent replication in the liver. Hepatitis A virus will be secreted in the bile, and finally high concentrations are found in stool specimens. It causes a potentially severe but controllable loss of liver function and general malaise. Proper medical care will generally result in full recovery of liver function and full clearance of virus from the host, with effective and lifelong immunity against reinfection. The mortality rate stands at about 0.02 per 100,000 of reported cases. If people above 65 years are considered than the mortality rate rises to 0.15 per 100,000 (CDC, 2024b). There are significant regional differences in the proportion of hepatitis A cases that are attributed to food due to endemic prevalence and vaccine utilization. In developed countries, with good sanitary infrastructures, the infection rate is low, and some of these countries have instituted an immunization schedule for children or recommend vaccination for adults traveling to endemic regions. However, there might be concern as in non-endemic countries one has seen seroprotection decline among adults as those who were exposed naturally to hepatitis A in the past are replaced by an adult population that is neither vaccinated nor previously exposed. In US, the number of hepatitis A cases in 2022 remained almost 2 times as high as in 2015. There is a need to identify also in non-endemic countries restaurant patrons and get them vaccinated when an HAV-infected food handler is discovered. Moreover, this type of approach for vaccination of persons in developed countries is increasingly being counterbalanced by the importation of foods from highly endemic areas of the world (Holmberg, 2012) .

Similarly with Norovirus, Hepatitis A virus can be transmitted person-to-person, or through contaminated water and food, particularly fruit or ready-to-eat foods that have not been sufficiently heated to inactivate the virus. International trade of foods plays an important role in transmission to susceptible populations and since some of these products have a long shelf-life especially if frozen or preserved, the potential is created for extended Hepatitis A outbreaks. A typical example was an outbreak in 2009 that was related to contaminated sun-dried tomatoes and affected individuals in several Australian states and further hepatitis A outbreaks linked to semidried tomatoes in France and the Netherlands. The outbreak was extensive (562 cases reported in Australia) and sustained for more than 12 months despite public health interventions, the design and implementation of which were complicated by limitations in food testing capability for Hepatitis A virus and complex supply chains (Donnan et al., 2012). Wider compliance with international standards, e.g. good agricultural and hygiene practices, is likely to reduce global transmission.

5.3 Hepatitis E Virus

Hepatitis E virus is currently classified in the family Hepeviridae. It is a positive-strand RNA virus with 7.2-kb genome. Although Hepatitis E virus is a non-enveloped virus in bile and feces, it circulates in the bloodstream wrapped in cellular membranes. Hepatitis E is a liver infection, usually self-limited but may persist and cause chronic hepatitis in immunocompromised patients. Currently, the virus is classified into eight genotypes, with G1–4 and genotype 7 (G7) capable of infecting humans. Genotype 3 (G3) HEV has been found to be the most geographically diverse of the viruses thus far and is the genotype which has emerged in the past 2 decades in many developed countries. It raised public health concerns for zoonosis and food safety (Treagus et al., 2021).

In contrast to Norovirus and hepatitis A virus which are linked to human reservoir, it is also possible for the Hepatitis E virus to be transmitted through foodstuffs, due to the ability for some HEV genotypes to infect non-human animals. The G3 HEV is known to be a zoonotic pathogen, with a broad host range. The primary source of HEV within more economically developed countries is considered to be pigs, and consumption of undercooked pork products is a significant risk factor and known transmission route for the virus to humans. In scientific literature HEV was detected frequently in both commercial livers at slaughterhouse (1.3–20.8%) or pork sausages or pork pâtés sampled in retail (11.7–68.9%) (Treagus et al., 2021). However, it should be considered that the HEV prevalence in these studies may be over- or underestimated by the fact that the sample sizes for some of these studies are relatively small and different methods have been used between studies (including RNA based detection protocols that might not be unambiguously linked to viability or infectivity of the HEV). Zoonotically transmitted hepatitis E virus can also enter the food chain by meat products derived from wild animals (wild boar, deer, etc).

Control measures can relate to preventing cross-contamination during slaughter and further processing, avoiding the use of high-risk tissues e.g. liver and blood in product formulations and virus inactivation by sufficient heat treatment. In meat, hepatitis E virus is highly resistant to heat. For example, it was reported that it took 20 minutes in a pâté-like product to obtain the similar inactivation as was observed for a relatively pure virus suspension treated at the same temperature (70–72 °C) for 2 minutes (FAO and WHO, 2024). Hepatitis E virus is definite an emerging pathogen for which further research is required to better understand how HEV is transmitted to humans through food. People with liver diseases or immune deficiencies and pregnant women should be discouraged from eating under-cooked meat and liver from wild boar and pork.

6 Opportunistic Pathogens

6.1 *Cronobacter sakazakii*

Before 1980, *Cronobacter sakazakii* was considered a variant of *Enterobacter cloacae* and was called 'yellow-pigmented *E. cloacae*' after the ability of this bacterium to form yellow-pigmented colonies on Nutrient Agar. In 1980, this bacterium was renamed *E. sakazakii*, which was eventually found to consist of five different species and it was proposed to classify these species into a new genus, *Cronobacter*. *C. sakazakii* is a Gram-negative rod and is classified in the family *Enterobacteriaceae*. *Cronobacter* strains exhibit α-glucosidase activity and this property is used to distinguish *Cronobacter* from *Enterobacter* spp. and other *Enterobacteriaceae* on chromogenic culture media. It is an opportunistic pathogen that mainly causes life-threatening infections (mortality rate 50–80%) in immunocompromised newborns, such as necrotizing enterocolitis, bloodstream infections and meningitis. However, infections have recently been reported in immunocompromised adults as well (Healy et al., 2010).

Cronobacter spp. is ubiquitous in the environment and can be isolated from a wide range of foodstuffs such as milk, cheese, dry food, raw meat, vegetables, rice, bread, etc. However, several *Cronobacter* foodborne infections in newborns have been associated with contaminated infant formula (Henry & Fouladkhah, 2019).

This typically involves post-contamination from the production environment or from ingredients added after pasteurization. *C. sakazakii* is more resistant to drying than other *Enterobacteriaceae* and can therefore survive well in milk powder (or its production environment). In addition, they might multiply well after rehydration of the milk powder and prolonged storage at elevated temperatures.

FAO (2008) developed a code of hygienic practice for powdered formulae for infants and young children. *C. sakazakii* is a hazard in Powdered Infant Formula (PIF) products that is hard to detect due to localized and low-level contamination and thus testing may not be the most useful control measure or means of verification of good practices. Still, if used it is important to reflect on the most appropriate or relevant sampling plan (Kim et al., 2023). Moreover, specific recommendations on the proper use of powdered infant formula in neonatal intensive care units or for mothers on the preparation of infant formula for their newborn babies are of interest (Kalyantanda et al., 2015).

6.2 Other Opportunistic Pathogens

Many other opportunistic pathogens naturally occur in foodstuffs. Table 7 provides some examples of clinically important opportunistic pathogens and their taxonomic status, morphology, associated infections and associated foodstuffs. These microorganisms are ubiquitous in the environment, naturally occurring in air, water, plants and soil. From these reservoirs they can enter the food chain, where they are mainly known as typical spoilage organisms. In addition, a number of these organisms (such as *Enterococcus, Klebsiella, Enterobacter, Pantoea, Pseudomonas, Aeromonas* etc.) are also part of the normal intestinal microbiota of humans, animals, plants or water where they behave as commensals.

As opportunistic pathogens, these microorganisms are primarily associated with hospital infections. It remains unclear whether and to what extent they are transmitted via food and are therefore relevant as foodborne pathogens. The reported infections are typically wound infections or hospital-acquired infections transmitted by e.g. infected catheters (Mirtella et al., 2021; Scales et al., 2014). The natural presence of these opportunistic pathogens in high concentrations on various foodstuffs, for example fresh vegetables, without any associated foodborne outbreak, suggests they pose a low food safety risk for the general population. Consumers should therefore not be overly concerned by the scientific description of many bacterial species as potential human pathogens. Similar to the lack of a clear definition of healthy or unhealthy food, there is a grey area regarding which germs are primarily pathogenic or opportunistic pathogenic. The key is to focus on epidemiological evidence: have foodborne infections been reported (repeatedly) in individuals or groups of people? Therefore, monitoring and thoroughly reporting foodborne infections and foodborne outbreaks remain vital sources of information.

Mainly for (highly) immunocompromised individuals, such as patients who use immunosuppressive drugs for organ transplants, chemotherapy, etc., extra attention

Table 7 Non-exhaustive list of examples of clinically relevant opportunistic pathogens

	Family	Morfology	Associated (clinical) infections	Associated foodstuffs as indigenous microbiota or spoilage organisms
Enterococcus spp. (*E. faecalis, E. faecium*)	*Enterococcaceae*	G^+ cocci	Hospital-acquired infections, mainly related to intensive-care: Bloodstream infections, endocarditis, urinary tract infections, etc.	Raw meat, (fermented) meat products, raw milk, cheese
Burkholderia cepacia complex (Bcc)	*Burkholderiaceae*	G^- rods	Life-threatening lung infections, especially in patients with underlying lung disease such as cystic fibrosis	Mainly cereals, also in fresh (leafy) vegetables. Used as biological control to suppress soil-borne plant pathogens.
Stenotrophomonas maltophilia	*Xanthomonadaceae*	G^- rods	Hospital-acquired infections such as bloodstream infections, endocarditis, respiratory infections, etc.	Raw vegetables, mainly Brassicaceae
Aeromonas spp. (*A. hydrophila, A. caviae, A. sobria*)	*Aeromonadaceae*	G^- rods	Gastroenteritis, bloodstream infections	Drinking water, fish, shellfish, raw meat, fresh vegetables, raw milk
Pseudomonas aeruginosa	*Pseudomonadaceae*	G^- rods	Mainly respiratory infections, but also hospital-acquired infections in the bloodstream, urinary tract, etc.	Water, fresh (leafy) vegetables
Klebsiella pneumoniae, Enterobacter cloacae, Serratia marcescens, Citrobacter freundii, …	*Enterobacteriaceae*	G^- rods	Severe pneumonia, bloodstream infections, urinary tract infections, postoperative wound infections, etc.	Fresh (leafy) vegetables

Table with information compiled from the following references: Berg et al. (2005), Berg et al. (2014), Giraffa (2002), Moradali et al. (2017) and Neyts et al. (2000)

should be paid to opportunistic pathogens. Healthcare providers and especially the dieticians involved typically recommend a neutropenic diet, as well as extra attention to respecting certain time/temperature combinations to reduce the number of germs in food and to prevent (re)contamination and further growth (De Bock et al., 2023). Healthcare providers and dieticians often give such vulnerable groups the simple rule 'cook it, peel it or forget it'. However, for the majority of the population, opportunistic pathogens cannot cause disease and are most relevant as spoilage organisms, possibly causing taste or odour deviations in a food product. Most microorganisms are benign and are an inherent part of humans, animals or plants and the wider environment, and adhering to good hygiene practices is typically sufficient to ensure safe food.

7 Conclusion

As we learn more and more about foodborne pathogens, they come under control. But more research activities into communities of microbes that are essential in plant, animal, human and environmental health may come with impact in terms of food quality and safety (e.g. refer to https://www.microbiomesupport.eu/resources/). Better insights in the mechanisms of pathogenicity, improved diagnostics and monitoring may lead to recognition of new pathogens, new transmission routes or better reporting of large, dispersed outbreaks identified through whole genome sequencing. There are a number of factors that can contribute to emergence of re-emergence of foodborne pathogens. Pathogens can emerge due to factors like changing consumer preferences such as the trend towards more sustainable plant-based foods including the introduction of alternative protein sources and associated changes in microbial ecology. But also introduction of alternative food production systems such as increased attention for a circular food economy seeking to reduce waste and supplant traditional linear supply chains, or higher interest in unprocessed foods or low or unregulated production and distribution routes in the food supply chain leads to new microbial challenges. Globalization, and climate change are also identified as factors that increase the risk of pathogen contamination in food systems. Moreover, the decrease in food literacy may impact safe eating habits. The raise in immunocompromised groups within our population may also result in re-emergence of foodborne pathogens. Combating foodborne disease and foodborne pathogens is a never-ending story and will need our continuous attention. Encouraging collaboration between microbiologists, epidemiologists, molecular biotechnology, food technology but also interdisciplinary food system thinking can lead to more comprehensive research and innovative solutions. But also increasing public awareness about food safety practices, educational campaigns or dialogue can play a crucial role to inform stakeholders in the farm to fork continuum about the dynamics of food safety and (re)-emerging pathogens.

References

ANSES. (2008). Fiche de description de danger biologique transmissible par les aliments/*Yersinia enterocolitica, Yersinia pseudotuberculosis*. (https://www.anses.fr/fr/system/files/BIORISK2016SA0266Fi.pdf)

Berg, G., Eberl, L., & Hartmann, A. (2005). The rhizosphere as a reservoir for opportunistic human pathogenic bacteria. *Environmental Microbiology, 7*(11), 1673–1685. https://doi.org/10.1111/j.1462-2920.2005.00891.x

Berg, G., Erlacher, A., Smalla, K., & Krause, R. (2014). Vegetable microbiomes: Is there a connection among opportunistic infections, human health and our 'gut feeling'? *Microbial Biotechnology, 7*(6), 487–495. https://doi.org/10.1111/1751-7915.12159

Boerlin, P. (1999). Evolution of virulence factors in Shiga-toxin-producing *Escherichia coli*. *Cellular and Molecular Life Sciences, 56*(9–10), 735–741. https://doi.org/10.1007/s000180050020

Boysen, L., Rosenquist, H., Larsson, J. T., Nielsen, E. M., Sørensen, G., Nordentoft, S., & Hald, T. (2014). Source attribution of human campylobacteriosis in Denmark. *Epidemiology and Infection, 142*(8), 1599–1608. https://doi.org/10.1017/s0950268813002719

Brands, D. A. (2006). Chapter 2 *Salmonella* and food-borne illness. In *Salmonella*. Chelsea House Publishers. ISBN 0-7910-8500-7.

Brauge, T., Mougin, J., Ells, T., & Midelet, G. (2024). Sources and contamination routes of seafood with human pathogenic *vibrio* spp.: A farm-to-fork approach. *Comprehensive Reviews in Food Science and Food Safety, 23*(1), 1–25. https://doi.org/10.1111/1541-4337.13283

Burtscher, J., Etter, D., Biggel, M., Schlaepfer, J., & Johler, S. (2021). Further insights into the toxicity of Bacillus cytotoxicus based on toxin gene profiling and vero cell cytotoxicity assays. *Toxins, 13*, 234. https://doi.org/10.3390/toxins13040234

Caprioli, A., Morabito, S., Brugere, H., & Oswald, E. (2005). Enterohaemorrhagic *Escherichia coli*: Emerging issues on virulence and modes of transmission. *Veterinary Research, 36*(3), 289–311. https://doi.org/10.1051/vetres:2005002

Casadevall, A. (2017). The pathogenic potential of a microbe. *Msphere, 2*(1), e00015-17. https://doi.org/10.1128/mSphere.00015-17

Casadevall, A., & Pirofski, L. A. (2002). What is a pathogen? *Annals of Medicine, 34*(1), 2–4. https://doi.org/10.1080/078538902317338580

Casadevall, A., & Pirofski, L. A. (2018). What is a host? Attributes of individual susceptibility. *Infection and Immunity, 86*(2). https://doi.org/10.1128/IAI.00636-17

CDC. (2024a). *Norovirus outbreaks*. https://www.cdc.gov/norovirus/outbreak-basics/index.html

CDC. (2024b). *Hepatitis A basics*. https://www.cdc.gov/hepatitis-a/about/index.html

Ceuppens, S., Boon, N., & Uyttendaele, M. (2013). Diversity of *Bacillus cereus* group strains is reflected in their broad range of pathogenicity and diverse ecological lifestyles. *FEMS Microbiology Ecology, 84*(3), 433–450. https://doi.org/10.1111/1574-6941.12110

Ceuppens, S., Van Boxstael, S., Westyn, A., Devlieghere, F., & Uyttendaele, M. (2016). The heterogeneity in the type of shelf life label and storage instructions on refrigerated foods in supermarkets in Belgium and illustration of its impact on assessing the *listeria monocytogenes* threshold level of 100 CFU/g. *Food Control, 59*, 377–385. https://doi.org/10.1016/j.foodcont.2015.06.009

Chaidoutis, E., Keramydas, D., Papalexis, P., Migdanis, A., Migdanis, I., Lazaris, A. C., & Kavantzas, N. (2022). Foodborne botulism: A brief review of cases transmitted by cheese products (review). *Biomedical Reports, 16*(5), 41. https://doi.org/10.3892/br.2022.1524

Chhabra, P., de Graaf, M., Parra, G. I., Chan, M. C., Green, K., Martella, V., Wang, Q., White, P. A., Katayama, K., Vennema, H., Koopmans, M. P. G., & Vinje, J. (2020). Corrigendum: Updated classification of norovirus genogroups and genotypes. *The Journal of General Virology, 101*(8), 893. https://doi.org/10.1099/jgv.0.001475

Clavel, T., Carlin, F., Lairon, D., Nguyen-The, C., & Schmitt, P. (2004). Survival of *Bacillus cereus* spores and vegetative cells in acid media simulating human stomach. *Journal of Applied Microbiology, 97*(1), 214–219. https://doi.org/10.1111/j.1365-2672.2004.02292.x

Conrad, A. R., Tubach, S., Cantu, V., Webb, L. M., Stroika, S., Moris, S., Davis, M., Hunt, D. C., Bradley, K. K., Kucerova, Z., Strain, E., Doyle, M., Fields, A., Neil, K. P., Gould, L. H., Jackson, K. A., Wise, M. E., Griffin, P. M., & Jackson, B. R. (2023). *Listeria monocytogenes* illness and deaths associated with ongoing contamination of a multiregional brand of ice cream products, United States, 2010–2015. *Clinical Infectious Diseases, 76*(1), 89–95. https://doi.org/10.1093/cid/ciac550

Daelman, J., Jacxsens, L., Lahou, E., Devlieghere, F., & Uyttendaele, M. (2013). Assessment of the microbial safety and quality of cooked chilled foods and their production process. *International Journal of Food Microbiology, 160*(3), 193–200. https://doi.org/10.1016/j.ijfoodmicro.2012.10.010

De Bock, T., Zhao, X., Jacxsens, L., Devlieghere, F., Rajkovic, A., Spanoghe, P., Höfte, M., & Uyttendaele, M. (2021). *Evaluation of B. thuringiensis-based biopesticides in the primary production of fresh produce as a food safety hazard and risk* (Vol. 130, p. 108390). Food Control. https://doi.org/10.1016/j.foodcont.2021.108390

De Bock, T., Jacxsens, L., Maes, F., Van Meerhaeghe, S., Reygaerts, M., & Uyttendaele, M. (2023). Microbiological profiling and knowledge of food preservation technology to support guidance on a neutropenic diet for immunocompromised patients. *Frontiers in Microbiology, 14*, 1136887. https://doi.org/10.3389/fmicb.2023.1136887

Devriese, P. P. (2000). De Gentse hoogleraar Emile van Ermengem en de ontdekking van botuline. *Ghendtsche Tydinghen, 29*(1). https://doi.org/10.21825/gt.v29i1.6237

Donnan, E. J., Fielding, J. E., Gregory, J. E., Lalor, K., Rowe, S., Goldsmith, P., Antoniou, M., Fullerton, K. E., Knope, K., Copland, J. G., Bowden, D. S., Tracy, S. L., Hogg, G. G., Tan, A., Adamopoulos, J., Gaston, J., & Vally, H. (2012). A multistate outbreak of hepatitis A associated with semidried tomatoes in Australia, 2009. *Clinical Infectious Diseases, 54*(6), 775–781. https://doi.org/10.1093/cid/cir949

EFSA. (2007). Monitoring and identification of human enteropathogenic *Yersinia* spp. – scientific opinion of the panel on biological hazards. *EFSA Journal, 5*(12), 595. https://doi.org/10.2903/j.efsa.2007.595

EFSA. (2008). Overview of methods for source attribution for human illness from food-borne microbiological hazards – scientific opinion of the panel on biological hazards. *EFSA Journal, 6*(7), 764. https://doi.org/10.2903/j.efsa.2008.764

EFSA. (2009). Technical specifications for the monitoring and reporting of verotoxigenic *Escherichia coli* (VTEC) on animals and food (VTEC surveys on animals and food). *EFSA Journal, 7*(11), 1366. https://doi.org/10.2903/j.efsa.2009.1366

EFSA. (2010). Scientific opinion on quantification of the risk posed by broiler meat to human campylobacteriosis in the EU. *EFSA Journal, 8*(1), 1437. https://doi.org/10.2903/j.efsa.2010.1437

EFSA. (2011). *EFSA provides up-to-date information on food-borne viruses*. https://www.efsa.europa.eu/en/news/efsa-provides-date-information-food-borne-viruses. Accessed 14 July 2011.

EFSA. (2013a). *Manual for reporting of food-borne outbreaks in accordance with directive 2003/99/EC from the year 2012*. Supporting publication 2013:EN-410 [49 pp]. https://doi.org/10.2903/sp.efsa.2013.en-410

EFSA. (2013b). Analysis of the baseline survey on the prevalence of *Listeria monocytogenes* in certain ready-to-eat foods in the EU, 2010-2011 part A: *Listeria monocytogenes* prevalence estimates. *EFSA Journal, 11*(6), 3241. https://doi.org/10.2903/j.efsa.2013.3241

EFSA. (2016). Risks for public health related to the presence of *Bacillus cereus* and other *Bacillus* spp. including *Bacillus thuringiensis* in foodstuffs. *EFSA Journal, 14*(7), e04524. https://doi.org/10.2903/j.efsa.2016.4524

EFSA. (2020a). Pathogenicity assessment of Shiga toxin-producing *Escherichia coli* (STEC) and the public health risk posed by contamination of food with STEC. *EFSA Journal, 18*(1), e05967. https://doi.org/10.2903/j.efsa.2020.5967

EFSA. (2020b). Update and review of control options for *Campylobacter* in broilers at primary production. *EFSA Journal, 18*(4). https://doi.org/10.2903/j.efsa.2020.6090

EFSA. (2024). Public health aspects of *Vibrio* spp. related to the consumption of seafood in the EU. *EFSA Journal, 22*(7). https://doi.org/10.2903/j.efsa.2024.8896

EFSA and ECDC. (2023). The European Union one health 2022 Zoonoses report. *EFSA Journal, 21*(12), p211202. https://doi.org/10.2903/j.efsa.2023.8442

Etesami, H., Jeong, B. R., & Glick, B. R. (2023). Potential use of *Bacillus* spp. as an effective biostimulant against abiotic stresses in crops—a review. *Current Research in Biotechnology, 5*, 100128. https://doi.org/10.1016/j.crbiot.2023.100128

FAO. (2008). *Code of hygienic practice for powdered formulae for infants and young children*. https://www.fao.org/input/download/standards/11026/CXP_066e.pdf

FAO and WHO. (2021a). *Microbiological risk assessment – Guidance for food*. Guidance. Microbiological Risk Assessment Series No. 36. Rome. https://doi.org/10.4060/cb5006en

FAO and WHO. (2021b). Advances in science and risk assessment tools for *Vibrio parahaemolyticus* and *V. vulnificus* associated with seafood: Meeting report. Microbiological risk assessment series N°35. Rome. https://iris.who.int/handle/10665/343548

FAO and WHO. (2024). *Summary report of the joint FAO/WHO expert meeting on the microbiological risk assessment of viruses in foods part 2: prevention and intervention measures*. https://www.who.int/publications/m/item/jemra-of-viruses-in-foods-part-2-prevention-and-intervention-measures

Feaugas, T., & Sauvonnet, N. (2021). Organ-on-chip to investigate host-pathogens interactions. *Cellular Microbiology, 23*(7), e13336. https://doi.org/10.1111/cmi.13336

Fredriksson-Ahomaa, M., & Korkeala, H. (2003). Low occurrence of pathogenic *Yersinia enterocolitica* in clinical, food, and environmental samples: A methodological problem. *Clinical Microbiology Reviews, 16*(2), 220–229. https://doi.org/10.1128/CMR.16.2.220-229.2003

Giraffa, G. (2002). Enterococci from foods. *FEMS Microbiology Reviews, 26*(2), 163–171. https://doi.org/10.1111/j.1574-6976.2002.tb00608.x

Goulet, V., Hebert, M., Hedberg, C., Laurent, E., Vaillant, V., De Valk, H., & Desenclos, J. C. (2012). Incidence of listeriosis and related mortality among groups at risk of acquiring listeriosis. *Clinical Infectious Diseases, 54*(5), 652–660. https://doi.org/10.1093/cid/cir902

Granum, P. E., & Lund, T. (1997). *Bacillus cereus* and its food poisoning toxins. *FEMS Microbiology Letters, 157*(2), 223–228. https://doi.org/10.1111/j.1574-6968.1997.tb12776.x

Guinebretiere, M. H., Thompson, F. L., Sorokin, A., Normand, P., Dawyndt, P., Ehling-Schulz, M., Svensson, B., Sanchis, V., Nguyen-The, C., Heyndrickx, M., & De Vos, P. (2008). Ecological diversification in the *Bacillus cereus* group. *Environmental Microbiology, 10*(4), 851–865. https://doi.org/10.1111/j.1462-2920.2007.01495.x

Guinebretiere, M. H., Velge, P., Couvert, O., Carlin, F., Debuyser, M. L., & Nguyen-The, C. (2010). Ability of *Bacillus cereus* group strains to cause food poisoning varies according to phylogenetic affiliation (groups I to VII) rather than species affiliation. *Journal of Clinical Microbiology, 48*(9), 3388–3391. https://doi.org/10.1128/JCM.00921-10

Gyawali, P., Fletcher, G. C., McCoubrey, D.-J., & Hewitt, J. (2019). Norovirus in shellfish: An overview of post-harvest treatments and their challenges. *Food Control, 99*, 171–179. https://doi.org/10.1016/j.foodcont.2018.12.049

Hanning, I. B., Nutt, J. D., & Ricke, S. C. (2009). Salmonellosis outbreaks in the United States due to fresh produce: Sources and potential intervention measures. *Foodborne Pathogens and Disease, 6*(6), 635–648. https://doi.org/10.1089/fpd.2008.0232

Harrison, L. M., Balan, K. V., Hiett, K. L., & Babu, U. S. (2022). Current methodologies and future direction of *Campylobacter* isolation and detection from food matrices, clinical samples, and the agricultural environment. *Journal of Microbiological Methods, 201*, 106562. https://doi.org/10.1016/j.mimet.2022.106562

Healy, B., Cooney, S., O'Brien, S., Iversen, C., Whyte, P., Nally, J., Callanan, J. J., & Fanning, S. (2010). Cronobacter (Enterobacter sakazakii): An opportunistic foodborne pathogen. *Foodborne Pathogens and Disease, 7*(4), 339–350. https://doi.org/10.1089/fpd.2009.0379

Hennekinne, J. A., De Buyser, M. L., & Dragacci, S. (2012). *Staphylococcus aureus* and its food poisoning toxins: Characterization and outbreak investigation. *FEMS Microbiology Reviews, 36*(4), 815–836. https://doi.org/10.1111/j.1574-6976.2011.00311.x

Henry, M., & Fouladkhah, A. (2019). Outbreak history, biofilm formation, and preventive measures for control of *Cronobacter sakazakii* in infant formula and infant care settings. *Microorganisms, 7*(3), 77. https://doi.org/10.3390/microorganisms7030077

Holmberg, S. D. (2012). Hepatitis a epidemiology goes global. *Clinical Infectious Diseases, 54*(6), 782–783. https://doi.org/10.1093/cid/cir945

HPA. (2009). *Guidelines for assessing the microbiological safety of ready-to-eat foods placed on the market*. Available at: https://assets.publishing.service.gov.uk/government/uploads/system/uploads/attachment_data/file/363146/Guidelines_for_assessing_the_microbiological_safety_of_ready-to-eat_foods_on_the_market.pdf. Accessed 31 Mar 2022

Hyytia, E., Hielm, S., & Korkeala, H. (1998). Prevalence of clostridium botulinum type E in Finnish fish and fishery products. *Epidemiology and Infection, 120*(3), 245–250. https://doi.org/10.1017/s0950268898008693

Jaaskelainen, E. L., Teplova, V., Andersson, M. A., Andersson, L. C., Tammela, P., Andersson, M. C., Pirhonen, T. I., Saris, N. E., Vuorela, P., & Salkinoja-Salonen, M. S. (2003). In *vitro* assay for human toxicity of cereulide, the emetic mitochondrial toxin produced by food poisoning *Bacillus cereus*. *Toxicology In Vitro, 17*(5–6), 737–744. https://doi.org/10.1016/s0887-2333(03)00096-1

Jacxsens, L., Stals, A., De Keuckelaere, A., Deliens, B., Rajkovic, A., & Uyttendaele, M. (2017). Quantitative farm-to-fork human norovirus exposure assessment of individually quick frozen raspberries and raspberry puree. *International Journal of Food Microbiology, 242*, 87–97. https://doi.org/10.1016/j.ijfoodmicro.2016.11.019

Kalyantanda, G., Shumyak, L., & Archibald, L. K. (2015). *Cronobacter* species contamination of powdered infant formula and the implications for neonatal health. *Frontiers in Pediatrics, 3*, 56. https://doi.org/10.3389/fped.2015.00056

Karlsson, P. A., Tano, E., Jernberg, C., Hickman, R. A., Guy, L., Jarhult, J. D., & Wang, H. (2021). Molecular characterization of multidrug-resistant *Yersinia enterocolitica* from foodborne outbreaks in Sweden. *Frontiers in Microbiology, 12*, 664665. https://doi.org/10.3389/fmicb.2021.664665

Kim, M., Reyes, G. A., Cheng, X., & Stasiewicz, M. J. (2023). Simulation evaluation of power of sampling plans to detect Cronobacter in powdered infant formula production. *Journal of Food Protection, 86*(8), 100115. https://doi.org/10.1016/j.jfp.2023.100115

Kuhn, K. G., Nygård, K. M., Guzman-Herrador, B., Sunde, L. S., Rimhanen-Finne, R., Trönnberg, L., Jepsen, M. R., Ruuhela, R., Wong, W. K., & Ethelberg, S. (2020). *Campylobacter* infections expected to increase due to climate change in Northern Europe. *Scientific Reports, 10*(1). https://doi.org/10.1038/s41598-020-70593-y

Kumari, S., & Sarkar, P. K. (2016). *Bacillus cereus* hazard and control in industrial dairy processing environment. *Food Control, 69*, 20–29. https://doi.org/10.1016/j.foodcont.2016.04.012

Lake, R. J., Campbell, D. M., Hathaway, S. C., Ashmore, E., Cressey, P. J., Horn, B. J., Pirikahu, S., Sherwood, J. M., Baker, M. G., Shoemack, P., Benschop, J., Marshall, J. C., Midwinter, A. C., Wilkinson, D. A., & French, N. P. (2021). Source attributed case-control study of campylobacteriosis in New Zealand. *International Journal of Infectious Diseases, 103*, 268–277. https://doi.org/10.1016/j.ijid.2020.11.167

Lechner, S., Mayr, R., Francis, K. P., Pruss, B. M., Kaplan, T., Wiessner-Gunkel, E., Stewart, G. S., & Scherer, S. (1998). *Bacillus weihenstephanensis* sp. *nov.* is a new psychrotolerant species of the *Bacillus cereus* group. *International Journal of Systematic Bacteriology, 48 Pt 4*, 1373–1382. https://doi.org/10.1099/00207713-48-4-1373

Leclercq, A., Tourdjman, M., Mattheus, W., Friesema, I., van Sorge, N. M., Halbedel, S., Wilking, H., & Lecuit, M. (2024). Outbreak of listeriosis associated with consumption of vegan cheese. *The New England Journal of Medicine, 390*(15), 1439–1440. https://doi.org/10.1056/NEJMc2400665

Liu, Y., Lai, Q., & Shao, Z. (2018). Genome analysis-based reclassification of *Bacillus weihenstephanensis* as a later heterotypic synonym of *Bacillus mycoides*. *International Journal of Systematic and Evolutionary Microbiology, 68*, 106–112. https://doi.org/10.1099/ijsem.0.002466

Lopez-Jimenez, A. T., & Mostowy, S. (2021). Emerging technologies and infection models in cellular microbiology. *Nature Communications, 12*(1), 6764. https://doi.org/10.1038/s41467-021-26641-w

Mawer, S. L. (1988). The pathogenicity of environmental *Campylobacters*--a human volunteer experiment. *Epidemiology and Infection, 101*(2), 295–300. https://doi.org/10.1017/s0950268800054212

McLane, B. A., Robertson, S. L., & Li, J. (2012). Chapter 18. *Clostridium perfringens*. In M. P. Doyle & R. L. Buchanan (Eds.), *Food microbiology: Fundamentals and frontiers* (4th ed., pp. 465–489). ASM Press. https://doi.org/10.1128/9781555818463.ch18. ISBN:9781119739135.

McLauchlin, J., Mitchell, R. T., Smerdon, W. J., & Jewell, K. (2004). *Listeria monocytogenes* and listeriosis: A review of hazard characterisation for use in microbiological risk assessment of foods. *International Journal of Food Microbiology, 92*(1), 15–33. https://doi.org/10.1016/S0168-1605(03)00326-X

Methot, P. O., & Alizon, S. (2014). What is a pathogen? Toward a process view of host-parasite interactions. *Virulence, 5*(8), 775–785. https://doi.org/10.4161/21505594.2014.960726

Miller, R. A., Beno, S. M., Kent, D. J., Carroll, L. M., Martin, N. H., Boor, K. J., & Kovac, J. (2016). *Bacillus wiedmannii* sp. nov., a psychrotolerant and cytotoxic *Bacillus cereus* group species isolated from dairy foods and dairy environments. *International Journal of Systematic and Evolutionary Microbiology, 66*, 4744–4753. https://doi.org/10.1099/ijsem.0.001421

Mirtella, D., Fedeli, P., Scendoni, R., Cannovo, N., & Cingolani, M. (2021). A case of nosocomial outbreak of pantoea agglomerans related to parenteral nutrition procedures. *Healthcare (Basel), 9*(6). https://doi.org/10.3390/healthcare9060684

Moradali, M. F., Ghods, S., & Rehm, B. H. (2017). *Pseudomonas aeruginosa* lifestyle: A paradigm for adaptation, survival, and persistence. *Frontiers in Cellular and Infection Microbiology, 7*, 39. https://doi.org/10.3389/fcimb.2017.00039

Morasi, R. M., Rall, V. L. M., Dantas, S. T. A., Alonso, V. P. P., & Silva, N. C. C. (2022). *Salmonella* spp. in low water activity food: Occurrence, survival mechanisms, and thermoresistance. *Journal of Food Science, 87*(6), 2310–2323. https://doi.org/10.1111/1750-3841.16152

Nestec Ltd. (2016). *Booklet 'Minimizing the risk of microbial contamination in primary production of berries.'* https://www.nestle.com/sites/default/files/asset-library/documents/library/documents/suppliers/best-practices-suppliers-minimizing-risk-microbial-contamiation-berries-2016.pdf

Nevas, M., Lindström, M., Hautamäki, K., Puoskari, S., & Korkeala, H. (2005). Prevalence and diversity of *Clostridium botulinum* types A, B, E and F in honey produced in the Nordic countries. *International Journal of Food Microbiology, 105*(2), 145–151. https://doi.org/10.1016/j.ijfoodmicro.2005.04.007

Neyts, K., Huys, G., Uyttendaele, M., Swings, J., & Debevere, J. (2000). Incidence and identification of mesophilic *Aeromonas* spp. from retail foods. *Letters in Applied Microbiology, 31*(5), 359–363. https://doi.org/10.1046/j.1472-765x.2000.00828.x

Nichols, G. L. (2005). Fly transmission of *Campylobacter*. *Emerging Infectious Diseases, 11*(3), 361–364. https://doi.org/10.3201/eid1103.040460

Nummer, B. A., Schaffner, D. W., Fraser, A. M., & Andress, E. L. (2011). Current food safety issues of home-prepared vegetables and herbs stored in oil. *Food Protection Trends, 31*(6), 336–342. Jun-11-Nummer.pdf

Palmer, A. D., & Slauch, J. M. (2017). Mechanisms of *Salmonella* pathogenesis in animal models. *Human and Ecological Risk Assessment, 23*(8), 1877–1892. https://doi.org/10.1080/10807039.2017.1353903

Pillsbury, A., Chiew, M., Bates, J., & Sheppeard, V. (2013). An outbreak of staphylococcal food poisoning in a commercially catered buffet. *Communicable Diseases Intelligence Quarterly Report, 37*(2), E144–E148.

Pope, J. E., Krizova, A., Garg, A. X., Thiessen-Philbrook, H., & Ouimet, J. M. (2007). *Campylobacter* reactive arthritis: A systematic review. *Seminars in Arthritis and Rheumatism, 37*(1), 48–55. https://doi.org/10.1016/j.semarthrit.2006.12.006

Public Health England. (2020). Investigation into an outbreak of *Listeria monocytogenes* infections associated with hospital-provided pre-prepared sandwiches. https://assets.publishing.service.gov.uk/media/5fbd0f898fa8f559e9ea363e/2019-05-Listeria-CC8-Outbreak-Report.pdf

Roussel, C., De Paepe, K., Galia, W., De Bodt, J., Chalancon, S., Leriche, F., Ballet, N., Denis, S., Alric, M., Van de Wiele, T., & Blanquet-Diot, S. (2020). Spatial and temporal modulation of enterotoxigenic *E. coli* H10407 pathogenesis and interplay with microbiota in human gut models. *BMC Biology, 18*(1), 141. https://doi.org/10.1186/s12915-020-00860-x

Rushton, S. P., Sanderson, R. A., Diggle, P. J., Shirley, M. D. F., Blain, A. P., Lake, I., Maas, J. A., Reid, W. D. K., Hardstaff, J., Williams, N., Jones, N. R., Rigby, D., Strachan, N. J. C., Forbes, K. J., Hunter, P. R., Humphrey, T. J., & O'Brien, S. J. (2019). Climate, human behaviour or environment: Individual-based modelling of *campylobacter* seasonality and strategies to reduce disease burden. *Journal of Translational Medicine, 17*(1), 34. https://doi.org/10.1186/s12967-019-1781-y

Sampers, I., Habib, I., De Zutter, L., Dumoulin, A., & Uyttendaele, M. (2010). Survival of *campylobacter* spp. in poultry meat preparations subjected to freezing, refrigeration, minor salt concentration, and heat treatment. *International Journal of Food Microbiology, 137*(2–3), 147–153. https://doi.org/10.1016/j.ijfoodmicro.2009.11.013

Sampers, I., Berkvens, D., Jacxsens, L., Ciocci, M.-C., Dumoulin, A., & Uyttendaele, M. (2012). Survey of Belgian consumption patterns and consumer behaviour of poultry meat to provide insight in risk factors for campylobacteriosis. *Food Control, 26*(2), 293–299. https://doi.org/10.1016/j.foodcont.2012.01.054

Scales, B. S., Dickson, R. P., LiPuma, J. J., & Huffnagle, G. B. (2014). Microbiology, genomics, and clinical significance of the Pseudomonas fluorescens species complex, an unappreciated colonizer of humans. *Clinical Microbiology Reviews, 27*(4), 927–948. https://doi.org/10.1128/CMR.00044-14

Schoeni, J. L., & Wong, A. C. (2005). *Bacillus cereus* food poisoning and its toxins. *Journal of Food Protection, 68*(3), 636–648. https://doi.org/10.4315/0362-028x-68.3.636

Sharapov, U. M., Wendel, A. M., Davis, J. P., Keene, W. E., Farrar, J., Sodha, S., Hyytia-Trees, E., Leeper, M., Gerner-Smidt, P., Griffin, P. M., Braden, C., & T. (2016). Multistate outbreak of *Escherichia coli* O157:H7 infections associated with consumption of fresh spinach: United States, 2006. *Journal of Food Protection, 79*(12), 2024–2030. https://doi.org/10.4315/0362-028X.JFP-15-556

Smith, A. M., Tau, N. P., Smouse, S. L., Allam, M., Ismail, A., Ramalwa, N. R., Disenyeng, B., Ngomane, M., & Thomas, J. (2019). Outbreak of *listeria monocytogenes* in South Africa, 2017–2018: Laboratory activities and experiences associated with whole-genome sequencing analysis of isolates. *Foodborne Pathogens and Disease, 16*(7), 524–530. https://doi.org/10.1089/fpd.2018.2586

Spagnoli, P., Jacxsens, L., & Vlerick, P. (2023). Towards a food safety culture improvement roadmap: Diagnosis and gap analysis through a conceptual framework as the first steps. *Food Control, 145*, 109398. https://doi.org/10.1016/j.foodcont.2022.109398

Stenfors Arnesen, L. P., Fagerlund, A., & Granum, P. E. (2008). From soil to gut: *Bacillus cereus* and its food poisoning toxins. *FEMS Microbiology Reviews, 32*(4), 579–606. https://doi.org/10.1111/j.1574-6976.2008.00112.x

Stevens, M. P., Humphrey, T. J., & Maskell, D. J. (2009). Molecular insights into farm animal and zoonotic *Salmonella* infections. *Philosophical Transactions of the Royal Society, 364*, 2709–2723. https://doi.org/10.1098/rstb.2009.0094

Tirloni, E., Stella, S., Celandroni, F., Mazzantini, D., Bernardi, C., & Ghelardi, E. (2022). *Bacillus cereus* in dairy products and production plants. *Food Security, 11*(17). https://doi.org/10.3390/foods11172572

Treagus, S., Wright, C., Baker-Austin, C., Longdon, B., & Lowther, J. (2021). The foodborne transmission of hepatitis E virus to humans. *Food and Environmental Virology, 13*(2), 127–145. https://doi.org/10.1007/s12560-021-09461-5

Uyttendaele, M. (2018). Chapter I.3 3. Food-borne pathogens. In M. Uyttendaele, A. De Loy Hendrickx, J. Debevere, F. Devlieghere, L. Jacxsens, & A. Vermeulen (Eds.), ISBN : 9782874035036 *Microbiological guidelines: Support for interpretation of microbiological test results of foods*. Die Keure publishers.

Uyttendaele, M., Busschaert, P., Valero, A., Geeraerd, A. H., Vermeulen, A., Jacxsens, L., Goh, K. K., De Loy, A., Van Impe, J. F., & Devlieghere, F. (2009). Prevalence and challenge tests of *Listeria monocytogenes* in Belgian produced and retailed mayonnaise-based deli-salads, cooked meat products and smoked fish between 2005 and 2007. *International Journal of Food Microbiology, 133*(1–2), 94–104. https://doi.org/10.1016/j.ijfoodmicro.2009.05.002

Van Damme, I., Berkvens, D., Vanantwerpen, G., Bare, J., Houf, K., Wauters, G., & De Zutter, L. (2015). Contamination of freshly slaughtered pig carcasses with enteropathogenic *Yersinia* spp.: Distribution, quantification and identification of risk factors. *International Journal of Food Microbiology, 204*, 33–40. https://doi.org/10.1016/j.ijfoodmicro.2015.03.016

Van Damme, I., De Zutter, L., Jacxsens, L., & Nauta, M. J. (2017). Control of human pathogenic *Yersinia enterocolitica* in minced meat: Comparative analysis of different interventions using a risk assessment approach. *Food Microbiology, 64*, 83–95. https://doi.org/10.1016/j.fm.2016.12.006

Van der Linden, I., Cottyn, B., Uyttendaele, M., Vlaemynck, G., Maes, M., & Heyndrickx, M. (2013). Long-term survival of *Escherichia coli* O157:H7 and Salmonella enterica on butterhead lettuce seeds, and their subsequent survival and growth on the seedlings. *International Journal of Food Microbiology, 161*(3), 214–219. https://doi.org/10.1016/j.ijfoodmicro.2012.12.015

Van Paepeghem, C., Taghlaoui, F., De Loy-Hendrickx, A., Vermeulen, A., Devlieghere, F., Jacxsens, L., & Uyttendaele, M. (2024). Prevalence and growth potential of *Listeria monocytogenes* in innovative, pre-packed, plant-based ready-to-eat food products on the Belgian market. *International Journal of Food Microbiology, 410*, 110506. https://doi.org/10.1016/j.ijfoodmicro.2023.110506

van Seventer, J. M., & Hochberg, N. S. (2017). Principles of infectious diseases: Transmission, diagnosis, prevention, and control. In *International encyclopedia of public health* (pp. 22–39). Elsevier.

Wauters, G., Kandolo, K., & Janssens, M. (1987). Revised biogrouping scheme of *Yersinia enterocolitica*. *Contributions to Microbiology and Immunology, 9*, 14–21.

Webb, M. D., Barker, G. C., Goodburn, K. E., & Peck, M. W. (2019). Risk presented to minimally processed chilled foods by psychrotrophic *Bacillus cereus*. *Trends in Food Science and Technology, 93*, 94–105. https://doi.org/10.1016/j.tifs.2019.08.024

WHO. (2006). *Five keys to safer food manual*. World Health Organization. https://iris.who.int/handle/10665/43546

WHO. (2015). *WHO estimates of the global burden of foodborne diseases: Foodborne diseases burden epidemiology reference group 2007–2015*. World Health Organization. https://iris.who.int/handle/10665/199350

Zhang, X., Wang, S., Chen, X., & Qu, C. (2021). Review controlling *Listeria monocytogenes* in ready-to-eat meat and poultry products: An overview of outbreaks, current legislations, challenges, and future prospects. *Trends in Food Science and Technology, 116*, 24–35. https://doi.org/10.1016/j.tifs.2021.07.014

Zhao, X., Zervas, A., Hendriks, M., Rajkovic, A., van Overbeek, L., Hendriksen, N. B., & Uyttendaele, M. (2022). Identification and characterization of *Bacillus thuringiensis* and other *Bacillus cereus* group isolates from spinach by whole genome sequencing. *Frontiers in Microbiology, 13*, 1030921. https://doi.org/10.3389/fmicb.2022.1030921

Zhao, X., Hendriks, M., Deleu, E., Spanoghe, P., Hofte, M., van Overbeek, L., & Uyttendaele, M. (2023). Prevalence, attachment ability and strength of the biological control agent *Bacillus thuringiensis* on tomato. *Food Microbiology, 112*, 104235. https://doi.org/10.1016/j.fm.2023.104235

Zweifel, C., & Stephan, R. (2012). Spices and herbs as source of *Salmonella* related foodborne diseases. *Food Research International, 45*(2), 765–769. https://doi.org/10.1016/j.foodres.2011.02.024

Zhao, X., Van Overbeek, L., Burgess, C. M., Holden, N., Brennan, F., Johannessen, G. S., Allende, A., Höfte, M., Cottyn, B., Pothier, J. F., Schikora, A., Uyttendaele M. (2025). Human pathogenic microorganisms in fresh produce production: lessons learned when plant science meets. Food Safety Journal of Food Protection 88(7):100551. https://doi.org/10.1016/j.jfp.2025.100551

Food Loss Reduction: The Pathway Towards Sustainability

O. M. Olanya, A. K. Hoshide, D. O. Ukuku, S. Mukhopadhyay, Z. He, T. Z. Jin, B. K. Sharma, K. Kohli, and O. A. Ijabadeniyi

1 Introduction

Food is important for nutrition, sustenance, and the survival of humanity and many other living organisms (FAO, 2016; 2020). For human beings, the sources of food can be diverse such as food derived from plants/crops as well as animals on land and in soil and water), marine organisms such as fish, shrimp, crabs, lobsters, and other invertebrates. Food can also be more industrially manufactured by the food industry to produce processed or semi-processed foods. Food can be consumed by households at home or food can be prepared by restaurants and sold to customers. These facets of food represent the broad categories of global food sources (FAO, 2020).

Food losses can occur due to the processes or actions that can result in loss, reduction, waste, or destruction of food during any of the stages of food production, processing, or utilization, rendering edible food unavailable and un-useable (Olajuyin et al., 2023). The loss of food can impact various resources used during the production of food. Reduction of food loss is important for eradication of poverty and hunger and is essential for the sustainability of food systems throughout the

O. M. Olanya (✉) · D. O. Ukuku · S. Mukhopadhyay · T. Z. Jin · B. K. Sharma
USDA-ARS, Eastern Regional Research Center, Wyndmoor, PA, USA
e-mail: modesto.olanya@usda.gov

A. K. Hoshide
Born Global Foundation, Portland, ME, USA

Z. He
USDA-ARS, Southern Regional Research Center, New Orleans, LA, USA

K. Kohli
CSIR Indian Institute of Petroleum, Separation Process Division, Dehradun, India

O. A. Ijabadeniyi
Department of Food Science & Technology, Durban University of Science and Technology, Durban, South Africa

D. Li (ed.), *Antimicrobial Strategies in the Food System: Updates, Opportunities, Challenges*, https://doi.org/10.1007/978-3-031-95056-8_2

world. Global estimates of food loss can vary a lot in low, medium, and high-income economies due to differences in technologies used for food production, processing and distribution, and utilization or consumption (FAO, 2020). It has been demonstrated that food loss can occur throughout the entire supply chain of the food production process from farm to fork. The rationale for minimizing food loss can be attributed to improvements in global sustainability. The availability of food in the domestic market or production of adequate amounts of food, access to food, or the potential to acquire or purchase it, as well as adequate food nutrition and safety can contribute to enhancing the sustainability of food systems.

This concept of food loss runs in tandem with food security as well as food safety, which are key elements in ensuring sustainable development as a sustainable future requires that food loss reduction should be optimized to the greatest extent possible. Intensification and enhancement of food production can be an essential component of sustainable development. Food loss reduction can reduce pressure to produce more food since food is used more effectively and efficiently. Similarly, strategies for management of antimicrobial resistance are also an important part of intensifying food production and it could simultaneously help reduce food loss. Although the exact quantity of global food loss is not exactly known, it has been estimated that approximately more than 30% of global food production (>1.4 billion metric tons) may be lost on an annual basis from the food production on farms to consumption of edible food by consumers (Gustavsson et al., 2011; HLPE, 2014). The lack of precise definitions for food loss and limited quantification methods may be a contributing factor to this precise estimate figures. It has been estimated that the food loss in developed countries is about 671 million metric tons while in developing countries this amount is approximately 600 million metric tons for a total of 1271 billion metric tons (Wang et al., 2021). The lack of precise definitions for food loss and limited quantification methods may be a contributing factor to verify these estimates more accurately. In addition, there may be tremendous variation in food loss between regions of the world.

The causes of food loss have been shown to vary depending on the geographical region and to be correlated to economic development. For example, in developing countries, most of the food loss occurs mainly during agricultural food production and post-harvest handling, storage, and processing. Meanwhile in developed economies, most of the food loss occurs at the distribution and consumption phase (Wang et al., 2021). Optimum food production can occur, provided conditions for agricultural production are adequate. However, pre-harvest food may be lost due to biotic factors such as insect pests and diseases as well as abiotic factors such as soil conditions and environmental conditions like temperature, humidity, and rainfall. Pre-harvest food can also just remain un-harvested in the field. Post-harvest food loss can occur in food systems via the principal mechanisms of post-harvest handling, processing, food storage, distribution, and following the consumption of food.

2 Food Loss Reduction Measures and Sustainable Pathways

Food losses may be driven by pre-harvest events, post-harvest losses, or losses that may occur in the food systems during production, processing, or utilization (Vagsholm et al., 2020). There are several pathways through which food losses could be reduced during food production and utilization. Among these pathways for reducing food loss are: pre-harvest measures in the field, greenhouse, aquaculture and animal production systems, (Adebowale et al., 2020; optimum harvest and post-harvest food processing technologies, (Aday et al., 2011; Adebisi, 2023); transportation, distribution, and storage of food or food products such as in retail and warehouse outlets, (Alvarez-Hendandez et al., 2020); Appendini & Hotchkiss, 2002); use of excess food for animal feed or food waste recycling (Appendini & Hotchkiss, 2002); An, 2016) excess or waste food for nutrient recycling and composts and (Arena, 2012) and food waste utilization for biofertilizer, energy conversion and biodiesel (Arena, 2012).

Food loss reduction measures can be framed in the context of how these translate to sustainable pathways. Sustainable directions in reducing food loss should have positive impacts on national and global economies, societies, and environments. Globally, the economic value of food wasted after harvest but before market is estimated at $400 billion, representing 14% of production with the capacity of feeding 1.26 billion people (Yue & Takács-György, 2023). While it is clear that the food loss reduction measures discussed can be economically beneficial, these measures can have both positive and negative impacts on society and the environment. Better managing both food loss and food waste can result in more sustainable food security translating to more stable societies (Yue & Takács-György, 2023). However, there can be potential adverse environmental impacts from increased resource use when using plastic packaging and climate control storages to better preserve food and reduce food loss (López-Gálvez et al., 2021). Food loss and waste as well as adverse impacts from reducing such loss/waste can be addressed through prevention, re-use, recycling, and disposal (O'Connor et al., 2023). In addition to private and public sector involvement, reducing food loss can also be the responsibility of consumers. For example, preparing and consuming appropriate amounts of food to reduce food loss involves consumer decisions to store food e.g., freezing, better plan for menu, and management of leftovers (Olajuyin et al., 2023).

2.1 *Food Loss Reduction and Diverse Production Systems*

Food loss reduction for crops in the field and greenhouse production, and aquaculture or animal production systems may be accomplished by using sustainable pest control measures that can reduce or eliminate insect populations below economic injury level in addition to pathogens and vertebrate pests that would otherwise damage food crops during preharvest, at harvest, or post-harvest. For example, the use of bio-pesticides or plant extracts are regarded as sustainable applications that could potentially reduce pest populations with limited environmental impact.

The use of optimum post-harvest food processing technologies can reduce food losses regardless of the production system used. In low-input systems in developing countries, post-harvest food processing is somewhat rudimentary where small-holder farmers conduct basic food crop harvests where processing is done with limited equipment (Manandhar et al., 2018). This may result in huge losses of harvested cereals (e.g., maize (*Zea mays*), millet (*Pennisetum glaucum*), sorghum (*Sorghum bicolor*), and rice (*Oryza sativa*)), root and tubers, (e.g., yams (*Dioscorea rotundata*), cocoyam (*Xanthosoma sagittifolium*), sweet potato (*Ipomoea batatas*), potato (*Solanum tuberosum*), cassava (*Manihot esculenta*) or leguminous crops (e.g., soybeans (*Glycine max*), pigeon peas (*Cajanus cajan*), common beans (*Phaseolus vulgaris*), lentils (*Lens culinaris*), and other pulses (*Fabaceae*)). There is a lack of mechanized harvesting equipment such as harvesters and mechanical hoppers, which would undoubtedly reduce pre-harvest and harvest losses.

These practices in developing nations are contrary to developed economies where the production practices, especially pre-harvest, post-harvest and food processing operations are highly mechanized including the use of cold chain technology, leading to limited food waste (Hodges et al., 2011; Balan et al., 2023). In aquaculture, poultry, and small and large animal production systems, food losses could also occur prior to animals being readied for slaughter facilities. In developed nations, there would be minimal losses if any at this stage. In poultry, fish, and animal production systems, use of antimicrobials administered during growth phases potentially limit food loss, enhancing animal weight gain (Banrie, 2012; Tang & Bondad-Reantaso, 2019).

2.2 Food Loss Reduction during Post-harvest, Storage and Transportation of Food

Similarly, food losses at post-harvest and during transportation may be reduced by sustainable post-harvest processing, storage, and transportation practices. For example, in low- and middle-income countries, post-harvest processing of grain could be considerably improved. Post-harvest processing activities such as drying, threshing/shelling, cleaning and winnowing are rudimentary, resulting in huge losses. Storage losses of cereals such as maize and rice could be considerably reduced by using hermetic storage technology for agricultural products (Manandhar et al., 2018). Hermetic bags and plastic storage silos which are airtight, sealed bags or containers, that restrict gas exchange (moisture, air, or water) have been successfully used to reduce food losses (e.g., grains) caused by insects, rodents, and pathogens during storage (Manandhar et al., 2018). In developed economies, steel alloy or aluminum storage silos which are well-ventilated with controlled atmosphere are often used to reduce food losses attributed to insects and other pests. Root and tuber food crops such as cassava in developing economies, are subject to rapid post-harvest deterioration once harvested and must be consumed immediately or processed. Such processing can produce value-added products (for industrial use such as flour or starch, or industrial byproducts such as glucose). Therefore, small or medium-size processors, for example for pigeon pea, cassava or tomato (*Solanum lycopersicum*) can be instrumental in improving the quality of value-chain products (Ojwang et al., 2016, 2021). In contrast, medium- to high-income countries have more advanced processing and cold chain storage technologies so transportation, processing and storage do not result in any significant post-harvest losses for fruits and vegetables (e.g., peaches (*Prunus persica*), apples (*Malus domestica*), tomatoes, cabbage (*Brassica oleracea* var. *capitata*) or roots and tubers such as potato (Hodges et al., 2011). Therefore, optimum post-harvest processing, storage, and transportation of food would considerably reduce food loss or waste.

2.3 Recycling Excess Food and Food Waste Utilization

The utilization of excess food as animal feed or conversion of food waste via recycling for nutrient production, composts, and value-added utilization for biofertilizer or energy sources are not readily practiced in low-income economies due to lack of processing facilities for such tasks. More often than not, most of the food losses in developing countries occur at smaller scales. In general, fragmented farm holders conduct operations individually with a lack of economies of scale. A greater degree of industrialization or having enough labor is required in order to conduct such post-harvest, value-added processing efficiently (Chrisendo et al., 2023). Under these conditions, excess food can be more effectively processed to reduce food loss such as the case when there is a glut of ripened mangoes or tomatoes in tropical countries (Beausang et al., 2017). In some situations, food waste may be combined with livestock manure in bio digesters for anaerobic digestion to produce biogas and the remnant digestate can also be used as a soil amendment for crops (Sarker et al., 2023).

3 Microbial Biopesticides and Plant Extracts—Direct/Indirect Effects on Pre-and Post-harvest Food Loss

3.1 Food Loss for Field or Horticultural Crops/Products

Food losses from field crops or horticultural plants may be caused by poor harvesting or post-harvest techniques, processing, and food storage. These food losses may result in serious consequences to global food security and human well-being. Food losses caused by plant pathogens and insects appear to be the most significant losses sustained in developing and developed countries (Savary et al., 2019). Food losses may be caused by biotic and abiotic agents or factors. Biotic factors include microorganisms (e.g., fungi, bacteria, viruses, and nematodes) mainly during the production stage of food from plants (Olanya et al., 2014). Post-harvest food losses especially during storage may be primarily attributed to insect pests and rodents (Cao et al., 2002). Biotic food losses or damage may occur for vegetables, cereals, root and tubers or other edible parts of plants or crops and may also be caused by insects (pre-harvest and post-harvest/storage), rodents, and birds.

3.2 Biotic and Abiotic Factors and Indirect Effects on Food Loss

Food losses during pre-harvest and post-harvest can result in fruit and vegetable rots and decay, cereal damage, root and tuber rots, and other types of damages on plant-derived foods. For vegetables and fruits, bacteria, yeasts, and molds are primarily responsible for decay and rot (Tournas, 2005). These organisms can contaminate the plant parts during growth and development in the field, during harvest, transport, processing, and/or storage. Once the degradation process begins, microorganisms spread rapidly, destroying plant tissue and rendering the affected food product unacceptable for consumption, resulting in great economic losses (Tournas, 2005).

Abiotic factors leading to food loss may be due to environmental conditions (e.g., soil, water, relative humidity/temperature) that can limit production of foods (Olanya et al., 2016a, 2016b). These food losses appear to vary among countries with most losses occurring in developing countries, while developed economies do have lower food losses comparatively (Balan et al., 2023). Overall, it appears that the damage attributed to the loss of market value of fruits, vegetables, and cereals during production, postharvest storage, shipment, and marketing can reach up to 40% in developed countries (Beausang et al., 2017; Magalhães et al., 2021). Food losses in the major food crops of wheat (*Triticum aestivum*), rice, maize, potatoes, and soybean range from 8–35% (Savary et al., 2019).

Food loss due to insect pests impacts approximately 20–30% of the world's potential harvested crops. Insect pests have direct effects on agricultural produce or foods due to chewing, sucking, and boring plant roots, stems and leaves. Food destruction by insects is particularly acute during storage where approximately 17% of the global food production is affected by insect pests during storage (FAO, 2020). Similarly, 7% of food products in storage are affected by mites, rodents, and post-harvest pathogens. Insects can result in the deterioration of food quality through their direct consumption or secretions as well as changes in the food products' chemical compositions due to presence of insect parts.

3.3 Food Loss and Processing Establishments

Food losses mostly in food production facilities, processing establishments, and restaurants can also occur due to foodborne pathogens contamination. Foodborne pathogens such as *Escherichia coli* O157:H7, *Listeria monocytogenes*, and *Salmonella enterica* among others can result in tremendous losses in processed foods that must be discarded due to safety precautions. For example, in 2017, outbreak of listeriosis in South Africa delis resulted in severe losses from contaminated food, plant closures, as well as human mortality (Olanya et al., 2019). Food losses in developing countries are not adequately quantified but vary immensely from country to country (Havelaar et al., 2015). In developed countries, food losses resulting from foodborne pathogens in the United States of America and the European Union can be over $240 million annually (Scallan et al., 2011; EFSA, 2013).

3.4 Biopesticides and Biotic Constraints as Indirect Effects on Reducing Food Loss

Biopesticides (biochemicals or pesticides) and microbial biopesticides may be categorized as natural products that can be applied to mitigate food losses from plant pathogens and insect pests during pre-harvest and post-harvest. Microbial biopesticides may also contain microorganisms and their metabolic products (microbial pesticides), as well as substances produced by plants containing added genetic material (plant-incorporated protectants) that are able to control pests and diseases. Microbial antagonists represent some of the common examples of biopesticides which are used for reducing food losses, through plant/human pathogen reduction. These biopesticides may contain microbial predators, mycoparasites, and antimicrobial compounds (Olanya et al., 2020; Vero et al., 2023). Therefore, crop damage and food losses during production, harvest, post-harvest and processing can be mitigated in several ways using specific biopesticides.

Biopesticides may refer to pesticides or compounds that may be obtained from natural substances (e.g., plants, bacteria, animals) or materials that have specific biological effects or actions and used for pest management or controls (Koul, 2023). This may include all natural products with negative effects on insects, rodents, fungi, and other plants (bioherbicides). Biopesticides may be used to reduce food loss by preserving food from contamination or degradation or directly control pests that destroy or degrade food. There are numerous examples of biopesticides that have been utilized for pest management. Examples of the commercial use of biopesticides include *Bacillus thuringiensis* (Bt) that have been used against insect populations, thereby reducing damage to food crops (Rad et al., 2016). Similarly, cyanobacterium (*Nostoc piscinale*) and green algae (*Chlorella vulgaris*) have shown efficacy against some pathogens (Goncalves, 2021). Some pathogen strains such as *Streptomyces* sp. and *Bacillus subtilis* have been used to control bacterial and fungal (oomycetes) pathogens such as *Phytophthora infestans* resulting in a reduction food loss in potato which would otherwise occur (Olanya & Larkin, 2006). Other fungal organisms such a s *Trichoderma harzianum* have been shown to be effective against fungal pathogens or as an insecticide (Tyśkiewicz et al., 2022).

3.5 Plant Extracts (Indirect Effects and Mitigation of Food Loss)

Plant extracts from various plant species may be used either by themselves or in combination with other micro-organisms (non-pathogenic bacteria) to reduce or suppress biotic agents that directly or indirectly affect food loss by impacting crop development or maturity (Olanya and Larkin, 2006). For example, neem extract has been used to guard against insect infestation of potato tubers in storage (Kumar et al., 2022). In other cases, essential oils, and phenolic and flavonoid compounds have demonstrated some potency in reducing food losses. Similarly, fungal entomopathogens in combination with plant extracts have been used to control whitefly on sweet potato (Jaber et al., 2018).

3.6 Mitigating Grain Loss at Post-harvest and in Storage

Minimizing food loss during harvest, transport, and storage is perhaps one of the most important activities that can be done by small-holder and large-scale farmers to reduce food loss. For grain-based foods (cereal crops) such as maize, wheat, rice, sorghum, millet, oats (*Avena sativa*), or barley (*Hordeum vulgare*), food loss reduction can be minimized during harvest, threshing and cleaning, drying, storage, milling and transportation (Kumar & Kalita, 2017). These operations differ in developed and developing countries, primarily due to the differences in technologies used. Hand harvesting or use of rudimentary harvest equipment often practiced in developing countries, compared to mechanical harvests with elaborate holding containers which can reduce grain loss significantly in developed nations with substantially less grain lost during combining, spillage, or the use of improper containers (Maiti et al., 2018). Therefore, losses during threshing and cleaning, drying, or transportation tend to be considerably lower in developed versus developing countries, due to mechanized efficiencies and/or technological advantages.

Overall, farm operations in many developing countries are manual which can involve cutting, manual threshing, sun drying/air drying, low level of storage and milling. Meanwhile in developed economies, farm operations may be moderately or highly mechanized such as combine harvest, machine threshing, mechanical drying, sealed storage/good humidity control and pest management, and commercial milling, as well as adequate transportation and storage facilities (Hodges et al., 2011).

Maintenance of proper storage facilities is crucial to prevent post-harvest losses in cereals from insect pests, rodents, and pathogen infections. This is because of a lack of proper humidity control can lead to damage if inoculum or any pests get into the grain storage area or facility. Therefore, improvement in the above-mentioned operations and storage facilities) can reduce food losses (Maiti et al., 2018).

3.7 Minimizing Food Loss of Fresh Produce (Vegetables, Fruits, Root, and Tubers), Meat/Fish and Dairy Products

Unlike grains or cereals, food losses of fresh produce, fruits, meat, and dairy products are perhaps, one of the most significant losses sustained by many farmers in the world. The high-water content of these foods combined with high moisture/temperature and lack of proper cold storage facilities with adequate temperature and humidly controls contribute to food losses particularly in developing countries. The perishable nature of food in developing countries is indicative of the ease of such deterioration that is often sustained (Hodges et al., 2011). Most of the loss of perishable food items and commodities are associated with logistical hurdles in developing nations as opposed to developed countries. Similarly, meat and fish products are subject to losses due to their perishability as cold storage and rapid transportation are often a limitation in low-income countries, even though smoke-based processing may be utilized in some regions of developing countries.

4 Effect of Antimicrobials in Mitigating Food Loss by Pathogen Suppression, Inactivation and Enhancement of Shelf Life of Food Products

4.1 Food Loss or Deterioration Due to Microbial Contaminants

Food loss due to microbial contamination, browning reaction, or changes in physical and other chemical attributes can be a problem for both consumers and food industries. Knowledge of the level and the behavior of microflora causing spoilage in foods should help processors understand storage conditions needed to maintain longer shelf life and at the same time maintain microbial safety of food products. For health reasons, people are consuming fresh foods or minimally processed foods, thereby exposing themselves to the risk of foodborne illness if such foods are contaminated with bacteria pathogens (Dewey-Mattia et al., 2018). *Escherichia coli* O157:H7, *Listeria monocytogenes*, and *Salmonella* are all recognized foodborne pathogens that have been linked to food borne illnesses (Ukuku et al., 2019; Olanya et al., 2018).

Most food surfaces are not free from natural contaminants, especially the surfaces of fruits and vegetables which are in frequent contact with the soil, insects, animals, and humans during the growing and harvesting in the field and during post-harvest processing (Ukuku et al., 2016; Olanya et al., 2018). By the time fresh produce reaches the packing house, it can retain populations of 10^4 to 10^6 microorganisms per gram of produce (FDA, 2015). However, physical and chemical treatments are used in food processing to eliminate or at least reduce the presence of pathogenic and spoilage microorganisms in foods (Olanya et al., 2023; Ukuku et al., 2013; Ukuku, 2023). Prepared foods, especially fresh-cut salads in the supermarket, are becoming very popular with consumers in the United States of America due to the benefits of a diet rich in fruits and vegetables. Fresh-cut salads are prepared in the supermarket or by regional distributors leading to quality and safety concerns.

The presence of any one of the bacterial pathogens in food is a safety hazard for both consumers and the food industries as this has led to several food borne outbreaks (Ukuku & Fett, 2002, 2006; Gorski et al., 2011; Jackson et al., 2013), costly recalls (Olanya et al., 2019), and food wastage. These bacterial pathogens can persist on a variety of foods and particularly in low acid foods. For example, *Escherichia coli* O157:H7 cells have been recovered in foods such as apple cider, potato, turkey roll, yogurt, raw milk, and raw fruits and vegetables (FDA, 2015; Ukuku et al., 2013). Contributing factors include poor hygiene and unsanitary procedures used by processing staff, inadequate cleaning of processing equipment, and the use of decayed or damaged parts (Ukuku et al., 2019). The ability of pathogenic bacteria to adhere to surfaces of foods continues to be a potential food safety problem of great concern to the produce industry. Surface structure and biochemical characteristics of bacteria and of a topography of produce play a major role in how and where bacteria may attach (Ukuku & Fett, 2006). Bacterial attachment to surfaces is influenced not only by cell surface charge and hydrophobicity but also by the presence of surface appendages such as flagella and fimbriae as well as extracellular charge of polysaccharides (Ukuku & Fett, 2002; Ukuku et al., 2019). Bacteria colonize surfaces and can migrate by responding to changes within their growth environment (Ukuku & Fett, 2002).

4.2 Antimicrobial Applications and Reduction of Food Loss

Nisin, a natural pentacyclic heterodetic subtype A lantibiotic peptide synthesized by *Lactococcus lactis* subsp. *lactis* has been used in foods to kill bacteria and extend the shelf life of foods (Ukuku and Shelef, 1996). Ukuku (1997) reported that inactivation of six strains of *L. monocytogenes* by nisin showed variation in sensitivity and concluded that efficacy was dependent on nisin concentration and the pH of the media. The authors also reported that presence of higher nisin concentration being soluble at high pH, but the stability and antimicrobial efficacy was impaired. In another study, it was reported rapid decrease in cell numbers and intracellular ATP of *L. monocytogenes* in the presence of 3–12 μg/ml of nisin (Ukuku & Shelef, 1996).

These authors reported an increase in the extracellular ATP, indicating membrane damage and cellular leakage. There are several reports that nisin used in combination with a chelating agent, such as ethylenediaminetetraacetic acid (EDTA), exhibits a bactericidal effect towards both gram-positive and gram-negative bacteria (Ukuku et al., 2006; 2010). Ukuku et al. (2009) reported changes in the lag phase and generation time of *Salmonella*, *E. coli* O157:H7, and *L. monocytogenes* populations in apple cider amended with nisin-EDTA. Also, a novel antimicrobial-antioxidant solution that contain organic acids (generally regarded as safe) that inactivated 99.99% bacterial pathogens on fresh-cut apples and at the same time inhibited browning of the fresh-cut apples was developed by scientist at the USDA-ARS-ERRC. This solution extended the shelf life of fresh-cut apples up to 10 days during refrigerated storage (Ukuku et al., 2013).

4.3 Mitigation of Water Activity, pH, Enzymatic and Storage Conditions to Reduce Food Loss

Other factors like water activity, pH of food system, storage container conditions, and temperatures lead to spoilage and waste. For example, a slight change in water activity and pH could lead to enzymatic dysfunction and or production of other metabolites by food microflora which subsequently can initiate changes in food samples leading to off odors. Generally, water activity <0.6 in a food system is required to inhibit effectively the deterioration of the food products (He et al., 2021). Thus, understanding of water (moisture) sorption phenomena in products and isotherms modeling is valuable for the controlling and optimization of the safe storage of the products to minimize food loss (Dushkova et al., 2022; He et al., 2021; Ondier et al., 2011). These off-odors may be a result of oxidative and/or enzymatic activity and not necessarily directly related to microbial populations. For example, rancid conditions can be classified as either hydrolytic or oxidative. Color deterioration of fresh-cut produce is caused by the oxidation of polyphenol oxidase and peroxidase which ultimately leads to the formation of dark brown polymers of quinoidal nature (Lopez-Serrano & Ros Barcelo, 1999; Cetin-Karaca, 2011; He et al., 2008; Mahajan et al., 2014). These changes in colors of minimally processed foods may lead to economic losses due to rejection by consumers.

A group of enzymes called cellulase and hemi-cellulase have all been implicated in the degradation of cellulose, a major structural component of plant cell walls. For example, fungal polyphenol oxidase has an antioxidant effect in juice systems by removing residual oxygen present which preserves flavor and color. Also, produce firmness is related to cellular characteristics such as cell walls, intracellular turgor pressure, and the number, and sizes of intercellular spaces, which are very susceptible to undesirable alterations such as changes in texture, flavor, and color because of injury suffered during storage, handling, and processing (Watada et al., 1990). Similarly, dehydration of plant cells results in a loss of cell turgor which provides textural "crispness" to fresh vegetables. Hydrolytic rancidity is mostly associated with dairy products, usually resulting from enzymatic activity (Ukuku, 2023). Hydrolytic rancidity warrants further studies to understand why and how this happens to food. It is also possible that these off-odors could be due to enzymatic activity related to the degradation of peroxides and/or scission of carbon chains in food which ultimately lead to the formation of carbonyl compounds that bring about foul-smelling volatile compounds.

Several researchers have reported that the level of sanitation and the microbiological load of finished products are of primary importance to for quality, shelf stability and microbial safety, and that the microflora of all food items are of practical significance to producers, processors, and consumers (Mukhopadhyay & Ramaswamy, 2012; Mukhopadhyay et al., 2018; Ukuku et al., 2011; CDC, 2016). Microorganisms require certain basic nutrients for growth and maintenance of metabolic functions. The amount and types of nutrients required range widely depending on the microorganism. Therefore, food preservation and extension of shelf life require complex treatments involving technological innovation given challenging conditions. For example, higher storage temperature and relative humidity as well as the type of bags used for food storage can affect the water activity, bacterial survival, and the pH of food samples.

5 Antimicrobials for Food Loss Reduction in Aquaculture, Poultry, and Animal Production Systems

5.1 Antimicrobials in Aquaculture and Food Loss Reduction

Antimicrobials are substances that have direct action on micro-organisms by killing or preventing them from reproduction. Antimicrobials may be used for prevention of pathogens or treatment of infections and infectious diseases. Many antimicrobials may include antibiotics, antivirals, antifungals, and anti-protozoans. Antibiotics often have direct effects or action on bacteria, either by killing or preventing bacteria from reproducing and are used for the treatment of infections or infectious diseases caused by bacteria (Banrie, 2012). Therefore, all antibiotics are antimicrobials, but not all antimicrobials are antibiotics. Antimicrobials that have no effect on bacteria are not antibiotics. Antimicrobials have been used to enhance production of aquaculture products, poultry, and animal production systems.

Globally, the use of fish as food may account for 17% of animal protein (Henriksson et al., 2015). Aquaculture production systems are quite varied as prior research has shown that in excess of 500 freshwater and marine animal species may be used for aquaculture using a variety of production systems (Troell et al., 2014). Intensive aquaculture systems previously used for production of salmon (*Salmo salar*) in Chile and tiger shrimp (*Penaus monodon*) in Vietnam use antimicrobials to combat fish pathogens and enhance fish growth. The production of these two types of fish types accounted for almost 23% of total antimicrobial use during 2002–2003 (Henriksson et al., 2015). Antimicrobials and antibiotics are used in aquaculture industries in order to reduce excessive losses of fish due to aquatic infectious diseases. For example, aquaculture industries in Vietnam, Malaysia, Mexico, and China took losses of about $40 million between 2010 and 2016 (Tang & Bondad-Reantaso, 2019).

In contrast to post-infection treatment using antibiotics, the use of antimicrobials as prophylactic treatments is relatively limited such as that for fish production in Norway. Antibiotic use has been considerably reduced in many aquaculture production systems globally over the years due to changes in practice, consumer demand for natural foods, and more efficient antibiotic or antimicrobial management. Similarly, antimicrobial use in aquaculture production systems in many developing countries is fairly limited due to the lack of funds available for such production systems. Here in many cases, aquaculture production systems are small-scale and limited to freshwater fish. Overall, antimicrobial use in aquaculture systems can reduce food loss and enhance sustainability.

5.2 Antimicrobials in Poultry Production to Reduce Food Loss

Poultry production is an important component of food on a worldwide basis with over 23 billion birds (FAOSTAT, 2016). On a global scale, the United States of America is the leading producer of chicken meat at roughly 21 million metric tons, while Brazil accounts for approximately 14.5 million metric tons (USDA-ERS, 2024; Mottet & Tiempio, 2017). In developing economies, poultry production contributes substantially to agri-business due to quick production cycles and a variety of conditions where production can occur. In both developed and developing nations, antimicrobials or antibiotics are used in poultry production due to perceived and real effects on poultry health and production. Therefore, its extensive use in various parts of the world has been viewed as a necessity for eradicating poultry diseases, particularly bacterial pathogens, and for increasing production efficiency. For many years, antimicrobial growth promoters were used as feed additives to prevent or minimize intestinal pathogens (Carrasco Diaz et al., 2016).

As the potential for antimicrobial resistance developed, options such as plant extracts with antimicrobial properties, particularly against *Clostridium perfringens* (necrotic enteritis), gained wider attention as alternative feed additives (Carrasco Diaz et al., 2016). Due to worldwide poultry mortality and contamination of chicken products by *C. perfringens*, global losses have been estimated at $2 billion annually. Therefore, therapeutic antimicrobials such as bacitracin, lincomycin, virginiamycin, penicillin, and tylosin have been used to control infections. Similarly, the same antibiotics have been used for control of Gram-positive bacterial pathogens in chicken. Additionally, several phytochemicals, such as plant extracts, have been used as additives to feed for chicken (*Gallus gallus domesticus*) and many of these have been used as additives to feed chicken and many of these have been used in Asia, Africa, and South America (Fernandez Miyakawa, 2016; Hashemi & Davoodi, 2011). Similarly, tannins and essential oils have also been used for their antimicrobial properties in poultry production. In the European Union, due to regulatory prohibition, many antibiotics may be used in poultry production on a limited scale. The only exception appears to be ionophore antibiotics for coccidiosis control in poultry which appears to be classified as feed additives, since they are not used in humans. Some of the anticoccidial drugs that have been used in poultry may include monensin compound (Coban, Rumensin, Rumensin CRC), lasalocid (Avatec, Bovatec), salinomycin (Bio-cox, Sacox), narasin (Monteban, Maxiban), maduramicin (Cygro), laidlomycin (Cattlyst) and semduramycin (Aviax) among others (Banrie, 2012).

5.3 Application of Antimicrobials During Animal Production to Reduce Food Losses

Antimicrobials in large and small animal production systems, including livestock, have been utilized extensively throughout the world for a long time, to reduce food losses and simultaneously enhance animal growth and production (Ma et al., 2021). This may seem contradictory due to antimicrobial resistance which is of considerable concern regarding animal and human health (Food and Drug Administration (FDA), 2022). One of the most notable uses of antimicrobials has been in the form of substances that can promote the growth and rapid development of animals. For example, growth promoter compounds such as sulfasuxidine, streptothricin, and streptomycin in swine or pig (*Sus scrofa domesticus*) feed have been shown to have beneficial attributes to animal growth and production (Moore et al., 1946; Stokstad & Jukes, 1950). The average global use of antimicrobials per kilogram of animal product was approximately between 45 mg/kg for cattle to 172 mg/kg for swine (i.e., pig) production (Ma et al., 2021).

These doses have been steadily increasing, primarily to reduce food losses, despite concerns for antimicrobial resistance. Approximately 30,000 tons of antimicrobials were used for animal production in China, which has used the greatest amount of antimicrobials in animal production systems, primarily promoting animal growth (Hu & Cheng, 2014). In the United States of America (U.S.A.), the amount of antimicrobials used for animal production has been estimated to about 14,000 tons (FDA, 2015), while in Europe, about 8000 tons of antimicrobials have been used for animal production (European Medicine Agency, 2025). In the U.S.A., the European Union, and most developed economies, most antibiotics used that are detected as antibiotic residues in pigs and cattle are mostly ciprofloxacin, Tetracyclines, chlortetracyclines, sulfonamides, and nitrofurans, mainly for cattle (*Bos taurus*) and pigs (Ma et al., 2021).

In developing countries primarily in Africa, Latin America, and Asia, antimicrobials or antibiotics have been used primarily as an essential component of veterinary medicine to enhance animal production by treating, controlling, and preventing infectious bacterial diseases mostly for large animals. Such antibiotic use can ensure excellent health and subsequently reduce losses in animal food production systems (Hosain et al., 2021). In sub-Saharan Africa, antimicrobial use is regarded as essential to produce healthy animals in the food chain with substantial economic impact even though environmental and public health concerns may be rampant. Due to high level of indiscriminate use of antimicrobials, precise data are missing from various countries (Mshana et al., 2021). In Nigeria's Ogun State, over 70% of pig farmers reported use of antimicrobials such as tetracycline, tylosin, and chloramphenicol (Adebowale et al., 2022). Similarly in other countries such as Tanzania, approximately 92% of farmers were reported to have deficiency in managing animal diseases. This results in a greater propensity to use antimicrobials in livestock or animal production systems, as proper veterinary services may be expensive or unavailable.

In Sub-Sahara Africa, economic returns appear to be the dominant rationale for widespread use of antimicrobials in animal production systems. Improvements in animal weight gain and feed conversion are considered to be key aspects for cost savings per unit of animal produced compared to more rudimentary production practices, in addition to considerable reduction in animal mortality (Odey et al., 2024). Conversely, antimicrobial or antibiotic use may be detected in composted manure generated from livestock wastes which can stabilize excess farm nutrients and reduce environmental impacts such as eutrophication (Odey et al., 2024). Consequently, the use of antibiotics aimed to reduce food losses from livestock mortality may unintentionally introduce detrimental pathogens into farm soil, water, agro-ecosystems, or the environment in general.

6 The Role of Active Food Packaging in Reduction of Food Loss and Waste

Globally, 1.3 billion tons of food is wasted each year, representing approximately one-third of the total food produced for human consumption (Gautam et al., 2022; Ranganathan et al., 2020). Food packaging plays a significant role in food loss and waste management and overall prevention. A survey for post-harvest losses in bulk-packaged fruits and vegetables transported from rural to urban centers in southern Asian countries indicated loss reduction (%) for tomatoes (87%), banana, *Musa acuminata* (61%), and cauliflower, *Brassica oleracea* var. *botrytis* (60%) if they were stored in plastic crates (FAO, 2017). Compared to traditional packaging, active and intelligent packaging plays a more active and important role in the preservation of the food. ReFED (2024) estimates that active packaging can help divert 1.1 million tons of food waste and save 314 billion gallons of water a year.

Active packaging is prepared by incorporating active agents into packaging materials and works by absorbing food-derived chemicals or releasing active agents into the surrounding environment. This is an advanced packaging technology and plays an important role in enhancing food safety, maintaining food quality, and extending food shelf life, hence, reducing food loss and waste from farm to fork (Guo et al., 2022). Typical active packaging systems include antimicrobial packaging, antioxidant packaging, carbon dioxide emitters, moisture absorbers, ethylene absorbers, etc. There have been numerical studies in these areas, and the following are just some of them.

6.1 Antimicrobial Packaging

The presence of spoilage and pathogenic microorganisms in foods following post-process contamination will lead to outbreaks of foodborne disease and food waste (Guo et al., 2014). Spoilage microorganisms include bacteria, mold, and yeast. The presence of these microorganisms in foods usually accelerate changes in the color, flavor, and texture of foods, resulting in a reduction in their quality and shelf life. The presence of pathogenic microorganisms such as *Listeria monocytogenes*, *Salmonella enterica*, enterohaemorrhagic *Escherichia coli*, and *Campylobacter jejuni* in foods generally results in foodborne illness and even death. Antimicrobial packaging was developed to eliminate or inhibit the growth of these microorganisms.

Jin et al. (2022) investigated the antimicrobial efficacy of antimicrobial coatings against populations of nalidixic acid-resistant Salmonella, *Listeria monocytogenes,* and native microorganisms on whole grape tomatoes. The coating treatments significantly reduced *Salmonella* and *Listeria* populations on Day 1, and to undetectable levels from Day 2 through Day 21. Similarly, the coating treatments reduced spoilage microorganism to an undetectable level after 2 days. The sample images clearly indicate that the tomatoes in the control group were moldy after 21 days while the treated tomatoes were still fresh-like. Use of broadleaf plantain (*Plantago major*) seed mucilage as a novel edible coating incorporated with dill (*Anethum graveolens*) essential oil, has served as a potent antimicrobial (Behbahani et al., 2017). Antimicrobial surface coating extended the shelf-life of meat from 6 to 18 days and inhibited bacterial growth and slowed down the oxidative changes. Antimicrobial surface coating extended the shelf life of fresh ginseng (*Panax ginseng*) up to 6 months (Jin et al., 2016, 2017). Jin et al. (2020) developed a special antimicrobial and anti-browning formula to coat fresh-cut apples. The treated apple slices had no-detectable pathogens and spoilage microorganisms and no color changes were observed for 35 days in storage at 4 °C. Antimicrobial packaging films that release essential oils into the headspace of packaged food were also developed and investigated in grape tomatoes and fresh-cut apples. The headspace film treatment effectively reduced both pathogenic and spoilage microbes in packaged tomatoes and fresh-cut apples, hence extending their shelf life (Jin et al., 2023).

6.2 *Oxygen Scavenging*

The gas composition in the headspace environment of food packaging has a significant effect on the growth of microorganisms and the sensory quality of food products. Oxidation reactions generally cause the formation of off-odors, off-flavors, toxic aldehydes, changes in the color and texture of foods, and degradation of nutrients in foods (Dominguez et al., 2018). Lipid oxidation usually results in quality deterioration and safety issues in foods that have a high content of polyunsaturated fatty acids and proteins such as meat or seafood (Han et al., 2018). Borzi et al. (2019) modified polyamide film with green tea extract and the active film had excellent antioxidant capacities and increased the shelf life of minced meat from 6 to 23 days. Similar results were also reported by Nisa et al. (2015) and Siripatrawan and Harte (2010), where they incorporated green tea extract into starch-based film or chitosan film to enhance the antioxidant properties of the films. The antioxidant activity of the optimum polyvinyl alcohol/ tannic acid-Fe^{+++} (II) fibrous mat was investigated by Yang et al. (2016) using sunflower oil. They found the PVA/TA-Fe^{+++} (II) mat was able to decrease the formation of conjugate dienes, especially for long term storage.

6.3 *Carbon Dioxide Emitters/Absorber*

Carbon dioxide (CO_2) is used due to its outstanding efficacy in inhibiting the growth of some aerobic bacteria and fungi by reducing the relative oxygen levels in food packaging and/or directly exerting an antibacterial effect (Lee et al., 2021). Hence, CO_2 emitters have been developed to increase the concentration of CO_2 in food packaging, which is generally applied in combination with modified atmosphere packaging (Hansen et al., 2016). This emitter can release CO_2 through the interactions of sodium bicarbonate with ferrous carbonate or ascorbic acid. Sodium sulfite deoxidizer can also be mixed with $NaHCO_3$ to generate CO_2 and deoxygenate at the same time (Soltani Firouz et al., 2021). Holck et al. (2014) demonstrated that chicken filets stored in pure CO_2 showed an enhanced shelf life in terms of microbiology and odor compared with the lower concentrations of CO_2 used commercially. Carbon dioxide emitters consisting of sodium bicarbonate ($NaHCO_3$) and citric acid have been tested on salmon (*Salmo salar*) filet packaging and cod (*Gadus* sp.) packaging (Hansen et al., 2007, 2009), resulting in similar or longer microbiological shelf life than vacuum packaging.

For many of the packaged fresh produce, CO_2 is produced via the respiration depending on the storage temperature. A CO_2 concentration above the tolerance limit causes physiological injury to the produce. The symptoms of CO_2 injury include discoloration, off-flavor development, and internal tissue breakdown. Therefore, CO_2 absorbers are needed to reduce the concentrations of CO_2 in the headspace of packaged food. The absorption or removal of CO_2 from a gaseous phase can be theoretically achieved by a chemical reaction with an alkaline solution, physical adsorption, membrane separation, and cryogenic condensation (Lee, 2016). Wang et al. (2015) developed a multi-functional agar-based biofilm for fresh produce packaging and used sodium carbonate (SC) and/or sodium glycinate (SG) as a CO_2 absorbent. They showed the package with CO_2 absorbent film had best quality preservation of shiitake mushrooms (*Agaricus bisporus*) in terms of color, firmness, flavor score, and bacterial growth after 5-day storage. Use of CO_2 absorbers in food packages was thus effective in preserving the quality of fresh produce products, such as strawberries, *Fragaria* × *ananassa* (Aday et al., 2011), inhibiting or delaying the internal browning of pears, *Pyrus communis* (Nugraha et al., 2015), preventing chilling injuries of eggplants, *Solanum melongena* (Veasna et al., 2012), and reducing level of decay of Shiitake mushrooms (An, 2016).

6.4 *Moisture Absorbers*

The presence of moisture in the food package can promote the growth of food spoilage microorganisms. In addition, the presence of moisture also leads to the softening of crispy foods and clumping of hygroscopic products such as milk powder and negatively affects consumer acceptance (Biji et al., 2015). Moisture absorbers work by physical adsorption of moisture from the surrounding environment or via a chemical reaction between two functional materials (Gaikwad et al., 2019). Commonly used materials to control moisture are porous clay, molecular sieves, silica gel, humectant salts, absorbent polymers and calcium oxide (Yildirim et al., 2018), which are applied in various forms including sachets, pads, sheets, trays, and polymeric films.

6.5 *Ethylene Absorbers*

Ethylene is a plant hormone which activates the enzymes that promote the ripening of fruits and vegetables. This causes changes in the color, flavor, texture, and nutritional value of fruits and vegetables, especially during storage and transportation, resulting in a shorter shelf life and postharvest losses. Several researchers used potassium permanganate as an ethylene absorber in the form of a sachet that was placed inside of plastic bag to absorb ethylene generated by packaged produce products. These sachets enhance shelf life of guava (*Psidium guajava*) up to 7 weeks (Singh & Giri, 2014), extended marketable life of sapota (*Manilkara achras* Mill., cv. Kallipatti) up to 13 days (Bhutia et al., 2011), and the shelf life of banana for an additional 7 days at 25 °C (Hassan et al., 2005). Similarly, decay of mango (*Magnifera indica*) was controlled for 21 days of refrigerated storage, papaya (*Carica papaya*) ripening was delayed and quality preserved for 25 days (Silva et al., 2009), shelf life of banana was extended up to 20 days at 25 °C (Illeperuma et al., 2000), blueberries (*Vaccinium* sect. *canococcus*) quality (texture and sweetness) was maintained after 21 days in storage (Álvarez-Hernández et al., 2019), and sensory quality of peach was maintained for 15 days (Álvarez-Hernández et al., 2020). Other researchers integrated zeolite as ethylene adsorber in plastic film and achieved broccoli (*Brassica oleracea* var. *italica*) shelf life of 20 days compared to 5 days for control samples (Esturk et al., 2014), and mango (*Mangifera indica*) shelf life of 40 days compared to 20 days for the control (Boonruang et al., 2012).

7 Pathways for Food Loss Reduction by Utilizing Excess Food and Food Waste for Nutrient Recycling and Composts

7.1 Nutrient Recycling

Nutrient recycling from green and animal manure is an important strategy for circular agriculture and environmental sustainability (Chimouriya et al., 2018; He et al., 2016). Composting and anaerobic digestion are the typical methods used for such nutrient cycling to help plant growth (He, 2012; Tejada et al., 2008). Such strategies should also be applicable for nutrient loss reduction due to excess food and food waste. In other words, a promising food waste management option would be the production of organic amendments, similar to composting manure, which recovers food nutrients. These nutrient-recycled resources are sometime referred to as "biofertilizers" (Sharma et al., 2023). While composting is an anerobic process of recycling organic matter materials into valuable fertilizers, anaerobic digestion of food waste can provide both fertilizers (as degassed biomass or digestate) and energy (as methane or storable hydrocarbon), and therefore be classified as either energy (refer to next section of this chapter) or as nutrient recycling (Teigiserova et al., 2020).

7.2 Composts

Local composting (household, community, and small-scale applications) could be considered a sustainable option for food waste recovery and is receiving increasing demand from society (Torrijos et al., 2021). While this type of biofertilizer is capable of improving chemical, biological, and physical properties of soil and agricultural productivity, its practical effectiveness depends not only on proper composting or conversion processes, but also on appropriate application techniques. Sharma et al. (2024) reported the use of food waste digestate as a biofertilizer for ornamental plant cultivation used as a soil amendment in growing media. They found that plant growth was enhanced with food waste digestate application, particularly growth parameters of chlorophyll content index, plant height, number of leaves, and leaf area. They further attributed this enhancing phenomenon to the increase in bacterial and archaeal communities in the rhizosphere which had been acting as plant growth promoters and nutrient solubilizers.

In another study, Song et al. (2021) evaluated the agronomic properties of both raw food-waste digestate and digestate-biochar residue as biofertilizers for four types of vegetables (Chinese spinach (*Amaranthus dubius*), water spinach (*Ipomoea aquatica*), Chinese cabbage (*Brassica rapa* subsp. *pekinensis*), and lettuce (*Lactuca sativa* var. *angustana*). These researchers reported that the raw food-waste digestate induce ammonia and salt stress in vegetables, while application of its 20–40% (v/v) dilution made the vegetables' shoot fresh weight comparable to that grown using a commercial fertilizer. Furthermore, they proposed that adoption of biochar to filter the food-waste digestate (100 g to 1 L) could be a better way to recover the nutrients in both the liquid and solid fractions of anaerobic digestate considering the difficulty and cost of large quantity handling and transporting of raw digestate. In addition, O'Connor et al. (2023) evalauted food waste-derived fertilizers (e.g. compost, liquid and solid anaerobic digestates) on soil chemistry, microbial function and crop productivity. They noted that the food compost and digestates are low nutrient value fertilizers, while food waste-derived fertilizers could be better used alongside synthetic fertilizers to optimize crop growth and yields.

7.3 Co-composting Food Waste with Organic Matter

Alternatively, a co-composting approach can be adopted to compost food waste with other organic substrates to take advantage of different composting substrates to optimize composting conditions and beneficial outcomes (Awasthi et al., 2020). Wang et al. (2024), evaluated the effects of reactor temperature and feeding ratio on the co-composting of household food waste and green waste. They observed that raised reactor temperature could promote composting efficiency and compost end-product maturity and the 2:1 feeding ratio of the two substrates made the fed-batch composting attained excellent performance regarding organic matter decomposition, weight loss rate, and carbon/nitrogen ratio. Optimal aeration management could speed up the composting process and characteristics of the composted products of food wastes, such as vegetable and fruit wastes (Jolanun et al., 2008; Lai et al., 2024).

In addition to aeration, maintening certain compost parameters such as soil or oil content are important. For example, the oil content of some food wastes can be high (e.g., 17–42%) which could inhibite the start of aerobic composting of such waste (Wang et al., 2023). The reduction or removal of oil in these food wastes may be needed by adding oil-degrading microorganisms in such composting ecosystem. Specifically, Wang et al. (2023) constructed a thermophilic cooking oil degrading bacterial alliance, and applied this to food waste undergoing aerobic composting. The data provided in this work demonstrated the feasibility of the application of microorganisms in the high-temperature aerobic composting of food waste materials with high-oil content. Vermicomposting could be another method of food waste treatment through the combination of earthworms (*Lumbricina* sp.) and microbes (Maharjan et al., 2022). While appropriate amounts of salt could promote earthworm growth, it is reported that the salinity of kitchen wastes should be pretreated to less than 0.2% before vermicomposting (Wu et al., 2019).

8 Food Loss Reduction by Conversion of Food Waste to Biofuels, Platform Chemicals and Biofertilizers

8.1 Bio-conversion Pathways for Food Loss and Waste

Food waste, if left as such or landfilled will produce a more potent greenhouse gas, methane, which is estimated to have a Global Warming Potential of 27–30 over 100 years compared to 1 for CO_2 (IPCC, 2013). Therefore, biorefining of both pre- and post-processed food waste is an increasingly popular sustainable development trend for waste treatment. This process involves converting different types of food waste into biomaterials, such as biofuels, platform chemicals, and biofertilizers (Tsegaye et al., 2021; Singh et al., 2022; Kohli et al., 2024). Therefore, energy can be recovered, and valuable components can be recycled while minimizing negative impacts on the environment. The products generated can help reduce dependency on fossil fuels.

The bio-conversion process is chosen based on the desired end product and can be classified as thermochemical (e.g., combustion, pyrolysis, torrefaction, gasification, and hydrothermal) and biological (e.g, anaerobic digestion and compositing) (Rajaeifar et al., 2017). In the thermochemical conversion processes, organic material is decomposed into products (e.g., gas, oil, and char) by a combination of thermal cracking and condensation reactions. For example, pyrolysis involves indirectly heating carbon-rich materials such as mixed organic wastes, agricultural wastes, sewage sludge, plastics, etc. at high temperatures without oxygen (Chen et al., 2014; Rajaeifar et al., 2017). This process produces gas stream, oil, char, as well as other inert materials and heavy metals. Gasification, on the other hand, involves heating waste in an atmosphere with slightly reduced oxygen concentration, mainly producing syngas (Arena, 2012).

Compositing is a natural biological process in which organic materials decompose into nutrient-rich soil under controlled aerobic conditions. Anaerobic digestion, in particular, is a widely used biological process that produces methane-enriched biogas and a nutrient-rich digestate that can be used for agricultural and horticultural purposes (Peng et al., 2020). Some of the potential products that can be produced from food waste biorefineries are discussed in the following sections.

8.2 Biofuels

Biodiesel, bioethanol, biohydrogen, and biogas are the biofuels that can be generated from food waste valorization. Food waste can be a viable source of raw materials for energy production. Enzymatic hydrolysis of food waste can produce food hydrolysate, which is rich in carbohydrates, and lipids. The hydrolysate can be used to produce bioethanol, while the lipids can be converted into biodiesel. For instance, the waste generated from the production of instant noodles was used as a feedstock for biodiesel (95.4% conversion) and bioethanol (98.5% conversion) production (Yang et al., 2014). Waste cooking oil has long been used for biodiesel production (Karmee et al., 2015). This household waste was used to produce bioethanol (Matsakas et al., 2014) using liquefaction and saccharification methods. Afterward, fermentation of the remaining solids was carried out to increase the bioethanol yield. The food processing waste and animal manure was first converted into biocrude oil using hydrothermal liquefaction, followed by distillation and esterification to produce diesel blend stock (Chen et al., 2018). This upgraded distillate was shown to perform well when blended at 10–120% level with petroleum diesel in engine tests.

Hydrogen is seen as a promising future fuel because it has a high energy content (12.07 MJ/kg) and produces zero carbon emissions. One sustainable and carbon-neutral way to produce hydrogen is through biological routes. The degradation of food waste primarily produces biohydrogen along with methane and bioethanol. In a recent study, it was found that the complete upgradation of discarded and inedible parts of date fruit produced 292 mL H_2/g and 235 mL methane/g (Yahmed et al., 2021). In another report, electro-fermentation of discarded food produces 4595 mg/L of volatile fatty acids, which generated 26% biohydrogen (Sravan et al., 2018).

Discarded food can also be processed through anaerobic digestion (AD) to produce biogas, which is mainly composed of methane. To achieve the highest yield of biogas, it is important to create optimal reaction conditions. Once the gas is generated, it undergoes a cleaning and dewatering process before it can be utilized in a biogas-based fuel. The by-products obtained in the AD process are used as organic manure for agricultural purposes. Singh et al. (2023) explored the potential for biogas generation from both pre- and post-processed food waste residues. Co-digestion of food waste and municipal waste can help improve the biogas yield by 40–50%, compared to using food waste alone (Lv et al., 2021). Another study by Miller et al. (2021) examined the feasibility of using mono- and co-digestion mixtures of food waste, grease waste, brewery waste, and agricultural residues of giant miscanthus (*Miscanthus giganteus*) in an AD system. The study found that co-digestion had a greater impact on biomethane production.

8.3 Platform Chemicals

Volatile fatty acids (VFAs) have significant industrial applications such as polymer production, flavoring agents, and other products in both the chemical and pharmaceutical industries. A variety of platform chemicals, include lactic acid, acetic acid (Fazzino et al., 2021), phenols (Leite et al., 2019), polyphenols (Zhou et al., 2017), pectin, carotenoid and antioxidants, succinic acid (Zhang et al., 2013), flavanols, and anthocyanin (Pataro et al., 2017) which can be extracted from different types of food waste. For example, co-fermentation of waste-activated sludge and food waste can produce VFAs, lactic acid, and carboxylic acid (Fazzino et al., 2021). Phosphorous, vivianite, and VFAs were recovered from the co-fermentation of waste-activated sludge and food waste (Wu et al., 2020). With conversion technologies rapidly advancing, it is crucial to conduct further research on different food wastes and appropriate waste conversion technologies for the production of platform chemicals, biofuels, and biomaterials to ensure the sustainability of life on our planet.

8.4 Biofertilizers

Biofertilizers are organic fertilizers derived from biological sources and are eco-friendly and avoid harming natural sources. This helps to improve soil quality and reduce the amount of waste disposed, benefiting the environment. Biofertilizers are used to replace chemical fertilizers in the agricultural sector. Chemical fertilizers can be detrimental to the environment and even deplete soil fertility over time and have much higher greenhouse gas potential compared to biofertilizers. The most practiced method for food waste recycling is compositing due to ease of handling, storage, and transportation. However, environmental factors like temperature, pH, and food waste type make it difficult to maintain a stable degradation process. Earthworms are used to accelerate food waste degradation in vermicomposting.

In the anerobic digestion (AD) process, although biogas is the primary product, the co-product digestate is used as a biofertilizer and is crucial as a revenue-generating strategy. AD digestate as a biofertilizer has various advantages such as adjusting the C/N ratio, enhancing pH, alleviating salinity, providing organic matter, and improving water-holding capacity in soils. The co-digestion (i.e., two-stage) anaerobic process followed by the aerobic process of fruit and vegetable waste mixed with slaughterhouse waste significantly enhanced biofertilizer formulations (Chakravarty & Mandavgane, 2020). Electrodialysis was used to recover a high-purity phosphorous (81%) from waste and 74% of nitrogen (in nitrate form) was recovered using gas-permeable membranes for biofertilizer production (Oliveira et al., 2021).

Yan et al. (2023) found that using food waste biofertilizers led to the successful growth of ornamental plants, specifically blue porterweed (*Stachytarpheta jamaicensis*). Siddiqui et al. (2023) evaluated food waste-derived organic liquid fertilizers for hydroponically grown cucumbers (*Cucumis sativus*), lettuce (*Lactuca sativa*), and cherry tomatoes (*Solanum lycopersicum* var. *ceraiforme*). Multiple studies, including those by Rady et al. (2016), Zheng et al. (2016), and Tartoura et al. (2014), have shown that food waste fertilizer can enhance the chemical, physical, and biological properties of crops such as tomato, rice, pakchoi (*Brasica rapa* var. *chinensis*), and beans. By converting food waste into biofertilizer, we can improve soil health and agricultural productivity, while also addressing waste management challenges, reducing environmental impacts, and supporting sustainable food systems.

9 The Role of Artificial Intelligence and Other Emerging Technologies in Food Loss Control

Artificial intelligence and other emerging technologies such as blockchain, internet of things, precision farming, big data analytics, 3D printing, Web 2 and Web 3 applications can play a role in food waste reduction and mitigation. The 3D food printing can enable certain food waste to be converted to customized and unique food products that will be acceptable to consumers and will they reduce cost of production. However, they will also contribute to a safer and sustainable food system (Rogers & Srivastava, 2021; Adebisi, 2023; Wong et al., 2022).

According to Sharma et al. (2022), artificial intelligence (AI) can assist in overcoming challenges associated with crop health, harvesting time, market demand, food waste accumulation in addition to waste disposal in the environment. AI also has the potential to adequately track food spoilage and limit its negative impact. Sonwani et al. (2022), developed a Convolutional Neural Network model that monitors the gas emission, humidity and temperature levels of fruits and veggies by using sensors and actuators to monitor food spoilage level. The model was found to increase shelf life of some foods by 2 days. Digital platforms such as Web 2 and 3 applications, assist individuals and organizations to share, donate and distribute their excess food thereby preventing food waste and contributing to food sustainability (Kor et al., 2022). Blockchain technology can aid traceability systems to be green, smart and very effective with an added advantage of tackling or reducing food loss and food recall (Bosona & Gebresenbet, 2023).

10 Conclusions

Food loss is a major constraint to food security in both developing and developed countries. Food loss can be mitigated through various measures such as both pre- and post-harvest management of food crops, application of biopesticides, plant extracts, or antimicrobials, and active packaging of plant-derived foods, or from aquaculture, poultry, and animal production systems. These measures are undertaken in small-holder, large-scale or commercial operations. Similarly, the remedies associated with post-harvest storage, processing and transportation of food products may vary immensely across locations and countries, depending on the technologies applied during food production or utilization. However, these measures, if utilized efficiently, can go a long way to reduce food losses associated with post-harvest handling and storage of food products, such as grains, fruits, and vegetables as well as root and tubers. Although fleshy fruits and vegetables are quite perishable, meat (animal-based) as well as fish (aquaculture-based) products may be subject to tremendous loss and food waste, unless put in cold storage or processed. The extent of food loss reduction measures will continue to vary across developing and developed countries unless applicable technologies for food production, transportation, processing, and storage are streamlined through upgrading or sharing of procedures and technologies. While utilization of food waste and biowaste for nutrient cycling and bioconversion platforms are common in developed countries, such processes are meager in developing countries. Ultimately, reduction of food loss will enhance sustainability.

Acknowledgements We thank Angela Burke and Edlyn Jusino for technical assistance. This research was supported by CRIS Project No. 8072-41420-022-00D, Mitigation of foodborne pathogens in water and fresh produce via application of biochar.

Disclaimer Mention of trade names or commercial products in this article is solely for the purpose of providing specific information and does not imply recommendation or endorsement by the U.S. Department of Agriculture. "USDA is an equal opportunity provider and employer."

References

Aday, M. S., Caner, C., & Rahvalı, F. (2011). Effect of oxygen and carbon dioxide absorbers on strawberry quality. *Postharvest Biology and Technology, 62*, 179–187. https://doi.org/10.1016/J.POSTHARVBIO.2011.05.002

Adebisi, J. (2023). Combating consumer food waste. An exploration of information communication technology approach. *Journal of Digital Food, Energy & Water Systems, V4i1.2439*, 10.36615/digital-food-energy-water-systems.

Adebowale, O. O., Bankole, M., Olasoju, M., et al. (2020). Farmers' perceptions and drivers of antimicrobial use and abuse in commercial pig production, Ogun State, Nigeria. *International Journal of Environmental Research and Public Health, 17*(10), 3579. https://doi.org/10.3390/ijerph17103579

An, D. S. (2016). Effect of active master packaging system on preservation of fresh shiitake mushrooms in supply chain. *Journal of the Korean Society of Food Science and Nutrition, 45*, 402–408. https://doi.org/10.3746/JKFN.2016.45.3.402

Appendini, P., & Hotchkiss, J. H. (2002). Review of antimicrobial food packaging. *Innovative Food Science & Emerging Technologies, 3*(2), 113–126. https://doi.org/10.1016/S1466-8564(02)00012-7

Arena, U. (2012). Process and technological aspects of municipal solid waste gasification. A review. *Waste Management, 32*(4), 625–639. https://doi.org/10.1016/j.wasman.2011.09.025

Awasthi, S. K., Sarsaiya, S., Awasthi, M. K., Liu, T., Zhao, J., Kumar, S., & Zhang, Z. (2020). Changes in global trends in food waste composting: Research challenges and opportunities. *Bioresource Technology, 299*, 122555. https://doi.org/10.1016/j.biortech.2019.122555

Balan, I. M., Trasca, T. I., Brad, I., Belc, N., Tulcan, C., Radoi, B. P., & Rinovetz, A. E. (2023). Approaches to limiting food loss and food waste. In D. I. Kiba (Ed.), *Transitioning to Zero Hunger* (pp. 215–244). https://doi.org/10.3390/books978-3-03897-863-3-9

Banrie. (2012). *Current antimicrobial use and resistance issues in the European Union: Reality and resolution*. The Poultry Site. https://www.thepoultrysite.com/articles/current-antimicrobial-use-and-resistance-issues-in-the-european-union-reality-and-resolution. Accessed 08 Apr 2024

Beausang, C., Hall, C., & Toma, L. (2017). Food waste and losses in primary production: Qualitative insights from horticulture. *Resources, Conservation and Recycling, 126*, 177–185. https://doi.org/10.1016/j.resconrec.2017.07.042

Behbahani, B. A., Shahidi, F., Yazdi, F. T., Mortazavi, S. A., & Mohebbi, M. (2017). Use of Plantago major seed mucilage as a novel edible coating incorporated with Anethum graveolens essential oil on shelf life extension of beef in refrigerated storage. *International Journal of Biological Macromolecules, 94*(Pt A), 515–526. https://doi.org/10.1016/j.ijbiomac.2016.10.055

Bhutia, W., Pal, R. K., Sen, S., & Jha, S. K. (2011). Response of different maturity stages of sapota (*Manilkara achras* Mill.) cv. Kallipatti to in-package ethylene absorbent. *Journal of Food Science and Technology, 48*(6), 763–768. https://doi.org/10.1007/s13197-011-0360-x

Biji, K. B., Ravishankar, C. N., Mohan, C. O., & Srinivasa Gopal, T. K. (2015). Smart packaging systems for food applications: A review. *Journal of Food Science and Technology, 52*(10), 6125–6135. https://doi.org/10.1007/s13197-015-1766-7

Boonruang, K., Chonhenchob, V., Singh, S. P., Chinsirikul, W., & Fuongfuchat, A. (2012). Comparison of various packaging films for mango export. *Packaging Technology and Science, 25*(2), 107–118. https://doi.org/10.1002/pts.954

Borzi, F., Torrieri, E., Wrona, M., & Nerín, C. (2019). Polyamide modified with green tea extract for fresh minced meat active packaging applications. *Food Chemistry, 300*, 125242. https://doi.org/10.1016/j.foodchem.2019.125242

Bosona, T., & Gebresenbet, G. (2023). The role of blockchain technology in promoting traceability systems in agri-food production and supply chains. *Sensors, 23*, 5342. https://doi.org/10.3390/s23115342

Cao, D., Pimentel, D., & Hart, K. (2002). Postharvest food losses (vertebrates). In D. Pimentel (Ed.), *Encyclopedia of pest management* (pp. 648–649). CRC Press. https://doi.org/10.1201/NOE0824706326

Carrasco Diaz, J. M., Redondo, L. M., Redondo, E. A., Dominguez, J. E., Chacana, A. P., Fernandez Miyakawa, M. E. (2016). Use of plant extracts as an effective manner to control *Clostridium perfringens* induced necrotic enteritis in poultry. *BioMed Research International*. Article ID 3278359, 15 pages. https://doi.org/10.1155/2016/3278359

Center for Disease Control and Prevention (CDC). (2016). Disinfection By-Products. https://www.cdc.gov/safewater/chlorination-byproducts.html. Accessed 4 Feb 2022.

Cetin-Karaca, H. (2011). Evaluation of natural antimicrobial phenolic compounds against foodborne pathogens. University of Kentucky Master's Theses. Paper 652. http://ukowledge.uky.edu/graschool_theses/652. 31 Jan 2014.

Chakravarty, I., & Mandavgane, S. A. (2020). Valorization of fruit and vegetable waste for biofertilizer and biogas. *Journal of Food Process Engineering, 44*(2), e13512, 1–8. https://doi.org/10.1111/jfpe.13512

Chen, D., Yin, L., Wang, H., & He, P. (2014). Pyrolysis technologies for municipal solid waste: A review. *Waste Management, 34*(12), 2466–2486. https://doi.org/10.1016/j.wasman.2014.08.004

Chen, W. T., Zhang, Y., Lee, T. H., Wu, Z., Si, B., Lee, C.-F., Lin, A., & Sharma, B. K. (2018). Renewable diesel blendstocks produced by hydrothermal liquefaction of wet biowaste. *Nature Sustainability, 1*(11), 702–710. https://doi.org/10.1038/s41893-018-0172-3

Chimouriya, S., Lamichhane, J., Gauchan, D. P., & Dhulikhel, K. (2018). Green manure for restoring and improving the soil nutrients quality. *International Journal of Research, 5*(20), 1064–1074. e-ISSN: 2348-6848.

Chrisendo, D., Piipponen, J., Heino, M., & Kummu, M. (2023). Socioecononomic factors of global food loss. *Agriculture & Food Security, 12*, 23. https://doi.org/10.1186/s40066-023-00426-4

Dewey-Mattia, D., Manikonda, K., Hall, A. J., Wise, M. E., & Crowe, S. J. (2018). Surveillance for foodborne disease outbreaks — United States, 2009–2015. *MMWR Surveillance Summaries, 67*(SS-10), 1–11. https://doi.org/10.15585/mmwr.ss6710a1

Dominguez, R., Barba, F. J., Gomez, B., Putnik, P., Kovačević, D. B., Pateiro, M., Santos, E. M., & Lorenzo, J. M. (2018). Active packaging films with natural antioxidants to be used in meat industry: A review. *Food Research International, 113*, 93–101. https://doi.org/10.1016/j.foodres.2018.06.073

Dushkova, M. A., Simitchiev, A. T., Kalaydzhiev, H. R., Ivanova, P., Menkov, N. D., & Chalova, V. I. (2022). Comparison and modeling of moisture sorption isotherms of deproteinized rapeseed meal and model extrudate. *Journal of Food Processing & Preservation, 46*(11), e16978.

Esturk, O., Ayhan, Z., & Gokkurt, T. (2014). Production and application of active packaging film with ethylene adsorber to increase the shelf life of broccoli (*Brassica oleracea* L. var. Italica). *Packaging Technology and Science, 27*(3), 179–191. https://doi.org/10.1002/pts.2023

European Food Safety Authority (EFSA). (2013). The European Union summary report on trends and sources of zoonoses, zoonotic agents and food-borne outbreaks in 2013. *EFSA Journal, 13*(1), 3991., p. 165. https://doi.org/10.2903/j.efsa.2015.3991

European Medicine Agency (2025). European sales and use of antimicrobials for veterinary medicine(ESUAvet). Annual Surveillance report for 2023 (EMA/CVMP/ESUAVET/80289/2025). https://doi.org/10.2809/4487470

FAO (2016) *World programme for the census of agriculture 2020.* FAO Statistical Development Series No. 1, FAO: Rome, Italy, pp. 1–204. ISBN: 978-92-5-108865-4. https://openknowledge.fao.org/items/3363e104-a665-4e57-a77c-ee23c1390853

FAO. (2017). *World programme for the census of agriculture 2020.* FAO Statistical Development Series No. 1. Rome.

FAO. (2020). *World Food and Agriculture - Statistical Yearbook 2020.* Rome. https://doi.org/10.4060/cb13292n

FAOSTAT. (2016). *FAO statistical database.* https://www.fao.org/faostat/en/#home. Accessed 1 July 2016.

Fazzino, F., Mauriello, F., Paone, E., Sidari, R., & Calabrò, P. S. (2021). Integral valorization of orange peel waste through optimized ensiling: Lactic acid and bioethanol production. *Chemosphere, 271*, 129602. https://doi.org/10.1016/j.chemosphere.2021.129602

FDA. (2015). Analysis and evaluation of preventive control measures for the control and reduction/elimination of microbial hazards on fresh and fresh-cut produce: Chapter IV. Outbreaks Associated with Fresh and Fresh-Cut Produce. Incidence, Growth, and Survival of Pathogens in Fresh and Fresh-Cut Produce. https://www.fda.gov/Food/FoodScienceResearch/ucm091363.htm. Accessed 4 Feb 2019.

Fernandez Miyakawa, M. E. (2016). Use of plant extracts as an effective manner to control *Clostridium perfringens* induced necrotic enteritis in poultry. *BioMed Research International, 2016*, 3278359., 15 pages. https://doi.org/10.1155/2016/3278359

Gaikwad, K. K., Singh, S., & Ajji, A. (2019). Moisture absorbers for food packaging applications. *Environmental Chemistry Letters, 17*(2), 609–628. https://doi.org/10.1007/s10311-018-0810-z

Gautam, K., Vishvakarma, R., Sharma, P., Singh, A., Kumar Gaur, V., Varjani, S., & Kumar Srivastava, J. (2022). Production of biopolymers from food waste: Constraints and perspectives. *Bioresource Technology, 361*, 127650. https://doi.org/10.1016/j.biortech.2022.127650

Goncalves, A. L. (2021). The use of microalgae and cynaobacteria in the improvement of agricultural practices: A review on their biofertilising, biostimulating and biopesticide roles. *Applied Sciences, 11*(2), 871. https://doi.org/10.3390/app11020871

Gorski, L., Parker, C. T., Liang, A., Cooley, M. B., Jay-Russell, M. T., Gordus, A. G., Atwill, E. R., & Mandrell, R. E. (2011). Prevalence, distribution, and diversity of *Salmonella* enterica in a major produce region of California. *Applied and Environmental Microbiology, 77*(8), 2734–2748. https://doi.org/10.1128/AEM.02321-10

Guo, M., Jin, T., Wang, L., Scullen, O. J., & Sommers, C. H. (2014). Antimicrobial films and coatings for inactivation of *Listeria innocua* on ready-to-eat deli Turkey meat. *Food Control, 40*, 64–70. https://doi.org/10.1016/j.foodcont.2013.11.018

Guo, M., Zhang, X., & Jin, T. Z. (2022). Active food packaging. In W. Geoffrey (Ed.), *Encyclopedia of food safety* (2nd ed.). Smithers Publisher: Elsevier Inc. https://doi.org/10.1016/B978-0-12-822521-9.00078-2

Gustavsson, J., C. Cederberg., U. Sonesson., R. Van Otterdijk & Meybeck, R. (2011). Global food losses and food waste-extent, causes and prevention. Food and Agriculture Organization of the United Nations, Rome, 2011, 29 pg

Han, J., Ruiz-Garcia, L., Qian, J., & Yang, X. (2018). Food packaging: A comprehensive review and future trends. *Comprehensive Reviews in Food Science and Food Safety, 17*(4), 860–877. https://doi.org/10.1111/1541-4337.12343

Hansen, A. A., Morkore, T., Rudi, K., Olsen, E., & Eie, T. (2007). Quality changes during refrigerated storage of MA-packaged prerigor fillets of farmed Atlantic cod (*Gadus morhua* L.) using traditional MAP, CO_2 emitter, and vacuum. *Journal of Food Science, 72*(9), M423–M430. https://doi.org/10.1111/j.1750-3841.2007.00561.x

Hansen, A. A., Morkore, T., Rudi, K., Rodbotten, M., Bjerke, F., & Eie, T. (2009). Quality changes of prerigor filleted Atlantic salmon (*Salmo salar* l.) packaged in modified atmosphere using CO_2 emitter, traditional map, and vacuum. *Journal of Food Science, 74*(6), M242–M249. https://doi.org/10.1111/j.1750-3841.2009.01233.x

Hansen, A. A., Moen, B., Rødbotten, M., Berget, I., & Pettersen, M. K. (2016). Effect of vacuum or modified atmosphere packaging (MAP) in combination with a CO_2 emitter on quality parameters of cod loins (*Gadus mor*hua). *Food Packaging and Shelf Life, 9*, 29–37. https://doi.org/10.1016/j.fpsl.2016.05.005

Hashemi, S. R., & Davoodi, H. (2011). Herbal plants and their derivatives as growth and health promoters in animal nutrition. *Veterinary Research Communications, 35*(3), 169–180. https://doi.org/10.1007/s11259-010-9458-2

Hassan, M. K., Shipton, W. A., Coventry, R. J., & Gardiner, C. P. (2005). Maintenance of fruit quality in organically-grown bananas under modified atmosphere conditions. *Asian Journal of Plant Sciences, 4*(4), 409–412. https://doi.org/10.3923/ajps.2005.409.412

Havelaar, A. H., Kirk, M. D., Torgerson, P. R., Gibb, H. J., Hald, T., Lake, R. J., et al. (2015). World Health Organization Global estimates and regional comparisons of the burden of foodborne disease in 2010. PLoS Medicine, 2, e1001923. https://doi.org/10.1371/journal.pmed.1001923

He, Z. (Ed.). (2012). *Applied research of animal manure: Challenges and opportunities beyond the adverse environmental concerns* (p. 325). Nova Science Publishers.

He, Q., Luo, Y., & Chen, P. (2008). Elucidation of the mechanism of enzymatic browning inhibition by sodium chlorite. *Food Chemistry, 110*(4), 847–851. https://doi.org/10.1016/j.foodchem.2008.02.070

He, Z., Pagliari, P. H., & Waldrip, H. M. (2016). Applied and environmental chemistry of animal manure: A review. *Pedosphere, 26*(6), 779–816. https://doi.org/10.1016/S1002-0160(15)60087-X

He, Z., Zhang, D., & Cheng, H. N. (2021). Modeling and thermodynamic analysis of the water sorption isotherms of cottonseed products. *Foundations, 1*(1), 32–44.

HLPE, (2014). Sustainable fisheries and acquaculture for food security and nutrition. A report by the High level Panel of Experts on Food Security and Nutrition of the Committee on World Food Security, Rome 2014, 118 pg. www.fao.org/cfs/cfs-hlpe

Henriksson, P. J. G., Troell, M., & Rico, A. (2015). Antimicrobial use in aquaculture: Some complementing facts – Letter. *Proceedings of the National Academy of Sciences of the United States of America, 112*(26), E3317. https://doi.org/10.1073/pnas.1508952112

Hodges, R. J., Buzby, J. C., & Bennet, B. (2011). Postharvest losses and waste in developed and less developed countries: Opportunities to improve resource use. *The Journal of Agricultural Science, 149*(S1), 37–45. https://doi.org/10.1017/S0021859610000936

Holck, A. L., Pettersen, M. K., Moen, M. H., & Sorheim, O. (2014). Prolonged shelf life and reduced drip loss of chicken filets by the use of carbon dioxide emitters and modified atmosphere packaging. *Journal of Food Protection, 77*(7), 1133–1141. https://doi.org/10.4315/0362-028X.JFP-13-428

Hosain, M. Z., Kabir, S. M. L., & Kamal, M. M. (2021). Antimicrobial uses for livestock production in developing countries. *Veterinary World, 14*(1), 210–221. https://doi.org/10.14202/vetworld.2021.210-221

Hu, Y., & Cheng, H. (2014). Research opportunities for antimicrobial resistance control in China's factory farming. *Environmental Science & Technology, 48*, 5364–5365. https://doi.org/10.1021/es502032c

Illeperuma, K., Chamara, D., & Galappatty, P. T. (2000). Effect of modified atmosphere and ethylene absorbers on extension of storage life of "Kolikuttu" banana at ambient temperature. *Fruits, 55*(6), 381–388. https://iifiir.org/en/fridoc/effect-of-modified-atmosphere-and-ethylene-absorbers-on-extension-of-115766

IPCC. (2013). Climate change 2013: The physical science basis. In T. F. Stocker, D. Qin, G. K. Plattner, M. Tignor, S. K. Allen, J. Boschung, A. Nauels, Y. Xia, V. Bex, & P. M. Midgley (Eds.), *Contribution of working group I to the fifth assessment report of the intergovernmental panel on climate change* (pp. 1–1535). Cambridge University Press. https://www.ipcc.ch/report/ar5/wg1/. Accessed 9 Apr 2024

Jaber, L. R., Araj, S. E., & Qasem, J. R. (2018). Compatibility of endophytic fungal entomopathogens with plant extracts for the management of sweetpotato whitefly *Bemisa tabaci* Gennadius (Homoptera: Aleyrodidae). *Biological Control, 117*, 164–171. https://doi.org/10.1016/j.biocontrol.2017.11.009

Jackson, B., Griffin, P., Cole, D., Walsh, K., & Chai, S. (2013). Outbreak-associated *Salmonella* enterica serotypes and food commodities, United States, 1998–2008. *Emerging Infectious Diseases, 19*, 1239–1244. https://doi.org/10.3201/eid1908.121511. [PubMed: 23876503].

Jin, T. Z., Huang, M., Niemira, B., & Cheng, L. (2016). Shelf life extension of fresh ginseng roots using sanitizer washing, edible antimicrobial coating and modified atmosphere packaging. *International Journal of Food Science and Technology, 51*(9), 2132–2139. https://doi.org/10.1111/ijfs.13201

Jin, T. Z., Huang, M., Niemira, B. A., & Cheng, L. (2017). Microbial reduction and quality preservation of fresh ginseng roots using nonthermal processing and antimicrobial packaging. *Journal of Food Processing & Preservation, 41*(1), e12871. https://doi.org/10.1111/jfpp.12871

Jin, T., Chen, W., Gurtler, J. B., & Fan, X. (2020). Effectiveness of edible coatings to inhibit browning and inactivate foodborne pathogens on fresh-cut apples. *Journal of Food Safety, 40*(4), e12802. https://doi.org/10.1111/jfs.12802

Jin, T. Z., Fan, X., & Mukhopadhyay, S. (2022). Antimicrobial coating with organic acids and essential oil for the enhancement of safety and shelf life of grape tomatoes. *International Journal of Food Microbiology, 378*, 109827. https://doi.org/10.1016/j.ijfoodmicro.2022.109827

Jin, T. Z., Yadav, M. P., & Qi, P. (2023). Antimicrobial and physiochemical properties of films and coatings prepared from bio-fiber gum and whey protein isolate conjugates. *Food Control, 148*, 109666. https://doi.org/10.1016/j.foodcont.2023.109666

Jolanun, B., Towprayoon, S., & Chiemchaisri, C. (2008). Aeration improvement in fed batch composting of vegetable and fruit wastes. *Environmental Progress, 27*(2), 250–256. https://doi.org/10.1002/ep.10259

Karmee, S., Patria, R., & Lin, C. (2015). Techno-economic evaluation of biodiesel production from waste cooking oil—A case study of Hong Kong. *International Journal of Molecular Sciences, 16*(30), 4362–4371. https://doi.org/10.3390/ijms16034362

Kohli, K., Prajapati, R., Shah, R., Das, M., & Sharma, B. K. (2024). Food waste: Environmental impact and possible solutions. *Sustainable Food Technology, 2*, 70–80. https://doi.org/10.1039/D3FB00141E

Kor, B., Krawezyk, A., & Wakkee, I. (2022). Addressing food loss and waste prevention. *British Food Journal, 124*, 2434–2460. https://doi.org/10.1108/BFJ-05-2021-0571

Koul, O. (2023). Biopesticides: Commercial opportunites and challenges. Chapter 1. In *Development and commercilaization of biopesticides: Costs and benefits* (pp. 1–23). Elsevier, Inc.. https://doi.org/10.1016/B978-0-323-95290-3.00009-1

Kumar, D., & Kalita, P. (2017). Reducing postharvest losses during storage of grain crops to strengthen food security in developing countries. *Food Security, 6*(1), 8, 22 pages. https://doi.org/10.3390/foods6010008

Kumar, S., Singh, N., Devi, L. S., Kumar, S., Kamle, M., Kumar, P., & Mukherjee, A. (2022). Neem oil and its nanoemulsion in sustainable food preservation and packaging: Current status and future prospects. *Journal of Agriculture and Food Research, 7*, 100254, 9 pages. https://doi.org/10.1016/j.jafr.2021.100254

Lai, J. C., Then, Y. L., Hwang, S. S., & Lee, C. S. (2024). Optimal aeration management strategy for a small-scale food waste composting. *Carbon Resources Conversion, 7*(1), 100190. https://doi.org/10.1016/j.crcon.2023.06.002

Lee, D. S. (2016). Review: Carbon dioxide absorbers for food packaging applications. *Trends in Food Science and Technology, 57*(Part A), 146–155. https://doi.org/10.1016/j.tifs.2016.09.014

Lee, D. S., Wang, H. J., Jaisan, C., & An, D. S. (2021). Active food packaging to control carbon dioxide. *Packaging Technology and Science, 35*(3), 213–227. https://doi.org/10.1002/pts.2627

Leite, P., Silva, C., Salgado, J. M., & Belo, I. (2019). Simultaneous production of lignocellulolytic enzymes and extraction of antioxidant compounds by solid-state fermentation of agro-industrial wastes. *Industrial Crops and Products, 137*, 315–322. https://doi.org/10.1016/j.indcrop.2019.04.044

López-Gálvez, F., Gómez, P. A., Artés, F., & Artés-Hernández, F. (2021). Interactions between microbial food safety and environmental sustainability in the fresh produce supply chain. *Food Security, 10*(7), 1655. https://doi.org/10.3390/foods10071655

Lopez-Serrano, M., & Ros Barcelo, A. (1999). H_2O_2-mediated pigment decay in strawberry as a model system for studying color alternations in processed plant foods. *Journal of Agricultural and Food Chemistry, 47*(3), 824–827. https://doi.org/10.1021/jf980576w

Lv, Y., Chang, N., Li, Y. Y., & Liu, Y. (2021). Anaerobic co-digestion of food waste with municipal solid waste leachate: A review and prospective application with more benefits. *Resources, Conservation and Recycling, 174*, 105832. https://doi.org/10.1016/j.resconrec.2021.105832

Ma, F., Xu, S., Tang, Z., Li, Z., & Zhang, L. (2021). Use of antimicrobials in food animals and impact of transmission of antimicrobial resistance on humans. *Biosafety and Health, 3*(1), 32–38. https://doi.org/10.1016/j.bsheal.2020.09.004

Magalhães, V. S. M., Ferreira, L. M. D. F., & Silva, C. (2021). Using a methodological approach to model causes of food loss and waste in fruit and vegetable supply chains. *Journal of Cleaner Production, 283*, 124574. https://doi.org/10.1016/j.jclepro.2020.124574

Mahajan, P. V., Caleb, O. J., Singh, Z., Watkins, C. B., & Geyer, M. (2014). Postharvest treatments of fresh produce. *Philosophical Transactions of the Royal Society A, 372*(2017), 20130309. https://doi.org/10.1098/rsta.2013.0309

Maharjan, K. K., Noppradit, P., & Techato, K. (2022). Suitability of vermicomposting for different varieties of organic waste: A systematic literature review (2012–2021). *Organic Agriculture, 12*(07), 581–602. https://doi.org/10.1007/s13165-022-00413-2

Maiti, R., Thakur, A. K., Gupta, A. I., & Debashis, M. (2018). Post-harvest management of agricultural produce. In *Research trends in bioresource management and technology* (pp. 137–166). American Academic Press.

Manandhar, A., Milindi, P., & Shah, A. (2018). An overview of the post-harvest grain storage practices of smallholder farmers in developing countries. *Agriculture, 8*(4), 57. https://doi.org/10.3390/agriculture8040057

Matsakas, L., Kekos, D., Loizidou, M., & Christakopoulos, P. (2014). Utilization of household food waste for the production of ethanol at high dry material content. *Biotechnology for Biofuels, 7*, 4. https://doi.org/10.1186/1754-6834-7-4

Miller, K. E., Herman, T., Philipinanto, D. A., & Davis, S. C. (2021). Anaerobic digestion of food waste, brewery waste, and agricultural residues in an off-grid continuous reactor. *Sustainable Forestry, 13*(12), 6509. https://doi.org/10.3390/su13126509

Moore, P. R., Eenson, A., Luckey, T. D., Mccoy, E., Elvehjem, C. A., & Hart, E. B. (1946). Use of sulfasuxidine, streptothricin, and streptomycin in nutritional studies with the chick. *The Journal of Biological Chemistry, 165*(2), 437–441.

Mottet, A., & Tiempio, G. (2017). Global poultry production: Current state and future outlook and challenges. *World's Poultry Science Journal, 73*(2), 245–256. https://doi.org/10.1017/S0043933917000071

Mshana, S. E., Sindato, C., Matee, M. I. N., & Mboera, L. E. G. (2021). Antimicrobial use and resistance in agriculture and food production systems in Africa: A systematic review. *Antibiotics, 10*(8), 976. https://doi.org/10.3390/antibiotics10080976

Mukhopadhyay, S., & Ramaswamy, R. (2012). Application of emerging technologies to control *Salmonella* in foods: A review. *Food Research International, 45*(2), 666–677. https://doi.org/10.1016/j.foodres.2011.05.016

Mukhopadhyay, S., Ukuku, D. O., Juneja, V. K., Nayak, B., & Olanya, M. (2018). Principles of food preservation. In V. Juneja, H. P. Dwivedi, & J. Sofos (Eds.), *Microbial control and food preservation: Theory and practice* (pp. 17–33). Springer. ISBN: 978-1-4939-7554-9. https://link.springer.com/book/10.1007/978-1-4939-7556-3#bibliographic-information. Accessed 09 Apr 2024

Nisa, B. A., Ashwar, A., Shah, A., Gani, A., Gani, F., & Masoodi, A. (2015). Development of potato starch based active packaging films loaded with antioxidants and its effect on shelf life of beef. *Journal of Food Science and Technology, 52*(11), 7245–7253. https://doi.org/10.1007/s13197-015-1859-3

Nugraha, B., Bintoro, N., & Murayama, H. (2015). Influence of CO_2 and C_2H_4 adsorbents to the symptoms of internal browning on the packaged 'silver bell' pear (Pyrus communis L.). *Agriculture and Agricultural Science Procedia, 3*, 127–131. https://doi.org/10.1016/j.aaspro.2015.01.025

Nunan, C. (2022). Achieving responsible farm antibiotic use through improving animal health and welfare in pig and poultry production: Ending routine farm antibiotic use in Europe. 2022. European Public Health Alliance (EPHA) (pp. 1–73). https://epha.org/wp-content/uploads/2022/02/report-ending-routine-farm-antibiotic-use-in-europe-final-2022.pdf. Accessed 09 Apr 2024.

O'Connor, J., Skeaff, S., Bremer, P., Lucci, G., & Mirosa, M. (2023). A critical review of on-farm food loss and waste: Future research and policy recommendations. *Renewable Agriculture and Food Systems, 38*(e24), 1–12. https://doi.org/10.1017/S1742170523000169

Odey, T. O. J., Tanimowo, W. O., Afolabi, K. O., Jahid, I. K., & Reuben, R. C. (2023). Antimicrobial use and resistance in food animal production: Food safety and associated concerns in Sub-Saharan Africa. *International Microbiology, 27*(1), 1–23.

Ojwang, D., Nyankanga, R., Imungi, J., Olanya, M., & Ukuku, D. (2016). Cultivar preference and sensory evaluation of vegetable pigeon pea (*Cajanus cajan*) in eastern Kenya. *Food Security, 8*, 757–767. https://doi.org/10.1007/s12571-016-0592-8

Ojwang, D., Nyankanga, R. O., Imungi, J., Ganga, R. N., Olanya, O. M., He, Z., & Juneja, V. K. (2021). Effects of processing and storage on nutrient composition of green vegetable pigeonpeas. *Journal of Food Processing & Preservation, 45*(9), e15714. https://doi.org/10.1111/jfpp.15714

Olajuyin, A. M., Raji, H., Iwu, A., Okunbor, H., Albert, K., Olajuyin, A. K., & Bamigbade, A. T. (2023). Food waste matters—The policy implications and sustainability of food waste practices. *Research, Society and Development, 12*(9), e2012942212. https://doi.org/10.33448/rsd-v12i9.42212

Olanya, O. M., & Larkin, R. P. (2006). Efficacy of essential oils and biopesticides on *Phytophthora infestans* suppression in laboratory and growth chamber studies. *Biocontrol Science and Technology, 16*(9), 112–119. https://doi.org/10.1080/09583150600827918

Olanya, O. M., Sites, J. E., & Hoshide, A. K. (2016a). Cost modelling of *Pseudomonas fluorescens* and *Pseudomonas chlororaphis* as biocontrol for competitive exclusion of *Salmonella enterica* on tomatoes. *Biocontrol Science and Technology, 26*(5), 651–664. https://doi.org/10.1080/09583157.2016.1142503

Olanya, O. M., Anwar, M., He, Z., Larkin, R. P., & Honeycutt, C. W. (2016b). Survival potential of *Phytophthora infestans* in relation to environmental factors and late blight occurrence. *Journal of Plant Protection Research, 56*(1), 73–81. https://doi.org/10.1515/JPPR-2016-0011

Olanya, O. M., Ukuku, D. O., Solaiman, D. S. Y., Ashby, R. D., Niemira, B. A., & Mukhopadhyay, S. (2018). Reduction of *Listeria monocytogenes, Salmonella enterica,* and *Escherichia coli* O157:H7 populations invitro and on tomato by sophorolipid and sanitizer as affected by temperature and storage time. *International Journal of Food Science and Technology, 53*(5), 1303–1315. https://doi.org/10.1111/ijfs.13711

Olanya, O. M., Hoshide, A. K., Oluwatosin, I. A., Ukuku, D. O., Mukhopadhyay, S., Niemira, B. A., & Ayeni, O. (2019). Cost estimation of listeriosis (*Listeria monocytogenes*) occurrence in South Africa in 2017-2018 and its food safety implications. *Food Control, 102*, 231–239. https://doi.org/10.1016/j.foodcont.2019.02.007

Olanya, O. M, Olanya B. A., Niemira J. M., Cassidy G., Boyd J., Uknalis (2020). Pathogen reduction by predatory bacteria and survival of Bdellovibrio bacteriovorus and Escherichia coli on produce and buffer treated with low-dose gamma radiation LWT 130109630. https://doi.org/10.1016/j.lwt.2020.109630

Oliveira, V., Dias-Ferreira, C., González-García, I., Labrincha, J., Horta, C., & García-González, M. C. (2021). A novel approach for nutrients recovery from municipal waste as biofertilizers by combining electro dialytic and gas permeable membrane technologies. *Waste Management, 125*, 293–302. https://doi.org/10.1016/j.wasman.2021.02.055

Ondier, G. O., Siebenmorgen, T. J., Bautista, R. C., & Mauromoustakos, A. (2011). Equilibrium moisture contents of pureline, hybrid, and parboiled rice. *Transactions of ASAE, 54*(3), 1007–1013.

Pataro, G., Bobinaite, R., Bobinas, Č., Šatkauskas, S., Raudonis, R., ˙ Visockis, M., Ferrari, G., & Viškelis, P, 2017. Improving the extraction of juice and anthocyanins from blueberry fruits and their by-products by application of pulsed electric fields. Food and Bioprocess Technology, 10, 1595–1605. https://doi.org/10.1007/s11947-017-1928-x

Peng, W., Lu, F., Hao, L., Zhang, L., & Shao, P. (2020). Digestate management for high-solid anaerobic digestion of organic wastes: A review. *Bioresource Technology, 297*, 122485. https://doi.org/10.1016/j.biortech.2019.122485

Rad, S. N., Shirazi, M. M. A., Kargari, A., & Marzban, R. (2016). Application of membrane separation technology in downstream processing of *Bacillus thuringiensis* biopesticide: A review. *Journal of Membrane Science and Research, 2*, 66–77. https://doi.org/10.22079/JMSR.2016.19154

Rady, M. M., Semida, W. M., Hemida, K. A., & Abdelhamid, M. T. (2016). The effect of compost on growth and yield of *Phaseolus vulgaris* plants grown under saline soil. *International Journal of Recycling of Organic Waste in Agriculture, 5*, 311–321. https://doi.org/10.1007/s40093-016-0141-7

Rajaeifar, M. A., Ghanavati, H., Dashti, B. B., Heijungs, R., Aghbashlo, M., & Tabatabaei, M. (2017). Electricity generation and GHG emission reduction potentials through different

municipal solid waste management technologies: A comparative review. *Renewable and Sustainable Energy Reviews, 79*, 414–439. https://doi.org/10.1016/j.rser.2017.04.109
Ranganathan, S., Dutta, S., Moses, J. A., & Anandharamakrishnan, C. (2020). Utilization of Food waste streams for the production of biopolymers. *Heliyon, 6*(9), e04891. https://doi.org/10.1016/j.heliyon.2020.e04891
ReFED. (2024). Food waste is a solvable problem—Here's how to do it. https://refed.org/food-waste/the-solutions#solutions. Accessed 9 Apr 2024.
Rogers, H., & Srivastava, M. (2021). Emerging sustainable supply chain models for 3D food printing. *Sustainable Forestry, 13*, 12085. https://doi.org/10.3390/su132112085
Sarker, A., Ahmmed, R., Ahsan, S. M., Rana, J., Ghosh, M. K., & Nandi, R. (2023). A comprehensive review of food waste valorization for the sustainable management of global food waste. *Sustainable Food Technology, 2*, 48–69.
Savary, S., Willocquet, L., Pethybrige, S. J., Esker, P., McRoberts, N., & Nelson, A. (2019). The global burden of pathogens and pests on major food crops. *Nature Ecology & Evolution, 3*(3), 430–439. https://doi.org/10.1038/s41559-018-0793-y
Scallan, E., Hoekstra, R. M., Angulo, F. J., Tauxe, R. V., Widdowson, M. A., Roy, S. L., Jones, J. L., & Griffin, P. M. (2011). Foodborne illness acquired in the United States-major pathogens. *Emerging Infectious Diseases, 17*(1), 7–15. https://doi.org/10.3201/eid1701.p11101
Sharma, P., Virmal, A., Vishvakarma, R., Kumar, P., Vandenberghe, L. P., Gaur, V. K., & Varjani, S. (2022). Deciphering the black box of omics approaches and artificial intelligence in food waste transformation and mitigation. *International Journal of Food Microbiology, 372*, 109691. https://doi.org/10.1016/j.ijfoodmicro.2022.109691
Sharma, P., Bano, A., Verma, K., Yadav, M., Varjani, S., Singh, S. P., & Tong, Y. W. (2023). Food waste digestate as biofertilizer and their direct applications in agriculture. *Bioresource Technology Reports, 23*, 101515. https://doi.org/10.1016/j.biteb.2023.101515
Sharma, P., Tiong, Y. W., Yan, M., Lam, H. T., Zhang, J., & Tong, Y. W. (2024). Plant growth enhancement under combined amendments of food waste digestate and rhizosphere microbial co-occurrence. *ACS Sustainable Resource Management, 1*(3), 563–576. https://doi.org/10.1021/acssusresmgt.4c00051
Siddiqui, Z., Hagare, D., Liu, M. H., Panatte, O., Hussain, T., Memon, S., Noorani, A., & Chen, Z. H. (2023). A food waste derived organic liquid fertiliser for sustainable hydroponic cultivation of lettuce, cucumber, and cherry tomato. *Food, 12*(4), 719.
Silva, D. F. P., Salomao, L. C. C., Lopes de Siqueira, D., Cecon, P. R., & Rocha, A. (2009). Potassium permanganate effects in postharvest conservation of the papaya cultivar sunrise golden. *Pesquisa Agropecuária Brasileira, 44*(7), 669–675. https://doi.org/10.1590/S0100-204X2009000700003
Singh, R., & Giri, S. K. (2014). Shelf life study of guava (*Psidium guajava L.*) under active packaging: An experiment with potassium permanganate salt as ethylene absorbent. *Journal of Food Safety and Food Quality, 65*(2), 32–39. https://doi.org/10.2376/0003-925X-65-32
Singh, P. K., Mohanty, P., Mishra, S., & Adhya, T. K. (2022). Food waste valorization for biogas-based bioeconomy production in circular bioeconomy: Opportunities, challenges, and future developments. *Frontiers in Energy Research, 10*, 903775. https://doi.org/10.3389/fenrg.2022.903775
Singh, D., Tembhare, M., Machhirake, N., & Kumar, S. (2023). Biogas generation potential of discarded food waste residue from ultra-processing activities at food manufacturing and packaging industry. *Energy, 263*(Part E), 126138. https://doi.org/10.1016/j.energy.2022.126138
Siripatrawan, U., & Harte, B. R. (2010). Physical properties and antioxidant activity of an active film from chitosan incorporated with green tea extract. *Food Hydrocolloids, 24*(8), 770–775. https://doi.org/10.1016/j.foodhyd.2010.04.003
Soltani Firouz, M., Mohi-Alden, K., & Omid, M. (2021). A critical review on intelligent and active packaging in the food industry: Research and development. *Food Research International, 141*, 110113. https://doi.org/10.1016/j.foodres.2021.110113

Song, S., Lim, J. W., Lee, J. T. E., Cheong, J. C., Hoy, S. H., Hu, Q., Tan, J. K. N., Chiam, Z., Arora, S., Lum, T. Q. H., Lim, E. Y., Wang, C.-H., Tan, H. T. W., & Tong, Y. W. (2021). Food-waste anaerobic digestate as a fertilizer: The agronomic properties of untreated digestate and biochar-filtered digestate residue. *Waste Management, 136*, 143–152. https://doi.org/10.1016/j.wasman.2021.10.011

Sonwani, E., Bansal, U., Alroobaca, R., Baqasah, A. M., & Hedabou, M. (2022). An artificial intelligence approach toward food spoilage detection and analysis. *Frontiers in Public Health, 9*, 816226. https://doi.org/10.3389/fpubh.2021.816226

Sravan, J. S., Butti, S. K., Sarkar, O., Krishna, K. V., & Mohan, S. V. (2018). Electrofermentation of food waste—Regulating acidogenesis towards enhanced volatile fatty acids production. *Chemical Engineering Journal, 334*, 1709–1718. https://doi.org/10.1016/j.cej.2017.11.005

Stokstad, E. L. R., & Jukes, T. H. (1950). Further observations on the "animal protein factor". *Experimental Biology and Medicine, 73*, 523–528. https://doi.org/10.3181/00379727-73-17731

Tang, K. F. J., & Bondad-Reantaso, M. G. (2019). Impacts of acute hepatopancreatic necrosis disease on commercial shrimp aquaculture. *Revue Scientifique et Technique, 38*(2), 477–490. https://doi.org/10.20506/rst.38.2.2999

Tartoura, K. A., Youssef, S. A., & Tartoura, E. S. A. (2014). Compost alleviates the negative effects of salinity via up-regulation of antioxidants in *Solanum lycopersicum* L. plants. *Plant Growth Regulation, 74*, 299–310. https://doi.org/10.1007/s10725-014-9923-y

Teigiserova, D. A., Hamelin, L., & Thomsen, M. (2020). Towards transparent valorization of food surplus, waste and loss: Clarifying definitions, food waste hierarchy, and role in the circular economy. *Science of the Total Environment, 706*, 136033. https://doi.org/10.1016/j.scitotenv.2019.136033

Tejada, M., Gonzalez, J., García-Martínez, A., & Parrado, J. (2008). Application of a green manure and green manure composted with beet vinasse on soil restoration: Effects on soil properties. *Bioresource Technology, 99*(11), 4949–4957. https://doi.org/10.1016/j.biortech.2007.09.026

Torrijos, V., Dopico, D. C., & Soto, M. (2021). Integration of food waste composting and vegetable gardens in a university campus. *Journal of Cleaner Production, 315*, 128175. https://doi.org/10.1016/j.jclepro.2021.128175

Tournas, V. H. (2005). Spoilage of vegetable crops by bacteria and fungi and related health hazards. *Critical Reviews in Microbiology, 31*(1), 33–44. https://doi.org/10.1080/10408410590886024

Troell, M., Naylor, R. L., Metian, M., Beveridge, M., Tyedmers, P. H., Folke, C., Arrow, K. J., Barrett, S., Crépin, A.-S., Ehrlich, P. R., Gren, A., Kautsky, N., Levin, S. A., Nyborg, K., Österblom, H., Polasky, S., Scheffer, M., Walker, B. H., Xepapadeas, T., & de Zeeuw, A. (2014). Does aquaculture add resilience to the global food system? *Proceedings of the National Academy of Sciences, 111*(37), 13257–13263. https://doi.org/10.1073/pnas.140406711

Tsegaye, B., Jaiswal, S., & Jaiswal, A. K. (2021). Food waste biorefinery: Pathway towards circular bioeconomy. *Food Security, 10*(6), 1174. https://doi.org/10.3390/foods10061174

Tyśkiewicz, R., Nowak, A., Ozimek, E., & Jaroszuk-Ściseł, J. (2022). Trichoderma: The current status of its application in agriculture for the biocontrol of fungal phytopathogens and stimulation of plant growth. *International Journal of Molecular Sciences, 23*, 2329. https://doi.org/10.3390/ijms23042329

U.S. Food Drug Administration (FDA). (2022). *Annual summary report on antimicrobials sold or distributed in 2021 for use in food-producing animals*. AASV News Archive https://www.fda.gov/animal-veterinary/cvm-updates/fda-releases-annual-summary-report-antimicrobials-sold-or-distributed-2021-use-food-producing. Accessed 09 Apr 2024

Ukuku, D. O. (1997). Sensitivity of six strains of *Listeria monocytogenes* to nisin. *Journal of Food Protection, 60*(7), 867–869. https://doi.org/10.4315/0362-028X-60.7.867

Ukuku, D. O. (2023). *Synergistic antibrowning-antimicrobial composition*. USDA DN: 0052.19, U.S. Patent Application No. 17/020,363, Patent#: 11,641,864.

Ukuku, D. O., & Fett, W. F. (2002). Relationships of cell surface charge and hydrophobicity with strength of attachment of bacteria to cantaloupe rind. *Journal of Food Protection, 65*(7), 1093–1099. https://doi.org/10.4315/0362-028X-65.7.1093

Ukuku, D. O., & Fett, W. F. (2006). Effect of cell surface charge and hydrophobicity on attachment of 16 *Salmonella* serovars to cantaloupe rind and decontamination with sanitizers. *Journal of Food Protection, 69*(8), 1835–1843. https://doi.org/10.4315/0362-028x-69.8.1835

Ukuku, D. O., & Shelef, L. A. (1996). Technical note: Bioluminescence measurements of the antilisterial activity of nisin: Comparison with ampicillin and streptomycin. *Journal of Bioluminescence and Chemiluminescence, 11*(3), 169–173. https://doi.org/10.1002/(SICI)1099-1271(199605)11:3<169::AID-BIO410>3.0.CO;2-R

Ukuku, D. O., Zhang, H. Q. and Huang, L. (2009). Growth parameters of Escherichia coli, Salmonella enteritidis and Listeria monocytogenes, and Aerobic Mesophilic Bacteria of Apple cider amended with nisin-EDTA. Foodborne Pathogens and Disease. 6:487-494. 2009.

Ukuku, D. O., Bari, M. L., & Kawamoto, S. (2011). Hydrogen peroxide (Chapter 11). In V. M. Gomez-Lopez (Ed.), *Decontamination of minimally processed produce*. Wiley and Blackwell: ,. ISBN: 9780813823843. https://doi.org/10.1002/9781118229187

Ukuku, D. O., Mukhopadhyay, S., & Olanya, M. (2013). Effect of organic acid treatments on microbial safety and overall acceptability of fresh-cut cantaloupe cubes. *Journal of Food, Agriculture and Environment, 11*(3), 340–345.

Ukuku, D. O., Mukhopadhyay, S., Geveke, D., Olanya, M., & Niemira, B. (2016). Effect of hydrogen peroxide in combination with minimal thermal treatment for reducing bacterial populations on cantaloupe rind surfaces and transfer to fresh-cut pieces. *Journal of Food Protection, 79*(8), 1316–1324. https://doi.org/10.4315/0362-028X.JFP-16-046

Ukuku, D. O., Niemira, B., & Uknalis, J. (2019). Nisin-based antimicrobial combination with cold plasma treatment inactivate *Listeria monocytogenes* on granny smith apples. *LWT - Food Science and Technology, 104*, 120–127. https://doi.org/10.1016/j.lwt.2018.12.049

USDA-ERS. (2024). *Livestock, dairy and poultry outlook: March 2024.* Situation and Outlook Report, LDP-M-357, 27 pages. https://www.ers.usda.gov/webdocs/outlooks/108794/ldp-m-357.pdf?v=4940.4. Accessed 09 Apr 2024.

Vagsholm, I., Arzoomand, N. S., & Boquit, S. (2020). Food security, safety, and sustainability—Getting the trade-offs right. *Frontiers in Sustainable Food Systems, 4*(16), 1–14. https://doi.org/10.3389/fsufs.2020.00016

Veasna, H., Hwang, Y.-S., Choi, J.-M., Ahn, Y.-J., Lim, B.-S., & Chun, J.-P. (2012). 1-Methylcyclopropene and carbon dioxide absorber reduce chilling injury of eggplant (*Solanum melongena* L.) during MAP storage. *Journal of Bio-Environment Control, 21*(1), 50–56. https://koreascience.kr/article/JAKO201217355622844.page. Accessed 09 Apr 2024

Vero, S., Garmendia, G., Allori, E., Sanz, J. M., Gonda, M., Alconada, T., Cavello, I., Dib, J. R., Diaz, M. A., Nally, C., Pimenta, R. S., da Silva, J. F. M., Vargas, M., Zaccari, F., & Wisniewski, M. (2023). Microbial biopesticides: Diversity, scope, and mechanisms involved in plant disease control. *Diversity, 15*(3), 457. https://doi.org/10.3390/d15030457

Wang, H. J., An, D. S., Rhim, J.-W., & Lee, D. S. (2015). A multi-functional biofilm used as an active insert in modified atmosphere packaging for fresh produce. *Packaging Technology and Science, 28*(12), 999–1010. https://doi.org/10.1002/pts.2179

Wang, Y., Yuan, Z., & Tang, Y. (2021). Enhancing food security and environmental sustainability: A critical review of food loss and waste management. *Resources, Environment and Sustainability, 4*(100023), 1–13. https://doi.org/10.1016/j.resenv.2021.100023

Wang, W., Zhao, Z., Yang, J., Lian, X., Xie, X., Chen, H., Wang, M., & Zheng, H. (2023). Application of oil-degrading agents consisted of thermophilic *Bacillus subtilis* and *Bacillus glycinifermentans* in food waste. *Environmental Technology, 45*, 4704. https://doi.org/10.1080/09593330.2023.2283064

Wang, F., Chen, C., Fu, D., & Singh, R. P. (2024). Effect of reactor temperature and feeding ratio on fed-batch composting of household food waste and green wastes. *Biomass and Bioenergy, 181*, 107040. https://doi.org/10.1016/j.biombioe.2023.107040

Watada, A. E., Abe, K., & Yamuchi, N. (1990). Physiological activities of partially processed fruits and vegetables. *Food Technology, 44*, 116–122. https://ucanr.edu/datastoreFiles/234-1893.PDF. Accessed 09 Apr 2024

Wong, G. H., Pant, A., Zhang, Y., Chua, C. C., Hashimoto, M., Leo, C. H., & Tau, U. (2022). 3D food printing sustainability through food waste upcycling. *Materials Today Proceedings, 70*, 627–630.

Wu, Z., Yin, B., Song, X., Qiu, J., Cao, L., & Zhao, Q. (2019). Effects of salinity on earthworms and the product during vermicomposting of kitchen wastes. *International Journal of Environmental Research and Public Health, 16*(23), 4737. https://doi.org/10.3390/ijerph16234737

Wu, Y., Cao, J., Zhang, T., Zhao, J., Xu, R., & Zhang, Q. (2020). A novel approach of synchronously recovering phosphorus as vivianite and volatile fatty acids during waste activated sludge and food waste co-fermentation: Performance and mechanisms. *Bioresource Technology, 305*, 123078. https://doi.org/10.1016/j.biortech.2020.123078

Yahmed, N. B., Dauptain, K., Lajnef, I., Carrere, H., Trably, E., & Smaali, I. (2021). New sustainable bioconversion concept of date by-products (Phoenix dactylifera L.) to biohydrogen, biogas and date-syrup. *International Journal of Hydrogen Energy, 46*(1), 297–305. https://doi.org/10.1016/j.ijhydene.2020.09.203

Yan, M., Tian, H., Song, S., Tan, H. T., Lee, J. T., Zhang, J., Sharma, P., Tiong, Y. W., & Tong, Y. W. (2023). Effects of digestate-encapsulated biochar on plant growth, soil microbiome and nitrogen leaching. *Journal of Environmental Management, 334*, 117481. https://doi.org/10.1016/j.jenvman.2023.117481

Yang, X., Lee, J. H., Yoo, H. Y., Shin, H. Y., Thapa, L. P., Park, C., & Kim, S. W. (2014). Production of bioethanol and biodiesel using instant noodle waste. *Bioprocess and Biosystems Engineering, 37*, 1627–1635. https://doi.org/10.1007/s00449-014-1135-3

Yang, W., Ana, M. M., Sousa, A. M. M., Thomas-Gahring, A., Fan, X., Jin, T. Z., Li, X., Tomasula, P., & Liu, L. S. (2016). Electrospun polymer nanofibers reinforced by tannic acid/Fe+++ complexes. *Materials, 9*(9), 757. 1–12. https://doi.org/10.3390/ma9090757

Yildirim, S., Röcker, B., Pettersen, M. K., Nilsen-Nygaard, J., Ayhan, Z., Rutkaite, R., Radusin, T., Suminska, P., Marcos, B., & Coma, V. (2018). Active packaging applications for food. *Comprehensive Reviews in Food Science and Food Safety, 17*(1), 165–199. https://doi.org/10.1111/1541-4337.12322

Yue, W., & Takács-György, K. (2023). Why does food loss and waste matter for food security—From the perspective of cause and magnitude. *Ecocycles 2023, 9*(3), 47–61. https://www.ecocycles.net/ojs/index.php/ecocycles/article/view/337

Zhang, A. Y., Sun, Z., Leung, C. C. J., Han, W., Lau, K. Y., Li, M., & Lin, C. S. K. (2013). Valorisation of bakery waste for succinic acid production. *Green Chemistry, 15*, 690–695. https://doi.org/10.1039/C2GC36518A

Zheng, S., Jiang, J., He, M., Zou, S., & Wang, C. (2016). Effect of kelp waste extracts on the growth and development of Pakchoi (*Brassica chinensis* L.). *Scientific Reports, 6*(1), 38683. https://doi.org/10.1038/srep38683

Zhou, Y., Zheng, J., Gan, R. Y., Zhou, T., Xu, D. P., & Li, H. (2017). Bin optimization of ultrasound-assisted extraction of antioxidants from the mung bean coat. *Molecules, 22*(4), 638. https://doi.org/10.3390/molecules22040638

Part II
What Are the Options?

Physical Antimicrobial Strategies Used in Food Systems

Jim Junhui Huang, Dan Li, and Weibiao Zhou

1 Introduction

Physical antimicrobial strategies are a broad category of disinfection approaches by applying thermal or non-thermal processing techniques, mainly based on electromagnetic waves, temperature controls, photodynamic inactivation, filtration, electric field, ultrasound (US), cold plasma and high hydrostatic pressure, to efficiently inactivate environmental microorganisms, especially those contaminated the foodstuffs, for safety and healthy purposes (Han et al., 2020; Qi et al., 2021; Pokhrel et al., 2022; Bai et al., 2023; Gallus et al., 2023; Jabin et al., 2023; Ahmed & Mohamed, 2024; Casado et al., 2024; Chong et al., 2024; Mamone et al., 2024; Spackman et al., 2024; Xu et al., 2024a). The advantages of physical antimicrobial techniques are obvious as they are a kind of clean technology that does not lead to contamination in the environment just as chemical antimicrobial techniques do. In addition, they normally do not require long processing duration, and their processing costs are relatively low, compared with the biological antimicrobial techniques that usually involve in the applications of expensive enzymic preparations (Al-Abri et al., 2019; Rashad et al., 2022; Fernandes et al., 2024). Thus, the applications of physical antimicrobial techniques in food systems have earlier been paid attention to and the advances in this field in recent years have been greatly recorded, as below reviewed.

Currently, the physical antimicrobial techniques for food systems could be divided into two aspects, e.g., thermal antimicrobial techniques (TATs) and non-thermal antimicrobial techniques (NTATs) (Rifna et al., 2019). TATs mainly employ heat energies from heat sources [e.g., sun, heating equipment or specific electromagnetic waves such as microwaves and infrared (IR) waves] to enhance

J. J. Huang · D. Li · W. Zhou (✉)
Department of Food Science & Technology, National University of Singapore, Singapore, Singapore
e-mail: weibiao@nus.edu.sg

D. Li (ed.), *Antimicrobial Strategies in the Food System: Updates, Opportunities, Challenges*, https://doi.org/10.1007/978-3-031-95056-8_3

environmental temperatures, subsequently denature the intracellular proteins, DNA and RNA of the involved microorganisms, and eventually kill these environmental microorganisms (e.g., foodborne pathogens) (Kim et al., 2021; Oguadinma et al., 2022; Chen et al., 2023; Saraeva et al., 2023), or remove heat energies from the environment by using refrigeration equipment to remarkably decrease environmental temperatures, thus deactivate the activities of microbial proteins, DNA and RNA, and finally lead to the hibernation of these microorganisms (Charpe et al., 2019; Zhou et al., 2022). While NTATs target on employing physical factors/entities/effects such as electromagnetic waves (excluding radio waves, microwaves, IR waves, red and green lights), photodynamic inactivation, electric fields, sound waves, active particles and pressure to exert the killing functions on microorganisms by destroying the structures of cells and cellular contents, followed by malfunctioning them, or utilizing the physical sizes of the targeted microorganisms to separate them from the foodstuffs by means of filtration techniques (Zhu et al., 2021; Jha et al., 2024; Kido et al., 2024; Ye et al., 2024). As noted in food systems, the typical features of TATs involve the significant enhancement or reduction of environmental and foodstuff temperatures, compared with room temperature, whereas those of NTATs are to maintain at around room temperature when processing the foodstuffs (Ulusoy et al., 2019; Tran et al., 2024). Therefore, TATs are preferred for the foodstuffs that are temperature-resistant or ready to cook, while NTATs are more fit for the temperature-sensitive foodstuffs such as juices and fresh produce.

In this chapter, the previous and the most recent findings on the physical antimicrobial techniques used in food systems are presented, with especially focusing on their mechanisms to inactivate pathogenic microorganisms, along with the concrete examples to demonstrate their advances recently, as well as critically reviewing their advantages and disadvantages when applied in food systems.

2 Physical Antimicrobial Strategies Applied in Food Systems

As above stated, physical antimicrobial strategies include TATs and NTATs. For TATs, the involved electromagnetic waves such as microwaves can result in the generation of heat energies for inactivating microorganisms during the processing, mainly due to the rotations and frictions among electric dipoles and ions presented in foodstuffs (Brittany & Hayes, 2002), while another kind of electromagnetic wave (e.g., IR waves) produces heat irradiation by itself, thus transferring heat energies to foodstuffs for executing antimicrobial operation (Xia et al., 2022a). Additionally, solar radiation could be other sources of heat because solar full-band electromagnetic spectra also contain IR waves in their profiles (Chaúque & Rott, 2021). Notably, TATs not only involve the temperature rise (heating) operation, but also associate with temperature reduction (refrigeration) manipulation. This is because refrigeration including cold storage and freeze can also contribute to the antimicrobial effects on foodstuffs (Ahmed & Mohamed, 2024; Suathong et al., 2024). For NTATs, their antimicrobial mechanisms are through the actions of short-wave electromagnetic radiations with wavelengths shorter than green light (500–600 nm), photo-driven photosensitizers, various electric field types, US waves, reactive nitrogen/oxygen species (RNS/ROS) and high hydrostatic pressure to disintegrate microorganisms, and destruct the organelles and molecular structures of the involved bioactive macromolecules (e.g., proteins, DNA and RNA) (Yogesh, 2016; Cho & Ha, 2019; Hyun & Lee, 2020; Zhu et al., 2021; Xia et al., 2022b; Bai et al., 2023; Zhang et al., 2023a), or to eliminate the targeted microorganisms from foodstuffs according to their actual physical sizes (Rosenqvist et al., 2024).

2.1 *Thermal Antimicrobial Techniques (TATs)*

In this section, TATs mainly include the techniques using microwaves, IR waves, solar radiation, artificial heat sources or refrigeration equipment that carry heat or can generate heat, to inactivate undesirable microorganisms contaminated in foodstuffs. Except for low temperature refrigeration technique, all the other TATs mentioned in this section involve in the temperature rise operations.

2.1.1 Electromagnetic-Wave-Based Thermal Antimicrobial Techniques

As noted, the electromagnetic waves involving in the generation of heat are microwaves that achieve this purpose through facilitating the frictions among the molecules inside the foodstuffs, while those in association with heat-carrying manner are the IR waves, no matter in the artificial or natural manners, in which solar radiation could provide the natural IR waves for executing thermal antimicrobial processing.

2.1.1.1 Microwaves

Microwaves, with wavelengths (frequencies/photon energy) ranging from one millimeter (300 GHz/1.24 meV) to one meter (300 MHz/1.24 μeV), belong to the family of electromagnetic spectrum involving in wavelengths longer than those of IR waves (700 nm to 1 millimeter), while shorter than those of radio waves (1 m and more) (Banik & Bandyopadhyay, 2018). In food industries, the application of microwaves is in a non-ionizing radiation manner, and it acts normally through converting microwave energy into thermal energy to heat the foodstuffs via frictions among molecules because the energy contained in microwave radiation is not enough to ionize molecules or induce the generation of free radicals (Ghazanfari et al., 2020; Shahi et al., 2021).

In detail, the principle of microwaves to play the heating roles in foodstuffs is subjected to two aspects. One aspect is to facilitate the rotations of electric dipoles (e.g., water molecules or polar molecules) in foodstuffs to follow the ever-oscillating electric field by microwave application, the different rotation speeds of various adjacent electric dipoles subsequently generate frictional forces among them, leading to the production of heat. While another is to achieve ionic conduction during microwave application, in this scenario, the free ion or ionic species in foodstuffs (e.g., various positive or negative ions contained in foods) would also like to align with the ever-oscillating electric field when they move translationally through space. Similarly to the rotations of electric dipoles, the relative motions of these species also generate frictional forces and then induce heat production (Fig. 1) (Galema, 1997; Brittany & Hayes, 2002). As can be seen, the more polar molecules and ionic species are presented in foodstuffs, the higher efficient heating effects of microwaves could be shown in elevating the temperatures of foodstuffs. By means of such property of microwave radiation, the antimicrobial mechanism of microwaves is mainly through the heat it generated to act on microbial cells and denature or dysfunction the intracellular proteins in their secondary, tertiary or quaternary structures, and eventually inactivate these microbial cells contained in foodstuffs (Fig. 1) (Shahi et al., 2021).

The 2450 MHz is normally the household microwave frequency (Sablani, 2024), the recent advances based on this frequency have achieved some valuable outcomes for application, as below reviewed. As found, this microwave frequency is excellent in sanitizing the solid surfaces of foodstuffs by efficiently removing the undesired microorganisms from them. For example, the sterilization treatment on red pepper (*Capsicum annuum* L.) using microwave (2450 MHz) for 10 W/g of microwave power density and up to 75 s could result in 8.85 and 7.37 $\log_{10}$ CFU/g reductions of total plate count, as well as yeast and mold count, respectively, on its surface (Table 1) (Shirkole et al., 2020).

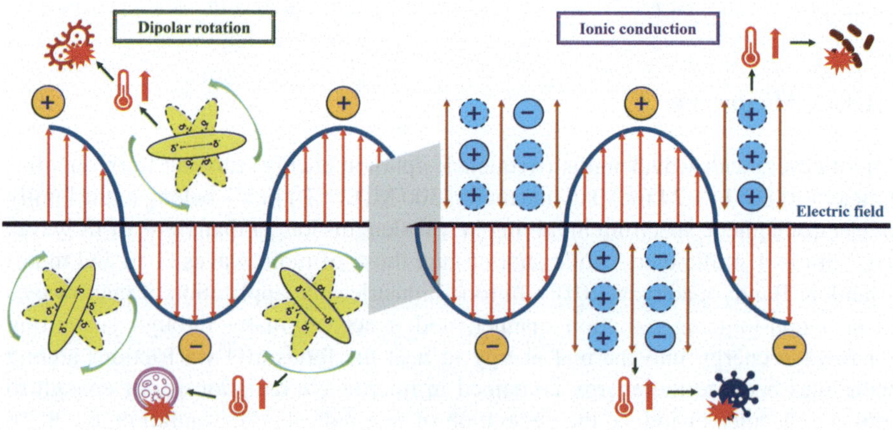

Fig. 1 Mechanism of microwave antimicrobial technique: Dipolar rotation and ionic conduction. (Adapted from Galema, 1997; Brittany & Hayes, 2002; Hay et al., 2022)

Table 1 Thermal antimicrobial techniques

TATs	Main physical conditions	Processed foodstuffs/ grown media	Disinfected microorganisms	Outcomes	Processed temperatures (°C)	Processed duration	Quality of foodstuffs after processed	References
Microwave	2450 MHz (10 W/g) + IR[a] heater (1000 W)	Paprika	*S.* Typhimurium *A. flavus*	8.85 log_{10} CFU[b]/g ↓[c] 7.37 log_{10} CFU/g ↓	Below 150	75 s	(1) a_w[d]↓; (2) color degradation ↑[e]; (3) peroxidase activity ↓	Shirkole et al. (2020)
	2450 MHz (4250 W)	Low-salt Sufu pehtze	Unspecific microorganisms	98.05% inactivated.	NM[f]	30 s	(1) Moisture content (58.45%); (2) a_w (0.848).	Fan et al. (2020)
	2450 MHz (950 W)	Jalapeño pepper Coriander leaves	*S.* Typhimurium	5.12 log_{10} CFU/g ↓ 4.45 log_{10} CFU/g ↓	63	25 s 10 s	(1) Color darkened; (2) Pepper greenness ↓; (3) Both firmness →[g].	De La Vega-Miranda et al. (2012)
	2450 MHz (360 W)	Whiting Salmon	*E. coli* O157:H7 *L. monocytogenes* *S. aureus*	4.00 log_{10} CFU/cm^2 ↓ 2.78 log_{10} CFU/cm^2 ↓ 2.78 log_{10} CFU/cm^2 ↓	50 70	2.2–5.2 min 1.5–3.1 min	ND[h]	Ulusoy et al. (2019)
	2450 MHz + PPA[i]	Medium agar Wheat grain Black peppercorn	*Salmonella* spp. *Enterococcus faecium* *Natural microbiota*	7.00 log_{10} CFU/g ↓ 2.10 log_{10} CFU/g ↓ 1.50 log_{10} CFU/g ↓	30–37	10 min 30 min 1 h	ND	Casado et al. (2024)
	2450 MHz (450/650 W)	Reconstituted PIFM[j]	*Cronobacter sakazakii*	1–2 log_{10} CFU/ml	53 71/88	5 s (650 W) 10/15 s (450–650 W)	89–94% bovine lactoferrin immunereactivity remained.	Harouna et al. (2020)
	2450 MHz (720 W)	Fresh apple juice	*E. coli* O157:H7 *S.* Typhimurium	7.00 log_{10} CFU/ml ↓	80–90	25 s	ND	Mendes-Oliveira et al. (2020)
	2455 MHz (600 W)	Raw cow's milk	Total plate count Yeast	5.00 log_{10} CFU/cm^3 ↓	40	47 s	ND	Kapcsándi et al. (2020)

(continued)

Table 1 (continued)

TATs	Main physical conditions	Processed foodstuffs/ grown media	Disinfected microorganisms	Outcomes	Processed temperatures (°C)	Processed duration	Quality of foodstuffs after processed	References
	2450 MHz (1000 W) + 3% ginger EO[k]	Mixed juices of vegetables (Celery, lettuce, parsley, spinach)	*S.* Typhimurium *L. monocytogenes*	9.82 s ↓ of inactivation initiation time 8.87 s ↓ of inactivation initiation time	Up to 70	30 s	Juice qualities →.	Hashemi et al. (2024)
	2450 MHz (700 W) + US[l] wave (60% amplitude)	Tangerine juice	*E. coli* *L. monocytogenes* *S. aureus* *S.* Enteritidis	5.00 $\log_{10}$ CFU/ml ↓ 7.19 $\log_{10}$ CFU/ml ↓ 5.27 $\log_{10}$ CFU/ml ↓ 5.12 $\log_{10}$ CFU/ml ↓	Up to 85	34 s (Microwave) 12 min (US)	Significant differences in colors found.	Demirok and Yıkmış (2022)
	915 MHz (18.7 kW) + HPP[m] (600 MPa)	Green beans	*Listeria innocua* ATCC 51742	9.00 $\log_{10}$ CFU/g ↓ (Microwave) 3.70 $\log_{10}$ CFU/g ↓ (HPP)	25 70	10 min 2 min	(1) Yellowness ↓, total color difference ↑; (2) Firmness →; (3) Total chlorophyll content →; (4) pH → .	Inanoglu et al. (2021)
	915 MHz (1.5 kW)	Hot chili sauce	*E. coli* O157:H7 *S.* Typhimurium *L. monocytogenes*	1.30 $\log_{10}$ CFU/ml ↓ 1.20 $\log_{10}$ CFU/ml ↓ 0.50 $\log_{10}$ CFU/ml ↓	100	65 s	(1) Color →; (2) a_w → .	Kim et al. (2021)
	915 MHz (1.5 kW) + Carvacrol (3 mM)	BPW[n] Hot chili sauce	*E. coli* O157:H7 *S.* Typhimurium *L. monocytogenes*	(1) Up to 6.00 $\log_{10}$ CFU/ml ↓ (BPW); (2) 758.5 of PI[o] uptake value (most damage to *E. coli*)	50–60 50–60	45 s 40 s	Hot chili sauce quality (color value and a_w) → .	Kim and Kang (2023)
IR	3.15 and 6.04 μm of resonant wavelengths	CaF_2 substrates	*Pseudomonas aeruginosa*	(1) Molecular vibrations of amide groups in protein (1500–1700 cm^{-1}); (2) C-H bond vibrations in membrane lipids and proteins (2800–3000 cm^{-1}).	25	3, 5 and 7 min	NI[p]	Saraeva et al. (2023)

	CIR[q] (3–5 μm)	Dry green Sichuan pepper	*P. Aeruginosa*	(1) Membrane dysfunction ↑; (2) Swimming motility ↓; (3) Bacterial DNA breakage ↑	70	4 min	(1) Alkaloid →; (2) Volatile oil →; (3) Aroma component →	Xia et al. (2022a)
	808 nm (2.0 W/cm^2) + (WO_3/PVA)$_4$ film	Pork	*E. coli* *S. aureus*	About 90% of *E. coli* and *S. aureus* ↓	21 → 50	45 min	Pork freshness monitoring achieved by (WO_3/PVA)$_4$ film	Wang et al. (2021)
	980 nm (7.5 W/cm^2) Totally 225 J/cm^2	Trypticase broth	*S. aureus* RN6390 biofilm	4.4 $\log_{10}$ CFU/ml ↓ (NIRT[r] + PDI[s]) 0.9 $\log_{10}$ CFU/ml ↓ (PDI) 0.8 $\log_{10}$ CFU/ml ↓ (NIRT)	46	30 s	NI	Mamone et al. (2024)
Solar radiation	SODIS[t]	Water	*Bacillus subtilis* (spores)	96% of spores inactivated	20–40	8 h	NI	Chaúque and Rott (2021)
	Sunlight	Deionized water	*Salmonella* spp. *E. coli*	Tolerance (*Salmonella* spp.) > Tolerance (*E. coli*)	26.5–30.0	Daily 180 min for 3 days	NI	Oguadinma et al. (2022)
	Sunlight + BTDA[u]-contained film	Mongolian cheese	*E. coli*	99.9% of antimicrobial effects	NM	3 h	Shelf stable for 8 days at 25 °C	Chen et al. (2023)
	Sunlight + *Curcuma longa* extract	Lawn cultures	Gram-negative and gram-positive bacteria	(1) The counts of bacteria ↓; (2) MIC[v] value of methicillin on *S. aureus* ↓	NM	10 min	NI	Bhadra et al. (2024)
High temperature pasteurization	HTST[w]	Bovine skim milk	Aerobic bacteria Coliform	(1) 2.3 $\log_{10}$ CFU/ml ↓; (2) Coliforms undetected; (3) Count of mold spores →	72	15 s	XO[x], IgG[y], LPO[z], LF[α], IgM[β], IgA[γ], native serum proteins and total sulfhydryl groups were 87, 78, 69, 57, 48, 41, 89 and 86% remained.	Li et al. (2021)

(continued)

Table 1 (continued)

TATs	Main physical conditions	Processed foodstuffs/ grown media	Disinfected microorganisms	Outcomes	Processed temperatures (°C)	Processed duration	Quality of foodstuffs after processed	References
	HTST	Materials from pastures	*Coxiella burnetii*	5.0 $\log_{10}$ CFU/ml ↓	68	15 s	NI	Wittwer et al. (2022)
	HTST	Raw milk	HPAIV[δ]	5.8 EID_{50}[ε] per ml of HPAIV ↓	72	15 s	NI	Spackman et al. (2024)
	HTST	Milk and dairy products	MRSA[ζ]	(1) ten of MRSA ↓ to 0 under LTLT[η] (63 °C for 30 min); (2) Two of MRSA ↓ to 0.	72	16 s	NI	Kanaan et al. (2023)
	HTLT[θ]	Orange and carrot nectar	*L. monocytogenes*	(1) *L. monocytogenes* ↓ to 0; (2) 12 days of stability of total mesophilic aerobes	88	32.7 min	(1) 85.87% vitamin C in nectar remained; (2) Acceptable flavor and texture achieved.	Wilma et al. (2024)
Low temperature pasteurization	Low temperature (2.5 °C) storage	Green pea sprouts	APCs[ι] *Listeria* spp.	(1) 2.5 °C (21 days of FP[κ]); (2) 5.0 °C (17 days of FP); (3) 10.0 °C (12 days of FP); (4) 15.0 °C (4 days of FP); (5) APCs ↓ (2.5 °C); (6) *Listeria* spp. (2.5 °C)	2.5	21 days	(1) Minimal changes in overall discoloration (ΔE^*), total ascorbic acid content and weight loss (%) achieved; (2) Longer freshness period.	Suathong et al. (2024)
	Low temperature (5.0 °C) storage	Broccoli and radish sprouts	Microorganisms	Pathogenic microorganisms undetected	5.0	14 days	(1) Freshness →; (2) TPC[λ] and glucosinolates →	Baenas et al. (2017)
	Low temperature (4.0 °C) storage	Chilled beef (CB) Frozen beef (FB)	AMB[μ]	(1) 5.3×10^4 CFU/g (CB); (2) 9.7×10^3 CFU/g (FB); (3) 10^7 CFU/g (room temperature).	4.0	NM	Hygienic quality still inadequate.	Ahmed and Mohamed (2024)

	Low temperature (1, 5 and 10 °C) storage	Buckwheat microgreens	APCs	APCs ↓ (1, 5 and 10 °C); APCs → (15 and 20 °C).	1, 5 and 10 15 and 20	10 days	Less tissue electrolyte leakage at 1, 5 & 10 °C than at 15 & 20 °C.	Kou et al. (2013)
	Low temperature (1, 4, 8 and 12 °C) storage	Baby spinach	APCs	APCs ↓ (1, 4, 8 and 12 °C); APCs → (16 and 20 °C).	1, 4, 8 and 12 16 and 20	16 days	Sensory quality inversely correlated with indigenous microorganisms.	Zhou et al. (2022)
	Low temperature (4 °C) storage	PIFM	*C. sakazakii*	*C. sakazakii* undetected	4	6 days	NI	Harouna et al. (2020)
	Low temperature (−18 °C) storage	FB	Molds Yeasts	(1) 28 out of 70 samples found; (2) 42 out of 70 samples found.	−18	NM	NI	Mohamed et al. (2023)
	Low temperature (4 and −18 °C) storage	CB FB 50 of each	*S. aureus* *E. coli* *Salmonella* spp. *L. monocytogenes*	(1) 24% and 16% in CB and FB; (2) 14% and 8% in CB and FB; (3) 5% and 3% in CB and FB; (4) 3% and 1% in CB and FB.	4 −18	NM	Hygienic quality still inadequate.	Ahmed and Mohamed (2024)

[a]IR, infrared; [b]CFU, colony forming unit; [c]↓, decreased; [d]a_w, water activity; [e]↑, increased; [f]NM, not mentioned; [g]→, unchanged; [h]ND, not determined; [i]PPA, plasma-processed air; [j]PIFM, powdered infant formula; [k]EO, essential oils; [l]US, ultrasound; [m]HPP, high-pressure processing; [n]BPW, buffered peptone water; [o]PI, propidium iodide; [p]NI, not involved; [q]CIR, catalytic infrared; [r]NIRT, near-infrared treatment; [s]PDI, photodynamic inactivation; [t]SODIS, solar disinfection; [u]BTDA, 3,3,4,4-benzophenone tetracarboxylic acid; [v]MIC, minimum inhibitory concentration; [w]HTST, high-temperature short-time; [x]XO, xanthine oxidase; [y]IgG, immunoglobulin G; [z]LPO, lactoperoxidase; [α]LF, lactoferrin; [β]IgM, immunoglobulin M; [γ]IgA, immunoglobulin A; [δ]HPAIV highly pathogenic avian influenza virus; [ε]EID_{50}, 50% egg infectious doses; [ζ]MRSA, multidrug-resistant *Staphylococcus aureus*; [η]LTLT, low-temperature long-time; [θ]HTLT, high-temperature long-time; [ι]APCs, aerobic plate counts; [κ]FP, freshness period; [λ]TPC, total phenolic compounds; [μ]AMB, aerobic mesophilic bacteria

For efficiently sanitizing the dense solid surfaces of foodstuffs, usually, the higher microwave power could significantly shorten the corresponding treatment duration, while the lower one might require longer time, nevertheless, it was also subjected to foodstuff types. As reported, it took 30 s for microwave (2450 MHz) at 4250 W to inactivate approximately 98.05% of microorganisms originally presented on the surfaces of low-salt sufu pehtze (Table 1) (Fan et al., 2020). In another trial, it took 25 and 10 s, respectively, for microwave (2450 MHz) at 950 W to heat the samples of fresh vegetables including Jalapeño pepper (*Capsicum annuum* L.) and Coriander leaves (*Coriandrum sativum* L.) till 63 °C, which could significantly decrease the counts of *Salmonella* Typhimurium on their surfaces by 5.12 and 4.45 $\log_{10}$ CFU/g, respectively (Table 1) (De La Vega-Miranda et al., 2012). However, it took 2.2–5.2 min and 1.5–3.1 min, respectively, for microwave (2450 MHz) at 360 W to cook fish fillets to reach the internal temperatures of 50 °C for whiting (*Merlangius merlangus*), as well as 70 °C for salmon (*Salmo salar*), which led to the significant count reductions of *Eschierichia coli* O157:H7, *Listeria monocytogenes*, and *Staphylococcus aureus* on their surfaces by 4.00, 2.78 and 2.78 $\log_{10}$ CFU/cm^2, respectively (Table 1) (Ulusoy et al., 2019). Recently, this frequency of microwave (2450 MHz) was reported to combine with the emerging technique such as non-thermal plasma by driving the plasma source to generate plasma-processed air (PPA), Through microwave-driven PPA, the count reductions of microorganisms (e.g., *Salmonella*, *Enterococcus faecium* and natural microbiota) on medium agar, the dry surfaces of wheat grains and black peppercorns were 7.0, 2.1 and 1.5 $\log_{10}$ CFU/g, respectively (Table 1) (Casado et al., 2024).

With regards to the loose solid surfaces such as powders, the microwave power and treatment duration can be markedly reduced to achieve the identical antimicrobial goal. This is because loose powder surface might contain less microorganisms per unit area due to bigger interval spaces presented on the powder surface. As tested, the microwave treatments (2450 MHz) at 650 W for 5 s, or 450–650 W for 10 and 15 s, respectively, could simultaneously fully sanitize *Cronobacter sakazakii* (1–2 $\log_{10}$ CFU/ml) in the reconstituted powdered infant formula (PIFM) (Table 1) (Harouna et al., 2020). Based on the above results, it can still be observed that higher microwave power helped to shorten the treatment duration to completely inactivate these foodborne pathogens.

Besides solid surface, microwave can also work well on liquid foodstuffs. As reported, the pasteurization process on fresh apple juice using microwave (2450 MHz, up to the treatments of 720 W and 25 s) could achieve up to 7 $\log_{10}$ CFU/ml reductions of pathogens in this liquid sample (Table 1) (Mendes-Oliveira et al., 2020). In another study, the heating process on cow's milk by employing microwave (2455 MHz and up to 600 W) for 47 s could also remarkably reduce the counts of total plate and yeast cell by around 5 $\log_{10}$ CFU/cm^3 (Table 1) (Kapcsándi et al., 2020).

Sometimes microwaves in cooperation with specific extraction techniques can act as the efficient auxiliary means to assist in improving the antimicrobial activities of some bioactive extracts. As revealed, the hydro-distillation extraction assisted by microwave (2450 MHz) at 500–800 W for 80 min significantly enhanced the antimicrobial activity of essential oil (EO) extracted from coriander seeds, especially on *Candida albicans* (Ghazanfari et al., 2020). This meant that the participation of microwave could help to extract more bioactive EO substances to display their antimicrobial activities. Interestingly, such extracts can in turn also play the auxiliary roles by helping microwave antimicrobial technique in its very beginning to kill the pathogenic microorganisms in the processed foodstuffs. A recent study showed that the treatment of microwave (2450 MHz, 1000 W & 30 s) in combination with 3% EO could significantly shorten the inactivation initiation time of *S.* Typhimurium and *L. monocytogenes* in a mixed vegetable juice containing celery, lettuce, parsley and spinach by 9.82 and 8.87 s, respectively, compared with those without the participation of EO, indicating that EO could be acted as an auxiliary substance to execute the inactivation process when the temperatures of samples did not reach the designated levels at the very beginning of microwave process (Table 1) (Hashemi et al., 2024).

Apart from cooperating with extraction technique or bioactive substances, microwave combining with another physical antimicrobial technique (e.g., US wave) could be an alternative solution to efficiently inactivate the common pathogenic microorganisms such as *E. coli*, *S. aureus*, *Salmonella* Enteritidis, and *L. monocytogenes* in beverage industries. Using response surface methodology (RSM), the optimal combination to achieve 5 $\log_{10}$ CFU/ml reduction of *E. coli* presented in tangerine juice was confirmed as microwave (2450 MHz, 700 W & 34 s) in cooperation with 60% of US amplitude for 12 min. Such optimal combination processing on the pathogen-inoculated tangerine juice led to 7.19, 5.27 and 5.12 $\log_{10}$ CFU/ml reductions of *L. monocytogenes*, *S. aureus* and *S.* Enteritidis, respectively, significantly higher than them at the RSM central points all achieving lower than 4.00 $\log_{10}$ CFU/ml reductions (Table 1) (Demirok & Yıkmış, 2022).

Besides the 2450 MHz, the 915 MHz is also the frequently used microwave frequency applied in food industries to inactivate pathogenic microorganisms contained in foodstuffs (Kim et al., 2024). Compared with 2450 MHz, the advantage of 915 MHz is that it possesses longer wavelength and thus strengthens its diffraction effect deeply into foodstuffs for heating (Bhunia et al., 2024). Therefore, besides superficial disinfection, the 915 MHz of microwave can theoretically be more efficient to be applied in sanitizing the specific foodstuffs such as the denser liquid-like sauces, compared with the 2450 MHz one (Hernandez-Gomez et al., 2021; Kim et al., 2021; Kim & Kang, 2023).

Using 915 MHz as the operation frequency, the microwave pasteurization treatment at 18.7 kW of total power on green beans led to 9.0 $\log_{10}$ CFU/g of pathogen reduction, compared with the technique of high hydrostatic pressure that led to 3.7 $\log_{10}$ CFU/g of reduction only (Table 1) (Inanoglu et al., 2021). In the sauce case, the application of 915 MHz microwave at 1.5 kW took 65 s to increase the temperature of hot chili sauce (25 g) by up to 100 °C (Table 1) (Kim et al., 2021), in which the treatment duration for 40 s resulted in 1.3, 1.2 and 0.5 $\log_{10}$ CFU/ml reductions of *E. coli* O157:H7, *S.* Typhimurium and *L. monocytogenes*, respectively, in this sauce (Table 1) (Kim & Kang, 2023).

Similarly to 2450 MHz, the cooperation between 915 MHz and EO could also perform the good antimicrobial results on foodstuffs. For example, the combined processes of microwave (915 MHz, 1.5 kW) heating and carvacrol (an EO) for 45 s were found to achieve up to 6 $\log_{10}$ CFU/ml decrement of the maximal inactivation efficiency for the pathogens (e.g., *E. coli* O157:H7, *S.* Typhimurium and *L. monocytogenes*) presented in the buffered peptone water, while the similar tendency was demonstrated in hot chili sauce. Using propidium iodide (PI) uptake study, such combination of microwave and carvacrol was shown to lead to the severe destruction to the cellular membrane of *E. coli* O157:H7 (758.5 of PI value) (Table 1) (Kim & Kang, 2023).

All in all, the main advantage of microwave antimicrobial technique is significant by displaying their higher heating efficiencies, compared with the traditional heating manners. This is because the process of microwaves can achieve deeply into centimeter of foodstuffs by directly co-acting with various polar molecules presented in the surfaces and interiors of foodstuffs simultaneously, while the traditional ones depend on thermal conductivity, by which heat energy can only be transferred from heat source, through the container, then via the surfaces of foodstuffs, and eventually reaching the interiors of foodstuffs. Obviously, these traditional manners are time-consuming and inefficient (Galema, 1997; Brittany & Hayes, 2002). From here, it indicated that microwave antimicrobial technique could inactivate the microorganisms in foodstuffs faster and more efficient, compared with the traditional heating one, because it could heat the foodstuffs not only from the surface, but also deep in centimeter simultaneously. As found in a comparison study, the heating process by microwave on the mixed vegetable juice significantly reduced the treatment duration from 240 to 30 s, and simultaneously saved 67% of the electrical power, compared with the traditional heating approaches (Table 1) (Hashemi et al., 2024). Additionally, another advantage of microwave antimicrobial technique is that it can help to maintain the appearance quality of foodstuffs (e.g., colors) after processing because microwave heating duration is usually short, which is hard to induce further destructions on the double bond structures of molecules, thus avoiding appearance color variations of foodstuffs. For instance, the processing of 915 MHz microwave at 1.5 kW for 40 s did not lead to significant variations of color (L*, a* and b* values) and water activity (a_w) of the hot chili sauce (Table 1) (Kim et al., 2021).

However, the potential disadvantage of microwave on displaying its antimicrobial effect on foodstuffs should be highlighted. As this technique can generate uneven heating due to the uneven distribution of the electromagnetic field at the different parts of foodstuffs, it might lower its efficiency in killing pathogens (e.g., invasive bacteria), compared with the other heating techniques. Additionally, its transitory heating duration followed by the low heat energy it produced might also discount its antimicrobial capability acting on foodstuffs (Principato & Spigno, 2024).

2.1.1.2 Infrared Waves

IR waves, with wavelengths (frequencies/photon energy) ranging from 700 nanometer (430 THz/1.7 eV) to one millimeter (300 GHz/1.24 meV), belong to the family of electromagnetic spectrum involving in wavelengths shorter than those of microwaves (1 millimeter to 1 meter), while longer than those of visible lights (400–700 nm) (Rumble et al., 2019; Yadav et al., 2020). Similarly to microwaves, the application of IR waves in food industries is also in a non-ionizing radiation manner to heat up foodstuffs and the IR waves are normally emitted through infrared lamps (e.g., quartz heaters) to radiate on the foodstuffs that can preferentially absorb the IR waves, compared with the surroundings, because the vibration and rotation of water molecules contained in foodstuffs are well at IR frequencies (Ling et al., 2016).

In detail, the mechanism of IR radiation to produce heat is that the rotation and vibration of molecules and atoms in the IR emitting materials of the infrared lamps are more intense when they absorb energy (e.g., thermal energy) from the electrical power, which might further motivate the transitions of electrons inside these molecules and atoms that can thus produce IR waves. When exposed to foodstuffs, these waves can be preferentially captured by the water molecules of foodstuffs, which lead to more exquisite rotations and vibrations of these water molecules. As well recognized, IR power radiation is positively related to the fourth power of the absolute temperature, therefore, the temperatures of foodstuffs are greatly enhanced after IR exposure (Ling et al., 2016).

Based on the above feature, the mechanism of IR waves to disinfect microorganisms (e.g., bacteria) presented on the foodstuff surfaces is similar to that of microwaves, by also resorting to the heat it produced to denature the functions of the intracellular proteins and genetic materials (e.g., DNA and RNA) of the targeted microorganisms, and finally kill them (Shahi et al., 2021). Besides, the capability of IR waves to induce the vibrations of molecules and chemical bonds inside microorganisms could be another cause to eliminate them. A recent study reported that IR waves based on the femtosecond IR laser radiation technique could efficiently disinfect *Pseudomonas aeruginosa* under 3.15 and 6.04 μm of the resonant wavelengths, which led to the molecular vibrations of amide groups in proteins (1500–1700 cm^{-1}), as well as C-H bond vibrations in the membrane lipids and proteins (2800–3000 cm^{-1}) contained in bacterial cells, thus killing the bacteria (Table 1) (Saraeva et al., 2023). The phenomena of bactericidal structural molecular changes after femtosecond IR laser radiation treatment could be observed by the stationary Fourier-transform IR spectroscopy, however, by means of scanning and transmission electron microscopy, it was interesting to found out that the cellular membrane of *P. aeruginosa* was not significantly damaged by this technique although the bacteria have been disinfected. These results meant that the damage of femtosecond IR laser radiation on bacterial cells was focused on the intracellular macromolecules, rather than on the cellular structure, due to its treatment duration that was very short, although the normal IR radiation has been shown to destruct the bacterial cell membrane (Xia et al., 2022a; Saraeva et al., 2023).

As a promising technique in inactivating foodborne pathogens, the advantages of IR antimicrobial technique show that it is especially suitable to sanitize pathogens (e.g., *P. aeruginosa*) presented on foodstuffs (e.g., dry green Sichuan pepper, *Zanthoxylum schinifolium*) with rough surfaces, this is because IR waves are incapable of fully penetrating dry materials and its direct effect on foodstuff surfaces does not require the participation of additive and medium, therefore diminishing its influence on the qualities of foodstuffs (Xia et al., 2022a). Furthermore, IR waves possess potent diffraction ability to allow them to arrive at the cracks and gaps of the targeted foodstuffs to completely decontaminate pathogens. Besides these, the more even heating manner, the shorter treatment duration, the cleaner and environmentally-friendly means, are also its advantages in inactivating foodborne pathogens, compared with the traditional heating techniques (Pei et al., 2018; Zavala-Franco et al., 2020; Xia et al., 2022a).

Apart from displaying the direct antimicrobial property, IR waves could also work well as an assistant tool to execute antimicrobial purposes in food packaging. As reported, with the help of near-IR light (NIR808, 808 nm, 2.0 W/cm^2), the antimicrobial activity of polyvinyl alcohol/WO_3 [$(WO_3/PVA)_4$] films, responding to H_2S by demonstrating color change and thus employed for pork freshness monitoring as the indicator, was greatly improved by showing approximately 90% of antibacterial efficiencies against *E. coli* and *S. aureus*, in which the physical damage, oxidative stress and temperature increment, due to the participation of NIR808, remarkably supported the elevation of antibacterial activity of $(WO_3/PVA)_4$ films (Table 1) (Wang et al., 2021).

Among all the IR spectra, near-infrared treatment (NIRT) can be an efficient pre-treatment tool in cooperation with photodynamic inactivation (PDI) technique to improve its antimicrobial capability, which mechanism is commonly associated with ROS production and thermal effects. As reported, the pre-treatment using NIR irradiation prior to visible light exposure to *S. aureus* biofilm could elevate biofilm's sensitivity to PDI, which usually employ a visible-light-activated photosensitizer to produce ROS for killing bacteria. This was achieved by using a laser NIRT (980 nm) before PDI exposure, which demonstrated a combined effect to result in 4.4 $\log_{10}$ CFU/ml reduction of bacterial viability, compared with those (0.9 and 0.8 $\log_{10}$ CFU/ml reduction) of PDI-exposure-only and NIRT-exposure-only treatment, respectively. Further investigation showed that NIRT helped to enhance the absorption of photosensitizer and trigger clefts in the matrix of biofilm, leading to the discharges of *S. aureus* cells from the biofilm (Table 1) (Mamone et al., 2024).

Looking into the recent advances on the antimicrobial effect of IR waves, the catalytic infrared (CIR) radiation is an emerging technique that should be introduced, it is based on IR radiation technique but using natural gas, rather than electric energy as the energy source. The advantages of CIR superior to electric IR are that CIR achieves higher heat transfer efficiency, and it has more concentrated working wavelengths ranging from 3 to 5 μm (Table 1) (Feng et al., 2019; Xia et al., 2022a). Recently, the mechanism of antimicrobial effect of CIR radiation on foodstuffs has been revealed due to its damage on bacterial cell membrane leading to membrane dysfunction, its suppression on the swimming motility of bacteria, and bacterial DNA breakage it caused. Delightfully, CIR radiation technique seems not to significantly affect the flavors (e.g., the contents of alkaloid and volatile oil, as well as aroma component) of dry green Sichuan pepper even after 4 min of treatment (Xia et al., 2022a).

Besides the above advantages mentioned, there are some disadvantages of IR antimicrobial technique that should be cautioned. Firstly, it should be careful to operate the IR devices/equipment as IR radiation might cause burns due to the high temperature it produced; Secondly, the ruptures and damage of skin tissues after IR radiation exposure could be occurred in the long run (Aboud et al., 2019); Thirdly, IR radiation cannot work well deeply into foodstuffs to disinfect the pathogens because its penetration capability is weak, thus, it can only be used for the surface sterilizations of foodstuffs (Aboud et al., 2019; Xia et al., 2022a).

2.1.1.3 Solar Radiation

Solar full waveband radiation to reach earth's surface mainly includes ultraviolet A and B (UVA and UVB), visible lights and IR waves (Huang et al., 2020), therefore, solar radiation has long been recognized as a kind of efficient and natural antimicrobial tool to sanitize foodstuffs (e.g., fish) and water (Nguyen et al., 2014; Barba et al., 2019), because IR waves, blue light in the visible light spectrum, as well as UVA and UVB lights include the excellent antimicrobial wavebands that have been presented in the above section and are to be introduced in the below sections. The antimicrobial mechanism of IR waves has been introduced in Section 2.1.1.2, while those of blue lights, UVA, and UVB to play their antimicrobial roles are to be demonstrated in Sections 2.2.1.1, 2.2.1.2, and 2.2.1.3, respectively. Besides solar radiation itself, the heat generated through the natural IR waves in solar radiation is also another cause that can lead to bacteria inactivation through denaturing the intracellular proteins and genetic materials of bacteria as above mentioned (Shahi et al., 2021).

The traditional solar radiation antimicrobial technique, or termed as solar drying technology, is capable of inactivating microorganisms attached to the surfaces of foodstuffs by reducing their water contents or water activities to safe levels, where the actions of the attached microorganisms are thus inhibited. This could be normally seen in the open places where seafoods, meats, vegetables and fruits were basked to obtain the dried products (Janjai & Bala, 2012; Nguyen et al., 2014). As noted, the advantage of solar radiation antimicrobial technique is that the solar energy acquired for basking these products is free, thus significantly decreasing its operation cost. However, its disadvantages are also outstanding because sometimes the treatment duration is quite long due to climatic instability, which therefore reduces its antimicrobial efficiency on treating the foodstuffs. Furthermore, the exposure to solar radiation outdoors might have the potential to attach dust and dirt onto the surfaces of foodstuffs, which would arouse hygienic problems. Thus, the recently-developed solar-radiation-involved techniques are chiefly focused on sterilizing water sources [e.g., the solar disinfection (SODIS) technique that has been proved for sanitizing water in a convenient, free and clean way] (Chaúque & Rott, 2021; Oguadinma et al., 2022), or playing the auxiliary tools to improve the antimicrobial effect of PDI technique (Chen et al., 2023; Bhadra et al., 2024).

The sterilizing role played by solar radiation on water disinfection has been shown in some recent studies (Chaúque & Rott, 2021; Oguadinma et al., 2022). For example, the above mentioned SODIS technique could be an alternative for large-scale public drinking water supply. This can be achieved by designing high-performance continuous flow systems for large-scale water supply. Additionally, the combination of SODIS with solar pasteurization (SOPAS) and nanomaterials could be a better solution to perform well, compared with single approach system. In this scenario, the bacterial spores in the water are the major subject to be tested for the availability of SODIS system (Table 1) (Chaúque & Rott, 2021). In fact, the reductions of bacterial counts in the sterile deionized water, inoculated by *Salmonella* and *E. coli* and exposed to sunlight on three consecutive days for daily 180 min, have been reported closely associated with sunlight spectra, especially UV doses, as well as temperature. It was found that *Salmonella* was more tolerant to sunlight, compared with *E. coli*, in which the fluctuations of water daily could markedly affect the survival rates of these foodborne pathogens in water (Table 1) (Oguadinma et al., 2022). From both trials, sunlight was still deemed to play the positive role in removing the common foodborne pathogens away from water supply, although different pathogens responded differently to the same solar radiation dose.

As the auxiliary role to support PDI technique, sunlight is an effective tool in inducing 3,3,4,4-benzophenone tetracarboxylic acid (BTDA), an organic photosensitizer to fabricate the modified poly (vinyl alcohol) film (through chemical grafting reaction) that can display excellent antibacterial features, to generate ROS including hydrogen peroxide, hydroxyl radical, superoxide anion, as well as singlet oxygen (Niu et al., 2021; Chen et al., 2023). These ROS are capable of oxidizing microbial lipids, nucleic acids and proteins, thus killing the involved microorganisms eventually. As reported, such sunlight-induced antibacterial film could achieve more than 99.9% of antimicrobial efficiencies on five pathogenic microorganisms such as *E. coli*. Under room temperature (25 °C), Mongolian cheese packaged by this film can be well stored by up to 8 days (Table 1) (Chen et al., 2023).

Besides the cooperation with organic photosensitizer, sunlight could also play an auxiliary tool to strengthen the antimicrobial effects of some plant-based extracts with antimicrobial potential. A case in point was the ethanolic extract of *Curcuma longa* plant, which powder mainly containing curcumin, is the major component of curry (Ng et al., 2024). When this extract was poured into the lawn cultures of the gram-negative and gram-positive bacteria and then exposed to sunlight, the counts of bacteria was found significantly decreased, compared with those under extract-free control and another extract-containing treatment exposed to artificial UV light. In addition, the micro broth dilution assay further demonstrated the reduction of minimum inhibitory concentration (MIC) value of methicillin on *S. aureus* after sunlight exposure, indicating that sunlight could help to elevate the antimicrobial effect of plant-based antimicrobial agents (Table 1) (Bhadra et al., 2024).

2.1.2 Temperature-Based Thermal Antimicrobial Techniques

Temperature-based thermal antimicrobial techniques usually refer to the technologies involving the transfers of thermal energies into or out of foodstuffs to inactivate the targeted microorganisms. For transferring thermal energies into foodstuffs, the conventional manner is cooking, which can completely disinfect microorganisms (e.g., *E. coli* O157:H7) in foodstuffs if thorough cooking is performed (Ježek et al., 2020). However, some specific foodstuffs such as milk and fruit/vegetable juices are not likely to be cooked before selling to the customers because it might affect the qualities of such foodstuffs, thus, the pasteurization method could be a better alternative (Peng et al., 2017). For transferring thermal energies out of the foodstuffs, refrigeration such as cold storage and freeze are the common methods. Through refrigeration, the reductions of environmental and foodstuff temperatures might slow down the metabolic activities of microorganisms and suppress their growth, thus extending the shelf lives of processed foodstuffs (Xu et al., 2024a).

2.1.2.1 High Temperature Pasteurization

The antimicrobial mechanism of high temperature is through seeking the assistance from heat energy (generated by high temperature more than 60 °C) to denature the bacterial proteins in their secondary, tertiary, or quaternary structures, by which the involved covalent bonds are disrupted, and the functions of proteins are thus lost, this finally destroys microorganisms (Acharya & Chaudhuri, 2021). For high-temperature-based thermal antimicrobial technique, the high-temperature short-time (HTST) pasteurization, also called flash pasteurization, is a widely applied technique in food industries nowadays, because high temperature in HTST is responsible for inactivating microorganisms, while short time is intended to minimize the negative effect on the qualities of foodstuffs during processing (Rupasinghe & Yu, 2012; Aganovic et al., 2016).

The HTST technique has been proved to efficiently inactivate bacteria and viruses in milk and materials from pastures that might contaminate dairy products. For example, the application of HTST pasteurization at 72 °C for 15 s remarkably decreased the bacteria count in bovine skim milk by 2.3 $\log_{10}$ CFU/ml. After HTST treatment, coliforms in bovine skim milk could not be detected (Table 1) (Li et al., 2021). For treating the materials from pastures, the combined HTST pasteurization of 68 °C and 15 s of holding time was found to result in 5 $\log_{10}$ CFU/ml reduction of *Coxiella burnetii* presented in six German field isolates obtained from the abortion or afterbirth materials of cattle, ovine and caprine. As known, *C. burnetii* is a kind of heat-resistant Gram-negative, obligate intracellular bacterium that might cause zoonosis Q fever and could be presented as the pathogen in the milks of cattle, sheep and goats (Table 1) (Wittwer et al., 2022). For the virus case, the combination of 72 °C and 15 s of HTST continuous flow pasteurization processing under two flow rates (e.g., 0.5 and 1.0 L/min) completely disinfected

H5N1 highly pathogenic avian influenza virus (HPAIV) in the contaminated raw milk, with an average 5.8 $\log_{10}$ 50% egg infectious dose (EID_{50}) per ml of HPAIV reduction (Table 1) (Spackman et al., 2024).

With regards to the disadvantages of HTST pasteurization technique, it was found incapable of inactivating fungi, this could be seen in the above case where the counts of mold spores in bovine skim milk were not significantly affected by HTST pasteurization treatment at 72 °C for 15 s (Table 1) (Li et al., 2021). Furthermore, high temperature keeps possibility to exert some heat damages, which might, more or less, affect the qualities of foodstuffs after processing although its treatment duration is quite short. As noticed, the levels of bioactive xanthine oxidase (XO), immunoglobulin G (IgG), lactoperoxidase (LPO), lactoferrin (LF), IgM, IgA and native serum proteins were 87, 78, 69, 57, 48, 41 and 89% remained in bovine skim milk, compared with the control without treatment. In addition, the concentrations of total sulfhydryl groups of milk involved in protein's molecular oxidative modifications were decreased by 14%, owing to the thermal-oxidation process acting on protein molecules (Table 1) (Xiong & Guo, 2020; Li et al., 2021).

Moreover, HTST also induces the generations of heat-resistant bacterial species. For instance, ten of multidrug-resistant *S. aureus* (MRSA) isolates from milk and dairy products, processed under low temperature and long time (LTLT, 63 °C for 30 min, relative to HTST), were found completely disinfected, in which only two of them were observed also adaptable to HTST (72 °C for 16 s) processing (Table 1) (Khairullah et al., 2020; Kanaan et al., 2023). These results indicated that most of isolates could survive after short time (16 s) of heating under higher temperature (72 °C) due to the quorum sensing structures they have, to allow them to withstand heat damage induced by HTST, thus, the careful investigations are required for identifying the capabilities of thermal tolerances of these pathogens isolates from different sources when carrying out the actual HTST treatment in the contaminated milk (Table 1) (Kanaan et al., 2023).

Besides HTST, high-temperature long-time (HTLT) pasteurization technique might sometimes work well to inactivate bacteria in some specific foodstuffs (e.g., nectar). For instance, vacuum pasteurization under the combinations of different high temperature and treatment durations could display their potent antimicrobial effects on foodstuffs. When orange (*Citrus sinensis*) and carrot (*Daucus carota* L.) nectar were treated under the conditions of T1 (92°C/3.3 min), T2 (90°C/10.3 min), T3 (88°C/32.7 min), T4 (70°C/2.3 min), T5 (65°C/56.6 min), and T6 (60°C/56.6 min), *L. monocytogenes* in all the treatments were found absent, indicating the antimicrobial effectiveness of HTLT pasteurization technique on these foodstuffs, in which T3 was recognized as the best heat treatment among the six ones because it could prevent 85.87% of vitamin C in nectar from loss while maintaining a microbiological stability of 12 days in terms of total mesophilic aerobes, simultaneously, it could achieve the best flavor and texture, as well as the acceptable property of nectar, compared with the other treatments (Table 1) (Wilma et al., 2024).

2.1.2.2 Low Temperature Refrigeration

The mechanism of low temperature refrigeration technique to achieve antimicrobial purpose on foodstuffs is to significantly inhibit the activities of intracellular enzymes of microorganisms by means of low temperature environments (e.g., refrigerator), thus suppressing the growths of these associated microorganisms. The advantage of low temperature refrigeration technique is mainly to extend the storage lives of processed foodstuffs. Under low temperatures, the lag phase for bacteria would be prolonged before multiplication, which might thus inhibit the growth of bacteria attached to the foodstuff surfaces and extend the shelf lives of these stored foodstuffs (Charpe et al., 2019; Zhou et al., 2022). Additionally, low temperature refrigeration technique can also maintain the qualities of foodstuffs during and after processing. As found at 2.5 °C, the minimal changes in overall discoloration (ΔE^*), total ascorbic acid content, and weight loss (%) of the green pea sprouts were achieved, after 21 d of refrigeration (Table 1) (Suathong et al., 2024). In another case, the qualities of broccoli and radish sprouts including freshness and bioactive phytochemicals such as total phenolic compounds and glucosinolates could be greatly maintained, after stored under 5 °C for 14 d (Table 1) (Baenas et al., 2017).

Currently, low temperature refrigeration technique is mainly applied in frozen meats below 0 °C, or cold storages of chilled meats, vegetables and fruits above 0 °C, or the powder-type food products, to minimize the growths of undesired microorganisms on them (Harouna et al., 2020; Darwish et al., 2024; Suathong et al., 2024). For instance, the aerobic mesophilic bacteria (AMB) counts in the chilled beef samples (with store temperature not higher than 4 °C) could be as low as 5.3×10^4 CFU/g, while those in the frozen beef samples was touched down to 9.7×10^3 CFU/g, compared with those with more than 10^7 CFU/g of AMB counts under room temperature, indicating that the growths of such bacteria on the chilled/frozen meats can be greatly suppressed under low temperature refrigeration treatment (Table 1) (Ahmed & Mohamed, 2024).

For the cases of vegetables and fruits, the aerobic plate counts (APCs) levels of the buckwheat microgreens kept at 1, 5 and 10 °C were found significantly reduced after 10 days of preservation, compared with those preserved under 15 and 20 °C (Table 1) (Kou et al., 2013). Similarly, the baby spinach samples stored under 1, 4, 8 and 12 °C also demonstrated lower APCs, compared with those kept under 16 and 20 °C (Table 1) (Zhou et al., 2022). Stored under 5 °C for 14 d, no pathogenic microorganisms were found presented on the surfaces of broccoli and radish sprouts (Table 1) (Baenas et al., 2017). At 2.5 °C, the green pea sprouts could keep as long as 21 d of freshness period, significantly higher than those preserved at 5, 10 and 15 °C for 17, 12 and 4 d, respectively. Furthermore, under this temperature during the whole preservation period, the minimal APCs and *Listeria* spp. counts on the surfaces of sprouts were found (Table 1) (Suathong et al., 2024). For the powder-type products, it was found that the growth of *C. sakazakii* in the reconstituted infant formula (e.g., PIFM) could not be observed during a 6-day's storage at 4 °C (Table 1) (Harouna et al., 2020).

However, the disadvantage of low temperature refrigeration technique is also evident because the effect of low temperature is mainly to inhibit the growth of microorganisms, rather than eliminate them. Thus, there are still some risks for the pathogenic microorganisms to threaten public health after the foodstuffs are thawed or returned to room temperature (Darwish et al., 2024). Such risks are due to the possible existence of these undesired microorganisms even attached to the frozen foodstuffs when refrigerated. As reported, molds and yeasts could be found in 28 and 42 out of a total 70 frozen beef samples, respectively, purchased from the retail stores in Qena, Egypt (Table 1) (Mohamed et al., 2023). Astonishingly, 24% and 16% of chilled and frozen beef samples (50 of each), respectively, could detect *S. aureus*. Furthermore, *E. coli* could be found in 14% and 8% of chilled and frozen samples, respectively. Moreover, *Salmonella* spp. was in 5% and 3%, whereas *L. monocytogenes* was in 3% and 1%, respectively (Table 1) (Ahmed & Mohamed, 2024).

2.2 Non-thermal Antimicrobial Techniques (NTATs)

In this section, NTATs mainly include the techniques based on non-thermal-associated electromagnetic waves, photodynamic inactivation effect, filtration, pulsed electric field, US, cold plasma and high hydrostatic pressure that are not involved in the transfers of thermal energies, to eliminate undesirable microorganisms from foodstuffs.

2.2.1 Electromagnetic-Wave-Based Non-thermal Antimicrobial Techniques

Electromagnetic-wave-based non-thermal antimicrobial techniques normally contain the techniques using blue, UVA, UVB, and UVC lights, gamma ray, and X ray that do not lead to the significant enhancements of foodstuff temperatures, however, they are capable of, not only inactivating undesirable microorganisms from foodstuffs, but also maintaining the qualities of foodstuffs during and after processing.

2.2.1.1 Blue Light

Blue light, with wavelengths (frequencies/photon energy) ranging from 400 nanometer (750 THz/3.1 eV) to 500 nanometer (600 THz/2.5 eV), is included in the visible light family of electromagnetic spectrum involving in wavelengths shorter than that of ultraviolet (UV, 10 nm to 400 nm), whereas longer than that of green light (500–600 nm) (Rumble et al., 2019). The antimicrobial mechanism of blue light demonstrates that it is mainly capable of increasing the permeability and depolarization of microbial cell membrane, enhancing microbial DNA oxidation levels, as well as elevating intracellular levels of ROS and lipid peroxidation of microorganisms, all of which finally contribute to kill these treated microorganisms (Hyun & Lee, 2020).

In fact, although blue light exerts similar effects to disorder the intracellular operations of microorganisms as above stated (Hyun & Lee, 2020), it seems that different microorganisms (e.g., bacteria) have their own strategies to protect themselves, leading to their different responses to the same blue light treatment. As reported under up to 4080 J/cm^2 of dose at 4, 10 and 25 °C, the 460 nm blue-LED irradiation remarkably inhibited the growths and decreased the counts of six foodborne bacteria such as *Bacillus cereus*, *E. coli* O157:H7, *L. monocytogenes*, *P. aeruginosa*, *S.* Typhimurium and *S. aureus*. However, different bacteria responded differently to the same dose. As found, it took *S. aureus* population 7 h to achieve below 2 $\log_{10}$ CFU/ml (detection limit), while *B. cereus* required 9 h only to reach 3 $\log_{10}$ CFU/ml of count reduction (Table 2) (Kumar et al., 2017). The metabolic mechanism study by investigating and comparing the metabolic profiling of non-exposed and exposed cells showed that 11-deoxycortisol (inhibition of ketosteroid reductase) in all six bacterial cells, actinonin (inhibitor of peptide deformylase) in *L. monocytogenes*, *E. coli* and *P. aeruginosa*, as well as coformycin (an inhibitor in salvage pathway) in *B. cereus*, *E. coli* and *S.* Typhimurium, were down-regulated, while tyramine (H_2O_2 production) in *L. monocytogenes* and *E. coli*, as well as chitobiose (ROS scavenger) in *B. cereus* and *P. aeruginosa*, were up-regulated instead, under 460 nm blue-LED illumination at 25 °C for 1 h (Table 2) (Kumar et al., 2017). These could be the reasons why different bacteria have their respective sensitivities to 460 nm blue light. Meanwhile, these results also provided us information to adopt appropriate 460 nm blue light doses for fully disinfecting different species of bacteria.

As a non-antibiotic antimicrobial mean, blue light is also capable of fighting with multidrug-resistant bacteria. Recently, the genetic mechanisms of multidrug-resistant bacteria responding to blue light were investigated (Kruszewska-Naczk et al., 2024). In this case, how blue light (415 nm, 25 mW/cm^2) worked to eliminate these bacteria and how they responded to high blue light doses for survival were studied through genome-wide-based testing. Unfortunately, the current results showed that there was not a predominant mechanism that could clearly explain the sensitivities of these involved bacteria (e.g., *E. coli* BW25113) to photoinactivation. However, sixty-four single-gene-deleted mutants absent in certain protective genes were identified, these were the phenotypes that demonstrated the sensitivity to the increased doses of blue light by up to 64.8 J/cm^2. These results hinted that further studies on the full mechanism are necessary, while the resistant capabilities developed by these involved pathogens should be particularly concerned (Table 2) (Kruszewska-Naczk et al., 2024).

Currently, there are two waveband ranges of blue light (e.g., 405 and 460 nm) that are widely applied in inactivating microorganisms, in which the shorter wavelength of blue light (e.g., 405 nm) could have stronger antibacterial activities on foodborne bacteria, compared with the longer one (e.g., 460 nm). As reported, the exposure of *Vibrio parahaemolyticus* to 405 and 460 nm of blue-LED could receive more than

Table 2 Non-thermal antimicrobial techniques

NTATs	Main physical conditions	Processed foodstuffs/ grown media	Disinfected microorganisms	Outcomes	Processed temperatures (°C)	Processed duration/flow rate	Quality of foodstuffs after processed	References
Blue light	460 nm ($4080\ J/cm^2$)	Tryptic soy broth (TSB)	*Bacillus cereus* *E. coli* O157:H7 *L. monocytogenes* *P. aeruginosa* *S.* Typhimurium *S. aureus*	(1) *S. aureus* (below 2 log_{10} CFU[a]/ml); (2) *B. cereus* (3 log_{10} CFU/ml ↓[c]); (3) 11-deoxycortisol ↓, actinonin ↓, coformycin ↓, tyramine↑[d], chitobiose ↑	4, 10 and 25	7 h 9 h 1 h	NI[b]	Kumar et al. (2017)
	415 nm ($25\ mW/cm^2$)	Luria-Bertani (LB) broth/LB agar medium	*E. coli* BW25113	64 single-gene-deleted mutants sensitive to blue light (up to $64.8\ J/cm^2$) identified.	37	NM[e]	NI	Kruszewska-Naczk et al. (2024)
	405 nm 460 nm ($120\ mW/cm^2$)	TSB	*Vibrio parahaemolyticus* *Lactobacillus plantarum* *S. aureus*	For *V. parahaemolyticus*, (1) 4.0 log_{10} CFU/ml ↓; (2) 405 nm > 460 nm in antibacterial effects; (3) *V. parahaemolyticus* > *L. plantarum* and *S. aureus* in sensitivity	4, 10 and 25	7 h	NI	Kumar et al. (2016)
	405 nm ($87\ W/m^2$)	Tomato	*Botrytis cinerea* *Rhizopus stolonifer*	(1) 1.9 log_{10} CFU/g ↓ (*B. cinerea*); (2) 3.2 log_{10} CFU/g ↓ (*R. stolonifer*).	16.8	12 days	(1) TPC[f] (24.6% ↑); (2) TEAC[g] (28.2% ↑); (3) VC[h] (16.5% ↑); (4) mass loss ↑; (5) lycopene content ↓.	Chua et al. (2021)
	410 nm (28.8 mW/cm^2)	Paperboard carton	sessile *L. monocytogenes* (Planktonic *L. monocytogenes*)	For sessile *L. monocytogenes*, (1) 3.0 log_{10} CFU/cm^2 ↓ (120 min); (2) 5.0 log_{10} CFU/cm^2 ↓ (390 min).	7.0 (14.9)	120 (80) min 390 (84) min	NI	Ghate et al. (2019)

	455 nm (785.7 J/cm^2)	Pet food	powdered *S.* Typhimurium and *S.* Typhimurium on pet food	1.44 $\log_{10}$ CFU/g ↓ (powdered) 3.22 $\log_{10}$ CFU/g ↓ (on pet food)	20–23	45 min	(1) Weight ↓; (2) a_w[i] ↓; (3) Lipid oxidation ↑.	Prasad et al. (2021)
	460 nm (15–27 kJ/cm^2)	Fresh-cut pineapple slices	*Salmonella* spp.	(1) Under 7 and 16°C (disinfected); (2) Under 25°C (suppressed).	7, 16, 25	24, 13.91 and 8.66 h	Bleaching phenomenon found.	Ghate et al. (2017)
	460–470 nm (3V/1W)	Fresh-cut apples	*E. coli* O157:H7	By blue light + 0.025% carvacrol, *E. coli* count was significantly reduced.	4	5 days	Browning indexes down to (1) 39.85 (with 1% VC); (2) 38.30 (with 0.025% carvacrol) and (3) 37.67 (with 0.125% citric acid).	Hyun et al. (2022)
	405 nm (842mW/cm^2)	Paprika	*Salmonella* spp.	2.1–7.9 $\log_{10}$ CFU/g ↓	131.7	240 s	(1) Redness (a* value) ↓; (2) DPPH[j] ↑.	Minor et al. (2024)
UVA[k]	365 nm (1.6 mW/cm^2)	YEPG[l] and M90 media	*Candida auris* *Saccharomyces cerevisiae* *Cladosporium cladosporioides*	(1) For D90 dose for 1.0 $\log_{10}$ CFU/ml ↓, 365 nm < 400 and 450 nm; (2) For D90 dose, *C. cladosporioides* > *C. auris* > *S. cerevisiae.*	24–27	13.5 h 6.5 h 13.0 h	NI	Gierke and Hessling (2024)
	356 nm (0.03 mW/cm^2)	Spinach	*E. coli* O157:H7 *S.* Typhimurium *L. monocytogenes*	(1) 0.64 $\log_{10}$ CFU/ml ↓ (E. coli); (2) 0.23 $\log_{10}$ CFU/ml ↓ (*S.* Typhimurium); (3) 0.34 $\log_{10}$ CFU/ml ↓ (*L. monocytogenes*)	22	60 min	(1) Color →[m]; (2) TPC →; (3) Texture →	Jeong and Ha (2022)

(continued)

Table 2 (continued)

NTATs	Main physical conditions	Processed foodstuffs/ grown media	Disinfected microorganisms	Outcomes	Processed temperatures (°C)	Processed duration/flow rate	Quality of foodstuffs after processed	References
	365 nm (15.2 kJ/m^2)	BHI[n] culture medium	*Campylobacter jejuni* *E. coli* *V. parahaemolyticus* *S.* Enteritidis	(1) −6.4 of LSR[o] (*C. jejuni*); (2) −0.086 of LSR (*E. coli*); (3) −0.017 of LSR (*V. parahaemolyticus*); (4) −0.049 of LSR (*S.* Enteritidis).	37	4 min	NI	Kido et al. (2024)
	365 nm (Up to 114 kJ/m^2) (30 min)	Chicken meats	*C. jejuni*	(1) −0.33 of LSR (10 min); (2) −0.72 of LSR (20 min); (3) −1.34 of LSR (30 min).	37	10 min 20 min 30 min	Color →	Kido et al. (2024)
	365 nm (0.043 W/cm^2)	MHB[p]	*C. jejuni* NCTC 11168	Time to $4D_t$ by log linear model (132.5 s) < log linear and shoulder model (145.0 s), weibull model (145.0 s) and double weibull model (144.8 s).	NM	30, 45, 60, 90, 120, 135, and 150 s	NI	Soro et al. (2021)
	365 nm (183 mW/cm^2)	LB[q] broth + 0.9% NaCl	*S. aureus*	(1) Inactivation rate varied; (2) Bacteria sensitivity ↑; (3) Antibacterial effect ↑; (4) •OH ↑.	NM	15 min interval of fractionated exposure	NI	Zhao et al. (2023)
UVB[r]	312 nm (0.45 J/cm^2)	Coconut water Fructose solution	*E. coli* K12	(1) 4.5 log_{10} CFU/ml ↓ at 0.45 J/cm^2 (312 nm) > 4.0 log_{10} CFU/ml ↓ at 0.20 J/cm^2 (254 nm) or at 0.35 J/cm^2 (312 nm).	19–22	NM	8% of ascorbic acid ↓	Jha et al. (2024)

				DNA damage ↑; (2) Impaired metabolic activity ↑; (3) 4.0 log_{10} CFU/ml ↓.				
	313 nm (2.9 J/cm^2)	PDA[s]	*Geotrichum citri-aurantii* *Penicillium italicum* *Penicillium digitatum.*	(1) Fully inactivated by 2.9 J/cm^2; (2) Fully inactivated by 3.5 J/cm^2; (3) Fully inactivated by 5.2 J/cm^2.	25	NM	NI	Zuluaga-Acosta et al. (2023)
	280 and 300 nm (0.112 W/cm^2)	MHB	*C. jejuni* NCTC 11168	(1) For inactivation rate, 280+300 nm > 280, 300 and 365 nm; (2) 4.0 log_{10} CFU/ml ↓ (280+300 nm)	NM	5 s	NI	Soro et al. (2021)
	313 nm (0.1 J/cm^2)	Lemon	*Geotrichum citri-aurantii* *Penicillium italicum* *Penicillium digitatum.*	Disease incidences ↓ by up to 55% and severities ↓.	25	NM	Lemon qualities and H_2O_2 levels at lemon wound maintained.	Zuluaga-Acosta et al. (2023)
UVC[t]	254 nm (67 J/m^2)	Environmental surface	Coronavirus (SARS-CoV-2)	CoV inactivated range: 7–241 J/m^2, with an average of 67 J/m^2.	RT[u]	1–30 min	NI	Quevedo et al. (2020)
	254 nm (221.8 mW/cm^2)	Milk Flow rate: 3 L/min	Coliform	4.0 log_{10} CFU/ml ↓	38.5	9.9 s	Lipid peroxidation ↓ (after 72 h exposure)	Makararpong et al. (2024)
	254 nm (27 W of output)	Bovine skim milk	Bacterial counts Coliform Spores	(1) 2.3 log_{10} CFU/ml ↓ (Bacterial counts); (2) Below 2.0 log_{10} CFU/ml (Coliform); (3) 0.3 log_{10} CFU/ml ↓ (Spores).	50	NM	(1) XO[v], IgG[w], LPO[x], LF[y], IgM[z] and IgA[α] →; (2) Carbonyls 26% ↑; 3) Native serum proteins 5% ↓.	Li et al. (2021)

(continued)

Table 2 (continued)

NTATs	Main physical conditions	Processed foodstuffs/ grown media	Disinfected microorganisms	Outcomes	Processed temperatures (°C)	Processed duration/flow rate	Quality of foodstuffs after processed	References
	254 nm ($240\ mJ/cm^2$)	Carrots	*L. monocytogenes* *S. aureus*	(1) + 5.0 log_{10} CFU/cm^2 ↓; (2) + 72 h of anti-listerial effect last; (3) Such UVC pretreatment adapted to *S. aureus*.	RT	5 min	NI	Rajalingam et al. (2024)
	254 nm (29.4–882.0 mJ/cm^2)	Inshell walnuts	*Salmonella*, spp. *E. coli* O157:H7 *L. monocytogenes* *S. aureus*	(1) The tailing effects were found; (2) From 0.82–2.41 log_{10} CFU/walnut ↓.	22	1, 3, 5, 10, 20, and 30 min	Lipid oxidation → (up to 882.0 mJ/cm^2)	Yun et al. (2022)
	265 nm ($1680\ mJ/cm^2$)	Kimchi seasoning	White colony-forming yeast: *Candida sake*, *Kazachstania servazzii*	(1) Both suppressed; (2) For resistance to UVC, *K servazzii* > *C. sake*.	10	NM	Physicochemical properties →	Song et al. (2022)
	279 nm ($1420\ mJ/cm^2$)	Orange juice	*S. cerevisiae*	(1) 4.86 log_{10} CFU/ml ↓ (*S. cerevisiae*); (2) Plasma membrane destroyed; (3) 4.44 log_{10} CFU/ml ↓ (orange juice).	20	NM	Color parameters, pH, titratable acidity and total soluble solids →	Niu et al. (2021)
	280 nm ($7.26\ mJ/cm^2$)	Media Spinach Sliced deli meat	*E. coli* O157:H7 *S.* Typhimurium	(1) 3.00 log_{10} CFU/ml ↓ (Media); (2) 1.50 log_{10} CFU/ml ↓ (Spinach); (3) 3.00 log_{10} CFU/ml ↓ (deli meat).	21	1.35 s-*E. coli*; 3.60 s-S. Typhimurium	Quality →	Kim and Kang (2020)

	265, 285 nm or 255+265+285 nm (301.3 mJ/cm^2)	Beef broth Diced beef	*L. monocytogenes* *S.* Typhimurium	(1) 7.5 log_{10} CFU/ml ↓ (beef broth); (2) 0.9 log_{10} CFU/g ↓ (diced beef); (3) 7.4 log_{10} CFU/ml ↓ (beef broth); (4) 0.6 log_{10} CFU/g ↓ (diced beef).	RT	60 min	NI	McSharry et al. (2022)
	270/365 nm (128/ 12600 mJ/cm^2) (UVC-LED[β]) 254 nm UV mercury lamps (160 mJ/cm^2)	Plastic/Powdered seasoning ingredients (onion, garlic, chili, cheese)	*L. monocytogenes* *E. coli* O157:H7 *B. subtilis* *S.* Typhimurium	Antimicrobial effect: (1) 270 nm UVC-LED > the other two; (2) Plastic > seasoning powders; (3) 0.75–3 log_{10} CFU/g ↓ (powders)	RT	40 s	NI	Nyhan et al., 2021
	254 nm (1273.3 and 89.7 mJ/cm^2)	Apple juice Apple peel discs	*E. coli* *L. monocytogenes*	(1) First decimal reductions: Apple juice (1273.3 mJ/cm^2), Apple peel discs (89.7 mJ/cm^2). (2) 3.3 log_{10} CFU/ml ↓ (*E. coli*), 5.3 log_{10} CFU/ml ↓ (*L. monocytogenes*).	RT	NM	NI	Nicolau-Lapeña et al., 2022
X-ray	150 kV (125 Gy)	Stainless steel	*P. fluorescens* biofilm cells	EPS[γ], bacterial glucose uptake system and membrane integrity damaged.	RT	130 s	NI	Pang et al., (2022)
	150 kV of low-energy X-ray (300 and 350 Gy)	Dry cardamom	*E. coli* *S.* Typhimurium	(1) 5.0 log_{10} CFU/g ↓ (300 Gy); (2) 5–6 log_{10} CFU/g ↓ (350 Gy).	RT	300–350 s	(1) 2-ACBs[δ], dimethyl disulfide & 3-methyl-thiophene undetected; (2) No off-odor and metal odor generated; (3) 80% PUFAs[ε] remained (300 Gy).	Zhang et al. (2021a)

(continued)

Table 2 (continued)

NTATs	Main physical conditions	Processed foodstuffs/ grown media	Disinfected microorganisms	Outcomes	Processed temperatures (°C)	Processed duration/flow rate	Quality of foodstuffs after processed	References
	150 kV of low-energy X-ray (350 and 400 Gy)	Dry edible bird's nest (EBN)	*E. coli* O157:H7 *S.* Typhimurium	(1) 6.35–2.00 $\log_{10}$ CFU/g; (2) 5.84–2.00 $\log_{10}$ CFU/g; (3) Resistance to X-ray: *S.* Typhimurium > *E. coli* O157:H7	RT	350–400 s	(1) Sulfur-containing volatile compounds undetected; (2) Normal odor of EBN →; (3) Yellowness ↑.	Zhang et al. (2020)
	800 Gy	Sliced ham	*S.* Typhimurium *E. coli* O157:H7 *L. monocytogenes*	(1) 5.7 $\log_{10}$ CFU/g ↓ (*S.* Typhimurium); (2) 7.2 $\log_{10}$ CFU/g ↓ (*E. coli* O157:H7); (3) 6.9 $\log_{10}$ CFU/g ↓ (*L. monocytogenes*).	RT	800 s	(1) Textural properties →; (2) Color values →.	Cho and Ha (2019)
	160 kV (1.5 kGy)	Lettuce	*E. coli* O157:H7 *L. monocytogenes* *S.* Typhimurium	(1) 6.89 $\log_{10}$ CFU/g ↓ (*E. coli* O157:H7); (2) 3.22 $\log_{10}$ CFU/g ↓ (*L. monocytogenes*); (3) 4.48 $\log_{10}$ CFU/g ↓ (*S.* Typhimurium).	RT	60 min	(1) Texture →; (2) Color →.	Jee et al. (2023)
	160 kV (1.75–2.00 kGy)	Chilled chicken breast	Firmicutes	(1) Bacterial diversity ↓; (2) Microflora structure changed; (3) Firmicutes ↓.	RT	44–51 min	(1) Shelf life ↑ by 6 days; (2) Odor of irradiated chicken ↑; (3) Corruption indicators ↑.	Gao et al. (2024)
	2.5 MeV (3.10/7.00 kGy)	Pork cutlets	*E. coli* O157:H7 *L. monocytogenes* *S.* Typhimurium *S. aureus*	(1) 370 Gy ($D_{10}{}^{\zeta}$) (*E. coli* O157:H7); (2) 520 Gy (D_{10}) (*L. monocytogenes*);	RT	51/117 min	Color, pH and TBARS$^{\eta}$ → even up to 7.00 kGy of irradiation.	Yeom et al. (2024)

				(3) 620 Gy (D_{10}) (*S.* Typhimurium); (4) 360 Gy (D_{10}) (*S. aureus*).				
	150 kV (175–225 Gy)	Acrylic Polyethylene Stainless steel	*L. monocytogenes* biofilm*Pseudomonas fluorescens* biofilm	(1) 4 $\log_{10}$ CFU/cm^2 ↓; (2) Biofilm (*L. monocytogenes* and *P. fluorescens*) formation ability: Acrylic > Polyethylene and stainless steel; 3) Resistant abilities: *L. monocytogenes* + *P. fluorescens* > *L. monocytogenes* > *P. fluorescens*.	RT	233–300 s	NI	Zhang, Pang, et al. (2021b)
	160 kV (500 Gy)	PVC$^{\theta}$ STS 2B$^{\iota}$ Teflon	Biofilms of *E. coli* O157:H7, *L. monocytogenes*, *S.* Typhimurium	Biofilm removing ability: Teflon > PVC and STS 2B.	RT	20 min	NI	Cha and Ha (2022)
	150 kV (125 Gy)	Phosphate-buffered saline solution/stainless-steel	*P fluorescens* Planktonic cells Biofilm	(1) 4.60 $\log_{10}$ CFU/ml ↓ ($t_{R1}$$^{\kappa}$: 14.8 Gy) (Planktonic cells); (2) 4.21 $\log_{10}$ CFU/cm^2 ↓ (t_{R1}: 11.6 Gy) (Biofilm).	RT	130 s	NI	Pang et al. (2022)
Gamma ray	0.2, 0.5, 1.0, 5.0 and 10.0 kGy	Wheat	*B. cereus* *Enterobacteriaceae* Yeasts and molds	(1) The microorganism numbers negatively related to increased doses; (2) Up to 10.0 kGy, all microorganisms disinfected; (3) At 5.0 kGy, only yeasts and	RT	1.3, 3.3, 6.5, 32.6 and 65.2 min	Shelf life ↑.	Boshevska et al. (2024)

(continued)

Table 2 (continued)

NTATs	Main physical conditions	Processed foodstuffs/ grown media	Disinfected microorganisms	Outcomes	Processed temperatures (°C)	Processed duration/flow rate	Quality of foodstuffs after processed	References
				molds survived; (4) Above 1.0 kGy of dose can kill *Enterobacteriaceae*.				
	1.5, 2.0 and 4.0 kGy	Mutton	Coliform Yeasts and molds	TVC^{λ}, TCC^{μ} and $TYMC^{\nu}$ negatively related to increased doses.	RT	NM	(1) Dry matter, ether extracts, peroxide value, TBARS, free fatty acids ↑; (2) ash, crude protein and raw pH ↓; (3) tenderness, juiciness →; (4) color and flavor changed; (5) 4.0 kGy was the best.	Hashem et al. (2022)
	0.5, 1.0, 1.5 and 2.0 kGy	Mango (Amrapali and BARI-11)	Total microbial count	(1) 1.4 log_{10} CFU/ml ↓ (Amrapali at 2.0 kGy); (2) 1.5 log_{10} CFU/ml ↓ (BARI-11 at 2.0 kGy).	RT	5.2, 10.3, 15.5 and 20.7 min	(1) Carb, energy, fat, fiber, protein, soluble solids, sugar affected; (2) Antioxidant activity ↑ (1.5 kGy); (3) Textures kept firm (1.5 kGy with 5.811 kGy/h).	Jabin et al. (2023)
	250, 500, 750, 1000 and 1500 Gy (9.1, 3.9 and 0.22 kGy/h of dose rates)	Rice	*E. coli*, *S.* Typhimurium, *L. monocytogenes*	(1) Dose rate unrelated to D_{10}; (2) Lower D_{10} of 0.22 kGy/h than that of 0.76 kGy/h of X-ray.	RT	For 0.22 kGy/h, 68, 136, 205, 273 and 409 min	NI	Begum et al. (2020)

	0.5 kGy + 6 µL/ml of myrtle juice and 0.08 µL/ml of thyme EO[ξ]	Packaged ready-to-eat salads	*S. aureus*	3.0 $\log_{10}$ CFU/g ↓	22	25.5 min	Odor, taste and color were acceptable.	Zernadji et al. (2024)
PDI[π]	405 nm + CHL[ρ] (8.6 mW/cm^2)	DRBC[σ] agar	*Rhizopus stolonifera* *Penicillium expansum*	(1) 77.4% inactivation (*R. stolonifera*); (2) 52.1% inactivation (*P. expansum*).	25	0, 15 and 30 min	NI	Chong et al. (2024)
	405 nm + CHL (26.1 mW/cm^2)	Bread	*R. stolonifera* *P. expansum* *Aspergillus niger*	51–58% counts ↓	~26	30 min	(1) Moisture and texture →. (2) Colors changed.	Chong et al. (2024)
	405 nm + porphyrins (2.2 mW/cm^2) (5.46 fg/CFU)	DRBC agar	*Fusarium oxysporum*	*F. oxysporum* population ↓ below detection limit.	25.5	6 h	NI	Xu et al. (2024b)
	405 nm + porphyrins (2.2 mW/cm^2) (5.46 fg/CFU)	Cavendish bananas	*F. oxysporum*	70% count ↓	25.5	48 h	(1) Appearance, mass, hardness and peel color →; (2) polyphenol oxidase ↓.	Xu et al. (2024b)
Filtration	MF[τ]-1.4/ MF-0.8 (1.4/0.8 µm pore diameters)	Bovine skim milk	Aerobic bacteria Coliform Spores Somatic cells	(1) 3.0 $\log_{10}$ CFU/ml ↓; (2) Coliforms, spores and somatic cells undetected.	50	50, 100, 150 and 200 mL/min	(1) XO, IgG, LPO, LF, IgM, IgA and native serum proteins → (MF-1.4); 2) IgG and LPO —, IgA (93%), XO (92), LF (91) and IgM (85%) remained (MF-0.3).	Li et al. (2021)

(continued)

Table 2 (continued)

NTATs	Main physical conditions	Processed foodstuffs/ grown media	Disinfected microorganisms	Outcomes	Processed temperatures (°C)	Processed duration/flow rate	Quality of foodstuffs after processed	References
	CNC[υ]/Cu_9S_5	Bacterial suspension	*S. aureus*	(1) 4.6 $\log_{10}$ CFU/ml ↓ (4 nm size); (2) +99.9% disinfection rate; (3) Filtration flux ↑; (4) Retention capacity ↑.	RT	40 s	Good water dispersibility	Ye et al. (2024)
	SCF[φ]-ZnONPs[χ]	Water	*E. coli*	(1) Biofilm inhibited; (2) Anti-adhesion ↑; (3) Biofouling mitigation ↑; (4) No toxicity.	RT	NM	Wate purified.	Khalaf et al. (2024)
	MF tubular ceramic membranes	Water from El-Salam sluiceway	Coliform	(1) F3: 70 → 14 CFU/100 ml; (2) F2: 22.9 L/(m^2•h) of permeate flux; (3) 80% of bacterial removal ability.	25	45 s	Coliform in water: 14 CFU/100 ml.	Rashad et al. (2022)
	Nadir® MV020T/PA PVDF[ψ]-AgNPs[ω]	Bacterial suspension	*E. coli*	*E. coli* (3–4 mm inhibitory zones); Nadir® MV020T (0.47% AgNPs lost), PA PVDF (2.12% AgNPs lost).	RT	21 days	NM	Gallus et al. (2023)
	SSF[à] + HRF[á] (0.2 m/h)	Drinking water	Coliform	99.6% of total coliform removed	RT	0.2 m/h	100% free of total coli can not be reach.	Fitriani et al. (2020)
	Sand bed + 10% ZVI[â]	River water	*E. coli* *B. subtilis*	1.25×10^7 cells/ml of *E. coli* and *B. subtilis* retained.	RT	4 m/day	1.25×10^7 cells/ml of Bacteria and 4.0×10^6 cells/ml of algae removed.	Wang et al. (2022)

	SSF (16 cm layers) + Chitosan (10 mg/l)	Water	*E. coli* MS2[ã] coliphage	(1) Turbidity ↓; (2) $\log_{10}$ CFU/ml ↓.	RT	57 days	Meet the two-star WHO[ä] performance level of effectiveness.	Holmes et al. (2023)
	CFF[å] + 104 kPa	Sugarcane juice	Bacteria Yeast and mold	(1) 6.0 $\log_{10}$ CFU/ml ↓ (Bacteria); (2) 4.0 $\log_{10}$ CFU/ml ↓ (Yeast and mold).	RT	30 L/h	(1) 3-fold less in PPO[æ]; (2) 98% recovery of sucrose; (3) 80% recovery of polyphenol; (4) Juice storability ↑.	Panigrahi et al. (2020)
	CFF + spiral-shaped microchannel	Tap water	*E. coli*	Recovery rate: 40% → +90%	RT	Up to 260 mL/min	NI	Zhang et al. (2023b)
	TFUF[ç] + HA[è] TFUF + Try[é]	Water	Enteric virus (bacteriophage MS2)	Recovery rates: (1) 92.16% by TFUF + HA; (2) 81.50% by TFUF + Try.	RT	Up to 500 mL/min	NI	Wang, et al. (2024a)
	TFF[ê] + 20–25% carvacrol	Bacterial suspension	*S. aureus*	(1) Bacteria deposition ↓; (2) Membrane permeate fluxes ↑.	RT	0.007 m/s	NI	Nowak et al. (2023)
PEF[ë]	PEF (30 kV/cm, 200–500 μs, 17.5–87.6 kJ/l)	Sorbitol/ Nisin-unloaded/ loaded PecA nanoparticles solution	*S.* Typhimurium *L. innocua*	PEF treatment efficacy does not linearly depend on the energy provided to the bacterial cells.	40	2 min	NI	Novickij et al. (2020)
	HVAEF[ì], HVEF[í], HVPEF[î] (>10 kV)	Solid medium Food surfaces (Salmon, griskin, cheese & sausage)	*S. aureus*	(1) Sensitive sequences: HVPEF > HVEF > HVAEF; > 99.9% of mortality rate; (2) 98.3–99.6% of mortality rates	RT	15 min	NI	Qi et al. (2021)

(continued)

Table 2 (continued)

NTATs	Main physical conditions	Processed foodstuffs/ grown media	Disinfected microorganisms	Outcomes	Processed temperatures (°C)	Processed duration/flow rate	Quality of foodstuffs after processed	References
Ultrasound	LFUS[ï] (20 kHz, and 300W)	Bacterial suspension	*E. coli* O157:H7	(1) The structures of bacterial cell surface appendages destroyed; (2) cell membrane broken; (3) curli-formation genes down-regulated; (4) metabolic capacity ↓; (5) biofilm-forming capability ↓.	20	5 min	NI	Bai et al. (2023)
	LFUS (20 kHz, 60/100 W)	Lettuce	*P. fluorescens*	(1) The 27 metabolites differentially expressed; Under 15.79 W/cm^2 and 60 W, (2) EPS, energy level and genetic information of biofilm ↓; (3) lipid ↑; Under 26.32 W/cm^2 and 100 W, metabolic level and osmotic pressure of biofilm ↓, UDP-Gal[ð], ADPR[ñ] and 4-PA[ò] ↑; (4) biofilm ↓.	30	15 min	Under 26.32 W/cm^2 and 100 W, quality →.	Wang, et al. (2024b)
	LFUS (20 kHz, 573 W/cm^2)	Bacterial suspension	*E. coli*	3.52 $\log_{10}$ CFU/ml ↓	RT	2 s of pulsed mode, 27 min	NI	He et al. (2021)

	LFUS (20 kHz, 100%, 50% and 10% amplitudes)	Raw goat milk	TAMB[6], coliforms, *Streptococci*, *Lactobacilli*, yeast and mold, *Brucella melitensis* type 3, *E. coli*, *L. monocytogenes*, methicillin-resistant *S. aureus*, and *S.* Typhimurium	(1) Antimicrobial effects (20 kHz): 100% amplitudes > 50% and 10%; (2) TAMB, coliforms, *Streptococci*, *Lactobacilli*, yeast and mold ↓: 4.02–6.52 log_{10} CFU/ml; (3) 99% of *Brucella melitensis* type 3, *E. coli*, *L. monocytogenes*, methicillin-resistant *S. aureus*, and *S.* Typhimurium inactivated.	Initial 15	10 min	Color changed: brightness (L-value ↑), yellow (b-value) .	Tavsanli et al. (2022)
	LFUS (20 kHz and 300W)	Polystyrene sheets	*E. coli* biofilm	(1) 99.999% of biofilm removed (5 min); (2) bacterial cells still survived (30 min).	20	5 and 30 min	NI	Bai et al. (2023)
	US (20–100 kHz and 2–10 MHz)	Almond milk	*E. coli* O157:H7 *L. monocytogenes*	(1) Growth reduction (E. coli O157:H7); (2) Lag phase extended (*L. monocytogenes*).	RT	8 min (6s) 2 min (6s)	(1) Shelf life ↑; 2) food sensory properties →.	Onyeaka et al. (2023)
	LFUS (20 kHz and 300W)	Lettuce leaves, cantaloupe and cucumber epidermis	*E. coli* O157:H7	Anti-biofilm functions ↑;	20	NM	Colors →.	Bai et al. (2023)
Cold plasma	CP[6] (8 kV) Power: 51.7 W Power density: 71.5 W/cm^2.	Fresh vegetables, fruits, nuts, powdered food samples	*E. coli* O157:H7 *S.* Typhimurium *L. monocytogenes*	Similar inactivation rates of these three bacteria were found over time.	RT	20 min	Surface roughness negatively related to inactivation rate.	Han et al. (2020)

(continued)

Table 2 (continued)

NTATs	Main physical conditions	Processed foodstuffs/ grown media	Disinfected microorganisms	Outcomes	Processed temperatures (°C)	Processed duration/flow rate	Quality of foodstuffs after processed	References
	ACP[ö] (80 kV) Power: up to 112 W Power density: 0.089 W/cm^2.	Fresh shrimps Frozen shrimps	*L. monocytogenes* *V. parahaemolyticus*	(1) 0.94 (fresh) and 1.68 (frozen) $\log_{10}$ CFU/g; (2) 1.26 (fresh) and 2.00 (frozen) $\log_{10}$ CFU/g.	4	30 min	(1) Shelf life ↑ 2 days; (2) total viable count ↓, total volatile basic nitrogen ↓, pH ↓; (3) sensory scores ↑; (4) TBARS ↑.	Liu et al. (2024)
	CP + 170 V CP + 230 V	Black pepper Cinnamon Fennel	Yeast and mold	(1) 1.1 $\log_{10}$ CFU/g ↓ (170V + 5 min); (2) 1.3 $\log_{10}$ CFU/g ↓ (170V + 5 min); (3) 1.0 $\log_{10}$ CFU/g ↓(170V + 5 min); (4) Fully disinfected (230V+15 min).	RT	5 min 15 min	(1) Black pepper (a_w ↓ 0.12), Cinnamon (a_w ↓ 0.18), Fennel (a_w ↓ 0.17); (2) color and textures →; (3) Under 230 V and 15 min, TPC↑, EO →.	Kahar et al. (2024)
HHP[ö]	HHP (350 MPa)	Donor human milks	Group B *Streptococcus* *E. coli*	(1) Group B *Streptococcus* count ↓; (2) *E. coli* count ↓ (At early timepoints).	38	NM	(1) Qualities → up to 6 months; (2) HHP-treated lactoferrin →.	Tran et al. (2024)
	HHP (200–400 MPa)	Carrot-orange juice blends	*L. innocua* (ATCC 51742) AMB[ø]	(1) +5 $\log_{10}$ CFU/ml ↓ (200-400 MPa+5 min); (2) 6 $\log_{10}$ CFU/ml ↓ (300 MPa + pH 4 + 2min/400 MPa + pH 5 + 1 min/400 MPa + pH 6 + 3 min); (3) AMB below 2 $\log_{10}$ CFU/ml.	20.0–28.6	5 min	(1) ascorbic acid, total carotenoids, total phenolic compounds →; (2) shelf lives ↑ +28 days; (3) total soluble solids, pH and color →.	Pokhrel et al. (2022)
	HPP[ù] (600 MPa)	Apple juice (pH 4.5)	*E. coli* O157:H7 *S. enterica* *L. monocytogenes*	5 $\log_{10}$ CFU/ml ↓	5	3.2 min	pH, aw, total soluble solids and turbidity →.	Usaga et al. (2021)

[a]CFU, colony forming unit; [b]NI, not involved; [c]↓, decreased; [d]↑, increased; [e]NM, not mentioned; [f]TPC, total phenolic contents; [g]TEAC, trolox equivalent antioxidant capacity; [h]VC, vitamin C; [i]a_w, water activity; [j]DPPH, 1,1-diphenyl-2-picrylhydrazil; [k]UVA, ultraviolet band A; [l]YEPG, Yeast extract-peptone-glucose; [m]→, unchanged; [n]BHI, brain heart infusion; [o]LSR, log survival ratio; [p]MHB, Mueller Hinton Broth; [q]LB, Luria-Bertani; [r]UVB, ultraviolet band B; [s]PDA, potato dextrose agar; [t]UVC, ultraviolet band C; [u]RT, room temperature; [v]XO, xanthine oxidase; [w]IgG, immunoglobulin G; [x]LPO, lactoperoxidase; [y]LF, lactoferrin; [z]IgM, immunoglobulin M; [α]IgA, immunoglobulin A; [ϝ]LED, light-emittin diodes; [γ]EPS, extracellular polymeric substances; [δ]2-ACBs, 2-alkylcyclobutanones; [ε]PUFAs, polyunsaturated fatty acids; [ζ]D_{10}, the dose needed to lead to 90% microorganism reduction; [η]TBARS, thiobarbituric acid values; [θ]PVC, polyvinyl chloride; [ι]STS 2B, stainless steel with finish 2B; [κ]t_{R1}:, the dose for achieving 1-log microorganism reduction; [λ]TVC, total viable count; [μ]TCC, total coliform count; [ν]TYMC, total yeast-molds count; [ξ]EO, essential oils; [π]PDI, photodynamic inactivation; [ρ]CHL, sodium chlorophyllin; [σ]DRBC, dichloran rose-bengal chloramphenicol; [τ]MF, microfiltration; [υ]CNC, cellulose nanocrystal; [φ]SCF, scoured cotton fabrics; [χ]ZnONPs, zinc oxide nanoparticles; [ψ]PVDF, polyvinylidene fluoride; [ω]AgNPs, silver nanoparticles; [ȃ]SSF, slow sand filtration; [á]HRF, horizontal roughing filter; [ã]ZVI, zero-valent iron; [ā]MS2, Emesvirus zinderi; [ä]WHO, world health organization; [å]CFF, cross-flow filtration; [æ]PPO, polyphenol oxidase; [ς]TFUF, tangential flow ultrafiltration; [è]HA, humic acid; [é]Try, tryptophan; [ê]TFF, tangential flow filtration; [ë]PEF, Pulsed electric field; [ì]HVAEF, alternating electric field; [í]HVEF, high voltage electrostatic field; [î]HVPEF, prick electrostatic field; [ï]LFUS, low-frequency ultrasound; [ð]UDP-Gal, uridine diphosphate galactose; [ñ]ADPR, adenosine diphosphate ribose; [ò]4-PA, 4-pyridoxic acid; [ó]TAMB, total aerobic mesophilic bacteria; [ô]CP, cold plasma; [õ]ACP, atmospheric cold plasma; [ö]HHP, high hydrostatic pressure; [ø]AMB, aerobic mesophilic bacteria; [ù]HPP, high-pressure processing

4 $\log_{10}$ CFU/ml reduction in its population, where 405 nm blue-LED demonstrated higher antibacterial effect than 460 nm one under the same doses. In addition, the exposures of *Lactobacillus plantarum* and *S. aureus* to blue-LED also displayed significance, compared with the control (520 nm), but they showed less sensitive to blue-LED illumination than *V. parahaemolyticus* (Table 2) (Kumar et al., 2016). These results further confirmed that the different bacteria have their own sensitivities to blue light, similarly to those found in the study of Kumar et al. (2017).

Apart from bacteria, the 405 nm range blue light could also efficiently eliminate fungi attached to foodstuffs. Exposed to 87 W/m^2 of 405 nm blue-LED for 12 days, for example, the 405-nm-treated *Botrytis cinerea* and *Rhizopus stolonifer* (two fungi) on tomato surfaces could achieve 1.9 and 3.2 $\log_{10}$ CFU/g reductions, respectively, proving the antifungal capability of 405 nm blue light (Table 2) (Chua et al., 2021). Noteworthily, this antimicrobial property of 405 nm range blue light could be transferred to treat the sessile bacteria attached to the food packaging material, not just on the foodstuff surfaces. For instance, the 410 nm blue-LED has been reported to inactivate sessile *L. monocytogenes* on the surface of paperboard, a food packaging material. To achieve below the detection limit, the sessile counts sized 5 and 7 $\log_{10}$ CFU/cm^2 needed to take 120 and 390 min, respectively, which were significantly higher than the durations (80 and 84 min) for the respective planktonic counts of *L. monocytogenes* to reach the same goal. This indicated that the overlap of sessile bacterial cells could provide some protections on the populations, proved by scanning electron microscopy (SEM) technique. However, the illumination duration of 410 nm blue-LED was still positively associated with the increase of membrane-damaged cells, revealing that such wavelength of blue-LED could potentially decrease the risk of *L. monocytogenes* to contaminate paperboard cartons used for food packaging, in a dose-dependent manner (Table 2) (Ghate et al., 2019).

Although it seemed that 460 nm range of blue light was weaker than 405 nm one to display its antimicrobial activity, the 460 nm range of blue light can still demonstrate its efficiency in inactivating *S.* Typhimurium. In this case related to the treatment of pet food, the 460 nm range was more effective in eliminating *S.* Typhimurium attached to the surface of pet foods, than the powdered *S.* Typhimurium. For example, the 455 nm of LED in cooperation with 785.7 J/cm^2 of dose resulted in 1.44 log CFU/g of decrement in powdered *S.* Typhimurium count, which was significantly lower than 3.22 log CFU/g of decrement in *S.* Typhimurium count presented on pet food. In addition, blue light could also display its drying potential by decreasing the weight and a_w of the treated pet food (Table 2) (Prasad et al., 2021). Besides pet food, the 460 nm blue-LED could also show its potential against *Salmonella* on the surface of fresh-cut pineapple slices. In this scenario, temperature factor might affect the antibacterial effect of 460 nm blue-LED application because *Salmonella* under 7 and 16 °C could be fully disinfected, while its growth was just suppressed under 25 °C. Additionally, the irradiance (15–27 kJ/cm^2) of blue-LED achieved no significance to influence the antibacterial effect of 460 nm blue-LED (Table 2) (Ghate et al., 2017).

Looking into the advantages of blue light when displaying its antimicrobial properties on foodstuffs, some studies have pointed out that it could help to enhance the nutritive values, as well as maintain the appearance qualities of some treated foodstuffs as well (Chua et al., 2021; Hyun et al., 2022). For example, the 87 W/m^2 of 405 nm blue-LED exposure for 12 days could help to improve the physicochemical properties of postharvest tomatoes through enhancing their total phenolic content, Trolox equivalent antioxidant capacity (TEAC) and vitamin C by 24.6, 28.2 and 16.5%, respectively, compared with the non-illuminated samples (Table 2) (Chua et al., 2021). The cause might be that blue light could induce the enhanced intracellular levels of ROS and lipid peroxidation of, not only microorganisms, but also foodstuffs, which triggers the living cells in the foodstuffs to synthesize more antioxidants to deal with ROS as the defending strategies, thus increasing these antioxidant's concentrations and enhancing the antioxidant activities of the processed foodstuffs (e.g., postharvest amaranth) (Hyun & Lee, 2020; Jin et al., 2021). In another study, blue light could also help to prevent the color deterioration of fresh produce. As reported, the browning indexes of fresh-cut apples under the combined treatments of blue-LED (460–470 nm, 3 V, 1 W) with ascorbic acid (1%), carvacrol (0.025%), and citric acid (0.125%) for 5 d were remarkably decreased down to 39.85, 38.30 and 37.67, respectively, compared with the control (59.36) using distilled water, as well as the treatments of carvacrol-only (53.52) and citric-acid-only (50.33) (Table 2) (Hyun et al., 2022).

However, the disadvantages of blue light when demonstrating its antimicrobial properties on foodstuffs should also be noticed, in which the results and phenomena associated with lipid oxidation, bleaching or quality variations could be found in some treated foodstuffs. For instance, the remarkable lipid oxidation was found in the blue-LED-treated pet food (455 nm of LED with 785.7 J/cm^2 of dose), compared with the non-treated control, indicating that the double bonds of lipids in pet food could be destructed by 455 nm blue light under high dose and the quality of pet food could be negatively affected due to the application of strong blue light (Table 2) (Prasad et al., 2021). In the other studies, the bleaching phenomenon on the surfaces of pineapple slices was found after exposing them to 460 nm blue LED (15–27 kJ/cm^2), indicating that 460 nm of blue light under higher dose might, more or less, affect the appearance of fresh-cut pineapples (Table 2) (Ghate et al., 2017). Also, the 87 W/m^2 of 405 nm blue-LED exposure for 12 days could result in the remarkable mass losses and lycopene content decrements of the illuminated tomatoes (Table 2) (Chua et al., 2021). The cause could be that blue light might damage the double bonds of carotenoids or anthocyanin contained in the pineapple pulps, thus fading the pulp's colors. Based on the above results, it can be concluded that higher dose of blue light or short-wave blue light (e.g., 405 nm) are the main reasons to trigger the negative effects on the processed foodstuffs.

Although blue light antimicrobial technique is a non-thermal approach to inactivate microorganisms, the applications of short-wave blue light and its extremely higher dose could still generate some amounts of heats to increase the temperatures of foodstuffs. This could be seen through photo-thermal pasteurization method applying ultra-high irradiance blue light (405 nm) that was used to evaluate its capability of disinfecting *Salmonella* spp. in paprika powder (Minor et al., 2024). As reported, when employing blue LED light under various irradiances (e.g., 548, 697 and 842 mW/cm^2) followed by treatment durations (60, 120, 180, and 240 s) to expose to the *Salmonella*-inoculated paprika samples (about 8.7 $\log_{10}$ CFU/g), the best combined condition was showed to be 842 mW/cm^2 and 240 s, which resulted in the decreased levels of *Salmonella* counts from 2.1 to 7.9 $\log_{10}$ CFU/g, and the increased sterilization-like temperature by up to 131.7 °C. After the processing, the redness (e.g., a* values) in the exposed paprika powders was remarkably reduced, while a higher DPPH scavenging activity was found in such samples, compared with the nonexposed control (Table 2) (Minor et al., 2024).

Besides its own antimicrobial property, blue light could also play an important role in elevating the antibacterial effectiveness of other antimicrobials or photosensitizers in disinfecting *E. coli* O157:H7 on the surface of agar media and fresh produce such as cherry tomatoes and fresh-cut apples, thus prolonging their shelf lives (Hyun et al., 2022). This study was carried out under 4 °C for 5 days, the combined treatments of blue-LED (460–470 nm, 3 V, 1 W) and 0.025% of carvacrol (an antimicrobial) were found to significantly reduce the counts of *E. coli* O157:H7 on the surfaces of these fresh produce, compared with those treated only by carvacrol, blue-LED, curcumin (0.0078%, a photosensitizer) and riboflavin (0.0078%, a photosensitizer), respectively (Table 2) (Hyun et al., 2022).

2.2.1.2 UVA

UVA light, or called long-wave UV and with wavelengths (frequencies/photon energy) ranging from 320 nanometer (937 THz/3.9 eV) to 400 nanometer (750 THz/3.1 eV), is included in the UV light family of electromagnetic spectrum involving in wavelengths shorter than that of blue light (400 nm to 500 nm), whereas longer than that of UVB light (280–320 nm) (Rumble et al., 2019). The mechanism to use UVA light for inactivating microorganisms (e.g., bacteria/pathogens or microalgae) is to carry out the non-ionizing radiation triggered by UVA light to form 4-thymine dimerization (DNA kink) on the DNA chain of the microorganism, thus damage the replication of DNA in the microorganism, and eventually kill the microorganism (Huang et al., 2024). Simultaneously, UVA can also induce the formation of pyrimidine dimers and oxidative DNA damage in bacterial cells (Fuentes-León et al. 2024). All of which contribute to the antimicrobial mechanisms of UVA. Currently, the wavelength applied in studying the antimicrobial effect of UVA on foodstuffs has fallen on 365 nm, which is normally fixed as the waveband of the commercial UVA lamps.

As well recognized, UVA light could achieve better antimicrobial activity than blue light, because UVA light with higher frequency and photon energy can destroy microbial cells more efficiently, compared with blue light. This could be seen in a report showing that a lower D90 dose for 1 $\log_{10}$ CFU/ml reduction was achieved by 365 nm of UVA irradiation for two yeasts (*Candida auris* and *Saccharomyces cerevisiae*) and a mold (*Cladosporium cladosporioides*), compared with those by longer wavelengths (e.g., 400 and 450 nm) (Table 2) (Gierke & Hessling, 2024). This meant that UVA light was more powerful than blue light to eliminate fungi by only using less dose to achieve the same antimicrobial effect as blue light.

Similarly to blue light, different microorganisms (e.g., fungi and bacteria) also respond and resist differently to the same UVA dose. In the above case, it was found that UVA-irradiated *S. cerevisiae* was more photosensitive, compared with *C. auris*, while the UVA-irradiated *C. cladosporioides* required much higher D_{90} dose, compared with the two yeasts. Such different response mechanisms might be associated with the intracellular photosensitizers that participated. Further investigating the potential photosensitizers involved in this study, flavins and some porphyrins were detected, while thiamine was assumed according to fluorescence spectrum (Gierke & Hessling, 2024). In the bacteria case, the application of UVA (356 nm) at mean irradiation intensity of 0.03 mW/cm^2 on spinach surface led to significant $\log_{10}$ CFU/ml reductions of foodborne pathogens such as *E. coli* O157: H7, *S.* Typhimurium and *L. monocytogenes* by 0.64, 0.23 and 0.34, respectively, after 60 min of treatment, which were equal to 0.11 J/cm^2 of total light fluence (Table 2) (Jeong & Ha, 2022). For 365-nm UVA-LED-exposed *Campylobacter jejuni* that is a foodborne bacterium which can trigger gastroenteritis in humans, the significant inactivation with up to −6.4 of log survival ratio after 4 min of irradiation (15.2 kJ/m^2 of dose) was achieved, compared with the other foodborne pathogens such as *E. coli*, *V. parahaemolyticus* and *S.* Enteritidis with only −0.086, −0.017 and − 0.049 of which under the same conditions, respectively, showing that these three pathogens were not effectively inactivated by such irradiation, compared with *C. jejuni* (Table 2) (Kido et al., 2024).

The antimicrobial effect of UVA on foodstuffs is also subjected to treatment duration. When *C. jejuni*-inoculated chicken meats were exposed to 365-nm UVA-LED, the log survival ratio of *C. jejuni* on meats demonstrated a significant reduction from −0.33 and − 0.72 to −1.34 after 10, 20 and 30 min of irradiation, respectively, in a time-dependent manner. This indicated that there was a positive relation between UVA irradiation duration and antimicrobial effect on *C. jejuni*, furthermore, only longer 365 nm UVA-LED irradiation duration by up to 30 min (114 kJ/m^2 of dose) could remarkably inactivate *C. jejuni* attached to the surfaces of chicken meats, compared with shorter durations (Table 2) (Kido et al., 2024).

In the study of antimicrobial effect, an appropriate mathematical model to describe the survival rate curves of the treated microorganisms is necessary, because it can allow us to accurately estimate the antimicrobial effects of certain techniques (e.g., UVA antimicrobial technique) on foodstuffs. As reported by using log linear model, the time to reach a 4 $\log_{10}$ CFU/ml decrement ($4D_t$) by 365 nm of UVA-LED lights in the population of *C. jejuni* NCTC 11168 in Maximum Recovery Diluent (MRD) was 132.5 s. In addition, this log linear model was found the most fitted for the survival rate curve of *C. jejuni* because it achieved the smallest $4D_t$ value, compared with the other models such as log linear and shoulder mode (145.0 s), Weibull model (145.0 s) and double Weibull model (144.8 s) (Table 2) (Soro et al., 2021). Such results can make people more clearly about the actual antimicrobial effects of UVA on the designated microorganisms.

Recently, fractionated exposure has been introduced as an efficiently auxiliary manner to strengthen the inactivation effect of UVA-LED on bacteria. As found, the 15 min (not longer or shorter) interval of fractionated exposure during 365 nm UVA-LED illumination could significantly inactivate *S. aureus*. The mechanism involved pointed out that such operation varied the inactivation rate constant, as well as removed the shoulder in the fluence-response curves, which enhanced the sensitivity of bacteria to UVA, leading to higher damage of UVA to bacteria, thus improving the antibacterial effect of such operation with UVA-LED. Further investigation revealed that hydroxyl radical (•OH) induced by UVA played an essential role in interacting with bacterial cells during the interval without UVA-LED exposure, therefore supporting such antibacterial effect. Moreover, the longer interval might lead to the level decrement of •OH, along with the recovery of intracellular glutathione (GSH, an antioxidant), eventually bringing down such antibacterial effect (Table 2) (Zhao et al., 2023).

Observing the advantages of UVA antimicrobial technique, it is shown to be a clean and environmentally-friendly technology that does not produce wastes or chemicals to contaminate the environments during processing. Furthermore, the exposure duration for UVA to inactivate microorganisms is normally short (e.g., around up to 60 min), thus saving electrical energy and cost (Jeong & Ha, 2022). Moreover, UVA is excellent in achieving better superficial disinfection of foodstuffs. However, there are some disadvantages that should be learnt. Besides microorganism cells, UVA light can also attack the DNA of human skin cells, leading to the similar damages as that of microorganism cells by generating pyrimidine dimers, therefore, the appropriate protection on the skin of the operator is necessary during UVA exposure. In addition, as a superficial disinfection technique, UVA light could not efficiently penetrate deeply into the interiors of foodstuffs to carry out disinfection tasks.

2.2.1.3 UVB

UVB light, or termed medium-wave UV and with wavelengths (frequencies/photon energy) ranging from 280 nanometer (1.1 PHz/4.4 eV) to 320 nanometer (937 THz/3.9 eV), is included in the UV light family of electromagnetic spectrum involving in wavelengths shorter than that of UVA light (320 nm to 400 nm), whereas longer than that of UVC light (10–280 nm) (Rumble et al., 2019). The antimicrobial mechanism of UVB light is similar to that of UVA light by specially forming cyclobutane pyrimidine dimers (CPD), pyrimidine (6,4) pyrimidone (6–4 PP) dimers, and oxidized bases in DNA molecules that can result in mutation, through which the replication of bacterial DNA could be fully disordered, thus the treated microorganisms are eventually killed (Fuentes-León et al. 2024).

UVB application can work well on eliminating pathogenic bacteria from foodstuffs. The manipulation of UVB at 312 nm could lead to 4.5 $\log_{10}$ CFU/ml reduction of *E. coli* K12 in coconut water at 0.45 J/cm^2 dose. The associated mechanism involved in DNA damage, along with impaired metabolic activity, while the occurrences of membrane destruction and intracellular thiol oxidation were not triggered in this bacterium (Table 2) (Jha et al., 2024). Besides bacteria, UVB light can also show its antimicrobial activity on fungi. Nevertheless, different fungal phytopathogen conidia and mycelia might show different sensitivities to UVB light. As reported, the most sensitive fungus to UVB at 313 nm was *Geotrichum citri-aurantii*, which could be fully inactivated at 2.9 J/cm^2 of UVB radiation dose, while the lowest UVB dose to achieve the fully antimicrobial effect on *Penicillium italicum* was 3.5 J/cm^2. Notably, the 5.2 J/cm^2 of UVB dose was the threshold to totally disinfect *Penicillium digitatum*. As noted, these above-mentioned fungal phytopathogens could result in green and blue molds and sour rot which are postharvest diseases that might exert adverse influence on lemon quality (Table 2) (Zuluaga-Acosta et al., 2023). From the above cases, it was clear that the UVB radiation doses to disinfect fungi were normally higher than those to eliminate bacteria. The cause might be that fungi contain thick cell walls that could act as the shield to partially block UVB light, while bacteria that do not have such cell wall structure were directly exposed to UVB radiation, resulting that bacteria were more sensitive to UVB, compared with fungi.

As expected, the combined treatments of shorter UVB wavebands can achieve better antimicrobial activities than those of the other combined with longer UVA wavebands under similar intensities. In a recent study, the synergetic effects of 280 and 300 nm of UVB lights were found to reach the highest inactivation rate of *C. jejuni* NCTC 11168 in MRD, compared with the other combined effects of UV wavelengths (e.g., 280, 300 and 365 nm), where 4 $\log_{10}$ CFU/ml of reduction achieved for 5 s in a 5 $\log_{10}$ CFU/ml of *C. jejuni* suspension was observed in this strain. Modelling study showed that log linear models could be best fit to the survival curves of *C. jejuni* under such combined UVB exposure. These models can be used as valuable tools to predict the inactivation kinetics and UVB's efficiency in disinfecting food pathogens (Table 2) (Soro et al., 2021). These results indicated that UVB or their combinations with higher frequencies included more energies, therefore displaying more potent antimicrobial activities, compared with longer wavebands of UVA.

Noteworthily, when observing the appropriate mathematical model to describe the inactivation kinetics of *E. coli* K12 by 0.45 J/cm^2 of 312 nm UVB dose, a Weibull model was found the most suitable, compared with 0.20 J/cm^2 dose of 254 nm UVC which kinetics is in the log-linear relationship achieving 4 $\log_{10}$ CFU/ml reduction. In addition, the 0.35 J/cm^2 of 312 nm UVB dose was needed for reaching the same 4 $\log_{10}$ CFU/ml reduction (Table 2) (Jha et al., 2024). On one hand, this revealed that the inactivation kinetics of the same bacterium under different UV wavebands might respond differently, on the other hand, more UVB dose is required to arrive at the same antimicrobial goal, compared with UVC dose. This is understandable because UVC with high frequency contains more energy and can exert stronger damage on the bacterial cells, compared with UVB, thus, less UVC dose is equal to certain UVB dose to produce the same damage on these cells. However, although UVB was required more dose to achieve the same goal, it could facilitate less damage on the foodstuffs (Jha et al., 2024). Continuing with the above example, the 312 nm UVB irradiation induced lower ROS level in fructose solution, compared with UVC. In the further trial for the requirement of 4 $\log_{10}$ CFU/ml reduction of *E. coli* K12 exposed to UVB, only 8% of ascorbic acid was found degraded. This indicated that such UVB dose resulted in less photo-degradation of ascorbic acid mediated by fructose, compared to UVC treatment (Table 2) (Jha et al., 2024).

UVB can be an excellent pre-treatment tool to reduce fungi-induced diseases on fresh produce. As reported, UVB pre-treatment at 0.1 J/cm^2 of dose on lemons remarkably decreased the disease incidences induced by the fungal phytopathogens (e.g., *G. citri-aurantii*, *P. italicum*, and *P. digitatum*) by up to 55%, as well as lessened the severities caused by disease, simultaneously, it did not achieve negative effect on the quality of lemons. Noteworthily, UVB pre-treatment could also keep the levels of H_2O_2 at the wound location of lemons infected by *P. digitatum* that could inhibit the generation of H_2O_2 in the lemon tissue (Table 2) (Zuluaga-Acosta et al., 2023).

The advantages and disadvantages of UVB antimicrobial technique are similar to those of UVA one, by presenting that it is also a clean and environmentally-friendly technology. While its advantage superior to UVA is that the exposure duration for UVB technique to inactivate microorganisms is usually shorter than UVA one, which can save more electrical energy and cost, compared with UVA technique, because UVB with higher frequency might contain more energy to attack DNA under the same dose as UVA. However, such higher frequency also leads to the disadvantage of UVB technique to exert more severe damages on the skin of operator if inappropriately handled, compared with UVA technique. Additionally, similarly to UVA light, UVB light also could not efficiently penetrate deeply into the interiors of foodstuffs to sanitize pathogenic microorganisms.

2.2.1.4 UVC

UVC light, or called short-wave UV and with wavelengths (frequencies/photon energy) ranging from 10 nanometer (30 PHz/124 eV) to 280 nanometer (1.1 PHz/4.4 eV), is included in the UV light family of electromagnetic spectrum involving in wavelengths shorter than that of UVB light (280 nm to 320 nm), whereas longer than that of X ray (10 pm to 10 nm) (Rumble et al., 2019). Looking into the antimicrobial mechanisms of UVC, there are three ways that should be noticed. The first way is to facilitate UVC light to trigger the formation of dimers inside DNA, thus suppressing the further transcription and translation of DNA, and finally disinfecting microorganisms (Niu et al., 2021; Fuentes-León et al. 2024). The second way is to let UVC light induce the productions of various intracellular ROS such as hydrogen peroxide and hydrogen radicals, these photoproducts can result in oxidative damages of microbial macromolecules including DNA and proteins (Niu et al., 2021). The third way is that the peak absorbances of DNA and proteins (e.g., tryptophan and tyrosine) are approximately from 260 to 280 nm which are fallen inside the wavelength range of UVC, thus, all the UVC light within these wavelengths are most likely to be fully absorbed by these macromolecules, leading to their disintegration and eventually inactivating the microorganisms (Kim et al., 2017).

Currently, there are two waveband ranges of UVC (e.g., 250–265 nm and 275–285 nm) that are mainly applied in eliminating microorganisms from foodstuffs. For 250–265 nm waveband of UVC light, the recent studies were mostly focused on bacteria and fungi, however, their antimicrobial effects on the viruses were also significant. With regards to their actions on viruses as recently reported, the application of UVC at 254 nm for up to 30 min was found to markedly disinfect coronavirus (SARS-CoV-2) at an average dose of 67 J/m^2 (equal to 6.7 mJ/cm^2) (Table 2) (Quevedo et al. 2020). This meant that 250–265 nm waveband range of UVC light were capable of actively sanitizing virus-type microorganisms.

In the cases related to eliminate bacteria in liquid foodstuffs, the exposure of UVC, emitted via a 144 W thin-film 254 nm UVC reactor, to the flowing milk with a flow rate of 3 L/min was found to remarkably decrease bacterial counts by more than 4 log_{10} CFU/ml, in which the count of coliform was greatly reduced, compared with those of coagulase-negative staphylococci and lactic acid bacteria (Table 2) (Makararpong et al., 2024). In another case, the treatment of 254 nm UVC light (27 W of output) could remarkably decrease the bacteria counts in bovine skim milk by 2.3 log_{10} CFU/ml. After UVC treatment, coliforms were fully disinfected, furthermore, the counts of spores were decreased by 0.3 log_{10} CFU/ml (Table 2) (Li et al., 2021).

Apart from liquid foodstuffs, UVC (250–265 nm range) exposure could also work well on solid foodstuffs. As studied, *L. monocytogenes* on the surfaces of 254 nm UVC-pretreated carrots could be greatly inactivated, with more than 5 of log_{10} CFU/cm^2 reduction, after 24 h of UVC exposure (30 W of output). Notably, the anti-listerial effect of carrots pretreated by UVC could last for more than 72 h. Furthermore, the antimicrobial effects of all the investigated carrots could be remarkably elevated through UVC pretreatment. Moreover, such UVC pretreatment to generate antimicrobial effect could also be adapted to *S. aureus*, rather than *E. coli* O157:H7 and *S.* Typhimurium (Table 2) (Rajalingam et al., 2024). For the walnut samples, the effects of 254 nm UVC exposure (at doses from 29.4 to 882.0 mJ/cm^2) to the inoculated pathogens (e.g., *Salmonella*, *E. coli* O157:H7, *L. monocytogenes* and *S. aureus*) on the surfaces of inshell walnuts demonstrated a dose-dependent manner, where the tailing effects were found involving in these inhibited pathogens (Table 2) (Yun et al., 2022). For fungi case, the manipulation of 265 nm UVC-LED could significantly suppress the formation of white colony-forming yeast on the surface of kimchi seasoning. In addition, *Candida sake* and *Kazachstania servazzii* contained in kimchi could also be sanitized by UVC-LED (Table 2) (Song et al., 2022).

Besides 250–265 nm range of UVC, the 275–285 nm range of which can also bring forth good antimicrobial effect on foodstuffs. In a study related to fungi, the exposure of 279 nm UVC light emitted from UVC-LEDs to *S. cerevisiae* remarkably inactivated their populations in a dose-dependent manner, in which 1420 mJ/cm^2 of dose led to up to 4.86 log_{10} CFU/ml of reduction. After exposure and observed by

SEM technique, the plasma membrane of *S. cerevisiae* was found to be destroyed badly. A similar dose-dependent relation could be found when UVC-LEDs were used to treat this yeast in orange juice, where the same 1420 mJ/cm^2 of dose resulted in 4.44 $\log_{10}$ CFU/ml decrement of *S. cerevisiae* population (Table 2) (Niu et al., 2021). In another study involving pathogenic bacteria, the working of 280 nm UVC-LED on the surfaces of chosen media facilitated 3 $\log_{10}$ CFU/ml declines of *E. coli* O157:H7 and *S.* Typhimurium by no more than 1.35 and 3.60 s, respectively. Furthermore, the survival rates of these UVC-treated foodborne pathogens on the chosen media could be depicted well by log linear and Weibull models. Moreover, the inoculation experiments of these pathogens onto the surfaces of spinach and sliced deli meat showed that the 1.5–3 $\log_{10}$ CFU/ml reductions of their counts could be achieved within 34 s of UVC-LED treatment (Table 2) (Kim & Kang, 2020).

Sometimes, the combined exposures of different UVC waveband ranges seemed to achieve impressive antimicrobial effect on foodstuffs, which might reach higher bacterial count reductions while significantly decreasing the output of electrical power. As reported, after 301.3 mJ/cm^2 of irradiation dose by 265, 285 nm or the combined 255, 265 and 285 nm UV-LED for up to 60 min, the 7.5 $\log_{10}$ CFU/ml and up to 0.9 $\log_{10}$ CFU/g decrements of *L. monocytogenes* in beef broth and diced beef, respectively, were observed. Simultaneously, the 7.4 $\log_{10}$ CFU/ml and up to 0.6 $\log_{10}$ CFU/g decrements of *S.* Typhimurium in beef broth and diced beef, respectively, were also found under the same dose (Table 2) (McSharry et al., 2022).

Excitingly, UVC can not only efficiently eliminate viruses and bacteria from foodstuffs, but also maintain foodstuff qualities after exposure as much as possible. As reported, the occurrence probability of lipid peroxidation in 254 nm UVC-exposed milk was markedly lower than those in non-exposed and pasteurized milks, after 72-hour's exposure (Table 2) (Makararpong et al., 2024). These results showed that UVC treatment could well keep the quality of milk while guaranteeing sufficient decrement of microbial counts in milk, in a safe manner. In another liquid foodstuff case, the levels of 254 nm UVC-exposed bioactive XO, IgG, LPO, LF, lgM and IgA were 100% remained in bovine skim milk, while those of UVC-exposed native serum proteins were 95% remained. Moreover, the concentrations of carbonyls, involved in the protein's molecular oxidative modifications to reduce protein's activities, were found enhanced by 26%, compared with the control without treatment, due to the photo-oxidation induced by UVC (Table 2) (Li et al., 2021). Similarly for solid foodstuff case, no remarkable variations of lipid oxidation in walnuts were observed after 254 nm UVC application (dose by up to 882.0 mJ/cm^2) (Table 2) (Yun et al., 2022). In the above fungi cases, the physicochemical properties of kimchi seasoning were not remarkably varied after exposed to 265 nm UVC-LED (Table 2) (Song et al., 2022). Similarly, the treatment of 279 nm UVC light did not generate significant variations of orange juice in color parameters, pH, titratable acidity and total soluble solids, except for its total phenolic content (Table 2) (Niu et al., 2021). These results indicated that UVC treatment could still maintain many good qualities of orange juice even under the higher dose (1420 mJ/cm^2) applied.

In these UVC practices, the selection of appropriate UVC light source is crucial. Normally, UVC-LED lamp is the best choice presently for antimicrobial purpose on foodstuffs because the configuration of LED lamp has higher luminous efficacy that can emit more UVC lights under the same electrical power, compared with those of the other types such as mercury and fluorescent lamps. In a comparative experiment, the comparisons of UV mercury lamps at 254 nm, UVC-LEDs at 270 nm and UVA-LEDs at 365 nm on the surfaces of plastic and four powdered seasoning ingredients including those from onion, garlic, cheese and the combined onion and chili showed that UVC-LEDs could achieve the best performance, compared with the other two lamp-types. Although the effects of UVC-LEDs on the surfaces of the seasoning powders were inferior to that of plastic, about 0.75–3 $\log_{10}$ CFU/g reductions could also be reached after 40 s of exposure duration (Table 2) (Nyhan et al., 2021).

The advantages of UVC antimicrobial technique are similar to that of UVB one, because they are both the clean and environmentally-friendly technologies and normally require short exposure duration to efficiently disinfect microorganisms, which thus saves more electrical energy and operation cost. Nevertheless, such higher frequency also results in the disadvantage of UVC technique, like UVB one, to lead to more serious damages on the skins of operators if they are unluckily exposed. Furthermore, another disadvantage of UVC light, similar to UVA and UVB lights, is that UVC light could also inefficiently penetrate the interiors of foodstuffs to sanitize pathogenic microorganisms. Thus, to achieve the better antimicrobial effect on foodstuffs, the penetrating ability of UVC should be evaluated. It was reported that the maximal UVC does at 254 nm to lead to first decimal reductions in apple juice and apple peel discs were 1273.3 and 89.7 mJ/cm^2, respectively. From here, it was clear that the thickness of apple juice layer should be considered because the transmittance of UVC through juice was relatively low, indicating that a thinner juice layer is expected. Further application showed that the radiation on the water-immersed apple peel was better than that on the surface by more effectively disinfecting the pathogens. The above UVC application could eventually result in 3.3–5.3 $\log_{10}$ CFU/ml reductions in *E. coli* and *L. monocytogenes* (Table 2) (Nicolau-Lapeña et al., 2022).

2.2.1.5 X-Ray

X-ray, with wavelengths (frequencies/photon energy) ranging from 10 picometer (30 EHz/124 keV) to 10 nanometer (30 PHz/124 eV), is included in the family of electromagnetic spectrum involving in wavelengths shorter than that of UVC light (10–280 nm), whereas longer than that of gamma ray (less than 10 pm) (Rumble et al., 2019). X-ray antimicrobial technique is a kind of non-thermal approach that can efficiently eliminate pathogenic microorganisms from foodstuffs. The mechanism involved by X-ray to display its lethality to microorganisms is through its main effects to inactivate intracellular enzyme and destroy cellular DNA of these microorganisms (Cho & Ha, 2019). Besides the direct action on microorganism cells by killing them, X-ray can also effectively eliminate biofilms formed by these micro-organisms, which mechanism is via the removal of extracellular polymeric substances (EPS, e.g., exopolysaccharides) in biofilms by disintegrating them from different tested surfaces, where the removal rate was positively related to the irradiation dose of X-ray (Table 2) (Cha & Ha, 2022). This could be seen from a concrete example, it was found that the EPS structure secreted by *P. fluorescens* biofilm cells and the bacterial glucose uptake system were strongly damaged by X-ray irradiation (125 Gy of dose), which further led to the partial destruction of bacterial membrane integrity and eventually eliminated the bacteria (Table 2) (Pang et al., 2022).

Currently, the application of X-ray in food industries for antimicrobial purposes is mainly through using low-energy X-rays, which are also termed superficial X-rays that demonstrate less penetrating depth, compared with the high-energy ones. These low-energy X-rays are usually generated by means of X-ray tubes triggered at 10–100 kV range of voltages, which achieve a 10–100 keV range of peak energies (Podgorsak, 2005). Recently, the antimicrobial studies of low-energy X-ray technique (e.g., continuing polychromatic low-energy X-ray irradiation technique) on foodstuffs were chiefly focused on the radiation doses ranging from 300 Gy to 3.10 kGy [1 Gy (gray) = 1 Joule/kilogram] (Cho & Ha, 2019; Zhang et al., 2020; Zhang et al., 2021a; Jee et al., 2023; Gao et al., 2024; Yeom et al., 2024).

As reported, exposed to 150 keV of low-energy X-ray at 300 Gy, the counts of *E. coli* and *S.* Typhimurium in dry cardamom were significantly decreased by up to 5 $\log_{10}$ CFU/g, while exposed to 350 Gy, the counts of *S. aureus* and *L. monocytogenes* attached to this foodstuff demonstrated 5–6 $\log_{10}$ CFU/g reductions (Table 2) (Zhang et al., 2021a). In another study related to dry edible bird's nest (EBN), exposed to 150 keV of low-energy X-ray at 350 and 400 Gy, the counts of the above two foodborne pathogens (e.g, *E. coli* O157:H7 and *S.* Typhimurium) on EBN surface were remarkably reduced from 6.35 and 5.84 $\log_{10}$ CFU/g, respectively, to below 2.00 $\log_{10}$ CFU/g (Table 2) (Zhang et al., 2020). With the increment of X-ray irradiation dose, it seemed that the higher bacterial count reduction on foodstuff surface could be achieved. For instance, the exposure of pathogen-inoculated sliced ham to 800 Gy of X-ray irradiation led to significant count reductions of *S.* Typhimurium, *E. coli* O157:H7 and *L. monocytogenes* by 5.7, 7.2 and 6.9 $\log_{10}$ CFU/g, respectively, on its surface (Table 2) (Cho & Ha, 2019). When the dose kept going up to 1.5 kGy, it was reported to work well on the fresh produce surface. As found, the illumination of low-energy X-rays at 1.5 kGy on the leaf surfaces of lettuces could significantly decrease the counts of *E. coli* O157:H7, *L. monocytogenes*, *S.* Typhimurium attached to them by 6.89, 3.22 and 4.48 $\log_{10}$ CFU/g, respectively, which were remarkably superior to the wash treatment using sodium hypochlorite (200–400 ppm) that was inefficient in sanitizing the above-mentioned internalized cells on lettuce leaves (Table 2) (Jee et al., 2023). Notably, X-ray reaching higher irradiation dose could facilitate the decrement of bacterial diversity and simultaneously change the structures of microflora. For example, the 1.75–2.00 kGy of X-ray irradiation dose could induce significant reduction in Firmicutes on chilled chicken breast (Table 2) (Gao et al., 2024). In fact, although the D_{10} values (the dose needed to lead to 90% microorganism reduction in the irradiated foodstuffs) of X-ray-irradiated pathogenic bacteria were relatively low, the X-ray irradiation dose was required to be higher than 3.0 kGy, if all the bacteria on foodstuff surfaces were supposed to be completely disinfected. For instance, on the surfaces of pork cutlets, the D_{10} values of X-ray-irradiated *E. coli* O157:H7, *L. monocytogenes*, *S.* Typhimurium and *S. aureus* were only 370, 520, 620 and 360 Gy, respectively. For fully eliminating all the pathogens, however, the minimal X-ray irradiation dose of 3.10 kGy was requested. In this scenario, the X-ray irradiation to decrease the counts of all the pathogens was in a dose-dependent manner (Table 2) (Yeom et al., 2024).

For the effects of X-ray irradiation on the qualities (e.g., appearances, flavors and nutritional contents) of foodstuffs, some valuable facts were found, where a majority of findings showed that this technique could maintain the qualities of the exposed foodstuffs, while a minority of results demonstrated change (Cho & Ha, 2019; Zhang et al., 2020, 2021a; Jee et al., 2023; Gao et al., 2024; Yeom et al., 2024). For example, exposed to 150 keV of low-energy X-ray at 300 Gy and under the detection limit of 0.030 μg/g, the 2-alkylcyclobutanones (2-ACBs), marked as the product of triglycerides breakage, was not found in X-ray-treated cardamom. Moreover, the sulfur-containing compounds such as dimethyl disulfide and 3-methyl-thiophene in this irradiated foodstuff were not detected, indicating that no off-odor and metal odor were generated after this dry cardamom was treated by X-ray (Table 2) (Zhang et al., 2021a). The similar result was also found in EBN, where X-ray irradiation (350 and 400 Gy) did not achieve the release of sulfur-containing volatile compounds from the irradiated samples, revealing that the normal odor of EBN has not been markedly affected after exposure (Table 2) (Zhang et al., 2020). On sliced ham, the 800 Gy of X-ray irradiation on its surface did not remarkably change the textural properties and color values of this foodstuff, compared with those of the non-exposed control (Table 2) (Cho & Ha, 2019). Noteworthily, the qualities of lettuces such as texture and color were not negatively affected by even up to 1.5 kGy of irradiation dose of treatment (Table 2) (Jee et al., 2023). In fact, the higher X-ray irradiation (1.75–2.00 kGy) could significantly prolong the shelf life of chilled chicken breast by up to 6 days, doubling that of the control group, due to the quality sustainment of this foodstuff after processing (Table 2) (Gao et al., 2024). Astonishingly, even up to 7.00 kGy of dose still did not show negative effects on color, pH and thiobarbituric acid reactive substance (TBARS, which represents the concentration of malondialdehyde that is utilized for evaluating the degree of lipid oxidation) of pork cutlets. This study indicated that up to 7.00 kGy of X-ray irradiation is capable of completely disinfecting pathogens grown on the surfaces of pork cutlets and keeping meat's qualities simultaneously (Table 2) (Yeom et al., 2024).

Although many foodstuff qualities were maintained as above cited, there were still some variations that should be noticed after X-ray irradiation. For instance, the level of cardamom polyunsaturated fatty acids was found to be significantly reduced by 20% after X-ray (300 Gy) exposure treatment (Table 2) (Zhang et al., 2021a). In addition, X-ray irradiation (350 and 400 Gy) might increase EBN's yellowness, thus changing its appearance (Table 2) (Zhang et al., 2020). Furthermore, the enhanced irradiation doses (1.75–2.00 kGy) could induce the level enhancements of hexanal and 1,3-bis (1,1 dimethylethyl)-benzene derived from the oxidations of lipids and amino acids, this compound contributes to the odor of X-ray-irradiated chicken (Table 2) (Gao et al., 2024). This result was quite different from that of Zhang et al. (2021a), which foodstuffs were irradiated under 300 Gy. This seemed to indicate that higher X-ray irradiation dose was more likely to induce the generations of off-odor and metal odor from foodstuffs. Moreover, the corruption indicators (e.g., 1-pentanol, 2-pentanone and acetoin) in association with the activities of *Acinetobacter*, *Brochothrix* and *Pseudomonas* were observed increased in all the irradiated chilled chicken breast meats, indicating that it still has potential risk to result in meat spoilage after X-ray irradiation (1.75–2.00 kGy) (Table 2) (Gao et al., 2024).

Similarly to the other electromagnetic radiations, different bacteria might respond differently to the identical radiation. In the further trial, *S.* Typhimurium was found more resistant to this low-energy X-ray irradiation than *E. coli* O157:H7, this is because *S.* Typhimurium has 83.3 Gy of D_{10} value that is significantly higher than *E. coli* O157:H7 with 37.6 Gy of t_{R1} value (the dose for achieving 1-log microorganism reduction, i.e., 90% reduction). In this context, D_{10} value has the same meaning as t_{R1} value (Table 2) (Zhang et al., 2020).

Apart from efficiently working on the surfaces of foodstuffs, low energy X-ray (175–225 Gy) could efficiently sanitize some typical food contact surfaces such as acrylic, polyethylene and stainless steel. For instance, the count reduction of X-ray-exposed *L. monocytogenes* biofilm formed on these surfaces could be reached by up to 4 $\log_{10}$ CFU/cm^2. Additionally, the acrylic surface among the above three was found to be the most favorable place for the biofilm formation of *L. monocytogenes* and *Pseudomonas fluorescens* (Table 2) (Zhang et al., 2021b). This indicated that the adsorption abilities of the typical bacterial biofilm to different contact surfaces were varied. Similarly, the bacterial removal efficiencies of different contact surfaces to the typical bacterial film were dissimilar too. For example, up to 500 Gy of low-energy X-rays were efficient in sanitizing the biofilms of major pathogens such as *E. coli* O157:H7, *L. monocytogenes* and *S.* Typhimurium grown on three food contact surfaces, e.g., polyvinyl chloride (PVC), stainless steel with finish 2B (STS 2B) and Teflon. Among these three surfaces, biofilm cells on Teflon surface were found most easily removed by such X-ray irradiation, further investigation showed that Teflon demonstrated the lowest D_5d (the irradiation dose leading to a 5 $\log_{10}$ CFU/coupon reduction in cell count), compared with the other two surface materials (Table 2) (Cha & Ha, 2022).

As stated by the above mechanism, low-energy X-ray is capable of decomposing EPS secreted by bacteria attached to the food contact surfaces (Cha & Ha, 2022). However, the resistant abilities of different bacterial biofilms formed are dissimilar, for instance, the biofilm of *L. monocytogenes* was found more resistant to low-energy X-rays than that of *P. fluorescens*, based on the t_{R1} value (46.36–50.81 Gy) of *L. monocytogenes* which was significantly higher than that (25.61–31.33 Gy) of *P. fluorescens* (Table 2) (Zhang et al., 2021b). Interestingly, the synergetic effect of two or more bacteria formed in the biofilm could help to improve the resistant ability of this biofilm to X-ray. As found, the combined biofilm of *L. monocytogenes* and *P. fluorescens* could efficiently elevate their resistance to such irradiation, according to the respective higher tR1 values of *L. monocytogenes* (59.54–70.77 Gy) and *P. fluorescens* (32.73–45.13 Gy) that were achieved (Table 2) (Zhang et al., 2021b).

Noteworthily, no matter planktonic or biofilm cells of bacteria, X-ray irradiation can demonstrate similar inactivation effects in some cases. As noted, low-energy X-ray at 125 Gy of irradiation dose displayed its excellent inactivation effects on planktonic *P fluorescens* cells suspending in phosphate-buffered saline solution by reducing 4.60 $\log_{10}$ CFU/ml, as well as on *P. fluorescens* biofilm cells attached to stainless-steel food contact surface by decreasing 4.21 $\log_{10}$ CFU/cm^2, respectively. Although the achieved t_{R1} values for X-ray-irradiated *P. fluorescens* planktonic and biofilm cells were 14.8 and 11.6 Gy, respectively, this study showed that there was no significance ($p > 0.05$) of resistance between planktonic and biofilm cells of *P. fluorescens* to X-ray irradiation (Table 2) (Pang et al., 2022). These results gave us some hints that the penetrating power of X-ray to specific bacteria was strong enough for eliminating biofilm cells, which are normally recognized hard to be killed due to the overlapping effect (Ghate et al., 2019).

Similarly to those of UV light, the advantage of X-ray antimicrobial technique is that it is a clean and environmentally-friendly technology which does not bring contamination to food during processing. It also takes less treatment duration for efficiently eliminating microorganisms in an economical manner. Different from UV light and blue light, X-ray with shorter wavelength and higher frequency could penetrate the interatomic space of the processed substance more efficiently, which contributes to its another advantage to effectively kill the bacteria in the interior of biofilm (Pang et al., 2022). Nevertheless, although the short-term exposure to X-ray is relatively safe, it could still induce severe damage to the operators if they are exposed to X-ray in the long term. Therefore, the appropriate protection measure to keep the operators from X-ray is very necessary.

2.2.1.6 Gamma Ray

Gamma rays, with wavelengths (frequencies/photon energy) less than 10 picometer (30 EHz/124 keV), are included in the family of electromagnetic spectrum involving wavelengths shorter than that of X ray (10 pm to 10 nm). Gamma rays are generated due to the radioactive decay of atomic nuclei and is the known shortest electromagnetic wave so far (Rumble et al., 2019). Recently, gamma ray irradiation, as a potent ionizing radiation approach, has been well recognized as an efficient antimicrobial technique used in the preservation of various fruits, grains and vegetables for extending their shelf lives (Boshevska et al., 2024). The mechanisms of gamma ray to eliminate microorganisms in foodstuffs are through two effects, e.g., the direct and indirect ones. For direct effect, gamma ray directly breaks down microbial DNA, thus suppressing microbial division and eventually kill the microorganisms contained in foodstuffs. For indirect effect, gamma ray acts on the water molecules firstly to generate the radiolytic products such as •OH, e^{-} aq and H•, which can subsequently attack and oxidize microbial biomolecules like proteins, lipids, nucleic acids and DNA, and finally inactivate these microorganisms (da Silva Aquino, 2012).

No doubt, higher doses of gamma ray can eliminate more microorganisms, this forms the negative correlation between doses and microbial counts. For example, exposed to a gradually increasing doses of gamma ray radiation (0.2, 0.5, 1.0, 5.0 and 10.0 kGy), the numbers of microorganisms (including *B. cereus*, *Enterobacteriaceae*, as well as yeasts and molds) presented on wheat (*Triticum aestivum* L.), were found to negatively related to the above increased doses (Table 2) (Boshevska et al., 2024). This tendency could also be found in the gamma-ray-exposed mutton samples. As found, the total viable count (TVC), total coliform count (TCC) and total yeast-molds count (TYMC) values of fresh mutton after gamma ray exposure followed by 30 and 60 days of −20 °C storage, respectively, were negatively related to the gradually increasing doses of gamma ray irradiation (1.5, 2.0 and 4.0 kGy) (Table 2) (Hashem et al., 2022).

Similarly to the other electromagnetic wavebands (e.g., UV light and X-ray), different microorganisms are also sensitive to different doses of gamma ray. For instance, up to 10.0 kGy of gamma ray radiation dose was found to completely disinfect all the exposed microorganisms such as *B. cereus*, *Enterobacteriaceae*, as well as yeasts and molds on wheat, while at 5.0 kGy, only yeasts and molds could be detected. Moreover, the highest dose set in this study that *Enterobacteriaceae* could tolerate was 1.0 kGy (Table 2) (Boshevska et al., 2024). In another case, the total microbial counts of two 2-kGy-exposed mango varieties (e.g., Amrapali and BARI-11) after 1 week of storage were dropped down to 1.4 and 1.5 $\log_{10}$ CFU/ml, respectively, compared with 6.8 and 6.4 $\log_{10}$ CFU/ml of the control (Table 2) (Jabin et al., 2023). All these above results meant that the tolerances of different microorganisms to gamma ray were varied.

Dose rate is not a significant factor to influence the antimicrobial effect of gamma ray. This could be seen by a study showing that the dose rate of gamma-ray did not remarkably influence D_{10} values of the pathogens (e.g., *E. coli*, *S.* Typhimurium and *L. monocytogenes*) in rice exposed to cobalt-60 gamma ray (Table 2) (Begum et al., 2020). However, lower dose rate of gamma ray seemed to more efficiently than higher dose rate of X-ray in eliminating microorganisms. It was documented that the rice-inoculated pathogens (e.g., *E. coli*, *S.* Typhimurium and *L. monocytogenes*) treated by low energy X-ray (125 keV, 0.76 kGy/h of dose rate) contained higher D_{10} values (lower microbicidal effectiveness), compared with gamma-ray-treated ones (9.1, 3.9 and 0.22 kGy/h of dose rates) (Table 2) (Begum et al., 2020). The results indicated that gamma ray was more powerful than X-ray to disinfect microorganisms, even the dose rate of X-ray was higher than that of gamma ray. This could be due to the higher frequency and energy contained in gamma ray, compared with those of X-ray, leading to higher lethality of gamma ray than that of X-ray on microorganisms (Rumble et al., 2019).

As noted, gamma rays might affect the qualities of foodstuffs to some degree after exposure treatment. However, some good qualities could still be maintained. For instance, exposed to different doses of gamma ray irradiation (0.5, 1.0, 1.5, and 2.0 kGy), the major parameters of peels of two mango cultivars (e.g., Amrapali and BARI-11) including carb, energy, fat, fiber, protein, soluble solids, sugar, etc., more or less, were affected. However, the higher antioxidant activities could be achieved when these mango samples were exposed to 1.5 kGy. Additionally, with 5.811 kGy/h of dosage rate, the 1.5 kGy-exposed mangoes' textures could still be kept firm after 2 weeks of preservation under 19 °C, revealing that such gamma ray dosage could help to significantly prolong the shelf lives of these treated mangoes, the cause could be due to the killing effect of sufficient gamma ray dosage on those environmental microorganisms (Table 2) (Jabin et al., 2023). Besides fruits, the gamma ray irradiations at 1.5, 2.0 and 4.0 kGy were found to extend the shelf life of mutton and keep its nutritional qualities by increasing dry matter, ether extracts, peroxide value, TBARS and free fatty acid values, while decreasing ash, crude protein and raw pH. Additionally, the tenderness and juiciness of mutton were not significantly changed, compared with the non-irradiated mutton. However, remarkable changes in color and flavor after gamma ray irradiation were detected. Overall, the 4.0 kGy of irradiation dose could achieve the best effect of preservation for mutton by fulfilling the requirements of satisfied microbial control, longer shelf life and quality acceptability (Table 2) (Hashem et al., 2022).

Sometimes, the synergistic actions of gamma ray with some bioactive substances (e.g., EO and plant extracts) can achieve desired antimicrobial effect. As reported, the combined treatment of low dose gamma ray irradiation with *Thymus capitatus* EO and myrtle juice could significantly suppress the growth of *S. aureus*. In a recent trial, the synergistic effect of 0.5 kGy gamma irradiation and 6 μL/mL myrtle juice, as well as 0.08 μL/mL of thyme EO (mainly containing 79.55% of carvacrol and 7.93% of *p*-cymene) was found to lead to more than 3 $\log_{10}$ CFU/g of decrement in *S. aureus*, which was fully inactivated in this scenario (Table 2) (Zernadji et al., 2024).

As a pollution free technology, the advantage of gamma ray antimicrobial technique is outstanding by achieving no contamination (e.g., residuals and radioactivity) to foodstuffs during processing. Similarly to X-ray, it also disinfects the foodstuffs by applying short treatment duration, thus saving costs. Likewise, as X-ray, gamma ray with the shortest wavelength and highest frequency of electromagnetic waves could fully penetrate the interatomic space of the processed substance to exert killing tasks on microorganisms. However, as gamma rays are normally emitted by the radioactive sources (e.g., cobalt-60), the associated safeguard procedures are thus very crucial, not only to prevent the operators from the direct exposure to gamma ray, but also to avoid them to contact the radioactive sources (da Silva Aquino, 2012).

2.2.2 Photodynamic-Inactivation-Based Non-thermal Antimicrobial Technique

Based on the interaction of non-thermal photophysical and photochemical reactions, the antimicrobial mechanism of PDI is through employing lights (e.g., ultraviolet light, visible light or near-IR light) and a photosensitizer (PS, e.g., dye) to produce ROS under oxygen and illumination environments, to eventually oxidize and inactivate microbial cells, in which the critical point is that different photosensitizers can be activated by the specific light wavelengths, to generate potent ROS for antimicrobial purpose (Jiang et al., 2014; Zhu et al., 2021). As summarized, PDI is potential to be applied in the food categories such as aquatic products, fresh fruits and vegetables, juices, meat, as well as seeds and grains for antimicrobial objectives (Zhu et al., 2021).

In terms of antimicrobial mechanism, two major categories of PDI reactions are involved (Alves et al., 2015). In the first category, the excited singlet state ($^1PS^*$) of PS can be achieved by PS under ground state ($^1PS^0$) through absorbing photons from light source. By means of fluorescence emission, some parts of $^1PS^*$ will lose energy and regress back to $^1PS^0$, while the other parts of $^1PS^*$ can change to the long-lived triplet state ($^3PS^*$) via the intersystem crossing (ISC) process (Fig. 2). Such $^3PS^*$ also has two ways to go. One is to return to $^1PS^0$ by means of phosphorescent

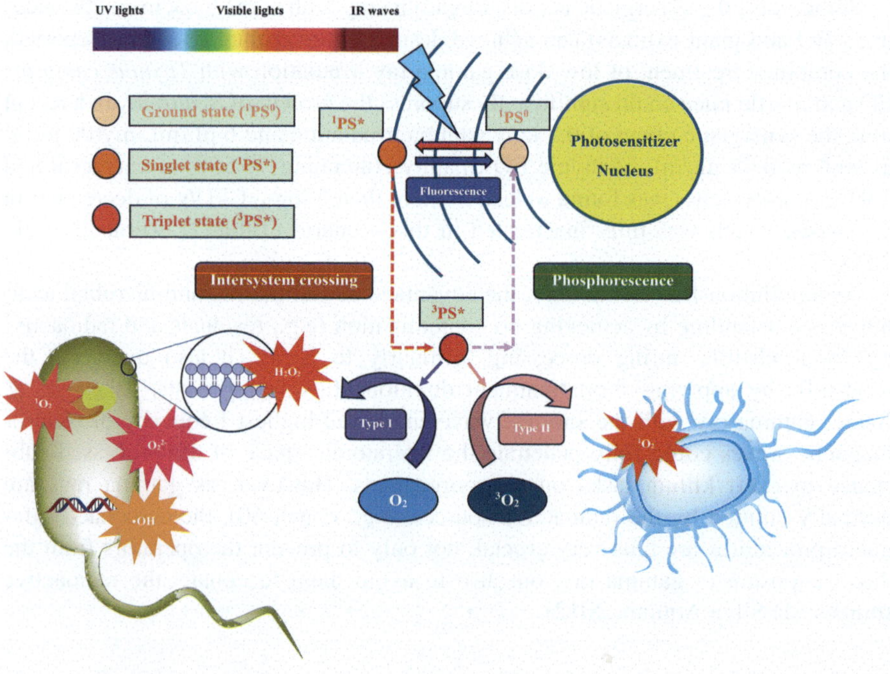

Fig. 2 Mechanisms of PDI-based non-thermal antimicrobial technique. (Referred to Chatterjee et al., 2008; Zhu et al., 2021)

emission process, another is to react with the environmental organic matters or ground state oxygen to generate ROS including hydrogen peroxide, hydroxyl radical, singlet oxygen and superoxide anion that can inactivate the microbial cells efficiently, via combining with proteins, nucleic acids and lipids, as well as interacting with organic molecules such as acids, alcohols, amines, carbohydrates, cytokines, hormone, nitrogen heterocyclic molecules, pyrrole and vitamins, to disable their functions and eventually lead to cell apoptosis/necrosis. Through this category (Type I), the loss of PS is recorded (Fig. 2) (Chatterjee et al., 2008; Zhu et al., 2021). In the second category (Type II), a portion of $^3PS^*$ may bond with the triplet oxygen molecule ($^3O_2$) to produce singlet oxygen ($^1O_2$) that can deoxidize the substrate or microorganisms (Fig. 2). Obviously, PS is not involved in this category and thus is not consumed. It can finally regress back to the ground state, ready for the next reaction cycle (Chatterjee et al., 2008).

Recently, PDI technique has been reported to eliminate the post-baking mold spoilage of bread. In this scenario, different PS have different antifungal effects against environmental mold spores. As compared, the 30-min combination of chlorophyllin (CHL, a food-grade PS) and 405 nm blue-LED efficiently inactivated two common bread spoilage molds, e.g., *R. stolonifer* and *Penicillium expansum* grown on dichloran rose bengal chloramphenicol (DRBC) agar by 77.4 and 52.1%, significantly higher than the combined riboflavin (another food-grade PS) and 445 nm blue-LED with 22.8 and 45.5%, respectively (Table 2) (Chong et al., 2024). In another study to achieve antifungal effect on DRBC-agar-grown *Fusarium oxysporum* that could lead to *Fusarium* crown rot in bananas, the minimal irradiance and treatment duration of 405 nm blue-LED were 2.2 mW/cm^2 and 6 h, respectively, where the endogenous porphyrins (5.46 fg/CFU) in *F. oxysporum*, as the PS performing PDI with blue light, contributes to the effect (Table 2) (Xu et al., 2024b).

Practically, the above-combined CHL and 405 nm blue-LED applied on the surface of bread remarkably reduced the counts of three molds including the two above-mentioned strains and *Aspergillus niger* by 51–58% (Table 2) (Chong et al., 2024). In the practice using Cavendish bananas which were *Fusarium*-spot-inoculated on their stem scar after stored under simulated warehouse conditions (15.0 °C and 80.0% RH) for 24 h, a remarkable 70% decrement of *F. oxysporum* count was found when these bananas were further exposed to 2.2 mW/cm^2 of blue-LED for 48 h under the simulated retail conditions (25.5 °C and 63.5% RH for a total of 54 h) (Table 2) (Xu et al., 2024b).

As for the influences of PDI technique on the qualities of foodstuffs, some positive and negative effects were found. The above CHL-based PDI technique could keep the moisture and texture of bread, however, significant changes of bread colors were also presented, which might affect the appearances of bread accordingly (Table 2) (Chong et al., 2024). For the banana case, the 405 nm blue-LED illumination using endogenous porphyrins as PS did not lead to remarkable differences in the physicochemical qualities of bananas such as appearance, hardness, mass and peel color. Furthermore, it could slightly inhibit the activity of polyphenol oxidase, an enzyme involving in browning, indicating that such illumination could also help to postpone the browning of bananas for maintaining their good properties (Table 2) (Xu et al., 2024b).

As could be seen, the disadvantages of PDI technique are aroused mainly by the characteristics of PS. So far, three generations of PS have been developed, in which the defects of the first-generation PS (e.g., hematoporphyrin and its derivatives) were high phototoxicity, low chemical purity and poor targeting, while the drawbacks of the second-generation PS (e.g., porphyrin derivatives, quinones and metal phthalocyanines) were their poor solubilities in water. Recently, the three generation of PS including natural and nano PS have been developed, with the advantages of high light efficiency, low toxic and side effects, reliable sources, as well as fast removal rate in foodstuffs, in which nano PS can be the promising ones because they apply various polymers that can enhance their water solubilities and bio-availabilities (Liu et al., 2023a).

2.2.3 Filtration-Based Non-thermal Antimicrobial Techniques

According to the concept of filtration, the filtration-based antimicrobial techniques are defined as the process in which solid microorganisms in the liquid foodstuffs or water are removed through the application of a filter medium with certain pore diameters that allows the liquid foodstuffs to transit, while holds back the solid microorganisms on the filter medium. Normally, the objective of filtration technique is to remove or separate the microorganisms from the foodstuffs, rather than kill them, which antimicrobial mechanism is thus showed to be different from those of the other antimicrobial techniques. Furthermore, as this process does not require the participation of heat energy, they are classified as non-thermal antimicrobial techniques. Currently, microfiltration (MF) membrane, slow sand filtration (SSF) and cross-flow filtration (CFF) are the three main filtration-based non-thermal antimicrobial techniques applied in food industries.

2.2.3.1 Microfiltration Membrane

In food industries, MF membranes are a kind of special structure materials that are usually employed in membrane filtration technology for effectively filtering away impurities, pathogenic microorganisms, and toxic pollutants from drinking water or beverages (Khalaf et al., 2024). The antimicrobial mechanism of these MF membranes mainly resorts to gravity or pressure, as well as their smaller pore sizes, compared with the microorganisms, for filtering out these microbes as far as possible. In a concrete study, for example, the operation of MF membrane with 1.4/0.8 μm pore diameters significantly decreased the bacteria count by 3.0 $\log_{10}$ CFU/ml. After MF treatment, coliforms, spores and somatic cells in bovine skim milk could not be detected (Table 2) (Li et al., 2021).

Recently, many kinds of novel MF membrane materials (non-commercial or commercial) have been developed and exemplary results have been achieved to highlight their antimicrobial properties. For example, the novel rod-shaped cellulose nanocrystal/Cu_9S_5 (CNC/Cu_9S_5) antibacterial nanocomposites with 4 nm of size has been developed, which can demonstrate, not only excellent water dispersibility, but also potent antibacterial function to achieve 4.6 $\log_{10}$ CFU/ml reduction of *S. aureus* under cellulose nanocrystal of middle aspect ratio (CNC-M) (Table 2) (Ye et al., 2024). For a newly invented scoured cotton fabrics (SCF) microfilter loaded with zinc oxide nanoparticles (ZnONPs), it was reported to demonstrate remarkable antibacterial effect on the growths of tested pathogenic bacteria (e.g., *E. coli*) (Table 2) (Khalaf et al., 2024). Undoubtedly, ceramic substances can also be acted as MF materials because they normally contain very small sizes of pores. Recently, three low-cost antibacterial MF tubular ceramic membranes, using materials from F1 (Aswan ball clay, pore size from 0.4 to 90 μm), F2 (Aswan kaolin, pore size from 0.09 to 45 μm) and F3 (a mixture of clay, potash feldspar, and quartz, pore size from 0.18 to 89 μm), were made for testing their antibacterial effects on water from El-Salam sluiceway (Table 2) (Rashad et al., 2022). For the commercial membranes (Nadir® MV020T and PA PVDF) loaded with silver nanoparticles (AgNPs) with sizes from 20 to 50 nm, they could demonstrate their antibacterial effects against *E. coli* in the size of 3–4 mm inhibitory zones (Table 2) (Gallus et al., 2023).

Concerning the above CNC/Cu_9S_5 material, the antibacterial glass fiber/CNC-M/Cu_9S_5 MF membrane was further fabricated, this membrane displayed good bactericidal effect with more than 99.9% disinfection rate for *S. aureus*, followed by high filtration flux and retention capacity, which is appropriate for MF of water purification (Table 2) (Ye et al., 2024). Additionally, the application of the above ZnONPs with sizes of around 55.91 nm on the SCF could greatly enhance the anti-adhesion and biofouling mitigation properties of this microfilter through suppressing the growth of biofilm on it. Moreover, this microfilter was proved to be safe via toxicity test. These findings seemed to support that ZnONPs-loaded SCF microfilter is qualified for applying in the water purification or disinfection industries (Table 2) (Khalaf et al., 2024). For the above ceramic membrane substances, F3 was found the best membrane to significantly decrease the count of total coliform in the water samples from 70 down to 14 CFU/100 ml, while F2 was confirmed as the optimal tubular ceramic membrane for antibacterial MF processing as it could achieve up to 22.9 L/(m^2•h) of permeate flux, 80% of bacterial removal ability, and up to 2 MPa of the highest compressive strength, by comprehensive considerations on the properties of these three membranes (Table 2) (Rashad et al., 2022). Sometimes, longer filtration operation duration might damage the membranes by running away some of their components. Related to the above commercial membranes, the lost percentages of these AgNPs on the two membranes, after 21 days of filtration, were 0.47 and 2.12%, respectively. As a solution, polyvinylpyrrolidone could be employed as a stabilizer to enhance the stability of these AgNPs (Table 2) (Gallus et al., 2023).

The major advantage of MF membrane is that it can largely maintain the levels of bioactive metabolites in foodstuffs. In the above case through filtering bovine skim milk, its bioactive XO, IgG, LPO, LF, IgM, IgA and native serum proteins could be 100% remained by using MF-1.4, whereas its IgG and LPO could be 100% kept by MF-0.8, but 93, 92, 91, and 86% of its IgA, XO, LF, and IgM could be retained through this MF-0.8 membrane (Table 2) (Li et al., 2021). However, the disadvantage of MF membrane is also significant because MF process is easier to accumulate and form biofilm on the membrane surfaces, leading to biofouling. To guarantee a high filtration efficiency, MF membrane is required to be cleaned or substituted frequently, which no doubt increases the operation cost of MF membrane (Kacprzyńska-Gołacka et al. 2020).

2.2.3.2 Slow Sand Filtration

SSF technique seems to be one of the oldest manners for making drinking water or solution through using gravity as the driving force. This could be achieved by facilitating the treated liquids to pass through a sand bed from top to bottom, through which, the microorganisms, fine suspended particulates, flocs and chemicals contained in the liquids can be eliminated via interacting with SSF biofilm produced by the resident microorganisms (Abdiyev et al., 2023; Rosenqvist et al., 2024).

As noted, SSF technique can filter about 99% of total coliform (as the water quality standard) through the aid of schmutzecke layer located in the filter. In a recent study, the treatment of SSF combined with horizontal roughing filter (HRF) on the water sample (4386 colonies/100 ml) was found to remove 99.60% of total coliform, under 0.2 m/h of filtration rate. Nevertheless, the purified water made through such combined SSF and HRF processing still could not meet the quality standard of drinking water (100% free of total coli) and further improvement was needed in the future study (Table 2) (Fitriani et al., 2020).

In fact, the combination of SSF technique with appropriate substances can significantly improve the efficacy of such filtration process (Wang et al., 2022; Holmes et al., 2023). For example, the addition of zero-valent iron (ZVI) into sand beds could be an efficient method to improve the performance of SSF technique. This is because ZVI possesses potent adsorption property that can efficiently capture bacteria in the sand, furthermore, ZVI can markedly inactivate almost all the captured bacteria through destructing the structures of EPS secreted by bacteria and bacterial cell membranes, thus increasing the levels of intracellular ROS, decreasing the content of ATP, and eventually leading to bacterial death (Wang et al., 2022; Liu et al., 2023b). It was reported that the combination of sand bed and 10 wt% ZVI could fully store both Gram-negative *E. coli* and Gram-positive *Bacillus subtilis* at their original concentrations of 1.25×10^7 cells/ml in the sand column (Table 2) (Wang et al., 2022). As documented, chitosan was another substance that could achieve better SSF performance. For instance, the combination of SSF (16 cm layers of sands) with chitosan (10 mg/l), an organic coagulant-flocculant, was found to achieve the turbidity decrements and $\log_{10}$ CFU/ml reduction of bacteria and viruses (e.g., *E. coli* and MS2 coliphage) in water that could meet the two-star WHO performance level of effectiveness, which was significantly superior to the sole sand filtration processing without chitosan pretreatment (Table 2) (Holmes et al., 2023).

Nevertheless, there are several disadvantages of SSF technique that should be noticed. Firstly, the particle size of sand filter material is usually small, which can easily lead to clogs on sand interspace and subsequently shorten the filtration path, thus affecting the filtration efficacy. Secondly, as the name of slow sand is called, the filtration speed is quite slow, which remarkably enhances the operation duration. Thirdly, the cover area of SSF is commonly large, which might push up the performance cost (Liu et al., 2020).

2.2.3.3 Cross-Flow Filtration

CFF is also termed as tangential flow filtration (TFF) (Puertas et al., 2023). In the food system, the antimicrobial principle of this technique is by directing liquid foodstuffs (e.g., juice) or water tangentially across the surface of filtration membrane, through which these liquid foodstuffs would pass through the membrane, while the microorganisms (e.g., bacteria, viruses and spores) in the liquid are blocked at the opposite side of the membrane. After the filtration, microorganisms will be fully separated from liquid foodstuffs, thus eventually achieving the elimination goal (Nowak et al., 2023). Nowadays, some CFF membranes (e.g., cellulose acetate MF membrane) have been commercial and widely applied (Nowak et al., 2023).

Regarding the concrete examples for CFF studies, the application of cross-flow hollow fiber ultrafiltration membrane with molecular weight cut off 30 kDa, in combination with the optimal conditions [e.g., 104 kPa of transmembrane pressure and 30 L/h of cross flow rate (Reynolds number 249)] were found to remarkably reduce the counts of bacteria, as well as those of yeast and mold in the sugarcane juice by 6 and 4 $\log_{10}$ CFU/ml, respectively (Table 2) (Panigrahi et al., 2020). Normally, the pores of membrane applied in the CFF technique are usually expected not to be blocked by microorganisms or impurities for maintaining the stable filtration efficiency. To solve this, a dean-vortex-assisted cross-flow microfilter with spiral shaped microchannel was recently designed and proved that it could increase the recovery rate of *E. coli* in the water samples from approximately 40% to more than 90%. Additionally, such Dean vortex generated in the microchannel could effectively avoid the obstructions of *E. coli* on the surface of membrane. When this technique was applied in aggregating *E. coli* spiked in tap water samples, the similar results were shown, indicating its potential to be used in monitoring the possible existence of pathogenic microorganisms in drinking water (Table 2) (Zhang et al., 2023b).

Noteworthily, the tangential flow ultrafiltration (TFUF) is a special TFF technique that is adapted for filtering away viruses from liquids. For example, TFUF has been reported to be a good enrichment approach for recovering enteric virus in the water samples, in which dissolved organic matters and suspended particles could execute the major effects on the virus recovery. As documented through the simulated reclaimed water experiments, the humic acid (HA) and tryptophan (Try) could increase the recovery rates of bacteriophage MS2 by 92.16 and 81.50%, respectively, while kaolin led to 63.13% of suppressive rate of MS2 instead (Table 2) (Wang et al., 2024a). To play the similar roles as the above matters and particles (e.g., HA and Try) for eluting viruses from membranes and suspending the viral lifecycles for keeping workable filtration efficiency, carvacrol at 20% and 25% of percentages could be a good choice to be loaded into cellulose acetate membranes to reduce the depositions of bacteria (e.g., *S. aureus*) on them and thus improve the permeate fluxes of the membranes (Table 2) (Nowak et al., 2023).

Looking back to the above mechanisms involved, the atomic force microscopy showed that the clusters of MS2-HA and MS2-Try have looser membrane surfaces with larger roughness values, facilitating it easier to be eluted, while MS2-Kaolin cluster has compact surfaces that made it hard to be eluted. Most importantly, the single hydrogen bond with S=O was used to make the connection between MS2-HA cluster and the membrane, which prompted viral elution easier and led to the highest recovery rate, compared with the clusters of MS2-Try and MS2-Kaolin connected to the membrane by means of two hydrogen bonds that were difficult to achieve MS2 eluting tasks (Wang et al., 2024a). For the case of carvacrol which is a potent antimicrobial agent, its presence on the cellulose acetate membranes might directly kill the microorganisms, thus decreasing the depositions of microbial cells on the membranes, for maintaining the better membrane permeate fluxes (Table 2) (Nowak et al., 2023).

As the major advantage, CFF technique can largely maintain the qualities of foodstuffs, because it does not involve heat transfer, or the other drastic damages induced by rays. This could be seen by applying the above cross-flow hollow fiber ultrafiltration membrane to make the clear sugarcane juice. This clarified juice could achieve three-fold less in the polyphenol oxidase (PPO), compared with the control, and facilitate 98% recovery of sucrose, as well as 80% recovery of polyphenol, all of which contribute to the extended juice storability. After CFF treatment, the shelf life of clarified sugarcane juice could be prolonged to 9 weeks under 4 °C (Table 2) (Panigrahi et al., 2020). However, the cost of filtration membrane is usually expensive, which shows to be the main disadvantage that limits the further development of CFF/TFF technique (Biswas & Bandyopadhyaya, 2017). In addition, biofouling can be easily accumulated on the membrane and leads to clogs, which might lead to higher operating pressures followed by shorter membrane life, therefore reducing the efficiency of membrane (Kang et al., 2004). To improve membrane efficacy, the procedures of chemical cleaning or membrane replacement are the routine solutions, nevertheless, these manners might further push up the operating costs when performing this technique.

2.2.4 Electric-Field-Based Non-thermal Antimicrobial Technique

Electric field antimicrobial technique is a kind of widely applied non-thermal approach for efficiently inactivating undesired microorganisms contaminated on foodstuffs, in which pulsed-electric-field (PEF), high-voltage alternating electric field (HVAEF), electrostatic field (HVEF) and prick electrostatic field (HVPEF) are the main electric field modes that have been utilized for performing antimicrobial purposes (Qi et al., 2021). As for the mechanism of PEF to inactivate microorganisms, it involves the use of repetitive transitory high voltage pulses (around 20–1000 μs) with several thousand electric field strength (kV/cm) to act on the foodstuffs that are set between two electrodes, during which electroporation is performed to puncture microbial cell membranes, thus disintegrating the membranes and eventually killing these microorganisms (Yogesh, 2016). Based on the recent findings, the efficacy of PEF treatment (30 kV/cm, 200 ns - 500 μs protocols, 17.5–87.6 kJ/l) was found to non-linearly associate with the energy transferred to the cells of foodborne pathogens such as *S.* Typhimurium and *L. innocua* (Table 2) (Novickij et al., 2020).

Interestingly, microorganisms might respond differently to dissimilar electric fields, which could thus achieve different antimicrobial effects by these electric fields. Recently, the comparison of the capabilities of the above three electric fields, e.g., HVAEF, HVEF and HVPEF to suppress the growth of *S. aureus* on the solid medium plates and food systems was carried out. The results showed that the sensitive sequences of this strain to the electric fields was from HVPEF down to HVEF and then HVAEF, where the mortality rate of this strain, positively related to processing time, was not lower than 99.9% while the voltage was adjusted to more than 10 kV (Table 2) (Qi et al., 2021). The mechanism study revealed that HVPEF, similarly to PEF, destructed the structure of bacterial cell membrane followed by enhancing the permeability of membrane, through which, the morphology of *S. aureus* was varied while the intracellular materials were leaked. Furthermore, HVPEF induced the decrease of nonspecific esterase activity whereas markedly increased the intracellular ROS levels, compared with the control when voltage rose. When HVPEF was worked on the *S. aureus*-inoculated food surfaces (e.g., salmon, griskin, cheese and sausage), the increment of bacterial mortality rates (from 98.3% to 99.6%) were found positively related to electric voltage and treatment time (Table 2) (Yogesh, 2016; Qi et al., 2021).

Checking about the disadvantages of electric field, PEF in particular, the high cost of the equipment to produce PEF could be the major hinderances to limit its development to execute antimicrobial tasks on foodstuffs (Arshad et al., 2020). Additionally, since PEF could puncture microbial cell membrane, it could also puncture the biological cells of foodstuffs too, which might lead to the run-off of essential nutrients in foodstuffs when carrying out antimicrobial performance, because the study on PEF to help to extract plant-based food pigments and colorings has proved the puncture effect of PEF on biological cell membranes of foodstuffs and the improved extraction efficiency (Bocker & Silva, 2022). As such, the balance between the antimicrobial performance on foodstuffs and the avoidance of nutrient loss of foodstuffs should be carefully manipulated.

2.2.5 Ultrasound-Based Non-thermal Antimicrobial Technique

US-based antimicrobial technique is a promising non-thermal approach applied in food processing for efficiently inactivating foodborne pathogens while maintaining the acceptable quality and sensory attributes of the processed foodstuffs. The mechanism of US to display its antimicrobial effect is through using electric energy to converse into extremely short wavelength of mechanical waves (e.g., US or acoustic energy, more than 16 kHz of frequency rate) to lead to the violent collisions of the condensed molecules to generate shock waves, which can further produce the regions containing up to 5500 °C of temperature and 50,000 kPa of pressure, the implosions of these regions might thus create pressure variations that can eventually eliminate the microorganisms (Juraga et al., 2011; Nora & Borges, 2017). Besides the above common effects, the antimicrobial mechanisms, especially the antibacterial mechanisms, of US are also shown to change the membrane microstructures and properties of bacteria. In these scenarios, the permeabilities of the bacterial inner and outer membranes are remarkably increased after US processing, which further leads to the decreases of membrane fluidity and membrane depolarization, as well as the conformational and compositional variations of a few membrane components (He et al., 2021). Additionally, US application [e.g., low-frequency US (LFUS) technique] could destroy the structures of bacterial cell surface appendages as well, while simultaneously break their cell membranes by means of acoustic cavitation, along with mechanical vibration (Bai et al., 2023).

From the viewpoint of molecular biology to explain the antimicrobial mechanism of US, it was revealed that US treatment (20 kHz and 300 W) for 5 min resulted in the remarkable down-regulations of bacterial genes involved in curli formation, which is a process that leads to cell aggregation and irreversible adhesion on the material surfaces (Table 2) (Bai et al., 2023). In addition to the effect based on molecular biology level, US treatment could significantly disturb the metabolic responses of the processed microorganisms as well. As documented, the 27 metabolites of US-treated *P. fluorescens* (20 kHz, 60/100 W for 15 min), mainly involving in the metabolisms of amino acid, carbohydrate, lipid, and nucleotide, were found differentially expressed through non-targeted metabolomics analysis (Table 2) (Wang et al., 2024b). For another bacterium, e.g., *E. coli* O157:H7, US treatment (20 kHz and 300 W) for 5 min could also weaken the metabolic capacity and destruct the biofilm-forming capability of its separated cells (Table 2) (Bai et al., 2023).

Interestingly, different US intensities and powers seemed to lead to dissimilar metabolic responses of the bacteria. Under lower US intensity at 15.79 W/cm^2 and power at 60 W, for example, the exopolysaccharides, energy level and genetic information of biofilm formed by *P. fluorescens* were severely damaged, however, the intracellular lipid metabolites could be significantly accumulated as a response of this strain to the stress (Table 2) (Wang et al., 2024b). Under higher US intensity at 26.32 W/cm^2 and power at 100 W, the metabolic level and osmotic pressure of *P. fluorescens* biofilm were greatly disturbed, which led to the suppression on protein synthesis and disrupted genetic information within bacterial cells, the stress response thus facilitated *P. fluorescens* to synthesize uridine diphosphate galactose

(UDP-Gal), ADP-ribose (ADPR) and 4-pyridoxic acid (4-PA) to save energy for survival purpose. As shown by this case, US-induced extremely momentary high temperature and potent oxidation seemed to contribute to the divergences of bacterial metabolic responses (Table 2) (Wang et al., 2024b).

Observing some concrete studies to show the antimicrobial effects of US on foodstuffs, frequency is the major influence factor. It was reported that the treatment of US sound waves between 20–100 kHz (low frequency) and 2–10 MHz (high frequency) on foods (e.g., almond milk) significantly suppressed the growth of foodborne pathogens (Table 2) (Onyeaka et al. 2023). In another related study, 20 kHz of low frequency and up to 573 W/cm^2 of ultrasound power intensity, along with 2 sec of pulsed mode, as well as 27 min of processing duration, were found to achieve up to 3.52 $\log_{10}$ CFU/ml reduction in *E. coli* populations (Table 2) (He et al., 2021).

Apart from frequency, amplitude, intensity and power are also crucial factors to affect the antimicrobial efficacy of US. As found, the most potent damage exerted by US practices on the selected pathogenic microorganisms was achieved by the combination of 20 kHz with 100% of amplitudes, compared with the combined 20 kHz with 50% and 10% of amplitudes, indicating that amplitude could affect the antimicrobial efficacy of US in a positively related manner. Under this scenario, the counts of total aerobic mesophilic bacteria (TAMB), coliforms, *Streptococci*, *Lactobacilli*, yeast and mold were reduced from at least 4.02 by up to 6.52 $\log_{10}$ CFU/ml in raw goat milk. Furthermore, such US application could inactivate about 99% of foodborne pathogens such as *Brucella melitensis* type 3, *E. coli*, *L. monocytogenes*, methicillin-resistant *S. aureus*, and *S.* Typhimurium inoculated in this milk sample (Table 2) (Tavsanli et al., 2022). For *P. fluorescens*, the intensities and powers of US at 15.79 W/cm^2 and 60 W, as well as those at 26.32 W/cm^2 and 100 W, could effectively eliminate its biofilm inoculated on the lettuce, in which the remained amounts of biofilm on the plant surface were negatively related to US intensities and powers, which meant that higher US intensity and powers were more efficient in removing biofilm from the surface of this vegetable (Table 2) (Wang et al., 2024b).

Besides directly working on the bacteria attached to foodstuffs, US can efficiently remove biofilm from the attached surface as well. As reported after LFUS treatment (20 kHz and 300 W) for 5 min, the 99.999% of biofilm removal rate from polystyrene sheets could be achieved. Nevertheless, the potent biofilm removal capacity of US does not also mean that it can more effectively to kill bacterial cells in the biofilm, for example, it was noticed that up to 30 min of LFUS treatment still could not fully eliminate the bacterial cells presented in the biofilm (Table 2) (Bai et al., 2023).

As noted, US treatment does not remarkably affect the major qualities of foodstuffs, which is demonstrated to be the major advantage to display its antimicrobial effect in the food system. This could be observed through the study showing that the treatment of US sound waves between 20–100 kHz and 2–10 MHz could prolong the shelf lives of foods (e.g., almond milk), as well as keep the desired food sensory properties simultaneously (Table 2) (Onyeaka et al. 2023). When applied in fresh produce (e.g., lettuce leaves, cantaloupe and cucumber epidermis), US technique (e.g., LFUS) could, not only demonstrate good anti-biofilm functions on their surfaces, but also exert no significant effects on the overall qualities (e.g., colors) of these fresh produces (Bai et al., 2023). Also, under 26.32 W/cm^2 of intensity and 100 W of power, the quality of US-treated lettuce was not significantly affected, compared with the untreated control (Table 2) (Wang et al., 2024b).

No doubt, high frequency US might display its disadvantage on processing foodstuffs, because such process can result in the rapid enhancement of foodstuff temperature, which negatively influences the qualities (e.g., nutritional values and sensory attributes) of the processed foodstuffs (Arvanitoyannis et al., 2017). This is why LFUS was chosen as the non-thermal antimicrobial technique, even applied in the fresh produce, as it does not lead to significant temperature increase during processing (Bai et al., 2023). However, some disadvantages still exist nowadays, regarding the application of the US in displaying its antimicrobial effect on foodstuffs. The first disadvantage comes to the fact that the developments of the appropriate US equipment for large-scale and continuous business applications are still on the way, while this equipment is commonly expensive, which might lead to higher operation costs. The second one is that US treatment still has potential to exert adverse effects on the properties of foodstuffs by degrading some essential metabolites such as phenolic compounds, vitamins and anthocyanin in the fresh produce, leading to the reductions of their nutritional values. Thirdly, US can also work well on eliminating chemical contaminants contained in foodstuffs, through which these contaminants can be degraded into other forms of toxic products after US treatment, which might further trigger potential safety risks in the processed foodstuffs (Yuan et al., 2021).

2.2.6 Cold-Plasma-Based Non-thermal Antimicrobial Technique

Cold plasma (CP), or called non-thermal plasma, is a promising and efficient technique recently applied in food processing industry for inactivating microbes and pathogens under relatively-low temperatures and within short time, in which its major parameters such as electrodes gap distance, frequency, gas composition, frequency, treatment time and voltage could affect the type and level of reactive plasma species, as well as the antimicrobial efficiency of CP eventually (Laroque et al., 2022).

As well recognized, plasma could be defined as an ionized gas that contains reactive nitrogen species (RNS) including NO, NO_2 and NO_x, ROS such as O, O_2, ozone (O_3) and •OH, along with charged particles, free radicals and ultraviolet photons (Bourke et al., 2018; Laroque et al., 2022). The occurrence of plasma is achieved through employing free electrons to collide with gas molecules static or flowing between two high-electrical-potential-difference-equipped electrodes to lead to the ionization of gas (Fig. 3) (Mandal et al., 2018). Normally, the combination of relatively low energy (1–10 eV) and electronic density (no more than 10^{10}/cm^3) is the requisite to generate CP (Roualdes & Roussac, 2017). In the scenario of CP, a

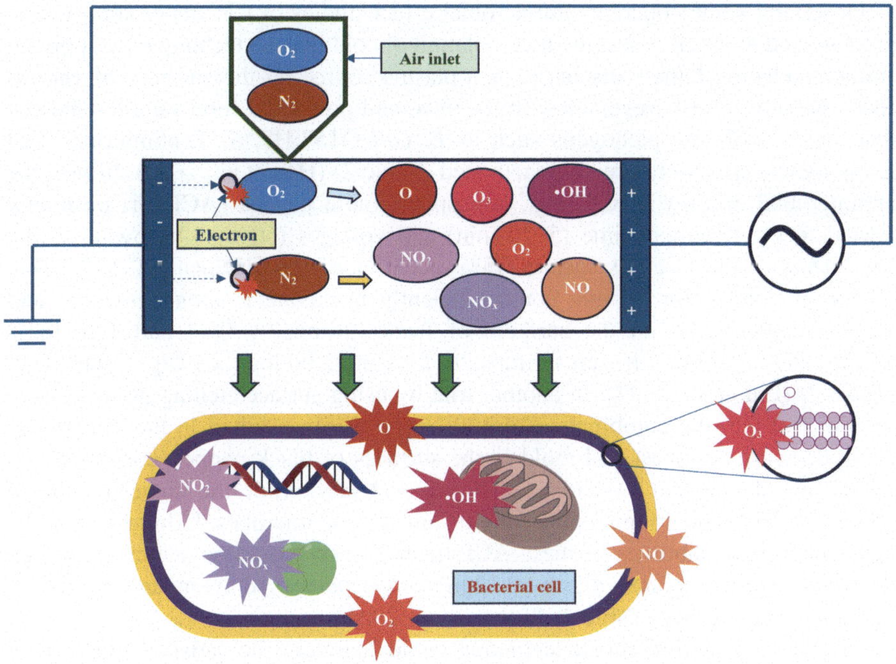

Fig. 3 Mechanisms of Cold-plasma-based non-thermal antimicrobial technique. (Adapted from Bourke et al., 2018; Mandal et al., 2018; Laroque et al., 2022)

thermodynamic non-equilibrium between electrons and those bigger species such as RNS, ROS, charged particles, free radicals and ultraviolet photons could be found, resulting that the temperatures of electrons are much higher than those of above-mentioned species because the masses of electrons are much lower than those of the species, this merely achieves the exchange of a little proportion of total energy (Misra et al., 2018). Through such process, the temperature reductions of the species are easier than those of electrons, which finally maintains the gas at a low temperature (Misra, 2016). As found, the average CP electron energy to excite the atomic and molecular species, and subsequently to break their chemical bonds to form ionic species and free electrons, is no more than 10 eV, thereout, the structures of all organic molecules with 3–6 eV of ionization and dissociation energies could be readily destructed by CP (Laroque et al., 2022).

Noteworthily, the above-mentioned reactive species produced by CP such as RNS, ROS, charged particles, free radicals and ultraviolet photons are also the excellent bactericidal species, expert in killing the microbes (Laroque et al., 2022). Thus, the antimicrobial mechanisms of CP mainly involve in the oxidation and damage of cell membrane, the degradation of proteins, the breakages of nucleic acids in microbes, as well as disrupting EPS secreted by microorganisms and disturbing quorum sensing to form biofilms (Fig. 3) (Zhang et al., 2023a).

Observing some concrete antimicrobial effect studies of CP, under atmospheric pressure and ambient gases, the manipulation of cold plasma technique by applying surface dielectric barrier discharge as a plasma source on the surfaces of various foods including fresh vegetables, fruits, nuts, and powdered food samples showed that three foodborne pathogens such as *E. coli* O157:H7, *S.* Typhimurium, and *L. monocytogenes* were greatly inactivated (Table 2) (Han et al., 2020). In fact, the antimicrobial efficacy of CP [e.g., atmospheric cold plasma (ACP)] is positively related to the treatment time (5–20 min) and voltages (60–80 kV) owing to the increments of reactive gas species. In a recent case, the treatment of ACP using 80 kV of voltage for 30 min could efficiently inactivate *L. monocytogenes* and *V. parahaemolyticus* on the surfaces of fresh shrimp, by 0.94 and 1.26 $\log_{10}$ CFU/g, and on those of frozen shrimps, by 1.68 and 2.00 $\log_{10}$ CFU/g, respectively (Table 2) (Liu et al., 2024). In another trial by using surface etching and oxidative reactions of CP, the combination of 170 V and 5 min resulted in the remarkable count decreases of yeast and mold in the surfaces of black pepper, cinnamon and fennel by 1.1, 1.3 and 1.0 $\log_{10}$ CFU/g, respectively, while the higher combined values of 230 V and 15 min could fully disinfect these microbes. Exhilaratingly, the better microbial stabilities of these ACP-treated spices could be achieved through decreasing their water activities by 0.12 in black pepper, 0.18 in cinnamon and 0.17 in fennel, respectively (Table 2) (Kahar et al., 2024).

Joyfully, CP antimicrobial technique could maintain the satisfied qualities of foodstuffs after processing. According to the above case, CP (e.g., ACP) treatment could significantly prolong the shelf life of shrimp by 2 days, thus improving the preservation quality of this seafood. In detail, the better qualities such as lower total viable count, total volatile basic nitrogen and pH, followed by higher sensory scores shown by these ACP-treated shrimps stored at 4 °C were achieved, compared with the controls, although the slightly higher TBARS levels were simultaneously detected (Table 2) (Liu et al., 2024). Similarly, the unchanged color and textural characteristics of the spices (e.g., black pepper, cinnamon and fennel) during ACP treatment further indicated the capability of this technique to maintain the freshness of these foodstuffs. Furthermore, ACP treatment under 230 V and 15 min could help to improve the qualities of these spices through increasing the total phenolic contents of black pepper, cinnamon and fennel by 0.11, 2.26 and 0.33%, respectively, contributed by the surface etching phenomenon of cold plasma. Moreover, the unvaried composition profile of volatile oil such as EO in these spices during ACP treatment indicated that such technique could also keep their features (Table 2) (Kahar et al., 2024).

However, several disadvantages regarding CP should be highlighted. Firstly, the reactive species induced by CP have potential to react with essential food components such as lipids and proteins, for example, those in dairy and meat products, leading to their oxidations and eventually affecting the whole qualities of foodstuffs (Jadhav & Annapure, 2021). Secondly, CP still can vary the qualities of foodstuffs under different treatment conditions. According to the previous report for treating pork loin, CP could achieve reductions in pH and L* values, it could also negatively affect its appearance, color, odor and acceptability (Kim et al., 2013; Pankaj et al.,

2018). Additionally, CP can reduce the firmness of blueberries, leading to fruit softening (Sarangapani et al., 2017; Pankaj et al., 2018). Thirdly, the treatment of CP is in a non-uniformity manner (Zhang et al., 2022). This also indicated that the antimicrobial effect of CP on foodstuffs is uneven and some regions inside foodstuffs can not be fully sanitized.

2.2.7 High-Hydrostatic-Pressure-Based Non-thermal Antimicrobial Tcchniquc

The antimicrobial technique of high hydrostatic pressure (HHP), or termed as high-pressure processing (HPP), is a non-thermal pasteurization approach that has been used in the food industries for efficiently disinfecting microorganisms and inactivating microbial enzymes contained in the foodstuffs, to enhance food safety and prolong their shelf lives (Xia et al., 2022b). The normal applied pressure ranges of HHP are from 100 to 1000 MPa, while its temperature ranges are from 0 to 120 ° C, in which pressure between 200 and 800 MPa and temperature between 20 and 60 °C are the industrial application ranges (Giannoulis & Karatzas, 2024). Additionally, the commercially applied pressure level of HPP can also be 600 MPa (Pokhrel et al., 2022).

The antimicrobial mechanism of HHP seems simple by facilitating high pressure to irreversibly damage the cell structures of microorganisms, thus varying the permeability of microbial cells and eventually killing them (Liu et al., 2022). However, some microorganisms are baro-resistant, which means that the antimicrobial effect of HHP is subjected to the microorganism species (Xia et al., 2022b). As noted, the resistance divergences and sensitivities of different pathogenic microorganisms (e.g., *E. coli*, *Salmonella* spp. and *L. monocytogenes*) to HPP are closely related to the genotypes and phenotypes of these strains, as well as the compositions of the contaminated foods. In acidic juices, the total soluble solids and pH values are the essential factors to control the high-pressure resistances of these pathogenic microorganisms. Meanwhile, the high acidic juices could support the commercial application of HHP in a safe manner (Gouvea et al., 2023).

Observing the cases relating to the potent antimicrobial effects of HHP on the processed foodstuffs (e.g., human milks and fruit juices), it was found that HHP technique (350 MPa at 38 °C) demonstrated its strong bactericidal influence against Group B *Streptococcus* and certain bacteriostatic effect against *E. coli* presented in the donor human milks. Most importantly, these HHP-treated donor milks could maintain their qualities for up to 6 months (Table 2) (Tran et al., 2024). In another case, HHP at 200–400 MPa for up to 5 min of treatment duration could achieve more than 5 $\log_{10}$ CFU/ml reduction of *L. innocua* (ATCC 51742) in the carrot-orange juice blends. In addition, HHP at 300 MPa and pH 4 for 2 min, at 400 MPa and pH 5 for 1 min, as well as at 400 MPa and pH 6 for 3 min, respectively, could reach at least 6 $\log_{10}$ CFU/ml reduction of this strain in the above-mentioned blends, indicating that higher pressure or lower pH value can help to shorten the processing duration, whereas achieved the same inactivation purpose in these blends (Table 2) (Pokhrel et al., 2022). In acid and acidified juices and beverages such as apple juice (pH 4.5) at 5 °C, the combined HPP conditions of 600 MPa and up to 3.2 min of processing duration led to 5 $\log_{10}$ CFU/ml reductions of three pathogenic microorganisms, e.g., *E. coli* O157:H7, *Salmonella enterica* and *L. monocytogenes* (Table 2) (Usaga et al., 2021).

As well recognized, the advantages of HHP are outstanding through keeping the physicochemical stability and maintaining the qualities of food matrices (Xia et al., 2022b). These could be proved via some studies. As found by above case, the level of HHP-treated lactoferrin under 350 MPa at 38 °C, the importantly healthy component in human milk, was observed similar with that of the raw donor human milk, whereas such level was remarkably reduced after conventional holder pasteurization (HoP, 62.5 °C for 30 min) treatment (Table 2) (Tran et al., 2024). Additionally, HHP processing did not significantly affect the contents of ascorbic acid, total carotenoids, and total phenolic compounds in the carrot-orange juice blends, compared with the untreated ones. Moreover, HHP was proved to prolong the shelf lives of these blends for more than 28 days through remaining natural microbiota such as aerobic mesophilic bacteria below 2 $\log_{10}$ CFU/ml. For the major qualities of these blends such as the levels of total soluble solids, pH values and color attributes (L*, a*, and b*), HHP did not lead to their variations after treatment followed by 28 days of storage, compared with those of the untreated blends that have been changed in 10 days of storage (Table 2) (Pokhrel et al., 2022).

Besides the advantages, some disadvantages of HHP application should also be noticed. The first disadvantage is associated with the operation cost of HHP that is relatively high, because HHP involves in the utilization of expensive equipment to produce high pressure, during which a lot of water are consumed (Wang et al., 2016). The second one is that HHP might not be effective against all microorganisms because some microorganisms (e.g., molds, spores and viruses) are baroresistant as above stated (Rivalain et al., 2010; Xia et al., 2022b). The third disadvantage comes to its adverse effect on the qualities of some foodstuffs by influencing the conformation of proteins and further resulting in their denaturation (Dehnad et al., 2024), or destructing the tissues of fruits (e.g., blueberries) and leading to the leakages of their intracellular bioactive metabolites (Paciulli et al., 2019). The fourth one states that HHP is ineffective in processing dry products and low water activity (Aw), which could reduce the efficacy of HHP treatment (Jofré et al., 2009).

3 Conclusion and Future Prospectives

This chapter showed that physical antimicrobial strategies including TATs and NTATs have been widely used in food systems and received great advances recently. As the concrete techniques, TATs and NTATs have their respective advantages and disadvantages. As summarized, their mutual advantages demonstrated that they were normally efficient in disinfecting the targeted microorganisms (e.g., pathogens) presented in foodstuffs, while their disadvantages on foodstuffs have their own features. On one hand, TATs usually lead to the increases of foodstuff's temperatures, which might not be favored by some foodstuffs that are sensitive to temperature. On the other hand, although NTATs do not result in the significant enhancements of foodstuff's temperatures during process, their other features such as the rays they provided (e.g., UVC, gamma ray and X ray) might induce the damages on the intracellular structures of foodstuffs at the same time when carrying out antimicrobial functions, these drawbacks require that the appropriate doses or intensities of NTATs are needed to process the foodstuffs, in order not to affect their qualities or appearances.

In views of the current situation, the future directions in this field should be focused on developing more efficient technologies to fully disinfect various pathogens presented on the foodstuffs, while simultaneously, the damages on the foodstuffs through these technologies should be minimized during process, so that the qualities and appearances of these processed foodstuffs are not remarkably influenced, thus affecting their selling prices. Additionally, the energy consumption issues are needed to be considered and well controlled as physical antimicrobial strategies normally apply electricity to execute the process, which might push up the operation costs and eventually elevate the selling prices of these foodstuffs if too much electrical powers are wasted to achieve the identical antimicrobial purpose on the foodstuffs.

References

Abdiyev, K., Azat, S., Kuldeyev, E., Ybyraiymkul, D., Kabdrakhmanova, S., Berndtsson, R., et al. (2023). Review of slow sand filtration for raw water treatment with potential application in less-developed countries. *Water, 15*(11), 2007.

Aboud, S. A., Altemimi, A. B., Al-Hilphy Asaad, R. S., Yi-Chen, L., & Cacciola, F. (2019). A comprehensive review on infrared heating applications in food processing. *Molecules, 24*(22), 4125.

Acharya, V. V., & Chaudhuri, P. (2021). Modalities of protein denaturation and nature of denaturants. *International Journal of Pharmaceutical Sciences Review and Research, 69*(2), 19–24.

Aganovic, K., Grauwet, T., Siemer, C., Toepfl, S., Heinz, V., Hendrickx, M., & Van Loey, A. (2016). Headspace fingerprinting and sensory evaluation to discriminate between traditional and alternative pasteurization of watermelon juice. *European Food Research and Technology, 242*, 787–803.

Ahmed, A. M., & Mohamed, F. S. (2024). Prevalence of some foodborne pathogens (FBPs) in imported chilled and frozen beef in Egypt. *Benha Veterinary Medical Journal, 46*(2), 80–84.

Al-Abri, M., et al. (2019). Chlorination disadvantages and alternative routes for biofouling control in reverse osmosis desalination. *npj Clean Water, 2*(1), 2.

Alves, E., Faustino, M. A., Neves, M. G., Cunha, A., Nadais, H., & Almeida, A. (2015). Potential applications of porphyrins in photodynamic inactivation beyond the medical scope. *Journal of Photochemistry and Photobiology C Photochemistry Reviews, 22*, 34–57.

Arshad, R. N., Abdul-Malek, Z., Munir, A., Buntat, Z., Ahmad, M. H., Jusoh, Y. M., et al. (2020). Electrical systems for pulsed electric field applications in the food industry: An engineering perspective. *Trends in Food Science and Technology, 104*, 1–13.

Arvanitoyannis, I. S., Kotsanopoulos, K. V., & Savva, A. G. (2017). Use of ultrasounds in the food industry–methods and effects on quality, safety, and organoleptic characteristics of foods: A review. *Critical Reviews in Food Science and Nutrition, 57*(1), 109–128.

Baenas, N., Gómez-Jodar, I., Moreno, D. A., García-Viguera, C., & Periago, P. M. (2017). Broccoli and radish sprouts are safe and rich in bioactive phytochemicals. *Postharvest Biology and Technology, 127*, 60–67.

Bai, M., Dai, J., Li, C., Cui, H., & Lin, L. (2023). Antibacterial and antibiofilm performance of low-frequency ultrasound against Escherichia coli O157: H7 and its application in fresh produce. *International Journal of Food Microbiology, 400*, 110266.

Banik, B. K., & Bandyopadhyay, D. (Eds.). (2018). *Advances in microwave chemistry*. CRC Press.

Barba, F. J., Gavahian, M., Es, I., Zhu, Z., Chemat, F., Lorenzo, J. M., & Khaneghah, A. M. (2019). Solar radiation as a prospective energy source for green and economic processes in the food industry: From waste biomass valorization to dehydration, cooking, and baking. *Journal of Cleaner Production, 220*, 1121–1130.

Begum, T., Follett, P. A., Hossain, F., Christopher, L., Salmieri, S., & Lacroix, M. (2020). Microbicidal effectiveness of irradiation from Gamma and X-ray sources at different dose rates against the foodborne illness pathogens *Escherichia coli*, *Salmonella* Typhimurium and *Listeria* monocytogenes in rice. *LWT, 132*, 109841.

Bhadra, K., Basu, A., Choudhuri, B. N., Guchhait, P., Dawn, A. K., & Das, S. (2024). *Direct and sunlight-induced photosensitizing antibacterial activities of curcuma longa extract* (Vol. 6, pp. 28–49).

Bhunia, K., Tang, J., & Sablani, S. S. (2024). Microwave-based sustainable in-container thermal pasteurization and sterilization technologies for foods. *Sustainable Food Technology, 2*, 926–944.

Biswas, P., & Bandyopadhyaya, R. (2017). Biofouling prevention using silver nanoparticle impregnated polyethersulfone (PES) membrane: E. coli cell-killing in a continuous cross-flow membrane module. *Journal of Colloid and Interface Science, 491*, 13–26.

Bocker, R., & Silva, E. K. (2022). Pulsed electric field assisted extraction of natural food pigments and colorings from plant matrices. *Food Chemistry: X, 15*, 100398.

Boshevska, M., Sandeva, I., Verde, S. C., Spasevska, H., & Jankuloski, Z. (2024). Effects of different irradiation doses and storage period on microbiological characteristics of wheat (*Triticum aestivum* L.). *Food Control, 158*, 110201.

Bourke, P., Ziuzina, D., Boehm, D., Cullen, P. J., & Keener, K. (2018). The potential of cold plasma for safe and sustainable food production. *Trends in Biotechnology, 36*(6), 615–626.

Brittany, L., & Hayes, D. (2002). *Microwave synthesis: Chemistry at the speed of light*. CEM.

Casado, E. G., Yao, Y., Zaffora, B., Battaggia, D., Schnabel, U., Zuber, S., & den Besten, H. M. (2024). Inactivation of *Salmonella*, *Enterococcus faecium* and natural microbiota on dry food matrices with microwave-driven plasma processed air. *Innovative Food Science & Emerging Technologies*, 103822.

Cha, M. Y., & Ha, J. W. (2022). Low-energy X-ray irradiation effectively inactivates major foodborne pathogen biofilms on various food contact surfaces. *Food Microbiology, 106*, 104054.

Charpe, A. M., Sedani, S., Murumkar, R., & Bhad, R. G. (2019). Effect of temperature on microbial growth in food during storage. *Multilogic in Science, 8*, 56–58.

Chatterjee, D. K., Fong, L. S., & Zhang, Y. (2008). Nanoparticles in photodynamic therapy: An emerging paradigm. *Advanced Drug Delivery Reviews, 60*(15), 1627–1637.

Chaúque, B. J. M., & Rott, M. B. (2021). Solar disinfection (SODIS) technologies as alternative for large-scale public drinking water supply: Advances and challenges. *Chemosphere, 281*, 130754.

Chen, S., Wang, P., Wu, L., Saldaña, M. D., Gao, Y., & Sun, W. (2023). Sunlight-induced antibacterial poly (vinyl alcohol) grafted 3, 3′, 4, 4′-benzophenone tetracarboxylic dianhydride film for the preservation of Mongolian cheese. *Food Packaging and Shelf Life, 40*, 101200.

Cho, G. L., & Ha, J. W. (2019). Application of X-ray for inactivation of foodborne pathogens in ready-to-eat sliced ham and mechanism of the bactericidal action. *Food Control, 96*, 343–350.

Chong, L., Ghate, V., Seah, C., & Zhou, W. (2024). Photosensitization can be an effective risk-reduction strategy against the post-baking mold spoilage of bread. *Food Microbiology, 117*, 104390.

Chua, A., Chong, L., Ghate, V., Yuk, H. G., & Zhou, W. (2021). Antifungal action of 405 nm light emitting diodes on tomatoes in a meso-scale system and their effect on the physicochemical properties. *Postharvest Biology and Technology, 172*, 111366.

Darwish, W. S., El Bayoumi, R. M., Mohamed, N. H., & Hussein, M. A. (2024). Microbial contamination of meat at a low temperature storage: A review. *Journal of Advanced Veterinary Research, 14*(2), 322–325.

da Silva Aquino, K. A. (2012). Sterilization by gamma irradiation. *Gamma Radiation, 9*, 172–202.

Dehnad, D., Emadzadeh, B., Ghorani, B., Assadpour, E., Yang, N., & Jafari, S. M. (2024). The influence of high hydrostatic pressure on different properties of legume proteins with an emphasis on soy proteins; a comprehensive review. *Food Hydrocolloids, 146*, 109188.

De La Vega-Miranda, B., Santiesteban-López, N. A., López-Malo, A., & Sosa-Morales, M. E. (2012). Inactivation of *Salmonella* Typhimurium in fresh vegetables using water-assisted microwave heating. *Food Control, 26*(1), 19–22.

Demirok, N. T., & Yıkmış, S. (2022). Combined effect of ultrasound and microwave power in tangerine juice processing: Bioactive compounds, amino acids, minerals, and pathogens. *PRO, 10*(10), 2100.

Fan, X., Lv, X., Meng, L., Ai, M., Li, C., Teng, F., & Feng, Z. (2020). Effect of microwave sterilization on maturation time and quality of low-salt sufu. *Food Science & Nutrition, 8*(1), 584–593.

Feng, Y., Zhou, C., Yagoub, A. E. A., Sun, Y., Owusu-Ansah, P., Yu, X., et al. (2019). Improvement of the catalytic infrared drying process and quality characteristics of the dried garlic slices by ultrasound-assisted alcohol pretreatment. *Lwt, 116*, 108577.

Fernandes, S., Gomes, I. B., Simões, M., & Simoes, L. C. (2024). Novel chemical-based approaches for biofilm cleaning and disinfection. *Current Opinion in Food Science, 55*, 101124.

Fitriani, N., Ardiyanti, P. E., Kuncoro, E. P., Budiyanto, W. D., Isnadina, D. R. M., Wahyudianto, F. E., & Mohamed, R. M. S. R. (2020). Behavior of schmutzdecke with varied filtration rates of slow sand filter to remove total coliforms. *Heliyon, 6*(4), e03736.

Fuentes-León, F., et al. (2024). Genotoxicity of ultraviolet light and sunlight in the bacterium Caulobacter crescentus: Wavelength-dependence. *Mutation Research, Genetic Toxicology and Environmental Mutagenesis, 894*, 503727.

Galema, S. A. (1997). Microwave chemistry. *Chemical Society Reviews, 26*(3), 233–238.

Gallus, I. J., Boyraz, E., & Maryška, J. (2023). Antimicrobial properties of nanofiber membrane and commercial micromembrane by modification with diethylenetriamine (DETA) and attachment of silver nanoparticles. *Journal of Nanomaterials, 2023*(1), 8927774–8927713.

Gao, W., Li, X., Wan, J., & Yan, H. (2024). Influence of X-ray irradiation on quality and core microbiological characteristics of chilled chicken meat. *LWT, 206*, 116582.

Ghate, V., Kumar, A., Kim, M. J., Bang, W. S., Zhou, W., & Yuk, H. G. (2017). Effect of 460 nm light emitting diode illumination on survival of *Salmonella* spp. on fresh-cut pineapples at different irradiances and temperatures. *Journal of Food Engineering, 196*, 130–138.

Ghate, V., Zelinger, E., Shoyhet, H., & Hayouka, Z. (2019). Inactivation of *Listeria* monocytogenes on paperboard, a food packaging material, using 410 nm light emitting diodes. *Food Control, 96*, 281–290.

Ghazanfari, N., Mortazavi, S. A., Yazdi, F. T., & Mohammadi, M. (2020). Microwave-assisted hydrodistillation extraction of essential oil from coriander seeds and evaluation of their composition, antioxidant and antimicrobial activity. *Heliyon, 6*(9), e04893.

Giannoulis, N., & Karatzas, K. A. G. (2024). The combined effect of chitosan and high hydrostatic pressure on *Listeria* monocytogenes and Escherichia coli. *Innovative Food Science & Emerging Technologies, 94*, 103693.

Gierke, A. M., & Hessling, M. (2024). Photoinactivation by UVA radiation and visible light of *Candida auris* compared to other fungi. *Photochemical & Photobiological Sciences, 23*(4), 681–692.

Gouvea, F. S., Koutchma, T., Ferreira, E. H. R., Walter, E. H. M., & Rosenthal, A. (2023). Resistance of Escherichia coli, *Salmonella* spp., and *Listeria* monocytogenes in high and low-acidity juices processed by high hydrostatic pressure. *International Journal of Food Microbiology, 395*, 110189.

Han, J. Y., Song, W. J., Eom, S., Kim, S. B., & Kang, D. H. (2020). Antimicrobial efficacy of cold plasma treatment against food-borne pathogens on various foods. *Journal of Physics D: Applied Physics, 53*(20), 204003.

Harouna, S., Franco, I., Carramiñana, J. J., Blázquez, A., Abad, I., Pérez, M. D., et al. (2020). Effect of hydrolysis and microwave treatment on the antibacterial activity of native bovine milk lactoferrin against Cronobacter sakazakii. *International Journal of Food Microbiology, 319*, 108495.

Hashem, M. A., Hossain, M. A., Sadakuzzaman, M., Rahman, M. M., & Islam, M. A. (2022). Effect of gamma irradiation on the shelf life and quality of mutton. *Turkish Journal of Agriculture-Food Science and Technology, 10*(2), 117–124.

Hashemi, S. M. B., Akbari, M., Roohi, R., & Phimolsiripol, Y. (2024). Hurdle approaches using conventional thermal, microwave heating and ginger essential oil in vegetable juice mixture: Heat transfer modeling and microbial inactivation kinetics. *LWT, 198*, 115958.

Hay, N., Huy, L. Q., & Kien, P. V. (2022). A review of drying methods assisted by infrared radiation, microwave and radio frequency. In A comprehensive review of the versatile dehydration processes. Jovanovic, J. D. (Editor). Publisher: IntechOpen.

He, Q., Liu, D., Ashokkumar, M., Ye, X., Jin, T. Z., & Guo, M. (2021). Antibacterial mechanism of ultrasound against Escherichia coli: Alterations in membrane microstructures and properties. *Ultrasonics Sonochemistry, 73*, 105509.

Hernandez-Gomez, E. S., Olvera-Cervantes, J. L., Sosa-Morales, M. E., Corona-Vazquez, B., Corona-Chavez, A., Lujan-Hidalgo, M. C., & Kataria, T. K. (2021). Dielectric properties of Mexican sauces for microwave-assisted pasteurization process. *Journal of Food Science, 86*(1), 112–119.

Holmes, E. B., Oza, H. H., Bailey, E. S., & Sobsey, M. D. (2023). Evaluation of chitosans as coagulants—Flocculants to improve sand filtration for drinking water treatment. *International Journal of Molecular Sciences, 24*(2), 1295.

Huang, J. J., Huang, W., Li, J., Li, P., & Cheung, P. C. K. (2020). Potential advancement of ultraviolet-free solar radiation technology in enriching the nutrient composition and biodiesel feedstock production in marine green microalga *Platymonas subcordiformis*. *Bioresource Technology Reports, 11*, 100534.

Huang, J. J., Xu, W., Lin, S., & Cheung, P. C. K. (2024). The bioactivities and biotechnological production approaches of carotenoids derived from microalgae and cyanobacteria. *Critical Reviews in Biotechnology, 45*, 276–304.

Hyun, J. E., & Lee, S. Y. (2020). Blue light-emitting diodes as eco-friendly non-thermal technology in food preservation. *Trends in Food Science and Technology, 105*, 284–295.

Hyun, J. E., Moon, S. K., & Lee, S. Y. (2022). Application of blue light-emitting diode in combination with antimicrobials or photosensitizers to inactivate Escherichia coli O157: H7 on fresh-cut apples and cherry tomatoes. *Food Control, 131*, 108453.

Inanoglu, S., Barbosa-Cánovas, G. V., Patel, J., Zhu, M. J., Sablani, S. S., Liu, F., et al. (2021). Impact of high-pressure and microwave-assisted thermal pasteurization on inactivation of *Listeria* innocua and quality attributes of green beans. *Journal of Food Engineering, 288*, 110162.

Jabin, T., Kamal, S., Islam, S., Razu, M. H., Paul, G. K., Karmaker, P., et al. (2023). Effect of gamma irradiation on chemical composition, antioxidant activity, antibacterial activity, shelf life, and cytotoxicity in the peels of two mango varieties grown in Bangladesh. *Arabian Journal of Chemistry, 16*(6), 104708.

Jadhav, H. B., & Annapure, U. (2021). Consequences of non-thermal cold plasma treatment on meat and dairy lipids–a review. *Future Foods, 4*, 100095.

Janjai, S., & Bala, B. K. (2012). Solar drying technology. *Food Engineering Reviews, 4*, 16–54.

Jee, D. Y., Cha, S. Y., & Ha, J. W. (2023). Evaluation of low-energy x-rays as an alternative to chlorine washing to control internalized foodborne pathogens in lettuce. *Journal of Food Safety, 43*(6), e13094.

Jeong, Y. J., & Ha, J. W. (2022). Synergistic antimicrobial effect of UV-A irradiation and malic acid combination treatment against foodborne pathogens on spinach and the underlying mechanism. *Food and Bioprocess Technology, 15*(2), 379–390.

Ježek, F., Kameník, J., Macharáčková, B., Bogdanovičová, K., & Bednář, J. (2020). Cooking of meat: Effect on texture, cooking loss and microbiological quality–a review. *Acta Veterinaria, 88*(4), 487–496.

Jha, A., Montemayor, A. M., & Tikekar, R. V. (2024). Investigation of antimicrobial activity of ultraviolet B radiation and its potential to generate fructose-mediated reactive oxygen species in coconut water. *Innovative Food Science & Emerging Technologies, 92*, 103577.

Jiang, Y., Leung, A. W., Hua, H., Rao, X., & Xu, C. (2014). Photodynamic action of LED-activated curcumin against Staphylococcus aureus involving intracellular ROS increase and membrane damage. *International Journal of Photoenergy, 2014*(1), 637601.

Jin, S., Ding, Z., & Xie, J. (2021). Study of postharvest quality and antioxidant capacity of freshly cut amaranth after blue LED light treatment. *Planning Theory, 10*(8), 1614.

Jofré, A., Aymerich, T., Grèbol, N., & Garriga, M. (2009). Efficiency of high hydrostatic pressure at 600 MPa against food-borne microorganisms by challenge tests on convenience meat products. *LWT- Food Science and Technology, 42*(5), 924–928.

Juraga, E., Sobota Šalamon, B., Herceg, Z., & Režek Jambrak, A. (2011). Application of high intensity ultrasound treatment on Enterobacteriae count in milk. *Mljekarstvo, 61*(2), 125–134.

Kacprzyńska-Gołacka, J., Łożyńska, M., Barszcz, W., Sowa, S., Wieciński, P., & Woskowicz, E. (2020). Microfiltration membranes modified with composition of titanium oxide and silver oxide by magnetron sputtering. *Polymers, 13*(1), 141.

Kahar, S. P., Shelar, A., & Annapure, U. S. (2024). Effect of pin-to-plate atmospheric cold plasma (ACP) on microbial load and physicochemical properties in cinnamon, black pepper, and fennel. *Food Research International, 177*, 113920.

Kanaan, M. H. G., Sheriff, N. A., & Abdullah, S. S. (2023). Analysis of the responses to pasteurization temperatures of methicillin resistant *Staphylococcus aureus* derived from milk items. *Microbiology Research Journal International, 33*(7), 9–16.

Kang, S. T., Subramani, A., Hoek, E. M., Deshusses, M. A., & Matsumoto, M. R. (2004). Direct observation of biofouling in cross-flow microfiltration: Mechanisms of deposition and release. *Journal of Membrane Science, 244*(1–2), 151–165.

Kapcsándi, V., Cserpán, M., & Lakatos, E. H. (2020). Impact assessment of microwave treatment of raw cow's milk on its microbiological properties. *Analecta Technica Szegedinensia, 14*(2), 60–76.

Khairullah, A. R., Sudjarwo, S. A., Effendi, M. H., Harijani, N., Tyasningsih, W., Rahmahani, J., et al. (2020). A review of methicillin-resistant Staphylococcus aureus (MRSA) on milk and milk products: Public health importance. *Systematic Reviews in Pharmacy, 11*(8), 59–69.

Khalaf, M. M., Gouda, M., Abou Taleb, M. F., Abdelaziz, M. A., & Abd El-Lateef, H. M. (2024). Modification of microfiltration process using cellulosic fabric treated with zinc oxide nanoparticles as a pioneering gravity-driven filter for biofouling mitigation in water purification. *Journal of Environmental Chemical Engineering, 12*, 113286.

Kido, J., Shimohata, T., Aihara, M., Tsunedomi, A., Hatayama, S., Amano, S., et al. (2024). Reduction of *Campylobacter jejuni* contamination by using UVA-LED and sodium hypochlorite on the surface of chicken meat. *Journal of Microorganism Control, 29*(2), 91–97.

Kim, H. J., Yong, H. I., Park, S., Choe, W., & Jo, C. (2013). Effects of dielectric barrier discharge plasma on pathogen inactivation and the physicochemical and sensory characteristics of pork loin. *Current Applied Physics, 13*(7), 1420–1425.

Kim, D. K., Kim, S. J., & Kang, D. H. (2017). Bactericidal effect of 266 to 279 nm wavelength UVC-LEDs for inactivation of gram positive and gram negative foodborne pathogenic bacteria and yeasts. *Food Research International, 97*, 280–287.

Kim, D. K., & Kang, D. H. (2020). Inactivation efficacy of a sixteen UVC LED module to control foodborne pathogens on selective media and sliced deli meat and spinach surfaces. *LWT, 130*, 109422.

Kim, W. J., Kim, S. H., & Kang, D. H. (2021). Combination effect of 915 MHz microwave heating and carvacrol for inactivation of *Escherichia coli* O157: H7, *Salmonella* Typhimurium and *Listeria monocytogenes* in hot chili sauce. *Food Control, 121*, 107578.

Kim, W. J., & Kang, D. H. (2023). Synergistic effects of 915 MHz microwave heating and essential oils on inactivation of foodborne pathogen in hot-chili sauce. *International Journal of Food Microbiology, 398*, 110210.

Kim, W. J., Lee, J. G., & Kang, D. H. (2024). Inactivation of foodborne pathogens on fruit juice with different solid contents by using pilot-scale 915 MHz microwave heating system. *LWT, 213*, 116924.

Kou, L., Luo, Y., Yang, T., Xiao, Z., Turner, E. R., Lester, G. E., et al. (2013). Postharvest biology, quality and shelf life of buckwheat microgreens. *LWT- Food Science and Technology, 51*(1), 73–78.

Kruszewska-Naczk, B., Grinholc, M., Waleron, K., Bandow, J. E., & Rapacka-Zdończyk, A. (2024). Can antimicrobial blue light contribute to resistance development? Genome-wide analysis revealed aBL-protective genes in Escherichia coli. *Microbiology Spectrum, 12*(1), e02490–e02423.

Kumar, A., Ghate, V., Kim, M. J., Zhou, W., Khoo, G. H., & Yuk, H. G. (2016). Antibacterial efficacy of 405, 460 and 520 nm light emitting diodes on *Lactobacillus plantarum*, *Staphylococcus aureus* and *Vibrio parahaemolyticus*. *Journal of Applied Microbiology, 120*(1), 49–56.

Kumar, A., Ghate, V., Kim, M. J., Zhou, W., Khoo, G. H., & Yuk, H. G. (2017). Inactivation and changes in metabolic profile of selected foodborne bacteria by 460 nm LED illumination. *Food Microbiology, 63*, 12–21.

Laroque, D. A., Seó, S. T., Valencia, G. A., Laurindo, J. B., & Carciofi, B. A. M. (2022). Cold plasma in food processing: Design, mechanisms, and application. *Journal of Food Engineering, 312*, 110748.

Li, Z., Liu, D., Xu, S., Zhang, W., Hemar, Y., Regenstein, J. M., & Zhou, P. (2021). Effects of pasteurization, microfiltration, and ultraviolet-c treatments on microorganisms and bioactive proteins in bovine skim milk. *Food Bioscience, 43*, 101339.

Ling, S. J., Moebs, W., & Sanny, J. (2016). 16.5 the electromagnetic spectrum. *University Physics Volume* 2.

Liu, J., Liu, L., Huang, Z., Fu, Y., & Huang, Z. (2020). Contaminant removal and optimal operation of bio-slow sand filtration water treatment based on nature-based solutions. *Polish Journal of Environmental Studies, 29*(2), 1703–1713.

Liu, X., Zhang, L., Pang, X., Wu, Y., Wu, Y., Shu, Q., et al. (2022). Synergistic antibacterial effect and mechanism of high hydrostatic pressure and mannosylerythritol lipid-A on *Listeria monocytogenes*. *Food Control, 135*, 108797.

Liu, D., Gu, W., Wang, L., & Sun, J. (2023a). Photodynamic inactivation and its application in food preservation. *Critical Reviews in Food Science and Nutrition, 63*(14), 2042–2056.

Liu, J., Lu, J., Li, Z., Fan, Y., & Liu, S. (2023b). An ultra-small fluorescence zero-valent iron nanoclusters selectively kill gram-positive bacteria by promoting reactive oxygen species generation. *Colloids and Surfaces B: Biointerfaces, 227*, 113343.

Liu, X., Chen, X., Ou, Y., Li, F., Jiao, Y., & Shi, H. (2024). Pathogens inactivation and preservation of Pacific white shrimp by atmospheric cold plasma treatment. *Innovative Food Science & Emerging Technologies, 93*, 103638.

Makararpong, D., Tantayanon, S., Gowanit, C., Jareonsawat, J., Samgnamnim, S., Wataradee, S., et al. (2024). Enhancing raw bovine milk quality using ultraviolet-C (UV-C) irradiation: A microbial and lipid peroxidation study. *Food Science of Animal Resources, 44*(2), 372–389.

Mamone, L., Tomás, R. S., Di Venosa, G., Gándara, L., Durantini, E., Buzzola, F., & Casas, A. (2024). Laser NIR irradiation enhances antimicrobial photodynamic inactivation of biofilms of Staphylococcus aureus. bioRxiv, 2024-05.

Mandal, R., Singh, A., & Singh, A. P. (2018). Recent developments in cold plasma decontamination technology in the food industry. *Trends in Food Science and Technology, 80*, 93–103.

McSharry, S., Koolman, L., Whyte, P., & Bolton, D. (2022). Inactivation of *Listeria monocytogenes* and *Salmonella* Typhimurium in beef broth and on diced beef using an ultraviolet light emitting diode (UV-LED) system. *LWT, 158*, 113150.

Mendes-Oliveira, G., Deering, A. J., San Martin-Gonzalez, M. F., & Campanella, O. H. (2020). Microwave pasteurization of apple juice: Modeling the inactivation of Escherichia coli O157: H7 and *Salmonella* Typhimurium at 80–90 C. *Food Microbiology, 87*, 103382.

Minor, M., Martínez-Monteagudo, S. I., & Sabillón, L. (2024). Photothermal pasteurization of paprika powder by ultra-high irradiance blue (405 nm) light: Impact on *Salmonella* inactivation and quality attributes. *Innovative Food Science & Emerging Technologies, 94*, 103672.

Misra, N. N., Martynenko, A., Chemat, F., Paniwnyk, L., Barba, F. J., & Jambrak, A. R. (2018). Thermodynamics, transport phenomena, and electrochemistry of external field-assisted non-thermal food technologies. *Critical Reviews in Food Science and Nutrition, 58*(11), 1832–1863.

Misra, N. N. (2016). Chapter 1: Plasma in food and agriculture. In *Cold plasma in food and agriculture*. Elsevier.

Mohamed, H. M., Aljasir, S. F., Moftah, R. F., & Younis, W. (2023). Mycological evaluation of frozen meat with special reference to yeasts. *Veterinary World, 16*(3), 571–579.

Ng, T. P., Nyunt, S. Z., Gao, Q., Gwee, X., Chua, D. Q. L., & Yap, K. B. (2024). Curcumin-rich curry consumption and life expectancy: Singapore longitudinal ageing study. *GeroScience, 46*(1), 969–980.

Nguyen, M. V., Arason, S., & Eikevik, T. M. (2014). *Drying of fish. Seafood processing: Technology, quality and safety* (pp. 161–175).

Nicolau-Lapeña, I., Colás-Medà, P., Viñas, I., & Alegre, I. (2022). Inactivation of Escherichia coli, *Salmonella* enterica and *Listeria monocytogenes* on apple peel and apple juice by ultraviolet C light treatments with two irradiation devices. *International Journal of Food Microbiology, 364*, 109535.

Niu, L., Wu, Z., Yang, L., Wang, Y., Xiang, Q., & Bai, Y. (2021). Antimicrobial effect of UVC light-emitting diodes against Saccharomyces cerevisiae and their application in orange juice decontamination. *Journal of Food Protection, 84*(1), 139–146.

Nora, F. M. D., & Borges, C. D. (2017). Ultrasound pretreatment as an alternative to improve essential oils extraction. *Ciência Rural, 47*(9), e20170173.

Nowak, M., Misic, D., Koter, S., Jankowski, W., Półbrat, T., Trusek, A., & Zizovic, I. (2023). The antimicrobial substance's content influences the membrane blocking mechanism in cross-flow filtration–flow cytometric assessment of the membrane's activity. *Journal of Water Process Engineering, 56*, 104371.

Novickij, V., Stanevičienė, R., Staigvila, G., Gruškienė, R., Sereikaitė, J., Girkontaitė, I., et al. (2020). Effects of pulsed electric fields and mild thermal treatment on antimicrobial efficacy of nisin-loaded pectin nanoparticles for food preservation. *LWT, 120*, 108915.

Nyhan, L., Przyjalgowski, M., Lewis, L., Begley, M., & Callanan, M. (2021). Investigating the use of ultraviolet light emitting diodes (UV-leds) for the inactivation of bacteria in powdered food ingredients. *Food Security, 10*(4), 797.

Oguadinma, I. C., Mishra, A., & Kumar, G. D. (2022). Sunlight parameters influence the survival and decline of *Salmonella* and *Escherichia coli* in water. *Journal of Food Protection, 85*(11), 1614–1624.

Onyeaka, H., Miri, T., Hart, A., Anumudu, C., & Nwabor, O. F. (2023). Application of ultrasound technology in food processing with emphasis on bacterial spores. *Food Reviews International, 39*(7), 3663–3675.

Paciulli, M., Medina Meza, I. G., Rinaldi, M., Ganino, T., Pugliese, A., Rodolfi, M., et al. (2019). Improved physicochemical and structural properties of blueberries by high hydrostatic pressure processing. *Food Security, 8*(7), 272.

Pang, X., Zhang, H., Seck, H. L., & Zhou, W. (2022). Inactivation effect of low-energy X-ray irradiation against planktonic and biofilm *Pseudomonas fluorescens* and its antibacterial mechanism. *International Journal of Food Microbiology, 374*, 109716.

Panigrahi, C., Mondal, M., Karmakar, S., Mishra, H. N., & De, S. (2020). Shelf life extension of sugarcane juice by cross flow hollow fibre ultrafiltration. *Journal of Food Engineering, 274*, 109880.

Pankaj, S. K., Wan, Z., & Keener, K. M. (2018). Effects of cold plasma on food quality: A review. *Foods, 7*(1), 4.

Pei, Y., Tao, T., Yang, G., Wang, Y., Yan, W., & Ding, C. (2018). Lethal effects and mechanism of infrared radiation on Sitophilus zeamais and Tribolium castaneum in rough rice. *Food Control, 88*, 149–158.

Peng, J., Tang, J., Barrett, D. M., Sablani, S. S., Anderson, N., & Powers, J. R. (2017). Thermal pasteurization of ready-to-eat foods and vegetables: Critical factors for process design and effects on quality. *Critical Reviews in Food Science and Nutrition, 57*(14), 2970–2995.

Podgorsak, E. B. (2005). Treatment machines for external beam radiotherapy. *IAEA radiation oncology physics: A handbook for teachers and students international atomic energy agency, Vienna, 38*.

Pokhrel, P. R., Boulet, C., Yildiz, S., Sablani, S., Tang, J., & Barbosa-Cánovas, G. V. (2022). Effect of high hydrostatic pressure on microbial inactivation and quality changes in carrot-orange juice blends at varying pH. *LWT, 159*, 113219.

Prasad, A., Gänzle, M., & Roopesh, M. S. (2021). Antimicrobial activity and drying potential of high intensity blue light pulses (455 nm) emitted from LEDs. *Food Research International, 148*, 110601.

Principato, L., & Spigno, G. (2024). Microwave heating in food processing. In *Food packaging and preservation* (pp. 299–329). Academic.

Puertas, G., Cazón, P., & Vázquez, M. (2023). Assessment of whole egg fractionation by tangential flow filtration: The problem of low-density lipoprotein aggregates. *Food and Bioproducts Processing, 140*, 99–109.

Qi, M., Zhao, R., Liu, Q., Yan, H., Zhang, Y., Wang, S., & Yuan, Y. (2021). Antibacterial activity and mechanism of high voltage electrostatic field (HVEF) against Staphylococcus aureus in medium plates and food systems. *Food Control, 120*, 107566.

Quevedo, R., Bastías, J. M., Espinoza, T., Ronceros, B., Balic, I., & Muñoz, O. (2020). Inactivation of coronaviruses in food industry: The use of inorganic and organic disinfectants, ozone, and UV radiation. *Scientia Agropecuaria, 11*(2), 257–266.

Rajalingam, N., Choi, S. Y., & Van Haute, S. (2024). Ultraviolet-C pretreatment enhances the antimicrobial efficacy of unpeeled carrots against subsequent contamination with *Listeria monocytogenes*. *International Journal of Food Microbiology, 421*, 110800.

Rashad, D., Amin, S. K., Mansour, M. S., & Abdallah, H. (2022). Fabrication of low-cost antibacterial microfiltration tubular ceramic membranes. *Ceramics International, 48*(8), 11489–11501.

Rifna, E. J., Singh, S. K., Chakraborty, S., & Dwivedi, M. (2019). Effect of thermal and non-thermal techniques for microbial safety in food powder: Recent advances. *Food Research International, 126*, 108654.

Rivalain, N., Roquain, J., & Demazeau, G. (2010). Development of high hydrostatic pressure in biosciences: Pressure effect on biological structures and potential applications in biotechnologies. *Biotechnology Advances, 28*(6), 659–672.

Rosenqvist, T., Chan, S., Ahlinder, J., Salomonsson, E. N., Suarez, C., Persson, K. M., et al. (2024). Inoculation with adapted bacterial communities promotes development of full scale slow sand filters for drinking water production. *Water Research, 253*, 121203.

Roualdes, S., & Roussac, V. (2017). Comprehensive membrane science and engineering II

Rumble, J. R., Lide, D. R., & Bruno, T. J. (2019). *CRC handbook of chemistry and physics: A ready-reference book of chemical and physical data* (100th ed.). CRC Press.

Rupasinghe, H. V., & Yu, L. J. (2012). Emerging preservation methods for fruit juices and beverages. *Food Additive, 22*, 65–82.

Sablani, S. S. (2024). Product and package testing in microwave processes. In *Microwave processing of foods: Challenges, advances and prospects* (p. 605).

Saraeva, I., Tolordava, E., Sheligyna, S., Nastulyavichus, A., Khmelnitskii, R., Pokryshkin, N., et al. (2023). FT-IR analysis of *P. aeruginosa* bacteria inactivation by femtosecond IR laser radiation. *International Journal of Molecular Sciences, 24*(6), 5119.

Sarangapani, C., O'Toole, G., Cullen, P. J., & Bourke, P. (2017). Atmospheric cold plasma dissipation efficiency of agrochemicals on blueberries. *Innovative Food Science & Emerging Technologies, 44*, 235–241.

Shahi, S., Khorvash, R., Goli, M., Ranjbaran, S. M., Najarian, A., & Mohammadi Nafchi, A. (2021). Review of proposed different irradiation methods to inactivate food-processing viruses and microorganisms. *Food Science & Nutrition, 9*(10), 5883–5896.

Shirkole, S. S., Jayabalan, R., & Sutar, P. P. (2020). Dry sterilization of paprika (*Capsicum annuum* L.) by short time-intensive microwave-infrared radiation: Establishment of process using glass transition, sorption, and quality degradation kinetic parameters. *Innovative Food Science & Emerging Technologies, 62*, 102345.

Song, H., Dang, Y. M., Ha, S., & Ha, J. H. (2022). Effect of ultraviolet-C light-emitting diode irradiation on inactivation of white colony-forming yeast in kimchi seasoning. *Food Control, 140*, 109157.

Soro, A. B., Whyte, P., Bolton, D. J., & Tiwari, B. K. (2021). Modelling the effect of UV light at different wavelengths and treatment combinations on the inactivation of *Campylobacter jejuni*. *Innovative Food Science & Emerging Technologies, 69*, 102626.

Spackman, E., Anderson, N., Walker, S., Suarez, D. L., Jones, D. R., McCoig, A., et al. (2024). Inactivation of highly pathogenic avian influenza virus with high-temperature short time continuous flow pasteurization and virus detection in bulk milk tanks. *Journal of Food Protection, 87*(10), 100349.

Suathong, W., Ongkunaruk, P., & Trevanich, S. (2024). The shelf life and quality of green pea (*Pisum sativum*) sprouts during storage at different refrigerated temperatures and durations of indirect-sunlight exposure. *Postharvest Biology and Technology, 207*, 112590.

Tavsanli, H., Aydin, M., Ede, Z. A., & Cibik, R. E. C. E. P. (2022). Influence of ultrasound application on the microbiota of raw goat milk and some food pathogens including Brucella melitensis. *Food Science and Technology International, 28*(7), 634–640.

Tran, L. C., Marousez, L., Micours, E., De Lamballerie, M., Thys, L., Gottrand, F., et al. (2024). High hydrostatic pressure is similar to holder pasteurization in preserving donor milk antimicrobial activity. *Pediatric Research, 95*, 1–5.

Ulusoy, Ş., Üçok Alakavuk, D., Mol, S., & Coşansu, S. (2019). Effect of microwave cooking on foodborne pathogens in fish. *Journal of Food Processing & Preservation, 43*(8), e14045.

Usaga, J., Acosta, Ó., Churey, J. J., Padilla-Zakour, O. I., & Worobo, R. W. (2021). Evaluation of high pressure processing (HPP) inactivation of Escherichia coli O157: H7, *Salmonella* enterica, and *Listeria monocytogenes* in acid and acidified juices and beverages. *International Journal of Food Microbiology, 339*, 109034.

Wang, C. Y., Huang, H. W., Hsu, C. P., & Yang, B. B. (2016). Recent advances in food processing using high hydrostatic pressure technology. *Critical Reviews in Food Science and Nutrition, 56*(4), 527–540.

Wang, Q., Wang, H., Zhang, T., Hu, Z., Xia, L., Li, L., et al. (2021). Antibacterial activity of polyvinyl alcohol/WO3 films assisted by near-infrared light and its application in freshness monitoring. *Journal of Agricultural and Food Chemistry, 69*(3), 1068–1078.

Wang, S., Zhang, M., He, L., Li, M., Zhang, X., Liu, F., & Tong, M. (2022). Bacterial capture and inactivation in sand filtration systems with addition of zero-valent iron as permeable layer under both slow and fast filtration conditions. *Journal of Hazardous Materials, 436*, 129122.

Wang, Z., Zhang, C. M., & Li, Y. F. (2024a). Influence of suspended particles and dissolved organic matters on virus enrichment in reclaimed water by two-step tangential flow ultrafiltration: Phenomena and mechanisms. *Journal of Hazardous Materials, 472*, 134494.

Wang, M., Jiang, L., Liu, M., Yuan, S., Guo, Y., Yao, W., & Yu, H. (2024b). Effects of ultrasound on disrupting metabolite profiles of *Pseudomonas fluorescens* biofilms cultured on the surface of lettuce. *Food Control, 155*, 110103.

Wilma, L. S., Burgos, J., Ortiz, J., Samaniego, I., Marcia, J., José, M., et al. (2024). Effects of vacuum pasteurization on the nutritional, sensory and microbiological properties of orange (citrus× sinensis) and carrot (Daucus carota L.) Nectar. *Applied Microbiology, 4*(2), 731–744.

Wittwer, M., Hammer, P., Runge, M., Valentin-Weigand, P., Neubauer, H., Henning, K., & Mertens-Scholz, K. (2022). Inactivation kinetics of *Coxiella burnetii* during high-temperature short-time pasteurization of milk. *Frontiers in Microbiology, 12*, 753871.

Xia, G., Li, Y., Tao, H., Zhang, L., Zhang, J., Yang, H., et al. (2022a). Inactivation mechanism of catalytic infrared against *Pseudomonas aeruginosa* and its decontamination application on dry green Sichuan pepper (*Zanthoxylum schinifolium*). *Food Control, 132*, 108483.

Xia, Q., Liu, Q., Denoya, G. I., Yang, C., Barba, F. J., Yu, H., & Chen, X. (2022b). High hydrostatic pressure-based combination strategies for microbial inactivation of food products: The cases of emerging combination patterns. *Frontiers in Nutrition, 9*, 878904.

Xiong, Y. L., & Guo, A. (2020). Animal and plant protein oxidation: Chemical and functional property significance. *Food Security, 10*(1), 40.

Xu, Z., Li, Y., Liu, Z., Soteyome, T., Li, X., & Liu, J. (2024a). Current knowledge on cryogenic microorganisms and food safety in refrigerators. *Trends in Food Science and Technology, 104382*, 104382.

Xu, A., Ghate, V., Chong, L., Wang, F., Zou, Y., Yuk, H. G., & Zhou, W. (2024b). 405 nm LED illumination for the reduction of Fusarium spoilage risk in Cavendish bananas in a simulated retail environment. *Food and Bioprocess Technology, 17*, 1–14.

Yadav, G., Gupta, N., Sood, M., Anjum, N., & Chib, A. (2020). Infrared heating and its application in food processing. *The Pharma Innovation Journal, 9*(2), 142–151.

Ye, J., Wang, H., Wang, L., Wang, D., Li, X., & Wang, X. (2024). Rod-shaped cellulose nanocrystal/Cu9S5 antibacterial nanocomposites for microfiltration. *Industrial Crops and Products, 216*, 118742.

Yeom, S. J., Kim, S. S., Yoon, K. N., Cho, M. J., Song, B. S., Park, J. H., & Kim, J. K. (2024). Effects of X-ray irradiation on pathogen inactivation, quality changes, and hydrocarbon profiles of pork cutlets. *International Journal of Food Science and Technology, 59*(11), 8372–8382.

Yogesh, K. (2016). Pulsed electric field processing of egg products: A review. *Journal of Food Science and Technology, 53*, 934–945.

Yuan, S., Li, C., Zhang, Y., Yu, H., Xie, Y., Guo, Y., & Yao, W. (2021). Ultrasound as an emerging technology for the elimination of chemical contaminants in food: A review. *Trends in Food Science and Technology, 109*, 374–385.

Yun, Y. S., Bae, S. J., & Park, S. H. (2022). Inactivation of foodborne pathogens on inshell walnuts by UV-C radiation. *Journal of Food Protection, 85*(8), 1172–1176.

Zavala-Franco, A., Arámbula-Villa, G., Ramírez-Noguera, P., Salazar, A. M., Sordo, M., Marroquín-Cardona, A., et al. (2020). Aflatoxin detoxification in tortillas using an infrared radiation thermo-alkaline process: Cytotoxic and genotoxic evaluation. *Food Control, 112*, 107084.

Zernadji, W., Jebri, S., Rahmani, F., Amri, I., Aissaoui, D., Trabelsi, M. H., et al. (2024). Effect of gamma irradiation on pathogenic *Staphylococcus aureus* in packaged ready-to-eat salads treated with biological extracts. *Journal of Food Protection, 87*(3), 100232.

Zhang, H., Ha, T. M. H., Seck, H. L., & Zhou, W. (2020). Inactivation of Escherichia coli O157: H7 and *Salmonella Typhimurium* in edible bird's nest by low-energy X-ray irradiation. *Food Control, 110*, 107031.

Zhang, H., Seck, H. L., & Zhou, W. (2021a). Inactivation of *Salmonella Typhimurium, Escherichia coli* O157: H7, *Staphylococcus aureus*, and *Listeria monocytogenes* in cardamom using 150 KeV low-energy X-ray. *Innovative Food Science & Emerging Technologies, 67*, 102556.

Zhang, H., Pang, X., Seck, H. L., & Zhou, W. (2021b). Low-energy X-ray inactivation of *Listeria monocytogenes* in mono−/co-culture biofilms with *Pseudomonas fluorescens* on food contact surfaces. *Food Microbiology, 100*, 103841.

Zhang, B., Tan, C., Zou, F., Sun, Y., Shang, N., & Wu, W. (2022). Impacts of cold plasma technology on sensory, nutritional and safety quality of food: A review. *Food Security, 11*(18), 2818.

Zhang, H., Zhang, C., & Han, Q. (2023a). Mechanisms of bacterial inhibition and tolerance around cold atmospheric plasma. *Applied Microbiology and Biotechnology, 107*(17), 5301–5316.

Zhang, B., Zhao, W., Wu, W., Shang, X., Thanh Nguyen, B. T., Zhang, J. B., et al. (2023b). Dean vortex trapping for concentrating *Escherichia coli* in a cross-flow microfilter. *ACS Applied Engineering Materials, 1*(10), 2809–2816.

Zhao, N., Lv, L. P., Ma, P., Zhang, Y. Y., Deng, J., & Zhang, Y. Y. (2023). A novel exposure mode based on UVA-LEDs for bacterial inactivation. *Journal of Photochemistry and Photobiology B: Biology, 239*, 112641.

Zhou, B., Luo, Y., Huang, L., Fonseca, J. M., Yan, H., & Huang, J. (2022). Determining effects of temperature abuse timing on shelf life of RTE baby spinach through microbial growth models and its association with sensory quality. *Food Control, 133*, 108639.

Zhu, S., Song, Y., Pei, J., Xue, F., Cui, X., Xiong, X., & Li, C. (2021). The application of photodynamic inactivation to microorganisms in food. *Food Chemistry: X, 12*, 100150.

Zuluaga-Acosta, J., Volentini, S. I., Debes, M. A., Hilal, M., Cerioni, L., & Rapisarda, V. A. (2023). Application of UV-B light and low-toxicity compounds to prevent postharvest spoilage on lemons. *Food and Bioprocess Technology, 16*, 804–823.

Chemical and Biochemical Antimicrobial Strategies in the Food Environment

Doris H. D'Souza

Pathogenic and spoilage organisms in the food environment need to be eliminated in order to ensure food safety, extend shelf-life and reduce food waste. Traditionally used systems though effective have drawbacks that can damage equipment and may not be user- or environmentally-friendly. Novel approaches that overcome some of these drawbacks are therefore being researched that maintain food safety, quality, and maintain or increase shelf-life. Some of these approaches including chemical (ozone, electrolyzed water, hydrogen peroxide, organic acids) and plant-based antimicrobials that are reported to have antibacterial effects against foodborne bacterial pathogens as well as antiviral activity against human foodborne viruses are highlighted and summarized.

1 Ozone

Both aqueous and gaseous ozone continue to gain increasing interest as antimicrobial agents for application in the food industry due to their strong oxidative properties. The energization and splitting of diatomic O_2 into monatomic (O) oxygen that bonds with other O_2 molecules results in the formation of the powerful oxidant, ozone (O_3), typically generated by corona discharge using dry oxygen (Choi, 2020; Choi and D'Souza, 2023; Xue et al., 2023; CDC, 2016a). Aqueous ozone can cause oxidation through hydroxyl radical mediated oxidation and generation of reactive oxygen species that destroys microbial cells (Premjit et al., 2022). In addition to cell wall (peptidoglycan layer), membrane (unsaturated lipid oxidation) and nucleic acid damage (DNA double bond breakage by singlet state oxygen),

D. H. D'Souza (✉)
Department of Food Science, University of Tennessee, Knoxville, TN, USA
e-mail: ddsouza@utk.edu

D. Li (ed.), *Antimicrobial Strategies in the Food System: Updates, Opportunities, Challenges*, https://doi.org/10.1007/978-3-031-95056-8_4

further damage to cytoplasmic components of bacteria including enzymes and proteins, and inhibition of gene expression can occur with cell component leakage that causes eventual cell death. The factors that play a role in the stability or decay of ozone in water include pH (increased pH causes rapid decrease), temperature (lower is favorable), and the presence of dissolved anions, organic matter and carbonates (Premjit et al., 2022). Ozone is reported to have a half-life in distilled water at 20 °C between 20 and 30 min, where higher temperatures decrease its solubility and accelerates degradation (Premjit et al., 2022).

With regards to foodborne viruses, ozone that readily oxidizes molecules is known to destroy the capsid proteins and nucleic acid of viruses (Predmore et al., 2015; Fuzawa et al., 2020). Aqueous ozone at 0.49 mg/L (0.49 ppm) after 15 s treatment was reported to cause 2.63 log MPN/mL inactivation of adenovirus type 40 (AD40; ~10^3 to 10^5 MPN/mL) and 3.55 log MPN/mL inactivation with 0.30 mg/L (0.30 ppm) ozone after 10 min treatment (Thurston-Enriquez et al., 2005). These researchers also demonstrated that the human norovirus (HNoV) surrogate, feline calicivirus, FCV (~10^3 to 10^5 MPN/mL) treated with 0.06 mg/L aqueous ozone showed inactivation of 1.85 log MPN/mL after 15 s and 1.00 mg/L ozone caused 4.74 log MPN/mL inactivation after 1.20 min. Other researchers also showed that treatment of FCV at 7 log $TCID_{50}$/mL in water with 6.25 ppm aqueous ozone for 30 s caused 1.39 ± 0.08 log $TCID_{50}$/mL reduction, while 10 min treatments caused FCV titer reductions of 6.79 ± 0.06 log $TCID_{50}$/mL (limit of detection ≤100 viral units) (Hirneisen et al., 2011). This research study also showed that MNV titers of 6 log PFU/mL were lowered by 2.0 ± 1.18 log PFU/mL after treatment with 6.25 ppm aqueous ozone for 30 s, while 10 min treatments caused 5.31 ± 0.81 log PFU/mL decreases in MNV titers (Hirneisen et al., 2011).

Aqueous ozone and also gaseous ozone has been applied to decontaminate fruits and vegetables with varying inactivation rates depending on the contact time, surface properties, and levels of contamination with organic matter (Premjit et al., 2022; Strano et al., 2021; Sun et al., 2020; Zhou et al., 2018). Mixing with higher transfer rate/contact efficiency and spraying have shown improved antimicrobial effects due to friction and shear with increased microbial detachment and oxidation.

Strawberries dipped for 2 min in 0.1 mg/L aqueous ozone were reported to have 21% lower weight loss, 16% increased/higher firmness, and 15% lower fruit color change during 2-day room temperature storage (Nayak et al. 2020). FDA's approval of ozone as a direct additive to food in 2001 has increased research in this technology for microbial inactivation (U.S. FDA 2004). Peach juice treated with 10 and 18 mg/L aqueous ozone in a bubble column resulted in 3.9 and 4.9 log reduction of *L. innocua* ATCC 33090 after 12-min treatment, respectively, with ~4 log reduction of *E. coli* ATCC 11229 after 12 min by 18 mg/L aqueous ozone, but only 1 log reduction of *S. cerevisiae* KE162 (Loredo et al., 2015). Orange juice (pH 3.5 to 4) spiked with 6 log CFU/mL *E. coli* ATCC 25922 and NCTC 12900 strains after treatment with 75–78 μg/mL ozone for 60 s at the optimum flow rate of 0.12 L/min showed reduction to non-detectable levels (Patil et al., 2009). Thus, by considering pH, turbidity, organic content, and with use of appropriate concentration that maintain the sensory attributes and nutritive value, aqueous ozone has potential as a processing approach to inactivate target microorganisms in juices.

With regards to liquid or fluid foods and the juice industry, sanitation practices are important to ensure food safety and extend shelf-life. Typically for pipes, nozzles, and equipment that process the fluids, cleaning-in-place (CIP) processes, using hot or cold-water flushes, alkaline and acidic cleaners, detergents, and disinfectants are used. However, aqueous ozone can have advantages of sustainability using less water, energy (heat requirement) and less chemicals for CIP. During bottling of wine, the treatment of the bottling machines with 3.5 mg/L aqueous ozone for 30 min was found to be more effective in cleaning pipes than 1% peracetic acid for 15 min (Englezos et al. 2019; Premjit et al., 2022).

Aqueous ozone as well as gaseous ozone has been used also against human enteric viruses and cultivable surrogates on fresh produce including alfalfa seeds, lettuce, green onions, strawberries and raspberries (Hirneisen et al., 2011; Wang et al., 2015; Predmore et al., 2015; Zhou et al., 2018). Green onion plants sprayed with 6.25 ppm ozone for 10 min showed reduction of 2 log PFU of MNV-1/plant (Hirneisen et al., 2011). Similarly, lettuce and green onions treated for 4 min with 6.25 ppm ozone showed decreases of 2 log TCID50/mL titers of FCV-F9, while >6 log $TCID_{50}$/mL reduction was observed in water (Hirneisen & Kniel, 2013). Alfalfa seeds spiked with the cultivable human norovirus surrogates, Tulane virus (TV), murine norovirus (MNV-1) and HNoV genogroup II showed reductions of 1.66 to 5.6 log PFU/g seeds, and reduction of 1.5 to 3 log genomic copies/g seeds after treatment with 6.25 ppm aqueous ozone for 30 min (Wang et al., 2015). Similar reduction of the genomic copy numbers of TV to HNoV GII was also reported (Wang et al., 2015).

Studies with 6% gaseous ozone treatments for 40 min indicated decreases of MNV titers to levels that could not be detected by the plaque assay, with only 0.6 log PFU/mL of TV remaining detectable (Predmore et al., 2015). Furthermore, on strawberry surfaces a 3.3 log PFU/g reduction of MNV-1 was reported, while only 1.5 log PFU/g reduction of internalized MNV-1 was obtained after a 40 min treatment with 6% gaseous ozone. Similar 6% gaseous ozone treatments by these researchers showed that TV was non-detectable on strawberry surfaces, while only 2 log reduction was reported to be observed on internalized strawberries. On lettuce surfaces, both TV and MNV-1 titers decreased by 2 to 2.5 log PFU/g after 40 min treatments with 6% gaseous ozone (Predmore et al., 2015). Others showed that raspberries treated with 3 ppm gaseous ozone for 1 min could decrease MNV-1 titers by >3.3 log TCID50, however hepatitis A virus (HAV) titers did not show any reduction even after 3 min treatments with 5 ppm gaseous ozone, the highest concentration level tested that maintained the attributes of raspberries after storage for 3 days (Brié et al., 2018). Thus, these research results demonstrate that higher concentration and longer treatment times with gaseous ozone are needed to have optimal antiviral effects.

Other alternate forms of using ozone as antimicrobials for decontamination include catalytic ozonation and microbubbles. Catalytic ozonation using a low-cost volcanic rock, with increased disinfection efficiency has been researched for human enteric viral inactivation. HNoV genotype I and II were reported to be completely removed/inactivated to non-detectable levels using catalytic ozonation after 150 min treatment of municipal effluent water (from initial 3 log or 5 log GC/L, respectively (Gomes et al., 2019).

Ozone micro bubbles (that range from 10 to 50 μm in diameter), nanobubbles (that range from less than a micron size to within nanometer diameters) or combined micro-nano bubbles (diameters ranging from nano-meters to micro-meters) are being researched for decontamination and disinfection. Their advantages are attributed to their large interfacial area, longer residence time in water, lower bubble rising velocity, high internal pressure with increased ozone mass transfer and enhanced oxidation capabilities leading to the use of lower volumes and concentrations for effective inactivation (Tekile et al., 2017). The hydroxyl radical concentration generated from ozone plays a key role, as it is a more powerful oxidant than ozone. The effective generation of hydroxyl radicals (·OH) by the collapse of ozone micro-nanobubbles under a broad pH spectrum combined with hydrodynamic cavitation has great potential for decontamination and virus inactivation (Tekile et al., 2017). Overall, the use of various forms of ozone as an alternate to aqueous chlorine needs further consideration.

Biofilms of *Pseudomonas fluorescens*, *Staphylococcus aureus*, and *L. monocytogenes* on AISI 304 stainless steel coupons treated with 0.5 ppm ozonated water in static conditions showed an estimated decrease in viable cells of 1.61–2.14 log CFU/cm^2 after 20 min, with higher reduction of 3.26–5.23 log CFU/cm^2 with similar treatments under dynamic conditions (Marino et al., 2018). This study showed the potential application of aqueous ozone for daily sanitation to prevent or control formation of microbial biofilms in the food environment. Researchers showed that aqueous ozone concentrations of greater than 1 mg/L could damage *V. parahaemolyticus* cell membranes as observed using transmission electron microscopy and also as determined by lactate dehydrogenase release into the extracellular space, causing inactivation of enzymes, gene expression inhibition, and degradation of genomic DNA and total RNA that led to cell death (Feng et al., 2018).

In addition, aqueous ozone (11 μg/mL) was reported to inactivate spores and could decrease the spore counts of *B. cereus*, *B. subtilis* and *B. stearothermophilus* by 1.3 to 6.1 log CFU/mL, with higher resistance shown by spores of *B. stearothermophilus* (Khadre, 2001). Furthermore, *B. subtilis* spores treated with ozone showed damage to the spore coat suggesting that the mode of sporicidal action of ozone is by targeting the outer spore coat layers (Khadre, 2001).

The surface of carrots, sweet potatoes, and butternut squash (commonly used in raw meat-based diets for companion animals) inoculated with two strains of *L. monocytogenes* and three serovars of *Salmonella enterica* subsp. *enterica* upon treatment with 5 ppm aqueous ozone for either 20 s or 60 s in a batch wash ozone sanitation system (BWOSS) showed mean reduction of 1.56 log CFU/mL or 0.63–1.66 log CFU/mL by 20 s in a spray wash ozone sanitation system, with highest reductions at lower temperatures (freeze-thawed samples) compared to ambient temperature (Chandran et al., 2023).

Boston bibb lettuce and cherry tomatoes after 40 min treatment with 0.86–0.99 ppm ozonated water wash or water alone showed 2 to 4 log reduction of *Enterobacter cloacae* and *Bacillus cereus* that was similar to that obtained with water-only wash (Dawley et al., 2021). However, this study showed that when MNV was contaminated together with *B. cereus* a higher/increased reduction of 1-log reduction (2.60 log PFU/mL) compared to MNV contained by itself/alone (1.63 log PFU/mL) was obtained (Dawley et al., 2021). This study showed that the low concentration of ozonated water was not sufficient to cause significant/impactful reduction compared to washing with water alone.

Treatment of fresh-cut cabbage with 1.4 mg/L aqueous ozone for 10 min showed significant reduction of bacteria, coliforms, and molds after 12-day storage at 4 °C compared to the control (Liu et al., 2020). These researchers showed that the coliform counts in the control were 2.1 log CFU/g on day 4, while coliforms were initially non-detectable (<1 log CFU/g) after 5 and 10 min treatments, that further resulted in 0.8 log CFU/g reduction in coliforms, and 1.1 to 1.4 log reduction in yeast count, and non-detectable levels of molds on day 12 with 10 min treatments compared to control. Furthermore, 5 min treatments with 1.4 mg/L aqueous ozone caused greater removal of trichlorfon, chlorpyrifos, methomyl, dichlorvos, and omethoate from fresh-cut cabbage ($p < .05$) than the control, indicating an economic and effective approach for pesticide residue removal as well as shelf-life extension of fresh-cut cabbage (Liu et al., 2020).

Washing of cherry tomatoes with 0.85 ± 0.2 mg/L of aqueous ozone and mono-mode frequency irradiation was reported to cause <1 log CFU/g reduction of spoilage microorganisms (mesophilic bacteria on standard plate count agar and yeasts and molds on potato dextrose agar), while dual-mode frequency irradiation (DMFI) caused higher reduction of 1.3–2.61 log CFU/g, with further increase in reduction by >3 log CFU/g using the combined treatment system (20/40 kHz + aqueous ozone) (Taiye et al. 2020). These researchers noted that this combined treatment did not result in any adverse effect on the physicochemical properties (except firmness), antioxidant or bioactive compounds during 21 day refrigerated storage indicating its potential application to extend the shelf life of cherry tomatoes (Taiye et al. 2020).

Alfalfa seeds inoculated with *Salmonella* and STEC showed significant reduction of 3.6 and 2.9 log CFU/g, ~~respectively,~~ and 3.1 and 3.0 log CFU/g reduction on alfalfa sprouts, respectively after immersing for 15 min or 20 min into ozonated water followed by immersion for 15 min in 1 L of acidic electrolyzed water (pH 3.0), without any significant changes in color, weight, quality, or shelf life compared to controls (Mohammad et al., 2020).

Gaseous ozone (800 or 1600 ppm) treatments for 30 min were reported to reduce populations of a two-strain cocktail of *S.* Typhimurium on tomatoes by <0.6 log CFU/grape tomato smooth surface and stem scar area (Fan et al., 2020).

In purified bottled water manufacturing, per the US-FDA regulations, residual ozone should not exceed 0.4 mg/L (Lou et al., 2024). Researchers showed that *Enterococcus faecium*, two serotypes of *E. coli*, *L. monocytogenes* (three strains), *P. aeruginosa*, and *S. enterica* (three serovars) could be decreased by ≥5-log by applying aqueous ozone at 0.832 mg/L for 1 min, 0.416 mg/L for 2 min, or other combinations at 21 °C for decontamination of purified water and also showed that

E. faecium ATCC 8459 was a suitable surrogate strain for the tested pathogens to validate water decontamination by aqueous ozone (Lou et al., 2024).

Aqueous ozone at 4 ppm after 2 min treatment was shown to decrease dairy cattle manure-based pathogens (MBP; namely *Bacillus cereus*, *Citrobacter freundii*, *Enterococcus italicus*, *E. coli*, and *Proteus hauseri*) by ≥5-log on smooth plastic (6.1 log) and metal surfaces (5.1 log), and 9 ppm aqueous ozone could decrease MBP by 6.0 log on nylon after 2 min and 5.4-log on rubber surfaces after 8 min (Megahed et al., 2018). However, this study showed that on wood, both concentrations of aqueous ozone at up to 9 ppm could decrease MBP by only 3.6 log.

Sequential soaking (10×) and spraying chicken drumsticks contaminated with 6.9 log CFU/cm^2 of a two strain *Salmonella* cocktail of avirulent *S.* Typhimurium and vaccine strain of *S. choleraesuis* (aSTC) with 8 ppm aqueous ozone and aqueous ozone-lactic acid blend resulted in reductions of 1.6 and 1.2 log/cm^2 on the skin surface, and showed 1.1 and 0.9 log/cm^2 reduction on subcutaneous skin by soaking and spraying, respectively (Megahed et al. 2020). Thus, this research showed the application of aqueous ozone use in the poultry processing facilities to reduce microbial loads.

Other researchers studied the combination of 9.00 ± 0.25 mg/L aqueous ozone with 1% sodium citrate solution for 35 s pretreatment of vacuum-package beef for cold temperature storage at 4 °C and showed at least 1 log reduction (7.30 log CFU/g to 6.16 log CFU/g on day 21) after treatment with aqueous ozone compared to the control (Zhang et al. 2020; Premjit et al., 2022).

Taken together, these data suggest the potential application of aqueous ozone in the food environment to enhance food safety.

2 Electrolyzed Water

Electrolyzed water (EW) or electrochemically activated solution is emerging as a broad-spectrum oxidative antimicrobial that is gaining increased interest in the food industry. EW is typically generated by electrolysis (with an electrolytic cell) of a dilute sodium chlorine solution, with reasonably low cost and low environmental impact (Moorman et al., 2017). The dilute sodium chloride solution through electrolysis "dissociated into acidic electrolyzed water (AEW) with a pH of 2 to 3, an oxidation-reduction potential (ORP) of >1,100 mV, and an active chlorine content of 10 ppm to 90 ppm, while basic/alkaline electrolyzed water results in a pH range of 10 to 13 and has an oxidation-reduction potential (ORP) in the range of -800 to -900 mV" (Hricova et al., 2008). The antimicrobial effects of EW are associated with pH, oxidation-reduction potential (ORP) and also the chlorine content "(balance of hypochlorous acid [HOCl], chlorine gas [Cl_2], and hypochlorite ion [OCl^-])" (Moorman et al., 2017). Hypochlorous acid, HOCl is known to be more effective as an antimicrobial (that is predominant at neutral pH) compared to the active ingredient of bleach, hypochlorite, that is its dissociated form (OCl^-) (Moorman et al. 2017). Factors that influence the antimicrobial activity of EW include treatment temperature, mixing and agitation (improves antimicrobial activity), and presence of organic matter (decreases activity) (Hricova et al., 2008). Gram-negative bacteria are generally more susceptible to AEW at higher temperatures as their cell membranes become more fluid, allowing for cell entry of AEW, while the residual chlorine is stabilized at low temperatures of storage.

EW is being researched for applications as a fresh produce wash and for clean-in-place systems in dairy manufacturing against foodborne pathogens (Kim et al., 2001; Kim et al., 2003; Tian et al., 2011; Moorman et al., 2017). Treatments with 250 ppm free available chlorine (FAC) of neutral electrolyzed water (NEW, pH 7 to 8) for 10, 15, and 30 min was shown to reduce HNoV GII.4 Sydney strain inoculated on stainless steel coupons (without soil or organic load) by 1.6, 2.4, and 5.0 log RNA copy numbers and caused 3.0, 3.8. and 4.1 log PFU/mL reduction of TV titers, respectively (Moorman et al., 2017). This same study also showed that NEW with levels of 250 ppm free available chlorine after a 1 min treatment caused HNoV RNA copy numbers to be decreased by 4.8 log in suspension tests (Moorman et al., 2017). NEW at 5 ppm was also shown to cause almost complete loss of HNoV GII.4 receptor binding after treatment for 30 s, while the VP1 major capsid protein was degraded after treatment with 250 ppm for 5 min, and virus particle aggregation was observed after 30 min with 150 ppm NEW. Therefore, NEW at 250 ppm FAC showed promise as an alternate to bleach and as an antinoroviral disinfectant for use under clean conditions.

Strawberries that were spot-inoculated with 6.6 log PFU/g of bacteriophage MS2 and rinsed with EW (50 to 55 ppm free chlorine; ProduceMaxx, Sterilox™) or 50 ppm chlorine for 90 s before and upon storage caused an increased 1 log reduction compared to washing with water alone after 30 day frozen storage (Huang et al., 2019). However, acidic EW, basic EW and NEW were not effective in removing HNoV bound to raspberries or lettuce compared to washing with tap water alone that removed >95% HNoV from raspberries (Tian et al., 2011). This research study also demonstrated that acidic EW wash caused only ~7.5% and ~4% removal from raspberries and lettuce, respectively indicating an increased HNoV binding to these food matrices (Tian et al., 2011). The binding and electrostatic interaction of the virus with the plants plays a key function and role in the choice of decontaminants. Thus, it appears that the use of NEW is limited to clean fomite surfaces such as stainless steel.

Slightly acidic electrolyzed water (SAEW, with an available chlorine concentration (ACC) of 0.5 mg/L) was shown to decrease *E. coli* levels from 8.4 log CFU/mL to <3.9 log CFU/mL, (>4.5 log decrease) while NaOCl at the same ACC of 0.5 mg/L resulted in only a 1.3 log decrease in water (Naka et al., 2020). Similarly, SAEW decreased *P. aeruginosa* from 8.5 log CFU/mL to 4.1 log CFU/mL, but was decreased to only 6.2 log CFU/mL by NaOCl at the same ACC of 0.5 mg/L (Naka et al., 2020).

AEW was reported to reduce *L. monocytogenes* in biofilms on stainless steel to nondetectable levels within 5 min, potentially by destabilizing the glycocalyx and allowing for penetration of the oxidizing agents (Kim et al., 2001). *L. monocytogenes* Scott A and *S. aureus* biofilm cells were found to be inactivated by ~2.45 and 2.57 log CFU/mL after treatments for 5-min with 25 mg/L slightly AEW, with lower reduction of mixed-species biofilms by 1.95 and 1.43 log CFU/mL, respectively (Yan et al., 2022). Furthermore, these researchers using atomic force microscopy and confocal laser scanning microscopy showed that changes in the cell membrane permeability of the mixed-species biofilm occurred after treatment with SAEW.

The biofilms of *L. monocytogenes* containing 7.57 log CFU/mL on glass dishes after treatment with alkaline electrolyzed water (BEW) containing available chlorine concentration (ACCs) of 7 mg/L for 1 min and acidified electrolyzed water with 60 mg/L ACC for 4 min showed complete removal of *L. monocytogenes* biofilms, though showed improved effects with combined treatments requiring less time to decrease surviving populations (Huang et al., 2023). This study also showed its dissociative effect on ribosomes, with inhibitory effects on ATP binding, sugar and amino acid transport, signal transduction and metabolism of sugar and amino acids. Washing biofilms of *L. innocua* formed on food contact surfaces (rubber, glass, propylene and stainless steel) with 70 ppm of SAEW for 5 min was also shown to significantly reduce the bacterial populations during food processing (Jeon et al., 2018).

The biofilm biomass of *S. aureus* formed on flat-bottom 96-well polystyrene microtiter plates was shown to be effectively removed using basic EW, where EW at pH 10.8 reduced 42% of the biofilm mass after treatment for 2 min, while increasing the pH of BEW to pH 11.6 caused 78% reduction of the biofilm biomass that was confirmed using SEM (Sun et al., 2012). This study also showed that AEW at pH 3.5 and 2.5 did reduce the *S. aureus* biofilm biomass.

C. sakazakii biofilm counts were shown to decrease from initial 8.98 log CFU/mL to 5.34 log CFU/mL after treatment for 2 min with 25% slightly AEW and to 3.11 log CFU/mL with 100% slightly AEW (80 mg/mL available chlorine) (Tao et al., 2022). These researchers also showed that SAEW damaged the cell integrity and changed membrane permeability causing leakage of nucleic acids, intercellular proteins, and potassium ions with reduced enzymatic activity of superoxide dismustase and catalase. Furthermore, they showed that biofilm formation was inhibited with decrease in gene expression of *C. sakazakii* response regulators (Tao et al., 2022).

AEW treatment of strawberries for 10 min was shown to decrease naturally present aerobic bacteria, coliforms, and fungi by 1.6, 2.4, and 1.6 log CFU per strawberry, respectively with similar results on cucumbers (Koseki et al., 2004), while *E. coli* O157: H7, *L. monocytogenes*, and *Salmonella* Enteritidis were decreased by 7.5 log CFU per tomato (Bari et al., 2003).

One kg of golden delicious apples (6 to 8 apples) or cherry tomatoes (18 to 20 tomatoes) dipped in 2 L or 1 L suspensions of *E. coli* O157:H7 and *L. monocytogenes* to obtain 6.52 ± 0.13 and 6.39 ± 0.01 log CFU/surface, respectively resulted in reductions of 2.28 ± 0.10 and 2.30 ± 0.10 log CFU/apple after a 3 min dip treatment with SAEW (Tango et al., 2017). On cherry tomatoes, the initial surface counts were 6.80 ± 0.01 and 6.58 ± 0.11 log CFU/tomato that was reduced to 2.73 ± 0.08 and 2.35 ± 0.12 log CFU/tomato, respectively after the same 3 min dip treatment with SAEW (Tango et al., 2017).

Sunflower seed contaminated with *Salmonella* and treated with AEW (pH 2.76, oxidation-reduction potential 1093 mV, ACC 48 mg/L) for 15 min showed 2.3 log reduction without any negative impact on sunflower seed germination (Inpitak & Udompijitkul, 2022). Alfalfa seeds soaked with SAEW at varied ACC of 25, 35 and 45 mg/L and pH levels of 5.0, 5.7 and 6.4 for 0.5 and 6 h showed a significant reduction of 0.97–1.50 log of *Enterobacteriaceae* populations from initial 7.05–7.61 log CFU/g on day 1 to 5.55–6.64 log CFU/g on day 9 without any adverse effects on sprout quality (Zhang et al., 2021).

High-pressure spray treatment with BEW (52 °C for 10 s at pH 11.2) and AEW (60 °C for 10 s at pH 2.4 and an ACC of 70 ppm) on cattle hides resulted in decrease of total microbial counts by 3.5 log CFU/100 cm^2 and *Enterobacteriaceae* counts by 4.3 log CFU/100 cm^2 (Bosilevac et al., 2005). Dipping of beef in SAEW (ACC of 30 mg/L) for 5 min caused reduction in *S.* Enteritidis populations of 3.36 ± 0.27 log CFU/g, while 3.29 ± 0.21 CFU/g reduction was observed after dipping in AEW (Bing et al., 2022). Pork chops that were artificially contaminated with 6 log CFU/mL *L. monocytogenes* and treated with 58 ppm NEW showed reduction of 80.19% and 90.35% after 24 h and 8 days, while *in vitro* NEW caused >99.98% reduction (Torres-Rosales et al., 2020).

Total microbial reduction of 23.48×10^3 CFU/egg was observed after spraying eggshells of hatching eggs with 100 mg/L SAEW for 180 s (Zang et al., 2019). Another study reported that chicken carcasses sprayed with 30 mg/L SAEW for 15 s resulted a 1.0 log CFU/cm^2 reduction of bacterial populations (Zang et al., 2019).

Electrolyzed reduced water (ERW) spray treatment of eggs for 16 s showed *S.* Enteritidis population reduction of 1.70 log CFU/egg compared to the reduction of 1.51 log CFU/egg by distilled water alone (Yuan et al., 2022). Significantly higher reduction of 1.89–4.69 log CFU/egg by SAEW alone at ACC of 5 and 15 mg/L was reported after 16 to 34 s treatments compared to the 1.77 to 4.41 log CFU/egg reduction obtained by the combined treatment of ERW and SAEW (Yuan et al., 2022).

Chicken breast meat samples treated with near-neutral electrolyzed water (NEW) at 100 and 200 μg/mL, peroxyacetic acid (PAA) at 200 and 400 μg/mL, and their combination (NEW 100 μg/mL + PAA 200 μg/mL) were reported to effectively decrease microbial populations after storage at 4 °C for 2, 4, and 6 days, with the best antimicrobial effects observed using combined treatments (Moghassem Hamidi et al., 2021). This study reported that the aerobic plate counts, psychrophilic plate count, *Enterobacteriaceae*, lactic acid bacteria, and *Pseudomonas* counts in the NEW 100 μg/mL + PAA 200 μg/mL group were 1.33, 1.40, 1.45, 1.01, and 1.45 log CFU/g, respectively, significantly lower than the control group with the highest sensory evaluation score (Moghassem Hamidi et al., 2021).

Neutral electrolyzed water was reported to cause greater than 3 log reduction of *S.* Typhimurium, *E. coli* O157:H7, *Campylobacter jejuni, L. monocytogenes* and *S. aureus* on polypropylene boards, with <3 log reduction on wooden surfaces after 1 min treatment (Al-Qadiri et al., 2016). However, increasing the treatment time to 5 min caused greater than 5 log reduction of all tested bacteria on polypropylene surfaces. Using NEW and QUAT solutions within 5 min reduced Gram-negative bacteria by 4.58 to 4.85 log10 compared to more than 5 log10 reductions in Gram-positive bacteria inoculated onto wooden surfaces (Al-Qadiri et al., 2016).

B. cereus ATCC 14579 and RB223 spores were reported to be reduced by an additional 2.44 and 1.64 log CFU/mL and 1.89 and 1.38 log CFU/mL after treatment with SAEW and mild heat for 1 and 5 min, respectively, in comparison to SAEW (ACC level: 60 mg/L) treatment alone that showed about 3 log reduction after 5 min (Yang et al., 2021).

The advantages of AEW includes its environment friendly nature and low operational costs as it uses only salt and is finally converted to water, without the need for special handling or storage, and without adversely affecting the sensory properties of produce and fish, with continued research applications in the poultry environment as well (Hricova et al., 2008; Cichoski et al., 2019). However, certain disadvantages need to be noted that include the initial cost of the equipment, corrosive nature and formation of chlorine gas that may pose a hazard to the operator, with its high oxidation reduction potential and free chlorine content, and reduced antimicrobial activity in the presence of organic matter or long storage time (Hricova et al., 2008).

3 Hydrogen Peroxide

Hydrogen peroxide (H_2O_2) (as well as peracetic acid (PAA)) are being researched as alternates to household bleach (sodium hypochlorite) for their antimicrobial activity associated with their powerful oxidizing potential and maintenance of quality of water used in washing (Wessels & Ingmer, 2013). Their mechanism of action implies a lower risk for developing resistance in microbial populations, and this is supported by the lack of reports in the literature about development of antimicrobial resistance after use (Nicolau-Lapeña et al., 2019). Hydrogen peroxide (H_2O_2) produces hydroxyl-free radicals and is a strong oxidant with antimicrobial activity against bacteria and viruses, by attacking the viral capsid and genome. As H_2O_2 breaks down to water and oxygen, its application in food processing poses a low environmental risk (Nasheri et al., 2021).

In terms of decontamination of rooms, viable FCV and MNV could not be detected in the test room/facility after hydrogen peroxide vapor (HPV) treatment, where FCV reduction of at least 3.65 log TCID50 and 3.67 log TCID50 MNV reduction were reported, while MNV reduction by plaque assays was at least 2.85 log PFU (Holmdahl et al., 2016). Using maximum 860 ppm hydrogen peroxide, with 205 g H_2O_2 at 100% humidity, a 33 min gassing time, and a 50 min dwell time at 20 °C in a test facility, treatment of fecal suspensions of HNoV genogroup I and II and MNV-1 dried on plastic plates, showed 4.5 to 5 log reduction of MNV in viability assays but 0.4 log reduction of HNoVs by qPCR (Holmdahl et al., 2019). With regards to foods, spraying of lettuce with 2% (vol/vol) H_2O_2 at 50 °C for around 60 s caused a 3-log reduction of MS2 but with lowered reduction at 10 °C or 25 °C (Xie et al., 2008).

Furthermore, treatment of strawberries by immersion in water at 43 °C containing 0.5% H_2O_2 resulted in 99.2% reduction of MS2, 96.9% reduction of phiX174, 99.6% reduction of phage PRD1, and 98.2% reduction of poliovirus-1 (Lukasik et al., 2003). Chocolate contaminated with 6 log PFU/mL of FCV was inactivated by 2 log after 30 s treatment with an advanced oxidative process (AOP) maintained at 48 °C in a UV-C box that used 6% v/v hydrogen peroxide (830 μL), 16.5 mg ozone (ozone generating185 nm lamps) and 1.9 J/cm^2 of UV-C at 254 nm, and by 4 log (complete) after 60 s (with corresponding double concentrations/volume of 1.67 mL hydrogen peroxide, 33 mg ozone and 3.8 J/cm^2 UV-C at 254 nm (Nasheri et al., 2021). Pistachios contaminated with similar titers of FCV showed reduction of 1.1 log after 30 s and 2 log after 60 s treatments and contaminated cornflakes showed reductions of 0.77 log and 0.97 log after similar 30 and 60 s treatments, respectively (Nasheri et al., 2021). These researchers also showed that MNV at 6 log PFU/mL in chocolate was decreased to non-detectable levels (4 log reduction) after similar treatments for 30 and 60 s, and in pistachios was reduced by 0.9 log and 1.9 log, with lower reduction of 0.4 log and 0.6 log in cornflakes after 30 and 60 s treatments. These researchers found that HAV at 5 log PFU/mL was the most resistant in chocolate to the AOP treatment with reduction of 1.7 log after 30 s and 1.9 log after 60 s treatment, with 1.1 log and 1.25 log reduction in pistachios and 1.3 and 1.15 log reduction in cornflakes after treatments for 30 and 60 s, respectively (Nasheri et al., 2021). Thus, in complex food systems, aqueous hydrogen peroxide in combination with other chemicals or processing methods show promise to lower the risk of viral foodborne associated illness.

Black pepper treated with 21.67 mg/L vacuum hydrogen peroxide vapor was shown to decrease population levels of *E. coli* O157:H7 and *S.* Typhimurium by >6.12 and 4.52 log CFU per gram, respectively, without negative effects on the color, while they were decreased by 4.35 and 2.36 log CFU per gram on red pepper, respectively (Song & Kang, 2022a). Treatment of alfalfa seeds inoculated with *Salmonella* Typhimurium, *Escherichia coli* O157:H7 and *Listeria monocytogenes* sequentially with 1 mL of up to 30% vacuumed hydrogen peroxide vapor (VHPV) for 1 min followed by dry heat (DH) or vacuum sealed dry heat (VH) for 2 h was shown to decrease the levels of the pathogens by >5 log CFU/g without decreasing germinability (Song & Kang, 2022b).

Grape tomatoes inoculated with a two-strain cocktail of S. Typhimurium showed decreases of 2.1 log CFU/tomato on the smooth surface area and 0.8 log CFU/tomato on stem scar area after 30 min treatments with aerosolized hydrogen peroxide (5% and 10%) (Fan et al., 2020). These researchers showed a synergistic antimicrobial effect using combination of gaseous ozone (800 ppm and 1600 ppm) and aerosolized hydrogen peroxide (H_2O_2) that caused decreases of 5.2 log CFU/fruit on smooth surface area and 4.2 log CFU/fruit on the stem scar (Fan et al., 2020).

Cabbage slices within a closed polyethylene terephthalate container with a H_2O_2-soaked polypropylene pad attached to the lid and treated with this H_2O_2-atmospheric dielectric barrier discharge cold plasma (ADCP) device (integrated treatment) was shown to decrease *E. coli* O157:H7 by 1.8 log/g, *L. monocytogenes* by 2.0 log CFU/g, *B. cereus* spores by 1 log spores/g and indigenous bacteria by 1.3 log CFU/g without altering the moisture content or texture of cabbage slices (Kim et al., 2024).

Treatments of tomatoes with 7.8% aerosolized hydrogen peroxide (droplet diameters of 40 nm and 3.0 μm) was shown to decrease populations of *S.* Typhimurium by 5.0 log CFU/piece, *E. coli* O157:H7 and *L. innocua* to non-detectable levels from initial levels of 2.9 and 6.3 log CFU/piece, respectively, (detection limit 0.6 log CFU/piece) on smooth surfaces (Jiang et al., 2017). These researchers showed lower reductions on the stem scar area with reductions of 1, 1.3 and 1.3 log of *E. coli* O157: H7, *S.* Typhimurium, and *L. innocua*, respectively. Similar treatments of cantaloupe caused decreases of 4.9, 1.3, and 3.0 log CFU/piece of *E. coli* O157:H7, *S.* Typhimurium and *L. innocua*, respectively and caused 1.5, 4.2 and 4.0 log reduction of *E. coli* O157:H7, *S.* Typhimurium and *L. innocua* on spinach, respectively without affecting their firmness or color (Jiang et al., 2017). Another study showed that treatment of cantaloupes and melons with hydrogen peroxide (5%) for 2 min could cause 1.5 to 2.5 log reductions of *E. coli* O157:H7, *L. monocytogenes*, *S.* Typhimurium and *S.* Enteritidis, as well as non-O157 Shiga toxin-producing *E. coli* (O26, O45, O103, O111, O121, and O145) (Svoboda et al., 2016). *E. coli* O157:H7 and *S.* Enteritidis showed a 4 log reduction on iceberg lettuce treated at 22 °C for 5 min with 1.5% lactic acid and 2% H_2O_2 combination and also after treatment with 2% H_2O_2 at 50 °C for 60 s, with a 3 log reduction of *L. monocytogenes*, without negatively affecting the sensory attributes of lettuce (Lin et al., 2002).

Suspensions of 8.19 log CFU/mL *S.* Enteritidis and 7.98 log CFU/mL *L. monocytogenes* treated with dielectric barrier discharge atmospheric CP at 75 kV for 150 s was shown to cause a 7 log reduction of both bacteria associated with the generated reactive oxygen species including hydroxyl radicals and hydrogen peroxide (Qian et al., 2022). This study reported damage to the membrane of *S.* Enteritidis with leakage of intracellular material while there was not much effect on the enzyme activity of *L. monocytogenes* (Qian et al., 2022).

A combined treatment of 60 °C hot water (10 min) followed by 3% or 4%H2O2/ 0.2% acetic acid treatment for 10 min was shown to result in >3 log reduction of bacterial populations when mung beans were co-inoculated with 6 log CFU/g *E. coli* O157:H7, *S.* enterica and *L. monocytogenes*, similar to the 3 log reduction obtained for *E. coli* and *S. enterica* with individual 2% and 4% H_2O_2 treatment after a 10 min hot water wash immediately after storage (Trząskowska et al., 2018).

Chicken frames used to prepare mechanically separated chicken upon treatment with 7% H_2O_2 and UV-C light (254 nm; 8 to 12 mW/cm^2) for 40 s were shown to decrease 1.5 log CFU of aerobic bacteria/frame from initial 3.6 log/frame while also causing reduction of cocktails of *S.* Enteritidis and *S.* Typhimurium inoculated on these frames (Jones-Ibarra et al., 2019). This study also showed that 5% H_2O_2 caused reductions of 1.1 log CFU *Salmonella* per frame similar to the 7% H_2O_2 treatments from initial 5.6 log CFU/frame (Jones-Ibarra et al., 2019).

White Leghorn hen egg shell surfaces after treatment with UV-C at 254 nm (8 to 12 mW/cm^2) followed by 3.5% H_2O_2 solution of UV-C (254 nm) with residence time (spray and UV) of 5 and 38 s travel time through the entire sanitization device showed >2 log reduction of aerobic counts and *S.* Enteritidis from a recovered control mean counts for both of ~3 log CFU/egg (Al-Ajeeli et al., 2016).

Biofilms of *E. coli* O157:H7 and three serovars of *S. enterica* (namely Anatum (FSIS1500-574), Dublin (FSIS1500-519), and Montevideo (FSIS1500-570)) on stainless steel surfaces were shown to be decreased by >3 log CFU/chip after treatment for 60 min with a 5 to 10% multicomponent sanitizer containing hydrogen peroxide and (300 ppm) quaternary ammonium chloride, with increased reduction with multiple consecutive treatments and higher exposure time (Wang et al., 2021). This study also used scanning electron microscopy to show that 5% strength sanitizer removed the biofilm and dissolved the connecting extracellular polysaccharide matrix, and with only few visible bacteria that had altered cell morphology indicating cellular injury and death using the 10% sanitizer (Wang et al., 2021).

Treatment of *S. enterica* and *E. coli* O157:H7 in suspension for 120 s with a 0.4% concentration of a newly developed mixed organic acid wash (lactic acid and one or more fruit acids) and hydrogen peroxide showed reduction of 7.5 and 7.1 log CFU/mL, respectively, while 0.8% solution decreased *L. monocytogenes* by 5 log CFU/mL (Gurtler et al., 2022). Cherry tomatoes dip inoculated and treated with 1.0% peroxyacid solution showed 3.8 and 3.4 log CFU/g reduction of *S. enterica* and *E. coli* O157:H7, respectively, and a 3.5 log CFU/g reduction of *L. monocytogenes* (Gurtler et al., 2022). While hydrogen peroxide at 10%, w/w was shown to decrease *B. cereus*, *B. subtilis*, and *B. stearothermophilus* spores by only 0.32 to 1.6 log CFU/mL (Khadre, 2001).

Thus, hydrogen peroxide shows broad spectrum of antimicrobial activity that can have improved synergistic effects together with organic acids, mild heat, and oxidizing agents for use in hurdle approaches for the more resistant microorganisms including spore-formers.

4 Acidic Treatments

Organic acids that are considered 'Generally Recognized as Safe' (GRAS) by the U.S. FDA are increasingly being studied as antimicrobials and surface sanitizers in the food environment, and can persist with residual effects on produce to prevent microbial attachment (Ortiz-Solà et al., 2020). Their mode of action involves lowering the internal cellular pH, anion accumulation, and alteration of membrane transport and increase of cell permeability. Some of the commonly used organic acids include citric acid, acetic acid, and lactic acid.

Citric Acid Citric acid (CA) is commonly used for washing of fruit and vegetables and in wash solutions (Ortiz-Solà et al., 2020). A new (alcohol based hand rub) formulation that comprised of 2% citric acid, 2% urea, 3.69% w/w 2-propanol, and 69.39% w/w ethanol was reported to cause 4.08 log decrease in MNV titers and 5.21 log decrease of adenovirus, AdV (Ionidis et al., 2016), and showed activity in a quantitative suspension tests as well as on dried viruses (Ionidis et al., 2016). In addition, citrate buffer could drastically alter the HNoV genogroup II.10 VLP morphology causing enlargement of VLPs after 10 min at room temperature, where the VLP diameters were ~42 to 44 nm in citrate buffer up to concentrations of 7.81 mM, however, the VLP diameters were ~46 to 49 nM in the presence of citrate buffer at concentrations of 62.50 and 125 mM (Koromyslova et al., 2015). Citrate buffer could also alter the ability to bind at the HBGA-binding pocket after 10 min treatment at room temperature (Koromyslova et al., 2015). It was also shown that in 15.63 mM citrate buffer, a few of the VLPs had a new ring-like structure on their outer spikes, and the ring-like structure was prevalent in the majority of the VLPs in 62.50 mM citrate buffer (Koromyslova et al., 2015). Research also showed that MNV-1 had <1 log PFU reduction at pH 2 after 2 h in citrate buffer (Cannon et al., 2006), while FCV showed 4.4 log reduction in citrate buffer at pH of 2 at 37 °C after 30 min, and other buffers tested at pH from 2 to 10 for 2 h included phosphate buffer and carbonate buffer (Cannon et al., 2006). Other researchers showed that FCV incubated with citric acid buffer for 30 min at 37 °C showed a 2 log decrease in infectivity at pH 6 and > 5 log decrease in titers at pH ≤ 2 and pH ≥ 10 (Duizer et al., 2004). A 4·0 log decrease in RNA copy number of HNoV GI.6 and GII. 4 after 5 min treatment with an antimicrobial containing silver dihydrogen citrate (SDC) was reported that meets the US EPA standard for antinorovirus efficacy (Manuel et al., 2017; US EPA 2015). However, this research reported that in the presence of 5% soil, SDC treatment for 30 min caused approximately 2·5 log reduction in RNA copy number, while carrier test assays also indicated a 2 to 3 log reduction in 30 min. These researchers also showed that treated particles were deformed with aggregate formation, reducing viral capsid protein band intensity by 50%, and histo-blood group antigen receptor-binding ability by 80% (Manuel et al., 2017). Thus, citric acid shows potential to decrease and inactivate viruses on surfaces. Furthermore, strawberries inoculated with 3.8 log $TCID_{50}$/mL of MNV-1 after washing with 1, 2.5 and 5% citric acid (pH 1.8 to 2.2) for 2 to 5 min resulted in >1.7 log$TCID_{50}$/mL reduction of MNV-1 (Ortiz-Solà et al., 2020). Thus, citric aid alone has some ability to decrease HNoV surrogate titers on produce, that may have improved effects in combination with other hurdles.

Upon treatment of various *L. monocytogenes* strains for 1 min with a commercial "eco-friendly" citric acid-based sanitizer (CAB) containing 0.66% w/v citric acid and 0.036%w/v sodium dodecylbenzene sulfonate (manufacturer's recommended concentration (MRC) and treatment time), a 5-log reduction in a standard broth assay at 4 °C, 15 °C, 23 °C, or 30 °C for 24 h was reported (Boucher et al., 2021). However, it is important to mention that these researchers also reported that this CAB did not consistently reduce *L. monocytogenes* by 5-log on stainless steel or plastic surfaces using the MRC and treatment time at 4 °C (Boucher et al., 2021). Further studies may therefore be warranted.

In food systems, *L. monocytogenes* was reported to be decreased by 3.0–3.3 log CFU/g at 4 °C, 1.7–3.9 log CFU/g at 10 °C, and 0.9–1.4 log CFU/g of hummus at 24 °C using citric acid at 0.5% to 1.5%, while *S. enterica* levels were decreased by 0.6–1.7, 4.1–4.9 and <1.5 log CFU/g, at 4, 10 and 24 °C, respectively, compared to the control during 10 day storage of hummus using 0.5 to 1.5% w/w citric acid (Olaimat et al., 2022). Citric acid at 0.5% and 1% and 4 °C was reported to decrease

S. sonnei by 1 log, though 2% citric acid showed reduction to non-detectable levels (from initial levels of 4 to 6 log/g) in hummus, with 3.0 log reduction at 10 °C (Olaimat et al., 2017). This study also showed that 0.5% and 1% citric acid had a bacteriostatic effect *on S. flexneri* (Olaimat et al., 2017). Spinach leaves treated with 1% citric acid for approximately 11 min decreased *E. coli* O157:H7 and *L. monocytogenes* by 0.45 and 1.23 log CFU/mL, respectively from initial 6 to 7 log CFU/leaf (Jeon & Ha, 2020). However, these researchers showed that the combination treatment of 0.3 kGy X-ray with 1% citric acid for 11 min decreased these bacterial levels by 4.23 and 3.69 log CFU/mL, respectively, while 0.3 kGy X-ray irradiation alone decreased the levels of *E. coli* O157:H7 and *L. monocytogenes* by only 2.83 and 1.32 log CFU/mL, respectively (Jeon & Ha, 2020). Thus, this hurdle technology shows synergistic effects for antimicrobial reduction on spinach leaves, which was attributed to the generation of intracellular reactive oxygen species in the pathogens without causing any negative effects on spinach leaf color or texture.

Irradiation of spinach leaves with xenon light (XL, 200 to 1500 nm, with integral intensity of 0.025 W/cm^2 and dose of 12.0 J/cm^2 during the maximum 8 min treatment) for 8 min was shown to decrease *E. coli* O157:H7, *S.* Typhimurium, and *L. monocytogenes* levels by 3.96, 4.36, and 4.62 log CFU/cm^2, respectively (Cho & Ha, 2021). However, this study showed that treatment with 1% citric acid caused only <1 log reduction of the three tested foodborne pathogens, while the combined treatment with 1% CA and XL irradiation for 8 min showed increased reduction of 5.25, 5.05, and 5.02 log CFU of *E. coli* O157:H7, *S.* Typhimurium, and *L. monocytogenes*, respectively that did not negatively impact the color, texture or total phenolic content of the spinach leaves (Cho & Ha, 2021). They also demonstrated that the combined treatment resulted in cellular damage.

A disinfectant comprising of 0.06% citric acid, 0.08% lactic acid, and 0.02% sodium dodecyl sulfate (SDS) after 15 s could decrease *Salmonella* levels by >4.7 log CFU/mL, with 0.02% citric acid, 0.025 lactic acid, and 0.0125% SDS needed for similar reduction of *Campylobacter* using suspension tests from initial populations of 7 to 8 log CFU/mL (Bai et al., 2022). The disinfectant was reported to continue to retain activity in the presence of 15% fetal bovine serum (Bai et al., 2022). This study also showed the utility of using this disinfectant in poultry slaughterhouses on carcasses, knives and gloves.

A blend of 5% citric acid/hydrochloric acid (v/v) treatment for 30 s at 4 °C with mechanical agitation was shown to decrease the populations of a five-serovar cocktail of *Salmonella enterica* after storage for 3 days by 1.32 ± 0.12 log CFU/g in chicken hearts and by 0.96 ± 0.27 log CFU/g in chicken livers (Nakimera et al., 2024). This study also showed that treatment with 0.05% peroxyacetic acid (PAA) for 90 s showed reductions of 1.33 ± 0.25 log CFU/g in chicken hearts and 1.10 ± 0.12 log CFU/g in chicken livers, and treatment with 2% v/v sulfuric acid for 30 s showed reductions of 1.40 ± 0.04 log CFU/g in chicken hearts and 1.09 ± 0.19 log CFU/g in chicken livers (Nakimera et al., 2024). Furthermore, the mesophilic aerobic plate count showed minimal growth after 3 day of this refrigerated storage without any changes in the color of the chicken hearts and liver. Broiler carcasses treated with Chicxide (buffered blend of natural 2.5% lactic acid-citric acid) for 5 s as a spray was reported to cause 1.3 log CFU/mL reduction of *Salmonella* and 2.3 log CFU/mL by immersion for 5, 10, or 20 s from an initial level of 5 log CFU/mL (Laury et al., 2009).

Beef brisket tissue samples inoculated with 6 log CFU/cm^2 of a 5-strain mixture of *E. coli* O157:H7, 12-strain mixture of non-O157:H7 STEC, 6-strain mixture of *Salmonella,* and 5-strain mixture of nonpathogenic *E. coli* after treatment in a spray cabinet, with combinations of 1.9% and 2.5% lactic acid-citric acid (LCA) blend solutions at 43 °C and 60 °C and application pressure of 15 and 30 lb./square inch, was reported to cause microbial reduction of 0.9 to 1.5 log CFU/cm^2 irrespective of inoculum type (Scott et al., 2015). In a commercial facility when beef carcasses were inoculated with 6 log CFU/cm^2 of *E. coli* surrogates to validate hot water treatments (92.2–92.8 °C, 13–15 lb/in^2) followed by the LCA blend (1.9%, 50 to 51.7 °C, 13 to 15 lb./in^2), 10 s), a 3.7 log CFU/cm^2 reduction was obtained (Scott et al., 2015). This showed the use of two acids and hot water as a hurdle technology to inactivate pathogens in the beef industry.

Fresh, unpasteurized carrot juice at pH 6.41 inoculated with 7 to 8 log CFU/mL of *E. coli* O157:H7 showed reductions of 7.46 log CFU/mL after a combination treatment of 5 mM caprylic acid and 5 mM citric acid at 50 °C for 5 min, and endogenous microbial populations were decreased by 3.07 log CFU/mL, without any significant reduction observed when individual treatments were used (Kim & Rhee, 2015). Furthermore, these researchers performed a validation study using 3.32 log CFU/mL *E. coli* O157:H7 that was decreased to non-detectable levels with the combined treatment of 5 mM caprylic acid and 2.5 mM citric acid at 50 °C for >5 min and also with 5 mM caprylic acid and 5 mM citric acid combination at 45 °C and 50 °C for >5 min (Kim & Rhee, 2015).

The addition of the food-grade antimicrobial, citric acid at 30 mM to commercially available powdered infant formula milk spiked with *C. sakazakii* at 5 log CFU/mL after incubation at room temperature for 3 h caused a ~0.5-log reduction, while combinations of 60 μM nisin and citric acid caused an increased reduction of 1 log, while Nisaplin, a commercial formulation containing nisin A combined with 30 mM citric acid caused about a significant >3 log reduction (Campion et al., 2017).

Commercial pasteurized milk treated with 16 μg/mL nisin showed only a slight growth inhibition of 6 log CFU/mL *S. aureus* or *L. monocytogenes*, while 0.25 mg/mL citric acid alone inhibited multiplication of *S. aureus* and *L. monocytogenes* in pasteurized milk and maintained the initial loads (Zhao et al., 2017). The study showed that combined nisin-citric acid at sub-inhibitory concentrations decreased *S. aureus* and *L. monocytogenes* by ≥2 log CFU/mL after 24 h compared to nisin or citric acid alone that had MIC values in the ranges of 32–64 μg/mL and 0.5–1 mg/mL, respectively (Zhao et al., 2017).

Thus, citric acid in combination with other hurdles has improved antimicrobial effects against microorganisms in the food environment.

Lactic Acid (LA) L-(+)-Lactic acid (2-hydroxypropanoic acid) is recognized as a GRAS organic acid in the family of carboxylic acids (U.S. Environmental Protection Agency 2009; Al-Qadiri et al., 2016). LA that is naturally produced by lactic acid bacteria has applications in the food industry as a flavoring agent, pH regulator and also as a preservative to control pathogens (Straube et al., 2011). The undissociated form of lactic acid is known to passively diffuse into the cytoplasm and decreases the cytoplasmic internal pH, further leading to protein denaturation, and proton motive force disruption (Al-Qadiri et al., 2016; Cherrington and others 1990; Culver and others 2014). Lactic acid solutions have been applied as sanitizing washes to control microbial contamination in meat carcasses and minimally processed or ready-to-eat produce (Al-Qadiri et al., 2016).

Lactic acid at 0.5%, 1% and 2% did not show any reduction of the 6 log PFU/mL TV after treatment for 5 min in solution, however incubation with 2% lactic acid (LA) combined with 2.8% SDS (v/v, pH 2.6), 1% LA-4%SDS (v/v, pH 2.8), 0.5% LA-0.7%SDS (v/v, pH 3.5) resulted in ≥4.5 log PFU/mL reduction of TV (Lacombe et al., 2018). MNV-S99 titers were reduced between pH values of 5.0 and 6.2 by 3.25 log TCID50 units to 0.5 log TCID50 after 3 days exposure to 90% D-L lactic acid (10:1 v/v, acid to virus) at 24 °C (Lange-Starke et al., 2014). FCV upon exposure to 0.3% and 0.4% D,L-lactic acid (pH 3.2) at 4 °C for 3 h was reduced by 0.4 to 0.9 log TCID50, with increased reduction of 0.6 to 1.5 log TCID50 after 3 h at 20 °C (Straube et al., 2011). Increased treatment time of 7 days with 0.3% and 0.4% lactic acid (pH < 3.6) caused >6 log reduction of FCV at 20 °C and 2 log reduction at 4 °C (Straube et al., 2011).

Treatment of murine norovirus (MNV) and influenza A virus A/WSN/33 (H1N1) in a 1:10 ratio of cell-free supernatants of *Lactobacillus curvatus* and its related metabolites (sakacin A, nisin, and lactic acid) for 3 days at 24 °C using simulated in vitro pH (5 to 6.2) and temperature of raw sausage fermentations was shown to decrease MNV titers by 1.25 log while D,L-lactic acid decreased MNV titers by 3.25 log and H1N1 titers by 2.5 log (Lange-Starke et al., 2014).

Washing with lactic acid at 2.5% for 2 and 5 min was shown to reduce MNV-1 titers by 1.7 log TCID50/mL on strawberries inoculated with 3.8 log TCID50/mL of this virus (Ortiz-Solà et al., 2020). HNoV GII.4 at 2.57 log genomic copies/μL were reduced to 1.58 log genomic copies/μL at 4 °C (pH 4.26) and reduced to 1.04 log genomic copies/μL at 10 °C (pH 4.06) in experimentally contaminated cabbage kimchi fermented (which corresponded well with the increased levels of lactic acid and acetic acid that occurred gradually) that was stored for 28 days (pH 6.02 at day 0 for both temperatures) as determined by using the combination of monoclonal antibody-conjugated immuno-magnetic capture and separation procedure together with real-time RT-qPCR (Lee et al., 2017). The study showed that only acetic acid at 2530.5 mg/kg was present at 4 °C and 2498.6 mg/kg acetic acid was present at 10 °C, that gradually increased to 3537.7 mg/kg at 4 °C and 3705.9 mg/kg at 10 °C after 28 days (Lee et al., 2017). However, they reported that storage at both temperatures caused lactic acid amounts to increase from day 0 to day 28 to 5739.3 mg/kg at 4 °C and 9971.0 mg/kg at 10 °C (Lee et al., 2017). Thus, both lactic acid and acetic acid could play a role in the reduction of HNoV GII.4 titers. Dongchimi (Korean water kimchi) spiked with FCV or MNV after overnight fermentation at room temperature and storage at 4 °C showed increased acidity over 20 days (pH decreased from 5.99 ± 0.06 to 3.49 ± 0.03) and caused reductions of FCV by 4.12 log PFU and MNV by 1.47 log PFU (Lee et al., 2012). They also reported that FCV titers were reduced from 5.69 ± 0.09 log PFU/g to 4.02 ± 0.31 log PFU/g by after 10 days, and to 3.30 ± 0.30 log PFU/g and 1.57 ± 0.51 log PFU/g after 15 and 20 days, respectively (Lee et al., 2012). Similarly, MNV titers were shown to decrease from 5.63 ± 0.05 log PFU/g to 5.09 ± 0.10 log PFU/g after 5 days, and to 4.16 ± 0.28 log PFU/g after 20 days (Lee et al., 2012). Raw *Lactococcus lactis* subsp. *lactis* LM0230 bacterial growth medium (BGMF) and also its bacterial cell suspension (BCS) was shown to cause FCV reductions of 1.3 log $TCID_{50}$ and 1.8 log $TCID_{50}$, respectively after FCV was pretreated with this media (Aboubakr et al., 2014). Co-treatments of CRFK host cells with FCV and pH 7-adjusted BGMF as well also FCV with BCS caused higher FCV titer reductions of 7.5 log $TCID_{50}$ and 6.0 log $TCID_{50}$, respectively (Aboubakr et al., 2014). Raw BGMF (pH 3 to 4) with high concentration of lactic acid could cause reductions of 0.7 to 1.3 log $TCID_{50}$ after 30 min to 24 h pretreatment (Aboubakr et al. 2014). Thus, lactic acid treatments have applications to inactivate foodborne viruses (based on surrogate data) in foods associated with low pH.

Long chain branched poly lactic acid at 0.3% concentration was reported to cause almost 100% inhibition of 8 log CFU/mL of *S. aureus* and *V. parahaemolyticus* (around 80% inhibition) after 3 and 7 day storage, while polylactic acid caused lower inhibition from 45% to 90% (Li et al., 2019). Treatment of wooden surfaces (cutting boards) with lactic acid at 2% for 5 min were shown to cause reductions of 2 log CFU/100 cm^2 of *S.* Typhimurium, 1.91 log CFU/100 cm^2 of *E. coli* O157:H7, 1.96 log CFU/100 cm^2 of *C. jejuni*, 2.38 log CFU/100 cm^2 of *L. monocytogenes* and 2.45 log CFU/100 cm^2 of *S. aureus* (Al-Qadiri et al., 2016). Similar treatments of propylene boards showed somewhat higher reduction of 2.29, 2.23, 2.2, 2.75, and 2.62 log CFU/100 cm^2 for *S.* Typhimurium, *E. coli* O157:H7, *C. jejuni*, *L. monocytogenes*, and *S. aureus*, respectively (Al-Qadiri et al., 2016).

Researchers impregnated bacterial nanocellulose (BNC produced by Komagataeibacter xylinus (*Gluconacetobacter xylinus*)) with lyophilized postbiotics of *Lactobacillus plantarum* (21.2%) for use as an antimicrobial ground meat wrapping nanopaper and showed that *L. monocytogenes* could be decreased by 5 log in ground meat after storage for 9-days at 4 °C (Yordshahi et al. 2020). Furthermore, they showed that the total mesophilic and psychrophile count and TBA values decreased significantly compared to the untreated controls (Yordshahi et al. 2020). Treatment of beef brisket tissue samples inoculated with 6 log CFU/cm^2 of a 5-strain *E. coli* O157:H7 mixture, 12-strain non-O157 Shiga toxin-producing *E. coli* mixture, 6-strain *Salmonella* mixture, or 5-strain nonpathogenic *E. coli* mixture with 1.9% and 2.5% of a commercially available lactic acid-citric acid blend (LCA) was shown to decrease the tested bacterial populations by 0.9 to 1.5 log CFU/cm^2 at temperatures of 43 °C and 60 °C and 15 and 30 lb./in^2 pressure (Scott et al., 2015). These researchers then inoculated 6 log CFU/cm^2 *E. coli* surrogates on beef carcasses in a commercial facility for validation of the use of a hot water treatment (92.2 °C to 92.8 °C, 13 to 15 lb./in^2) followed by 1.9% LCA treatment at 50 °C to 51.7 °C and 13 to 15 lb./in^2 for 10 s that caused 3.7 log CFU/cm^2 reduction of these *E. coli* (Scott et al., 2015).

Raw pork loins (4 cm thick, 200 to 300 g samples) artificially contaminated with 6 log CFU/mL of a cocktail of *E. coli* O157:H7 after treatment with 5% phage solution (PhageGuard) combined with 2.5% lactic acid was reported to cause decreases of 1.89 log on pre-tenderized loin surfaces and > 1.75 log in post-tenderized loin surfaces upon storage at 4 C for 1 h (Li et al., 2023).

Lactic acid at 0.5% used for washing of iceberg lettuce for 2 min was shown to cause 2.7 log CFU/g *E. coli* and 2 log CFU/g of *L. monocytogenes*, while washing with lactic acid at 1% for 5 min caused increased reduction of only 3 log CFU/g of *E. coli* and 2.2 log CFU/g of *L. monocytogenes* (Akbas and Olmez 2007). Others showed that 1% lactic acid washes of iceberg lettuce for 5 min caused 1.45 log CFU/g *E. coli* O157:H7, 1.39 log CFU/g *S.* Typhimurium, and 1.17 log CFU/g *L. monocytogenes* (Sagong et al. 2011; Al-Qadiri et al., 2016).

The MIC of lactic acid against *B. cereus* was shown to be 0.625 mg/mL, while the MIC of phenyllactic acid (PLA, produced by lactic acid bacteria) against *B. cereus* was reported to be 1.25 mg/mL (Ning et al., 2021). This study showed that *B. cereus* treated with 0.625 mg/mL of lactic acid caused 1.9 log reduction similar to the treatment with 1.25 mg/mL PLA, while the combination of PLA at 1/4MIC with LA at 1/2MIC showed decrease in population levels by 1.7 log, indicating a synergistic effect (Ning et al., 2021). Furthermore, the study showed that treatment of *B. cereus* with lactic acid caused down-regulation of periplasmic phosphorus-binding proteins, altered ribosome functions, and caused inhibition of phosphorus transport, TCA cycle, and also purine and pyrimidine metabolism (Ning et al., 2021).

Chicken thighs (skin on) inoculated with *Salmonella* cocktails treated for 15 s by spraying with 5.11% lactic acid and 5.85% buffered lactic acid (on both sides of chicken samples) at 38 °C showed 0.53 log and 0.59 log reductions, respectively, while skin-off chicken thighs treated at 21 °C showed 0.69 log and 0.66 log reductions, respectively compared to the water control (Ramirez-Hernandez et al., 2018). There were no significant differences between the skin-on and skin-off samples among acid treatments and treatment concentrations after spraying the chicken samples at 15.2 cm above and 5.1 cm below and going through a stainless steel mesh conveyor belt at 0.421 L/min flow rate and 138 kPa pressure (Ramirez-Hernandez et al., 2018).

Thus, lactic acid has potential for use as an antimicrobial agent in the food industry.

Acetic Acid (AA) Acetic acid is commonly used in the form of vinegar for preservation. TV titers at 6 log PFU/mL was lowered by 2.6 log PFU/mL upon 5 min treatment with 0.5:0.7% (v/v) acetic acid (AA)-SDS at pH 4.0 without any observable reduction with individual 0.5% AA or combined AA-SDS at pH 7 (Lacombe et al., 2018). Thus, the pH of the solution should be lower than the pKa of the organic acid to have antiviral effects. TV individually treated with AA, LA, tannic acid (TA), gallic acid (GA), *p*-coumaric acid, or SDS did not show any significant reduction after 5 min treatment at room temperature, indicating that inclusion of surfactants including SDS aid in the antiviral effects of these tested acids (Lacombe et al., 2018).

MNV-1 titers were much higher in 0% vinegar-containing (edible seaweed) lavers at 3.6 log PFU/mL than the 3.3 to 3.1 log PFU/mL in 5–15% vinegar-containing lavers during the entire 7 days of storage, where a 1-log reduction occurred in 0% vinegar-containing laver samples after storage for 5 days, while a 1 log reduction occurred after a shorter time of 3 days in 5–15% vinegar-containing laver samples (Park et al., 2016).

In terms of applications in food systems, fresh strawberries had a 1.7 log TCID50/mL reduction of MNV after treatment with 2.5% acetic acid for 2 to 5 min (Ortiz-Solà et al., 2020). Apple cider vinegar at 25% concentration was shown to have antifungal activity with *Candida albicans* being less susceptible, though *B. subtilis, E. coli* and *P. aeruginosa* were found to be susceptible showing complete inhibition (Gopal et al., 2019). Chitosan films (1% w/v) containing ~50% pomegranate or hawthorne vinegar (~50% acetic acid) was reported to cause ~2.5 log CFU/mL reduction of *S. aureus* and *E. coli* from an initial 6 log CFU/mL inoculum after 4 h (Adımcılar et al., 2023).

The application of acetic acid has been also studied for poultry applications. Chicken meat (drumsticks, breast, skins) surface inoculated with 6.91 log CFU/cm^2 of *S.* Typhimurium showed decreases of by 0.17, 1.05, and 2.33 log CFU/cm^2 after 10 min immersion treatment with 0.2%, 0.6%, and 0.8% of plasma activated acetic acid (PAAA), respectively in comparison to decreases of 0.07, 0.93, and 1.23 log CFU/cm^2 by treatments for 10 min with 0.2%, 0.6%, and 0.8% acetic acid, respectively (Kang et al., 2022). The researchers generated PAAA by exposing the plasma discharge (2.2 kHz and 8.4 kVpp) to 20 mL acetic acid (v/v) for 30 min. This study also showed that the pH and 2-thiobarbituric acid reactive substances (TBARS) of PAAA-treated chicken meat decreased significantly compared to the control group, while the lightness (L*) values of treated chicken breasts increased compared to the control, indicating utility in chicken products (Kang et al., 2022).

Others used single postchill treatments of raw chicken wings with 5% buffered vinegar fermentate (Verdad N6) with high CO_2 (60/40% CO_2/N_2), and high O_2 (75/25% O_2/CO_2) modified atmosphere (MA) storage that showed microbial growth inhibition to non-detectable levels during the first seven days of storage with reductions of potential spoilage organisms and lower scores for spoilage associated odor attributes (Heir et al., 2022).

In terms of application to produce for antibacterial purposes, ultrasound (50 kHz) combined with 1000 mg/L acetic acid or 2000 mg/L acetic acid was shown to decrease *S.* Typhimurium population levels on kale by 2.8 log CFU/g and 3.8 log CFU/g, respectively as well as caused 2.6 log CFU/g reduction of aerobic mesophiles, 2.4 log CFU/g of yeasts and molds and 2.6 log CFU/g of coliforms (de Moraes Motta Machado et al., 2022). This study also showed that microbial morphology was altered by these treatments without causing any changes in mass, vitamin C, titratable total acidity, or total phenolic compounds (Moraes Motta

Machado et al., 2022). Treatments of cucumber and watermelon seeds with 100 mg/L acetic and propionic acid vapors at 50 °C and 85% relative humidity (RH) for 2 h caused inactivation of *S. enterica* and *L. monocytogenes* on honeydew melon and watermelon seeds, while the phytopathogen *Acidovorax citrulli* was decreased within 1 h (Shin et al., 2020). This study also showed that *E. coli* O157:H7 was inactivated by the acetic acid vapor at 85% RH, while *A. citrulli* was reduced to non-detectable levels on honeydew melon seeds by the acetic acid vapor (Shin et al., 2020). Furthermore, these treatments did not affect the germination percentages of the seeds.

Treatments of serrano peppers and jalapeno peppers with 0.5% acetic acid for 5 min was shown to result in <2 log CFU of attached *L. monocytogenes*, *S.* Typhimurium, *S. flexneri*, *E. coli* O157:H7 and *Vibrio cholerae* (Rangel-Vargas et al., 2017). Raw enoki mushrooms treated with 5% vinegar at 22 °C for 10 min showed reductions of 1.23 log CFU/g of *L. monocytogenes*, higher than rinsing or dipping in water for 10 min (Chung et al., 2023). *E. coli* O157:H7 population levels were found to be decreased with increasing concentrations of acetic acid in cucumber puree samples stored at 22 °C than at 5 °C, and the antimicrobial effect was shown to decrease with combined 3% salt and acetic acid treatments (Lee et al., 2010b).

Challenge studies on strawberries contaminated with *E. coli* showed that treatment for 5 min at 15 ± 1 °C with 10 mL/L acetic acid solution (Vetec) could decrease the counts on strawberries by 1.26 log CFU/g from 5.38 log CFU/g, and 2 mL/L peracetic acid (Nippon Chemical) decreased counts by 1.08 log CFU/g with further decreases using combination of Ultrasound (40 kHz/110 W) at 15 °C and 2 mL/L peracetic of 2.21 log CFU/g (Donatti Leão Alvarenga et al., 2020). However, this study reported that no increased reduction was obtained by combining acetic acid and ultrasound that caused fragmentation of *E. coli* cells treated with peracetic acid-ultrasound combinations. Though, the combined ultrasound-peroxyacetic acid treatment also showed a reduction of up to 2.48 log CFU/g aerobic mesophiles and reduction of coliforms ranging between 0.89 and 1.45 log CFU/g on strawberries upon storage for 9 days at 6 °C (Donatti Leão Alvarenga et al., 2020). This study also showed that the anthocyanin content stayed constant at 13.47 mg/100 g strawberries. Gaseous acetic acid treatments of 4.7 mmol/L at 55 °C were shown to cause ~5 log reduction of *E. coli* O157:H7 and *Salmonella* Enteritidis after a 3 h treatment of black pepper and fenugreek seeds as well as caused 4 log CFU/g reduction of *B. subtilis* spores on fenugreek seeds and 3.5 log CFU/g reduction on black pepper, after 3 h of treatment (Nei et al., 2015). Similar to citric acid, combination of acetic aids with hurdle approaches may be the path forward to maintain safety of produce and food systems.

Peroxy acetic acid (PAA), a strong oxidizer, is also used as an antimicrobial in the food environment and the food processing industry (Baert et al., 2009; Cano et al., 2021). Chicken wings immersed with100 ppm PAA (pH adjusted to 8.5, 10 or 10.5) for 60 min showed 1.68 and 1.42 log CFU/mL reduction of Salmonella using the commercially available PAA stock solution (SaniDate) and 1.82 and 1.83 log CFU/mL reduction using the on-site PAA generating system (OxyFusion), while

SaniDate caused 1.59 and 1.36 log CFU/mL reduction of Campylobacter and OxyFusion caused reduction of 1.63 and 1.71 log CFU/mL (Vaddu et al., 2021). This study also showed that pH or type of PAA solution did not affect the antimicrobial efficacy and the treatments did not cause any significant color differences between control and treated chicken wings. Peracetic acid solution at 0.5% was reported to cause >4 log CFU/mL reduction of *E. coli* and also against *Pseudomonas aeruginosa* strain, in the absence of any artificial residue substance (Thomas et al., 2020). However, this effect was shown to be decreased for both lactic acid at 5 and 6.67% and 0.5% peracetic acid with reduction of <1.4 log CFU/mL when the artificial residue substance was applied. In the poultry slaughterhouse, 0.5% peracetic acid was shown cause >4 log CFU/mL decrease of the total aerobic bacteria and >3 log CFU/mL decrease in *Enterobacteriaceae* (Thomas et al., 2020).

Treatments of stainless steel [SS], ultra-high molecular weight polyethylene [UHMWPE]) and silicone rubber [SR] with 50–500 μg/mL PAA in combination with UV-C for 5 and 10 min was shown to cause 3.10–6.41 log CFU/cm^2 reduction of *S.* Enteritidis biofilms, while 0.5–2.0% lactic acid with UV-C for 5 and 10 min caused 3.35–6.41 log CFU/cm^2 reduction of *S.* Enteritidis biofilms (Byun et al., 2022). Biofilms of *S.* Enteritidis on chicken skin were decreased by ~2 log CFU/g with minor color and textural changes of chicken using combinations of 500 μg/mL PAA or 2.0% LA combined with UV-C for 10 min (Byun et al., 2022). Thus, PAA at relatively lower concentrations together with other hurdles may be useful in food applications that can preserve the quality attributes of the food commodity.

Gallic Acid (GA) Gallotannins upon hydrolysis by acid, bases or enzymes releases gallic acid that is naturally occurring in many plants (Zhang et al., 2012). Gallic acid (GA) and SDS showed a virucidal concentration of 12.5 mM GA-SDS (0.2:0.3% GA-SDS) at pH 7.0 against TV causing a 0.8-log PFU/mL decrease in titer, while higher concentrations of 50 mM GA-SDS (0.8:1.4% GA-SDS) at similar pH 7.0 of caused increased 1.6 log PFU/mL decrease in TV titer after treatments for 5 min (Lacombe et al., 2018). These researchers reported that gallic acid alone did not result in any significant reduction of TV (Lacombe et al., 2018). However, treatment with 0.1, 0.2, and 0.4 mg/mL gallic acid for 2 h at room temperature was reported to lower FCV-F9 titers at 5 log PFU/mL by 0.86 log PFU/mL to 2.36 log PFU/mL, respectively, and had minimal antiviral effects against at 7 log PFU/mL (high titers) of FCV-F9 (Su and D'Souza 2012). Moreover, these same concentrations of gallic acid and conditions did not result in any detectable decreases in titers of MNV-1 or bacteriophage MS2 even at low titers of 5 log PFU/mL (Su and D'Souza 2012). Other researchers showed that gallic acid (at 1.3 mg/g; ~8 μM), one of the minor polyphenolic components of black raspberry seed extract at 100 μM caused 50–65% inhibition of FCV-F9 and MNV-1 (Lee et al., 2016). Thus, the antimicrobial effects of gallic acid is dependent on the virus type and also its concentration.

The minimum inhibitory concentrations (MIC) of gallic acid was reported to be 2.5 mg/mL for *Pseudomonas putida* DSMZ 291, *P. fragi* DSMZ 3456 and *Pseudomonas* spp. P30-4, while a higher MIC of 5 mg/mL GA was reported for *P. fluorescens* DSMZ 50090, with a minimum bactericidal concentration (MBC) of 10 mg/mL for the tested *Pseudomonas* strains (Sorrentino et al., 2018). These researchers using SEM showed that the cells were injured with formation of pores, absence of flagella and lysis after treatment with 2.5 mg/mL gallic acid for 1 h. Furthermore, storage of fresh black truffles for 28 day at 4 °C showed decrease in population levels of *Pseudomonas spp.* (by ~1.2 log CFU/g), with minimal reduction *of Enterobacteriaceae* and *Eumycetes* that also did not affect the sensory properties after treatment with 2.5 mg/mL GA for 10 min (Sorrentino et al., 2018).

Encapsulated gallic acid at 0.88% in hamburgers was shown to decrease the growth of spoilage microorganisms (yeasts and molds) associated with cell wall damage and alteration of metabolic processes that did not negatively affect hamburger flavor (Asghari-Varzaneh et al., 2024). Gallic acid-grafted chitosan films were shown to possess photoactivated bactericidal effects after irradiation with 400 nm UVA light causing inactivation of 99.9% of *S. aureus* after treatment for 60 min, that showed degradation of the peptidoglycan layer of the cell wall, followed by altered membrane functions and cell death (Zhang et al., 2024). Gallic acid-modified agarose (GaAg) was reported to inhibit the growth of *E. coli* and *S. aureus* by 100% and also inhibited bacterial growth on grass carp (*Ctenopharyngodon idellus*) fillets during refrigeration that extended its shelf-life (Fu et al., 2024). Octyl gallate at 0.075 mM together with 200 mM ascorbic acid activated by 212 mW/cm^2 420 nm blue light was reported to decrease *E. coli* and *S. aureus* populations by >6 log after 5 min with eradication of biofilms due to the formation of hydroxyl radicals (Zheng et al., 2024).

The phenolic extracts of pine bark were shown to decrease growth of *E. sakazakii* and *E. faecalis* by 90% due to the presence of 0.15 mg/mL to 1.25 mg/mL caffeic acid and ferulic acid, while *Listeria* species was shown to be decreased by 98% due to 1.25 mg/mL taxifolin, and *S. aureus* by 92% with 1.25 mg/mL gallic acid after 16 h using a 96-well plate assay (Sánchez-Moya et al., 2024). Dried mango peel extracts that contain high levels of bioactive compounds (gallic acid, caffeic acid, catechin, chlorogenic acid, hesperidin/neohesperidin, and total phenolic content of 25 GAE/g DW), were shown to have antibacterial activity against *E. coli*, *Pseudomonas aeruginosa*, *B. cereus* and *S. aureus* as determined after 12 and 24 h incubation with ethanolic ultrasonic process and Soxhlet process extracts (Kučuk et al., 2024). Methanolic extracts of *Camellia annamensis* leaves and flowers containing 43.72 to 81.89 mg/g gallic acid and 67.31 to 70.68 mg/g (−)-epigallocatechin gallate were shown to have antimicrobial activity against *Enterococcus faecalis* ATCC299212, *S. aureus* ATCC25923, and *S. enterica* ATCC13076. (Linh et al., 2024). Phenolics and gallic acid from natural sources continue to be researched as sustainable antimicrobial approaches that show antimicrobial effects by damaging the cell wall and cell membrane functions. However, combinations with surfactants and other hurdles may have improved or synergistic effects that can be further explored.

Tannic Acid (TA) Tannic acid is a gallic acid derived polyphenol found in high quantities in various fresh fruits, beans, and teas, known for its health benefits including anti-inflammatory, anti-oxidant, neuroprotective, and anticancer and antimicrobial properties (Kaczmarek, 2020). Being FDA approved as a food additive, TA continues to be researched for applications in food systems.

Tannic acid as an antiviral agent against HNoVs has been researched. TA at 0.2 mg/mL was reported to lower high-titer FCV-F9 by 0.98 log PFU/mL but showed increased reduction of FCV-F9 at lower titers by 1.95 log PFU/mL in suspension after treatment for 2 h at RT, but did not cause any noticeable decreases in titers of MNV-1 or bacteriophage MS2 even at low titers of 5 log PFU/mL (Su and D'Souza 2011). Other researchers also similarly demonstrated that TV treated with 50 mM tannic acid (TA) alone for 5 min did not show any significant titer decrease, but TA at 50 mM with 50 mM SDS at pH 7 could reduce TV by 0.9 log upon a treatment time of 5 min (Lacombe et al., 2018). Commercial highly purified tannic acid could inhibit NoV P protein binding to both type A and B saliva with values of IC(50) ≈ 0.1 μM (Zhang et al., 2012). In addition, chinese gall wall rich in tannic acid could also block NoV P dimer binding to type A saliva with a reported IC(50) of 5.35 μg/mL and block binding to type B saliva with an IC(50) of 21.7 μg/mL (Zhang et al., 2012). Pomegranate rich tannic-acid was also shown to cause blocking of NoV P dimer binding to type A saliva with an IC(50) of15.59 μg/mL and to type B saliva with an IC(50) of 66.67 μg/m by these same group of researchers (Zhang et al., 2012).

Pomegranate peel, is typically obtained as a waste product during food processing, though it is known for health effects and broad-spectrum antimicrobial activity associated with its the high content of phenolic compounds and tannins (Živković et al., 2021). Pomegranate peel extracts (PPE, 5 mg/mL) was shown to decrease HuNoV GII from 5 log genomic equivalents (GE)/uL to 2.60 ± 0.04, 2.18 ± 0.05, and 1.62 ± 0.03 log GE after 1, 2 and 6 h at 37 °C, respectively, indicating 99% reduction in PBS (Živković et al., 2021). Furthermore, HuNoV GII capsid damage and disruption was observed by TEM using 5 mg/mL PPE. In addition, this study showed that HuNoV in simulated gastric fluid (SGF) was reduced to 2.7 ± 0.09 log GE after 2 h and in simulated intestinal fluid (SIF) decreased to 3.61 ± 0.09 and 2.24 ± 0.02 log after 2 and 6 h respectively (Živković et al., 2021). Spraying of contaminated cherry tomatoes and green onions with 5 mg/mL pomegranate peel extract was shown to decrease 5 log genomic equivalents (GE)/uL of HuNoV GII to 2.58 ± 0.08 log on cherry tomatoes and 2.49 ± 0.15 log on green onions after 1 h storage (Živković et al., 2021). These researchers also showed that reduction to 1.18 ± 0.06 log on cherry tomatoes and 1.30 ± 0.12 log on green onions were obtained by pre-treating these produce items before contamination with HuNoV.

Tannic acid has also been researched for its antibacterial effects. Lignocellulosic handsheets (60 and 130 g/m^2) impregnated with tannin-rich extracts from industrial bark of Norway spruce (TW), or commercial tannic acid (2–4%) was shown to have antibacterial activity after 50 min at room temperature against bioluminescent *Staphylococcus aureus* RN4220 + pAT19 (using 0.16–2.5 mg/mL TW) and *Escherichia coli* K12 + pCGLS11 (using 1.245 to 2.5 mg/mL TW) as well as the Enterovirus coxsackievirus B3 (Jyske et al., 2023). These researchers showed that tannin-rich extracts obtained from spruce bark side streams with green chemistry can be used to prepare substitute materials for plastic hygienic products, food packaging materials to extend shelf life of foods and in personal protection materials including surgical face masks to prevent the spread of illnesses (Jyske et al., 2023). Hibiscus extract rich in tannins, phenols, and flavonoids at 2.5 mg/mL was shown to inhibit swarming motility of *P. aeruginosa* PAO1 and had antimicrobial effects against yeast and molds (Hamrita et al., 2022).

Tannin-derived components, gallic acid (GA) and tannic acid (TA), was reported to display antibacterial effects against *S.* Typhimurium and inhibited invasion, adherence, and intracellular growth in macrophages (Reyes et al., 2017). Reduced *Salmonella* virulence was observed in mice treated with GA- or TA-treated as determined by their reduced mortality rates (Reyes et al., 2017). Tannic acid was also shown to inhibit *E. coli* biofilm formation through the repression of curli genes, involved in the extracellular matrix and biofilm backbone development. Thus, tannic acid is an alternative sustainable acid that can be used as antimicrobial agents in foods that can be derived from plants.

Levulinic Acid (LVA) Levulinic acid (4-oxo pentanoic acid, pKa: 4.61) is a US FDA Generally Recognized as Safe (21 CFR, 172.515) organic acid for direct addition as a flavoring agent or adjunct to foods and has been researched for its antimicrobial effects (Webb et al., 2013; Wang et al. 2012; Zhou et al. 2020). Initial studies with the organic acid, LVA showed that neither LVA at 0.5% to 5% nor SDS at 0.05% to 2% alone were could decrease MNV-1 or FCV titers, where $\leq$0.51-log reductions were reported (Cannon et al., 2012). However, this research indicated that the combination of 0.5% LVA with 0.5% SDS after 1 min treatments decreased MNV-1 titers and FCV titers by 3 to 4.21 log PFU/mL, and that MNV-1 was decreased by >1.50 log PFU/mL after treatments for 1 min when inoculated onto stainless steel and by >3.3 log PFU/mL after treatments for 5 min with a liquid or foam of 5% LVA with 2% SDS (Cannon et al., 2012). This same study also showed that the reduction or efficacy of the combined treatment did not decrease in the presence of 10% organic matter. Human noroviruses GI.1 and GII.4 were shown to be decreased by 8.97 and 8.13 relative differences in quantification cycle (Cq) values after treatment with 0.5% LVA combined with 0.1% SDS (Afolayan et al., 2016). Others also reported that combinations of 0.5% levulinic acid (LVA) with 0.5% sodium dodecyl sulfate (SDS) wash for 2 min on strawberries showed a 1.40 log decrease in titers of MNV-1, similar to the reductions observed with chlorine washes of 1.5 log reduction suitable for use as produce washes (Zhou et al., 2017).

Contaminated dried strawberries and blackberries (50 g samples) containing 4·4 log PFU HAV showed reduction of 1.7 log PFU after three successive 1-L distilled water washes (90 s each), while formulations of 0.5% w/v SDS alone or in combination with 0.5% v/v lactic acid, 0.5% v/v LVA or 3 ppm aqueous chlorine dioxide (2 washes followed by 1 wash in water) showed reduction of 2·1 to 2·9 log PFU of HAV (Kingsley & Annous, 2021). Washing of strawberries for 2 min with 0.5% LVA plus 0.5% SDS was shown to decrease HAV, MNV-1 and MS2 by 2.7, 1.4 and 2.4-log similar to the reductions obtained by chlorine washes (Zhou et al., 2017).

These generally recognized as safe food additives as stated before demonstrate potential for use in various food industry applications to decrease the risk of transmission of foodborne viral outbreaks.

With regards to antibacterial effects, the minimum inhibitory concentration (MIC) of LVA was reported to be 3.125 mg/mL and for SDS was reported to be 0.039 mg/mL against *S. aureus*, with an additive bacteriostatic effect between the two test compounds with the MIC of the mix being 1.562 and 0.020 mg/mL (Yan et al., 2023). The bactericidal effect was observed where LVA (3.125 mg/mL) and SDS (0.039 mg/mL) resulted in >3-log CFU/mL reduction of *S. aureus* within 6 h, and to below the detection limit (1.0 log CFU/mL) after12 h of at 37 °C, but when used alone had only bacteriostatic effects (Yan et al., 2023). These researchers also showed that SDS and LVA + SDS treatments had antimicrobial effects associated with the expression of cell surface proteins, virulence factors, adhesins, and capsular polysaccharides with SDS being the major antimicrobial (Yan et al. 2023).

Nonwoven fabrics dip-coated with 1% LVA + 0.5% SDS solution demonstrated >6 log reduction of *S. aureus* (from initial 7 log CFU) and >3.6-log reduction of aerosols of *S. aureus* and influenza A virus subtype H1N1 after exposure of 10 min, with coated fabrics being stable for 12 months for use to prevent transmission of pathogens in the food environment (Liu et al., 2022). Washing nitrile gloves used in food environments with LVA concentrations above 1.0% was shown to decrease *Salmonella* contamination with completely inactivation after washing with 0.5% LV/0.1% SDS (Erickson et al., 2018).

Washing of strawberries for 2 min with 0.5% LVA plus 0.5% SDS was shown to decrease *Enterococcus faecium*, *L. monocytogenes* and *Salmonella* by 2 log and *E. coli* O157:H7 and *E. coli* P1 were reduced by 1.9 and 1.8-log CFU/mL, respectively without altering the taste of strawberries (Zhou et al., 2017). A four-strain *Salmonella enterica* cocktail was shown to be reduced by 1.52–1.90 log CFU/fruit on the stem scar of grape tomatoes and by 3.54 log/fruit on the stem scar of round tomatoes after treatment for 2 min with 2% lactic acid +2% acetic acid +2% levulinic acid, 1.7% lactic acid +1.7% acetic acid +1.7% levulinic acid, and 3% lactic acid +3% acetic acid, while only .23–0.30 log CFU/fruit was removed after a water rinse (Fan et al., 2018). Cantaloupe surface tissue contaminated with 100 μL of 9 log CFU/mL of a five-strain mixture of *S.* Poona and treated with 2% LVA plus 0.2% SDS for 6 min followed by an additional sanitizer (2 mL) and brushing (to simulate tumbling over processing line brushes) was shown to cause decrease of *S.* Poona by 1.53 log CFU/g on the stem scar tissue and by 4.47 log CFU/g on the netted rind (Webb et al., 2013).

Ground beef packaged with food absorbent pads containing LVA with SDS was reported to show reduction of 6 log CFU/g of total aerobic counts (TAC), 4.3 CFU/g *Enterobacteriaceae*, 8.6 log CFU/g *Pseudomonas*, and 8.5 log CFU/g psychrotrophs after storage at 4 °C for 1 week showing application as antimicrobial packaging material (Wang et al., 2022).

Biofilms of 5 to 6-strain mixtures (9 log CFU/mL) of *L. monocytogenes*, *S.* Typhimurium, and Shiga toxin-producing *E. coli* (STEC) formed on stainless steel surfaces after 72 h after treatment with 3% LVA + 2% SDS for 10 min showed significant log reduction along with cell detachment and destruction of cellular integrity (Chen et al., 2015). Food slicer surfaces that were contaminated with 8.5 log CFU/blade of *L. monocytogenes, S.* Typhimurium, and *Escherichia coli* O157: H7 after sprayed with 1% LA plus 0.1% SDS as a foam (45–55 psi) for 1 min showed to decrease by 6.0 to 8.0 log CFU/blade (Chen et al., 2014).

Thus, LVA in combination with SDS has broad spectrum of application in the food industry to prevent cross-contamination from fomites, food equipment surfaces as surface sanitizers as well as produce washes and for use in coatings and packaging materials.

5 Salt

Sodium chloride salt: Sodium chloride has been used historically for preservation of food. NaCl at 5% and 10% when added to oysters fermentation at 18 °C for 15 days showed decrease in FCV-F9 titers from initial 6.39 ± 0.1 log PFU/g titer to a lower titer of 5.17 ± 0.27 log PFU/g after 10 days, though after 15 days the titer was decreased to 3.38 ± 0.17 log PFU/g (Seo et al., 2014). This study also showed that infectious MNV titers were lowered after 3 days from initial 6.44 ± 0.04 log PFU/g to 5.76 ± 0.24 log PFU/g and after 15 days was lowered to 4.84 ± 0.31 log PFU/g (Seo et al., 2014).

6 Other Natural Products

Table 1 summarized studies controlling enteric viruses and viral surrogates by select representative biobased/plant compounds.

Blueberries Blueberries (genus *Vaccinium* including *V. corymbosum*) contain mainly anthocyanins (anthocyanidin glycosides), proanthocyanidins (condensed tannins) and other phenolics with known health beneficial properties including antimicrobial effects (Joshi et al., 2019). The cultivable human norovirus surrogate, FCV-F9 when treated with 1 and 2 mg/mL blueberry proanthocyanidins (B-PAC) at 37 °C was reported to show reduction to levels that could not be detected by plaque assays after treatments for 5 min, and similarly after 3 h with commercial blueberry juice (BJ) (Joshi et al., 2016). On the other hand, infectious MNV-1 titers were reported to be decreased to levels that could not detected by the plaque assay after a longer incubation time of 3 h with 1 mg/mL B-PAC, while incubation with BJ even after 6 h did not result in any noticeable decrease in MNV-1 titers (Joshi et al., 2016). This study also demonstrated that infectious HAV could be decreased to levels that were not detected after incubation with 2 mg/mL B-PAC for 30 min, while decreased concentrations of 1 mg/mL B-PAC required incubation for 3 h for HAV titers to be reduced to levels that were non-detectable by the plaque assay (Joshi et al., 2016;

Table 1 Control of enteric viruses and viral surrogates by select representative biobased/plant compounds

Type of antiviral	Treatments	Virus	Reduction	References
American cranberry extracts (*Vaccinium macrocarpum*)	0.6 mg/mL C-PAC at RT for 1 h	FCV-F9 MNV-1 MS2	≥5 log PFU (non-detectable) ~3 log PFU ~1 log PFU	Su, Howell, and D'Souza (2010a, b)
Cranberry juice (*Vaccinium macrocarpum*)	50% juice for 30 min	Simian rotavirus Reovirus	Non-detectable from 9 log PFU/mL	Lipson et al. (2007a, b)
Blueberry (*Vaccinium* species)	Juice, 1, 2, 5 mg/mL B-PAC for up to 24 h at 37 C	FCV-F9 MNV-1 HAV AiV	Non-detectable Non-detectable with 1 mg/mL BPAC Non-detectable with 1 mg/mL BPAC ≥ 3 log PFU/mL	Joshi et al. (2016, 2019)
Chitosan	0.7% 53 kDa	FCV-F9 MS2 MNV-1	~3 log PFU/mL 1.4 log PFU/mL Not reduced	Su et al. (2009)
Clove (*Syzygium aromaticum*)	29.5% eugenol	FCV	6 log TCID50/0.1 mL	Aboubakr et al. (2016)
Curcumin (*Curcuma longa*)	2 mg/mL at 4 °C for 3 days	MNV-1	90.43%	Yang et al. (2016)
Black raspberry (BR; *Rubus coreanus*)	Juice	FCV-F9 MNV-1	Reduction after pretreatment or co-treatment of host cells	Oh et al. (2012)
BR seed exract	1 mg/mL	FCV-F9 MNV-1	Co-treatment of host cells: 84% inhibition and 87% inhibition by plaque assays	Lee et al. (2016)
Grapeseed extract (GSE; *Vitis vinifera*)	0.2 mg/mL GSE at 37 °C for 1 h	MNV-1	3 log PFU/mL	Li et al. (2012a, b)
GSE	1 mg/mL GSE at 37 °C for 2 h	FCV-F9	4.61 log PFU/mL	Su and D'Souza (2011)
		HAV	3.2 log PFU/mL	
		MNV-1	1.73 log PFU/mL	
		AiV	>4 log PFU/mL	Joshi and D'Souza (2021)
Ginger (*Zinziber officinale*)	10.7% R-(−)-1, 2, -propanediol for 24 h	FCV-F9	2.7 log TCID50	Aboubakr et al. (2016)
Green tea extract (GTE)	1 to 3 mg/mL aged GTE at 37 °C	HNoVGII.4 (HIE systems)	Non-detectable	Randazzo et al. (2020)
Analytical grade PAC	1 mg/mL and 10 mg/mL PAC for 10 s	FCV-F9, coxsackie virus A7, reovirus	>2 log	Iwasawa et al. (2009)

(continued)

Table 1 (continued)

Type of antiviral	Treatments	Virus	Reduction	References
Oregano oil (*Origanum vulgare*)	4% v/v for 15 min	MNV-1	1 log	Gilling et al. (2014a)
Carvacrol	0.5% for 1 h 1% for 2 h at 37 °C	MNV-1 HAV	3.87 log TCID50/mL 1 log	Gilling et al. (2014a) and Sánchez et al. (2015)
Thymol	1%	FCV MNV-1 HAV	Non-detectable 1.66 log TCID50/mL 1 log	Sánchez and Aznar (2015)
Pomegranate (*Punica granatum*)	32 mg/mL pomegranate polyphenol at RT for 1 h	FCV-F9 MNV-1 MS2	≥5 log PFU (non-detectable) ~3.6 log PFU ~1 log PFU	Su, Sangster, and D'Souza (2010); Su et al. (2011)

As adapted from D'Souza (2014)

Joshi et al., 2017). However, HAV titers were found to be lowered by ~2 log PFU/mL after incubation with BJ at 37 °C for 24-h. Earlier studies also showed that FCV-F9 titers were lowered to non-detectable levels in blueberry juice (BJ) after 24 storage at 4 °C, while MNV-1 titers showed no noticeable reduction even after storage at refrigerated temperatures of 4 °C after 14 days (Horm et al., 2012). The infectious titers of the foodborne Aichi virus (AiV) were found to be lowered by 2.13 log PFU/mL after treatment with 1 mg/mL B-PAC for 24 h and further decreased by ≥3 log PFU/mL (at levels that could not show plaque detection by the plaque assay) after treatment for 24 and 3 h with higher concentrations of 2 mg/mL and 5 mg/mL B-PAC, respectively (Joshi et al., 2019). Moreover, AiV titers were lowered by 0.17 log PFU/mL, 1.27 log PFU/mL, and 1.73 log PFU/mL after treatments with BJ for 1, 3, and 6 h, with ≥3 log PFU/mL reduction after treatment for 24 h at 37 °C (Joshi et al., 2019). Hepatitis C virus (though not a foodborne enteric virus) could also be inhibited by methanolic extracts of blueberry leaves (epicatechin) (Takeshita et al., 2009). Thus, blueberry bio-actives also show promise to help control human enteric viral spread. However, the mode of action needs to be further elucidated

Cranberries The beneficial components of cranberries (*Vaccinium macrocarpon*) include various phenolics along with considerable levels of condensed non-hydrolyzable tannins or proanthocyanidins that have anti-inflammatory and antimicrobial properties (Nogueira et al. 2003; Iwasawa et al., 2009). Cranberry proanthocyanidins (CPAC) contain mainly epicatechin tetramers and pentamers with a minimum of one A-type linkage averaging in content of 410 mg/100 g fresh cranberry (Côté et al., 2010; Iwasawa et al., 2009). Cranberry juice (CJ) available in grocery stores and CPAC at 0.15 mg/mL, 0.3 mg/mL, and 0.6 mg/mL after 30 min and 1 h treatments was reported to cause decreases of 5 log PFU/mL FCV-F9 to levels that were not detected by the plaque assay, while

MNV-1 titer levels were lowered/diminished by 2.06 log PFU/mL with CJ, and by 2.63 log PFU/mL, 2.73 log PFU/mL and 2.95 log PFU/mL after 1 h treatment with 0.15 mg/mL, 0.3 mg/mL and 0.6 mg/mL CPAC, respectively at RT (Su et al., 2010a). This study also showed that CJ reduced bacteriophage MS2 levels by 1.14 log PFU/mL, and 0.15 mg/mL C-PAC, 0.30 mg/mL C-PAC, and 0.60 mg/mL C-PAC lowered bacteriophage MS2 titers by 0.55 log PFU/mL, 0.80 log PFU/mL, and 0.96 log PFU/mL, respectively, after treatments for 1 h at RT. Another study reported that commercial analytical grade proanthocyanidin at 1 mg/mL concentrations for 10 s treatment times could also cause FCV-F9 titer reductions to levels that could not be detected, while for coxsackievirus A7 strain (Cox.A7) titer reductions of 2 logs or more could be obtained with 10 mg/mL CPAC after merely 10 s treatment times (Iwasawa et al., 2009).

CJ at 50%, 30%, and 10% concentrations could also lower initial titers of 9 log PFU/mL of bacteriophages T2 and T4 titers as well as simian rotavirus SA-11 titers to non-detectable levels after incubation at 37 °C, 23 °C, and 4 °C for 30 min treatment times (Lipson et al., 2007a). MA-104 host cells after pretreatment with cranberry cocktail extracts, 5% NutriCran-90 TM, NC 90% and 5% NutriCran-100 TM, NC 100 extracts were shown to decrease infectious reovirus to ~50% compared to untreated controls (Lipson et al., 2007b). Polio virus type 1 treatments with a cranberry drink at its natural pH of 2.6 as well as with the cranberry drink at pH 7.0 showed 21% survival and less than 1% survival, respectively (Konowalchuk & Speirs, 1978). Thus, these results indicate that cranberry bioactives show promise for use to control the spread of human enteric viruses and also human viral gastroenteritis.

Grape Seed Extract (GSE) Grape Seed Extract (GSE) is a value-added byproduct of the grape (*Vitis*) juice and wine industry that contains high levels of various polyphenols with antimicrobial properties (Jayaprakasha et al., 2003). The infectivity of the human norovirus surrogate, FCV-F9 was reported to be lowered by 3.64 log PFU/mL, 4.10 log PFU/mL, and 4.61 log PFU/mL after 2 h treatments with 0.25 mg/mL GSE, 0.5 mg/mL GSE, and 1 mg/mL GSE at 37 °C, respectively (Su & D'Souza, 2011). This study also demonstrated that MNV-1 titers upon 2 h treatments with 0.25 mg/mL GSE, 0.5 mg/mL GSE and 1 mg/mL GSE could be lowered by 0.82 log PFU/mL to 1.73 log PFU/mL, while infectious HAV titers could be decreased by 1.81 log PFU/mL, 2.66 log PFU/mL, and 3.20 log PFU/mL, respectively (Su & D'Souza, 2011). AiV titers of ~4.75 log PFU/mL were also shown to be lowered to non-detectable levels after 30 s treatments with 0.5 mg/mL GSE treatments at both 37 °C and RT in suspension (Joshi & D'Souza 2021).

When used as a produce wash, 0.25 mg/mL GSE, 0.50 mg/mL GSE, and 1 mg/mL GSE after 1 min treatment was reported to decrease FCV-F9 at ~7 log PFU/mL inoculated on lettuce by 2.33, 2.58, and 2.71 log PFU while on peppers decreases of 2.20 log PFU, 2.74 log PFU, and 3.05 log PFU were reported, respectively (Su & D'Souza, 2013a). Similar GSE treatments of 5 log PFU/mL MNV-1 on lettuce resulted in titer decreases of 0.2–0.3 log PFU and titer decreases of 0.8 log PFU were obtained on peppers after 1 min treatment (Su & D'Souza, 2013a). HAV titers were reported to be decreased to a lower extent than FCV on the tested produce samples, where 0.71 to 1.1 log PFU and 1 to 1.3 log PFU reductions after 0.25 mg/mL GSE to 1 mg/mL GSE wash treatments of lettuce and peppers were reported (Su & D'Souza, 2013a). Notably, these treatments did not show any differences in color or appearance when compared to the untreated control produce as determined by instrumental color analysis.

When 1 mg/mL GSE was used against AiV as a surface wash, AiV titers were decreased to titers that could not be detected under clean conditions after 30 s treatments; however, under soiled conditions titers were lowered by only 1.13 log PFU/mL and 1.71 log PFU/mL after exposure to 2 mg/mL GSE and 4 mg/mL GSE treatments, respectively (Joshi & D'Souza 2021). Similarly, in the presence of 0.02% and 0.2% dried milk as well as lettuce extract, the antiviral efficacy of GSE decreased requiring at least 2 mg/mL GSE for inactivation of 1.5 log PFU/mL to around 2 log PFU/mL of MNV-1 titers compared to that observed without the presence of milk or lettuce extracts (Li et al., 2012a). Additionally, 0.2 mg/mL GSE reduced MNV-1 titers by >3 log PFU/mL after 1 h at 37 °C, as well as diminished the binding of HNoV GII.4 to Caco-2 cells (>1 log nucleic acid genomic copies/mL), and blocked binding of noroVirus-like P particles (VLPs) to human salivary histo-blood group antigen receptors (Li et al., 2012b). Overall, GSE may be a sustainable approach for produce washes but the organoleptic attributes in food systems after treatments need to be evaluated as indicated earlier (D'Souza, 2014).

Pomegranates Pomegranate (*Punica granatum*) fruit extract also contains flavonoids as well as hydrolyzable tannins (Su, Sangster, & D'Souza, 2010; Howell and D'Souza, 2013; Lee et al., 2010a). Commercial pomegranate juice was shown to decrease MNV-1, FCV-F9, and bacteriophage MS2 titers at ~5 log PFU/mL by 0.32 log PFU/mL, 2.56 log PFU/mL and1.32 log PFU/mL, respectively after treatments for 1 h at RT (Su, Sangster, & D'Souza, 2010). FCV-F9 titers at 5 log PFU/mL could be decreased to non-detectable levels, while MNV-1 titers could be decreased/lowered by 1.30 log PFU/mL to 3.61 log PFU/mL after 1 h treatments with 8 to 32 mg/mL concentrations of commercially available pomegranate polyphenols (PP) at RT, respectively (Su, Sangster, & D'Souza, 2010). On the other hand, bacteriophage MS2 titers of 5 log PFU/mL could be decreased by only 0.41 log PFU/mL to 0.93 log PFU/mL after 1 h treatments with 4 mg/mL PP to 16 mg/mL PP, respectively (Su, Sangster, & D'Souza, 2010).

Punicalagin showed antiviral activity against human enterovirus 71 (EV71) infection using human rhadbomyosarcoma (RD) cells after 2 h with an inhibitory concentration (corresponding to an IC50 value) of 15 μg/mL (Yang et al., 2016). Furthermore, this research study demonstrated that punicalagin treatment of mice reduced the mortality of mice that were challenged by inoculation of a lethal dose of EV71 (Yang et al., 2016). Reports also show that poliovirus at 3 log PFU/0.05 mL was reduced to <1% after storage at 4 °C for 24 h in PJ (Konowalchuk & Speirs, 1978; Su et al., 2011). Thus, pomegranate polyphenols are also alternate naturally occurring compounds that can potentially help alleviate disease symptoms or decrease foodborne viral disease transmission risk. Further studies on the application of pomegranate polyphenols to elucidate the mode of action and against other foodborne viruses are needed.

Black Raspberry* (Rubus coreanus*) Black raspberry (BR) also contains high levels of polyphenols (gallic acid and quercetin), and their seeds are value-added products of the juice industry (Oh et al., 2012; Lee et al., 2016). The juice of black raspberry (BRB) was shown to cause reduction of infectious FCV-F9 and MNV-1 by the plaque assays and blocked attachment to the host cell receptors, while gallic acid or quercetin alone individually did not result in observable decreases in infectivity of these two viruses (Oh et al., 2012). Black raspberry seed extract (RCS) and its component of molecular weight less than 1 kDa (RCS-F1) at 0.1 to 1 mg/mL showed highest inhibition of MNV-1 and FCV-F9 when added simultaneously to host cells along with MNV-1 (87% inhibition with 1 mg/mL) or FCV-F9 (84% inhibition with 1 mg/mL) (Lee et al., 2016). Furthermore, treatment with the RCS-F1 component revealed enlarged capsids (from 35 to 100 nm) and caused their disruption as determined by visualization using Transmission electron microscopy (TEM) (Lee et al., 2016). These researchers identified two bioactive antiviral polyphenolic compounds that were derived from RCS-F1, namely cyanidin-3-glucoside (C3G) and gallic acid, using liquid chromatography-tandem mass spectrometry (Lee et al., 2016). Thus, this value added component from BRB seeds also shows promise as an antiviral agent that destroys the viral capsid and structure, though further research is needed on its effects against HAV, AiV and other foodborne viruses

7 Other Plant Flavonoids, Tannins, and Catechins Against Human Enteric Viruses

Flavonoids are also found in citrus fruits that include myricetin, naringenin, L-epicatechin, and tangeretin among others (D'Souza, 2014; Cowan, 1999). Treatments of FCV-F9 with 0.5 mM myricetin and 1 mM myricetin caused reductions to non-detectable levels, while 0.5 mM L-epicatechin caused 1.4 log PFU/mL reduction after 2 h at 37 °C (Su & D'Souza, 2013b). Higher titers of ~7 log PFU/mL FCV-F9 were reportedly lowered by 3.17 log PFU/mL and 0.72 log PFU/mL upon treatment with 1 mM with myricetin and 1 mM L-epicatechin in this same study, while no observable lowering of MNV-1 titers were reported upon treatment with these tested compounds (Su & D'Souza. 2013b). Further studies to elucidate their mode of action against enteric viruses are needed.

Persimmon (*Diospyros kaki*) tannins were reported to decrease FCV, MNV, poliovirus, coxsackie virus, adenovirus and rotavirus, by greater than 4 logs (Ueda et al., 2013). Persimmon tannin solutions at ≥0.11 mg/mL could lower HNoV GII. 4 genome copy numbers by >70% and bacteriophage MS2 titers by greater than 2.5 log PFU/mL (Kamimoto et al., 2014).

Theaflavins are polyphenols contained in black tea (Ohba et al. 2017). Crude theaflavin extracts from U. S. black tea inhibited 50% infection (EC_{50}) of bovine coronavirus (ATCC P2 strain) at 34.7 ± 9 µg/mL as examined by tissue culture infectious dose (TCID50) assays, with EC50 values of 0.353 µg/mL for the NCDV-Lincoln bovine rotavirus strain as determined in immunofluorescent focus-forming units (IFFU) (Clark et al., 1998). Reduction of FCV titers by theaflavins (TF2A, theaflavin-3-O-gallate) was reported to be concentration and time-dependent (Ohba et al., 2017). FCV titers were decreased from 6.2 ± 0.4 log TCID50/mL to 3.9 ± 0.3 log TCID50/mL after treatment with 225 µM of a mixture of theaflavin monogallates, TF2A and TF2B for 60 min (Ohba et al., 2017). These researchers also showed that FCV-F9 titers from 5.3 log TCID50/mL were reduced to 3.8, 3.8 and 3.2 log TCID50/mL after 5, 30 and 60 min treatments with 1.6 mg/mL theaflavins (2.25 mM), respectively. A dose dependent reduction of 6.2 log TCID50/mL FCV to 5.5, 4.5 and 3.2 log TCID50/mL after treatments with 0.016, 0.16 and 1.6 mg/mL theaflavin for 60 min was obtained, respectively (Ohba et al., 2017). The hydroxyl groups of the benzocycloheptenone ring of the theaflavins were associated with antiviral activity before the viral entry step (Ohba et al., 2017). Tannins that are derived from green tea, acacia and gallnuts showed antiviral activity against some of the tested viruses causing protein aggregation that decreased viral infectivity, however this study showed that coffee extracts did not decrease infectivity of any of the tested viruses (Ueda et al., 2013).

Green Tea Extract (GTE) (*Camellia sinensis* L.): from evergreen plants are rich in polyphenols as well. They have been associated with the inhibition of HAV, FCV-F9 and MNV-1 replication (Randazzo et al., 2020). HIE monolayers showed total inhibition of HNoV GII.4 replication after pre-treatment with 1.0 mg/mL aged GTE at 37 °C, 1.75 mg/mL aged GTE at 21 °C, and 2.5 mg/mL aged GTE for 7 °C (Randazzo et al., 2020). However, decreased infectivity of Tulane virus of 0.85 log $TCID_{50}$/mL, 0.75 log $TCID_{50}$/mL, and 0.65 log $TCID_{50}$/mL was obtained only after treatment with the higher concentration of 2.5 mg/mL aged-GTE at the three tested temperatures of 37 °C, 21 °C, and 7 °C, respectively (Randazzo et al., 2020). Others showed that controlled storage conditions at 25 °C for 24 h caused improved antiviral activity of GTE potentially linked to the increased catechin derivative accumulation (Falcó et al., 2019). Treatments with aged-GTE combined with heating at 50 °C for 30 min was reported to lower MNV-1 by more than 5 log PFU in PBS, while addition of aged GTE with heat at 40 °C for 30 min decreased HNoV binding to PGM to 4.7% compared to reduced binding of 41.5% by heating at 40 °C for 30 min alone (Falcó et al., 2020). Aged-GTE caused MNV-1 in juices heated at 50 °C for 30 min to be decreased by >4 log without showing any synergistic effect against HAV (Falcó et al., 2020). Extracts of green tea and green tea infusions could play major roles in decreasing the risk of enteric viral illness.

Hibiscus sabdariffa extracts (HE) have been gaining popularity for use in beverages and are rich in polyphenols. FCV-F9 titers at 5 log PFU/mL were decreased to non-detectable levels after aqueous HE treatments of 40 and 100 mg/mL for 15 min, while MNV-1 titers were decreased to levels that could not be detected after 24 h, and by 1.77 log PFU/mL and 1.88 log PFU/mL after treatments for 6 h with 40 mg/mL HE and 100 mg/mL HE, respectively (Joshi et al., 2015). This study also showed that HAV titers could be lowered to non-detectable levels by both HE concentrations after longer incubation times of 24 h. In addition, these researchers showed that protocatechuic acid (PCA) at 3 mg/mL lowered FCV-F9 titers to non-detectable levels, MNV-1 titers by 0.53 log PFU/mL and with no observable reduction of HAV titers after treatment for 6 h with 3 mg/mL protocatechuic acid (PCA), while ferulic acid could also decrease FCV-F9 titers to non-detectable levels after 3 h and MNV-1 and HAV titers could be decreased similarly after 24 h treatment (Joshi et al., 2015). In addition to green tea infusions, hibiscus extract infusions also show promise to alleviate symptoms and decrease the risk of enteric viral illness. Their mode of action needs to be determined as these sustainable approaches continue to be researched for their application in the food environment and in food systems.

Houttuynia cordata is a Southeast Asian herbal medicinal plant rich in polyphenols, flavonoids, sterols, fatty acids, and polysaccharides whose roots and leaves are used in beverages (Cheng et al., 2019). Treatments of FCV-F9 with a ~43 kDa, water-soluble pectin-like acidic polysaccharide extract of *H. cordata* polysaccharide (HP) at 500 μg/mL decreased MNV-1 titers at 8 log PFU/mL by 3.48 log PFU/mL and also MNV-1 at initial titers of 4.38 log PFU/mL were lowered to non-detectable levels (Cheng et al., 2019). Its water extract (HWE) at 250 μg/mL could decrease MNV-1 at 4.38 log PFU/mL to non-detectable levels, and could decrease MNV-1 at 8.09 log PFU/mL by 4.12 log PFU/mL (Cheng et al., 2019). Furthermore, this report also demonstrated that the spherical shaped 30 to 35 nM in size MNV-1 increased to 80 to 100 nM in size to after treatment with 250 μg/mL HP, along with denatured and disrupted particles (correlating well with the 2.75 log PFU reduction). Thus, the antiviral effects are based on destruction of the viral capsid structure. Overall this medicinal plant shows activity against the two tested cultivable HNoV surrogates. Further studies on its application against HAV, AiV and other enteric viruses are needed.

Artemisia princeps* var. *orientalis is an edible, aromatic, medicinal plant associated with temperate climate regions that contains borneol (12.1%, oxidized to camphor with anti-influenza properties), α-thujone (8.7%, present in essential oils), T-cadinol (6.7%), and 1, 8-cineole (6.2%) (Chung, 2017). FCV-F9 titers at 6 log PFU/mL were reported to be reduced moderately by 48% with 0.1% (v/v) of the above essential oils at RT after incubation times of 1 h, while 200 μM 2-Thiouridine, 2TU, a positive control that inhibits RNA-dependent RNA polymerase caused 25% reduction (Chung, 2017). This study showed that cotreatment with 0.1% (v/v) *A. orientalis* essential oil caused only 17% reduction of FCV-F9, while posttreatment resulted in 41% inhibition (Chung, 2017). Furthermore, pre-treatment of MNV-1 with 0.01% (v/v) essential oil caused 64% inhibition, while 50 μM 2TU showed 24% reduction, where cotreatment caused 53% inhibition, and with no observable reduction after post-treatment. Only 25 mM α-thujone when used individually could decrease FCV-F9 titers by 0.94 log PFU and MNV-1 titers by 0.50 log PFU after incubation at RT for 1 h treatment times (Chung, 2017). The mode of action could be postulated to be due to blocking of host cell receptors that prevent initial attachment and entry. However, the effects on the virus structure and nucleic acid need to be further determined and understood. Additional studies with these essential oils against HNoV, HAV, AiV and other enteric viruses are also needed.

Korean Ginseng (KRG) (*Panax ginseng*) widely utilized in oriental medicine at 5 μg/mL, 6.7 μg/mL, and 10 μg/mL of total saponin was shown to lower FCV-F9 infectious titers from 5.2 ± 0.3 log $TCID_{50}$/mL to 0.13 ± 0.51 log $TCID_{50}$/mL, 0.42 ± 0.42 log $TCID_{50}$/mL, and 0.83 ± 0.36 log $TCID_{50}$/mL, respectively after pretreatment of host cells, that also reduced 6.7 ± 0.2 log $TCID_{50}$/mL MNV-1 titers to 0.38 ± 0.41, 0.73 ± 0.19 and 1.48 ± 0.27 log $TCID_{50}$/mL, respectively ($p < 0.01$) (Lee et al., 2011). While HAV titers were reported to be reduced after both, pre-treatment and co-treatment causing 0.23 to 0.57 log PFU/mL reductions using up to 10 μg/mL red ginseng, and 0.45 to 0.66 log PFU/mL reductions were obtained with purified ginsenoside (Rb1 and Rg1) (Lee et al., 2013). Studies with cultivable HNoV genotypes will be useful to understand their role in decreasing infectious HNoV titers, once the HIE model cultivation systems are more easily available for all HNoV genogroups and routinely used in all labs.

Quillaja saponaria **(QS)**: extracts are rich in polyphenols, saponins and tannins, saponins, that are extracted from the Chilean soap-bark tree (D'Souza, 2014). These QS extracts at 0.015 mg/mouse reduced diarrhea induced by infection with rhesus rotavirus (RRV) from 79% to 11% after challenging mice with 500 PFU of RV for 5 consecutive days, that was reported to inhibit attachment to host cells and disrupt membrane proteins (Tam & Roner, 2011). FCV-F9 at titers of ~5 log PFU/mL after a 15 min treatment at 37 °C and RT with commercially available aqueous *Quillaja saponaria* extract (1:1 v/v QS; Desert King, USA) was shown to be reduced to undetectable levels, but 1:5 diluted QE caused FCV-F9 titer reductions only at 37 °C compared to the untreated controls (Joshi et al., 2023). Thus, QS effects against FCV-F9 were both concentration and temperature dependent, where 1:5 diluted aqueous QS at RT was less effective that lowered FCV-F9 titers by 2.05, 2.14, and 2.74 log PFU/mL after treatments for 30 min, 1 h and 2 h, respectively (Joshi et al., 2023). Moreover, this study showed that MNV-1 infectious titers were decreased minimally by <1 log PFU/mL after treatment with 1:1 or 1:5 aqueous QE for 15 min to 2 h, without any noticeable reduction at RT (Su & D'Souza 2012). Thus, QS shows antiviral activity against these viruses and further studies on their mechanisms of action and against other enteric viruses are warranted.

8 Other Spices

Curcumin Curcumin is a polyphenol obtained from turmeric spice (*Curcuma longa* L.) rhizomes (Yang et al., 2016). Curcumin (CCM) at 0.25 mg/mL, 0.5 mg/mL, 0.75 mg/mL, 1 mg/mL, and 2 mg/mL after treatment at 4 °C for 3 days neutralized 6 log PFU/mL MNV-1 titers by 52.85%, 68%, 68.96%, 89.99%, and 90.4%, respectively (Yang et al., 2016). In this same study, human norovirus, HNoV

(Norwalk) replicon-bearing HG23 cells were also treated with 2 μg/mL CCM for 72 h, that did not result in any inhibition of replication as determined by real-time RT-PCR of the RNA extracts (Yang et al., 2016). On the other hand, blue LED (dose of 3 J/cm^2) with 50 μg/mL CCM could cause 5 log $TCID_{50}$/mL reduction of FCV at 37 °C after treatments for 30 min, but caused only 0.73 log $TCID_{50}$/mL reduction of MNV-1 (Randazzo et al., 2016, 2017). Blue LED light (470 nM; 0.06 W/cm^2 for 60 s) at 3 J/cm^2 with 5 μM CCM was shown to lower MNV-1 by 1.32-log PFU/mL with increased reduction to >3-log PFU/mL using 20 μM curcumin in buffer (Wu et al. 2015). Curcumin-mediated photodynamic inactivation (PDI) was then used to cause 0.76-log PFU/mL and 1.15-log PFU/mL reduction of MNV-1 with 10 and 20 μM curcumin in oysters, without any noticeable color changes after bioaccumulation (Wu et al. 2015). TEM images demonstrated that viral structural integrity was destroyed by CCM-mediated PDI, and RNA genome degradation was demonstrated using RT-PCR in this study (Wu et al. 2015). Furthermore, 0.015 mg/mL CCM decreased the D-values (time to cause a log reduction of virus) of Tulane virus (TV, the most recently cultivable HNoV surrogate as reported before) from 4.55 ± 0.28 min to 1.08 ± 0.16 min at 52–68 °C in PBS to 4.32 ± 0.25 min to 0.62 ± 0.17 min with CCM (Patwardhan et al., 2020). This study also showed that D-values for HAV that ranged from to 9.21 ± 0.24 min to 0.67 ± 0.19 min at 52–68 ° C in PBS were lowered to 8.12 ± 0.18 to 0.55 ± 0.08 min with 0.015 mg/mL CCM (Patwardhan et al., 2020). Thus, curcumin shows potential for application in the food industry to decrease risk of viral disease transmission when used in hurdle approaches such as in combination with heat or light.

Clove (*Syzygium aromaticum*) FCV titers at ~7 log TCID 50/0.1 mL were significantly decreased by 6 log TCID50/0.1 mL after treatment for 24 h at 4 °C with 3.3% clove (containing 29.5% eugenol) (Aboubakr et al., 2016). In addition, pretreatment of FCV for 24 h with concentrated eugenol showed 3.78 log TCID50/0.1 mL reduction, while cotreatment of host cells with FCV and eugenol caused reduction of 3.34 log TCID50/0.1 mL, with a lowered titer of 2.56 log TCID50/0.1 mL after post-infection treatment (Aboubakr et al., 2016). Other researchers reported that FCV inoculated onto baby-leaf salads could still be detected during refrigerated storage (7 °C) for 9 days even after spraying with 0.8 mL emulsions of essential oils (EOs) from clove (*Eugenia caryophyllus*), zataria (*Zataria multiflora* Boiss) and oregano (*Origanum compactum*) (Azizkhani et al., 2013). *Zataria multiflora* Boiss. (zataria) at 0.1% essential oil was shown to lower HAV titers by <0.5 log TCID50/mL after incubation for 2 h at 37 °C (Sánchez & Aznar, 2015). Thus, only concentrated eugenol shows promise against the tested enteric viruses and their mode of action needs to be further explored.

Ginger (Zingiber officinale) Ginger is commonly used in a variety of foods and ginger extracts (10.7% R-(−)-1,2-propanediol) at 20% were shown to cause 2.7 log TCID50 reduction of FCV-F9 after treatment for 24 h (Aboubakr et al., 2016). Lowered D-values (52–58 °C) for TV with 0.1 mg/mL gingerol from 4.09 ± 0.18 to 0.72 ± 0.09 min and similarly for HAV reduction in time needed to cause 1 log reduction (D-value) were reported (Patwardhan et al., 2020). These studies indicate the ability of ginger extracts and its components to decrease infectivity of enteric viruses and their surrogates. However, the mode of action against foodborne viruses has not been clearly elucidated.

Oregano oil (***Origanum vulgare***, 4% v/v; an essential oil): and its bioactive component carvacrol at 0.25% and 0.5% were reported to decrease MNV (strain S7-PP3) titers by 1 log TCID50/mL after 15 min and showed detection of 1.07 log TCID50/mL even after 24 h incubation, while carvacrol caused reductions of 3·87 log after 1 h (Gilling et al., 2014a). In this study, the RNase I protection assay and TEM analysis implied that the virus capsid was damaged (increased in size from ≤35 to 75 nM in diameter with 800 nm carvacrol) followed by damage to the RNA (Gilling et al., 2014a). In another study, *Origanum vulgare* (oregano)'s active component thymol at 0.5% and 1% was shown to decrease FCV infectious titers to levels that could not be detected, while 1% and 2% thymol caused 1.66 and 2.45 log $TCID_{50}$/mL reductions of MNV-1 titers, respectively, without any noticeable effect on HAV titers after 2 h at 37 °C (Sánchez & Aznar, 2015). Carvacrol at 0.5% was shown to decrease 6 to 7 log TCID50/mL of FCV and MNV infectious titers to levels that were non-detectable after 2 h at 37 °C, however, carvacrol at 1% could cause only 1 log reduction of HAV after 2 h at 37 °C (Sánchez et al., 2015). When carvacrol was used in lettuce wash water, FCV was reduced by 4 log which was independent of the chemical oxygen demand (COD), but MNV did not show any reduction at COD of >300 ppm. However, this study showed that carvacrol at 1% after 30 min treatment decreased FCV and MNV on spiked lettuce by 1 log (Sánchez et al., 2015). These essential oils show promise for use in hurdle approaches to decrease enteric viral loads that act by destroying the virus capsid and nucleic acid material.

Furthermore, allspice oil, lemongrass oil, and citral were reported to cause reduction of 0.9 to 1.8 log MNV titers after 6 h of exposure, with increased reduction of 3.41, 3.0 and 2.74 log reductions after 24 h treatment, respectively (Gilling et al., 2014b). This study indicated that allspice oil effects against MNV titers were time and also concentration dependent; while lemongrass oil and citral were only time dependent (Gilling et al., 2014b). This study also showed that allspice oil acted directly on the capsid of the virus and also the RNA as determined by RNaseI protection assays and TEM analysis where the capsids were found to be enlarged to 75 nM after treatment (Gilling et al., 2014b). However, this study showed that lemongrass oil and citral did not have any effect on the viral capsid but only caused coating of the viral particle to prevent binding to host cells. These three tested compounds have the potential ability to be used as surface sanitizers to decrease HNoV spread.

Chitosan Chitosan is a biodegradable and non-toxic carbohydrate polymer found in fungi and exoskeletons of crustaceans (Su et al., 2009; Devlieghere et al., 2004; Li et al. 2008), with approval as a biopesticide on horticultural crops by the U.-S. Environmental Protection Agency (US EPA) (Docket No. EPA-HQ-OPP-2007-0566, Chitin Case 6063). Earlier work with 0.7% water-soluble chitosan (53 kDa) showed that FCV-F9 at 7 log PFU/mL could be decreased to 2.8 log PFU/mL, while

7 log PFU/mL MS2 could be decreased by 1.18 to 1.41 log PFU/mL using 0.35 to 1.4% water-soluble 53 kDa chitosan at 37 °C after 3 h, however, without any effect against MNV-1 (Su et al., 2009). Low molecular weight 5 kDa chitosan oligosaccharide at 1.4% was shown to cause reduction of 1.4 log PFU/mL FCV-F9, 1 log PFU/mL MS2 and again without any noticeable decrease in MNV-1 titers (Su et al., 2009). Further studies showed that 0.7% water-soluble chitosan (53 kDa) could decrease infectious titers of MNV-1, bacteriophage phiX174, bacteriophage MS2, and FCV-F9 by 0.34 log PFU/mL (minimal of MNV-1 as also shown in other studies), 0.59 log PFU/mL, 2.44 log PFU/mL, and 3.36 log PFU/mL, respectively, while higher molecular weight chitosans (222, 307, 421 and ~1150 kDa) in 0.5% acetic acid could decrease infectious titers of MNV-1 by 0.09–0.28 log PFU/mL (negligible reduction), infectious titers of bacteriophage phiX174 by 1.19–1.29 log PFU/mL, infectious titers of bacteriophage MS2 by 1.88–5.37 log PFU/mL, and infectious titers of FCV-F9 titers by 2.27–2.94 log PFU/mL after 3 h treatment at 37 °C (Davis et al., 2012). Thus, chitosan in combination with other hurdle approaches would be needed for adequate decrease in enteric virus surrogate titers that could cause potentially cause damage to the viral capsid (Davis et al., 2015).

Similarly, other studies showed that stock solutions of chitosan microparticles (CM) in 0.04% acetic acid and 1% Tween 80 (90% degree of acetylation) at 0.001% treatments for 1 h at room temperature did not result in any noticeable decreases in initial 7 log TCID50/mL infectious MNV-1 titers, though increased concentrations of 0.01%, 0.1% and 0.3% (w/v CM) caused 1.1, 1.3, and 1.5 log reductions, respectively (Zhu et al., 2020). Immediate contact with 0.01% CM was shown to cause 2.9 log PFU reduction of MS2 titers, while decreases to below the 1.85 log PFU/mL detection limit was observed after 24 h (Zhu et al., 2020). Chitosan applications as a coagulant in water filtration against pathogens including viruses have been explored as well as chitosan films with antimicrobials for enhanced antiviral activity against enteric viruses are being researched (Sánchez et al., 2015). Infectious titers of MNV-1 at 7 log PFU/mL were lowered by 1.7 log PFU, 2.5 log PFU, and 3.3 log PFU after treatment with 1, 1.5, and 2.5% aqueous green tea extract (GTE), respectively, while film forming solutions (FFS) of chitosan containing 2.5% and 5.0% GTE after incubation for 3 h lowered titers of MNV-1 by 2.5 log PFU and 4.0 log PFU, respectively (Amankwaah et al., 2020a, b). Dried chitosan films containing 5% and 10% GTE caused decreases of 1.6 and 4.5 log in MNV-1 infectious titers, respectively, while those containing 15% GTE decreased MNV-1 titers to levels that could not be detected after 24 h at 23 ± 1 °C (Amankwaah et al., 2020a, b).

Additionally, chitosan solutions (2% w/w) containing 2.5% GSE and 5% GSE were reported to decrease MNV-1 titers after 3 h treatment by 2.68 log PFU/mL and 4 log PFU/mL, respectively (Amankwaah et al., 2020b). Furthermore, this study also showed that chitosan films containing GSE at 5%, 10%, and 15% GSE after an incubation period of 4 h resulted in MNV-1 titer decreases of 0.92 log PFU/mL, 1.89 log PFU/mL, and 2.27 log PFU/mL, respectively. With increased incubation time of 24 h, MNV-1 titers were decreased by 1.9 and 3.26 log PFU/mL with the 5% GSE and 10% GSE containing chitosan films and the higher concentration of 15% GSE containing chitosan films resulted in non-detectable levels of MNV-1 after 24 h (Amankwaah et al., 2020b). Overall, chitosan in solution and films along with combinations of antimicrobials for synergistic effects shows potential for use in coatings, packaging materials, and films to decrease the risk of enteric virus contamination and transmission (Fabra et al., 2016).

9 Future Directions

On-going foodborne outbreaks highlight the need for the application of improved methods to control and prevent the transmission of foodborne diseases. Even though chlorine and chlorine based decontaminants are effective, their disadvantages and regulations for use in various countries drives the need for alternate strategies. Research on alternates that include ozone and ozonated products, electrolyzed water, hydrogen peroxide, various organic acids and natural antimicrobials remain on-going that show promise for use in hurdle approaches (alternate washing strategies along with improved storage conditions and use of improved packaging materials are being studied including but not limited to chitosan films with GSE or GTE, curcumin or gingerol with warm water treatments, etc). With the on-going focus on sustainable systems and biobased products, improved delivery systems and encapsulation methods for effective inactivation that can also maintain longer product shelf-life are needed. When used on produce or directly in food products, further research may be warranted to evaluate the effects of novel biobased products and alternate technologies on the sensory attributes of the food and also retention of nutrient content. These novel approaches should not only be effective but also sustainable and cost-effective for use in the food and agricultural environment.

Acknowledgements The author gratefully acknowledges the support provided by the University of Tennessee Institute of Agriculture (Multistate Project S1077)

References

Aboubakr, H. A., El-Banna, A. A., Youssef, M. M., Al-Sohaimy, S. A., & Goyal, S. M. (2004). Antiviral effects of Lactococcus lactis on feline Calicivirus, a human norovirus surrogate. *Food and Environmental Virology, 6*(4), 282–289. https://doi.org/10.1007/s12560-014-9164-2. Epub 2014 Aug 17. PMID: 25129102; PMCID: PMC7090733.

Aboubakr, H. A., El-Banna, A. A., Youssef, M. M., Al-Sohaimy, S. A., & Goyal, S. M. (2014, December). Antiviral effects of Lactococcus lactis on feline calicivirus, a human norovirus surrogate. *Food and Environmental Virology, 6*(4), 282–289. https://doi.org/10.1007/s12560-014-9164-2

Aboubakr, H. A., Nauertz, A., Luong, N. T., Agrawal, S., El-Sohaimy, S. A., Youssef, M. M., & Goyal, S. M. (2016). In vitro antiviral activity of clove and ginger aqueous extracts against feline calicivirus, a surrogate for human norovirus. *Journal of Food Protection, 79*(6), 1001–1012. https://doi.org/10.4315/0362-028X.JFP-15-593

Adımcılar, V., Kalaycıoğlu, Z., Akın-Evingür, G., Torlak, E., & Erim, F. B. (2023, July 1). Comparative physical, antioxidant, and antimicrobial properties of films prepared by dissolving chitosan in bioactive vinegar varieties. *International Journal of Biological Macromolecules, 242*(Pt 1), 124735. https://doi.org/10.1016/j.ijbiomac.2023.124735

Afolayan, O. T., Webb, C. C., & Cannon, J. L. (2016, March). Evaluation of a porcine gastric mucin and RNase a assay for the discrimination of infectious and non-infectious GI.1 and GII.4 norovirus following thermal, ethanol, or levulinic acid plus sodium dodecyl sulfate treatments. *Food and Environmental Virology, 8*(1), 70–78. https://doi.org/10.1007/s12560-015-9219-z

Al-Ajeeli, M. N., Taylor, T. M., Alvarado, C. Z., & Coufal, C. D. (2016, May). Comparison of eggshell surface sanitization technologies and impacts on consumer acceptability. *Poultry Science, 95*(5), 1191–1197. https://doi.org/10.3382/ps/pew014

Al-Qadiri, H. M., Ovissipour, M., Al-Alami, N., Govindan, B. N., Shiroodi, S. G., & Rasco, B. (2016, May). Efficacy of neutral electrolyzed water, quaternary ammonium and lactic acid-based solutions in controlling microbial contamination of food cutting boards using a manual spraying technique. *Journal of Food Science, 81*(5), M1177–M1183. https://doi.org/10.1111/1750-3841.13275

Akbas, M. Y., & Olmez, H. (2007). Inactivation of Escherichia coli and listeria monocytogenes on iceberg lettuce by dip wash treatments with organic acids. *Letters in Applied Microbiology, 44*, 619–624.

Amankwaah, C., Li, J., Lee, J., & Pascall, M. A. (2020a, July). Antimicrobial activity of chitosan-based films enriched with green tea extracts on murine norovirus, *Escherichiacoli*, and *Listeria innocua. International Journal of Food Science, 2*, 3941924. https://doi.org/10.1155/2020/3941924

Amankwaah, C., Li, J., Lee, J., & Pascall, M. A. (2020b). Development of antiviral and bacteriostatic chitosan-based food packaging material with grape seed extract for murine norovirus, Escherichia coli and Listeria innocua control. *Food Science & Nutrition, 8*(11), 6174–6181. https://doi.org/10.1002/fsn3.1910

Asghari-Varzaneh, E., Sharifian-Mobarakeh, S., & Shekarchizadeh, H. (2024). Enhancing hamburger shelf life and quality using gallic acid encapsulated in gelatin/tragacanth gum complex coacervate. *Heliyon, 10*(2), e24917. https://doi.org/10.1016/j.heliyon.2024.e24917

Azizkhani, M., Elizaquível, P., Sánchez, G., Selma, M. V., & Aznar, R. (2013, September 2). Comparative efficacy of Zataria multiflora Boiss., Origanum compactum and Eugenia caryophyllus essential oils against E. coli O157:H7, feline calicivirus and endogenous microbiota in commercial baby-leaf salads. *International Journal of Food Microbiology, 166*(2), 249–255. https://doi.org/10.1016/j.ijfoodmicro.2013.07.020

Baert, L., Vandekinderen, I., Devlieghere, F., Coillie, E., Debevere, J., & Uyttendaele, M. (2009, May). Efficacy of sodium hypochlorite and peroxyacetic acid to reduce murine norovirus 1, B40-8, Listeria monocytogenes, and Escherichia coli O157:H7on shredded iceberg lettuce and in residual wash water. *Journal of Food Protection, 72*(5), 1047–1054. https://doi.org/10.4315/0362-028x-72.5.1047

Bai, Y., Ding, X., Zhao, Q., Sun, H., Li, T., Li, Z., Wang, H., Zhang, L., Zhang, C., & Xu, S. (2022, June). Development of an organic acid compound disinfectant to control food-borne pathogens and its application in chicken slaughterhouses. *Poultry Science, 101*(6), 101842. https://doi.org/10.1016/j.psj.2022.101842

Bari, M. L., Sabina, Y., Isobe, S., Uemura, T., & Isshiki, K. (2003). Effectiveness of electrolyzed acidic water in killing Escherichia coli O157:H7, Salmonella Enteritidis, and Listeria monocytogenes on the surface of tomatoes. *Journal of Food Protection, 66*, 542–548.

Bing, S., Zang, Y., Li, Y., Zhang, B., Mo, Q., Zhao, X., & Yang, C. (2022, January). A combined approach using slightly acidic electrolyzed water and tea polyphenols to inhibit lipid oxidation and ensure microbiological safety during beef preservation. *Meat Science, 183*, 108643. https://doi.org/10.1016/j.meatsci.2021.108643

Bosilevac, J. M., Shackelford, S. D., Brichta, D. M., & Koohmaraie, M. (2005). Efficacy of ozonated and electrolyzed oxidative waters to decontaminate hides of cattle before slaughter. *Journal of Food Protection, 68*, 1393–1398.

Boucher, C., Waite-Cusic, J., Stone, D., & Kovacevic, J. (2021, August). Relative performance of commercial citric acid and quaternary ammonium sanitizers against Listeria monocytogenes under conditions relevant to food industry. *Food Microbiology, 97*, 103752. https://doi.org/10.1016/j.fm.2021.103752

Brié, A., Boudaud, N., Mssihid, A., Loutreul, J., Bertrand, I., & Gantzer, C. (2018). Inactivation of murine norovirus and hepatitis a virus on fresh raspberries by gaseous ozone treatment. *Food Microbiology, 70*, 1–6. https://doi.org/10.1016/j.fm.2017.08.010

Byun, K. H., Na, K. W., Ashrafudoulla, M., Choi, M. W., Han, S. H., Kang, I., Park, S. H., & Ha, S. D. (2022, April). Combination treatment of peroxyacetic acid or lactic acid with UV-C to control Salmonella enteritidis biofilms on food contact surface and chicken skin. *Food Microbiology, 102*, 103906. https://doi.org/10.1016/j.fm.2021.103906

Campion, A., Morrissey, R., Field, D., Cotter, P. D., Hill, C., & Ross, R. P. (2017, August). Use of enhanced nisin derivatives in combination with food-grade oils or citric acid to control Cronobacter sakazakii and Escherichia coli O157:H7. *Food Microbiology, 65*, 254–263. https://doi.org/10.1016/j.fm.2017.01.020

Cannon, J. L., Papafragkou, E., Park, G. W., Osborne, J., Jaykus, L. A., & Vinjé, J. (2006, November). Surrogates for the study of norovirus stability and inactivation in the environment: A comparison of murine norovirus and feline calicivirus. *Journal of Food Protection, 69*(11), 2761–2765. https://doi.org/10.4315/0362-028x-69.11.2761

Cannon, J. L., Aydin, A., Mann, A. N., Bolton, S. L., Zhao, T., & Doyle, M. P. (2012, August). Efficacy of a levulinic acid plus sodium dodecyl sulfate-based sanitizer on inactivation of human norovirus surrogates. *Journal of Food Protection, 75*(8), 1532–1535. https://doi.org/10.4315/0362-028x-69.11.2761

Cano, C., Meneses, Y., & Chaves, B. D. (2021, October). Application of peroxyacetic acid for decontamination of raw poultry products and comparison to other commonly used chemical antimicrobial interventions: A review. *Journal of Food Protection, 84*(10), 1772–1783. https://doi.org/10.4315/JFP-21-107

CDC. (2016a, September 18). *Chemical disinfectants | Disinfection & sterilization guidelines | Guidelines library | Infection control*. CDC. https://www.cdc.gov/infectioncontrol/guidelines/disinfection/disinfection methods/chemical.html

CDC. (2016b, September 18) *Other sterilization methods | Disinfection & sterilization guidelines | Guidelines library | Infection control*. CDC. https://www.cdc.gov/infectioncontrol/guidelines/disinfection/sterilization/other-methods.html

Chandran, S., Baker, C. A., Hamilton, A. N., Dhulappanavar, G. R., Jones, S. L., & Gibson, K. E. (2023, November). Aqueous ozone efficacy for inactivation of foodborne pathogens on vegetables used in raw meat based diets for companion animals. *Journal of Food Protection, 86*(11), 100175. https://doi.org/10.1016/j.jfp.2023.100175

Choi, J. (2020). *Inactivation of human Norovirus Surrogates, Feline Calicivirus and Tulane virus*, by Ozonated Water. UT Trace. M.S. thesis, University of Tennessee, Knoxville.

Choi, J. M., & D'Souza, D. H. (2023). Inactivation of Tulane virus and feline calicivirus by aqueous ozone. *Journal of Food Science, 88*(10), 4218–4229. https://doi.org/10.1111/1750-3841.16755. Epub 2023 Sep 7. PMID: 37680092.

Chen, D., Zhao, T., & Doyle, M. P. (2014, April). Transfer of foodborne pathogens during mechanical slicing and their inactivation by levulinic acid-based sanitizer on slicers. *Food Microbiology, 38*, 263–269. https://doi.org/10.1016/j.fm.2013.10.004

Chen, D., Zhao, T., & Doyle, M. P. (2015, August 17). Control of pathogens in biofilms on the surface of stainless steel by levulinic acid plus sodium dodecyl sulfate. *International Journal of Food Microbiology, 207*, 1–7. https://doi.org/10.1016/j.ijfoodmicro.2015.04.026

Cheng, D., Sun, L., Zou, S., Chen, J., Mao, H., Zhang, Y., Liao, N., & Zhang, R. (2019, May 13). Antiviral effects of *Houttuynia cordata* polysaccharide extract on murine norovirus-1

(MNV-1)—A human norovirus surrogate. *Molecules, 24*(9), 1835. https://doi.org/10.3390/molecules24091835

Cherrington, C. A., Hinton, M., & Chopra, I. (1990). Effect of short-chain organic acids on macro molecular synthesis in Escherichia coli. *The Journal of Applied Bacteriology, 68*, 69–74.

Cho, G. L., & Ha, J. W. (2021, April). Synergistic effect of citric acid and xenon light for inactivating foodborne pathogens on spinach leaves. *Food Research International, 142*, 110210. https://doi.org/10.1016/j.foodres.2021.110210

Chung, M. S. (2017, August 28). Antiviral activities of *Artemisia princeps* var. *orientalis* essential oil and its α-thujone against norovirus surrogates. *Food Science and Biotechnology, 26*(5), 1457–1461. https://doi.org/10.1007/s10068-017-0158-3

Chung, S. Y., Cho, T. J., Yu, H., Park, S. G., Kim, S. R. K., Rhee, S. A., & M.S. (2023). Efficacy of combined caprylic acid and thymol treatments for inactivation of Listeria monocytogenes on enoki mushrooms in household and food-service establishments. *Food Research International, 166*, 112601. https://doi.org/10.1016/j.foodres.2023.112601

Cichoski, A. J., Flores, D. R. M., Menezes, C. R., Jacob-Lopes, E., Zepka, L. Q., Wagner, R., Barin, J. S., Moraes Flores, É. M., Cruz Fernandes, M., & Campagnol, P. C. B. (2019, April). Ultrasound and slightly acid electrolyzed water application: An efficient combination to reduce the bacterial counts of chicken breast during pre-chilling. *International Journal of Food Microbiology, 166*, 112601. https://doi.org/10.1016/j.ijfoodmicro.2019.05.004

Clark, K. J., Grant, P. G., Sarr, A. B., Belakere, J. R., Swaggerty, C. L., Phillips, T. D., & Woode, G. N. (1998). An in vitro study of theaflavins extracted from black tea to neutralize bovine rotavirus and bovine coronavirus infections. *Veterinary Microbiology, 63*(2–4), 147–157.

Côté, J., Caillet, S., Doyon, G., Sylvain, J. F., & Lacroix, M. (2010). Bioactive compounds in cranberries and their biological properties. *Critical Reviews in Food Science and Nutrition, 50*, 666–679.

Cowan, M. M. (1999). Plant products as antimicrobial agents. *Clinical Microbiology Reviews, 12*(4), 564–582.

Culver, A., Geiger, C., Simon, D. (2014). *Safer products and practices for disinfecting and sanitizing surfaces*. SF Environment.org. Available from: http://www.sfenvironment.org/sites/default/files/fliers/files/sfe_th_safer_products_and_practices_for_disinfecting.pdf. Accessed 1 Oct 2015.

D'Souza, D. H. (2014, February 4). Phytocompounds for the control of human enteric viruses. *Current Opinion in Virology*, 44–49. https://doi.org/10.1016/j.coviro.2013.12.006

Davis, R., Zivanovic, S., D'Souza, D. H., & Davidson, P. M. (2012, October). Effectiveness of chitosan on the inactivation of enteric viral surrogates. *Food Microbiology, 32*(1), 57–62. https://doi.org/10.1016/j.fm.2012.04.008

Davis, R., Zivanovic, S., Davidson, P. M., & D'Souza, D. H. (2015, December). Enteric viral surrogate reduction by chitosan. *Food and Environmental Virology, 7*(4), 359–365. https://doi.org/10.1007/s12560-015-9208-2

Dawley, C. R., Lee, J. A., & Gibson, K. E. (2021, September). Reduction of norovirus surrogates alone and in association with bacteria on leaf lettuce and tomatoes during application of aqueous ozone. *Food and Environmental Virology, 13*(3), 390–400. https://doi.org/10.1007/s12560-021-09476-y

Devlieghere, F., Vermeiren, L., & Debevere, J. (2004). New preservation technologies: Possibilities and limitations. *International Dairy Journal, 14*, 273–285.

Donatti Leão Alvarenga, P., Mileib Vasconcelos, C., & São José, J. F. B. (2020, December 22). Application of ultrasound combined with acetic acid and peracetic acid: Microbiological and physicochemical quality of strawberries. *Molecules, 26*(1), 16. https://doi.org/10.3390/molecules26010016

Duizer, E., Bijkerk, P., Rockx, B., Groot, A., Twisk, F., & Koopmans, M. (2004, August). Inactivation of caliciviruses. *Applied and Environmental Microbiology, 70*(8), 4538–4543. https://doi.org/10.1128/AEM.70.8.4538-4543.2004

Englezos, V., Rantsiou, K., Cravero, F., Torchio, F., Giacosa, S., Segade, S. R., et al. (2019). Minimizing the environmental impact of cleaning in winemaking industry by using ozonefor cleaning-in-place (CIP) of wine bottling machine. *Journal of Cleaner Production, 233*, 582–589. https://doi.org/10.1016/j.jclepro.2019.06.097

EPA. (2011). *Water treatment manual: Disinfection*. Environmental Protection Agency.

Erickson, M. C., Liao, J. Y., Habteselassie, M. Y., & Cannon, J. L. (2018, August). Inactivation of Escherichia coli O157:H7 and salmonella during washing of contaminated gloves in levulinic acid and sodium dodecyl sulfate solutions. *Food Microbiology, 73*, 275–281. https://doi.org/10.1016/j.fm.2018.01.024

Fabra, M. J., Castro-Mayorga, J. L., Randazzo, W., Lagarón, J. M., López-Rubio, A., Aznar, R., & Sánchez, G. (2016, January). Efficacy of cinnamaldehyde against enteric viruses and its activity after incorporation into biodegradable multilayer systems of interest in food packaging. *Food and Environmental Virology, 8*(2), 125–132. https://doi.org/10.1007/s12560-016-9235-7

Falcó, I., Randazzo, W., Rodríguez-Díaz, J., Gozalbo-Rovira, R., Luque, D., Aznar, R., & Sánchez, G. (2019, March 2). Antiviral activity of aged green tea extract in model food systems and under gastric conditions. *International Journal of Food Microbiology, 292*, 101–106. https://doi.org/10.1016/j.ijfoodmicro.2018.12.019

Falcó, I., Díaz-Reolid, A., Randazzo, W., & Sánchez, G. (2020, December 2). Green tea extract assisted low-temperature pasteurization to inactivate enteric viruses in juices. *International Journal of Food Microbiology, 334*, 108809. https://doi.org/10.1016/j.ijfoodmicro.2020.108809

Fan, X., Gurtler, J. B., & Sokorai, K. J. B. (2018, September 2). Tomato type and post-treatment water rinse affect efficacy of acid washes against salmonella enterica inoculated on stem scars of tomatoes and product quality. *International Journal of Food Microbiology, 280*, 57–65. https://doi.org/10.1016/j.ijfoodmicro.2018.05.012

Fan, X., Sokorai, K. J. B., & Gurtler, J. B. (2020. January 2). Advanced oxidation process for the inactivation of salmonella typhimurium on tomatoes by combination of gaseous ozone and aerosolized hydrogen peroxide. *International Journal of Food Microbiology, 312*, 108387. https://doi.org/10.1016/j.ijfoodmicro.2019.108387

Feng, L., Zhang, K., Gao, M., Shi, C., Ge, C., Qu, D., Zhu, J., Shi, Y., & Han, J. (2018, August 28). Inactivation of *Vibrio parahaemolyticus* by aqueous ozone. *Journal of Microbiology and Biotechnology, 28*(8), 1233–1246. https://doi.org/10.4014/jmb.1801.01056

U.S. FDA, U. S. Food and Drug Administration (2004). *Juice HACCP hazards and controls guidance*. United States Food, Drug Administration, USFDA.

Fu, L., Xiao, Q., Ru, Y., Hong, Q., Weng, H., Zhang, Y., Chen, J., & Xiao, A. (2024, January). Bio-based active packaging: Gallic acid modified agarose coatings in grass carp (Ctenopharyngodon idellus) preservation. *International Journal of Biological Macromolecules, 255*, 128196. https://doi.org/10.1016/j.ijbiomac.2023.128196

Fuzawa, M., Bai, H., Shisler, J. L., & Nguyen, T. H. (2020, September 17). The basis of peracetic acid inactivation mechanisms for rotavirus and Tulane virus under conditions relevant for vegetable sanitation. *Applied and Environmental Microbiology, 86*(19), e01095–e01020. https://doi.org/10.1128/AEM.01095-20

Gilling, D. H., Kitajima, M., Torrey, J. R. B., & K.R. (2014a, May). Antiviral efficacy and mechanisms of action of oregano essential oil and its primary component carvacrol against murine norovirus. *Journal of Applied Microbiology, 116*(5), 1149–1163. https://doi.org/10.1111/jam.12453

Gilling, D. H., Kitajima, M., Torrey, J. R. B., & K.R. (2014b, August). Mechanisms of antiviral action of plant antimicrobials against murine norovirus. *Applied and Environmental Microbiology, 80*(16), 4898–4910. https://doi.org/10.1111/jam.12453

Gomes, J., Quinta-Ferreira, D. F. R. M., Matos, A., & Rui, C. (2019). Removal of enteric pathogens from real wastewater using single and catalytic ozonation. *Water, 11*, 127–127. https://doi.org/10.3390/w11010127

Gopal, J., Anthonydhason, V., Muthu, M., Gansukh, E., Jung, S., Chul, S., & Iyyakkannu, S. (2019, March). Authenticating apple cider vinegar's home remedy claims: Antibacterial, antifungal, antiviral properties and cytotoxicity aspect. *Natural Product Research, 33*(6), 906–910. https://doi.org/10.1080/14786419.2017.1413567

Gurtler, J. B., Dong, X., Zhong, B., & Lee, R. (2022, March 1). Efficacy of a mixed peroxyorganic acid antimicrobial wash solution against Salmonella, Escherichia coli O157:H7, or Listeria monocytogenes on cherry tomatoes. *Journal of Food Protection, 85*(5), 773–777. https://doi.org/10.4315/JFP-21-368

Hamrita, B., Emira, N., Papetti, A., Badraoui, R., Bouslama, L., Ben Tekfa, M. I., Hamdi, A., Patel, M., Elasbali, A. M., Adnan, M., Ashraf, S. A., & Snoussi, M. (2022, May 20). Phytochemical analysis, antioxidant, antimicrobial, and anti-swarming properties of *Hibiscus sabdariffa* L. calyx extracts: *In vitro* and *in silico* modelling approaches. *Evidence-based Complementary and Alternative Medicine, 2022*, 1252672. https://doi.org/10.1155/2022/1252672

Heir, E., Solberg, L. E., Jensen, M. R., Skaret, J., Grøvlen, M. S., & Holck, A. L. (2022, February 2). Improved microbial and sensory quality of chicken meat by treatment with lactic acid, organic acid salts and modified atmosphere packaging. *International Journal of Food Microbiology, 362*, 109498. https://doi.org/10.1016/j.ijfoodmicro.2021.109498

Hirneisen, K. A., & Kniel, K. E. (2013). Inactivation of internalized and surface contaminated enteric viruses in green onions. *International Journal of Food Microbiology, 166*(2), 201–206. https://doi.org/10.1016/j.ijfoodmicro.2013.07.013

Hirneisen, K. A., Markland, S. M., & Kniel, K. E. (2011). Ozone inactivation of norovirus surrogates on fresh produce. *Journal of Food Protection, 74*(5), 836–839. https://doi.org/10.4315/0362-028X.JFP-10-438

Holmdahl, T., Walder, M., Uzcátegui, N., Odenholt, I., Lanbeck, P., Medstrand, P., & Widell, A. (2016, May). Hydrogen peroxide vapor decontamination in a patient room using feline calicivirus and murine norovirus as surrogate markers for human norovirus. *Infection Control and Hospital Epidemiology, 37*(5), 561–566. https://doi.org/10.1017/ice.2016.15

Holmdahl, T., Odenholt, I., Riesbeck, K., Medstrand, P., & Widell, A. (2019, March). Hydrogen peroxide vapour treatment inactivates norovirus but has limited effect on post-treatment viral RNA levels. *Infect Dis (Lond), 51*(3), 197–205. https://doi.org/10.1080/23744235.2018.1546056

Horm, K. M., Davidson, P. M., Harte, F. M., & D'Souza, D. H. (2012). Survival and inactivation of human norovirus surrogates in blueberry juice by high-pressure homogenization. *Foodborne Pathogens and Disease, 9*(11), 974–979.

Howell, A. B., & D'Souza, D. H. (2013). The pomegranate: Effects on bacteria and viruses that influence human health. *Evidence-based Complementary and Alternative Medicine, 2013*, 606212.

Hricova, D., Stephan, R., & Zweifel, C. (2008, September). Electrolyzed water and its application in the food industry. *Journal of Food Protection, 71*(9), 1934–1947. https://doi.org/10.4315/0362-028x-71.9.1934

Huang, L., Luo, X., Gao, J., & Matthews, K. R. (2019). Influence of water antimicrobials andstorage conditions on inactivating MS2 bacteriophage on strawberries. *International Journal of Food Microbiology, 291*, 67–71. https://doi.org/10.1016/j.ijfoodmicro.2018.11.009

Huang, H., He, J., Gao, X., Lei, J., Zhang, Y., Wang, Y., Liu, X., & Hao, J. (2023, August 30). Mechanism of acid and alkali electrolyzed water on the elimination of Listeria monocytogenes biofilm based on proteomic analysis. *Journal of Proteomics, 286*, 104952. https://doi.org/10.1016/j.jprot.2023.104952

Inpitak, P., & Udompijitkul, P. (2022, June 2). Effect of household sanitizing agents and electrolyzed water on Salmonella reduction and germination of sunflower and roselle seeds. *International Journal of Food Microbiology, 370*, 109668. https://doi.org/10.1016/j.ijfoodmicro.2022.109668

Ionidis, G., Hübscher, J., Jack, T., Becker, B., Bischoff, B., Todt, D., Hodasa, V., BrillFh, S., & Steinmann, E. J. (2016, February 11). Development and virucidal activity of a novel alcohol-based hand disinfectant supplemented with urea and citric acid. *BMC Infectious Diseases, 16*, 77. https://doi.org/10.1186/s12879-016-1410-9

Iwasawa, A., Niwano, Y., Mokudai, T., & Kohno, M. (2009). Antiviral activity of proanthocyanidin against feline calicivirus used as a surrogate for noroviruses, and coxsackievirus used as a representative enteric virus. *Biocontrol Science, 14*(3), 107–111.

Jayaprakasha, G. K., Selvi, T., & Sakariah, K. K. (2003). Antibacterial and antioxidant activities of grape (Vitis vinifera) seed extracts. *Food Research International, 36*, 117–122.

Jeon, M. J., & Ha, J. W. (2020, October). Synergistic bactericidal effect and mechanism of X-ray irradiation and citric acid combination against food-borne pathogens on spinach leaves. *Food Microbiology, 91*, 103543. https://doi.org/10.1016/j.fm.2020.103543

Jeon, H. R., Kwon, M. J., & Yoon, K. S. (2018, April). Control of Listeria innocua biofilms on food contact surfaces with slightly acidic electrolyzed water and the risk of biofilm cells transfer to duck meat. *Journal of Food Protection, 81*(4), 582–592. https://doi.org/10.4315/0362-028X.JFP-17-373

Jiang, Y., Sokorai, K., Pyrgiotakis, G., Demokritou, P., Li, X., Mukhopadhyay, S., Jin, T., & Fan, X. (2017, May 16). Cold plasma-activated hydrogen peroxide aerosol inactivates Escherichia coli O157:H7, Salmonella typhimurium, and Listeria innocua and maintains quality of grape tomato, spinach and cantaloupe. *International Journal of Food Microbiology, 249*, 53–60. https://doi.org/10.1016/j.ijfoodmicro.2017.03.004

Jones-Ibarra, A. M., Alvarado, C. Z., Coufal, C. D., & Taylor, T. M. (2019, November). Sanitization of chicken frames by a combination of hydrogen peroxide and UV light to reduce contamination of derived edible products. *Journal of Food Protection, 82*(11), 1896–1900. https://doi.org/10.4315/0362-028X.JFP-19-175

Joshi, S. S., Dice, L., & D'Souza, D. H. (2015, December). Aqueous extracts of Hibiscus sabdariffa calyces decrease hepatitis a virus and human norovirus surrogate titers. *Food and Environmental Virology, 7*(4), 366–373. https://doi.org/10.1007/s12560-015-9209-1

Joshi, S. S., Howell, A. B., & D'Souza, D. H. (2016, December). Reduction of enteric viruses by blueberry juice and blueberry proanthocyanidins. *Food and Environmental Virology, 8*(4), 235–243. https://doi.org/10.1016/j.fm.2019.02.001

Joshi, S., Howell, A. B., & D'Souza, D. H. (2017, May). Blueberry proanthocyanidins against human norovirus surrogates in model foods and under simulated gastric conditions. *Food Microbiology, 63*, 263–267. https://doi.org/10.1016/j.fm.2016.11.024

Joshi, S. S., Howell, A. B., & D'Souza, D. H. (2019, September). Antiviral effects of blueberry proanthocyanidins against Aichi virus. *Food Microbiology, 82*, 202–208. https://doi.org/10.1016/j.fm.2019.02.001

Joshi, S.S., Dice, L., Ailavadi, S., D'Souza, D.H. (2023, January). Antiviral Effects of Quillaja saponaria Extracts Against Human Noroviral Surrogates. *Food and Environmental Virology, 15*(2),167–175. https://doi.org/10.1007/s12560-023-09550-7. Epub 2023 Mar 15. PMID: 36920726

Joshi, S. S., & D'Souza, D. H. (2021). Grape seed extract against Aichi virus infectivity in model foods and contact surfaces. *Food Microbiology, 98*, 103784. https://doi.org/10.1016/j.fm.2021.103784. Epub 2021 Mar 9. PMID: 33875212.

Jyske, T., Liimatainen, J., Tienaho, J., Brännström, H., Aoki, D., Kuroda, K., Reshamwala, D., Kunnas, S., Halmemies, E., Nakayama, E., Kilpeläinen, P., Ora, A., Kaseva, J., Hellström, J., Marjomäki, V. S., Karonen, M., & Fukushima, K. (2023, April 19). Inspired by nature: Fiber networks functionalized with tannic acid and condensed tannin-rich extracts of Norway spruce bark show antimicrobial efficacy. *Frontiers in Bioengineering and Biotechnology, 11*, 1171908. https://doi.org/10.3389/fbioe.2023.1171908

Kaczmarek, B. (2020, July 20). Tannic acid with antiviral and antibacterial activity as a promising component of biomaterials—A minireview. *Materials (Basel), 13*(14), 3224. https://doi.org/10.3390/ma13143224

Kamimoto, M., Nakai, Y., Tsuji, T., Shimamoto, T., & Shimamoto, T. (2014, May). Antiviral effects of persimmon extract on human norovirus and its surrogate, bacteriophage MS2. *Journal of Food Science, 79*(5), M941–M946. https://doi.org/10.1111/1750-3841.12462

Kang, T., Yim, D., Kim, S. S., Baek, K. H., Kim, H. J., & Jo, C. (2022, May). Effect of plasma-activated acetic acid on inactivation of Salmonella typhimurium and quality traits on chicken meats. *Poultry Science, 101*(5), 101793. https://doi.org/10.1016/j.psj.2022.101793

Khadre, Y. A. E. (2001, December 30). Sporicidal action of ozone and hydrogen peroxide: A comparative study. *International Journal of Food Microbiology, 71*(2–3), 131–138. https://doi.org/10.1016/s0168-1605(01)00561-x

Kim, S. A., & Rhee, M. S. (2015, August). Synergistic antimicrobial activity of caprylic acid in combination with citric acid against both Escherichia coli O157:H7 and indigenous microflora in carrot juice. *Food Microbiology, 49*, 166–172. https://doi.org/10.1016/j.fm.2015.02.009

Kim, C., Hung, Y. C., Brackett, R. E., & Frank, J. F. (2001). Inactivation of Listeria monocytogenes biofilms by electrolyzed oxidizing water. *Journal of Food Processing & Preservation, 25*, 91–100.

Kim, C., Hung, Y. C., Brackett, R. E., Lin, C. S., Vincent, E. M. D. H., Meade, G. K., Watson, C. L., & Fan, X. (2003). Efficacy of electrolyzed oxidizing water in inactivating Salmonella on alfalfa seeds and sprouts. *Journal of Food Protection, 66*, 208–214. https://doi.org/10.1016/j.ijfoodmicro.2013.11.018

Kim, Y. E., Myung, G. E., Jeon, Y. J., & Min, S. C. (2024, March 16). Integrated in-package treatment of hydrogen peroxide and cold plasma for microbial inactivation of cabbage slices. *Food Science and Biotechnology, 33*(7), 1633–1640. https://doi.org/10.1007/s10068-024-01536-w

Kingsley, D. H., & Annous, B. A. (2021, November). Evaluation of SDS and GRAS liquid disinfectants for mitigation of hepatitis A virus contamination of berries. *Journal of Applied Microbiology, 131*(5), 2586–2591. https://doi.org/10.1111/jam.15123

Konowalchuk, J., & Speirs, J. I. (1978). Antiviral effect of commercial juices and beverages. *Applied and Environmental Microbiology, 35*(6), 1219–1220.

Koromyslova, A. D., White, P. A., & Hansman, G. S. (2015, November). Treatment of norovirus particles with citrate. *Virology, 485*, 199–204. https://doi.org/10.1016/j.virol.2015.07.009

Koseki, S., Yoshida, Y., Isobe, S., & Itoh, K. (2004). Efficacy of acidic electrolyzed water for microbial decontamination of cucumbers and strawberries. *Journal of Food Protection, 67*, 1247–1251.

Kučuk, N., Primožič, M., Kotnik, P., Knez, Ž., & Leitgeb, M. (2024, February 11). Mango peels as an industrial by-product: A sustainable source of compounds with antioxidant, enzymatic, and antimicrobial activity. *Food Security, 13*(4), 553. https://doi.org/10.3390/foods13040553

Lacombe, A., Niemira, B. A., Gurtler, J. B., Kingsley, D. H., Li, X., & Chen, H. (2018, February). Surfactant-enhanced organic acid inactivation of Tulane virus, a human norovirus surrogate. *Journal of Food Protection, 81*(2), 279–283. https://doi.org/10.4315/0362-028X.JFP-17-330

Lange-Starke, A., Petereit, A., Truyen, U., Braun, P. G., Fehlhaber, K., & Albert, T. (2014, March). Antiviral potential of selected starter cultures, bacteriocins and D,L-lactic acid. *Food and Environmental Virology, 6*(1), 42–47. https://doi.org/10.1007/s12560-013-9135-z

Laury, A. M., Alvarado, M. V., Nace, G., Alvarado, C. Z., Brooks, J. C., Echeverry, A., & Brashears, M. M. (2009, October). Validation of a lactic acid- and citric acid-based antimicrobial product for the reduction of Escherichia coli O157: H7 and salmonella on beef tips and whole chicken carcasses. *Journal of Food Protection, 72*(10), 2208–2211. https://doi.org/10.4315/0362-028x-72.10.2208

Lee, C. J., Chen, L. G., Liang, W. L., & Wang, C. C. (2010a). Anti-inflammatory effects of Punica granatum Linne in vitro and in vivo. *Food Chemistry, 118*, 315–322.

Lee, S. Y., Rhee, M. S., Dougherty, R. H., & Kang, D. H. (2010b, April). Antagonistic effect of acetic acid and salt for inactivating Escherichia coli O157:H7 in cucumber puree. *Journal of Applied Microbiology, 108*(4), 1361–1368. https://doi.org/10.1111/j.1365-2672.2009.04543.x

Lee, M. H., Lee, B. H., Jung, J. Y., Cheon, D. S., Kim, K. T., & Choi, C. (2011). Antiviral effect of Korean red ginseng extract and ginsenosides on murine norovirus and feline calicivirus as surrogates for human norovirus. *Journal of Ginseng Research, 35*(4), 429–435.

Lee, M. H., Yoo, S. H., Ha, S. D., & Choi, C. (2012, September). Inactivation of feline calicivirus and murine norovirus during Dongchimi fermentation. *Food Microbiology, 31*(2), 210–214. https://doi.org/10.1016/j.fm.2012.04.002

Lee, M. H., Lee, B. H., Lee, S., & Choi, C. (2013, August 9). Reduction of hepatitis a virus on FRhK-4 cells treated with Korean red ginseng extract and ginsenosides. *Journal of Food Science, 78*, 9. https://doi.org/10.1111/1750-3841.12205

Lee, J. H., Bae, S. Y., Oh, M., Seok, J. H., Kim, S., Chung, Y. B., Gowda, K. G., Mun, J. Y., Chung, M. S., & Kim, K. H. (2016, June). Antiviral effects of black raspberry (Rubus coreanus) seed extract andits polyphenolic compounds on norovirus surrogates. *Bioscience, Biotechnology, and Biochemistry, 80*(6), 1196–1204. https://doi.org/10.1080/09168451.2016.1151337

Lee, H. M., Lee, J. H., Kim, S. H., Yoon, S. R. L., Ha, J. Y., & J.H. (2017, November 28). Correlation between changes in microbial/physicochemical properties and persistence of human norovirus during cabbage kimchi fermentation. *Journal of Microbiology and Biotechnology, 27*(11), 2019–2027. https://doi.org/10.4014/jmb.1707.07041

Li, Q., Mahendra, S., Lyon, D. Y., Brunet, L., Liga, M. V., Li, D., & Alvarez, P. J. (2008). Antimicrobial nanomaterials for water disinfection and microbial control: Potential applications and implications. *Water Research, 42*(18), 4591–4602. https://doi.org/10.1016/j.watres.2008.08.015. Epub 2008 Aug 27. PMID: 1880483.

Li, D., Baert, L., Xia, M., Zhong, W., Jiang, X., & Uyttendaele, M. (2012a). Effects of a variety of food extracts and juices on the specific binding ability of norovirus GII.4 P particles. *Journal of Food Protection, 75*(7), 1350–1354.

Li, D., Baert, L., Zhang, D., Xia, M., Zhong, W., Coillie, E., Jiang, X., & Uyttendaele, M. (2012b). Effect of grape seed extract on human norovirus GII.4 and murine norovirus 1 in viral suspensions, on stainless steel discs, and in lettuce wash water. *Applied and Environmental Microbiology, 78*(21), 7572–7578.

Li, Z., Liu, L., Chen, B., Zhao, T., Ran, L., Yuan, X., Cao, Z., & Wu, T. (2019, November). Structure and antimicrobial properties of long-chain branched poly (lactic acid). *Journal of Biomedical Materials Research. Part A, 107*(11), 2458–2467. https://doi.org/10.1002/jbm.a.36752

Li, S., Konoval, H. M., Marecek, S., Lathrop, A. A., Feng, S., & Pokharel, S. (2023, February). Control of Escherichia coli O157:H7 using lytic bacteriophage and lactic acid on marinated and tenderized raw pork loins. *Meat Science, 196*, 109030. https://doi.org/10.1016/j.meatsci.2022.109030

Lin, C. M., Moon, S. S., Doyle, M. P., & McWatters, K. H. (2002, August). Inactivation of Escherichia coli O157:H7, Salmonella enterica serotype enteritidis, and Listeria monocytogenes on lettuce by hydrogen peroxide and lactic acid and by hydrogen peroxide with mild heat. *Journal of Food Protection, 65*(8), 1215 1220. https://doi.org/10.4315/0362 028x 65.8.1215

Linh, N. N., Huong, N. T. M., Dai, D. N., Giang, N. T. A., Huong, L. T., Thuy, T. T. T., & Son, N. T. (2024, January). *Camellia annamensis* (Theaceae): Phytochemical analysis, cytotoxic, antioxidative, and antimicrobial activities. *Natural Product Research, 15*, 1–7. https://doi.org/10.1080/14786419.2024.2302900

Lipson, S. M., Cohen, P., Zhou, J., Burdowski, A., & Stotzky, G. (2007a). Cranberry cocktail juice, cranberry concentrates, and proanthocyanidins reduce reovirus infectivity titers in African green monkey kidney epithelial cell cultures. *Molecular Nutrition & Food Research, 51*, 752–758.

Lipson, S. M., Sethi, L., Cohen, P., Gordon, R. E., Tan, I. P., Burdowski, A., & Stotzky, G. (2007b). Antiviral effects on bacteriophages and rotavirus by cranberry juice. *Phytomedicine, 14*(1), 23–30.

Liu, C., Chen, C., Jiang, A., Zhang, Y., Zhao, Q., & Hu, W. (2020, December 16). Effects of aqueous ozone treatment on microbial growth, quality, and pesticide residue of fresh-cut cabbage. *Food Science & Nutrition, 9*(1), 52–61. https://doi.org/10.1002/fsn3.1870

Liu, Z., Long, H., Wang, Y., Shen, C., & Chen, D. (2022, August 7). Antimicrobial nonwoven fabrics incorporated with levulinic acid and sodium dodecyl sulfate for use in the food industry. *Food Security, 11*(15), 2369. https://doi.org/10.3390/foods11152369

Loredo, A. B. G., Guerrero, S. N., & Alzamora, S. M. (2015). Inactivation kinetics and growth dynamics during cold storage of Escherichia coli ATCC 11229, Listeria innocua ATCC 33090 and Saccharomyces cerevisiae KE162 in peach juice using aqueous ozone. *Innovative Food Science & Emerging Technologies, 29*, 271–279.

Lou, Y., Kasler, D. R., Hawkins, Z. L., Li, Z., Sannito, D., Fritz, R. D., & Yousef, A. E. (2024, January 15). Inactivation kinetics of selected pathogenic and non-pathogenic bacteria by aqueous ozone to validate minimum usage in purified water. *Frontiers in Microbiology, 14*, 1258381. https://doi.org/10.3389/fmicb.2023.1258381

Lukasik, J., Bradley, M. L., Scott, T. M., Dea, M., Koo, A., Hsu, W. Y., Bartz, J. A., & Farrah, S. R. (2003, February). Reduction of poliovirus 1, bacteriophages, Salmonella montevideo, and Escherichia coli O157:H7 on strawberries by physical and disinfectant washes. *Journal of Food Protection, 66*(2), 188–193. https://doi.org/10.4315/0362-028x-66.2.188

Manuel, C. S., Moore, M. D., & Jaykus, L. A. (2017, January). Efficacy of a disinfectant containing silverdihydrogen citrate against GI.6 and GII.4 human norovirus. *Journal of Applied Microbiology, 122*(1), 78–86. https://doi.org/10.1111/jam.13331

Marino, M., Maifreni, M., Baggio, A., & Innocente, N. (2018, August 28). Inactivation of foodborne bacteria biofilms by aqueous and gaseous ozone. *Frontiers in Microbiology, 9*, 2024. https://doi.org/10.3389/fmicb.2018.02024

Megahed, A., Aldridge, B., & Lowe, J. (2018, May 14). The microbial killing capacity of aqueous and gaseous ozone on different surfaces contaminated with dairy cattle manure. *PLoS One, 13*(5), e0196555. https://doi.org/10.1371/journal.pone.0196555

Megahed, A., Aldridge, B., & Lowe, J. (2020). Antimicrobial efficacy of aqueous ozone and ozone–lactic acid blend on salmonella-contaminated chicken drumsticks using multiple sequential soaking and spraying approaches. *Frontiers in Microbiology, 11*, 3121.

Moghassem Hamidi, R., Shekarforoush, S. S., Hosseinzadeh, S., & Basiri, S. (2021, September). Evaluation of the effect of neutral electrolyzed water and peroxyacetic acid alone and in combination on microbiological, chemical, and sensory characteristics of poultry meat during refrigeration storage. *Food Science and Technology International, 27*(6), 499–507. https://doi.org/10.1177/1082013220968713

Mohammad, Z., Kalbasi-Ashtari, A., Riskowski, G., Juneja, V., & Castillo, A. (2020, October). Inactivation of Salmonella and Shiga toxin-producing Escherichia coli (STEC) from the surface of alfalfa seeds and sprouts by combined antimicrobial treatments using ozone and electrolyzed water. *Food Research International, 136*(109488), 109488. https://doi.org/10.1016/j.foodres.2020.109488

Moorman, E., Montazeri, N., & Jaykus, L. A. (2017, August 1). Efficacy of neutral electrolyzed water for inactivation of human norovirus. *Applied and Environmental Microbiology, 83*(16), e00653–e00617. https://doi.org/10.1128/AEM.00653-17

Moraes Motta Machado, M. C., Lepaus, B. M., Bernardes, P. C., & São José, J. F. B. (2022, October 18). Ultrasound, acetic acid, and peracetic acid as alternatives sanitizers to chlorine compounds for fresh-cut kale decontamination. *Molecules, 27*(20), 7019. https://doi.org/10.3390/molecules27207019

Naka, A., Yakubo, M., Nakamura, K., & Kurahashi, M. (2020, February 18). Effectiveness of slightly acidic electrolyzed water on bacteria reduction: In vitro and spray evaluation. *PeerJ, 8*, e8593. https://doi.org/10.7717/peerj.8593

Nakimera, E., Cancio, L. P. M., Sullivan, G. A., Sadat, R., & Chaves, B. D. (2024, May). Antimicrobial efficacy of a citric acid/hydrochloric acid blend, peroxyacetic acid, and sulfuric acid against salmonella and background microbiota on chicken hearts and livers. *Journal of Food Science, 89*(5), 2933–2942. https://doi.org/10.1111/1750-3841

Nasheri, N., Harlow, J., Chen, A., Corneau, N., & Bidawid, S. (2021, March). Survival and inactivation by advanced oxidative process of foodborne viruses in model low-moisture foods. *Food and Environmental Virology, 13*(1), 107–116. https://doi.org/10.1007/s12560-020-09457-7

Nayak, S. L., Sethi, S., Sharma, R. R., Sharma, R. M., Singh, S., & Singh, D. (2020). Aqueous ozone controls decay and maintainsquality attributes of strawberry (Fragaria × ananassa Duch.).

Journal of Food Science and Technology, 57(1), 319–326. https://doi.org/10.1007/s13197-019-04063-3
Nei, D., Enomoto, K., & Nakamura, N. (2015, August). A gaseous acetic acid treatment to disinfect fenugreek seeds and black pepper inoculated with pathogenic and spoilage bacteria. *Food Microbiology, 49*, 226–230. https://doi.org/10.1016/j.fm.2015.02.011
Nicolau-Lapeña, I., Abadias, M., Bobo, G., Aguiló-Aguayo, I., Lafarga, T., & Viñas, I. (2019, October). Strawberry sanitization by peracetic acid washing and its effect on fruit quality. *Food Microbiology, 83*, 159–166. https://doi.org/10.1016/j.fm.2019.05.004
Ning, Y., Fu, Y., Hou, L., Ma, M., Wang, Z., Li, X., & Jia, Y. (2021, January). iTRAQ-based quantitative proteomic analysis of synergistic antibacterial mechanism of phenyllactic acid and lactic acid against Bacillus cereus. *Food Research International, 139*, 109562. https://doi.org/10.1016/j.foodres.2020.109562
Nogueira, M. C., Oyarzábal, O. A., & Gombas, D. (2003). Inactivation of Escherichia coli O157: H7, listeria monocytogenes, and salmonella in cranberry, lemon, and limejuice concentrates. *Journal of Food Protection, 66*(9), 1637–1641.
Oh, M., Bae, S. Y., Lee, J. H., Cho, K. J., & Kim, K. H. (2012). Chung MS: Antiviral effects of black raspberry (Rubus coreanus) juice on foodborne viral surrogates. *Foodborne Pathogens and Disease, 9*(10), 915–921.
Ohba, M., Oka, T., Ando, T., Arahata, S., Ikegaya, A., Takagi, H., Ogo, N., Zhu, C., Owada, K., Kawamori, F., Wang, Q., Saif, L. J., & Asai, A. (2017, April). Antiviral effect of theaflavins against caliciviruses. *Journal of Antibiotics (Tokyo), 70*(4), 443–447. https://doi.org/10.1038/ja.2016.128
Olaimat, A. N., Al-Holy, M. A., Abu-Ghoush, M. H., Osaili, T. M., Al-Nabulsi, A. A., & Rasco, B. A. (2017, August). Inhibition of Shigella sonnei and Shigella flexneri in hummus using citric acid and garlic extract. *Journal of Food Science, 82*(8), 1908–1915. https://doi.org/10.1111/1750-3841.13803
Olaimat, A. N., Al-Holy, M. A., Abu Ghoush, M. H., Al-Nabulsi, A. A., Osaili, T. M., Ayyash, M., Al-Degs, Y. S., & Holley, R. A. (2022, February 2). Use of citric acid and garlic extract to inhibit Salmonella enterica and Listeria monocytogenes in hummus. *International Journal of Food Microbiology, 362*, 109474. https://doi.org/10.1016/j.ijfoodmicro.2021.109474
Ortiz-Solà, J., Abadias, M., Colás-Medà, P., Sánchez, G., Bobo, G., & Viñas, I. (2020, December 2). Evaluation of a sanitizing washing step with different chemical disinfectants for the strawberry processing industry. *International Journal of Food Microbiology, 334*, 108810. https://doi.org/10.1016/j.ijfoodmicro.2020.108810
Park, S. Y., Kang, S., & Ha, S. D. (2016, December 5). Antimicrobial effects of vinegar against norovirus and Escherichia coli in the traditional Korean vinegared green laver (Enteromorpha intestinalis) salad during refrigerated storage. *International Journal of Food Microbiology, 238*, 208–214. https://doi.org/10.1016/j.ijfoodmicro.2016.09.002
Patil, S., Bourke, P., Frías, J. M., Tiwari, B. K., & Cullen, P. J. (2009). Inactivation of Escherichia coli in orange juice using ozone. *Innovative Food Science & Emerging Technologies, 10*, 551–557.
Patwardhan, M., Morgan, M. T., Dia, V., & D'Souza, D. H. (2020, September). Heat sensitization of hepatitis A virus and Tulane virus using grape seed extract, gingerol and curcumin. *Food Microbiology, 90*, 103461. https://doi.org/10.1016/j.fm.2020.103461
Predmore, A., Sanglay, G., Li, J., & Lee, K. (2015). Control of human norovirus surrogates in fresh foods by gaseous ozone and a proposed mechanism of inactivation. *Food Microbiology, 50*, 118–125. https://doi.org/10.1016/j.fm.2015.04.004
Premjit, Y., Sruthi, N. U., Pandiselvam, R., & Kothakota, A. (2022, March). Aqueous ozone: Chemistry, physiochemical properties, microbial inactivation, factors influencing antimicrobial effectiveness, and application in food. *Comprehensive Reviews in Food Science and Food Safety, 21*(2), 1054–1085. https://doi.org/10.1111/1541-4337.12886
Qian, J., Ma, L., Yan, W., Zhuang, H., Huang, M., Zhang, J., & Wang, J. (2022, February). Inactivation kinetics and cell envelope damages of foodborne pathogens Listeria

monocytogenes and Salmonella enteritidis treated with cold plasma. *Food Microbiology, 101*, 103891. https://doi.org/10.1016/j.fm.2021.103891

Ramirez-Hernandez, A., Brashears, M. M., & Sanchez-Plata, M. X. (2018, January). Efficacy of lactic acid, lactic acid acetic acid blends, and peracetic acid to reduce Salmonella on chicken parts under simulated commercial processing conditions. *Journal of Food Protection, 81*(1), 17–24. https://doi.org/10.4315/0362-028X.JFP-17-087

Randazzo, W., Aznar, R., & Sánchez, G. (2016). Curcumin-mediated photodynamic Inactivationof norovirus surrogates. *Food and Environmental Virology, 8*(4), 244–250. https://doi.org/10.1007/s12560-016-9255-3. Epub 2016 Aug 5. PMID: 27496054.

Randazzo, W., Falcó, I., Aznar, R., & Sánchez, G. (2017). Effect of green tea extract on enteric viruses and its application as natural sanitizer. *Food Microbiology, 66*, 150–156. https://doi.org/10.1016/j.fm.2017.04.018

Randazzo, W., Costantini, V., Morantz, E. K., & Vinjé, J. (2020, August 18). Human intestinal enteroids to evaluate human norovirus GII.4 inactivation by aged-green tea. *Frontiers in Microbiology, 11*, 1917. https://doi.org/10.1111/1541-4337.12886

Rangel-Vargas, E., Gómez-Aldapa, C. A., Falfan-Cortes, R. N., Rodríguez-Marín, M. L., Godínez Oviedo, A., Acevedo-Sandoval, O. A., & Castro-Rosas, J. (2017, March). Attachment of 13 types of foodborne bacteria to jalapeño and serrano peppers and antibacterial effect of roselle calyx extracts, sodium hypochlorite, colloidal silver, and acetic acid against these foodborne bacteria on peppers. *Journal of Food Protection, 80*(3), 406–413. https://doi.org/10.4315/0362-028X.JFP-16-269

Reyes, A. W. B., Hong, T. G., Hop, H. T., Arayan, L. T., Huy, T. X. N., Min, W., Lee, H. J., Lee, K. S., & Kim, S. (2017, August). The in vitro and in vivo protective effects of tannin derivatives against Salmonella enterica serovar Typhimurium infection. *Microbial Pathogenesis, 109*, 86–93. https://doi.org/10.1016/j.micpath.2017.05.034

Sánchez, G., & Aznar, R. (2015, January 31). Evaluation of natural compounds of plant origin for inactivation of enteric viruses. *Food and Environmental Virology, 7*, 183–187. https://doi.org/10.1007/s12560-015-9181-9

Sánchez, C., Aznar, R., & Sánchez, G. (2015, January 2). The effect of carvacrol on enteric viruses. *International Journal of Food Microbiology, 192*, 72–76. https://doi.org/10.1016/j.ijfoodmicro.2014.09.028

Sánchez-Moya, T., López-Nicolás, R., Peso-Echarri, P., González-Bermúdez, C. A., & Frontela-Saseta, C. (2024, March 27). Effect of pine bark extract and its phenolic compounds on selected pathogenic and probiotic bacterial strains. *Frontiers in Nutrition, 11*, 1381125. https://doi.org/10.3389/fnut.2024.1381125

Sagong, H. G., Lee, S. Y., Chang, P. S., Heu, S., Ryu, S., Choi, Y. J., & Kang, D. H. (2011). Combined effect of ultrasound and organic acids to reduce Escherichia coli O157:H7, Salmonella Typhimurium, and Listeria monocytogenes on organic fresh lettuce. *International Journal of Food Microbiology, 145*, 287–292.

Scott, B. R., Yang, X., Geornaras, I., Delmore, R. J., Woerner, D. R., Adler, J. M., & Belk, K. E. (2015, December). Antimicrobial efficacy of a lactic acid and citric acid blend against Shiga toxin-producing Escherichia coli, Salmonella, and nonpathogenic Escherichia coli biotype I on inoculated prerigor beef carcass surface tissue. *Journal of Food Protection, 78*(12), 2136–2142. https://doi.org/10.4315/0362-028X.JFP-15-194

Seo, D. J., Lee, M. H., Seo, J., Ha, S. D., & Choi, C. (2014, December). Inactivation of murine norovirus and feline calicivirus during oyster fermentation. *Food Microbiology, 44*, 81–86. https://doi.org/10.1016/j.fm.2014.05.016

Shin, H. J., Kim, H., Beuchat, L. R., & Ryu, J. H. (2020, December). Antimicrobial activities of organic acid vapors against Acidovorax citrulli, Salmonella enterica, Escherichia coli O157:H7, and listeria monocytogenes on Cucurbitaceae seeds. *Food Microbiology, 92*, 103569. https://doi.org/10.1016/j.fm.2020.103569

Song, W. J., & Kang, D. H. (2022a, January). Inactivation of Escherichia coli O157:H7 and Salmonella typhimurium in black and red pepper by vacuumed hydrogen peroxide vapour. *Journal of Applied Microbiology, 132*(1), 290–297. https://doi.org/10.1111/jam.15230

Song, W. J., & Kang, D. H. (2022b, June). Inactivation of Salmonella typhimurium, Escherichia coli O157:H7 and Listeria monocytogenes on alfalfa seeds by the combination treatment of vacuumed hydrogen peroxide vapour and vacuumed dry heat. *Journal of Applied Microbiology, 74*(6), 909–915. https://doi.org/10.1111/lam.13678

Sorrentino, E., Succi, M., Tipaldi, L., Pannella, G., Maiuro, L., Sturchio, M., Coppola, R., & Tremonte, P. (2018, February 2). Antimicrobial activity of gallic acid against food-related Pseudomonas strains and its use as biocontrol tool to improve the shelf life of fresh black truffles. *International Journal of Food Microbiology, 266*, 183–189. https://doi.org/10.1016/j.ijfoodmicro.2017.11.026

Straube, J., Albert, T., Manteufel, J., Heinze, J., Fehlhaber, K., & Truyen, U. (2011, November 15). In vitro influence of D/L-lactic acid, sodium chloride and sodium nitrite on the infectivity of feline calicivirus and of ECHO virus as potential surrogates for foodborne viruses. *International Journal of Food Microbiology, 151*(1), 93–97. https://doi.org/10.1016/j.ijfoodmicro.2011.08.010

Strano, M. C., Timpanaro, N., Allegra, M., Foti, P., Pangallo, S., & Romeo, F. V. (2021). Effect of ozonated water combined with sodium bicarbonate on microbial load and shelf life of cold stored clementine (Citrus Clementina Hort. Ex tan.). *Scientia Horticulturae, 276*, 109775.

Su, X., & D'Souza, D. H. (2011). Grape seed extract for control of human enteric viruses. *Applied and Environmental Microbiology, 77*(12), 3982–3987.

Su, X., & D'Souza, D. H. (2013a). Grape seed extract for foodborne virus reduction on produce. *Food Microbiology, 34*(1), 1–6.

Su, X., & D'Souza, D. H. (2013b). Naturally occuring flavonoids against human norovirus surrogates. *Food and Environmental Virology, 5*(2), 97–102.

Su, X., Zivanovic, S., & D'Souza, D. H. (2009, December). Effect of chitosan on the infectivity of murine norovirus, feline calicivirus, and bacteriophage MS2. *Journal of Food Protection, 72*(12), 2623–2628. https://doi.org/10.4315/0362-028x-72.12.2623

Su, X., Howell, A. B., & D'Souza, D. H. (2010a). Antiviral effects of cranberry juice and cranberry proanthocyanidins on foodborne viral surrogates—A time dependence study in vitro. *Food Microbiology, 27*(8), 985–991.

Su, X., Howell, A. B., & D'Souza, D. H. (2010b). The effect of cranberry juice and cranberry proanthocyanidins on the infectivity of human enteric viral surrogates. *Food Microbiology, 27*(4), 535–540.

Su, X., Sangster, M. Y., & D'Souza, D. H. (2010). In vitro effects of pomegranate juice and pomegranate polyphenols on foodborne viral surrogates. *Foodborne Pathogens and Disease, 7*(12), 1473–1479.

Su, X., Sangster, M. Y., & D'Souza, D. H. (2011). Time-dependent effects of pomegranate juice and pomegranate polyphenols on foodborne viral reduction. *Foodborne Pathogens and Disease, 8*(11), 1177–1183.

Su, X., & D'Souza, D. H. (2012). Inactivation of human norovirus surrogates by benzalkonium chloride, potassium peroxymonosulfate, tannic acid, and gallic acid. *Foodborne Pathogens and Disease, 9*(9), 829–834. https://doi.org/10.1089/fpd.2012.1155. Epub 2012 Aug 16. PMID: 22897744.

Sun, J. L., Zhang, S. K., Chen, J. Y., & Han, B. Z. (2012, April). Efficacy of acidic and basic electrolyzed water in eradicating Staphylococcus aureus biofilm. *Canadian Journal of Microbiology, 58*(4), 448–454. https://doi.org/10.1139/w2012-005

Sun, X., Ji, J., Gao, Y., Zhang, Y., Zhao, G., & Sun, C. (2020). Fateof deoxynivalenol and degradation products degraded by aqueous ozone in contaminated wheat. *Food Research International, 137*, 109357.

Svoboda, A., Shaw, A., Dzubak, J., Mendonca, A., Wilson, L., & Nair, A. (2016, April). Effectiveness of broad-spectrum chemical produce sanitizers against foodborne pathogens as in vitro planktonic cells and on the surface of whole cantaloupes and watermelons. *Journal of Food Protection, 79*(4), 524–530. https://doi.org/10.4315/0362-028X.JFP-15-490

Takeshita, M., Ishida, Y., Akamatsu, E., Ohmori, Y., Sudoh, M., Uto, H., Tsubouchi, H., & Kataoka, H. (2009). Proanthocyanidin from blueberry leaves suppresses expression of subgenomic hepatitis C virus RNA. *The Journal of Biological Chemistry, 284*(32), 21165–21176.

Taiye Mustapha, A., Zhou, C., Wahia, H., Amanor-Atiemoh, R., Otu, P., Qudus, A., Abiola Fakayode, O., & Ma, H. (2020). Sonozonation: Enhancing the antimicrobial efficiency of aqueous ozone washing techniques on cherry tomato. *Ultrasonics Sonochemistry*. https://doi.org/10.1016/j.ultsonch.2020.105059

Tam, K. I., & Roner, M. R. (2011). Characterization of in vivo anti-rotavirus activities of saponin extracts from Quillaja saponaria Molina. *Antiviral Research, 90*(3), 231–241.

Tango, C. N., Khan, I., Ngnitcho Kounkeu, P. F., Momna, R., Hussain, M. S., & Oh, D. H. (2017, October). Slightly acidic electrolyzed water combined with chemical and physical treatments to decontaminate bacteria on fresh fruits. *Food Microbiology, 67*, 97–105. https://doi.org/10.1016/j.fm.2017.06.007

Tao, H., Liao, Q., Xu, Y. I., & Wang, H. L. (2022, March 1). Efficacy of slightly acidic electrolyzed water for inactivation of Cronobacter sakazakii and biofilm cells. *Journal of Food Protection, 85*(3), 511–517. https://doi.org/10.4315/JFP-21-263

Tekile, A., Ilho, K., & Lee, J. L. (2017). Applications of ozone micro- and nanobubble technologies in water and wastewater treatment: Review. *Journal of the Korean Society of Water and Wastewater, 31*(6), 481–490.

Thomas, C., Schönknecht, A., Püning, C., Alter, T., Martin, A., & Bandick, N. (2020, April 1). Effect of peracetic acid solutions and lactic acid on microorganisms in on-line reprocessing systems for chicken slaughter plants. *Journal of Food Protection, 83*(4), 615–620. https://doi.org/10.4315/0362-028X.JFP-19-350

Thurston-Enriquez, J. A., Haas, C. N., Jacangelo, J., & Gerba, C. P. (2005). Inactivation of enteric adenovirus and feline calicivirus by ozone. *Water Research, 39*(15), 3650–3656. https://doi.org/10.1016/j.watres.2005.06.006

Tian, P., Yang, D., & Mandrell, R. (2011, August). Differences in the binding of human norovirus to and from romaine lettuce and raspberries by water and electrolyzed waters. *Journal of Food Protection, 74*(8), 1364–1369. https://doi.org/10.4315/0362-028X.JFP-10-494

Torres-Rosales, E., Rivera-Garcia, A., Rosario-Perez, P. J., Ramirez-Orejel, J. C., Paez-Esquiliano, D., Martinez-Vidal, S., Guzman-Olea, E., & Cano-Buendia, J. A. (2020, November 16). Application of neutral electrolyzed water on pork chops and its impact on meat quality. *Scientific Reports, 10*(1), 19910. https://doi.org/10.1038/s41598-020-76931-4

Trząskowska, M., Dai, Y., Delaquis, P., & Wang, S. (2018, December). Pathogen reduction on mung bean reduction of Escherichia coli O157:H7, Salmonella enterica and Listeria monocytogenes on mung bean using combined thermal and chemical treatments with acetic acid and hydrogen peroxide. *Food Microbiology, 76*, 62–68. https://doi.org/10.1016/j.fm.2018.04.008

Ueda, K., Kawabata, R., Irie, T., Nakai, Y., Tohya, Y., & Sakaguchi, T. (2013). Inactivation of pathogenic viruses by plant-derived tannins: Strong effects of extracts from persimmon (Diospyros kaki) on a broad range of viruses. *PLoS One, 8*(1), e55343.

U.S. Environmental Protection Agency. (2009). Biopesticides Registration Action Document, L-Lactic Acid. *Technical Document for L-lactic acid*. https://www3.epa.gov/pesticides/chem_search/reg_actions/registration/decision_PC-128929_1-Jun-09.pdf.

US EPA, United States Environmental Protection Agency. (2015). *Initial virucidal effectiveness test using feline calicivirus as a surrogate for norovirus*. Available at www.epa.gov/sites/production/files/2015-09/documents/fcv1_initial_surf_pcol.pdf.

Vaddu, S., Kataria, J., Belem, T. S., Sidhu, G., Moller, A. E., Leone, C., Singh, M., & Thippareddi, H. (2021, July). On-site generated peroxy acetic acid (PAA) technology reduces Salmonella and Campylobacter on chicken wings. *Poultry Science, 100*(7), 101206. https://doi.org/10.1016/j.psj.2021.101206

Wang, B. Y., Hong, J., Ciancio, S. G., Zhao, T., & Doyle, M. P. (2012). A novel formulation effective in killing oral biofilm bacteria. *Journal of the International Academy of Periodontology, 14*(3), 56–61.

Wang, Q., Markland, S., & Kniel, K. E. (2015). Inactivation of human norovirus and its surrogates on alfalfa seeds by aqueous ozone. *Journal of Food Protection, 78*(8), 1586–1591. https://doi.org/10.4315/0362-028X.JFP-15-029

Wang, R., Zhou, Y., Kalchayanand, N., Harhay, D. M., & Wheeler, T. L. (2021, March 1). Consecutive treatments with a multicomponent sanitizer inactivate biofilms formed by Escherichia coli O157:H7 and Salmonella enterica and remove biofilm matrix. *Journal of Food Protection, 84*(3), 408–417. https://doi.org/10.4315/JFP-20-321

Wang, X., Yan, X., Xu, Y., Liu, J., & Chen, D. (2022, September 2). Changes in the quality and microbial compositions of ground beef packaged on food absorbent pads incorporated with levulinic acid and sodium dodecyl sulfate. *International Journal of Food Microbiology, 376*, 109771. https://doi.org/10.1016/j.ijfoodmicro.2022.109771

Webb, C. C., Davey, L. E., Erickson, M. C., & Doyle, M. P. (2013, October). Evaluation of levulinic acid and sodium dodecyl sulfate as a sanitizer for use in processing Georgia-grown cantaloupes. *Journal of Food Protection, 76*(10), 1767–1772. https://doi.org/10.4315/0362-028X.JFP-13-057

Wessels, S., & Ingmer, H. (2013). Modes of action of three disinfectant active substances: A review. *Regulatory Toxicology and Pharmacology, 67*, 456–467.

Wu, J., Hou, W., Cao, B., Zuo, T., Xue, C., Leung, A. W., Xu, C., & Tang, Q. J. (2015). Virucidal efficacy of treatment with photodynamically activated curcumin on murine norovirus bio-accumulated in oysters. *Photodiagnosis and Photodynamic Therapy, 12*(3), 385–392. https://doi.org/10.1016/j.pdpdt.2015.06.005. Epub 2015 Jun 25. PMID: 26117199.

Xie, Y., Hajdok, C., Mittal, G. S., & Warriner, K. (2008, May). Inactivation of MS2 F(+) coliphage on lettuce by a combination of UV light and hydrogen peroxide. *Journal of Food Protection, 71*(5), 903–907. https://doi.org/10.4315/0362-028x-71.5.903

Xue, W., Macleod, J., & Blaxland, J. (2023, February 14). The use of ozone technology to control microorganism growth, enhance food safety and extend shelf life: A promising food decontamination technology. *Food Security, 12*(4), 814. https://doi.org/10.3390/foods12040814

Yan, P., Chelliah, R., Jo, K. H., Selvakumar, V., Chen, X., Jo, H. Y., & Oh, D. H. (2022, May 12). Stability and antibiofilm efficiency of slightly acidic electrolyzed water against mixed-species of *Listeria monocytogenes* and *Staphylococcus aureus*. *Frontiers in Microbiology, 13*, 865918. https://doi.org/10.3389/fmicb.2022.865918

Yan, X., Xu, Y., Shen, C., & Chen, D. (2023). Inactivation of *Staphylococcus aureus* by Levulinic acid plus sodium dodecyl sulfate and their antibacterial mechanisms on S. aureus biofilms by transcriptomic analysis. *Journal of Food Protection, 86*(3), 100050. https://doi.org/10.1016/j.jfp.2023.100050. Epub 2023 Jan 24. PMID:36916557.

Yang, M., Lee, G., Si, J., Lee, S. J., You, H. J., & Ko, G. (2016, October 20). Curcumin shows antiviral properties against norovirus. *Molecules, 21*(10), 1401. https://doi.org/10.3390/molecules21101401

Yang, G., Shi, Y., Zhao, Z., Zhong, M., Jin, T., Shi, C., Zhang, C., & Xia, X. (2021). Comparison of inactivation effect of slightly acidic electrolyzed water and sodium hypochlorite on *Bacillus cereus* spores. *Foodborne Pathogens and Disease, 18*(3), 192–201. https://doi.org/10.1089/fpd.2020.2811

Yordshahi, A. S., Shafipour, Y., Moradi, M., Tajik, H., & Molaei, R. (2020). Design and preparation of antimicrobial meat wrapping nanopaper with bacterial cellulose and postbiotics of lactic acid bacteria. *International Journal of Food Microbiology, 321*, 108561. https://doi.org/10.1016/j.ijfoodmicro.2020.108561. Epub 2020 Feb 11. PMID: 32078868.

Yuan, X., Li, Y., Mo, Q., Zhang, B., Shu, D., Sun, L., Yang, H., Xie, X., Liu, Y., & Zang, Y. (2022, September 30). A combined approach using slightly acidic electrolyzed water spraying and chitosan and pectin coating on the quality of the egg cuticle, prevention of bacterial invasion, and extension of shelf life of eggs during storage. *Food Chemistry, 389*, 133129. https://doi.org/10.1016/j.foodchem.2022.133129

Zang, Y. T., Bing, S., Li, Y. J., Shu, D. Q., Huang, A. M., Wu, H. X., Lan, L. T., & Wu, H. D. (2019, November 1). Efficacy of slightly acidic electrolyzed water on the microbial safety and shelf life of shelled eggs. *Poultry Science, 98*(11), 5932–5939. https://doi.org/10.3382/ps/pez373

Zhang, X. F., Dai, Y. C., Zhong, W., Tan, M., Lv, Z. P., Zhou, Y. C., & Jiang, X. (2012). Tannic acid inhibited norovirus binding to HBGA receptors, a study of 50 Chinese medicinal herbs. *Bioorganic & Medicinal Chemistry, 20*(4), 1616–1623.

Zhang L., Wang, Z., Song, Y., Li, M., & Yu, Q. (2020). Quality of vacuum packaged beef as affected by aqueous ozone and sodiumcitrate treatment. *International Journal of Food Properties, 23*(1),1475–1489. https://doi.org/10.1080/10942912.2020.1814322

Zhang, C., Zhao, Z., Yang, G., Shi, Y., Zhang, Y., Shi, C., & Xia, X. (2021, August). Effect of slightly acidic electrolyzed water on natural Enterobacteriaceae reduction and seed germination in the production of alfalfa sprouts. *Food Microbiology, 97*, 103414. https://doi.org/10.1016/j.fm.2020.103414

Zhang, X., Qiu, H., Ismail, B. B., He, Q., Yang, Z., Zou, Z., Xiao, G., Xu, Y., Ye, X., Liu, D., & Guo, M. (2024, January). Ultrasonically functionalized chitosan-gallic acid films inactivate Staphylococcus aureus through envelope-disruption under UVA light exposure. *International Journal of Biological Macromolecules, 255*, 128217. https://doi.org/10.1016/j.ijbiomac.2023.128217

Zhao, X., Zhen, Z., Wang, X., & Guo, N. (2017, December). Synergy of a combination of nisin and citric acid against Staphylococcus aureus and Listeria monocytogenes. *Food Additives & Contaminants. Part A, Chemistry, Analysis, Control, Exposure & Risk Assessment, 34*(12), 2058–2068. https://doi.org/10.1080/19440049.2017.1366076

Zheng, M. Z., Chen, W. X., Zhao, Y. X., Fang, Q., Wang, L. G., Tian, S. Y., Shi, Y. G., & Chen, J. S. (2024, August 1). Ascorbic acid potentiates photodynamic inactivation mediated by octyl gallate and blue light for rapid eradication of planktonic bacteria and biofilms. *Food Chemistry, 448*, 139073. https://doi.org/10.1016/j.foodchem.2024.139073

Zhou, Z., Zuber, S., Cantergiani, F., Butot, S., Li, D., Stroheker, T., Devlieghere, F., Lima, A., Piantini, U., & Uyttendaele, M. (2017, September 18). Inactivation of viruses and bacteria on strawberries using a levulinic acid plus sodium dodecyl sulfate based sanitizer, taking sensorial and chemical food safety aspects into account. *International Journal of Food Microbiology, 257*, 176–182. https://doi.org/10.1016/j.ijfoodmicro.2017.06.023

Zhou, Z., Zuber, S., Cantergiani, F., Sampers, I., & Devlieghere, F. M. (2018). Uyttendaele inactivation of foodborne pathogens and their surrogates on fresh and frozen strawberries using gaseous ozone. *Frontiers in Sustainable Food Systems, 2*, 51. https://doi.org/10.3389/fsufs.2018.00051

Zhou, M., Doyle, M. P., & Chen, D. (2020). Combination of levulinic acid and sodium dodecyl sulfate on inactivation of foodborne microorganisms: A review. *Critical Reviews in Food Science and Nutrition, 60*(15), 2526–2531. https://doi.org/10.1080/10408398.2019.1650249

Zhu, S., Barnes, C., Bhar, S., Hoyeck, P., Galbraith, A. N., Devabhaktuni, D., Karst, S. M., Montazeri, N., & Jones, M. K. (2020, April 19). Survival of human norovirus surrogates in water upon exposure to thermal and non-thermal antiviral treatments. *Viruses, 12*(4), 461. https://doi.org/10.3390/v12040461

Živković, I., Šavikin, K., Živković, J., Zdunić, G., Janković, T., Lazić, D., & Radin, D. (2021, June). Antiviral effects of pomegranate peel extracts on human norovirus in food models and simulated gastrointestinal fluids. *Plant Foods for Human Nutrition, 76*(2), 203–209.

The Biological Strategies

Xue Zhao and Lin Chen

1 Introduction

The use of synthetic chemicals effectively increases the durability and prolongs the shelf life of food products (Jongman et al., 2022). For example, fungicides prochloraz and bavistin have been effectively employed to preserve the freshness of avocados and mushrooms (Shimshoni et al., 2020). However, as synthetic chemicals gained traction in the early to mid-twentieth century, concerns arose regarding antimicrobial resistance (AMR), environmental pollution, and potential adverse health effects associated with their use (Fang et al., 2019; Gaston et al., 2020). Given the risks to human health and the increase in food safety incidents, it is not surprising that a shift from chemical to non-chemical applications, such as biological control, is necessary to ensure product quality and safety.

The term "biocontrol" refers to the use of biological agents or natural substances to prevent, reduce, or eliminate the growth of pathogenic microorganisms or pests in food products at the pre/post-harvest and processing stages (Jongman et al., 2022; Leneveu-Jenvrin et al., 2020). These biological strategies work in harmony with the environment, posing minimal risks to human health and leaving virtually no harmful residues on food. Embracing biocontrol approaches enables the food industry to bolster food safety and quality while reducing reliance on synthetic chemicals, thereby promoting sustainable and eco-friendly practices throughout the food supply

X. Zhao
School of Biological Sciences, Nanyang Technological University, Singapore, Singapore

Singapore Centre for Environmental Life Sciences Engineering, Nanyang Technological University, Singapore, Singapore

L. Chen (✉)
School of Chemistry, Chemical Engineering and Biotechnology, Nanyang Technological University, Singapore, Singapore
e-mail: chen.lin@ntu.edu.sg

D. Li (ed.), *Antimicrobial Strategies in the Food System: Updates, Opportunities, Challenges*, https://doi.org/10.1007/978-3-031-95056-8_5

chain (Oyetunji et al., 2024). Although various biological methods have been developed over the past few decades, their widespread adoption faces challenges due to limited applicability in existing food processing system (Jongman et al., 2022).

This chapter provides an overview of the latest biocontrol applications used at different points along the food supply chain. These applications include bacteriophages (phages), their endolysins and depolymerases, competitive bacterial and fungal species, as well as plant-derived products. We delve into the antimicrobial mechanisms of these biocontrol approaches and discuss their potential applications in the food industry.

2 Bacteriophage-Based Biocontrol Strategies

Phages are viruses that selectively infect and kill bacterial hosts, leaving other cells unaffected. This unique specificity makes them promising candidates for use as biopesticides or biocontrol agents in food production processes (Goodridge & Bisha, 2011). Phages were first described by William Twort in 1915, and further independently discovered by Félix d'Herelle in 1917. d'Herelle is widely considered a pioneer in applied phage science as he introduced the concept of phages in clinical medicine for both preventive and post-operative treatment (Kortright et al., 2019; Wittebole et al., 2014). Recognizing the selectivity and potency of phages, d'Herelle conducted lots of trials administering phages intravenously to combat invasive bacterial infections (Garvey, 2022). However, phage therapy became overlooked due to the challenges in quality control and reproducibility, compounded by the rapid progress in antibiotic discovery and development since the 1930s (Wittebole et al., 2014). On the other hand, the rise of AMR due to the overuse and misuse of antimicrobials has escalated into a global health crisis, leading to increasing treatment failures (Ferri et al., 2017). Therefore, novel strategies such as phages and their enzymes have recently received renewed attention (Garvey, 2022). Moreover, an increasing number of researchers are turning their focus to the application of phages as a promising strategy for pathogen control in food systems.

2.1 Bacteriophage

2.1.1 Antibacterial Mechanisms

As the most abundant life forms on Earth, phages outnumber bacteria by ten times, with an estimated total number of 10^{32} (Verheust et al., 2010). The extensive natural reservoir of phages presents myriad potential benefits in antibacterial contexts. These include broad-spectrum activity against diverse bacterial strains, including those exhibiting AMR, controlled antibacterial profiles preserving the indigenous

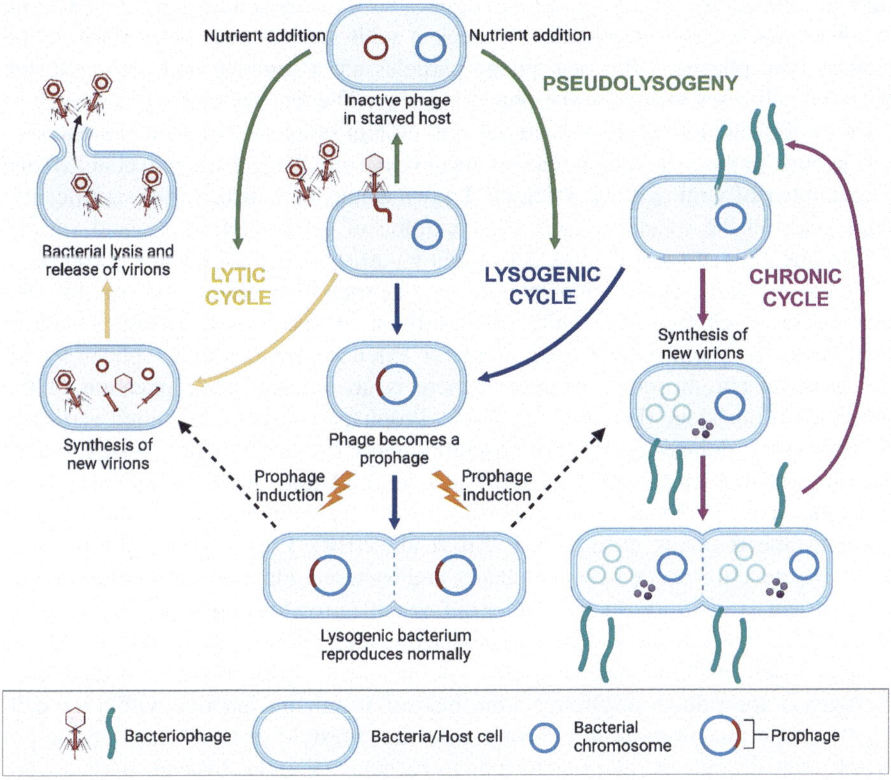

Fig. 1 Different life cycles of bacteriophages

microbiome, potential for minimal adverse effects, and cost-effectiveness (Helmy et al., 2023). As obligate intracellular parasites of bacterial cells, phages exhibit diverse life cycles within the bacterial host, including lysis, lysogeny, pseudolysogeny, and chronic infection (Fig. 1) (Huang et al., 2022; Wittebole et al., 2014). In phage therapy, primary focus is on lytic phages, which predominantly belong to three families within the *Caudovirales* order: *Myoviridae*, *Podoviridae*, and *Siphoviridae*. Additionally, there are reports on the applications of filamentous and cubic phages in bacterial inactivation (Villalpando-Aguilar et al., 2022; Wittebole et al., 2014).

Regardless of the phage life cycle, the initial stage involves attachment to host cells via different host receptors, including proteins, carbohydrates, or lipopolysaccharides, before penetrating the bacteria (Fig. 1) (Dunne et al., 2021). This process significantly influences the spectrum of potential phage-bacteria interactions. For example, phage λ specifically interacts with the LamB receptor of *Escherichia coli*. Spatiotemporal dynamics highlight the critical role of this event in successful bacterial invasion (Chatterjee & Rothenberg, 2012). After attaching to their specific receptor, phages create a pore in the bacterial cell wall and inject their nucleic acid into the cell. Subsequently, phage early genes are expressed, hijacking the bacterial synthetic machinery to replicate viral nucleic acids and proteins, particularly in the case of lytic phages. After new phage particles are assembled and packaged, the bacterial cell lyses to release the phage progeny. The late phage enzymes, such as lysins, holin, and inhibitors of bacterial wall protein synthesis, are then exploited for virion burst in the extracellular environment (Du Toit, 2017; Wittebole et al., 2014). The number of viral particles released, known as the burst size, varies significantly depending on the specific phage, the condition of the bacterial host, and various environmental factors, including surrounding nutrient levels (Weinbauer, 2004).

In the lysogenic cycle, temperate phages integrate their genetic material into the host bacterial chromosome either randomly or at specific sites (Fig. 1). Once integrated, the prophage remains silent for extended periods and replicates with the bacterial chromosome. Therefore, there is no independent replication of the prophage at this stage (Feiner et al., 2015). Prophage DNA is transmitted vertically to its progeny along with the entire bacterial genome until induction occurs, due to various environmental stress factors, such as oxidation, heating, and ultraviolet radiation. This induction results in excision of the viral genome from the host chromosome (Ładziak et al., 2024; Xu et al., 2018). Pseudolysogeny represents another interaction mode between phages and bacteria, initiated under conditions of low nutrient availability, such as under unfavorable growth conditions (Fig. 1). Ripp and Miller (1997) described pseudolysogeny as a phage-host interaction which the phage nucleic acid neither integrates into the host chromosome as a prophage (lysogeny) nor induces a lytic reaction. Instead, it remains inactive within the cell. Lastly, some phages exhibit a chronic infection lifestyle in which host cell is not lysed upon the release of progeny phage particles (Fig. 1). Instead, particles are continuously excreted to the extracellular environment through the membrane. Depending on the phage, its genome may be integrated into the host genome or remain in the cytoplasm (Mäntynen et al., 2021). Therefore, lysogenic, pseudolysogenic, and chronic infection life cycles are less effective at causing immediate death of the host bacteria. On the other hand, phages with a lytic life cycle are superior to temperate phages in terms of therapeutic potential.

2.1.2 Phage Biocontrol in Pre/Post-harvest Applications

The use of phages is gaining recognition as a sustainable biocontrol technology, noted for its remarkable specificity in targeting bacterial pathogens across various environments, covering each stage of the food processing from farm to table. Recently, with Generally Recognized As Safe status for the control of harmful pathogens in the food industry, some phage products have been introduced into the commercial market to address common foodborne pathogens, including *E. coli*, *Listeria monocytogenes*, and various *Salmonella* serotypes (de Melo et al., 2018). This advancement is highly promising, highlighting the industry's confidence in the effectiveness and safety of phage-based formulations. Many commercial phage companies, including FINK TEC GmbH, Phagelux, Intralytix, Passport Food Safety

Solutions, and Micreos Food Safety, have received FDA approval for their food safety products (Vikram et al., 2022).

Phage therapy has emerged as a promising pre-harvest intervention strategy for controlling pathogens in various meat-producing animals, including sheep, cattle, poultry, and swine, before slaughter (Goodridge & Bisha, 2011; Wittebole et al., 2014). To date, extensive studies on phage therapy aimed at controlling foodborne pathogens in poultry have been conducted, demonstrating significant efficacy in pathogen reduction. For instance, in a study conducted by Clavijo et al. (2019), the effectiveness of SalmoFREE®, a newly patented phage preparation designed to target *Salmonella*, was assessed throughout the production cycle of broiler chickens. Cloacal swab results indicated that the phage product successfully decreased *Salmonella* counts to 0% by day 34. These *in vivo* studies investigated the efficacy of employing phages either individually or in combination (cocktail) to combat foodborne pathogens. Generally, using a single phage in certain experiments leads to the development of phage resistance. Conversely, the use of multiple phages in a cocktail reduces the likelihood of resistance emergence (Bach et al., 2003; Raya et al., 2011).

Foods offer a nutrient-rich environment that supports the survival and proliferation of various bacterial pathogens. Numerous studies have documented intervention strategies utilizing phages to reduce pathogens in post-harvest food materials. The literature shows that carefully selected phages can effectively reduce the prevalence of specific harmful pathogens. In a recent study focusing on meat systems, Vikram et al. (2020) conducted a comprehensive investigation of the Intralytix phage cocktail EcoShield PX™, specifically targeting Shiga toxin-producing *E. coli*. Their results showcased the effectiveness of these phages in reducing *E. coli* levels (by 3.0 log CFU/g) across various food products, including ground beef, beef chuck roast, cooked chicken, chicken breast, salmon and cheese. In fruits and vegetables, Wong et al. (2020) explored the efficacy of a five-component phage cocktail in controlling seven *S. enterica* strains across four different serovars, including Enteritidis, Newport, Javiana, and Thompson. The study involved inoculating cantaloupe and romaine lettuce leaves with the phage cocktail 24 hours prior to bacterial inoculation. Results varied significantly across different *Salmonella* targets, highlighting that while the phages demonstrated potential for *Salmonella* biocontrol, their success was influenced by the strain specificity.

2.2 *Phage Enzymes*

2.2.1 Endolysin

The use of endolysins to control the proliferation of pathogenic microorganisms in food represents a relatively novel concept that has gained increasing interest in recent years. Endolysins are enzymes encoded by all phages, synthesized within the infected host bacterium at the end of the lytic cycle. These enzymes efficiently cleave peptidoglycan, the major component of the bacterial cell wall, resulting in cell lysis and subsequent release of progeny phages (Chaudhary et al., 2024). Gram-positive bacteria, unlike their Gram-negative counterparts, lack an outer membrane structure outside the cell wall. This characteristic allows endolysins to exert bacteriolytic activity with high specificity and efficacy when externally applied, without harming non-target bacteria (Zheng & Zhang, 2024). Nonetheless, significant progress has been made in using endolysins specific to Gram-negative pathogens like *Pseudomonas aeruginosa*, *Campylobacter* spp., *E. coli*, *Neisseria gonorrhoeae*, *Salmonella* spp., *Vibrio* spp., *Klebsiella pneumoniae*, and *Acinetobacter baumannii* (Endersen & Coffey, 2020). In addition, recent advancements in evaluating the bactericidal efficacy of endolysins have been demonstrated in various food-related applications. For example, the endolysin LysP53 has shown the capability to lyse *S.* enteritidis found on fresh romaine lettuce, indicating its potential as a biocontrol agent for reducing bacterial loads in fresh vegetable produce (Khan et al., 2023). Another study investigated the combined action of LysSA97 and carvacrol oil to combat *Staphylococcus aureus* in milk and beef. The results showed that the combined treatment resulted in an average reduction of 4.5 log CFU/mL in *S. aureus* counts across the tested food samples (Chang et al., 2017).

2.2.2 Depolymerase

Depolymerases, categorized as tail spike proteins, can degrade extracellular polysaccharides, including those forming capsules or biofilm matrices. Additionally, they can cleave structural polysaccharides such as lipopolysaccharide or peptidoglycan strands (Fernandes & São-José, 2018). Phages that express depolymerase activity exhibit improved access to host receptors due to their ability to break down capsular polysaccharides. This feature is believed to enhance efficacy against bacterial biofilms (Pires et al., 2016). Therefore, in devising phage or phage cocktail biocontrol strategies, incorporation of phages encoding polysaccharide depolymerases could bolster efficacy, broaden specificity, and facilitate biofilm removal. Additionally, employing depolymerases sourced directly from phages introduces a novel strategy for combating pathogenic bacteria and their biofilms. For example, the depolymerase LKA1gp49, derived from the *Pseudomonas* phage LKA1, effectively attaches to and cleaves the B-band lipopolysaccharide, thereby reducing the virulence of *P. aeruginosa* (Olszak et al., 2017). Furthermore, the depolymerase has shown effectiveness in eliminating *Klebsiella* sp. biofilms from food contact surfaces, achieving an 80% reduction after 4 h (Garvey, 2022). In practice, depolymerases offer greater advantage and flexibility compared to phages, avoiding certain disadvantages like the necessity for purification and endotoxin removal. Beyond endolysins and depolymerases, other phage-related enzymes involved in peptidoglycan degradation, such as virion-associated peptidoglycan hydrolases, have also received attention in recent years (Gutiérrez et al., 2018).

3 Bacteria-Based Biocontrol Strategies

Another biocontrol strategy involves the utilization of bioprotective microorganisms and/or their metabolites, which exhibit antagonistic effects against spoilage and pathogenic microorganisms, to enhance microbial safety and extend the shelf life of food products (Rathod et al., 2022). In addition to demonstrating antagonistic activity, these protective microorganisms must be safe for human consumption and must not negatively affect the nutritional or sensory quality of the product. Moreover, certain antagonistic microorganisms, known as probiotics, provide additional benefits. These include regulating the intestinal microflora balance, alleviating the impacts of allergens, easing lactose intolerance, lowering cholesterol levels, fostering immune response, and enhancing resistance to infections (Hossain et al., 2017; Kaur et al., 2015). Most probiotics are bacteria, with lactic acid bacteria (LAB) being the most prevalent type. Nonetheless, certain molds and yeasts are also utilized as probiotics. Among the frequently employed probiotic species are *Lactobacillus*, *Bifidobacterium*, *Carnobacterium*, *Lactococcus*, *Streptococcus*, *Enterococcus*, *Pediococcus*, *Bacillus*, *Leuconostoc*, and *Propionibacterium* (Hossain et al., 2017). Of these, *Lactobacillus* and *Bifidobacterium* are the most extensively researched genera.

3.1 Probiotics

3.1.1 Mechanism of Probiotic Action

The mechanisms underlying bacterial antagonism or antimicrobial activity are pivotal for maintaining equilibrium between beneficial and potentially pathogenic microorganisms (Hossain et al., 2017). The inhibitory ability results from multiple mechanisms, including competition for colonization sites, and nutrients. Additionally, probiotic microorganisms produce a variety of antimicrobial metabolites, such as bacteriocins, hydrogen peroxide, organic acids, biosurfactants, and enzymes, that can proficiently inhibit the growth of pathogens (Oliveira et al., 2018). For instance, the probiotic bacterium *L. reuteri* releases antimicrobial reuterin, which exhibits broad-spectrum activity against a range of pathogens, including bacteria, fungi, protozoa, and viruses (Spinler et al., 2008). Probiotic LAB produce organic acids, primarily acetic and lactic acids, to reduce the likelihood of intestinal colonization by pathogens. These organic acids create a favorable environment for the resident microbiota by lowering intestinal pH, thereby establishing an acidic milieu that inhibits the proliferation of pathogens (Servin, 2004). Moreover, some probiotics utilize enzymatic mechanisms, releasing enzymes capable of hydrolyzing bacterial toxins, modifying toxin receptors, and inhibiting toxin-mediated diseases. The LAB strain *Ped. parvulus* can biodegrade the mycotoxin ochratoxin A by hydrolyzing the amide group using specific peptidases, subsequently releasing non-toxic moieties (Ghadaksaz et al., 2022).

By releasing antimicrobial compounds, probiotics also exhibit the capacity to inhibit the formation of biofilms by pathogens, serving as a defense mechanism against infections. For example, the bacteriocin produced by probiotics significantly reduced the biofilm of numerous human pathogens, such as *S. aureus*, *S. pyogenes*, *K. pneumoniae*, *Shigella flexneri*, *Candida albicans*, and *Aspergillus fumigatus*, through structural deformation and disruption of bacterial membrane integrity (Krishnamoorthi et al., 2022). Some probiotic candidates prevent pathogen colonization, competitively inhibiting the pathogen adhesion to host cell surfaces (Hossain et al., 2017). This mechanism relies on specific proteins present on the surface of probiotic agents, such as S-layer (surface-layer) proteins found in *L. helveticus* and *L. crispatus*, which can inhibit the adhesion of pathogenic *E. coli* O157: H7 to T84, HEp-2, and HeLa cells (Johnson-Henry et al., 2007). It's important to note that the action mechanism of each probiotic depends on the specific strain, emphasizing the need to carefully select and scientifically evaluate individual probiotics for their intended use. These mechanisms typically involve the production of bioactive molecules or direct cell-to-cell interactions (Hossain et al., 2017).

3.1.2 Probiotics Applications

Similar to phage products, probiotics can be applied both pre-harvest on farms and post-harvest in foods. In livestock farming, probiotics are used to improve growth performance and control enteric pathogens (Hung et al., 2012). Therefore, probiotics could be a promising alternative to antibiotic feed additives. For example, some studies have highlighted poultry and their products as significant reservoirs of *Campylobacter* spp. Campylobacteriosis, caused by various species such as *C. jejuni*, *C. upsaliensis*, *C. coli*, and *C. lari*, is a prevalent zoonotic disease commonly associated with poultry production (Guyard-Nicodème et al., 2015). Some *in vitro* studies have successfully identified anti-*Campylobacter* probiotics, including strains of *L. salivarius* (Messaoudi et al., 2011). Additionally, *in vivo* trials have also reported the effective role of probiotics in inhibiting or reducing the shedding of *C. jejuni* load. Nishiyama et al. (2014) demonstrated the effectiveness of *L. gasseri* SBT2055 in inhibiting *C. jejuni* adherence and invasion of in vitro, as well as reducing the *C. jejuni* load in chickens in vivo (Table 1). Similarly, Caggia et al. (2015) showed that two probiotic strains, *L. rhamnosus* FS10 and *L. paracasei* PM8, isolated from cheese, exhibited strong antimicrobial activity in vitro against various foodborne pathogens, including *E. coli*, *S. aureus*, *L. monocytogenes*, and *E. faecalis*. The authors suggest these strains as promising candidates for incorporation into functional foods aimed at enhancing food safety.

Additionally, supplementing probiotics in other livestock, including cattle, sheep, and swine, has shown effective control and reduction of pathogen load (Table 1). Peterson et al. (2007) documented that daily administration of a feed supplement containing *L. acidophilus* NP51 to cattle led to a notable decrease in fecal *E. coli*

Table 1 The biocontrol use of probiotics in different pre/postharvest applications

Foods	Probiotics	Target pathogen	Observed effect	References
Chicken	*L. gasseri* SBT2055	*C. jejuni*	1–2 log CFU/well reduction	Nishiyama et al. (2014)
Cattle	*L. acidophilus* NP51	*E. coli* O157:H7	Reduction of counts in fecal sample	Peterson et al. (2007)
Sheep	*L. alimentarius* DDL 48, *L. reuteri* DDL 19, *B. bifidum* DDBA, and *E. faecium* DDE 39	*Shigella* & *Salmonella*	1.21 log and 1.09 log CFU/g reduction, respectively	Apás et al. (2010)
Swine	*L. fermentum I5007*	*Clostridium* spp. & *Escherichia* spp.	Decreased Clostridium spp. & *Escherichia* spp. levels	Liu et al. (2014)
Lettuce	*L. curvatus B.67, L. sakei D.7, L. plantarum I.60, L. mesenteroides J.27, L. plantarum M.2*, or *L. plantarum M.21*	*L. monocytogenes*	1.07–1.62 log CFU/cm^2 reduction	Hossain et al. (2020)
Cheese	*L. acidophilus La-5 and B. longum ATCC15707*	*E. coli* O157:H7 and *S. aureus*	1.7–1.88 log and 0.88–3 log cycle reductions, respectively	El-Kholy et al. (2014)
Sausage	*L. rhamnosus* E-97800, *L. rhamnosus* LC-705, *L. rhamnosus* GG, and *P. pentosaceus*	*E. coli* O157:H7	Decreased counts from around 5 to 2 log CFU/g	Erkkilä et al. (2000)
Mung bean sprouts	*B. subtilis LCA1*	EHEC cocktails	About 2.0 log CFU/g EHEC reduction	Shen et al. (2017)

O157:H7 counts. Moreover, the inclusion of *L. fermentum* I5007 in feed inhibited the growth of *Clostridium* spp. and *Escherichia* spp. in piglets (Liu et al., 2014). Apás et al. (2010) demonstrated that the administration of a probiotic mixture, containing *E. faecium*, *B. bifidum*, *L. reuteri*, and *L. alimentarius*, resulted in a significant reduction in *Shigella* and *Salmonella* in sheep feces by 1.21 and 1.09 log CFU/g, respectively. These findings demonstrate the potential antimicrobial activity of probiotics against significant foodborne pathogens, effectively limiting their colonization in animals. As a result, the use of probiotics can enhance animal growth rates and improve the quality and yield of meat, milk, and egg products.

Fresh produce, such as fruits and vegetables, is susceptible to microbial contamination. Table 1 also outlines various biocontrol applications of probiotics in foods. According to Hossain et al. (2020), six LAB strains isolated from kimchi demonstrated promising antimicrobial efficacy against *L. monocytogenes* on lettuce leaves. Vescovo et al. (1996) and Torriani et al. (1997) pioneered the use of *L. casei* to combat pathogens in ready-to-eat vegetables, highlighting its inhibitory properties against pathogens. Torriani et al. (1997) noted that the incorporation of 3% culture permeates of *L. casei* IMPC LC34 in salads led to decreased total mesophilic bacteria counts and inhibited the growth of enterococci, coliforms, and *A. hydrophila*. These probiotics are also widely used in fermented foods such as cheese, yogurt, kimchi, and sausage. For instance, the application of probiotics *B. longum* ATCC15707 and *L. acidophilus* La-5 has shown remarkable effectiveness in preventing the growth of *E. coli* O157:H7 and *S. aureus* during the production and storage of domiati cheese (El-Kholy et al., 2014). Erkkilä et al. (2000) reported that the probiotic *L. rhamnosus* E-97800, *L. rhamnosus* LC-705, *L. rhamnosus* GG, and commercial *P. pentosaceus* effectively suppressed the proliferation of pathogenic *E. coli* O157:H7 in dry sausage.

3.2 Other Antagonistic Bacteria

In addition to probiotics, alternative bacteria have been explored for pathogen control in food. Janisiewicz et al. (1999) successfully applied the antagonistic *P. syringae* L-59-66 to suppress the growth of *E. coli* O157:H7 on wounded apple tissue. The observed antagonistic effect is believed to stem from competition for both space and nutrients. Moreover, Alegre et al. (2013) showcased the antagonistic properties of *P. graminis*, a Gram-negative bacterium isolated from 'Golden Delicious' apples, against *E. coli* O157:H7 present in fresh-cut peach and apple. Similarly, *B. subtilis* isolates, collected from mung bean seeds and lettuce stems, were utilized in another study to surpress *E. coli* O157:H7 contamination in mung bean sprouts (Shen et al., 2017). The effectiveness of antagonistic approaches is subject to several factors, such as storage temperature, nutrient availability, inoculation concentration, and biofilm formation. For instance, *P. fluorescens* served as a biocontrol agent to regulate the growth of *E. coli* O157:H7 on spinach. Notably, a reduction in pathogen viability ranged from 0.5 to 2.1 log CFU/g, was recorded. Crucially, the efficacy of biocontrol was notably influenced by storage temperature, with more pronounced inhibitory effects observed at lower temperatures (Olanya et al., 2013). In addition, Kim et al. (2013) reported that the biofilm-forming strain *Paenibacillus polymyxa* T5 displayed potent antimicrobial activity against *E. coli* O157:H7, in comparison to antagonistic cells lacking biofilm formation.

4 Fungi-Based Biocontrol Strategies

A diverse range of fungal species has been employed as biocontrol agents against pathogens in food. The primary factors driving the use of fungal biocontrol are their notably high reproductive rates, both sexually and asexually, along with their short generation times and target-specific properties (Thambugala et al., 2020). Additionally, their efficacy in eradicating host organisms, extensive metabolic diversity facilitating the discovery of suitable isolates for biocontrol, and relative environmental safety due to their role as primary decomposers, all contribute to their expanded usage (Thomas & Read, 2007). Among the numerous fungal species demonstrating potential and applied in biocontrol practices, filamentous fungi (i.e., moulds) and yeasts stand out as the most researched and utilized.

4.1 Mechanisms Underlying the Fungal Biocontrol Activity

Similar to antagonistic bacteria, fungal species exhibit multiple mechanisms to control food and crop pathogens (Fig. 2). These include the release of antifungal metabolites, enzymes, and volatile organic compounds (VOCs), competition for nutrients and space, mycoparasitism, and induction of resistance and growth enhancement (Freimoser et al., 2019). Certain filamentous fungi and yeast species can impede fungal pathogens through competitive mechanisms, resulting in nutrient starvation, a primary cause of microbial mortality and reduced spore generation (Savita & Sharma, 2019). Iron is considered one of the most sought-after nutrients, and competition for it is acknowledged as a significant mechanism of action (Wachowska & Borowska, 2014). When faced with iron deficiency, filamentous fungi produce siderophores, aiding in its iron uptake from the environment (Eisendle

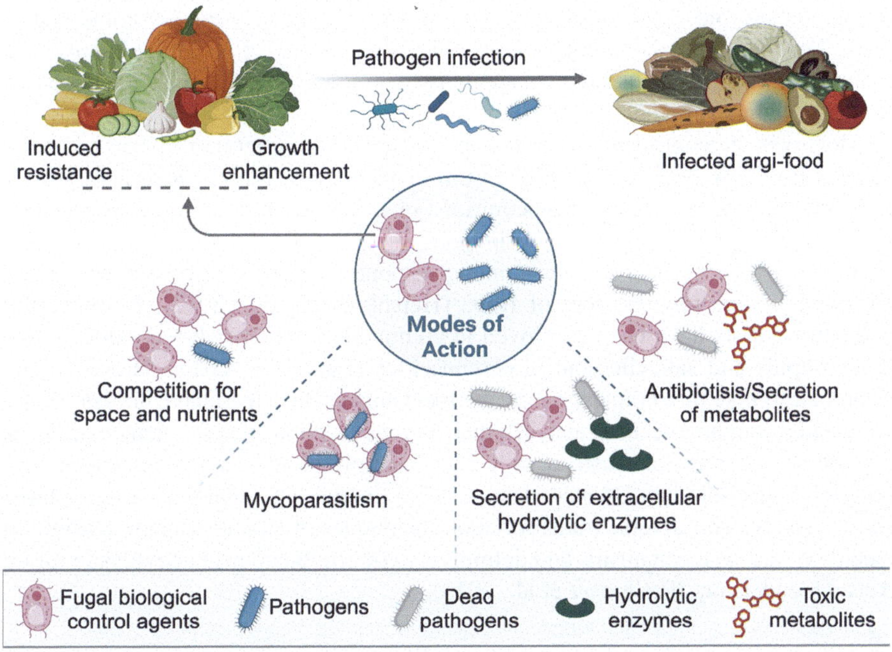

Fig. 2 Multiple action modes of fungi-based biocontrol strategies in agri-foods

et al., 2004). Regarding competition for space, the formation of biofilms stands out as a particularly effective strategy. In fungi, biofilm formation exhibits significant similarities to bacterial counterparts in terms of structure, function, and mechanisms, especially evident in budding yeasts (Bartholomew et al., 2021). The initiation of yeast biofilm formation typically involves individual cell adhesion to a surface, accompanied by cell wall modifications, extracellular matrix secretion, and frequently the development of hyphae or pseudohyphae (Cavalheiro & Teixeira, 2018).

The efficacy of beneficial fungi in biocontrol also relies on the production of secondary metabolites and/or lytic enzymes (Savita & Sharma, 2019). Among the most notable filamentous species exhibiting this trait are *Trichoderma* and *Gliocladium*, known for synthesizing and releasing potent antifungal antibiotics such as gliotoxin, viridin, and gliovirin (Mendoza et al., 2015). The production and release of enzymes that degrade cellular components is a general phenomenon in various interactions between fungi and pathogens. Typically, these enzymes are upregulated under conditions of nutrient scarcity to facilitate the liberation of nutrients from "prey" cells, potentially leading to their demise (Calderón et al., 2019). On the other hand, proteinaceous killer toxins, initially discovered in *S. cerevisiae*, are particularly notable metabolites produced by biocontrol yeasts. These toxins, primarily composed of proteins, play a crucial role in the antagonistic interactions between fungi and other microorganisms, often serving as potent weapons in the competition for resources and ecological dominance. Research on yeast killer toxins has predominantly focused on controlling spoilage yeasts in the food and beverage industry or exploring medical uses (Liu et al., 2015). Recent findings also highlight the significant role of the yeast volatilome in interactions between yeast and pathogens. For instance, the ability of specific food yeasts (e.g., *A. pullulans* and *S. cerevisiae*) to act as biocontrol agents is mainly attributed to their formation of 2-phenylethanol (Parafati et al., 2015).

Moreover, biocontrol fungi are capable of inducing systemic resistance in agricultural products against a wide spectrum of pathogens, a trait thought to greatly enhance their effectiveness in biocontrol. For instance, species like *S. cerevisiae*, *Rhodosporidium paludigenum*, *C. saitoana*, *C. oleophila*, and *Metschnikowia* have been shown to trigger an innate immune response, resulting in resistance against pathogens in the phyllosphere of fruits (Hershkovitz et al., 2012). Additionally, biocontrol yeasts have been employed in conjunction with resistance inducers like rhamnolipids and salicylic acid in certain cases (Yan et al., 2014). However, it's important to note that in many cases, the mechanisms described have not been fully validated through molecular analyses such as gene deletion and complementation, or heterologous expression. Instead, they have been proposed based on analogies with other biological systems. Nonetheless, with the increasing availability of annotated yeast genomes and diverse transformation techniques, it should become feasible to elucidate various mechanisms and definitively confirm biocontrol activities in future research endeavors (Freimoser et al., 2019).

4.2 *Application of Fungal Biocontrol Agents in Argi-foods*

4.2.1 Application of Biocontrol Filamentous Fungi

Filamentous fungi have emerged as highly effective biocontrol agents in safeguarding agricultural products. These fungal species include both basal and higher fungi. Among the basal fungi, representatives of the subphylum *Entomophthoromycotina* stand out, with species from genera like *Conidiobolus*, *Erynia*, and *Entomophaga* being particularly notable (Baron et al., 2019). Among higher filamentous fungi, numerous species within the phylum Ascomycota have shown promise, with several already in commercial use and application. These include *Purpureocillium*, *Aureobasidium pullulans*, *Curvularia pallescens*, *Metarhizium*, *Beauveria*, *Fusarium*, *Clonostachys*, *Epicoccum*, and various *Trichoderma* species (Faria et al., 2017). Filamentous fungi have been recognized as sustainable biological control agents suitable for mitigating many post-harvest pathogens affecting fruits, such as guava, mango, strawberry, apple, banana, jujube, citrus, and kiwifruit (Adetunji & Varma, 2020). *A. pullulans* was utilized in an effort to control *C. acutatum* in apples; however, this biocontrol presented limited effectiveness, achieving only a 44% inhibition rate (Di Francesco et al., 2015). In another study, *E. dendrobii* exhibited the ability to penetrate apple fruit through stomatal cells, inhibiting the germination of conidia and the formation of appressoria by *C. gloeosporioides* (Bian et al., 2021). Its antifungal effectiveness was significantly higher when used preventatively compared to post-infection, completely halting pathogen advancement. Compared to bacterial strains such as *Amycolatopsis* sp., *B. subtilis*, and *P. polymyxa*, fungal *E. dendrobii* exhibited superior control efficacy against *C. gloeosporioides* in apples. Filamentous fungi have also been extensively employed to control *C. musae* in bananas. Alvindia and Natsuaki (2008) and Sangeetha et al. (2009) showed the exceptional biocontrol capabilities of *Trichoderma* in managing crown rot in postharvest bananas, a disease caused by several pathogenic fungi, such as *F. verticillioides*, *Lasiodiplodia theobromae*, and *Colletotrichum musae*.

4.2.2 Application of Biocontrol Yeast

Among the biocontrol yeasts valued by the food and industrial sectors, *Saccharomyces* yeasts are particularly noteworthy. Among them, *S. cerevisiae* stands out prominently. This yeast has played a pivotal role in bread, beer, and wine production for centuries and holds the distinction of being the first genetically manipulated eukaryote (Di Canito et al., 2021). In the last two decades, researchers have explored the use of non-*Saccharomyces* as biocontrol solutions in the agri-food industry, recognizing their critical antagonistic activity against undesirable microorganisms (Jolly et al., 2003). This realm may encompass the utilization of killer toxins (zymocins or mycocins), antimicrobial peptides (e.g., Lactoferricin B), or VOCs generated by non-*Saccharomyces* yeasts as natural control strategies to combat contamination (Liu et al., 2015). For example, the increasing interest in minimizing the use of sulfur dioxide in the winemaking industry has spurred scientific investigation into the antimicrobial attributes of non-*Saccharomyces* yeasts as a substitute for conventional chemical additives (Agarbati et al., 2023). Some non-*Saccharomyces* yeasts utilized in winemaking exhibit the ability to manage filamentous fungi as well as spoilage yeasts in the vineyard and during the initial fermentation stage (Sizzano et al., 2023). They can specifically mitigate the presence of spoilage species like *Brettanomyces bruxellensis* and *Zygosaccharomyces rouxii* in bulk wine, through the release of active antimicrobial extracellular molecules (Berbegal et al., 2018).

Research into the antimicrobial characteristics of non-*Saccharomyces* yeasts has expanded beyond winemaking, indicating a broader exploration of their potential for biocontrol purposes across the food industry. Practical studies have showcased their effectiveness in cheesemaking and breadmaking contexts. Additionally, non-*Saccharomyces* yeasts have been widely applied to ensure the safety of pre-/post-harvest fruits and plants. They are used as potential antagonists against phytopathogenic fungi like *Penicillium*, *Aspergillus* genera, as well as *B. cinerea*, which affect table grapes, wine grapes, and raisins (Di Canito et al., 2021). For instance, *R. fluviale* and *R. paludigenum* have been employed to inhibit the growth of *A. japonicas*, *A. uvarum*, and *A. aculeatus* on post-harvest table grapes, through the secretion of lytic enzymes (Tilocca et al., 2019). In recent studies, yeast species from the *Pichia*, *Wickerhamomyces*, *Metschnikowia*, *Dekkera*, and *Rhodotorula* genera have emerged as effective inhibitors. This diversity offers a range of species with potential as antagonists against pathogenic fungi affecting grapes and raisins (Tryfinopoulou et al., 2019). Moreover, Leverentz et al. (2006) clearly indicated the effectiveness of *M. pulcherrima*, isolated from apple surfaces, against various foodborne pathogens, such as *S. enterica* and *L. monocytogenes*, on fresh-cut apple tissue.

5 Plant-Derived Antimicrobial Products

Plants harbor diverse natural compounds endowed with antimicrobial properties. Many studies have demonstrated the antimicrobial properties of plant extracts derived from herbs and spices. Furthermore, phytochemicals in plants, including saponins, tannins, steroids, alkaloids, and flavonoids, significantly contribute to their antimicrobial properties. The structural arrangement of these compounds greatly influences their effectiveness (Gyawali et al., 2015; Quinto et al., 2019). Indeed, the antimicrobial attributes of these natural products are often complemented by their antioxidant capacity. When both properties coexist within a molecule, the compound becomes more potent in combating microorganisms (Pisoschi et al., 2018). Gram-positive bacteria typically exhibit a stronger response to plant extracts than Gram-negative counterparts. The discrepancy primarily results from the presence of lipopolysaccharide cell wall in Gram-negative bacteria, which obstructs the diffusion of hydrophobic compounds (Nakamura & Hatanaka, 2002). Furthermore, the rise in antibiotic resistance in pathogens has prompted research to explore plant extracts as potential alternatives. These extracts hold the promise of solving the challenge of antibiotic resistance and may yield superior results compared to synthetic preservatives. Additionally, research shows that plant extracts can decrease antibiotic resistance through fostering synergistic interactions between natural antibiotics and antimicrobials (Li et al., 2023). This chapter provides an overview of the action mechanisms of plant-derived antimicrobials, their efficacy, and potential usage in mitigating pathogens in foods.

5.1 Plant Derived Compounds

5.1.1 Antimicrobial Mechanisms of Plant Extracts

The exact action mechanisms of plant extracts are often undisclosed or not fully elucidated. Table 2 summarizes some commonly used plant extracts and compounds along with their respective antimicrobial action against target pathogens. It's worth noting that the type of damage induced may vary among bacteria, even when exposed to the same antimicrobial substance. Additionally, some action modes have been reviewed for different antimicrobial groups, and some potential modes are listed below:

- Essential oils (EOs) cause functional and structural injury to bacterial cell membranes, and the hydrophobicity range plays a crucial role in their toxicity.
- Flavonoids exert their antimicrobial effects primarily by penetrating the bacterial cell membrane, leading to membrane damage and inducing alterations in membrane potential, intracellular pH, and adenosine triphosphate (ATP) synthesis.
- Membrane-disrupting compounds are able to induce cellular content leakage, disrupt active transport or metabolic enzyme function, and dissipate cellular energy metabolism.
- Certain extracts can directly lower the pH of the growth medium or substrate by increasing proton concentration, depressing intracellular pH through ionization of undissociated acid molecules, or disrupting substrate transport.
- Organic acids in plant extracts might inhibit NADH oxidation, thereby disrupting the supply of reducing agents to electron transport systems.
- Organic acids, particularly short-chain organic acids, can also alter membrane permeability and disrupt energy metabolism.
- Antimicrobial compounds from plants target the phospholipid bilayer, disrupt cell enzyme proteins, compromise the bacterial genetic materials, or generate fatty acid hydroperoxides through the oxygenation of unsaturated fatty acids.

The possible action modes mentioned above are commonly cited mechanisms to explain the antimicrobial effects of natural plant-derived products. Nevertheless, it's

Table 2 Antimicrobial mechanisms of selected plant-derived extracts

Plant extract/ compound	Target microorganism	Action mode	References
Grape seed extract	*L. monocytogenes* and *S. aureus*	Cell membrane permeability and intracellular unbalanced glycolysis and amino acid metabolism; partial disintegration of bacterial cell surfaces resulting in reduced residual cellular content	Al-Habib et al. (2010), Zhao et al. (2020a, b)
Thyme EO	*L. monocytogenes*	Cell wall disruption and increased roughness	Rasooli et al. (2006)
Cinnamon EO and cinnamaldehyde	*S. enteritidis*, *E. coli* O157:H7, *E. coli*, and *L. Monocytogenes*	Lower the content of adenosine triphosphate (ATP) and inhibit the activity of ATPase	Zhang et al. (2022)
Roselle, clove, and rosemary extract	*E. coli* and *S. aureus*	Cell wall disruption and decrease in cytoplasmic pH	Gonelimali et al. (2018)
Lindera glauca EO	*S. flexneri*	Cell membrane damage, metabolic function disruption, and disturbance of redox homeostasis	Chen et al. (2021)
Citrus EO	*Enterococcus* spp.	Morphological alternations	Fisher and Phillips (2008)
Garlic extract	*S.* Hadar	Loss of enzymatic activities, cell wall rupture, and nonhomogeneous distribution of cytoplasmic material	Belguith et al. (2009)
Green, oolong, black, and Fuzhuan tea extract	*S. typhimurium*, *S. aureus*, *E. coli*, and *E. faecalis*	Increased cell membrane permeability and changes in the relative electrical conductivity by catechins	Liu et al. (2022)

important to note that specific plant compounds may exhibit varying effects on different groups of target pathogens (Gyawali et al., 2015).

5.1.2 Applications of Plant Extracts in Foods

While research on plant extracts as antimicrobial agents in the food industry is limited, studies have shown their potential against microbial pathogens. Plant-derived antimicrobial agents are primarily utilized in food systems in the form of biofilms and edible coatings. These specialized films and coatings gradually permeate the food package, thereby prolonging the antimicrobial effect over time. For instance, Ouattara and Mafu (2003) documented an extension of the shelf life of shrimp by 20–21 days through the application of a protein coating containing 0.9% thyme oil and 1.8% trans-cinnamaldehyde, coupled with exposure to 3.0 kGy gamma rays. Moreover, grape seed extract has exhibited antimicrobial properties either independently or in conjunction with other hurdle technologies across various food applications, including cooked shrimp and fish fillets (Zhao et al., 2019, 2021; Zhao, Chen, Zhao, et al., 2020b). However, in the food industry, the use of plant extracts, like EOs, presents some challenges. Because the food matrix interferes with the active compounds in these extracts, their efficacy is often higher in laboratory media than in real food systems. In addition, most EOs derived from cloves, oregano, and thyme have poor water solubility, making it difficult to integrate them into complex food systems, thereby reducing antibacterial activity (Gyawali et al., 2015). Moreover, higher concentrations of these EOs may alter the sensory characteristics of food products. Therefore, it is important to consider the sensory impact and evaluate the appropriate amounts of these natural preservatives are used.

6 Conclusions

As consumer demand for food free of synthetic preservatives grows, the exploration of new alternatives becomes imperative. Biocontrol approaches offer promising avenues for curbing foodborne pathogens and extending the shelf life of foods. Additionally, they have the potential to increase the susceptibility of drug-resistant bacteria to antibiotics. Economically, the quest for natural antimicrobials must be cost-effective, with the integration of multiple approaches like hurdle technology emerging as a viable solution. Combining biocontrol agents from diverse groups or pairing them with other antimicrobial agents can synergistically reduce the required dosage for significant pathogen reduction. However, further research is essential to evaluate the safety and efficacy of these approaches.

References

Adetunji, C. O., & Varma, A. (2020). Biotechnological application of *Trichoderma*: A powerful fungal isolate with diverse potentials for the attainment of food safety, management of pest and diseases, healthy planet, and sustainable agriculture. In *Trichoderma: Agricultural applications and beyond* (pp. 257–285). Springer.

Agarbati, A., Canonico, L., Ciani, M., & Comitini, F. (2023). *Metschnikowia pulcherrima* in cold clarification: Biocontrol activity and aroma enhancement in Verdicchio wine. *Fermentation, 9*(3), 302.

Al-Habib, A., Al-Saleh, E., Safer, A.-M., & Afzal, M. (2010). Bactericidal effect of grape seed extract on methicillin-resistant *Staphylococcus aureus* (MRSA). *The Journal of Toxicological Sciences, 35*(3), 357–364.

Alegre, I., Viñas, I., Usall, J., Teixidó, N., Figge, M. J., & Abadias, M. (2013). Control of foodborne pathogens on fresh-cut fruit by a novel strain of *Pseudomonas graminis*. *Food Microbiology, 34*(2), 390–399.

Alvindia, D. G., & Natsuaki, K. T. (2008). Evaluation of fungal epiphytes isolated from banana fruit surfaces for biocontrol of banana crown rot disease. *Crop Protection, 27*(8), 1200–1207.

Apás, A. L., Dupraz, J., Ross, R., González, S. N., & Arena, M. E. (2010). Probiotic administration effect on fecal mutagenicity and microflora in the goat's gut. *Journal of Bioscience and Bioengineering, 110*(5), 537–540.

Bach, S. J., McAllister, T. A., Veira, D. M., Gannon, V. P., & Holley, R. A. (2003). Effect of bacteriophage DC22 on *Escherichia coli* O157: H7 in an artificial rumen system (Rusitec) and inoculated sheep. *Animal Research, 52*(2), 89–101.

Baron, N. C., Rigobelo, E. C., Zied, D. C., Baron, N. C., Rigobelo, E. C., & Zied, D. C. (2019). Filamentous fungi in biological control: Current status and future perspectives. *Chilean Journal of Agricultural Research, 79*(2), 307–315.

Bartholomew, H. P., Bradshaw, M., Jurick, W. M., & Fonseca, J. M. (2021). The good, the bad, and the ugly: Mycotoxin production during postharvest decay and their influence on tritrophic host–pathogen–microbe interactions. *Frontiers in Microbiology, 12*, 611881.

Belguith, H., Kthiri, F., Ben Ammar, A., Jaafoura, H., Ben Hamida, J., & Landoulsi, A. (2009). Morphological and biochemical changes of *Salmonella hadar* exposed to aqueous garlic extract. *International Journal of Morphology, 27*(3), 705–713.

Berbegal, C., Spano, G., Fragasso, M., Grieco, F., Russo, P., & Capozzi, V. (2018). Starter cultures as biocontrol strategy to prevent *Brettanomyces bruxellensis* proliferation in wine. *Applied Microbiology and Biotechnology, 102*(2), 569–576.

Bian, J.-Y., Fang, Y.-L., Song, Q., Sun, M.-L., Yang, J.-Y., Ju, Y.-W., Li, D.-W., & Huang, L. (2021). The fungal *Endophyte Epicoccum* dendrobii as a potential biocontrol agent against *Colletotrichum gloeosporioides*. *Phytopathology, 111*(2), 293–303.

Caggia, C., De Angelis, M., Pitino, I., Pino, A., & Randazzo, C. (2015). Probiotic features of *Lactobacillus* strains isolated from Ragusano and Pecorino Siciliano cheeses. *Food Microbiology, 50*, 109–117.

Calderón, C. E., Rotem, N., Harris, R., Vela-Corcía, D., & Levy, M. (2019). *Pseudozyma aphidis* activates reactive oxygen species production, programmed cell death and morphological alterations in the necrotrophic fungus *Botrytis cinerea*. *Molecular Plant Pathology, 20*(4), 562–574.

Cavalheiro, M., & Teixeira, M. C. (2018). *Candida* biofilms: Threats, challenges, and promising strategies. *Frontiers in Medicine, 5*, 28.

Chang, Y., Yoon, H., Kang, D.-H., Chang, P.-S., & Ryu, S. (2017). Endolysin LysSA97 is synergistic with carvacrol in controlling *Staphylococcus aureus* in foods. *International Journal of Food Microbiology, 244*, 19–26.

Chatterjee, S., & Rothenberg, E. (2012). Interaction of bacteriophage λ with its *E. coli* receptor, LamB. *Viruses, 4*(11), 3162–3178.

Chaudhary, V., Kajla, P., Lather, D., Chaudhary, N., Dangi, P., Singh, P., & Pandiselvam, R. (2024). Bacteriophages: A potential game changer in food processing industry. *Critical Reviews in Biotechnology, 44*, 1–25.

Chen, F., Miao, X., Lin, Z., Xiu, Y., Shi, L., Zhang, Q., Liang, D., Lin, S., & He, B. (2021). Disruption of metabolic function and redox homeostasis as antibacterial mechanism of *Lindera glauca* fruit essential oil against Shigella flexneri. *Food Control, 130*, 108282.

Clavijo, V., Baquero, D., Hernandez, S., Farfan, J., Arias, J., Arévalo, A., Donado-Godoy, P., & Vives-Flores, M. (2019). Phage cocktail SalmoFREE® reduces *Salmonella* on a commercial broiler farm. *Poultry Science, 98*(10), 5054–5063.

de Melo, A. G., Levesque, S., & Moineau, S. (2018). Phages as friends and enemies in food processing. *Current Opinion in Biotechnology, 49*, 185–190.

Di Canito, A., Mateo-Vargas, M. A., Mazzieri, M., Cantoral, J., Foschino, R., Cordero-Bueso, G., & Vigentini, I. (2021). The role of yeasts as biocontrol agents for pathogenic fungi on postharvest grapes: A review. *Food Security, 10*(7), 1650.

Di Francesco, A., Ugolini, L., Lazzeri, L., & Mari, M. (2015). Production of volatile organic compounds by *Aureobasidium pullulans* as a potential mechanism of action against postharvest fruit pathogens. *Biological Control, 81*, 8–14.

Du Toit, A. (2017). Viral infection: The language of phages. *Nature Reviews. Microbiology, 15*(3), 134–136.

Dunne, M., Prokhorov, N. S., Loessner, M. J., & Leiman, P. G. (2021). Reprogramming bacteriophage host range: Design principles and strategies for engineering receptor binding proteins. *Current Opinion in Biotechnology, 68*, 272–281.

Eisendle, M., Oberegger, H., Buttinger, R., Illmer, P., & Haas, H. (2004). Biosynthesis and uptake of siderophores is controlled by the PacC-mediated ambient-pH regulatory system in *Aspergillus nidulans*. *Eukaryotic Cell, 3*(2), 561–563.

El-Kholy, A., El-Shinawy, S., Meshref, A., & Korany, A. (2014). Microbiological quality of Domiati cheese and the influence of probiotics on the behavior of *Staphylococcus aureus* and *Escherichia coli* O157: H7 in Domiati cheese. *Journal of Food Safety, 34*(4), 396–406.

Endersen, L., & Coffey, A. (2020). The use of bacteriophages for food safety. *Current Opinion in Food Science, 36*, 1–8.

Erkkilä, S., Venäläinen, M., Hielm, S., Petäjä, E., Puolanne, E., & Mattila-Sandholm, T. (2000). Survival of *Escherichia coli* O157: H7 in dry sausage fermented by probiotic lactic acid bacteria. *Journal of the Science of Food and Agriculture, 80*(14), 2101–2104.

Fang, W., Peng, Y., Muir, D., Lin, J., & Zhang, X. (2019). A critical review of synthetic chemicals in surface waters of the US, the EU and China. *Environment International, 131*, 104994.

Faria, M., Martins, I., Souza, D. A., Mascarin, G. M., & Lopes, R. B. (2017). Susceptibility of the biocontrol fungi *Metarhizium anisopliae* and *Trichoderma asperellum* (Ascomycota: Hypocreales) to imbibitional damage is driven by conidial vigor. *Biological Control, 107*, 87–94.

Feiner, R., Argov, T., Rabinovich, L., Sigal, N., Borovok, I., & Herskovits, A. A. (2015). A new perspective on lysogeny: Prophages as active regulatory switches of bacteria. *Nature Reviews. Microbiology, 13*(10), 641–650.

Fernandes, S., & São-José, C. (2018). Enzymes and mechanisms employed by tailed bacteriophages to breach the bacterial cell barriers. *Viruses, 10*(8), 396.

Ferri, M., Ranucci, E., Romagnoli, P., & Giaccone, V. (2017). Antimicrobial resistance: A global emerging threat to public health systems. *Critical Reviews in Food Science and Nutrition, 57*(13), 2857–2876.

Fisher, K., & Phillips, C. (2008). Potential antimicrobial uses of essential oils in food: Is citrus the answer? *Trends in Food Science and Technology, 19*(3), 156–164.

Freimoser, F. M., Rueda-Mejia, M. P., Tilocca, B., & Migheli, Q. (2019). Biocontrol yeasts: Mechanisms and applications. *World Journal of Microbiology and Biotechnology, 35*(10), 154.

Garvey, M. (2022). Bacteriophages and food production: Biocontrol and bio-preservation options for food safety. *Antibiotics, 11*(10), 1324.

Gaston, S. A., Birnbaum, L. S., & Jackson, C. L. (2020). Synthetic chemicals and cardiometabolic health across the life course among vulnerable populations: A review of the literature from 2018 to 2019. *Current Environmental Health Reports, 7*(1), 30–47.

Ghadaksaz, A., Nodoushan, S. M., Sedighian, H., Behzadi, E., & Fooladi, A. A. I. (2022). Evaluation of the role of probiotics as a new strategy to eliminate microbial toxins: A review. *Probiotics and Antimicrobial Proteins, 14*(2), 224–237.

Gonelimali, F. D., Lin, J., Miao, W., Xuan, J., Charles, F., Chen, M., & Hatab, S. R. (2018). Antimicrobial properties and mechanism of action of some plant extracts against food pathogens and spoilage microorganisms. *Frontiers in Microbiology, 9*, 389103.

Goodridge, L. D., & Bisha, B. (2011). Phage-based biocontrol strategies to reduce foodborne pathogens in foods. *Bacteriophage, 1*(3), 130–137.

Gutiérrez, D., Fernández, L., Rodríguez, A., & García, P. (2018). Are phage lytic proteins the secret weapon to kill *Staphylococcus aureus*? *MBio, 9*(1), e01923-17.

Guyard-Nicodème, M., Rivoal, K., Houard, E., Rose, V., Quesne, S., Mourand, G., Rouxel, S., Kempf, I., Guillier, L., & Gauchard, F. (2015). Prevalence and characterization of *Campylobacter jejuni* from chicken meat sold in French retail outlets. *International Journal of Food Microbiology, 203*, 8–14.

Gyawali, R., Hayek, S., & Ibrahim, S. A. (2015). Plant extracts as antimicrobials in food products: Mechanisms of action, extraction methods, and applications. In *Handbook of natural antimicrobials for food safety and quality* (Vol. 49, pp. 49–62). Woodhead Publishing.

Helmy, Y. A., Taha-Abdelaziz, K., Hawwas, H. A. E.-H., Ghosh, S., AlKafaas, S. S., Moawad, M. M., Saied, E. M., Kassem, I. I., & Mawad, A. M. (2023). Antimicrobial resistance and recent alternatives to antibiotics for the control of bacterial pathogens with an emphasis on foodborne pathogens. *Antibiotics, 12*(2), 274.

Hershkovitz, V., Ben-Dayan, C., Raphael, G., Pasmanik-Chor, M., Liu, J., Belausov, E., Aly, R., Wisniewski, M., & Droby, S. (2012). Global changes in gene expression of grapefruit peel tissue in response to the yeast biocontrol agent *Metschnikowia fructicola. Molecular Plant Pathology, 13*(4), 338–349.

Hossain, M. I., Sadekuzzaman, M., & Ha, S.-D. (2017). Probiotics as potential alternative biocontrol agents in the agriculture and food industries: A review. *Food Research International, 100*, 63–73.

Hossain, M. I., Mizan, M. F. R., Ashrafudoulla, M., Nahar, S., Joo, H.-J., Jahid, I. K., Park, S. H., Kim, K.-S., & Ha, S.-D. (2020). Inhibitory effects of probiotic potential lactic acid bacteria isolated from kimchi against *Listeria monocytogenes* biofilm on lettuce, stainless-steel surfaces, and MBEC™ biofilm device. *LWT, 118*, 108864.

Huang, Y., Wang, W., Zhang, Z., Gu, Y., Huang, A., Wang, J., & Hao, H. (2022). Phage products for fighting antimicrobial resistance. *Microorganisms, 10*(7), 1324.

Hung, A. T., Lin, S.-Y., Yang, T.-Y., Chou, C.-K., Liu, H.-C., Lu, J.-J., Wang, B., Chen, S.-Y., & Lien, T.-F. (2012). Effects of *Bacillus coagulans* ATCC 7050 on growth performance, intestinal morphology, and microflora composition in broiler chickens. *Animal Production Science, 52*(9), 874–879.

Janisiewicz, W., Conway, W., & Leverentz, B. (1999). Biological control of postharvest decays of apple can prevent growth of *Escherichia coli* O157: H7 in apple wounds. *Journal of Food Protection, 62*(12), 1372–1375.

Johnson-Henry, K. C., Hagen, K. E., Gordonpour, M., Tompkins, T. A., & Sherman, P. M. (2007). Surface-layer protein extracts from Lactobacillus helveticus inhibit enterohaemorrhagic *Escherichia coli* O157: H7 adhesion to epithelial cells. *Cellular Microbiology, 9*(2), 356–367.

Jolly, N. P., Augustyn, O. P. R., & Pretorius, I. S. (2003). The effect of non-*Saccharomyces* yeasts on fermentation and wine quality. *South African Journal of Enology and Viticulture, 24*(2), 55–62.

Jongman, M., Carmichael, P., Loeto, D., & Gomba, A. (2022). Advances in the use of biocontrol applications in preharvest and postharvest environments: A food safety milestone. *Journal of Food Safety, 42*(2), e12957.

Kaur, S., Kaur, P., & Nagpal, R. (2015). *In vitro* biosurfactant production and biofilm inhibition by lactic acid bacteria isolated from fermented food products. *International Journal of Probiotics and Prebiotics, 10*(1), 17.

Khan, F. M., Chen, J.-H., Zhang, R., & Liu, B. (2023). A comprehensive review of the applications of bacteriophage-derived endolysins for foodborne bacterial pathogens and food safety: Recent advances, challenges, and future perspective. *Frontiers in Microbiology, 14*, 1259210.

Kim, S., Bang, J., Kim, H., Beuchat, L. R., & Ryu, J.-H. (2013). Inactivation of *Escherichia coli* O157: H7 on stainless steel upon exposure to *Paenibacillus polymyxa* biofilms. *International Journal of Food Microbiology, 167*(3), 328–336.

Kortright, K. E., Chan, B. K., Koff, J. L., & Turner, P. E. (2019). Phage therapy: A renewed approach to combat antibiotic-resistant bacteria. *Cell Host & Microbe, 25*(2), 219–232.

Krishnamoorthi, R., Srinivash, M., Mahalingam, P. U., Malaikozhundan, B., Suganya, P., & Gurushankar, K. (2022). Antimicrobial, anti-biofilm, antioxidant and cytotoxic effects of bacteriocin by *Lactococcus lactis* strain CH3 isolated from fermented dairy products—An *in vitro* and *in silico* approach. *International Journal of Biological Macromolecules, 220*, 291–306.

Ładziak, M., Prochwicz, E., Gut, K., Gomza, P., Jaworska, K., Ścibek, K., Młyńska-Witek, M., Kadej-Zajączkowska, K., Lillebaek, E. M., & Kallipolitis, B. H. (2024). Inactivation of lmo0946 (sif) induces the SOS response and MGEs mobilization and silences the general stress response and virulence program in *Listeria monocytogenes*. *Frontiers in Microbiology, 14*, 1324062.

Leneveu-Jenvrin, C., Charles, F., Barba, F. J., & Remize, F. (2020). Role of biological control agents and physical treatments in maintaining the quality of fresh and minimally-processed fruit and vegetables. *Critical Reviews in Food Science and Nutrition, 60*(17), 2837–2855.

Leverentz, B., Conway, W. S., Janisiewicz, W., Abadias, M., Kurtzman, C. P., & Camp, M. J. (2006). Biocontrol of the food-borne pathogens *Listeria monocytogenes* and *salmonella enterica* Serovar Poona on fresh-cut apples with naturally occurring bacterial and yeast antagonists. *Applied and Environmental Microbiology, 72*(2), 1135–1140.

Li, S., Jiang, S., Jia, W., Guo, T., Wang, F., Li, J., & Yao, Z. (2023). Natural antimicrobials from plants: Recent advances and future prospects. *Food Chemistry, 432*, 137231.

Liu, H., Zhang, J., Zhang, S., Yang, F., Thacker, P. A., Zhang, G., Qiao, S., & Ma, X. (2014). Oral administration of *Lactobacillus fermentum* I5007 favors intestinal development and alters the intestinal microbiota in formula-fed piglets. *Journal of Agricultural and Food Chemistry, 62*(4), 860–866.

Liu, G.-L., Chi, Z., Wang, G.-Y., Wang, Z.-P., Li, Y., & Chi, Z.-M. (2015). Yeast killer toxins, molecular mechanisms of their action and their applications. *Critical Reviews in Biotechnology, 35*(2), 222–234.

Liu, S., Zhang, Q., Li, H., Qiu, Z., & Yu, Y. (2022). Comparative assessment of the antibacterial efficacies and mechanisms of different tea extracts. *Food Security, 11*(4), 620.

Mäntynen, S., Laanto, E., Oksanen, H. M., Poranen, M. M., & Díaz-Muñoz, S. L. (2021). Black box of phage–bacterium interactions: Exploring alternative phage infection strategies. *Open Biology, 11*(9), 210188.

Mendoza, J. L. H., Pérez, M. I. S., Prieto, J. M. G., Velásquez, J. D. Q., Olivares, J. G. G., & Langarica, H. R. G. (2015). Antibiosis of *Trichoderma* spp strains native to northeastern Mexico against the pathogenic fungus *Macrophomina phaseolina*. *Brazilian Journal of Microbiology, 46*, 1093–1101.

Messaoudi, S., Kergourlay, G., Rossero, A., Ferchichi, M., Prévost, H., Drider, D., Manai, M., & Dousset, X. (2011). Identification of lactobacilli residing in chicken ceca with antagonism against *campylobacter*. *International Microbiology, 14*(2), 103–110.

Nakamura, S., & Hatanaka, A. (2002). Green-leaf-derived C6-aroma compounds with potent antibacterial action that act on both gram-negative and gram-positive bacteria. *Journal of Agricultural and Food Chemistry, 50*(26), 7639–7644.

Nishiyama, K., Seto, Y., Yoshioka, K., Kakuda, T., Takai, S., Yamamoto, Y., & Mukai, T. (2014). *Lactobacillus gasseri* SBT2055 reduces infection by and colonization of *Campylobacter jejuni*. *PLoS One, 9*(9), e108827.

Olanya, M. O., Ukuku, D. O., Annous, B. A., Niemira, B. A., & Sommers, C. H. (2013). Efficacy of *Pseudomonas fluorescens* for biocontrol of *Escherichia coli* O157: H7 on spinach. *Journal of Food, Agriculture and Environment, 11*(2), 86–91.

Oliveira, M., Ferreira, V., Magalhães, R., & Teixeira, P. (2018). Biocontrol strategies for Mediterranean-style fermented sausages. *Food Research International, 103*, 438–449.

Olszak, T., Shneider, M. M., Latka, A., Maciejewska, B., Browning, C., Sycheva, L. V., Cornelissen, A., Danis-Wlodarczyk, K., Senchenkova, S. N., & Shashkov, A. S. (2017). The O-specific polysaccharide lyase from the phage LKA1 tailspike reduces *Pseudomonas* virulence. *Scientific Reports, 7*(1), 16302.

Ouattara, B., & Mafu, A. (2003). Natural antimicrobials in combination with gamma irradiation. In *Natural antimicrobials for the minimal processing of foods* (pp. 263–271). Woodhead.

Oyetunji, O. E., Ojuederie, O. B., Thonda, O. A., Kotun, B., Glick, B. R., & Babalola, O. O. (2024). The expediency of fungi as biocontrol agents for the enhancement of food security. In *Biocontrol agents for improved agriculture* (pp. 1–28). Academic.

Parafati, L., Vitale, A., Restuccia, C., & Cirvilleri, G. (2015). Biocontrol ability and action mechanism of food-isolated yeast strains against *Botrytis cinerea* causing post-harvest bunch rot of table grape. *Food Microbiology, 47*, 85–92.

Peterson, R., Klopfenstein, T. J., Erickson, G. E., Folmer, J., Hinkley, S., Moxley, R. A., & Smith, D. R. (2007). Effect of *Lactobacillus acidophilus* strain NP51 on *Escherichia coli* O157: H7 fecal shedding and finishing performance in beef feedlot cattle. *Journal of Food Protection, 70*(2), 287–291.

Pires, D. P., Oliveira, H., Melo, L. D., Sillankorva, S., & Azeredo, J. (2016). Bacteriophage-encoded depolymerases: Their diversity and biotechnological applications. *Applied Microbiology and Biotechnology, 100*, 2141–2151.

Pisoschi, A. M., Pop, A., Georgescu, C., Turcuş, V., Olah, N. K., & Mathe, E. (2018). An overview of natural antimicrobials role in food. *European Journal of Medicinal Chemistry, 143*, 922–935.

Quinto, E. J., Caro, I., Villalobos-Delgado, L. H., Mateo, J., De-Mateo-Silleras, B., & Redondo-Del-Río, M. P. (2019). Food safety through natural antimicrobials. *Antibiotics, 8*(4), 208.

Rasooli, I., Rezaei, M. B., & Allameh, A. (2006). Ultrastructural studies on antimicrobial efficacy of thyme essential oils on *Listeria monocytogenes*. *International Journal of Infectious Diseases, 10*(3), 236–241.

Rathod, N. B., Nirmal, N. P., Pagarkar, A., Özogul, F., & Rocha, J. M. (2022). Antimicrobial impacts of microbial metabolites on the preservation of fish and fishery products: A review with current knowledge. *Microorganisms, 10*(4), 773.

Raya, R. R., Oot, R. A., Moore-Maley, B., Wieland, S., Callaway, T. R., Kutter, E. M., & Brabban, A. D. (2011). Naturally resident and exogenously applied T4-like and T5-like bacteriophages can reduce *Escherichia coli* O157: H7 levels in sheep guts. *Bacteriophage, 1*(1), 15–24.

Ripp, S., & Miller, R. V. (1997). The role of pseudolysogeny in bacteriophage-host interactions in a natural freshwater environment. *Microbiology, 143*(6), 2065–2070.

Sangeetha, G., Usharani, S., & Muthukumar, A. (2009). Biocontrol with *Trichoderma* species for the management of postharvest crown rot of banana. *Phytopathologia Mediterranea, 48*(2), 214–225.

Savita, & Sharma, A. (2019). Fungi as biological control agents. In *Biofertilizers for sustainable agriculture and environment* (pp. 395–411). Springer.

Servin, A. L. (2004). Antagonistic activities of lactobacilli and bifidobacteria against microbial pathogens. *FEMS Microbiology Reviews, 28*(4), 405–440.

Shen, Z., Mustapha, A., Lin, M., & Zheng, G. (2017). Biocontrol of the internalization of *Salmonella enterica* and Enterohaemorrhagic *Escherichia coli* in mung bean sprouts with an endophytic *Bacillus subtilis*. *International Journal of Food Microbiology, 250*, 37–44.

Shimshoni, J. A., Bommuraj, V., Chen, Y., Sperling, R., Barel, S., Feygenberg, O., Maurer, D., & Alkan, N. (2020). Postharvest fungicide for avocado fruits: Antifungal efficacy and peel to pulp distribution kinetics. *Food Security, 9*(2), 124.

Sizzano, F., Blackford, M., Berthoud, H., Amiet, L., Bailly, S., Vuichard, F., Monnard, C., Bieri, S., Spring, J.-L., Barth, Y., Descombes, C., Lefort, F., Cléroux, M., Simonin, S., Chappuis, C., Bourdin, G., & Bach, B. (2023). Bioprospecting of a *Metschnikowia pulcherrima* Indigenous strain for Chasselas winemaking in 2022 vintage. *Food Security, 12*(24), 4485.

Spinler, J. K., Taweechotipatr, M., Rognerud, C. L., Ou, C. N., Tumwasorn, S., & Versalovic, J. (2008). Human-derived probiotic *Lactobacillus reuteri* demonstrate antimicrobial activities targeting diverse enteric bacterial pathogens. *Anaerobe, 14*(3), 166–171.

Thambugala, K. M., Daranagama, D. A., Phillips, A. J. L., Kannangara, S. D., & Promputtha, I. (2020). Fungi *vs.* fungi in biocontrol: An overview of fungal antagonists applied against fungal plant pathogens. *Frontiers in Cellular and Infection Microbiology, 10*, 60492.

Thomas, M. B., & Read, A. F. (2007). Fungal bioinsecticide with a sting. *Nature Biotechnology, 25*(12), 1367–1368.

Tilocca, B., Balmas, V., Hassan, Z. U., Jaoua, S., & Migheli, Q. (2019). A proteomic investigation of *Aspergillus carbonarius* exposed to yeast volatilome or to its major component 2-phenylethanol reveals major shifts in fungal metabolism. *International Journal of Food Microbiology, 306*, 108265.

Torriani, S., Orsi, C., & Vescovo, M. (1997). Potential of *Lactobacillus casei*, culture permeate, and lacti acid to control microorganisms in ready-to-use vegetables. *Journal of Food Protection, 60*(12), 1564–1567.

Tryfinopoulou, P., Fengou, L., & Panagou, E. Z. (2019). Influence of *Saccharomyces cerevisiae* and *Rhotodorula mucilaginosa* on the growth and ochratoxin A production of *Aspergillus carbonarius*. *LWT, 105*, 66–78.

Verheust, C., Pauwels, K., Mahillon, J., Helinski, D. R., & Herman, P. (2010). Contained use of bacteriophages: Risk assessment and biosafety recommendations. *Applied Biosafety, 15*(1), 32–44.

Vescovo, M., Torriani, S., Orsi, C., Macchiarolo, F., & Scolari, G. (1996). Application of antimicrobial-producing lactic acid bacteria to control pathogens in ready-to-use vegetables. *Journal of Applied Microbiology, 81*(2), 113–119.

Vikram, A., Tokman, J. I., Woolston, J., & Sulakvelidze, A. (2020). Phage biocontrol improves food safety by significantly reducing the level and prevalence of *Escherichia coli* O157: H7 in various foods. *Journal of Food Protection, 83*(4), 668–676.

Vikram, A., Callahan, M. T., Woolston, J. W., Sharma, M., & Sulakvelidze, A. (2022). Phage biocontrol for reducing bacterial foodborne pathogens in produce and other foods. *Current Opinion in Biotechnology, 78*, 102805.

Villalpando-Aguilar, J. L., Matos-Pech, G., López-Rosas, I., Castelán-Sánchez, H. G., & Alatorre-Cobos, F. (2022). Phage therapy for crops: Concepts, experimental and bioinformatics approaches to direct its application. *International Journal of Molecular Sciences, 24*(1), 325.

Wachowska, U., & Borowska, J. (2014). Antagonistic yeasts competes for iron with winter wheat stem base pathogens. *Gesunde Pflanzen, 66*(4), 141–148.

Weinbauer, M. G. (2004). Ecology of prokaryotic viruses. *FEMS Microbiology Reviews, 28*(2), 127–181.

Wittebole, X., De Roock, S., & Opal, S. M. (2014). A historical overview of bacteriophage therapy as an alternative to antibiotics for the treatment of bacterial pathogens. *Virulence, 5*(1), 226–235.

Wong, C. W., Delaquis, P., Goodridge, L., Lévesque, R. C., Fong, K., & Wang, S. (2020). Inactivation of *Salmonella enterica* on post-harvest cantaloupe and lettuce by a lytic bacteriophage cocktail. *Current Research in Food Science, 2*, 25–32.

Xu, J., He, H., Wang, Y.-Y., Yan, R., Zhou, L.-J., Liu, Y.-Z., Jiang, F.-L., Maskow, T., & Liu, Y. (2018). New aspects of the environmental risks of quantum dots: Prophage activation. *Environmental Science. Nano, 5*(7), 1556–1566.

Yan, F., Xu, S., Chen, Y., & Zheng, X. (2014). Effect of rhamnolipids on *Rhodotorula glutinis* biocontrol of *Alternaria alternata* infection in cherry tomato fruit. *Postharvest Biology and Technology, 97*, 32–35.

Zhang, Z., Zhao, Y., Chen, X., Li, W., Wang, L., Li, W., Du, J., & Zhang, S. (2022). Effects of cinnamon essential oil on the physiological metabolism of *Salmonella enteritidis*. *Frontiers in Microbiology, 13*, 1035894.

Zhao, X., Wu, J. E., Chen, L., & Yang, H. (2019). Effect of vacuum impregnated fish gelatin and grape seed extract on metabolite profiles of tilapia (*Oreochromis niloticus*) fillets during storage. *Food Chemistry, 293*, 418–428.

Zhao, X., Chen, L., Wu, J. E., He, Y., & Yang, H. (2020a). Elucidating antimicrobial mechanism of nisin and grape seed extract against *Listeria monocytogenes* in broth and on shrimp through NMR-based metabolomics approach. *International Journal of Food Microbiology, 319*, 108494.

Zhao, X., Chen, L., Zhao, L., He, Y., & Yang, H. (2020b). Antimicrobial kinetics of nisin and grape seed extract against inoculated *Listeria monocytogenes* on cooked shrimps: Survival and residual effects. *Food Control, 115*, 107278.

Zhao, X., Chen, L., Wongmaneepratip, W., He, Y., Zhao, L., & Yang, H. (2021). Effect of vacuum impregnated fish gelatin and grape seed extract on moisture state, microbiota composition, and quality of chilled seabass fillets. *Food Chemistry, 354*, 129581.

Zheng, T., & Zhang, C. (2024). Engineering strategies and challenges of endolysin as an antibacterial agent against Gram-negative bacteria. *Microbial Biotechnology, 17*(4), e14465.

Hurdle Technologies

Yufan Wu and Xiang Wang

1 Introduction

Throughout the various stages of food production—such as raw material preparation, formulation, processing, packaging, storage, transportation, sale, and consumption—foods inevitably undergo a decline in their original quality over time. This deterioration occurs at different rates, influenced by physical, chemical, and biological factors. Among these, microbial contamination is particularly significant for both the quality and safety of food products. The microbiological stability and sanitary safety of foods depend on the interaction of various antimicrobial factors present in the product. Therefore, effectively controlling microbial growth during processing and storage is critical to maintaining product quality and safety.

In the food industry, several preservation methods, or "hurdles," are used to control microorganisms and enhance food quality and safety. These include high-temperature sterilization, low-temperature inhibition, pH adjustment, moisture activity regulation, reduction of oxidation-reduction potential, and the use of preservatives. Relying on a single method for microbial control typically requires more intensive processing. Researchers generally agree that no single method is sufficient on its own, necessitating the combined use of multiple food processing techniques for optimal results. Introduced in the 1970s, hurdle technology combines multiple preservation parameters at optimal levels to maximize microbial inactivation while minimizing changes to the sensory qualities of the food. This approach often

Y. Wu
Centre of Analysis and Test, School of Chemistry & Molecular Engineering, East China University of Science and Technology, Shanghai, China

X. Wang (✉)
School of Health Science and Engineering, University of Shanghai for Science and Technology, Shanghai, China
e-mail: xiang.wang@usst.edu.cn

D. Li (ed.), *Antimicrobial Strategies in the Food System: Updates, Opportunities, Challenges*, https://doi.org/10.1007/978-3-031-95056-8_6

produces foods that are more appealing than those treated with a single method, as it reduces the negative impact on taste, texture, and appearance.

The growing demand for minimally processed convenience foods, along with efforts to reduce sugar, salt, and preservatives, has presented new challenges in maintaining product stability and safety. Hurdle technology addresses these challenges by revealing the fundamental principles of food preservation and quality maintenance. It provides a flexible framework for designing or adjusting hurdle factors using modern technological methods, tailored to the preservation needs of different products. This enables the optimization of processing techniques, improvement of product quality, extension of shelf life, and enhancement of food safety and processing efficiency. The research and application of hurdle technology are of significant importance for advancing food science and technology, and it is widely applied in modern food industries.

2 Hurdle Technology

Hurdle technology is a comprehensive approach to food preservation and safety that combines multiple preservation methods to create synergistic effects. This strategy involves the strategic application of various "hurdles" or barriers to inhibit microbial growth and prevent food contamination. These hurdles may be physical, chemical, or biological, and when used together, they create a more effective antimicrobial environment than any single method could achieve alone. The concept was first introduced by Lothar Leistner in 1978. Today, over 60 hurdles have been identified that influence food stability, quality, and microbial safety (Leistner, 2000). Not all hurdles are used together, nor are they all applied to preserve a single food product. Hurdle technology promotes the deliberate integration of both existing and novel preservation techniques to establish a series of preservative factors that microorganisms cannot surpass, as illustrated in Fig. 1 (Leistner, 1995).

The fundamental principle of hurdle technology is that by applying multiple preservation methods, food producers can create a series of obstacles that microorganisms must overcome to survive and grow. This multi-faceted approach not only enhances food safety but also helps maintain food quality and extend shelf life. Each hurdle or barrier targets different aspects of microbial growth, making it more challenging for harmful bacteria to proliferate.

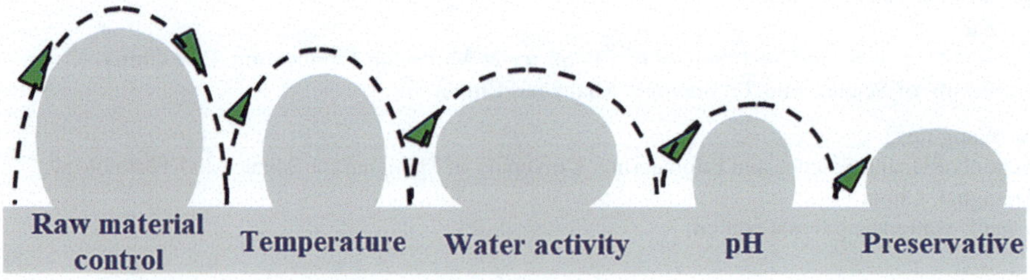

Fig. 1 An example of the hurdle effect in food preservation

3 Hurdles in Food Preservation

The concept of hurdle technology is elegantly simple yet highly effective: while some microorganisms may overcome individual hurdles, none can jump over all hurdles when combined. Traditionally, single hurdles, such as low pH (4.6) to inhibit *Clostridium botulinum* or high temperatures for canned foods, have been commonly used. Some hurdles, like low water activity (a_w), can be more intense than others, such as moderate pH levels. The microbial load and the food's nutrient content also play significant roles in the successful application of hurdle technology. Some important hurdles are outlined in Table 1. These hurdles can be physical (e.g., temperature control), chemical (e.g., pH adjustment, preservatives addition), or biological (e.g., use of beneficial microorganisms). The synergy created by multiple hurdles may result in a more effective antimicrobial environment than any single method would achieve alone. When multiple hurdles work together to create a synergistic effect, resulting in a more effective antimicrobial environment than any single method could achieve on its own, this is known as the "hurdle effect".

In recent decades, food processing technology has undergone significant advancements, driven by changing consumer demands, technological progress, and stricter food safety and quality standards. Consumers are increasingly concerned with health, nutrition, and convenience, motivating the food industry to develop innovative processing techniques. Technological advances have introduced groundbreaking methods such as irradiation, microwave processing, high-pressure processing, ultrasound treatment, and bio-preservatives (Table 1). These emerging technologies not only enhance food safety and extend shelf life but also help preserve the nutritional and sensory qualities of food products more effectively than traditional methods.

Table 1 Hurdles in food preservation

Hurdles	Application in food processing
High temperature	Heating
Low temperature	Chilling, freezing
Reduced a_w	Drying, curing, conserving
Increased acidity	Acid addition formations
Reduced redox potential	Modified-atmosphere packaging, addition of reducing agents
Preservatives	Sorbate, sulfite, nitrite
Competitive flora	Microbial fermentation
Natural antimicrobials	Bacteriocins, antibiotic peptides, essential oils, etc.
Novel methods	Irradiation, microwave processing, hydrostatic pressure, ultrasound, high-voltage pulse, etc.

3.1 Physical Hurdles

Physical hurdle techniques are widely used antimicrobial methods in the food industry due to their efficiency, making them essential for maintaining food quality and safety. These methods include thermal sterilization, non-thermal sterilization, freezing and refrigeration, and modified atmosphere preservation, and so on (Yousef & Balasubramaniam, 2012). Thermal sterilization methods range from far-infrared heating and microwave sterilization to ultra-high temperature, high-pressure, and atmospheric-pressure sterilization (pasteurization). Non-thermal sterilization, on the other hand, employs mechanisms like irradiation, high pressure, and ultrasound to kill microorganisms without heating the food. Freezing sterilization disrupts the cell structure and protein expression of pathogens by lowering the temperature, though it can also impact the food's quality. Modified atmosphere packaging (MAP) alters the gaseous environment around the food, isolating it from external contaminants and inhibiting microbial growth.

In recent years, physical hurdle techniques have evolved and expanded, becoming more widely applied in food safety and quality control. Gamma radiation is increasingly recognized as an effective preservation method, and the use of ultra-high pressure for preserving prepackaged, value-added foods is on the rise. Emerging techniques, such as pulsed electric fields, UV light, and ultrasound, are also being developed to ensure food safety while minimizing the negative effects of processing on product quality.

3.2 Chemical Hurdles

Chemical antimicrobial techniques include methods such as salt, sugar, vinegar, and alcohol preservation, as well as the use of chemical preservatives (Davidson et al., 2012). Preservatives are generally classified into inorganic and organic types based on their source and properties. Inorganic preservatives include substances like nitrates and nitrites, commonly used in processed meats to prevent the growth of *Clostridium botulinum* and enhance color stability. Salt (sodium chloride) is another widely used inorganic preservative, primarily functioning primarily by reducing water activity and inhibiting microbial growth. Additionally, carbon dioxide (CO_2) is often employed in modified atmosphere packaging to slow down spoilage and microbial proliferation in various food products. Organic preservatives consist of compounds such as sorbic acid and its salts (potassium sorbate), often used in baked foods, dairy products, and beverages to prevent mold and yeast growth. Benzoic acid and its salts (sodium benzoate) are also common organic preservatives, particularly in acidic foods like soft drinks, fruit juices, and pickles, where they inhibit bacterial and fungal growth. These organic preservatives work by disrupting microbial cell membranes, altering enzyme activity, and ultimately slowing or stopping microbial growth.

3.3 Biological Hurdles

Biological hurdle factors are natural substances derived directly from animals, plants, or microorganisms, or produced by the microorganisms themselves. These substances naturally occur in the environment, making them popular with consumers. Microorganisms and their natural products play a crucial role in food preservation and antimicrobial action. Bioprotective cultures, which are active microorganisms added directly to food, can extend shelf life and/or inhibit the growth of pathogens without affecting the sensory qualities of the food (Islam et al., 2024).

Plants can produce antimicrobial compounds, including certain aromatic components and peptides. Plant peptides are small molecules synthesized by plants to defend against microbial attacks from their environment. Essential oils extracted from plants contain a variety of secondary metabolites that have been shown to delay or inhibit the growth of bacteria, yeasts, and molds. As a major phytochemical, essential oils are widely found in various plants. The antimicrobial compounds in plant materials are commonly found in the essential oil components of branches, leaves, flowers or buds, roots, and seeds. Research has demonstrated that volatile substances, which are plant secondary metabolites, possess the ability to inhibit the growth of bacteria, yeasts, and molds (Manion & Widder, 2017). Generally, the antimicrobial effectiveness of plant essential oils depends on the chemical structure and concentration of their components. Most of the essential oil components with antimicrobial properties are phenols, terpenes, aliphatic alcohols, aldehydes, ketones, acids, and isoflavones.

Animal-derived antimicrobial compounds are abundant and often evolve from host defense mechanisms. Lysozyme is an enzyme that kills bacteria and the commercial lysozyme is derived from egg whites. Lysozyme can effectively inhibit various food spoilage microorganisms and can be successfully used to extend the shelf life of various food products, including raw and processed meats, cheeses, and other dairy products. Many inherent antimicrobial substances in animals exist in the form of antimicrobial peptides. These peptides are widely distributed in nature and serve as important components of the nonspecific host defense system. For example, Pleurocidin, isolated from the skin mucus layer of the Winter Flounder (*Pleuronectes americanus*), effectively controls pathogens including *Vibrio parahaemolyticus*, *Listeria monocytogenes*, *E. coli* O157:H7. Chitosan, a natural biopolymer, is obtained from the exoskeletons of crustaceans and arthropods, and its polycationic nature provides antifungal and antibacterial activity (Cole et al., 2000).

Certain microorganisms and their metabolic products can inhibit the growth and reproduction of other microbes, making them effective as preservatives. Currently, lactic acid bacteria are the primary microbial preservatives used (And & Hoover, 2003). Notable microbial metabolites used as food preservatives include bacteriocins produced by lactic acid bacteria and natamycin, which is produced by the bacterium *Streptomyces natalensis*. Lactic acid bacteria, widely recognized as safe for consumption, are among the most extensively studied microbial live cultures for food preservation. These bacteria are not only known for their positive effects on gut microbiota but also their antimicrobial properties. Many strains of lactic acid bacteria produce compounds that inhibit the growth of spoilage or pathogenic microorganisms. These compounds include fermentation end-products like organic acids, hydrogen peroxide, diacetyl, and bacteriocins, as well as other antimicrobial substances such as reuterin (And & Hoover, 2003). Bacteriocins are proteinaceous antimicrobial compounds synthesized by ribosomes as antibacterial peptides. Numerous bacteriocins have been isolated from lactic acid bacteria, and their properties have been extensively studied. Some of these bacteriocins demonstrate significant antimicrobial activity against pathogenic and spoilage organisms in food and are utilized as food preservatives. Notable examples of bacteriocins include nisin, diplococcin, acidophilin, bulgarican, helveticin, lactacin, and plantaricin. Nisin, produced by the *Lactococcus lactis* strain, is the most thoroughly researched and is used globally as a food additive (Field et al., 2023).

4 Mechanisms of Antimicrobial Effect of Hurdles

Hurdle technology is a strategic approach that effectively targets the homeostasis mechanisms microorganisms have evolved to survive environmental stresses. By disrupting these mechanisms, hurdles limit the organism's energy resources, hindering growth and reproduction. To achieve optimal effectiveness, multi-targeted hurdle technology is recommended. This approach involves combining various hurdles that target different microbial systems, such as the cell wall, membrane transport, receptor functions, signal transduction, gene expression, and enzyme systems. Synergistic effects can enhance the efficacy of hurdle technology. Some hurdles induce stress, making microorganisms more susceptible to subsequent hurdles, allowing for the use of lower-intensity hurdles to achieve the desired level of microbial control. Additionally, combining high-intensity microbiocidal hurdles can reduce the overall intensity required while maintaining the desired shelf life.

4.1 Homeostasis in Microorganisms

Homeostasis refers to the tendency of organisms to maintain internal stability. In microorganisms, this stability must be disturbed to prevent their proliferation in food products. When the homeostasis of microorganisms is disrupted by various preservative factors or hurdles in foods, the microorganisms may remain in a long lag phase or die before re-establishing their stable state (Leistner, 2000). For instance, the maintenance of a defined pH is a prerequisite and feature of living cells, and this applies to higher organisms as well as to microorganisms. Another crucial homeostasis mechanism involves the regulation of internal osmotic pressure, known as osmohomeostasis. In this process, cells ensure positive pressure by maintaining the osmolarity of the cytoplasm at a higher level than that of the surrounding environment. This regulation is achieved through the production of osmoprotective compounds, such as proline and betaine, which help the cells maintain their osmotic balance effectively (Erkmen & Bozoglu, 2016).

4.2 Metabolic Exhaustion

Metabolic exhaustion of microorganisms, leading to "auto sterilization" of foods (Leistner, 2000), is a practically significant phenomenon. This was initially observed in tests involving mildly heated liver sausage (with a core temperature of 95 °C), where salt and fat were added to adjust water activities, and the sausages were inoculated with *Clostridium sporogenes* before being stored at 37 °C. Despite some spores of *Clostridium* surviving the heat treatment, they disappeared during storage. The prevailing explanation is that spores, though capable of germinating, do so under conditions that are less favorable for vegetative bacterial growth (Leistner, 1992). Therefore, in stable hurdle-technology foods, spore counts decrease during storage, particularly in unrefrigerated conditions. This behavior can be explained by the fact that vegetative microorganisms that cannot grow will eventually die, and this process is expedited when conditions are marginally stable, temperatures are higher, antimicrobials are present, and the microorganisms are sublethally injured. In attempting to maintain homeostasis in such hostile environments, microorganisms exhaust their energy reserves and perish, resulting in auto sterilization. Consequently, the microbiological safety of these foods improves during storage, especially at room temperature.

4.3 Stress Reactions in Microorganisms

Some bacteria may exhibit increased resistance or virulence under stress by producing stress shock proteins (Ding et al., 2022). The synthesis of these protective proteins is triggered by conditions (hurdles) such as heat, pH changes, low water activity, ethanol, oxidative compounds, and starvation. Stress responses in microorganisms can pose challenges to food preservation and complicate the application of hurdle technology. Microorganisms may become more tolerant to other stresses due to a particular stress, i.e. "cross-protection" (Isohanni et al., 2013). However, the simultaneous induction of genes responsible for producing these proteins may be more challenging when multiple stresses are applied at once. Exposure to various stresses concurrently demands the energy-intensive production of numerous protective stress shock proteins, potentially leading to metabolic exhaustion in microorganisms. Consequently, utilizing multi-target food preservation approaches could be crucial to preventing the synthesis of stress shock proteins, thereby enhancing the microbial stability and safety of foods managed with hurdle technology (Leistner, 2000).

4.4 Multitarget Preservation

Multitarget preservation represents the pinnacle objective of hurdle technology, where employing multiple preservation factors induces a synergistic impact on microbial cells. By simultaneously targeting various cellular components like membranes, DNA, and enzyme systems, these hurdles effectively disrupt microbial homeostasis. This method not only complicates the repair of homeostasis and the synthesis of stress shock proteins but also significantly enhances the antimicrobial effectiveness of the preservation process.

5 Application of Hurdle Technology in Foods

Hurdle technology has seen widespread applied in various food systems, especially proving effective in sectors like meat and meat products, fruits and vegetables. This comprehensive approach combines multiple preservation techniques to tackle the distinct challenges of each food category. Choosing the right hurdles for a specific food product is crucial. For instance, bacterial heat resistance tends to increase when the a_w is low, but it decreases in the presence of certain preservatives. Fermented meat products can be considered safe and stable when both the a_w and pH levels are maintained within appropriate ranges. Integrating new hurdles with traditional ones offers the potential to enhance quality preservation and extend the shelf life of food products.

5.1 Application in Fruits and Vegetables

Vegetables and fruits are essential components of a healthy diet, offering a rich source of nutrients, vitamins, and minerals crucial for human health. However, their minimal processing and high moisture content make them particularly vulnerable to microbial contamination. In recent years, fresh-cut produce, such as peeled, sliced, or diced fruits and vegetables, has gained popularity due to its convenience. However, the mechanical processes involved in preparing fresh-cut products can disrupt the tissue structure, leading to nutrient leakage and creating conditions that favor microbial growth. To mitigate these risks, hurdle technology uses a combination of mild antimicrobial factors, or "hurdles," that work together to enhance food safety while maintaining a fresh-like quality. This approach results in shelf-stable, high-moisture products with minimal processing, which meet consumer demand for convenience and freshness.

Hurdle technology enables manufacturers to innovate by addressing safety and quality concerns while catering to the growing preference for cleaner labels and fewer synthetic additives. By understanding the specific microbial ecology of different foods, producers can customize their hurdle strategies to achieve the ideal balance of safety, quality, and consumer acceptance. Research shows that this flexible approach not only ensures compliance with food safety standards but also enhances the product's market appeal, opening up opportunities for growth in the competitive food industry.

The application of hurdle technology is continuously evolving, with ongoing research exploring new preservation methods, such as bio-preservation using natural antimicrobials derived from plants. Contemporary techniques involving plant extracts and essential oils are also being investigated to further improve food safety and extend shelf life. As consumer preferences shift toward natural, minimally processed, and health-conscious foods, hurdle technology is expected to play an increasingly important role in food preservation. This trend supports sustainable food production while maintaining high standards of safety and quality.

Fruit and vegetable preservation combines both conventional and innovative hurdles to extend shelf life while maintaining quality and safety. Traditional methods, such as short-time heat treatments, low-temperature storage, and chemical preservatives, have long been used in the food industry to ensure product safety and longevity. However, recent advancements in food preservation have introduced novel techniques that complement these conventional methods. Emerging technologies include bio-preservation, the use of natural antimicrobials like plant extracts and essential oils, and advanced packaging solutions that enhance food safety without compromising freshness. Additionally, novel physical processing technologies such as high-pressure processing (HPP), ultrasound, and pulsed light sterilization are now being utilized as new hurdle factors in the preservation of fruits and vegetables.

5.1.1 Temperature

Temperature is one of the most important factors influencing the growth and metabolism of microorganisms and plant cells, and it is also one of the most crucial hurdles in fruit and vegetable preservation. Low temperatures are highly effective at slowing down the growth and reproduction of microorganisms. In practical production processes, maintaining the correct temperature is critical for handling fruits and vegetables. Typically, for fresh-cut fruits and vegetables raw produce is stored at 5 ° C immediately after harvesting. During trimming and peeling, the ambient temperature is generally kept between 10 °C and 15 °C, and after processing, fresh-cut products are cooled to a storage temperature of 2 °C to 5 °C. However, despite low temperatures, some psychrophilic microorganisms can still grow and proliferate. Therefore, to effectively extend the shelf life of fresh-cut fruits and vegetables, low-temperature storage should be used in combination with other hurdle technologies. Soliva-Fortuny et al. (2004) investigated the effects of combining cold storage with strategies to reduce water activity, utilize modified atmosphere packaging, and/or incorporate preservatives can preserve avocado puree for 4 months (Soliva-Fortuny et al., 2004).

Mild heat treatments can effectively reduce the respiratory rate of fresh-cut fruits and vegetables, extending their shelf life while maintaining their sensory and nutritional qualities, and ensuring microbial inactivation. For instance, a study showed that treating pre-cut melons with 67 °C water followed by 0.5 kGy gamma irradiation significantly reduced the total microbial load without affecting their color, firmness, or other sensory properties (Fan et al., 2006).

5.1.2 Reduction of pH

Low pH limits for growth of some important food spoilage and pathogens. The pH value of most fruits ranges from 3.0 to 4.5 (Andrés-Bello et al., 2013). This lower pH is due to the presence of various organic acids in fruits, such as citric acid and malic acid. The naturally low pH is more favorable for fruit preservation. In contrast, most vegetables have a higher pH, typically ranging from 5.0 to 7.0 (Parry, 2012). Typically, lowering the pH of food systems to the range of 3.0 to 5.0 can effectively limit the growth of various microorganisms. In fruits and vegetables, organic acid antimicrobial agents such as citric acid, benzoic acid, sorbic acid, and acetic acid are commonly used to reduce pH. Citric acid has been widely accepted as effective in reducing superficial pH of cut fruits such as orange, apple, peach, apricot and kiwifruit, avocado and bananas. Research has shown that soaking fresh-cut melons in a low concentration citric acid solution for 30 s before modified atmosphere packaging not only inhibits microbial growth but also prevents translucency and discoloration. It was found that gently scrubbing fresh-cut carrots with a 2% lactic acid solution for 1 min significantly reduced the total microbial count and had a notable inhibitory effect on inoculated Pseudomonas pathogens (Uyttendaele et al., 2004). Although adding organic acid to reduce pH can improve product stability, the amount of organic acid needed to achieve the desired preservation effect may negatively impact sensory quality. It is important to ensure that the taste of fruits and vegetables remains acceptable when applying organic acid treatments. The combination effects of citric acid, cold temperature and honey on preservation was evaluated in guava fruits and slices (Eman et al., 2015). The results indicated that under cold storage (5 ± 1 °C, with 85–90%) using 2% citric acid alone, or in combination with 15% honey, significantly reduced the respiration rate of guava slices. The treatment was more effective in extending the shelf life of the guava slices and fruits, maintaining quality for up to 22 days by delaying the ripening process.

5.1.3 Disinfectants Washing

Washing fruits and vegetables with disinfectants is effective in reducing the initial microbial load, which helps extend their shelf life. Disinfectants like hydrogen peroxide (H_2O_2) and sodium hypochlorite (NaClO) are effective in reducing microbial contamination on fruits and vegetables. Ukuku et al. (2005) compared the efficacy of different washing agents on cantaloupe (Ukuku et al., 2005). A combination of 2.5% H_2O_2 with 1% H_2O_2, 25 μg/mL nisin, 1% sodium lactate, and 0.5% citric acid demonstrated significantly better bactericidal effects than 2.5% H_2O_2 alone. Cantaloupes treated with this combined washing solution showed no detectable pathogenic bacteria after cutting and had lower levels of endogenous microorganisms. However, due to safety concerns, the U.S. Food and Drug Administration (FDA) has not approved H_2O_2 as a disinfectant for fresh-cut produce, although it is permitted in other food processing and packaging applications (Stearns et al., 2022). Additionally, many European countries prohibit the use of sodium hypochlorite for sanitizing fresh-cut fruits and vegetables (Gil et al., 2009). As a result, the future development of food washing disinfectants will prioritize safety, environmental sustainability, and efficiency.

Electrolyzed water (EW), a new environmentally friendly preservative and disinfectant, is showing promise for fruit and vegetable preservation. Recent studies have proven that EW not only removes microorganisms and pesticide residues from fruits and vegetables but also improves the quality of postharvest fruits and vegetables (Lu et al., 2022). EW has attracted widespread attention in the food industry for its advantages of environmental protection, high-performance sterilization, low price and simple production. Produced by electrolyzing a saline or dilute hydrochloric acid solution, this water has excellent bactericidal properties due to its low pH, high redox potential, and a certain concentration of active chlorine. The effects of combination of acidic electrolyzed water and storage temperatures on the microbial growth of fresh-cut lettuce and cabbage was evaluated (Koseki & Itoh, 2001). Their results indicated that acidic electrolyzed water significantly reduced the initial microbial load on the fresh-cut produce. At 1 °C, the total microbial count remained stable during storage, whereas at 10 °C, the microbial count increased rapidly, reaching the same level as that of tap water-washed samples after 3 days. This suggests that combining acidic electrolyzed water with low-temperature storage effectively inhibits microbial growth.

5.1.4 Modified Atmosphere Packaging

Modified Atmosphere Packaging (MAP) is a method that involves altering the composition of gases (such as oxygen, carbon dioxide, and nitrogen) in the packaging environment around food products. This technique helps in preserving the freshness and extending the shelf life of various perishable items, including fruits, vegetables, and ready-to-eat foods (Mattheis & Fellman, 2000). In MAP, CO_2, O_2, and N_2 are commonly used as modifying gases. CO_2 is important in inhibiting the growth of bacteria and fungi, which also helps slow down cellular respiration. However, excessive concentrations of CO_2 can negatively impact the appearance of the packaging. O_2 helps inhibit the growth of anaerobic bacteria in food. Reducing O_2 concentrations can weaken the respiratory activity of produce, thereby extending its shelf life. However, extremely low O_2 levels can lead to pigment degradation due to anaerobic microbial respiration, particularly affecting compounds such as anthocyanins and flavonoids (Jones et al., 2014). N_2 acts primarily as a filler and does not participate in chemical reactions. By adjusting the proportions of CO_2, O_2, and N_2, optimal preservation conditions can be achieved. These gases are cost-effective, safe, and do not alter the flavor of the food.

The combined use of MAP and other hurdle technologies in fruits and vegetables provides a synergistic effect, leading to significant reductions in microbial growth and substantial extensions in shelf life. Fresh-cut cantaloupe stored under MAP conditions of 4 kPa O_2 and 10 kPa CO2 at 5 °C demonstrated a significantly reduced the total microbial count and extended shelf life of 9 days (Bai et al., 2001). Microwave-treated semi-dried green asparagus stored under MAP of 30% CO_2 and 70% N_2 at 4 °C effectively retained its quality and sensory attributes (Sergio et al., 2018).

5.1.5 Irradiation

Irradiation, a non-thermal sterilization technique involving exposure to electromagnetic radiation or accelerated electrons, is a recognized safe method for killing microorganisms. In the food industry, UV-C and gamma irradiation are commonly used for fruit and vegetable decontamination. However, UV-C has limited penetration depth and is more suitable for surface sterilization. To enhance its efficacy, low-dose UV-C can be combined with other hurdle technologies. For instance, Hadjok et al. (2008) studied the combined effects of hydrogen peroxide, UV-C, and temperature on fresh lettuce (Hadjok et al., 2008). The optimal combination was found to be 15% hydrogen peroxide, 37.8 mJ/m^2 UV-C, and 50 °C, which significantly reduced *Salmonella* counts both on the surface and inside the lettuce leaves. Vibhakara et al. (2005) investigated the preservation of high-moisture spice-based kohlrabi and cauliflower using a combination of hurdles, including brine blanching, spice treatment, packaging, and irradiation (Vibhakara et al., 2005). By combining 4% sodium chloride, 0.5% citric acid, and potassium metabisulfite with irradiation doses of 3–8 kGy, they achieved microbiological stability for up to 4 months at ambient temperature. Irradiation effectively reduced microbial loads, including coliforms and Aspergillus flavus, and contributed to the overall stability of the products.

5.1.6 Preservatives

Chemical preservatives play a crucial role in extending the shelf life of fruits and vegetables by inhibiting microbial growth and preventing oxidation. These substances are added intentionally to foodstuffs to increase durability and enhance or modify properties, including appearance, flavor, or structure. Common chemical preservatives used in fruits and vegetables include sorbic acid and its salts, benzoic acid and benzoates, and sulphur dioxide and sulphites. In fruit preservation, chemical preservatives are often applied to control fungal growth and prevent spoilage. For instance, sorbic acid and potassium sorbate are widely used in dried fruits, jams, and fruit juices. In vegetable preservation, benzoic acid and its salts are effective against yeasts and molds, making them suitable for pickled vegetables. While chemical preservatives are effective in maintaining food quality and safety, their use is regulated to ensure consumer safety. The food industry continues to explore optimal combinations of preservatives and application methods to meet both safety standards and consumer preferences for minimally processed fruits and vegetables. Soliva-Fortuny et al. (2004) examined the impact of various preservation techniques, including controlled storage temperature, pH reduction, lowering water activity, modified atmosphere packaging, and/or the use of preservatives, on the microbial quality of avocado puree stored at 4 °C and 22 °C over a four-month period (Soliva-Fortuny et al., 2004). The findings showed that sorbic acid was particularly effective in maintaining microbial stability, with a concentration of 300 mg/kg typically sufficient to inhibit the growth of dominant yeasts and molds in avocado. Furthermore, vacuum packaging combined with storage at 4 °C was found to extend the shelf life of avocado puree for up to 4 months, even in the absence of sorbic acid. Barwal et al. (2005) preserved cauliflower (var. Botrytis) using a low-cost, energy-efficient process based on hurdle technology, which combined various potassium metabisulfite (KMS; 0.2%), salt (5%, 10%, and 15%), and citric acid (1%) after blanching (Barwal et al., 2005). The physicochemical, sensory, and microbiological qualities were assessed on the day of preparation and at 15, 45, 90, and 180 days of storage. After 90 and 180 days, the cauliflower was reconstituted in running water and evaluated for pickle and pakora preparation. Samples treated with 10% and 15% salt solutions containing 0.2% KMS consistently ranked highest in quality. The production cost for these formulations was INR 4.03 and 4.15 per jar, respectively. After reconstitution, both pickle and pakora made from the preserved cauliflower were rated above acceptable by sensory panels, highlighting the effectiveness of chemical preservation in combination with salt for long-term storage. Vibhakara et al. (2006) developed a shelf-stable preservation process for grated carrots using hurdle technology (Vibhakara et al., 2006). The process involved the infusion of additives, including 3.0% sodium chloride to reduce water activity (a_w to 0.94), 0.9% citric acid to lower pH below 4.5, 1.0% sodium citrate, and 0.2% sodium benzoate as an antimicrobial agent, followed by partial dehydration. The product was then stored in flexible polymer pouches and monitored for stability at ambient temperatures (19–33 °C). The partially dehydrated grated carrots, with a moisture content of 66.2%, remained acceptable for over 6 months, retaining 82.5% of their carotenoid content.

In recent years, the interest towards health benefits, together with the demand for products without chemical preservatives, has been increasing around the world. This trend promoted a great interest in natural compounds. Natural preservatives can be employed on fresh-cut fruits and vegetables through various methods such as dipping, fumigation, spraying, or integrating with preservation papers and coatings. Some natural preservatives are even used as edible coatings to manage microbial contamination. For instance, applying a chitosan coating to fresh-cut pineapple significantly reduced microbial counts during storage. Incorporating vanillin, a natural antimicrobial agent, into the chitosan coating further enhanced its antibacterial properties, though it also led to a reduction in nutritional value, with vitamin C levels dropping to just 10% of their initial amount after 8 days at 10 °C (Sangsuwan et al., 2008). Raybaudi-Massilia et al. (2008) examined the impact of combining malic acid with essential oils from cinnamon, palmarosa, and lemongrass, in conjunction with passive MAP, on fresh-cut 'Piel de Sapo' melon (Raybaudi-Massilia et al., 2008). Essential oils and their active components (eugenol, geraniol, and citral) were added to an alginate-based edible coating, and the melon pieces were inoculated with *Salmonella* enteritidis before the application of these coatings. The integration of passive MAP and malic acid effectively enhanced the microbiological shelf life of the melon to about 9.6 days and improved its physicochemical properties for over 14 days, in comparison to non-coated melon, which lasted only 3.6 days. Furthermore, incorporating essential oils into the coating extended the microbial shelf life beyond 21 days while notably reducing *Salmonella* enteritidis levels. Among the oils, palmarosa oil at 0.3% within the coating and passive MAP showed promising results, maintaining the fruit's quality, inhibiting native microbial growth, and reducing *Salmonella*, while also receiving positive feedback from taste panels. Ukuku et al. (2002) demonstrated that washing cantaloupe using a combination of nisin and EDTA at 20 °C for 5 min, followed by an additional 1-min wash with the same solution and storing at 5 °C, significantly lowered the microbial load and prolonged the shelf life (Ukuku & Fett, 2002). Moreover, the use of nisin together with EDTA, potassium sorbate, or sodium lactate successfully eradicated inoculated *Salmonella* and minimized its transfer to fresh-cut cantaloupe (Ukuku & Fett, 2004). Karagözlü et al. (2011) investigated the impact of soaking fresh-cut lettuce and purslane in varying concentrations of peppermint and basil essential oils, along with low-temperature storage, on reducing inoculated *Salmonella* and *E. coli* levels (Karagözlü et al., 2011). Their findings indicated that higher concentrations of essential oils were more effective, with peppermint oil exhibiting greater antibacterial activity than basil oil.

5.2 Application in Meat and Products

Hurdle technology was initially applied in meat processing, primarily in simple and experiential ways. Nowadays, hurdle technology is widely applied in multiple fields of the food industry. In the meat industry, the conscious use of hurdle technology can significantly extend product shelf life (Leistner, 2000).

Meat, with a_w above 0.90, is highly susceptible to microbial contamination (Turantaş et al., 2015). The microorganisms most commonly involved in meat spoilage are *Pseudomonas* spp., the Enterobacteriaceae family, lactic acid bacteria and the *Brochothrix thermosphacta* (Kameník, 2013). Common pathogens include Gram-positive strains like *Listeria monocytogenes*, *Staphylococcus aureus*, *Bacillus cereus*, *Clostridium botulinum*, *Clostridium difficile*, and *Clostridium perfringens*, as well as Gram-negative strains such as *Escherichia coli*, *Yersinia enterocolitica*, *Campylobacter jejuni*, *Salmonella* enteritidis, *Shigella*, *Enterobacter*, *Brucella*, and *Aeromonas* (Shaltout et al., 2022). To ensure the microbial quality and safety of meat and meat products, effective preservation techniques are crucial. Conventional methods include thermal processing, refrigerated storage, vacuum packaging, and the use of nitrites and other preservatives. However, with advancements in technology, non-thermal treatments are gaining increasing favor, as they avoid harmful residues and preserve product quality. Techniques such as high-pressure processing (HPP), X-rays, and γ-rays have been employed to eliminate surface bacteria on meat. The use of hurdles in the preservation of meat and meat products is summarized in Table 2. The use of various hurdles can significantly extend the shelf life or control the growth of foodborne pathogens.

Table 2 Overview of studies evaluating the hurdle technology in meat and meat products

Meat and meat products	Hurdles	Effects	References
Raw chicken fillets	Ozone (0.6 ppm for 10 min) + Lyophilisation (20.5 h of primary drying-12 h at 0 °C and 8.5 h at 10 °C at 30 Pa)	Combined treated showed a reduction of 4.77 log CFU/g of lactic acid bacteria and a reduction of 6.8 log CFU/g of total aerobic mesophilic bacteria. Extended the shelf-life of the treated sample for up to 8 months	Cantalejo et al. (2016)
Fresh beef, ground beef, beef burger	Microwave (20 s) + Gamma-irradiation (3 kGy)	Synergistic effect, a shelf-life of at least 2 weeks at 5 °C, while the untreated samples had a shelf-life of less than 7 days	Aziz et al. (2002)
Goat meat surfaces	1% lemongrass oil + Exposure to 200 μW UV per cm^2 for 2 min	A 6-log reduction of *Escherichia coli* K12 was achieved	Degala et al. (2018)
Cooked ham	High-pressure processing (535–580 MPa) and 0.3% InbacTm (organicacids and nisin formula)	Extend the shelf life	O'Neill (2018)
Chicken sausage	Textured soy protein + Lactic acid	Extend the shelf life of chicken sausage to 6 days	Rindhe et al. (2017)
Sausage	Osmotic dehydration using salt (a_w 0.80 ± 0.02) + Polyethylene packaging + Gamma-irradiation (10 kGy)	Safe shelf-stable casing for sausage making	Chawla et al. (2006)
Keema	Isolated soy protein (a_w 0.88, 0.90) + Lactic acid (pH 5.50, 5.65 and 5.80)	Decrease the growth rate of both aerobic and anaerobic organisms and eliminate *Staphylococcus aureus* in standardized keema	Karthikeyan et al. (2000)

(continued)

Table 2 (continued)

Meat and meat products	Hurdles	Effects	References
Vacuum-packaged frankfurter	*Lactobacillus curvatus* CRL705 and *Lactobacillus sakei* CRL1862 (Bact705/1862), Nisin and organic acids (lactic acid, LA; acetic acid, AA) + 10 °C	The combined addition of Bact705/1862 + LA/AA was the most effective approach for pathogen reduction below detection level from day 6 to final storage	Castellano et al. (2018)
Mutton meat emulsions	Gamma irradiation + Antimicrobial (sodium nitrate and nisin) and antioxidants	The combination of nisin and irradiation effectively reduced microbial spoilage, comparable to sodium nitrate-treated samples	Akhter et al. (2021)
Pork loin	Cold plasma (5000–-600 W) + Lemongrass essential oil (5–10 mg/mL)	A combination of cold plasma (120 s, 500 W) and essential oil (5 mg/mL, 30 min) reduced *L. monocytogenes* by 2.80 log CFU/g, showing a synergistic antimicrobial effect	Cui et al. (2018)
Beef burgers	HPP + Sage powder	Lipid oxidation was delayed, and microbial quality remained acceptable for 60 days	Mizi et al. (2019)
Sliced-dry cured ham	HPP + Thymol	A combination of HPP (450 MPa, 10 min) and thymol reduced *L. monocytogenes* by 1.30 and 1.45 log CFU compared to control in dry-cured ham at 4 °C and 12 °C	Pérez-Baltar et al. (2019)
Raw ground beef	HPP + *Melissa officinalis* extracts and chemical constituents	A reduction of 3–6 log CFU/g of *Escherichia coli* was achieved at 350 and 400 MPa using extracts and chemical constituents	Chien et al. (2017)
Chicken breast	HPP + Dried rosemary	Rosemary attenuated the lipid oxidation in chicken breast	Bragagnolo et al. (2005)
Low-sodium sliced Turkey breast ham	HPP + Carvacrol	Carvacrol helped prevent lipid oxidation. Reduced the growth of *L. monocytogenes*, lactic acid bacteria, and psychrotrophic bacteria	de Oliveira et al. (2015)
Fresh meat	MAP + Oregano essential oils +5 °C and 15 °C	The synergy effect extends meat shelf life by affecting its microbiological and physicochemical properties	Skandamis and Nychas (2002)

(continued)

Table 2 (continued)

Meat and meat products	Hurdles	Effects	References
Fresh meat patties	MAP + Thymol + 4 °C	The combination showed better effects on the product quality, with a consequent prolongation of the shelf life	Del Nobile et al. (2009)
Non-conventional poultry patties	MAP + Thymol, carvacrol + 0–18 °C	The combination resulted in higher reduction of Pseudomonal spp. and longer shelf life	Mastromatteo et al. (2009)
Fresh chicken meat	MAP + oregano oil + 4 °C	The oregano oil and MAP exhibited an additive preservation effect, ca. 3–4 days for samples containing 0.1% oregano oil, 2–3 days for samples under MAP and 5–6 days for samples under MAP containing 0.1% of oregano oil	Chouliara et al. (2007)
Fresh chicken meat	MAP + Nisin, EDTA + 4 °C	MAP in combination with nisin–EDTA resulted in an organoleptic extension of refrigerated meat	Economou et al. (2009)

5.2.1 Conventional Hurdles in Meat and Meat Products

Several classic meat products effectively utilize traditional hurdle technology to ensure microbial safety while preserving quality. For instance, traditional sausages such as German Brühwurst, Italian salami, and Spanish chorizo rely on a combination of methods like pH reduction, dehydration, and the addition of antimicrobial agents like nitrates (Allen, 2015; Lucke & Zangerl, 2014). These hurdles collectively inhibit the growth of spoilage organisms and pathogens. Additionally, in products like mortadella and Gelderse Rookworst, hurdle technology incorporates factors such as low water activity and controlled temperatures during fermentation and smoking, which contribute to extending shelf life and ensuring food safety (Noort & Visschers, 2015). Charqui meat, a tropical intermediate meat product, is formulated using hurdle technology (Shimokomaki et al., 1998). Salt, sodium nitrite, dehydration, and packaging are hurdles sequentially applied to inhibit deteriorating microorganisms with the possibility of selecting desirable microbiota.

Santos et al. (2002) developed methods for a sausage that remains shelf-stable at room temperature, employing hurdle technology (Santos et al., 2002). The sausage was crafted from pork, fat, soybean derivatives, wheat flour, starch, water, salts, and seasonings to achieve a fine paste with small meat chunks, all encased in artificial impermeable casings ensuring a_w of 0.95. Sausages were then thermally processed to either 75 °C or 80 °C at the core before being stored at room temperature for up to 3 months. Both fresh and stored sausages underwent evaluations for physicochemical attributes, sensory qualities, and microbiological safety. Heating to a core temperature of 80 °C, paired with the use of artificial casings, successfully produced a shelf-stable sausage with a three-month shelf life. With the exception of color, sensory characteristics remained consistent over the storage period. Cepero et al. (2006) evaluated the effectiveness of hurdle technology in extending the shelf life of fresh sausages (Cepero et al., 2006). Two formulations were stored at 4–8 °C with 93 ± 5% relative humidity under two conditions: hung on bars and vacuum packaged. The sausages were assessed weekly for physicochemical, sensory, and microbiological qualities through an acceptance-refusal test, pH measurements, and microbiological analysis. The results showed that the shelf life of Cuban fresh sausages could be extended from 4 to 32 days by using hurdle technology, which included vacuum packaging to reduce oxidation-reduction potential, potassium sorbate as a preservative, and low-temperature storage (4–8 °C). Karthikeyan et al. (2000) aimed to extend the shelf life of the highly perishable Indian meat product, chevon keema, using hurdle technology. Variable hurdles included a_w and pH, while constant hurdles were vacuum packaging, preservatives, and heat treatment. The a_w was adjusted to 0.90 and 0.88, with 0.90 being preferred for sensory quality. The pH was modified to 5.50, 5.65, and 5.80, with 5.80 being the most acceptable. The meat product with a_w 0.90 and pH 5.80 was stored at ambient temperature and evaluated. Over time, increases in moisture and tyrosine values were observed, while aerobic and anaerobic growth slowed, and *Staphylococcus aureus* was inhibited. Sensory scores decreased with storage, but hurdle-treated keema remained well accepted for 3 days and fairly accepted for 5 days, compared to traditional keema, which was only acceptable on the first day. Hurdle technology significantly extended the product's shelf life.

Nitrites are used in meat products to prevent bacterial growth (especially *Clostridium botulinum*), maintain the color, enhance flavor, and act as an antioxidant to extend meat shelf life. Rödel and Scheuer (2007) explored a model system to evaluate hurdle technology and its impact on microbial growth in meat products (Rödel & Scheuer, 2007). They studied the effects of different combinations of sodium nitrite, temperature and water activity on the growth of *Escherichia coli*. The findings revealed complex interactions between these factors, with microbial growth patterns varying significantly. However, the use of nitrites is regulated due to potential health risks. Jafari and Emam-Djomeh (2007) conducted a study aimed at reducing nitrite content in hot dogs using hurdle technology while maintaining product safety and quality (Jafari & Emam-Djomeh, 2007). In their approach, the a_w of the hot dogs was lowered to 0.95 by adding humectants. Although the pH was adjusted to 5.4 using glucono-delta-lactone, the product had the lowest acceptance ($p > 0.05$) due to organoleptic changes, particularly a sour taste. The hot dogs were heated to 80 °C for 1 h, targeting an internal temperature of 75 °C. After cooking, the temperature was reduced to 5–6 °C within 40–45 min, and the sausages were stored at chilled temperatures (between 3 °C and 10 °C) for their shelf life. Total aerobic counts in the hurdle-treated hot dogs (with 50 ppm nitrite) were lower compared to the control (with 120 ppm nitrite), while *Clostridium perfringens* counts and *Clostridium botulinum* detection remained similar ($p < 0.05$) in both treated and control samples. The study demonstrated that both the hurdle-treated and control samples had the same ($p < 0.05$) acceptability and sensory attributes.

5.2.2 Novel Hurdles in Meat and Meat Products

With increasing consumer demand for safe, naturally processed meats, the meat industry is exploring new methods to produce safe meat products without compromising their sensory qualities. Novel technologies can preserve the sensory and nutritional characteristics of processed meats while ensuring microbial safety, making them a promising solution to reduce the need for chemical preservatives in meat products.

The use of biopreservatives, particularly through combining multiple microbial hurdles, offers a potential solution for controlling spoilage microorganisms and foodborne pathogens in meat and meat products. In a systematic review by Karbowiak et al. (2023), the authors explore whether applying various microbial hurdles together can serve as an effective decontamination strategy for meat products. The studies reviewed show that common biopreservative methods include using a combination of different starter cultures and non-starter protective cultures. The findings generally indicate a positive impact, with most studies reporting that combining various bacterial antimicrobials effectively inhibits the growth of both spoilage and pathogenic bacteria in meat products. Katikou et al. (2005) applied *Lactobacillus* protective cultures to refrigerated vacuum-packaged beef slices can slow spoilage, especially with strains producing bacteriocin-like inhibitory substances (Katikou et al., 2005). *L. sakei* CECT 4808 effectively preserves vacuum-packaged beef by reducing spoilage microorganisms and preventing lipid oxidation, extending storage life. Castellano et al. (2018) evaluated the antilisterial activity of a combination of three LAB strains (Lactiguard®) against *Listeria monocytogenes* on frankfurters (Koo et al., 2012). After 8 weeks of refrigerated storage, the LAB combination reduced *L. monocytogenes* by 0.6 log on frankfurters without lactate/diacetate, and by 1.2 log when a cell-free extract was added. In frankfurters containing lactate/diacetate, the LAB and LAB plus cell-free extract reduced *L. monocytogenes* by 2 and 3.3 log, respectively. Rindhe et al. (2017) aimed to develop shelf-stable chicken sausages using textured soy protein as a humectant, lactic acid as an acidulant, and Nisin to extend shelf life (Rindhe et al., 2017). The incorporation of Nisin at 40 mg/kg did not significantly affect sensory quality but helped maintain overall acceptability for up to 6 days at ambient temperature (37 ± 1 °C). Nisin also effectively inhibited yeast and mold growth, prolonging the sausages' shelf life.

Gamma radiation is a nonthermal technology known as the best technology to destroy pathogenic and spoilage microbes in food products (Silva et al., 2021). Combining irradiation, reduced water activity, and vacuum packing effectively demonstrates the microbiological safety and shelf stability of meat products (Santos et al., 2002). A radiation dose of 2.5 kGy eliminated 10^6 CFU of *Clostridium sporogenes* and *Bacillus cereus*, but not *Clostridium sporogenes*. Maintaining a_w of 0.85 and using vacuum packaging prevented the growth of all three organisms during 3 months of room temperature storage. Irradiation also inactivated yeasts and molds, which otherwise grew after 2 months, and *B. cereus* numbers decreased over time. These findings that combining these hurdles results in microbiologically safe, shelf-stable meat products. Kanatt et al. (2002) developed ready-to-serve, shelf-stable spiced mutton and chicken products with intermediate moisture using hurdle technology, employing moisture reduction, vacuum packaging, and irradiation (Kanatt et al., 2002). The meats were grilled or hot-air dried to reach a_w of about 0.80, then vacuum packaged and irradiated at doses from 0 to 10 kGy. Microbiological analysis indicated that higher radiation doses resulted in greater reductions in total viable counts and *Staphylococcus* numbers. Without irradiation, the meat products developed visible fungal growth within 2 months. Chawla et al. (2006) created safe, shelf-stable natural casings by applying hurdle technology involving reduced water activity, packaging, and gamma irradiation (Chawla et al., 2006). Lamb intestines were salted to lower water activity to about 0.80, packed in polyethylene bags, and subjected to 5 or 10 kGy gamma irradiation. Nonirradiated samples exhibited high microbial counts, while irradiation significantly reduced these counts. A 5 kGy dose notably decreased viable microbes, and 10 kGy eliminated them entirely. Lower water activity prevented microbial growth during storage. Sausages using these casings retained good sensory and textural qualities, showing that hurdle technology successfully produces safe casings. Aziz et al. (2002) observed that the combination of gamma radiation and microwave treatments effectively improves the microbial safety of beef products (Aziz et al., 2002). Initially, a 2–3 log reduction of total bacteria count was observed after 5.0 kGy gamma irradiation, while microwave heating reduced bacteria by 1 log in 20 s and 2 logs in 30 s. Untreated samples had a shelf life of under 7 days, but those treated with 3 kGy and microwaved for 20 s lasted at least 2 weeks at 5 °C. Gamma irradiation had minimal effects on odor and flavor; the combined treatments did not significantly alter thiamine and free fatty acids but increased peroxide values during storage.

Hydrodynamic pressure processing (HPP) is a non-thermal food preservation technique that uses high-pressure shock waves generated by explosives or mechanical methods in water. These shock waves disrupt microbial cells and alter food structures, enhancing safety and extending shelf life without significant heat damage to the food's quality, texture, or nutrients. However, the effectiveness of spore inactivation is reduced. Thus, hurdle technology, which combines HPP with mild heat treatment, effectively inactivates bacterial spores, extending the shelf life of meat products while preserving their freshness. Silva (2016) studied the effects of combining HPP at 600 MPa with thermal treatment at 70 °C on the inactivation of *Bacillus cereus* spores in beef slurry (Silva, 2016). The study found that this combined approach achieved a 4.9 log reduction in spores after 20 min, compared to only a 0.5 log reduction with thermal treatment alone at 70–90 °C for the same duration. Another limitation of HPP is the lipid oxidation that occurs in meat products. Consequently, some studies have explored combining low-pressure treatments with natural preservatives (sage powder, thymol, dried rosemary, carvacrol) to preservation of fresh meat. The antimicrobial synergy between HPP and essential oils, their active compounds, or bacteriocins impacts microbial cell membranes. This suggests that antimicrobials partially damage bacterial membranes, making them more susceptible to pressure-induced inactivation (Pérez-Baltar et al., 2021). For example, throughout the shelf life of 60 days at 4 °C, carvacrol helped mitigate HPP-induced lipid oxidation in the breast samples (de Oliveira et al., 2015). The combination of carvacrol at 200 ppm with HPP (600 MPa for 180 s at 25 °C) effectively extended the shelf life of low-sodium sliced turkey breast ham. This natural hurdle approach significantly prolonged the product's shelf life by reducing microbial growth rates and extending the lag phases of *L. monocytogenes* and other spoilage organisms like lactic acid and psychrotrophic bacteria.

Cold plasma is an ionized gas composed of various antimicrobial agents, including reactive oxygen species, charged particles, electromagnetic fields, UV-C photons, and reactive nitrogen substances. Numerous studies have demonstrated that cold plasma effectively reduces the counts of pathogenic and spoilage microorganisms in both raw and processed meats (Han et al., 2016). To achieve effective microbial decontamination, a combined approach using cold plasma and safe antimicrobial substances, such as essential oils and lactic acid has been employed. It has been reported that higher concentrations of lemongrass essential oil (10 mg/mL) for longer treatment time (40 min) and high power cold plasma (600 W) for increased time (180 s) were required to achieve a higher inactivation rate of *L. monocytogenes* when these were used individually. In contrast, a combination of cold plasma (120 s, 500 W) and essential oil (5 mg/mL, 30 min) exerted a synergistic antimicrobial effect on the *L. monocytogenes* (4 log CFU/g) with a decrease of 2.80 logs CFU/g at a reduced concentration of essential oil, cold plasma power and treatment times. Additionally, physicochemical properties (color, TBARS, and texture) and sensory characteristics (off-flavor, appearance, taste, color, and overall acceptability) of pork loin were not affected (Cui et al., 2017).

Modified atmosphere packaging (MAP) is employed to complement low-temperature food preservation techniques. However, a significant challenge in MAP's commercial application is its varying effectiveness across different products. Additionally, MAP can alter physicochemical properties and potentially allow the growth of pathogenic microorganisms, rendering MAP alone potentially insufficient for ensuring food quality and safety. To address this, combining MAP with other preservation methods is essential. The properties of fresh meat at 5 °C and 15 °C were affected by oregano essential oil volatiles and modified atmosphere packaging (MAP), using different gas compositions (Skandamis & Nychas, 2002). Shelf life extension varied by packaging, in the order of air < vacuum <40% CO_2/30% N_2/30% O_2 < 80% CO_2/20% air <100% CO_2, with enhanced effects when combined with oregano essential oil. This combination inhibits microbial growth and spoilage more effectively than MAP alone. The essential oil especially delayed glucose and lactate consumption, indicators of spoilage, and affected other metabolites like formic acid.

5.3 Application in Aquatic Food Products

The selection of hurdle factors is based on the basic nutritional composition of the product and specific spoilage bacteria and pathogens. In aquatic food products, these are mainly aerobic bacteria spoilage organisms or pathogenic bacteria, such as *Pseudomonas*, *Alcaligenes*, *Flavobacterium*, *Bacillus*, and *Vibrio parahaemolyticus* (Salama & Chennaoui, 2024). The application of hurdle technology in the processing and storage of aquatic food products has been extensively researched and has shown significant effectiveness. The use of conventional hurdle technology can lead to the production of aquatic food products seasonings that can be stored for extended periods. In addition to traditional aquatic food products preservation techniques such as chilling, freezing, drying, smoking, salting, fermentation, and canning, there are some more recent methods being applied. These include the use of natural preservatives, high hydrostatic pressure treatment, ozonation, irradiation, pulse light technology, retort pouch processing, and packaging combined with refrigeration or freezing (Kontominas et al., 2021).

For perishable aquatic food products like fish and shrimp, maintaining low temperatures is often the primary, and sometimes the only, preservation hurdle utilized. However, when these fresh aquatic products are subjected to temperature abuse during distribution and storage, this preservation method fails. This failure leads to quality deterioration and heightened safety risks, such as histamine poisoning in scombroid fish and the growth of *Vibrio parahaemolyticus* in oysters. These risks are closely linked to elevated storage temperatures (Tsironi et al., 2017).

For fresh aquatic food products, it is important to use a combination of preservation methods such as low temperatures, salt addition, mild thermal treatment, and packaging. These products have high moisture content with a_w above 0.9 and are typically chilled or frozen. However, maintaining low temperatures in the cold chain is energy-intensive and costly. Furthermore, temperature fluctuations in the cold chain can significantly impact the quality and shelf life of these products at any stage in the food supply chain. Therefore, in addition to low temperatures, other hurdles should be implemented to preserve quality and extend the shelf life of fresh aquatic food products. Common hurdles used for preserving these products include reducing water activity, mild heat treatment and using preservatives (Erkmen & Bozoglu, 2016).

Achieving reduced a_w levels in fresh fish can be accomplished with the application of salt or sugar. The use of additives such as maltodextrin and NaCl has been shown to reduce the a_w of fish fillet to approximately 0.95, effectively inhibiting the growth of *Pseudomonas* spp., a primary spoilage factor in aerobically packed, chilled Mediterranean fish (Tsironi & Taoukis, 2014). Incorporating agents like glucono-δ-lactone in the osmotic solution can decrease the pH of osmotically treated fish, providing an additional hurdle that synergizes with low a_w and refrigerated storage to further inhibit microbial growth and extend shelf life (Tsironi & Taoukis, 2012). Similar synergistic effects have been observed with the inclusion of natural antimicrobials like carvacrol and nisin, as well as plant extracts, in the solution used for osmotic treatments of fish fillets (Sofra et al., 2018; Tsironi & Taoukis, 2012). These minimal processing techniques are both cost-effective and energy-efficient, offering the advantage of extending shelf life without significantly altering the initial sensory properties of the food products.

Thermal treatment of fish products often utilizes hurdle technology to achieve extended shelf life. A prominent example involves the use of 6% salt and 0.2% sorbate at a pH of 5.7, which, when combined with heat treatment for 10 min at 80 °C, extends the shelf life of minced pelagic fish to 15 days at 15 °C, compared to less than 3 days for untreated samples. Additionally, combining hot smoking with edible coatings that contain rosemary essential oil or extract has been assessed. The addition of these ingredients demonstrated moderate antimicrobial properties and significantly slowed the formation of oxidation products, showcasing a synergistic effect that enhances shelf life (Aguilera et al., 1992).

Due to the loss of heat-sensitive nutrients and negative impacts on the inherent sensory properties such as color, flavor, and texture of foods caused by heat treatment, there has been an increasing interest in non-thermal sterilization technologies that preserve the original characteristics of food as much as possible. Cold sterilization methods, which eliminate microorganisms through physical means like hydrostatic pressure, magnetic oscillation, and gamma-ray irradiation, include ultra-high-pressure sterilization, irradiation sterilization, magnetic sterilization, pulsed light sterilization, and titanium dioxide sterilization. The development of shelf-stable, ready-to-eat (RTE) shrimp using a combination of preservation hurdles marks a significant advancement in food preservation techniques (Kanatt et al., 2006). This approach employed specific hurdles such as reduced water activity (0.85 ± 0.02), packaging, and gamma irradiation at 2.5 kGy. By integrating these methods, the shelf life of RTE shrimp was significantly extended. At an ambient storage temperature of 25 ± 3 °C, the shelf life increased from 15 days to 8 weeks. In contrast, the control samples developed mold growth within 15 days. These innovations not only extend product longevity but also ensure acceptable quality for consumers. Ates et al. (2016) observed a synergistic effect was observed on the inactivation of *L. monocytogenes* by combining high-pressure processing with mild heat (Ates et al., 2016). The study showed a reduction of 6.62 log CFU/g in inoculated fish soup using 500 MPa at 44 °C for 5 min, compared to a reduction of 2.64 log CFU/g at the same pressure but a lower temperature of 16 °C for the same duration. A similar synergistic effect was also noted in a study by Amanatidou et al. (2000), who combined high-pressure processing with modified atmosphere packaging of Atlantic salmon (Amanatidou et al., 2000). Combining processing technologies can enhance the feasibility of methods like ultrasound for seafood treatment. Generally, ultrasonication requires long processing times to achieve sufficient bacterial inactivation, which can negatively affect the product. However, when combined with mild heat treatment (<53 °C), it becomes a more effective processing approach for shrimp. This combination increases bacterial reduction from 0.6 to 4 log CFU/g and shortens the necessary processing time (Wang et al., 2013). Aubourg et al. (2013) discovered that treating Atlantic mackerel with 150 MPa hydrostatic pressure before freezing could inhibit lipid oxidation and enhance the quality of frozen fish without altering its color (Aubourg et al., 2013). Sensory analysis and microwave cooking of fish fillets demonstrated that the flavor and texture were comparable to those of fresh fish fillets. Ramirez-Suarez & Morrissey (2006) investigated the effects of ultra-high pressure on the shelf life of tuna, finding that treatment with 310 MPa allowed tuna to be stored for more than 23 days at 4 °C and 93 days at −20 °C. Additionally, the combination of non-thermal sterilization with hurdle technology has been explored in seafood preservation. For instance, A study on the shelf life of rainbow trout fillets treated with high-dose irradiation and vacuum packaging, showing a significant increase in shelf life with higher doses, reaching 28, 42, 70, and 98 days at 3.5 °C (Křížek et al., 2012).

Packing or modified atmosphere packing (MAP) are one of the most often methods in combinations for aquatic products preservation. Antimicrobial packaging involves incorporating antimicrobial agents into packaging materials, allowing these agents (such as potassium sorbate, nisin, etc.) to be gradually released into the food through the permeability of the packaging film. Since the concentration of the antimicrobial agents in the packaging is much higher than in the food, the agents slowly migrate from the high-concentration packaging film to the lower-concentration food. This ensures that the antimicrobials do not exceed maximum permissible levels in the food, allowing for a prolonged antimicrobial effect and extended freshness. New antimicrobial agents are continually being developed and added to packaging materials, such as essential oils. Tunç & Duman (2011) used carvacrol as an antimicrobial in methylcellulose-based nanocomposite packaging material and found that it exhibited antibacterial properties against *E. coli* and *Staphylococcus aureus*. Additionally, the concentration of montmorillonite in the film matrix and storage temperature affected the release rate of carvacrol. Gómez-Estaca et al. (2007) studied the combination of high-pressure processing (300 MPa, 20 °C, 15 min) with functional edible films containing antimicrobials (oregano oil/rosemary oil/chitosan) for cold-smoked sardine packaging (Gómez-Estaca et al., 2007). The results indicated that edible films containing plant essential oils provided strong antioxidant properties to the fish muscle, while edible films made from gelatin-chitosan controlled release materials demonstrated more significant antibacterial effects. These results suggest that the combination of high pressure and active packaging offers both antioxidative and antibacterial benefits. MAP in combination of other hurdles also showed good synergy effect. Abel et al. (2019) explored the effects of mild heat treatment (40 °C, 50 °C, or 60 °C) combined with MAP on microbial growth and overall product quality (Abel et al., 2019). The results showed enhanced inhibition of *Listeria innocua* and emphasized the potential for achieving safe products with low-temperature sous vide treatments at 40 and 50 °C. Furthermore, combining MAP with superchilling extended the shelf life of Atlantic cod from 13 days in ice storage to over 32 days when superchilled in MAP packaging (13, 17, 23, and over 32 days for iced air, iced MAP, superchilled air, and superchilled MAP, respectively), demonstrating the synergistic effect of this hurdle combination (Sørensen et al., 2020).

6 Conclusion

Hurdle technology plays an important role in modern food preservation, offering a comprehensive approach to ensure both food quality and safety. By integrating various physical, chemical, and biological preservation methods, hurdle technology establishes multiple barriers that impede microbial growth, thereby extending the shelf life of food products. This approach is crucial in meeting the increasing consumer demand for minimally processed foods that maintain high nutritional value, freshness, and safety. The advent of new technologies, such as non-thermal treatments, provides enhanced flexibility in combining hurdles. The combination of non-thermal techniques with traditional preservation methods presents an exciting opportunity for future advancements. This allows producers to improve quality, safety while reducing costs and facilitating the development of innovative products. The application of multiple hurdles requires more than mere combination; it should be based on an understanding of the synergistic effects mechanism of these methods. It is also important to consider not only microbial control but also the impact of multiple hurdles on overall food quality. Furthermore, the integration of predictive microbiology, Hazard Analysis and Critical Control Points (HACCP), and Good Manufacturing Practices (GMP) into hurdle technology will be essential for its successful industrial application.

References

Abel, N., Rotabakk, B. T., & Lerfall, J. (2019). Effect of heat treatment and packaging technology on the microbial load of lightly processed seafood. *LWT- Food Science and Technology, 101*, 123–129.

Aguilera, J. M., Francke, A., Figueroa, G., Bornhardt, C., & Cifuentes, A. (1992). Preservation of minced pelagic fish by combined methods. *International Journal of Food Science and Technology, 27*(2), 171–177.

Akhter, R., Masoodi, F., Wani, T. A., Rather, S. A., & Hussain, P. R. (2021). Synergistic effect of low dose γ-irradiation, natural antimicrobial and antioxidant agents on quality of meat emulsions. *Radiation Physics and Chemistry, 189*, 109724.

Allen, G. (2015). *Sausage: A global history*. Reaktion Books.

Amanatidou, A., Schlüter, O., Lemkau, K., Gorris, L., Smid, E., & Knorr, D. (2000). Effect of combined application of high pressure treatment and modified atmospheres on the shelf life of fresh Atlantic salmon. *Innovative Food Science & Emerging Technologies, 1*(2), 87–98.

And, H. C., & Hoover, D. G. (2003). Bacteriocins and their food applications. *Comprehensive Reviews in Food Science and Food Safety, 2*(3), 82–100.

Andrés-Bello, A., Barreto-Palacios, V., García-Segovia, P., Mir-Bel, J., & Martínez-Monzó, J. (2013). Effect of pH on color and texture of food products. *Food Engineering Reviews, 5*, 158–170.

Ates, M. B., Rode, T. M., Skipnes, D., & Lekang, O.-I. (2016). Modeling of *Listeria monocytogenes* inactivation by combined high-pressure and mild-temperature treatments in model soup. *European Food Research and Technology, 242*, 279–287.

Aubourg, S. P., Torres, J. A., Saraiva, J. A., Guerra-Rodríguez, E., & Vázquez, M. (2013). Effect of high-pressure treatments applied before freezing and frozen storage on the functional and

sensory properties of Atlantic mackerel (*Scomber scombrus*). *LWT- Food Science and Technology, 53*(1), 100–106.
Aziz, N. H., Mahrous, S. R., & Youssef, B. M. (2002). Effect of gamma-ray and microwave treatment on the shelf-life of beef products stored at 5°C. *Food Control, 13*(6–7), 437–444.
Bai, J. H., Saftner, R., Watada, A., & Lee, Y. (2001). Modified atmosphere maintains quality of fresh-cut cantaloupe (*Cucumis melo L.*). *Journal of Food Science, 66*(8), 1207–1211.
Barwal, V., Rakesh Sharma, R. S., & Rajinder Singh, R. S. (2005). Preservation of cauliflower by hurdle technology. *Journal of Food Science and Technology - Mysore, 42*(1), 26–31.
Bragagnolo, N., Danielsen, B., & Skibsted, L. H. (2005). Effect of rosemary on lipid oxidation in pressure-processed, minced chicken breast during refrigerated storage and subsequent heat treatment. *European Food Research and Technology, 221*, 610–615.
Cantalejo, M. J., Zouaghi, F., & Pérez-Arnedo, I. (2016). Combined effects of ozone and freeze-drying on the shelf-life of broiler chicken meat. *LWT - Food Science and Technology, 68*, 400–407.
Castellano, P., Peña, N., Ibarreche, M. P., Carduza, F., Soteras, T., & Vignolo, G. (2018). Antilisterial efficacy of *Lactobacillus bacteriocins* and organic acids on frankfurters. Impact on sensory characteristics. *Journal of Food Science and Technology, 55*, 689–697.
Cepero, Y., Beldarrain, T., Campos, A., Santos, R., Bruselas, A., & Vergara, N. (2006). Use of the hurdle effect in a conservation of fresh sausage. *Ciência e Tecnologia de Alimentos, 16*(2), 7–13.
Chawla, S., Chander, R., & Sharma, A. (2006). Safe and shelf-stable natural casing using hurdle technology. *Food Control, 17*(2), 127–131.
Chien, S.-Y., Sheen, S., Sommers, C., & Sheen, L.-Y. (2017). Modeling the inactivation of *Escherichia coli* O157: H7 and uropathogenic *E. coli* in ground beef by high pressure processing and citral. *Food Control, 73*, 672–680.
Chouliara, E., Karatapanis, A., Savvaidis, I., & Kontominas, M. (2007). Combined effect of oregano essential oil and modified atmosphere packaging on shelf-life extension of fresh chicken breast meat, stored at 4 °C. *Food Microbiology, 24*(6), 607–617.
Cole, A. M., Darouiche, R. O., Legarda, D., Connell, N., & Diamond, G. (2000). Characterization of a fish antimicrobial peptide: Gene expression, subcellular localization, and spectrum of activity. *Antimicrobial Agents and Chemotherapy, 44*(8), 2039–2045.
Cui, H., Wu, J., Li, C., & Lin, L. (2017). Promoting anti-listeria activity of lemongrass oil on pork loin by cold nitrogen plasma assist. *Journal of Food Safety, 37*(2), e12316.
Cui, H., Bai, M., Yuan, L., Surendhiran, D., & Lin, L. (2018). Sequential effect of phages and cold nitrogen plasma against *Escherichia coli* O157: H7 biofilms on different vegetables. *International Journal of Food Microbiology, 268*, 1–9.
Davidson, P. M., Taylor, T. M., & Schmidt, S. E. (2012). Chemical preservatives and natural antimicrobial compounds. In *Food microbiology* (pp. 765–801). Royal Society of Chemistry.
de Oliveira, T. L. C., Junior, B. R. d. C. L., Ramos, A. L., Ramos, E. M., Piccoli, R. H., & Cristianini, M. (2015). Phenolic carvacrol as a natural additive to improve the preservative effects of high pressure processing of low-sodium sliced vacuum-packed Turkey breast ham. *LWT- Food Science and Technology, 64*(2), 1297–1308.
Degala, H. L., Mahapatra, A. K., Demirci, A., & Kannan, G. (2018). Evaluation of non-thermal hurdle technology for ultraviolet-light to inactivate *Escherichia coli* K12 on goat meat surfaces. *Food Control, 90*, 113–120.
Del Nobile, M. A., Conte, A., Cannarsi, M., & Sinigaglia, M. (2009). Strategies for prolonging the shelf life of minced beef patties. *Journal of Food Safety, 29*(1), 14–25.
Ding, T., Liao, X., & Feng, J. (2022). *Stress responses of foodborne pathogens*. Springer.
Economou, T., Pournis, N., Ntzimani, A., & Savvaidis, I. (2009). Nisin–EDTA treatments and modified atmosphere packaging to increase fresh chicken meat shelf-life. *Food Chemistry, 114*(4), 1470–1476.
Eman, A., Kamel, H., Zaki, Z. A., & Rehab, M. A. (2015). Effect of honey and citric acid treatments on postharvest quality of fruits and fresh-cut of guava. *World Journal of Agricultural Science, 11*(5), 255–267.

Erkmen, O., & Bozoglu, T. F. (2016). *Food microbiology: Principles into practice* (2 Vol. set). Wiley.

Fan, X., Annous, B. A., Sokorai, K. J., Burke, A., & Mattheis, J. P. (2006). Combination of hot-water surface pasteurization of whole fruit and low-dose gamma irradiation of fresh-cut cantaloupe. *Journal of Food Protection, 69*(4), 912–919.

Field, D., Fernandez de Ullivarri, M., Ross, R. P., & Hill, C. (2023). After a century of nisin research-where are we now? *FEMS Microbiology Reviews, 47*(3), fuad023.

Gil, M. I., Selma, M. V., López-Gálvez, F., & Allende, A. (2009). Fresh-cut product sanitation and wash water disinfection: Problems and solutions. *International Journal of Food Microbiology, 134*(1–2), 37–45.

Gómez-Estaca, J., Montero, P., Giménez, B., & Gómez-Guillén, M. C. (2007). Effect of functional edible films and high pressure processing on microbial and oxidative spoilage in cold-smoked sardine (*Sardina pilchardus*). *Food Chemistry, 105*(2), 511–520.

Hadjok, C., Mittal, G., & Warriner, K. (2008). Inactivation of human pathogens and spoilage bacteria on the surface and internalized within fresh produce by using a combination of ultraviolet light and hydrogen peroxide. *Journal of Applied Microbiology, 104*(4), 1014–1024.

Han, L., Boehm, D., Amias, E., Milosavljević, V., Cullen, P., & Bourke, P. (2016). Atmospheric cold plasma interactions with modified atmosphere packaging inducer gases for safe food preservation. *Innovative Food Science & Emerging Technologies, 38*, 384–392.

Islam, F., Saeed, F., Imran, A., Shehzadi, U., Ali, R., Nosheen, F., Chauhan, A., Asghar, A., & Ojukwu, M. (2024). Bio-preservatives and essential oils as an alternative to chemical preservatives in the baking industry: A concurrent review. *Journal of Food Science and Technology, 61*(4), 609–620.

Isohanni, P., Huehn, S., Aho, T., Alter, T., & Lyhs, U. (2013). Heat stress adaptation induces cross-protection against lethal acid stress conditions in *Arcobacter butzleri* but not in *Campylobacter jejuni*. *Food Microbiology, 34*(2), 431–435.

Jafari, M., & Emam-Djomeh, Z. (2007). Reducing nitrite content in hot dogs by hurdle technology. *Food Control, 18*(12), 1488–1493.

Jones, B. M., Kaiser, R. I., & Strazzulla, G. (2014). Carbonic acid as a reserve of carbon dioxide on icy moons: The formation of carbon dioxide (CO_2) in a polar environment. *The Astrophysical Journal, 788*(2), 170.

Kameník, J. (2013). The microbiology of meat spoilage: A review. *Maso International – Journal of Food Science and Technology, 2013*, 1–9.

Kanatt, S. R., Chawla, S., Chander, R., & Bongirwar, D. (2002). Shelf-stable and safe intermediate-moisture meat products using hurdle technology. *Journal of Food Protection, 65*(10), 1628–1631.

Kanatt, S., Chawla, S., Chander, R., & Sharma, A. (2006). Development of shelf stable, ready to eat (RTE) shrimps (Penaeus indicus) using γ-radiation as one of the hurdles. *LWT - Food Science and Technology, 39*(6), 621–626.

Karagözlü, N., Ergönül, B., & Özcan, D. (2011). Determination of antimicrobial effect of mint and basil essential oils on survival of *E. coli* O157: H7 and *S.* typhimurium in fresh-cut lettuce and purslane. *Food Control, 22*(12), 1851–1855.

Karthikeyan, J., Kumar, S., Anjaneyulu, A., & Rao, K. (2000). Application of hurdle technology for the development of Caprine keema and its stability at ambient temperature. *Meat Science, 54*(1), 9–15.

Katikou, P., Ambrosiadis, I., Georgantelis, D., Koidis, P., & Georgakis, S. (2005). Effect of *Lactobacillus*-protective cultures with bacteriocin-like inhibitory substances' producing ability on microbiological, chemical and sensory changes during storage of refrigerated vacuum-packaged sliced beef. *Journal of Applied Microbiology, 99*(6), 1303–1313.

Karbowiak, M., Szymański, P., & Zielińska, D. (2023) Synergistic effect of combination of various microbial hurdles in the biopreservation of meat and meat products—Systematic Review. *Foods*, 12(7),1–26.

Kontominas, M. G., Badeka, A. V., Kosma, I. S., & Nathanailides, C. I. (2021). Innovative seafood preservation technologies: Recent developments. *Animals, 11*(1), 92.

Koo, O.-K., Eggleton, M., O'Bryan, C. A., Crandall, P. G., & Ricke, S. C. (2012). Antimicrobial activity of lactic acid bacteria against *Listeria monocytogenes* on frankfurters formulated with and without lactate/diacetate. *Meat Science, 92*(4), 533–537.

Koseki, S., & Itoh, K. (2001). Prediction of microbial growth in fresh-cut vegetables treated with acidic electrolyzed water during storage under various temperature conditions. *Journal of Food Protection, 64*(12), 1935–1942.

Křížek, M., Matějková, K., Vácha, F., & Dadáková, E. (2012). Effect of low-dose irradiation on biogenic amines formation in vacuum-packed trout flesh (*Oncorhynchus mykiss*). *Food Chemistry, 132*(1), 367–372.

Leistner, L. (1992). Food preservation by combined methods. *Food Research International, 25*(2), 151–158.

Leistner, L. (1995). Principles and applications of hurdle technology. In *New methods of food preservation* (pp. 1–21). Springer.

Leistner, L. (2000). Basic aspects of food preservation by hurdle technology. *International Journal of Food Microbiology, 55*(1), 181–186.

Lu, L., Guo, H., Kang, N., He, X., Liu, G., Li, J., He, X., Yan, X., & Yu, H. (2022). Application of electrolysed water in the quality and safety control of fruits and vegetables: A review. *International Journal of Food Science and Technology, 57*(9), 5698–5711.

Lücke, F.-K., & Zangerl, P. (2014). Food safety challenges associated with traditional foods in German-speaking regions. *Food Control, 43*, 217–230.

Manion, C. R., & Widder, R. M. (2017). Essentials of essential oils. *American Journal of Health-System Pharmacy, 74*(9), e153–e162.

Mastromatteo, M., Lucera, A., Sinigaglia, M., & Corbo, M. R. (2009). Combined effects of thymol, carvacrol and temperature on the quality of non conventional poultry patties. *Meat Science, 83*(2), 246–254.

Mattheis, J., & Fellman, J. K. (2000). Impacts of modified atmosphere packaging and controlled atmospheres on aroma, flavor, and quality of horticultural commodities. *HortTechnology, 10*(3), 507–510.

Mizi, L., Cofrades, S., Bou, R., Pintado, T., López-Caballero, M., Zaidi, F., & Jiménez-Colmenero, F. (2019). Antimicrobial and antioxidant effects of combined high pressure processing and sage in beef burgers during prolonged chilled storage. *Innovative Food Science & Emerging Technologies, 51*, 32–40.

Noort, M., & Visschers, R. (2015). Herformuleren van voeding: lekker gezond. *Tijdschrift voor gezondheidswetenschappen, 93*, 248–249.

O'Neill, C. M. (2018). *High pressure processing as a hurdle technology for development of consumer-accepted, low-salt processed meat products with enhanced safety and shelf-life*. PhD thesis. University College Cork

Parry, R. (2012). *Principles and applications of modified atmosphere packaging of foods*. Springer.

Pérez-Baltar, A., Serrano, A., Bravo, D., Montiel, R., & Medina, M. (2019). Combined effect of high pressure processing with enterocins or thymol on the inactivation of *Listeria monocytogenes* and the characteristics of sliced dry-cured ham. *Food and Bioprocess Technology, 12*, 288–297.

Pérez-Baltar, A., Serrano, A., Medina, M., & Montiel, R. (2021). Effect of high pressure processing on the inactivation and the relative gene transcription patterns of *Listeria monocytogenes* in dry-cured ham. *LWT - Food Science and Technology, 139*, 110555.

Ramirez-Suarez, J. C., & Morrissey, M. T. (2006). Effect of high pressure processing (HPP) on shelf life of albacore tuna (*Thunnus alalunga*) minced muscle. *Innovative Food Science & Emerging Technologies, 7*(1–2), 19–27.

Raybaudi-Massilia, R. M., Mosqueda-Melgar, J., & Martín-Belloso, O. (2008). Edible alginate-based coating as carrier of antimicrobials to improve shelf-life and safety of fresh-cut melon. *International Journal of Food Microbiology, 121*(3), 313–327.

Rindhe, S., Shinde, A., Jadhav, B., & Kumbhar, V. (2017). Application of hurdle technology for preservation of chicken sausages at ambient temperature (37±10 °C). *Indian Journal of Pure & Applied Biosciences, 5*(4), 480–488.

Rödel, W., & Scheuer, R. (2007). Recent results on the hurdle technology: Measuring of combined hurdles. *Fleischwirtschaft -Frankfurt, 87*(9), 111–115.

Salama, Y., & Chennaoui, M. (2024). Microbial spoilage organisms in seafood products: Pathogens and quality control. *European Journal of Microbiology and Infectious Diseases, 1*(2), 66–89.

Sangsuwan, J., Rattanapanone, N., & Rachtanapun, P. (2008). Effect of chitosan/methyl cellulose films on microbial and quality characteristics of fresh-cut cantaloupe and pineapple. *Postharvest Biology and Technology, 49*(3), 403–410.

Santos, R., Mella, R., Ramos, M., Valladares, C., García, A., Casals, C., & Cordova, A. (2002). Preservation of a sausage at room temperature. *Alimentaria, 331*, 21–25.

Sergio, L., Cantore, V., Spremulli, L., Pinto, L., Baruzzi, F., Di Venere, D., & Boari, F. (2018). Effect of cooking and packaging conditions on quality of semi-dried green asparagus during cold storage. *LWT - Food Science and Technology, 89*, 712–718.

Shaltout, F., Barr, A. A. H., & Abdelaziz, M. E. (2022). Pathogenic microorganisms in meat products. *Biomedical Journal of Scientific & Technical Research, 41*(4), 32836–32843.

Shimokomaki, M., Franco, B. D. G. d. M., Biscontini, T., Pinto, M. F., Terra, N., & Zorn, T. M. T. (1998). Charqui meats are hurdle technology meat products. *Food Reviews International, 14*(4), 339–349.

Silva, F. V. (2016). Modeling the inactivation of psychrotrophic *Bacillus cereus* spores in beef slurry by 600 MPa HPP combined with 38–70 °C: Comparing with thermal processing and estimating the energy requirements. *Food and Bioproducts Processing, 99*, 179–187.

Silva, D. R., Haddad, G. B., de Moura, A. P., de Souza, P. M., Ramos, A. L., Hopkins, D. L., & Ramos, E. M. (2021). Safe cured meat using gamma radiation: Effects on spores of *Clostridium sporogenes* and technological and sensorial characteristics of low nitrite cooked ham. *LWT - Food Science and Technology, 137*, 110392.

Skandamis, P. N., & Nychas, G.-J. E. (2002). Preservation of fresh meat with active and modified atmosphere packaging conditions. *International Journal of Food Microbiology, 79*(1–2), 35–45.

Sofra, C., Tsironi, T., & Taoukis, P. S. (2018). Modeling the effect of pre-treatment with nisin enriched osmotic solution on the shelf life of chilled vacuum packed tuna. *Journal of Food Engineering, 216*, 125–131.

Soliva-Fortuny, R. C., Elez-Martínez, P., Sebastián-Calderó, M., & Martín-Belloso, O. (2004). Effect of combined methods of preservation on the naturally occurring microflora of avocado purée. *Food Control, 15*(1), 11–17.

Sørensen, J. S., Bøknæs, N., Mejlholm, O., & Dalgaard, P. (2020). Superchilling in combination with modified atmosphere packaging resulted in long shelf-life and limited microbial growth in Atlantic cod (*Gadus morhua L.*) from capture-based-aquaculture in Greenland. *Food Microbiology, 88*, 103405.

Stearns, R., Freshour, A., & Shen, C. (2022). Literature review for applying peroxyacetic acid and/or hydrogen peroxide to control foodborne pathogens on food products. *Journal of Agriculture and Food Research, 10*, 100442.

Tsironi, T., & Taoukis, P. (2012). Shelf-life extension of gilthead seabream fillets by osmotic treatment and antimicrobial agents. *Journal of Applied Microbiology, 112*(2), 316–328.

Tsironi, T. N., & Taoukis, P. S. (2014). Effect of processing parameters on water activity and shelf life of osmotically dehydrated fish filets. *Journal of Food Engineering, 123*, 188–192.

Tsironi, T., Ronnow, P., Giannoglou, M., & Taoukis, P. (2017). Developing suitable smart TTI labels to match specific monitoring requirements: The case of *Vibrio* spp. growth during transportation of oysters. *Food Control, 73*, 51–56.

Tunç, S., & Duman, O. (2011). Preparation of active antimicrobial methyl cellulose/carvacrol/montmorillonite nanocomposite films and investigation of carvacrol release. *LWT - Food Science and Technology, 44*(2), 465–472.

Turantaş, F., Kılıç, G. B., & Kılıç, B. (2015). Ultrasound in the meat industry: General applications and decontamination efficiency. *International Journal of Food Microbiology, 198*, 59–69.

Ukuku, D. O., & Fett, W. F. (2002). Effectiveness of chlorine and nisin-edta treatments of whole melons and fresh-cut pieces for reducing native microflora and extending shelf-life. *Journal of Food Safety, 22*(4), 231–253.

Ukuku, D. O., & Fett, W. F. (2004). Effect of nisin in combination with EDTA, sodium lactate, and potassium sorbate for reducing *Salmonella* on whole and fresh-cut cantaloupe. *Journal of Food Protection, 67*(10), 2143–2150.

Ukuku, D. O., Bari, M., Kawamoto, S., & Isshiki, K. (2005). Use of hydrogen peroxide in combination with nisin, sodium lactate and citric acid for reducing transfer of bacterial pathogens from whole melon surfaces to fresh-cut pieces. *International Journal of Food Microbiology, 104*(2), 225–233.

Uyttendaele, M., Neyts, K., Vanderswalmen, H., Notebaert, E., & Debevere, J. (2004). Control of Aeromonas on minimally processed vegetables by decontamination with lactic acid, chlorinated water, or thyme essential oil solution. *International Journal of Food Microbiology, 90*(3), 263–271.

Vibhakara, H., Manjunatha, S., Radhika, M., Gupta, D., & Bawa, A. (2005). Effect of gamma-irradiation in combination preservation technique for stabilizing high moisture spice based vegetables. *Journal of Food Science and Technology, 42*(5), 434–438.

Vibhakara, H. S. J., Gupta, D. K. D., Jayaraman, K. S., & Mohan, M. S. (2006). Development of a high-moisture shelf-stable grated carrot product using hurdle technology. *Journal of Food Processing & Preservation, 30*(2), 134–144.

Wang, W., Li, M., & Li, Y. (2013). Modeling the thermoultrasound inactivation of *Vibrio parahaemolyticus* in raw peeled shrimps. *Journal of Food Protection, 76*(10), 1712–1718.

Yousef, A. E., & Balasubramaniam, V. M. (2012). Physical methods of food preservation. In *Food microbiology* (pp. 735–763). Royal Society of Chemistry.

Delivery Systems for Antimicrobials

Xiuqin Chen and Tian Ding

1 Introduction

Food-borne illnesses are very common around the world and have a significant effect on both the public health and social economy (Scallan et al., 2011). The majority of food-borne illnesses are infectious and caused by pathogens that contaminate food (Kumar et al., 2024). According to the World Health Organization (WHO), consuming tainted food causes illness in around 10% of the world's population each year, resulting in over 420,000 deaths worldwide (Hoffmann et al., 2017). Therefore, it is very important to ensure the microbiological safety of food, several preservation techniques have been developed to keep food secure from pathogens.

To increase food safety and quality, food antimicrobial compounds are either added on purpose or found naturally in food systems (Hector et al., 2023). They work by slowing down the growth of microorganisms or killing them off. Several significant obstacles may arise when directly added antimicrobial chemicals are used in food products: (1) low solubility in aqueous systems; (2) limited stability against physical or chemical degradation; (3) uncontrolled release; and (4) potential negative effects on food sensory attributes. Using delivery systems is one way to tackle these issues (Fu et al., 2016). Antimicrobial delivery systems provide an approach to maintaining food quality, increasing shelf life, and guaranteeing consumer safety in the field of food technology. These systems may provide a harmonious balance of flavor, nutrition, and protection by combining the potency of antimicrobials with delivery techniques (Grumezescu & Oprea, 2017).

X. Chen
Department of Food Science and Nutrition, Zhejiang University, Hangzhou, Zhejiang, China

Zhongyuan Institute, Zhejiang University, Zhengzhou, China

T. Ding (✉)
Department of Food Science and Nutrition, Zhejiang University, Hangzhou, Zhejiang, China
e-mail: tding@zju.edu.cn

D. Li (ed.), *Antimicrobial Strategies in the Food System: Updates, Opportunities, Challenges*, https://doi.org/10.1007/978-3-031-95056-8_7

Delivery systems can enhance the efficacy of antimicrobial agents by preventing them from interacting with food ingredients (Khaneghah et al., 2018). Nanotechnology, with its unique ability to manipulate matter at the nanoscale, has emerged as a promising approach in this regard (Keskin et al., 2021). Nanomaterials, such as nanoemulsions, liposomes, and nanofibers, offer unique properties that can improve bioavailability, and reduce toxicity associated with traditional antimicrobial agents (Donsì et al., 2012; Yousefi et al., 2019; Huang et al., 2022). The different kinds of antimicrobial agents used in food and the delivery mechanisms be examined in detail in this chapter, discussing its potential applications, and associated challenges.

2 Food Antimicrobials

Antimicrobials are widely utilized techniques in ensuring the microbial safety of food products, and they primarily encompass both synthetic and natural compounds (Dhanasekaran et al., 2015). Both types of antimicrobials play crucial roles in ensuring food safety and are widely used in the food industry, thus preserving the quality and extending the shelflife of food products (Liu et al., 2022).

2.1 Synthetic Antimicrobials

Synthetic antimicrobials are typically produced through chemical synthesis and have a broad spectrum of antimicrobial activity. However, the widespread use of synthetic antimicrobials may lead to concerns about their potential adverse effects, such as development of antimicrobial resistance (Bennani et al., 2020). They can also interact with human cells and tissues, leading to unwanted side effects, such as allergic reactions, liver or kidney damage, and disruptions of gut microbiota (Pisoschi et al., 2018; Cao et al., 2020). Nowadays, the growing concern among consumers regarding the use of naturally occurring antimicrobials derived from plants, microorganisms, and animal sources is primarily due to the increased knowledge of the dangers connected to the use of artificially manufactured additives and preservatives in the food industry.

2.2 Natural Antimicrobials

Natural antimicrobials on the other hand, are derived from natural sources such as plants, animals, or microorganisms, and they often have specific antimicrobial activities against certain microorganisms. Plant-based antimicrobials, such as essential oil, can get depleted in the food system. They may adhere to hydrophobic food components, such as lipids and proteins, reducing their effectiveness against the

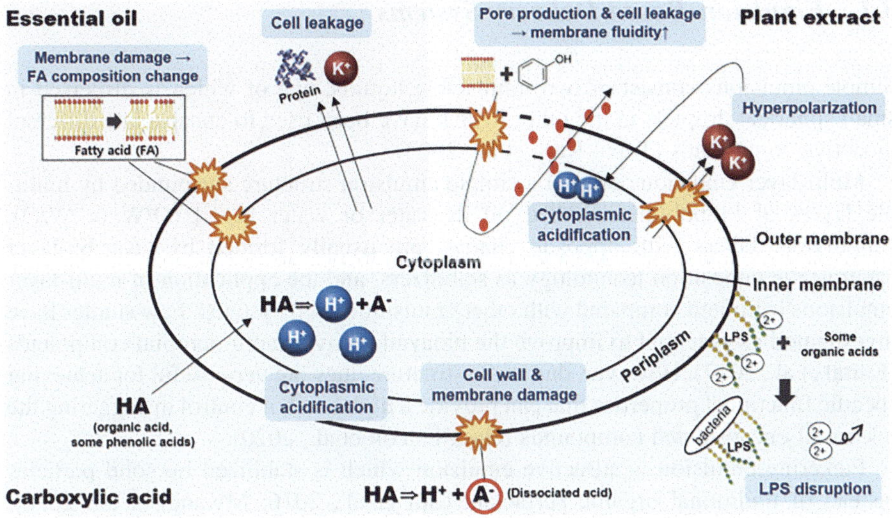

Fig. 1 Antimicrobial mechanisms of natural compounds (Kim & Lee, 2022)

target pathogens (Shah et al., 2012). Certain plant compounds, such as carvacrol and thymol are very volatile, with minimal impact on sensory attributes, they can exhibit antibacterial action in the gas phase at substantially lower concentrations than in the liquid phase. However, due to fast evaporation and oxidation, these chemicals' high volatility frequently, makes them unable to meet the requirements of regulated release in food preservation (Karam et al., 2020). Pleurocidin, lactoferrin, defensins, chitosan, lipids, and protamine are examples of antimicrobials that belong to a broad class of compounds derived from animals (Hoskin & Ramamoorthy, 2008). Microorganism source antimicrobials primarily consist of acid food preservatives (nisin and natamycin), enzymatic preservatives (lysozyme), and microbial metabolite preservatives (antimicrobial peptides and acids) (Hondrodimou et al., 2011; Rajendran et al., 2011; Elsser-Gravesen & Elsser-Gravesen, 2014; Hugo & Hugo, 2015). The known and putative mechanisms of natural antimicrobials are presented in Fig. 1.

3 Types of Delivery Systems for Food Antimicrobials

Systems including microemulsions, nanoemulsions, liposomes, nanoparticles and nanofibers are widely used in the delivery of antimicrobials in food (Fu et al., 2016). The delivery system is a carrier that protects the active compounds from oxidation, precipitation, crystallization, denaturation, light, temperature, pH, moisture, changes in ionic strength, adsorption, diffusion and biochemical interactions (McClements et al., 2009).

3.1 Emulsion-Based Delivery Systems

Simple emulsions consist of two immiscible liquids, one of which is dispersed in small spherical droplets in the other, which have been used to encapsulate different bioactive compounds (Teixé-Roig et al., 2023).

Multi-layer emulsions present a simple emulsion structure surrounded by multiple layers of biopolymers, called oil in water or water in oil (O/W or W/O). Biopolymer layers with opposite charges are usually formed by layer-by-layer electrostatic deposition technology as stabilizers, and the application of multi-layer emulsions is limited compared with other emulsion-based systems. Few studies have investigated its potential to improve the bioavailability of antimicrobial compounds (Kartal et al., 2017). However, due to the structure, they are very useful for achieving specific functional properties that can provide a high level of control in triggering the release of encapsulated compounds (Gasa-Falcon et al., 2020).

Pickering emulsion is attractive emulsion which is stabilized by solid particles instead of traditional organic surfactant (Shi et al., 2016; Mwangi et al., 2020). Compared with the surfactant molecule stable emulsion, pickering emulsion has good stability and good structure, so it has a wide range of applications in the food field. In the pickering emulsions delivery system, the solid particles form a film in the emulsion layer between the oil-water phase to stabilize the droplets, thereby preventing the accumulation of droplets. Such delivery systems can be used to encapsulate and deliver lipophilic or hydrophilic bioactive substances (Wu & Ma, 2016). By wrapping these substances in Pickering emulsions, they can be protected from environmental conditions and improve their stability and bioavailability.

Solid lipid nanoparticles and nanostructured lipid carriers are two types of nanoparticle systems consisting of a lipid core formed from a solid lipid or a mixture with a liquid lipid (Müller et al., 2002). Moreover, compared with conventional emulsifiers, lipid carriers have an inherently adjustable microstructure, acting as an active carrier and providing emulsion stability, it has been successfully used for the dual purposes of encapsulating or controlling the release of hydrophobic active substances and stabilizing O/W or W/O (Wolska et al., 2020; Sakellari et al., 2021).

3.2 Nanosized Carrier-Based Delivery Systems

At present, nanocarrier food delivery system has been applied in many fields, such as nanoemulsion, nanocapsule, nanofiber (Shishir et al., 2018; Koshani & Jafari, 2019). These nanocarriers can be prepared by different methods, such as emulsification, solvent volatilization, nanoprecipitation, self-assembly, electrospinning, spray drying, supercritical fluid, and microfluidics. However, the nanocarrier food delivery system also faces some challenges and problems, such as the safety of nanomaterials, the stability of nanocarriers, and the dispersion of nanomaterials in food. Therefore, when studying and applying nanocarrier food delivery systems, it is

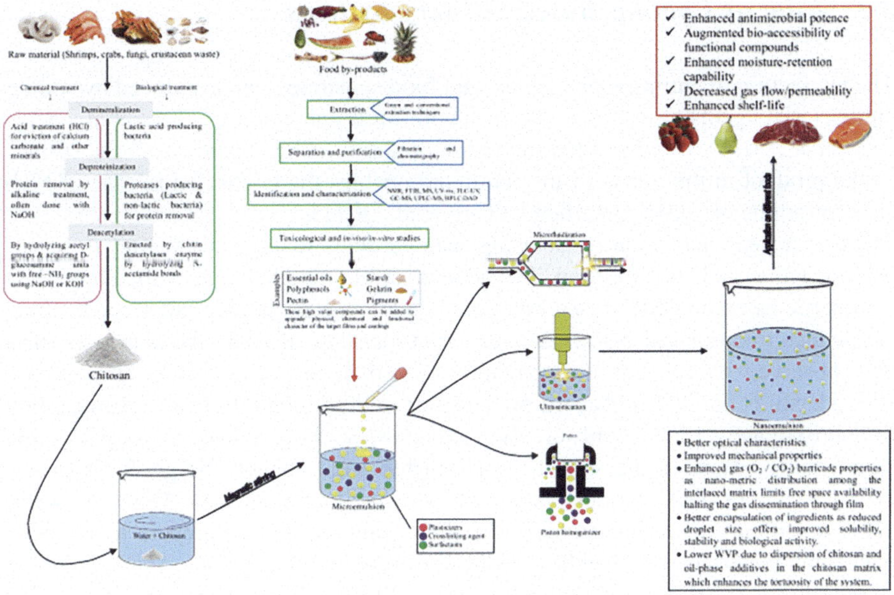

Fig. 2 Schematic representation of fabrication, application, and effect of chitosan nanoemulsion on food systems (Chaudhary, 2024)

necessary to consider their advantages, challenges, and problems comprehensively to ensure their safety and effectiveness.

Nanoemulsions, which are emulsions containing 50–500 nm nanoscale oil droplets, have been shown to have higher package stability during storage compared to conventional emulsions due to their more stable structure (Salvia-Trujillo et al., 2013; Teixé-Roig et al., 2020). Nanoemulsions for antimicrobials administration have demonstrated efficacy in treating infectious infections in both in vitro and in vivo settings. Nanoemulsions offer unique properties such as excellent stability, transparency, and bioavailability, which make them particularly appealing for biological applications. Figure 2 depicts an outline of the key components, fabrication methodologies, and influence of chitosan nanoemulsions loaded with byproducts and derivatives for food preservation.

Nanofibers, typically produced by electrospinning, are nanoscale fibers with diameters ranging from tens to hundreds of nanometers (Hu et al., 2014). They have been explored for the delivery of antimicrobial agents due to their high surface area, porosity, and ability to target specific sites (Shetty et al., 2022). Nanofibers can be loaded with antimicrobials and released slowly, providing sustained antimicrobial activity.

3.3 Film or Coating-Based Delivery Systems

The interaction of antimicrobial agents and food substrates can be reduced by adding antibacterial agents to the food interface with films and coatings (Ture et al., 2011; Resa et al., 2014). It reduces the migration of antimicrobial agents from the surface of the product to the inside of the food, and enables the controlled release of high concentrations of active ingredients (Ture et al., 2011).

Many researchers found that using nanoparticles to food packaging materials extended the shelf-life of food. Priyadarshi et al. reported that embedding ZnO nanoparticles into a CMC-based functional membrane containing grape seed extract can reduce microorganisms in packaged beef meat. In addition, the composite film was observed to prevent the oxidation of lipids in meat during refrigeration (Priyadarshi et al., 2021). According the study carried by Singh et al., impregnation of approximately 15.3 mg/mL of silver nanoparticles into cellulose food packaging significantly extended the shelf-life of tomatoe and cabbage (Singh & Sahareen, 2017). Ag, ZnO and CuO metal nanoparticles were mixed into low density polyethylene at different concentrations in different combinations to prepare active food packaging materials, which enhanced the antibacterial effect against *Staphylococcus aureus* and *Escherichia coli* (Dehghani et al., 2019). Films containing alginate clay and mixed with marjoram essential oil inhibited the growth of *Listeria monocytogenes* on rainbow trout fillets (Alboofetileh et al., 2016). In another study, a coating based on psyllium seed extract and rich in Anethum Gravolens essential oil inhibited the growth of microorganisms such as *E. coli*, *S. aureus*, *Streptococcus pyogenes*, *Bacillus subtilis*, *Bacillus cereus* and *Candida albicans* (Behbahani et al., 2017). The common types of antimicrobial packaging systems are summarized in Fig. 3.

The application of films and coatings in food technology provides new opportunities for the development of new types of food biodegradable packaging, and the addition of antibacterial nanomaterials to packaging films can achieve the controlled release of high concentrations of active ingredients. Therefore, films and coatings with antibacterial activity constitute a promising form of antimicrobial delivery for food preservation.

4 Selection Criteria for Antimicrobials Delivery Systems

Delivery method selection must consider the distinct structural and functional properties, target application environment, and the required release kinetics of various antimicrobial agents. Microcapsules or nanoemulsions may be used as delivery methods for complicated combinations, such as essential oils, to enhance their stability and controlled release capabilities (Weisany et al., 2022). Liposomes or nanoparticles can be thought of as delivery systems for single substances like

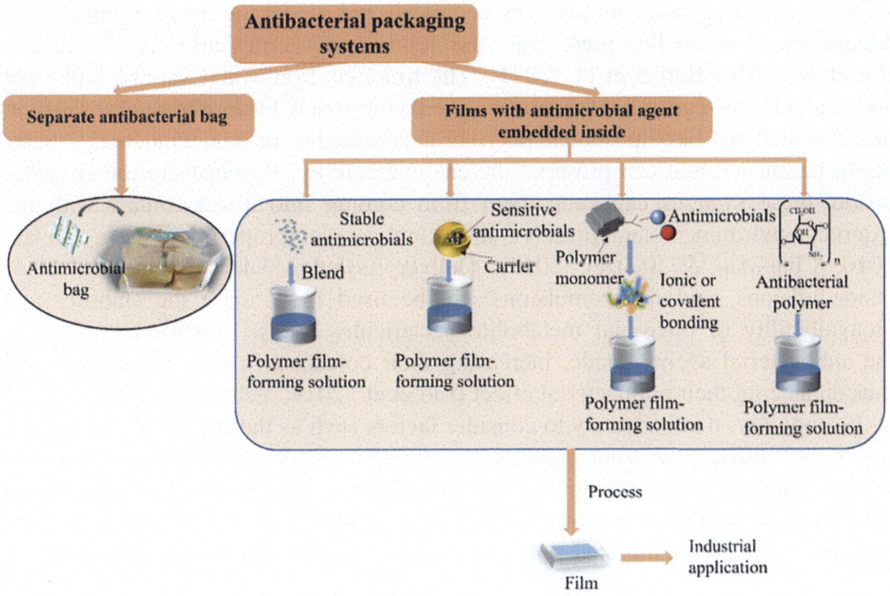

Fig. 3 Five types of antibacterial packaging systems (Wang et al., 2023)

thymol and carvacrol, which can boost their concentration and bioavailability at the target site (Doost et al., 2020).

Pleurocidin has antibacterial properties, but due to its relatively large molecular weight, it requires carrier systems such as nanoparticles or liposomes to achieve effective delivery (Sultana et al., 2021). Lactoferrin, an iron-binding glycoprotein, may require a delivery mechanism that can sustain its activity and stability, such as emulsions or microspheres, due to its iron-binding capacity (Kazemi-Taskooh & Varidi, 2021). Defensins are cationic peptides with disulfide bonds, duo to their cationic characteristics and low molecular weight, they are ideal for combining with negatively charged delivery methods to improve stability and targetability (Bakare et al., 2023). Protamine is an alkaline protein that is high in arginine, due to its large molecular weight and alkaline characteristics, it can be combined with delivery systems like nanoparticles or microspheres that can shield it from enzymatic hydrolysis and degradation (Cao et al., 2019). Chitosan is a linear copolymer produced by the combination of glucose and deacetylation, with excellent biocompatibility and gel characteristics. Because of its cationic and gel qualities, it is excellent to produce hydrogels, microspheres, and other delivery systems (Hejazi & Amiji, 2001). Lipids perform crucial structural and functional roles in microorganism, such as forming cell membranes and facilitating signal transduction. Lipid-based antibacterial agents could be combined with liposomes or emulsions to boost stability and bioavailability (Motsoene et al., 2023).

Acidic food preservatives require a low pH to be effective against bacteria, delivery techniques such as microcapsules or nanocapsules may be used. These systems can encapsulate antibacterial chemicals and, under the right circumstances, release them gradually, preserving the food's long-term antibacterial efficacy (Fu et al., 2016; Batiha et al., 2021). The time, environmental variables like pH and temperature can quickly deactivate enzyme-based bactericides. As a result, delivery methods like liposomes, polymer nanoparticles, or biodegradable particles might be chosen that can preserve the enzyme activity. By encasing the enzymes inside, these systems can keep them from coming into direct contact with the external environment and preserve their antibacterial properties (Chorny et al., 2010; Chapman & Stenzel, 2019). Delivery systems such polymer micelles, nanoemulsions, and microemulsions can be used to increase the stability and bioavailability of microbial metabolite bactericides. These systems can disperse the antibacterial agents inside, increasing their contact area with target cells and thus enhancing their antibacterial effect (Liu et al., 2019; Yeh et al., 2020).

In summary, it is necessary to consider factors such as the structural characteristics of the antibacterial agent, stability, target application environment, and required release kinetics when selecting a delivery system. By choosing an appropriate delivery system, the antibacterial effect of microbial antibacterial agents can be improved while reducing their potential risks to the environment and humans.

5 Toxicity Associated with Antimicrobials Delivery Systems

While nanomaterials offer significant advantages in antimicrobial delivery, concerns about their potential toxicity remain. Nanomaterials' toxicity is determined by a variety of parameters, including size, shape, surface chemistry, and dose. Studies have indicated that some nanomaterials can interact with biological systems in ways that may have detrimental effects (McClements & Xiao, 2017). As a result, it is critical to thoroughly assess the safety of nanoparticles before they are widely used. The current chapter digs into the complex world of antimicrobials and their delivery systems, particularly in relation to food-borne infections. This subject is critical given the prevalence of food-related illnesses and their significant impact on global public health and the socioeconomic fabric.

In conclusion, the delivery system for antimicrobials is a crucial aspect of food safety and public health. Its importance lies in its ability to protect food from contamination, prevent the spread of infections, and treat food-borne illnesses effectively. As the global demand for safe and secure food continues to grow, so does the need for innovative and effective antimicrobial delivery systems.

References

Alboofetileh, M., Rezaei, M., Hosseini, H., & Abdollahi, M. (2016). Efficacy of activated alginate-based nanocomposite films to control Listeria monocytogenes and spoilage flora in rainbow trout slice. *Journal of Food Science and Technology, 53*, 521–530.

Bakare, O. O., Gokul, A., Niekerk, L.-A., Aina, O., Abiona, A., Barker, A. M., Basson, G., Nkomo, M., Otomo, L., & Keyster, M. (2023). Recent progress in the characterization, synthesis, delivery procedures, treatment strategies, and precision of antimicrobial peptides. *International Journal of Molecular Sciences, 24*, 11864.

Batiha, G. E.-S., Hussein, D. E., Algammal, A. M., George, T. T., Jeandet, P., Al-Snafi, A. E., Tiwari, A., Pagnossa, J. P., Lima, C. M., & Thorat, N. D. (2021). Application of natural antimicrobials in food preservation: Recent views. *Food Control, 126*, 108066.

Behbahani, B. A., Shahidi, F., Yazdi, F. T., Mortazavi, S. A., & Mohebbi, M. (2017). Use of Plantago major seed mucilage as a novel edible coating incorporated with Anethum graveolens essential oil on shelf life extension of beef in refrigerated storage. *International Journal of Biological Macromolecules, 94*, 515–526.

Bennani, H., Mateus, A., Mays, N., Eastmure, E., Stärk, K. D., & Häsler, B. (2020). Overview of evidence of antimicrobial use and antimicrobial resistance in the food chain. *Antibiotics, 9*, 49.

Cao, S.-j., Xu, S., Wang, H.-m., Ling, Y., Dong, J., Xia, R.-d., & Sun, X.-h. (2019). Nanoparticles: Oral delivery for protein and peptide drugs. *AAPS PharmSciTech, 20*, 1–11.

Cao, Y., Liu, H., Qin, N., Ren, X., Zhu, B., & Xia, X. (2020). Impact of food additives on the composition and function of gut microbiota: A review. *Trends in Food Science and Technology, 99*, 295–310.

Chapman, R., & Stenzel, M. H. (2019). All wrapped up: Stabilization of enzymes within single enzyme nanoparticles. *Journal of the American Chemical Society, 141*, 2754–2769.

Chaudhary, S. (2024). Sustainable chitosan nanoemulsion coatings/films with agri-food byproducts: Advances, composition, production methods and applications in food preservation. *Journal of Food Measurement and Characterization, 18*, 1627–1649.

Chorny, M., Hood, E., Levy, R. J., & Muzykantov, V. R. (2010). Endothelial delivery of antioxidant enzymes loaded into non-polymeric magnetic nanoparticles. *Journal of Controlled Release, 146*, 144–151.

Dehghani, S., Peighambardoust, S. H., Peighambardoust, S. J., Hosseini, S. V., & Regenstein, J. M. (2019). Improved mechanical and antibacterial properties of active LDPE films prepared with combination of Ag, ZnO and CuO nanoparticles. *Food Packaging and Shelf Life, 22*, 100391.

Dhanasekaran, D., Thajuddin, N., & Panneerselvam, A. (2015). *Antimicrobials: Synthetic and natural compounds*. CRC Press.

Donsì, F., Annunziata, M., Vincensi, M., & Ferrari, G. (2012). Design of nanoemulsion-based delivery systems of natural antimicrobials: Effect of the emulsifier. *Journal of Biotechnology, 159*, 342–350.

Doost, A. S., Nasrabadi, M. N., Kassozi, V., Nakisozi, H., & Van der Meeren, P. (2020). Recent advances in food colloidal delivery systems for essential oils and their main components. *Trends in Food Science and Technology, 99*, 474–486.

Elsser-Gravesen, D., & Elsser-Gravesen, A. (2014). Biopreservatives. In *Biotechnology of food and feed additives* (pp. 29–49). Springer.

Fu, Y., Sarkar, P., Bhunia, A. K., & Yao, Y. (2016). Delivery systems of antimicrobial compounds to food. *Trends in Food Science and Technology, 57*, 165–177.

Gasa-Falcon, A., Acevedo-Fani, A., Oms-Oliu, G., Odriozola-Serrano, I., & Martín-Belloso, O. (2020). Development, physical stability and bioaccessibility of β-carotene-enriched tertiary emulsions. *Journal of Functional Foods, 64*, 103615.

Grumezescu, A., & Oprea, A. E. (2017). *Nanotechnology applications in food: Flavor, stability, nutrition and safety*. Academic.

Hector, G.-L., Isabel, F.-S., Edgar, P.-E., Susana, R., Alejandro, R., Maria, R.-R., & Jose, M. B. (2023). Immobilization of natural antimicrobial compounds on food-grade supports as a new strategy to preserve fruit-derived foods. *Foods, 12*, 2060.

Hejazi, R., & Amiji, M. (2001). Chitosan-based delivery systems: Physicochemical properties and pharmaceutical applications. In *Polymeric biomaterials, revised and expanded* (pp. 227–252). CRC Press.

Hoffmann, S., Devleesschauwer, B., Aspinall, W., Cooke, R., Corrigan, T., Havelaar, A., Angulo, F., Gibb, H., Kirk, M., Lake, R., et al. (2017). Attribution of global foodborne disease to specific foods: Findings from a World Health Organization structured expert elicitation. *PLoS One, 12*, e0183641.

Hondrodimou, O., Kourkoutas, Y., & Panagou, E. (2011). Efficacy of natamycin to control fungal growth in natural black olive fermentation. *Food Microbiology, 28*, 621–627.

Hoskin, D. W., & Ramamoorthy, A. (2008). Studies on anticancer activities of antimicrobial peptides. *Biochimica et Biophysica Acta, 1778*, 357–375.

Hu, X., Liu, S., Zhou, G., Huang, Y., Xie, Z., & Jing, X. (2014). Electrospinning of polymeric nanofibers for drug delivery applications. *Journal of Controlled Release, 185*, 12–21.

Huang, H., Song, Y., Zhang, Y., Li, Y., Li, J., Lu, X., & Wang, C. (2022). Electrospun nanofibers: Current progress and applications in food systems. *Journal of Agricultural and Food Chemistry, 70*, 1391–1409.

Hugo, C. J., & Hugo, A. (2015). Current trends in natural preservatives for fresh sausage products. *Trends in Food Science and Technology, 45*, 12–23.

Karam, L., Chehab, R., Osaili, T. M., & Savvaidis, I. N. (2020). Antimicrobial effect of thymol and carvacrol added to a vinegar-based marinade for controlling spoilage of marinated beef (Shawarma) stored in air or vacuum packaging. *International Journal of Food Microbiology, 332*, 108769.

Kartal, C., Unal, M. K., & Otles, S. (2017). Production and stabilization of a flaxseed oil multi-layer emulsion containing sodium caseinate and pectin. *International Journal of Food Properties, 20*, 833–844.

Kazemi-Taskooh, Z., & Varidi, M. (2021). Food-based iron delivery systems: A review. *Trends in Food Science and Technology, 116*, 75–89.

Keskin, D., Zu, G., Forson, A. M., Tromp, L., Sjollema, J., & van Rijn, P. (2021). Nanogels: A novel approach in antimicrobial delivery systems and antimicrobial coatings. *Bioactive Materials, 6*, 3634–3657.

Khaneghah, A. M., Hashemi, S. M. B., & Limbo, S. (2018). Antimicrobial agents and packaging systems in antimicrobial active food packaging: An overview of approaches and interactions. *Food and Bioproducts Processing, 111*, 1–19.

Kim, J.-Y., & Lee, S.-Y. (2022). Application of food-grade natural antimicrobials for the control of crop disease caused by phytopathogens. *Food Science and Biotechnology, 31*, 275–284.

Koshani, R., & Jafari, S. M. (2019). Ultrasound-assisted preparation of different nanocarriers loaded with food bioactive ingredients. *Advances in Colloid and Interface Science, 270*, 123–146.

Kumar, Y., Kaur, I., & Mishra, S. (2024). Foodborne disease symptoms, diagnostics, and predictions using artificial intelligence-based learning approaches: A systematic review. *Archives of Computational Methods in Engineering, 31*, 553–578.

Liu, Y., Shi, L., Su, L., van der Mei, H. C., Jutte, P. C., Ren, Y., & Busscher, H. J. (2019). Nanotechnology-based antimicrobials and delivery systems for biofilm-infection control. *Chemical Society Reviews, 48*, 428–446.

Liu, G., Nie, R., Liu, Y., & Mehmood, A. (2022). Combined antimicrobial effect of bacteriocins with other hurdles of physicochemic and microbiome to prolong shelf life of food: A review. *Science of the Total Environment, 825*, 154058.

McClements, D. J., & Xiao, H. (2017). Is nano safe in foods? Establishing the factors impacting the gastrointestinal fate and toxicity of organic and inorganic food-grade nanoparticles. *npj Science of Food, 1*, 6.

McClements, D. J., Decker, E. A., Park, Y., & Weiss, J. (2009). Structural design principles for delivery of bioactive components in nutraceuticals and functional foods. *Critical Reviews in Food Science and Nutrition, 49*, 577–606.

Motsoene, F., Abrahamse, H., & Kumar, S. S. D. (2023). Multifunctional lipid-based nanoparticles for wound healing and antibacterial applications: A review. *Advances in Colloid and Interface Science, 321*, 103002.

Müller, R., Radtke, M., & Wissing, S. (2002). Nanostructured lipid matrices for improved microencapsulation of drugs. *International Journal of Pharmaceutics, 242*, 121–128.

Mwangi, W. W., Lim, H. P., Low, L. E., Tey, B. T., & Chan, E. S. (2020). Food-grade Pickering emulsions for encapsulation and delivery of bioactives. *Trends in Food Science and Technology, 100*, 320–332.

Pisoschi, A. M., Pop, A., Georgescu, C., Turcuş, V., Olah, N. K., & Mathe, E. (2018). An overview of natural antimicrobials role in food. *European Journal of Medicinal Chemistry, 143*, 922–935.

Priyadarshi, R., Kim, S.-M., & Rhim, J.-W. (2021). Carboxymethyl cellulose-based multifunctional film combined with zinc oxide nanoparticles and grape seed extract for the preservation of high-fat meat products. *Sustainable Materials and Technologies, 29*, e00325.

Rajendran, K., Nagappan, R., & Ramamurthy, K. (2011). Short communication a study on the bactericidal effect of nisin purified from Lactococcus lactis. *Ethiopian Journal of Biological Sciences, 10*, 95–102.

Resa, C. P. O., Jagus, R. J., & Gerschenson, L. N. (2014). Natamycin efficiency for controlling yeast growth in models systems and on cheese surfaces. *Food Control, 35*, 101–108.

Sakellari, G. I., Zafeiri, I., Pawlik, A., Kurukji, D., Taylor, P., Norton, I. T., & Spyropoulos, F. (2021). Independent co-delivery of model actives with different degrees of hydrophilicity from oil-in-water and water-in-oil emulsions stabilised by solid lipid particles via a Pickering mechanism: A-proof-of-principle study. *Journal of Colloid and Interface Science, 587*, 644–649.

Salvia-Trujillo, L., Qian, C., Martín-Belloso, O., & McClements, D. J. (2013). Influence of particle size on lipid digestion and β-carotene bioaccessibility in emulsions and nanoemulsions. *Food Chemistry, 141*, 1472–1480.

Scallan, E., Hoekstra, R. M., Angulo, F. J., Tauxe, R. V., Widdowson, M.-A., Roy, S. L., Jones, J. L., & Griffin, P. M. (2011). Foodborne illness acquired in the United States—Major pathogens. *Emerging Infectious Diseases, 17*, 7.

Shah, B., Davidson, P. M., & Zhong, Q. (2012). Nanocapsular dispersion of thymol for enhanced dispersibility and increased antimicrobial effectiveness against Escherichia coli O157: H7 and Listeria monocytogenes in model food systems. *Applied and Environmental Microbiology, 78*, 8448–8453.

Shetty, K., Bhandari, A., & Yadav, K. S. (2022). Nanoparticles incorporated in nanofibers using electrospinning: A novel nano-in-nano delivery system. *Journal of Controlled Release, 350*, 421–434.

Shi, W.-J., Tang, C.-H., Yin, S.-W., Yin, Y., Yang, X.-Q., Wu, L.-Y., & Zhao, Z.-G. (2016). Development and characterization of novel chitosan emulsion films via Pickering emulsions incorporation approach. *Food Hydrocolloids, 52*, 253–264.

Shishir, M. R. I., Xie, L., Sun, C., Zheng, X., & Chen, W. (2018). Advances in micro and nano-encapsulation of bioactive compounds using biopolymer and lipid-based transporters. *Trends in Food Science and Technology, 78*, 34–60.

Singh, M., & Sahareen, T. (2017). Investigation of cellulosic packets impregnated with silver nanoparticles for enhancing shelf-life of vegetables. *LWT, 86*, 116–122.

Sultana, A., Luo, H., & Ramakrishna, S. (2021). Antimicrobial peptides and their applications in biomedical sector. *Antibiotics, 10*, 1094.

Teixé-Roig, J., Oms-Oliu, G., Ballesté-Muñoz, S., Odriozola-Serrano, I., & Martín-Belloso, O. (2020). Improving the in vitro bioaccessibility of β-carotene using pectin added nanoemulsions. *Food Security, 9*, 447.

Teixé-Roig, J., Oms-Oliu, G., Odriozola-Serrano, I., & Martín-Belloso, O. (2023). Emulsion-based delivery systems to enhance the functionality of bioactive compounds: Towards the use of ingredients from natural, sustainable sources. *Foods, 12*, 1502.

Ture, H., Eroglu, E., Ozen, B., & Soyer, F. (2011). Effect of biopolymers containing natamycin against Aspergillus niger and Penicillium roquefortii on fresh kashar cheese. *International Journal of Food Science and Technology, 46*, 154–160.

Wang, M., Wei, Z., & Zhang, Z. (2023). Antimicrobial edible films for food preservation: Recent advances and future trends. *Food and Bioprocess Technology, 17*, 1391–1411.

Weisany, W., Yousefi, S., Tahir, N. A.-R., Golestanehzadeh, N., McClements, D. J., Adhikari, B., & Ghasemlou, M. (2022). Targeted delivery and controlled released of essential oils using nanoencapsulation: A review. *Advances in Colloid and Interface Science, 303*, 102655.

Wolska, E., Sznitowska, M., Krzemińska, K., & Ferreira Monteiro, M. (2020). Analytical techniques for the assessment of drug-lipid interactions and the active substance distribution in liquid dispersions of solid lipid microparticles (SLM) produced de novo and reconstituted from spray-dried powders. *Pharmaceutics, 12*, 664.

Wu, J., & Ma, G.-H. (2016). Recent studies of Pickering emulsions: Particles make the difference. *Small, 12*, 4633–4648.

Yeh, Y.-C., Huang, T.-H., Yang, S.-C., Chen, C.-C., & Fang, J.-Y. (2020). Nano-based drug delivery or targeting to eradicate bacteria for infection mitigation: A review of recent advances. *Frontiers in Chemistry, 8*, 286.

Yousefi, M., Ehsani, A., & Jafari, S. M. (2019). Lipid-based nano delivery of antimicrobials to control food-borne bacteria. *Advances in Colloid and Interface Science, 270*, 263–277.

Part III
Safety at Every Step

Combatting Microbial Hazards Associated with Fresh Produce

A. Allende, P. Truchado, and M. I. Gil

1 Introduction

Safety of fresh produce is a critical concern in public health due to the high incidence of foodborne illnesses linked to the consumption of contaminated fruits and vegetables (EFSA and ECDC, 2023). Unlike processed foods, fresh produce is often consumed raw, without any step that would eliminate microbial hazards. Increasingly, health-conscious consumers are incorporating more fruits and vegetables into their diets, heightening the need for stringent safety measures. Reports attribute ~40% of *Salmonella*, 65% of *E. coli*, and 60% of *Listeria* illnesses in the U.S. to consumption of fruits or vegetables (Interagency Food Safety Analytics Collaboration 2021). In Europe, *Salmonella*, *Bacillus cereus*, norovirus, *Cryptosporidium* and *Clostridium botulinum* were the main causative agents associated with food of non-animal origin (EFSA and ECDC, 2023). All these pathogens can contaminate produce at any point from farm to table, including during growth, harvest, processing, and distribution.

The unique challenges presented by fresh produce require a comprehensive understanding of microbial contamination sources and the implementation of effective mitigation strategies throughout the production and supply chain (Devarajan et al., 2023). Good Agricultural Practices (GAPs) are fundamental in minimizing the risk of microbial contamination. GAPs encompass a range of practices designed to ensure food safety and suitability from pre-planting to post-harvest handling. The sources of microbial contamination in fresh produce are diverse, encompassing environmental, agricultural, and human vectors (FAO/WHO, 2023). Environmental sources include contaminated water, soil, and animal activity near cultivation sites. Agricultural practices also play a critical role, as an example, the use of manure as

A. Allende (✉) · P. Truchado · M. I. Gil
CEBAS-CSIC, Murcia, Spain
e-mail: aallende@cebas.csic.es

D. Li (ed.), *Antimicrobial Strategies in the Food System: Updates, Opportunities, Challenges*, https://doi.org/10.1007/978-3-031-95056-8_8

fertilizer can introduce pathogens unless properly treated. Human handling during harvesting, processing, and packaging can further introduce or spread contamination if proper hygiene practices are not followed (EC, 2017). It is important to understand that implementing GAPs requires not only adherence to technical recommendations but also active management and monitoring to adapt practices as new risks are identified.

Post-harvest handling of fresh fruit and vegetables presents its own set of challenges. Once harvested, the handling, storage, and transportation of fresh produce must be managed to prevent damage and microbial contamination. This includes maintaining the cold chain, using clean and sanitized processing equipment, and implementing effective wash and sanitation processes (FAO/WHO, 2023). One critical step during post-harvest handling of fresh fruit and vegetables is all the operations were water is used. Water used through the post-harvest handling of fresh fruit and vegetables, even that which has been conventionally treated and disinfected, may potentially contain human pathogens, albeit at low concentrations. A risk assessment, appropriate to the national or local production context, should be conducted to assess the potential risks associated with a specific water source or supply in order to devise the appropriate risk mitigation strategies. The water used in the processing of fresh fruits and vegetables, even when it has been disinfected, might still harbor human pathogens at low concentrations. To formulate effective risk mitigation strategies, it is essential to conduct a risk assessment tailored to the specific national or local production context, evaluating the potential hazards associated with a particular water source or supply (FAO/WHO, 2021). It should be considered that contaminated fresh produce cannot be effectively decontaminated once it has been compromised. This is because current technologies and methods for decontamination, such as washing and disinfecting, are not reliable in completely removing pathogens from produce (FAO/WHO, 2023). In response to the limitations of washing methods, recent advancements in post-process treatments have been developed to enhance the microbial safety of fresh produce (Truchado et al., 2020). Post-process treatments based on the use of bacteriophages and bacteriocin-producing cultures as 'biopreservatives' have been successfully tested to avoid growth of *L. monocytogenes* in different fruit and vegetables (Gómez-Galindo et al., 2023, 2024). Other alternative such as endolysins and essential oils have been suggested, but they are not jet implemented by the industry.

Environmental monitoring plays a crucial role in identifying and managing potential sources of contamination in fresh produce processing environments (FAO/WHO, 2022). By systematically sampling and testing surfaces within production facilities, operators can detect the presence of pathogens and implement corrective actions. This proactive approach is supported by risk assessment methodologies that evaluate the likelihood and severity of contamination based on specific environmental and operational factors (Gil et al., 2023).

Consumer awareness and education are critical components of reducing the risk of foodborne illnesses associated with fresh produce. Consumers must be informed about proper storage, handling, and preparation practices to minimize the risk of contamination in the home. Public health campaigns and labeling initiatives can enhance consumer knowledge and contribute to safer consumption practices.

This chapter summarize current knowledge on microbial hazards of fresh produce, acknowledging that ensuring the microbial safety of fresh fruit and vegetables is a complex challenge that requires a multifaceted approach. From farm-level interventions to advanced processing technologies and consumer education, each element plays a vital role in the overarching goal of public health protection. Continuous research and adaptation of best practices is essential as new threats emerge and consumer habits evolve. The commitment to maintaining stringent safety standards not only supports public health but also enhances consumer confidence in the fresh produce market.

2 Good Agricultural and Hygiene Practices

Agricultural and hygiene practices play a crucial role in combatting microbial hazards associated with fresh produce mostly because they do not undergo a "killing step" like cooking or pasteurization, which are common methods used to eliminate microbial hazards in other types of food. This is the reason why "the first line of battle" is to prevent contamination throughout the entire process and implementation of good hygiene and agricultural practices provide an effective means of establishing farming practices to minimize microbial hazards (FAO and WHO, 2023). Good agricultural and hygiene practices can be grouped in five categories including: (1) Environmental factors including animal reservoirs; (2) Fertilizers and plant protection products; (3) Agricultural water; (4) Staff hygiene and health status; (5) Hygiene conditions at primary production (EC, 2017). Figure 1 represent and illustration of microbial risk factors in pre-harvest operations. Among them, ensuring the microbiological quality of irrigation water is crucial for sustainable agriculture, food safety, and public health. It requires collaboration among farmers, agricultural stakeholders, regulators, and researchers to develop and implement effective strategies for managing water quality throughout the agricultural production process.

Conventionally, primary production of fresh fruit and vegetables relied on outdoor farming techniques for agriculture mostly due to the land availability and agreeable climate. However, climate change is reshaping the landscape of agriculture, including the establishment and operation of new production systems like controlled environmental agriculture (CEA). To effectively address these challenges, CEA operators must implement adaptive strategies, invest in resilient infrastructure and technologies, and collaborate with stakeholders across the agricultural supply chain to build a more sustainable and climate-resilient food system. Controlled Environment Agriculture (CEA) production has evolved from simple shade structures and hoop houses to full indoor vertical highly sophisticated greenhouse facilities with controlled lighting, water, and ventilation. These protected systems promote the efficient production of high-value crops at maximum productivity in an

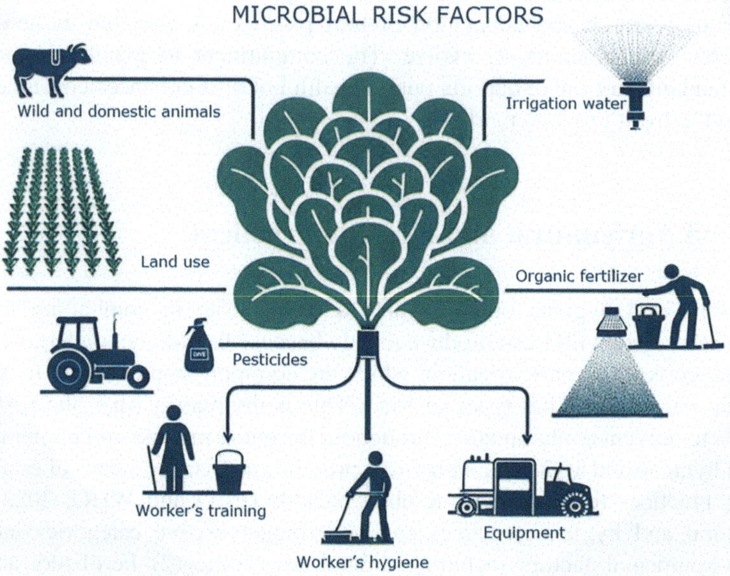

Fig. 1 Scheme summarizing the most relevant microbial risk factors in the primary production of fruit and vegetables

environmentally friendly way but they are not inherently safer than open systems as contamination can occur through production practices and procedures that introduce hazards into the environment (Motzer et al., 2024).

Prevalence studies of foodborne pathogens have been performed mostly in the production environment of open fields (Harrand et al., 2020; Belias et al., 2020) as well as packing houses and processing environment of fresh-cut facilities (Sullivan & Wiedmann, 2020; Belias et al., 2022; Gil et al., 2023). However, more research is needed to obtain science-based guidance for systematically assessing food safety risks in the growing and harvesting of leafy greens under CEA systems. Particularly, the industry needs more research describing field-based assessments and data-based development of corrective actions. Identification of potential risk sources and routes of contamination in CEA systems including a comprehensive understanding of the behaviour of persisting bacterial forms in controlled environmental niches.

There are already specific guidelines describing Good Agricultural Practices (GAPs) and Good Hygiene Practices (GHPs) for CEA operations (FAO, 2017;

Controlled Environment Agriculture Alliance, 2023), which include general recommendations such as: (i) protected facility structures should be located, designed and constructed to avoid contamination and harborage of pests; (ii) worker training and sanitation practices are necessary in all operations; (iii) proper water management and soil amendment use are critical to controlling and reducing risks.

Overall, GAP relevant for open field and CEA are similar. Likely microbial food safety challenges in both production systems include: (i) environmental factors, (ii) pathogen contamination of inputs (e.g., seeds, substrates, fertilizers), (iii) pathogen contamination and persistence in water and water associated infrastructure, (iv) worker health and hygiene, and (v) equipment and tools. Environmental factors are very relevant in both production systems (i.e. open field and CEA) but the sources are quite different. While in open fields the presence of domestic and wild animals is a main concern, in CEA operations, pathogen contamination and persistence in the overall facility infrastructure such trays, beds, walls, automated systems employed for various tasks such (e.g. seeding, planting, plant movement, irrigation, harvesting, and post-harvest processing) coming from the production environment, is very relevant, making environmental monitoring programs necessary.

EM programs have been proven to be essential for preventing *Listeria monocytogenes* contamination in fresh produce processing and packing facilities, as they provide relevant information about pathogen sources, determine pathogen spread throughout the facility, and verify control strategies. Recently, scientific evidence has been generated demonstrating that *L. monocytogenes* is not only an issue during postharvest of fruits and vegetables, but also during primary production. EM programs involve the collection of numerous samples of strategically selected surfaces to test them for the presence of *L. monocytogenes* or *Listeria* spp. as an indicator. Based on the guidance from the CEA Alliance (Controlled Environment Agriculture Alliance, 2023), EM should prioritize areas with the highest risk of *L. monocytogenes* harborage and subsequent product contamination, based on relevant factors such as temperature, sanitation practices, and equipment/surface hygienic design.

A distinguishing feature between open-field farming and CEA facilities is the growing trend of incorporating processing and packaging tasks within CEA operations. Tasks like cutting, washing, and packing, traditionally conducted in separate fresh-cut processing facilities, are now often carried out within the confines of CEA operations, either in the same space or adjacent rooms where crops are cultivated (Hamilton et al., 2023). For this reason, in CEA operations, the focus of EM programs is typically from the harvest area onward (e.g. processing and packaging). This approach is supported by the environmental conditions in the growing area (e.g. warmer temperatures), where *Listeria* spp. may face competition from other microbiota. Conversely, the cooler environments of the harvest and storage areas are conducive to Listeria persistence and growth compared to other microbes (Controlled Environment Agriculture Alliance, 2023).

Apart from the above-mentioned GAPs, there are additional recommendations that should be equally applicable to all production systems. Among these relevant recommendations are: (i) Continuous training of farm workers on food safety practices is crucial and (ii) Traceability and record-keeping.

It is essential to understand that production of safe fruit and vegetables involved the establishment of a good food safety management system (FSMS) to control several potential hazards. A greater emphasis on a risk-based approach based on the assessment of the potential hazards should be articulated (FAO/WHO, 2019). Indeed, all GAPs should be risk-based, and any intervention should be system-based. This implies that interventions chosen should align with the identified hazards and the likelihood and severity of their occurrence.

3 Post-harvest Handling and Processing Practices

Contamination of fresh produce with human pathogens may occur after harvest, during handling, processing, packaging, distribution and final preparation. Once fruits and vegetables have been contaminated with bacterial, viral pathogens or parasites, there are no sanitation practices that can assure their safety (Harris et al., 2002; López-Gálvez et al., 2010). The attachment and survival of human pathogens on produce depend on a variety of factors, such as the type of microorganism, the ability to survive on the product, the characteristics of the produce, the attachment site, nutrient availability and also environmental factors such as temperature, humidity, and, gaseous atmosphere among others that can promote survival (Salvi, 2017; Alexa et al., 2024). Product contamination can be amplified at any point in the food chain by improper handling and storage conditions before consumption and generally, it is not confined to a single point in the supply chain. Figure 2 illustrate different risk factors linked to post-harvest handling and processing operations. Managing the handling and processing conditions and the supply chain is of paramount importance in preventing fresh produce contamination as fresh fruits and vegetables are consumed raw with minimal preparation. The key to addressing microbial contamination is to focus on reducing the risk of produce contamination where possible all along the food chain. This section will cover the sources and pathways of microbial contamination and explore effective post-harvest handling and processing practices to mitigate microbial hazards.

3.1 Postharvest Handling

In this operation, contamination can occur through various vectors such as contaminated harvesting equipment, workers' hands or gloves, and containers like bins, boxes, and buckets. To minimize this risk, regular cleaning of equipment, the use of potable water, and the sanitary condition of containers prior to use are recommended practices. The interaction between workers and fresh produce during handling presents a notable risk for microbial cross-contamination. The presence of

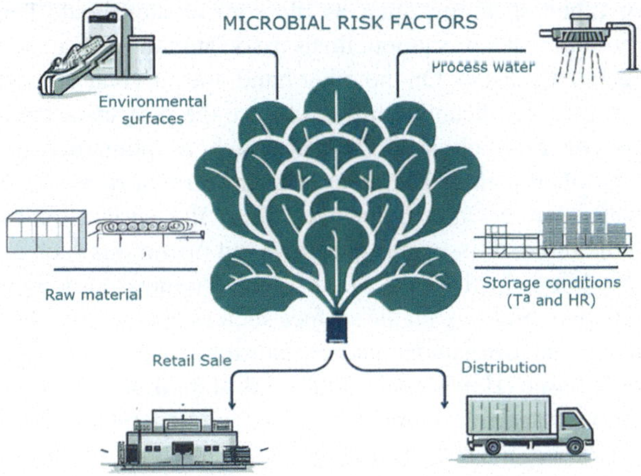

Fig. 2 Scheme summarizing the most relevant microbial risk factors in the postharvest handling and processing operations of fruit and vegetables

microorganisms on the surface of contaminated produce may facilitate their transfer to the hands of workers, or workers with pre-existing conditions may act as vectors for contaminating the crops. This risk is exacerbated by inadequate practices regarding the use of protective gloves, harvesting knives, and other utensils critical in both field and processing environments. Specifically, the infrequent replacement or disinfection of these items substantially elevates the potential for contamination. Moreover, the absence of essential sanitation infrastructure, such as toilets and handwashing facilities, coupled with a lack of disinfectants, significantly increases the likelihood of microbial transmission. Thus, maintaining stringent hygiene protocols and ensuring the availability of proper sanitation facilities are imperative measures to mitigate the risk of contamination (Julien-Javaux et al., 2019).

3.2 Intact Versus Fresh-Cut Products

Growth of pathogens on intact surfaces is not common because foodborne pathogens do not produce the enzymes necessary to break down the protective outer barriers on most produce that restricts the availability of nutrients and moisture (Harris et al., 2002). Under some circumstances, wash water may enter an intact fruit through the stem scar or other openings, such as blossom or stem end. This fact may have contributed to outbreaks of salmonellosis associated to fresh-market tomatoes eaten in restaurants (CDC, 2019). On the other hand, the survival of foodborne pathogens on produce can be significantly enhanced once the protective epidermal barrier has been broken either by physical damage, such as punctures or bruising, or by degradation by plant pathogens (bacteria or fungi) (Harris et al., 2003). Pathogenic microorganisms can be also transferred from the surface of the intact fruit or vegetable to the internal tissue through fresh-cut operations such as peeling, cutting, slicing or shredding. In the case of fresh-cut products, various enteric pathogens have been shown to multiply on the surface such as cut melons, shredded lettuce, and chopped parsley, and even under acidic conditions such as on chopped tomatoes and wounded apple tissue (Harris et al., 2002). Infiltration of wash water into intact fruit has been demonstrated in a number of fruits and vegetables. Wash water contaminated with microorganisms, including pathogens, can infiltrate the intercellular spaces through pores at any conditions during washing such as differences between product to water temperature, or washing under light or dark exposure, and also with different hydration-dehydration stages (Gómez-López et al., 2013).

3.3 Washing

During washing, microbial contamination present on the product can spread throughout the production batch, increasing the risk of illnesses. Sanitizers, including chlorine, are used to prevent water-mediated cross-contamination (Gombas et al., 2017; Murray et al., 2018; Wang & Ryser, 2020), but they have limited efficacy on the product itself. When fresh produce is contaminated by human pathogens there is no currently available process, other than cooking, that can completely eliminate them. While the acidity of most fruit juices prevents the multiplication of pathogens, survival in these products is better than has been previously assumed (Harris et al., 2003). Bacteria have been shown to preferentially adhere to stomata and cut surfaces of lettuce, making them less accessible to be remove and the reason that washing fails to eliminate pathogens (López-Gálvez et al., 2010).

While washing can remove some of the surface microorganisms, it cannot remove them all. Chlorinated wash water is usually no more or only slightly more effective than plain water at removing surface bacteria. Reductions observed over a wide range of conditions are typically from 0 to 1000-fold (Gehringer, 2017). When compare different sanitizers, including chlorine, the initial differences observed after washing disappeared after storage (Allende et al., 2008). However, improperly sanitized wash water itself become a major source of contamination and should be carefully monitored and controlled. If contaminated produce coming in from the field is washed, wash water can become contaminated and spread the contamination as the water is reused. This is known as water-mediated cross-contamination in which the pathogenic microorganisms are detached from contaminated surfaces into the wash water, where they may subsequently adhere to other uncontaminated products (Gombas et al., 2017). The presence and efficacy of antimicrobials in the wash system are crucial for mitigating this risk by inactivating pathogens before they can adhere to other produce. If not, pathogens can survive for relatively long times in water and can contaminate clean produce that passes through that water. Harris et al. (2002) mentioned a very good example such as cantaloupe cubes that were initially inoculated with a high concentration of *Salmonella* spp. (7 log CFU per piece), and immersed in chlorinated water with a concentration of 50 mg/L at a neutral pH (7.0) for 1 min. Despite the treatment, the level of *Salmonella* on the cantaloupe pieces remained constant, indicating that the chlorination at the specified conditions did not effectively reduce the *Salmonella* population on the cantaloupe. In contrast, *Salmonella* levels in the water demonstrated significant variation dependent on chlorination: in unchlorinated water, *Salmonella* concentrations increased to 5.6 log CFU/mL, while in chlorinated water, the population was effectively suppressed to less than 1 log CFU/mL (Harris et al., 2002). A subsequent experiment involved immersing cantaloupe cubes, which had not been pre-inoculated with *Salmonella*, in both the previously used chlorinated and unchlorinated soak solutions. The results showed a relevant evidence: those cantaloupe pieces soaked in chlorinated water did not show any *Salmonella* contamination, whereas those immersed in unchlorinated water exhibited a considerable acquisition of *Salmonella*, with levels reaching 4.2 log CFU per piece. These findings highlight the effectiveness of chlorinated water in preventing the transfer of *Salmonella* to the produce, underscoring the importance of chlorination in managing microbial risks in food processing environments (Harris et al., 2002). Thus, wash water sanitation is important to ensure that pathogens introduced into the water are inactivated and cannot act as a source of contamination.

Key factors influencing the likelihood and extent of cross-contamination include the type and concentration of antimicrobials used, environmental parameters such as pH, as well as operational conditions like water temperature and turbulence. Additionally, system overload, which reduces the distance between items and can lead to uneven antimicrobial distribution, significantly increases cross-contamination risks. Therefore, managing the load and ensuring optimal operation of the wash system are essential for minimizing the potential for cross-contamination (Gombas et al., 2017; López-Gálvez et al., 2021).

The reutilization of water for the purpose of washing can contribute microbial contamination attributable to the practice of reusing washing water multiple times prior to disposal, and frequently for sequential washes of diverse crop types. Moreover, the ratio of product to water, particularly with crops that bear debris and minor vegetative fragments, can precipitate an enhanced accumulation of organic material. This accumulation detrimentally impacts the efficacy of antimicrobial treatments designed for the decontamination of water (Machado-Moreira et al., 2019; Garrido et al., 2020). Consequently, the meticulous engineering of washing apparatuses, tanks, and processing infrastructures, alongside the implementation of rigorous quality control protocols, is imperative to mitigate the risk of cross-contamination. Different washing techniques (e.g., immersion, spray) (Bornhorst et al., 2018; Shen et al., 2020; Su et al., 2022) and the application of sanitizing agents (chlorine-based solutions, organic acids, UV light and cold plasma) have continued for fresh product and fresh-cut (Zhou et al., 2015; Huang et al., 2018; Singh & Thakur, 2024). The discussion on the effectiveness and limitations of these methods in reducing microbial contamination has been studied extensively and more recently by Sethi et al. (2020) and Raffo and Paoletti (2022). Cold plasma is one of the emerged promising postharvest solution as a dry treatment for products and plasma-activated water for the potential application in the sanitation of water (Singh & Thakur, 2024). Regardless of the sanitizing and monitoring system used, it is critical that the efficacy of the method be validated. Effective residual concentration and other important factors, such as pH, must be maintained in the wash water (Tudela et al., 2019; Marín et al., 2020). These properties need to be measured on a routine basis and if it is possible in a continuous monitoring way.

3.4 Modified Atmosphere Packaging and Cold Chain

Packaging is a critical step in protecting fresh produce from post-wash contamination. Packaging materials and technologies such as modified atmosphere packaging (MAP) can extend shelf-life and minimize microbial growth (Gorny, 2003). The role of packaging in maintaining the quality and extend the shelf life of fresh produce during transportation and storage are due to a reduction in the respiration rate, inhibition of enzymatic oxidation and inhibition of growth of aerobic microorganisms. The first two effects have been easily demonstrated but the inhibition of microbial growth is not always evident. Most of the authors conclude that MAP used for produce packaging does not significantly affect microbiota and microbial spoilage can be slow the growth in essence by refrigeration rather than the MAP. Several studies have found that MAP technologies commonly used in the fresh-cut industry have varying effects on the survival and growth of pathogens such as *E. coli* O157:H7, *Salmonella*, *Shigella* and *Listeria monocytogenes* while some strains may be inhibited, other unaffected, weakly inhibited or even stimulated (Caleb et al., 2013). Fresh-whole and fresh-cut produce packaged in gas-permeable films can modified its own atmosphere, creating a more favorable conditions for quality maintenance during storage. The reduces O_2 and elevated CO_2 typically found within produce MAP have been shown to slow the growth of some aerobic bacteria such as spoilage bacteria and increase the shelf life. However, MAP has a little effect in slowing the growth of most human pathogenic bacteria and may even stimulate the growth of some. This combination of a reduction in sensory spoilage microorganisms and an increase in time for pathogens to grow has been the cause of concern with fresh-cut produce in MAP (Harris et al., 2002).

Risk of human pathogens due to the use of MAP must be assessed on a per product basis due to the complex interactions between the produce, the pathogen and the environment. As an example, as *L. monocytogenes* can grow at refrigeration temperatures. Therefore, there is concern with extended shelf life by use of MAP that may allow *L. monocytogenes* to proliferate during shelf life (O'Beirne et al., 2015). Broad generalities cannot be drawn regarding the risk of specific human pathogens due to these complex interactions of the type of product and the interaction with MAP and storage conditions. The survival and/or growth of pathogens on fresh produce is influenced also by the conditions of storage. In general, pathogens will survive on the outer surface of fresh fruits or vegetables, especially if the humidity is high (Harris et al., 2003). Optimal storage conditions (temperature, humidity) for different types of fresh produce reduce or eliminate the ability of microorganisms to multiply. Ansah et al. (2018) provided comprehensive insights into the strategies and benefits of effective cold chain management and the implications in the maintenance on the proliferation of microorganisms in products. Multiplication of pathogens can be promoted once the protective barriers are broken and there is a degradation by plant pathogens, especially at higher temperatures (Harris et al., 2003). Temperature control becomes critical for preventing bacterial multiplication on any fresh-cut product. Maintaining low temperature throughout the distribution and cold chain is critical to preserve freshness and prevent the multiplication of most foodborne pathogens with the exception of *L. monocytogenes* that are capable of growing at refrigerated temperatures. Present case studies or examples where improper storage conditions led to outbreaks of foodborne illnesses have been described (Augustin et al., 2020). The preservation of cold chain conditions throughout the processing, shipping, retail, and final consumption phases is crucial for inhibiting the growth of microorganisms, particularly pathogenic ones, that might be present. Maintaining an unbroken cold chain ensures that perishable goods are kept at appropriate low temperatures, and significantly reducing the risk of microbial proliferation. This control is vital not only for preventing foodborne illnesses but also for maintaining the quality and shelf life of perishable products (Ansah et al., 2018). Thus, rigorous monitoring and quality assurance measures are indispensable in ensuring the integrity of the cold chain from production to the point of consumption.

4 Post-process Treatments

In the fresh produce industry, despite the widespread use of intervention strategies such as the application of sanitizers, preservatives, and non-thermal processing technologies to enhance food safety, foodborne outbreaks and recalls persist globally. This ongoing challenge highlights the need for alternative approaches to address food safety issues. Currently, there is a increasing interest in biological control methods as alternative strategies to combat pathogenic bacteria in fresh produce (Agriopoulou et al., 2020; Vikram et al., 2020). These methods encompass the use of bacteriophages, endolysins, lactic acid bacteria, bacteriocins, essential oils, and their derivatives. These treatments offer promising avenues for bolstering microbial safety and maintaining quality during the postharvest storage of raw and minimally processed fruits and vegetables (Pisoschi et al., 2018; Leneveu-Jenvrin et al., 2020; Gómez-Galindo et al., 2023, 2024). Moreover, the adoption of these post-processing treatments signifies a shift towards more sustainable and environmentally friendly practices in the food industry (Garvey, 2022). By diminishing reliance on traditional chemical interventions, these biological control methods align with the escalating consumer demand for safer, healthier, and more sustainable food choices. This section provides an overview of current and innovative post-processing treatments for fresh and minimally processed vegetables.

Bacteriophages, viruses that specifically target bacteria, offer potential as safe antimicrobial agents in food safety. Food and Drug Administration (FDA) approval for their use against *Listeria monocytogenes* in ready-to-eat meats was granted in 2006; however, they are still not approved for use in vegetable foods to date (Grigore-Gurgu et al., 2024). Although their effectiveness in decontaminating plant-derived foods has been demonstrated, bacteriophages commonly result in reductions in bacterial loads ranging from 1 to 3 log units (Vikram et al., 2022). Factors such as the physicochemical properties of the vegetable, initial concentrations of phages and bacteria, storage time, and vegetable type influence the efficacy of reducing pathogenic bacteria, and should be considered when the efficacy of bacteriophages treatment is studied (Shannon et al., 2020). However, the main challenge is the method of applying bacteriophages to fruit and vegetables, to ensure sufficient contact between the phages and bacteria across different matrices. Consumer acceptability is also a concern with using these strategies.

Currently, there are different commercial preparations based on bacteriophages for controlling pathogenic bacteria in food (Liu et al., 2023). The antimicrobial efficacy of some of them has been tested on fresh produce. For instance, the effectiveness of SalmoFresh was demonstrated in controlling Salmonella on romaine lettuce and sprouts (Zhang et al., 2019). Ecoshield (Intralytix) has been extensively used in studies to combat *E. coli* O157:H7 in tomatoes, broccoli, spinach, and lettuce (Abuladze et al., 2008; Sharma et al., 2009). Additionally, Listshield (Intralytix) and PhageGuard Listex (Micreos Food Safety) have been applied to the surfaces of artificially contaminated vegetables or fruits, effectively reducing *L. monocytogenes* levels (Gómez-Galindo et al., 2023, 2024; Truchado et al., 2020). The effectiveness of native bacteriophages isolated from wastewater, rivers, and soils against *L. monocytogenes* and *E. coli* O157:H7 inoculated in fresh celery, enoky mushroom and radish sprouts, respectively have bee also demonstrated in vegetables (Oh et al., 2024; Byun et al., 2022).

Phages specificity to certain bacterial hosts and potential for resistance development represent real challenges. Moreover, consumer preferences for additive-free products may conflict with phage incorporation in food processing. To address this, endolysin has emerged as a natural solution. Endolysins, or "enzybiotics," are enzymes produced by bacteriophages in the final stages of their lytic cycle. They target and degrade the cell walls of host bacteria, causing cell lysis and releasing new virions (Grigore-Gurgu et al., 2024). Endolysins are highly specific and have a low risk of inducing bacterial resistance. These enzymes show promise for food safety applications, being safe for human consumption and not altering food properties (Rahman et al. 2021). Limited literature is available on the use of endolysins as antimicrobial agents for controlling foodborne pathogenic bacteria in vegetables. However, some studies have already demonstrated their ability to reduce the presence of *C. perfringens*, *Salmonella*, and *L. monocytogenes* cells inoculated in lettuce (Zhao et al., 2023; Liu et al., 2019; Schmelcher et al., 2012). Moreover, bacteriophages and endolysins are proposed as disinfectants in food environments, albeit delivery strategies require refinement.

Lactic acid bacteria (LAB) are essential in food fermentation, improving taste and texture while also providing health benefits. They are commonly found in raw fruits and vegetables and are as safe (GRAS) microorganisms by the FDA and have been granted Qualified Presumption of Safety (QPS) status by the European Food Safety Authority (EFSA) (Agriopoulou et al., 2020). Their use in food preservation is gaining traction as an eco-friendly alternative to chemical methods (Zapaśnik et al., 2022). LAB outcompete pathogens and produce antimicrobial compounds like lactic acid, hydrogen peroxide, and bacteriocins, effectively inhibiting microbial growth (Cálix-Lara et al., 2014; Reis et al., 2012). In scientific literature, multiple studies have proposed LAB as an effective method for controlling pathogens such as *E. coli* O157:H7, *L. monocytogenes* and *Salmonella* in fresh vegetables and fruits (Truchado et al., 2020; Lokerse et al., 2016; Oliveira et al., 2015; Cálix-Lara et al., 2014).

Bacteriocins produced by LAB, are antimicrobial peptides, which are synthesized as defence mechanism against competing bacteria and pathogens/used as natural and safe food preservatives in a variety of foods, including fruits and vegetables, without altering the nutritional and sensorial properties or the physicochemical characteristics of the food. These peptides are categorized into three main classes: class I, which includes antibiotics characterized by post-translational modifications; and class II, comprising unmodified peptides further subdivided into class IIa, IIb, IIc, and IId (Kumariya et al., 2019). The most of study have evaluated the efficacy of bacteriocins, such as nisin, pediocin and enterocin, alone or combined with other hurdles regarding pathogenic bacteria growth inhibition in vegetables and vegetables based-products. Significant reductions have been reported when *L. monocytogenes, E. coli and Shigella* inoculated vegetables (e.g. fresh cut leafy green, tomatoes, mushroom and fruits) were treated with bacteriocins (Ramos et al., 2020; Mahajan et al., 2018; McManamon et al., 2019). The research of bacteriocins as potential alternatives for preserving fruits and vegetables is ongoing. While only a few, such as nisin or peodin PA-1/AcH, are FDA-approved as food additives, they have not yet received formal approval for use in fruits and vegetables (Agriopoulou et al., 2020). Several studies have also shown that they can be incorporated into packaging materials.

Essential oils (EOs), extracted from aromatic plants like thyme, rosemary, cinnamon, oregano, and clove, exhibit strong antimicrobial properties beneficial for food preservation (Falleh et al., 2020; Reis et al., 2022). Recognized as Safe (GRAS) status by the U.S. Food and Drug Administration (FDA) and are approved for food use under FDA (CFR, 2020). The European Commission also maintains a list of approved EO compounds for food applications due to their ability to inhibit bacteria and enhance flavour (EC, 2008). However, their potent aroma can affect food taste, requiring cautious use at low concentrations (Grigore-Gurgu et al., 2024; Prakash et al., 2023). Although in vitro experiments have been conducted to determine their antimicrobial capacity in vegetable products, their direct application is severely limited (Arellano et al., 2021; Kang & Song, 2018). Incorporating EOs into active food packaging emerges as a promising strategy to maintain food safety while minimizing sensory alterations, as the compounds migrate from packaging materials to food, providing ongoing protection against microbial contamination (López-Gómez et al., 2023; Sharma et al., 2021).

Although biological control agents have demonstrated promising antimicrobial effects, it is crucial to recognize that the outcomes observed in challenge studies, conducted both at laboratory and pilot scale, may not be reproducible in commercial applications. The transition to large-scale commercial production presents significant challenges, including concerns regarding the feasibility of achieving tested antimicrobial concentrations in the industrial setting, as well as considerations of cost-effectiveness and practicality. It is imperative to acknowledge that the implementation of biocontrol methods in fresh produce extends beyond simply extending shelf life and must also consider potential impacts on organoleptic quality. A recent investigation conducted by Gómez-Galindo et al. (2023, 2024) has assessed the use of bacteriophages on a commercial processing line of cut iceberg lettuce and its effects on product quality. Nevertheless, further research is required to comprehensively evaluate the effectiveness of post-processing treatments in commercial facilities. It is essential to underscore that while post-processing treatments serve as valuable tools for enhancing food safety, they should not be viewed as a replacement for proper food handling practices. Prevention should always remain the top priority. These treatments should be regarded as additional control measures aimed at improving the safety and quality of the food we consume.

5 Environmental Monitoring

The environmental monitoring (EM) is designed to be a proactive tool in identifying potential contamination risks within food processing environments (CFSAN, 2017). To this end, the program not only seeks to verify existing control measures but also to uncover and rectify any occurrences of pathogens and their harborage sites. For sampling food contact surfaces (FCS) and non-FCS, to collect the samples for identifying potential sources of contamination is important. Additionally, the use a systematic approach that covers all relevant areas of the plant, including hard-to-reach zones as hot spots that might be overlooked but could harbor microorganisms is essential. Once collected, the samples must be handled and transported under conditions that prevent any changes in the microbial population. The testing should be carried out after processing, at least 2 h latter than the start of the process to identify possible routes of entry and after cleaning and disinfection to detect any potential pathogens or indicator organisms that suggest a hygiene issue (Gil et al., 2023). In addition to detecting problems, the EM program should also guide the frequency and intensity of sanitation procedures, particularly when pathogens are detected. Once corrective actions are taken, subsequent rounds of testing should follow to ensure that these measures have been successful in eliminating the contamination sources.

EM programs for fresh produce are becoming increasingly crucial in packinghouses to ensure food safety by identifying potential contamination sources. These programs have historically been implemented in processing facilities, but their application in raw agricultural commodity packing operations is a more recent development. EM programs serve to verify the effectiveness of sanitation programs and to reduce the presence and proliferation of microorganisms that can lead to foodborne illnesses. About 62.5% of produce packers reported having an EM programs in place, with ATP testing being the most commonly used monitoring technique, especially for surfaces directly contacting the food (Zone 1 surfaces). Other targets for monitoring include generic *E. coli*, *Listeria* species, and aerobic plate counts, with an increasing emphasis on pathogen targets like *Salmonella* for Zones 2–4 (Critzer et al., 2024).

The importance of an EM program in fresh produce packinghouses is underscored by its ability to alert management about environmental contaminants before they come into contact with produce, thereby verifying the effectiveness of the facility's sanitation program. Correct zone identification within the facility, from direct food contact surfaces (Zone 1) to areas outside the finished product or food handling area (Zone 4), is critical for targeted and efficient monitoring (3M, 2019). This zonation helps in allocating resources appropriately for monitoring and cleaning, ensuring that the highest priority is given to zones with direct food contact. Moreover, implementing an EM program can provide fresh produce packers with confidence that they are minimizing the risk of microbial contamination in their facilities (Dunn et al., 2023). However, it's important to note that without proper implementation, an EM program could lead to false confidence and the wastage of resources. This highlights the necessity for fresh produce packers to thoughtfully design and execute their EM programs, considering the specific needs and risks associated with their operations. For further details on setting up and maintaining effective EM program in fresh produce environments, resources such as the guidance from United Fresh are invaluable (UFPA, 2018). In fresh-cut facilities, EM programs are essential for ensuring food safety and maintaining environmental sustainability. These programs help manage food safety concerns linked to the processing of fresh-cut vegetables, like the use of disinfectants, and address environmental impacts. EM programs in such settings are crucial for monitoring and controlling potential contaminants before they reach the produce, thereby reducing the likelihood of foodborne pathogens establishing in the packing environment (Gil et al., 2023; Critzer et al., 2024).

6 Consumer Awareness

Consumer awareness plays a crucial role in addressing microbiological hazards associated with fresh produce. As consumers increasingly prioritize high-quality, nutritious foods for a healthy lifestyle, they are consuming more produce, potentially increasing their exposure to hazards like *L. monocytogenes*. Fresh produce has become a significant source of foodborne illness outbreaks linked to *L. monocytogenes*, presenting a global public health concern (EFSA and ECDC, 2023). However, there may be insufficient recognition among consumers of the association between produce consumption and outbreaks. Additional risk communication efforts may be necessary to educate consumers about this link and empower them to make informed food choices (CDC, 2021).

The FAO/WHO highlights the need for retail-to-consumer risk assessments, particularly for minimally processed products like fresh produce (FAO/WHO, 2004). Leafy greens, cantaloupe/rock melon, and frozen vegetables are identified as priority areas for risk assessments due to their involvement in outbreaks. Educating consumers about the meaning of date labels, the importance of shelf-life limitations, temperature control, proper food storage, and safe handling methods can help minimize food waste and reduce the risk of foodborne illnesses associated with fresh produce (Newsome et al., 2014).

In general, there is a lack of comprehensive studies examining consumers' food safety knowledge, risk perceptions, and handling practices concerning fresh-cut produce. Available information indicate that consumers generally lack knowledge and safe handling practices concerning fresh-cut produce, particularly regarding storage hierarchy, surface cleaning and sanitizing, and time and temperature control. Men and millennial consumers were identified as exhibiting lower levels of safe handling practices for fresh-cut produce. Additionally, a significant interaction was observed between food safety knowledge and risk perceptions, indicating that consumers with higher levels of risk perception are more likely to translate their food safety knowledge into practice effectively (Yu et al., 2022). These results can inform the development of targeted consumer food safety education tools aimed at improving the safe handling of fresh-cut produce among specific demographic groups (Huang et al., 2020).

7 Conclusions

Combating microbial hazards in fresh produce stands as a critical concern, particularly as the consumption of raw fruits and vegetables continues to rise globally. The absence of a "killing step" in the processing of these foods elevates the importance of preventing contamination from the farm to the consumer. This chapter explored the different strategies and practices that aim to mitigate the risks associated with microbial contaminants in fresh produce.

The extensive examination of strategies to combat microbial hazards in fresh produce underscores the complexity and necessity of integrated approaches for ensuring food safety. GAPs and proper post-harvest handling emerge as foundational elements that significantly reduce the risk of microbial contamination. Maintaining the microbiological quality of water use in the primary production as well as during processing has been shown as a key element to avoid contamination of the final product. Environmental monitoring in CEA operations and fresh-cut processing plants is critical to avoid contamination from the production and processing environment. On the hand, the use of post-process treatments, including the use of bacteriophages and endolysins, reveals promising new strategies for enhancing microbial safety. In addition to these important pillars, the role of consumer awareness has been considered essential to reduce safety issues. Educating consumers about safe handling and storage practices can dramatically decrease the risk of foodborne illnesses. All the intervention strategies summarized in this chapter are highly interconnected, suggesting that the safety of fresh produce can only be managed through a comprehensive, multi-layered approach. The ongoing research developed within the frame of the Projects from the MxQ research group such as: 1) MCIN/AEI/10.13039/501100011033 (Project PID 2022-1417080B-I00) "ERDF A way of making Europe", 2) TED2021-131427B-C21 and TED2021-131427B-C22 Financed by the European Union—NextGeneration EURecovery, and 3) Transformation and Resilience Plan from Ministry of Science and Innovation), reflect the dynamic challenges that climate change, global supply chains, and evolving pathogens present to food safety professionals worldwide.

References

3M. (2019). *Environmental monitoring handbook for the food and beverage industries – Terms and definitions* (1st ed.). Cornell University. Available at: https://multimedia.3m.com/mws/media/1684575O/environmental-monitoring-handbook.pdf

Abuladze, T., Li, M. Y., Menetrez, T., Dean, A., Senecal, A., & Sulakvelidze, A. (2008). Bacteriophages reduce experimental contamination of hard surfaces tomato spinach broccoli and ground beef by *Escherichia coli* O157: H7. *Applied and Environmental Microbiology, 74*(20), 6230–6238. https://doi.org/10.1128/AEM.01465-08

Agriopoulou, S., Stamatelopoulou, E., Sachadyn-Król, M., & Varzakas, T. (2020). Lactic acid bacteria as antibacterial agents to extend the shelf life of fresh and minimally processed fruits and vegetables: Quality and safety aspects. *Microorganisms, 8*(6), 952. https://doi.org/10.3390/microorganisms8060952

Alexa, E. A., Papadochristopoulos, A., O'Brien, T., & Burgess, C. M. (2024). Microbial contamination of food. In *Food packaging and preservation*. https://doi.org/10.1016/B978-0-323-90044-7.00001-X

Allende, A., Selma, M. V., López-Gálvez, F., Villaescusa, R., & Gil, M. I. (2008). Role of commercial sanitizers and washing systems on epiphytic microorganisms and sensory quality of fresh-cut escarole and lettuce. *Postharvest Biology and Technology, 49*, 155–163. https://doi.org/10.1016/j.postharvbio.2007.12.010

Ansah, F. A., Amodio, M. L., & Colelli, G. (2018). Quality of fresh-cut products as affected by harvest and postharvest operations. *Journal of the Science of Food and Agriculture, 98*, 3614–3626. https://doi.org/10.1002/jsfa.8885

Arellano, S., Law, B., Friedman, M., & Ravishankar, S. (2021). Essential oil microemulsions inactivate antibiotic-resistant *Salmonella* Newport and spoilage bacterium Lactobacillus casei on Iceberg lettuce during 28-day storage at 4 °C. *Food Control, 130*. https://doi.org/10.1016/j.foodcont.2021.108209

Augustin, J. C., Kooh, P., Bayeux, T., Guillier, L., Meyer, T., Jourdan-Da Silva, N., Villena, I., Sanaa, M., & Cerf, O. (2020). Contribution of foods and poor food-handling practices to the burden of foodborne infectious diseases in France. *Food Security, 9*(11), 1644. https://doi.org/10.3390/foods9111644. PMID: 33187291; PMCID: PMC7697675.

Belias, A. M., Sbodio, A., Truchado, P., Weller, D., Pinzon, J., Skots, M., Allende, A., Munther, D., Suslow, T., Wiedmann, M., & Ivanek, R. (2020). Effect of weather on the die-off of *Escherichia coli* and attenuated *Salmonella typhimurium* on preharvest baby lettuce and spinach following a simulated irrigation event with contaminated water in California, New York, and Spain. *Applied and Environmental Microbiology*. https://doi.org/10.1128/AEM.00899-20

Belias, A., Sullivan, G., Wiedmann, M., & Ivanek, R. (2022). Factors that contribute to persistent listeria in food processing facilities and relevant interventions: A rapid review. *Food Control, 133*. https://doi.org/10.1016/j.foodcont.2021.108579

Bornhorst, E. R., Luo, Y., Park, E., Vinyard, B. T., Nou, X., Zhou, B., Turner, E., & Millner, P. D. (2018). Immersion-free single-pass commercial fresh-cut produce washing system: An alternative to flume processing. *Postharvest Biology and Technology, 146*, 124–133. https://doi.org/10.1016/j.postharvbio.2018.08.008

Byun, K. H., Han, S. H., Choi, M. W., Park, S. H., & Ha, S. D. (2022). Isolation, characterization and application of bacteriophages to reduce and inhibit *Listeria monocytogenes* in celery and enoki mushroom. *Food Control, 135*. https://doi.org/10.1016/j.foodcont.2022.108826

Caleb, O. J., Mahajan, P. V., Al-Said, F. A., & Opara, U. L. (2013). Modified atmosphere packaging technology of fresh and fresh-cut produce and the microbial consequences—A review. *Food and Bioprocess Technology, 6*, 303–329. https://doi.org/10.1007/s11947-012-0932-4

Cálix-Lara, T. F., Rajendran, M., Talcott, S. T., Smith, S. B., Mille, R. K., Castillo, A., Sturino, J. M., & Taylor, T. M. (2014). Inhibition of *Escherichia coli* O157: H7 and *Salmonella enterica* on spinach and identification of antimicrobial substances produced by a commercial lactic acid bacteria food safety intervention. *Food Microbiology, 38*, 192–200.

Center for Food Safety and Applied Nutrition (CFSAN). (2017). *Control of Listeria monocytogenes in ready-to-eat food: Guidance for industry*. Available at: https://www.fda.gov/files/food/published/Draft-Guidance-for-Industry%2D%2DControl-of-Listeria-monocytogenes-in-Ready-To-Eat-Foods-%28PDF%29.pdf. Accessed 14 Jan 2024.

Centers for Disease Control and Prevention (CDC). (2019). *Multistate outbreaks of Salmonella infections associated with raw tomatoes eaten in restaurants*. Available at: https://search.cdc.gov/search/?query=tomato&sitelimit=&utf8=%E2%9C%93&affiliate=cdc-main. Accessed 14 Jan 2024.

Centers for Disease Control and Prevention (CDC). (2021). *Listeria (Listeriosis)*. Retrieved from https://www.cdc.gov/listeria/index.html

CFR. (2020). *Code of Federal Regulations Title 21: Food and Drugs; Sec. 182.20: Essential oils, oleoresins (solvent-free), and natural extractives (including distillates)*. 21CFR182.20. https://www.accessdata.fda.gov/scripts/cdrh/cfdocs/cfcfr/cfrsearch.cfm?fr=182.20. Accessed 24 Jan 2024.

Controlled Environment Agriculture (CEA) Alliance. (2023). *Commodity specific food safety guidelines for controlled environment agriculture focused on leafy greens and herbs*. Available at: https://ceaalliance.com/wp-content/uploads/2023/11/cea_guide_oct23-2.pdf. Accessed 14 Jan 2024.

Critzer, F., Hamilton, A. M., Melendez, M., Danyluk, M. D., & Strawn, L. K. (2024). Survey of environmental monitoring practices in fresh produce packinghouses. *Food Protection Trends, 44*(2), 102–110. https://doi.org/10.4315/FPT-23-022

Devarajan, N., Weller, D. L., Jones, M., Adell, A. D., Adhikari, A., Allende, A., Arnold, N. L., et al. (2023). Evidence for the efficacy of pre-harvest agricultural practices in mitigating food-safety risks to fresh produce in North America. *Frontiers in Sustainable Food Systems, 7*, 1101435. https://doi.org/10.3389/fsufs.2023.1101435

Dunn, L. L., White, E., & Cimowsky, S. (2023). *Packinghouse environmental monitoring programs: Identifying packinghouse zones.* UGA Cooperative Extension bulletin 1524-1. Available at: extension.uga.edu/publications/detail.html?number=B1524-1. Accessed 14 Jan 2024.

EC. (2008). *European Commission (EC) Regulation. Off J Eur Union No 1334/2008 of the European Parliament and of the Council of 16 December 2008 on flavourings and certain food ingredients with flavouring properties for use in and on foods and amending council regulation (EEC) No. 1601/91 Regulations (EC) No 2232/96 and (EC) No 110/2008 and Directive 2000/13/EC; EUR-Lex.* European Commission

EC. (2017). Commission notice on guidance document on addressing microbiological risks in fresh fruits and vegetables at primary production through good hygiene (2017/C 163/01). *Official Journal of the European Union, 163*, 1–40.

EFSA and ECDC. (2023). The European Union one health 2022 Zoonoses report. *EFSA Journal, 21*, e8442. https://doi.org/10.2903/j.efsa.2023.8442

Falleh, H., Ben Jemaa, M., Saada, M., & Ksouri, R. (2020). Essential oils: A promising eco-friendly food preservative. *Food Chemistry, 30*, 127268. https://doi.org/10.1016/j.foodchem.2020.127268

FAO. (2017). *Good agricultural practices for greenhouse vegetable production in the South East European countries.* Available at: http://www.fao.org/3/a-i6787e.pdf. Accessed 24 Jan 2024.

FAO & WHO. (2004). *Guidance to governments on the application of HACCP in small and/or less-developed food businesses.* Available at: https://iris.who.int/handle/10665/43598. Accessed 24 Jan 2024.

FAO & WHO. (2019). *Safety and quality of water used in food production and processing—Meeting report* (Microbiological risk assessment series no. 33). FAO and WHO.

FAO & WHO. (2021). *Safety and quality of water used with fresh fruits and vegetables* (Microbiological risk assessment series no. 37). FAO and WHO. https://doi.org/10.4060/cb7678en

FAO & WHO. (2022). *Listeria monocytogenes in ready-to-eat (RTE) foods: Attribution, characterization, and monitoring—Meeting report* (Microbiological risk assessment series no. 38). FAO and WHO. https://doi.org/10.4060/cc2400en

FAO & WHO. (2023). *Prevention and control of microbiological hazards in fresh fruits and vegetables—Parts 1 & 2: General principles. Meeting report* (Microbiological risk assessment series no. 42). FAO and WHO. https://doi.org/10.4060/cc8490en

Food and Agriculture Organization of the United Nations/World Health Organization (FAO/WHO). (2004). *FAO/WHO guidance to governments on the application of HACCP in small and/or less-developed food businesses.* Available at: http://www.fao.org/3/y8702e/y8702e00.htm. Accessed 14 Jan 2024.

Garrido, Y., Tudela, J. A., Marín, A., Allende, A., & Gil, M. I. (2020). Reconditioning of wash water for the fresh-cut industry. *Advances in Food Processing and Technology, 1*, 126. https://doi.org/10.29011/2639-3387.100126

Garvey, M. (2022). Bacteriophages and food production: Biocontrol and bio-preservation options for food safety. *Antibiotics (Basel), 11*, 1324. https://doi.org/10.3390/antibiotics11101324

Gehringer, R. A. (2017). *Chlorinated water sanitation of leafy green vegetables for fresh produce processors. Ohioline.* Ohio State University Extension, Agriculture and Natural Resources AEX-262. Available at: https://ohioline.osu.edu/factsheet/aex-262. Accessed 14 Jan 2024.

Gil, M. I., Truchado, P., Tudela, J. A., & Allende, A. (2023). Environmental monitoring of three fresh-cut processing facilities reveals harborage sites for *Listeria monocytogenes*. *Food Control*, 110093. https://doi.org/10.1016/j.foodcont.2023.110093

Gombas, D., Luo, Y., Brennan, J., Shergill, G., Petran, R., Walsh, R., Hau, H., Khurana, K., Zomorodi, B., Rosen, J., et al. (2017). Guidelines to validate control of cross-contamination during washing of fresh-cut leafy vegetables. *Journal of Food Protection, 80*, 312–330.

Gómez-Galindo, M., Truchado, P., Allende, A., & Gil, M. I. (2023). Optimization of the use of a commercial phage-based product as a control strategy of *Listeria monocytogenes* in the fresh-cut industry. *Food Security, 12*(17), 3171. https://doi.org/10.3390/foods12173171

Gómez-Galindo, M., Truchado, P., Volpi, M., Elsser-Gravesen, A., Gil, M. I., & Allende, A. (2024). Inactivation efficacy of four commercial post-process treatments against *Listeria monocytogenes* and impact on the commercial quality of leafy greens. *Food Control*. https://doi.org/10.1016/j.foodcont.2024.110549

Gómez-López, V. M., Marín, A., Allende, A., Beuchat, L. R., & Gil, M. I. (2013). Postharvest handling conditions affect internalization of *Salmonella* in baby spinach during washing. *Journal of Food Protection, 76*, 145–1151. https://doi.org/10.4315/0362-028X.JFP-12-539

Gorny, J. R. (2003). *Packaging design for fresh-cut produce*. International Fresh-Cut Produce Association.

Grigore-Gurgu, L., Bucur, F. I., Mihalache, O. A., & Nicolau, A. I. (2024). Comprehensive review on the biocontrol of *Listeria monocytogenes* in food products. *Food Security, 13*(5), 734. https://doi.org/10.3390/foods13050734

Hamilton, A. N., Gibson, M. A., Amalaradjou, C. W. A., Callahan, P. D., Millner, S., Ilic, M. L., Lewis Ivey, A., & Shaw, A. M. (2023). Cultivating food safety together: Insights about the future of produce safety in the U.S. controlled environment agriculture sector. *Journal of Food Protection, 86*, 100190.

Harrand, A. S., Strawn, L. K., Illas-Ortiz, P. M., Wiedmann, M., & Weller, D. L. (2020). *Listeria monocytogenes* prevalence varies more within fields than between fields or over time on conventionally farmed New York produce fields. *Journal of Food Protection, 83*, 1958–1966. https://doi.org/10.4315/JFP-20-120

Harris, L. J., Zagory, D., & Gorny, J. R. (2002). Safety factors. In A. A. Kader (Ed.), *Postharvest technology of horticultural crops* (Agriculture and natural resources, publication 3311) (p. 301). University of California.

Harris, L. J., Farber, J. N., Beuchat, L. R., Parish, M. E., Suslow, T. V., Garrett, E. H., & Busta, F. F. (2003). Outbreaks associated with fresh produce: Incidence, growth, and survival of pathogens in fresh and fresh-cut produce. *Comprehensive Reviews in Food Science and Food Safety, 2*(1 SUPPL), 78–141. https://doi.org/10.1111/j.1541-4337.2003.tb00031.x

Huang, R., de Vries, D., & Chen, H. (2018). Strategies to enhance fresh produce decontamination using combined treatments of ultraviolet washing and disinfectants. *International Journal of Food Microbiology, 283*, 37–44. https://doi.org/10.1016/j.ijfoodmicro.2018.06.014

Huang, L., Cao, X., & You, X. (2020). Consumer food safety knowledge practices and risk perception associated with fresh-cut produce: A nationwide survey in the United States. *Food Control, 110*, 107031. https://doi.org/10.1016/j.foodcont.2019.107031

Interagency Food Safety Analytics Collaboration (2021). Foodborne illness source attribution estimates for 2019 for Salmonella, Escherichia coli O157, Listeria monocytogenes, and Campylobacter using multi-year outbreak surveillance data, United States. U.S. Department of Health and Human Services, CDC, FDA, USDA/ FSIS, Atlanta, Georgia and Washington, District of Columbia.

Julien-Javaux, F., Gérard, C., Campagnoli, M., & Zuber, S. (2019). Strategies for the safety management of fresh produce from farm to fork. *Current Opinion in Food Science, 27*, 145–152. https://doi.org/10.1016/j.cofs.2019.01.004

Kang, J. H., & Song, K. B. (2018). Inhibitory effect of plant essential oil nanoemulsions against *Listeria monocytogenes*, *Escherichia coli* O157:H7, and *Salmonella typhimurium* on red mustard leaves. *Innovative Food Science and Emerging Technologies, 45*, 447–454. https://doi.org/10.1016/j.ifset.2017.09.019

Kumariya, R., Garsa, A. K., Rajput, Y. S., Sood, S. K., Akhtar, N., & Patel, S. (2019). Bacteriocins: Classification, synthesis, mechanism of action, and resistance development in food spoilage causing bacteria. *Microbial Pathogenesis, 128*, 171–177.

Leneveu-Jenvrin, C., Charles, F., Barba, F. J., & Remize, F. (2020). Role of biological control agents and physical treatments in maintaining the quality of fresh and minimally-processed fruit and vegetables. *Critical Reviews in Food Science and Nutrition, 60*, 2837–2855.

Liu, A., Wang, Y., Cai, X., Jiang, S., Cai, X., Shen, L., Liu, Y., Han, G., Chen, S., & Wang, J. (2019). Characterization of endolysins from bacteriophage LPST10 and evaluation of their potential for controlling *Salmonella typhimurium* on lettuce. *LWT, 114*, 108372. https://doi.org/10.1016/j.lwt.2019.108372

Liu, S., Quek, S. Y., & Huang, K. (2023). Advanced strategies to overcome the challenges of bacteriophage-based antimicrobial treatments in food and agricultural systems. *Critical Reviews in Food Science and Nutrition*. https://doi.org/10.1080/10408398.2023.2254837

Lokerse, R. F. A., Maslowska-Corker, K. A., van de Wardt, L. C., & Wijtzes, T. (2016). Growth capacity of *Listeria monocytogenes* in ingredients of ready-to-eat salads. *Food Control, 60*, 338–345. https://doi.org/10.1016/j.foodcont.2015.07.041

López-Gálvez, F., Gil, M. I., Truchado, P., Selma, M. V., & Allende, A. (2010). Cross-contamination of fresh-cut lettuce after a short-term exposure during pre-washing cannot be controlled after subsequent washing with chlorine dioxide and sodium hypochlorite. *Food Microbiology, 27*, 199–204. https://doi.org/10.1016/j.fm.2009.09.009

López-Gálvez, F., Allende, A., & Gil, M. I. (2021). Recent progress on the management of the industrial washing of fresh produce with a focus on microbiological risks. *Current Opinion in Food Science, 38*, 46–51.

López-Gómez, A., Navarro-Martínez, A., Garre, A., Iguaz, A., Maldonado-Guzmán, P., & Martínez-Hernández, G. B. (2023). Kinetics of carvacrol release from active paper packaging for fresh fruits and vegetables under conditions of open and closed package. *Food Packaging and Shelf Life, 37*. https://doi.org/10.1016/j.fpsl.2023.101081

Machado-Moreira, B., Richards, K., Brennan, F., Abram, F., & Burgess, C. M. (2019). Microbial contamination of fresh produce: What, where, and how? *Comprehensive Reviews in Food Science and Food Safety, 18*(6), 1727–1750. https://doi.org/10.1111/1541-4337.12487

Mahajan, B. V. C., Tandon, R., Kapoor, S., & Sidhu, M. K. (2018). Natural coatings for shelf life enhancement and quality maintenance of fresh fruits and vegetables—A review. *Journal of Postharvest Technology, 6*, 12–26.

Marín, A., Tudela, J. A., Garrido, Y., Albolafio, S., Hernández, N., Andújar, S., Allende, A., & Gil, M. I. (2020). Chlorinated wash water and pH regulators affect chlorine gas emission and disinfection by-products. *Innovative Food Science and Emerging Technologies, 66*, 102533. https://doi.org/10.1016/j.ifset.2020.102533

McManamon, O., Kaupper, T., Scollard, J., & Schmalenberger, A. (2019). Nisin application delays growth of *Listeria monocytogenes* on fresh-cut iceberg lettuce in modified atmosphere packaging while the bacterial community structure changes within one week of storage. *Postharvest Biology and Technology, 147*, 185–195. https://doi.org/10.1016/j.postharvbio.2018.10.002

Motzer, C., El-Moghazy, A., Allende, A., Gil, M., Weesepoel, Y., van Overbeek, L., Liu, C., Van De Zedde, R., Bouzembrak, Y., Nitin, N., Ivanek, R., & Wiedmann, M. (2024). Food safety related data analytics, digital, and artificial intelligence needs and opportunities in controlled environment agriculture. *Food Protection Trends, 44*(6), 400–408.

Murray, K., Aldossari, H., Wu, F., & Warriner, K. (2018). Dynamic changes in free-chlorine levels within a commercial postharvest wash and prevention of cross-contamination between shredded lettuce batches. *Food Control, 85*, 127–134. https://doi.org/10.1016/j.foodcont.2017.09.029

Newsome, R., Balestrini, C. G., & Baum, M. D. (2014). Examining applications and perceptions of date labeling of food: A global view. *Waste Management, 34*(3), 612–620. https://doi.org/10.1016/j.wasman.2013.11.015

O'Beirne, D., Gomez-Lopez, V., Tudela, J. A., Allende, A., & Gil, M. I. (2015). Effects of oxygen-depleted atmospheres on survival and growth of *Listeria monocytogenes* on fresh-cut Iceberg lettuce stored at mild abuse commercial temperatures. *Food Microbiology, 48*, 17–21. https://doi.org/10.1016/j.fm.2014.11.012
Oh, M., Cevallos-Urena, A., & Kim, B. S. (2024). Bacteriophages PECP14, PECP20, and their endolysins as effective biocontrol agents for *Escherichia coli* O157:H7 and other foodborne pathogens. *International Journal of Food Microbiology, 409*, 110460. https://doi.org/10.1016/j.ijfoodmicro.2023.110460
Oliveira, M., Abadias, M., Colás-Medà, P., Usall, J., & Viñas, I. (2015). Biopreservative methods to control the growth of foodborne pathogens on fresh-cut lettuce. *International Journal of Food Microbiology, 214*, 4–11.
Pisoschi, A. M., Pop, A., Georgescu, C., Turcus, V., Olah, N. K., & Mathe, E. (2018). An overview of natural antimicrobials role in food. *European Journal of Medicinal Chemistry, 143*, 922–935.
Prakash, B., Singh, P. P., Gupta, V., & Raghuvanshi, T. S. (2023). Essential oils as green promising alternatives to chemical preservatives for agri-food products: New insight into molecular mechanism, toxicity assessment, and safety profile. *Food and Chemical Toxicology, 183*, 114241. https://doi.org/10.1016/j.fct.2023.114241
Raffo, A., & Paoletti, F. (2022). Fresh-cut vegetables processing: Environmental sustainability and food safety issues in a comprehensive perspective. *Frontiers in Sustainable Food Systems, 5*, 681459. https://doi.org/10.3389/fsufs.2021.681459
Rahman MU, Wang W, Sun Q, Shah JA, Li C, Sun Y, Li Y, Zhang B, Chen W, Wang S. (2021). Endolysin, a Promising Solution against Antimicrobial Resistance. Antibiotics (Basel), 10(11), 1277. https://doi.org/10.3390/antibiotics10111277. PMID: 34827215; PMCID: PMC8614784
Ramos, B., Brandão, T. R. S., Teixeira, P., & Silva, C. L. M. (2020). Biopreservation approaches to reduce *Listeria monocytogenes* in fresh vegetables. *Food Microbiology, 85*, 103282. https://doi.org/10.1016/j.fm.2019.103282
Reis, J. A., Paula, A. T., Casarotti, S. N., & Penna, A. L. B. (2012). Lactic acid bacteria antimicrobial compounds: Characteristics and applications. *Food Engineering Reviews, 4*, 124–140. https://doi.org/10.1007/s12393-012-9051-2
Reis, D. R., Ambrosi, A., & Luccio, M. D. (2022). Encapsulated essential oils: A perspective in food preservation. *Future Foods, 5*, 100126.
Salvi, D. (2017). *Understanding produce contamination*. Available at: https://www.aiche.org/resources/publications/cep/2017/may/understand-produce-contamination. Accessed 14 Jan 2024.
Schmelcher, M., Donovan, D. M., & Loessner, M. J. (2012). Bacteriophage endolysins as novel antimicrobials. *Future Microbiology, 7*, 1147–1171. https://doi.org/10.2217/fmb.12.97
Sethi, S., Nayak, S. L., Joshi, A., & Sharma, R. R. (2020). Sanitizers for fresh-cut fruits and vegetables. In *Fresh-cut fruits and vegetables*. https://doi.org/10.1016/B978-0-12-816184-5.00005-7
Shannon, R., Radford, D. R., & Balamurugan, S. (2020). Impacts of food matrix on bacteriophage and endolysin antimicrobial efficacy and performance. *Critical Reviews in Food Science and Nutrition, 60*(10), 1631–1640. https://doi.org/10.1080/10408398.2019.1584874
Sharma, M., Patel, J. R., Conway, W. S., Ferguson, S., & Sulakvelidze, A. (2009). Effectiveness of bacteriophages in reducing *Escherichia coli* O157: H7 on fresh-cut cantaloupes and lettuce. *Journal of Food Protection, 72*(7), 1481–1485. https://doi.org/10.4315/0362-028X-72.7.1481
Sharma, S., Barkauskaite, S., Jaiswal, A. K., & Jaiswal, S. (2021). Essential oils as additives in active food packaging. *Food Chemistry*. https://doi.org/10.1016/j.foodchem.2020.128403
Shen, X., Su, Y., Hua, Z., Cong, J., Dhowlaghar, N., Sun, Q., Lin, S., Green, T., Perrault, M., Galeni, M., Hanrahan, I., Suslow, T., & Zhu, M.-J. (2020). Verification of peroxyacetic acid treatment against L. *monocytogenes* on fresh apples using E. faecium NRRL B-2354 as a surrogate in commercial spray-bar operations. *Food Microbiology, 92*, 103590. https://doi.org/10.1016/j.fm.2020.103590

Singh, S. P., & Thakur, R. (2024). Postharvest applications of cold plasma treatment for improving food safety and sustainability outcomes for fresh horticultural produce. *Postharvest Biology and Technology, 209*, 112694. https://doi.org/10.1016/j.postharvbio.2023.112694

Su, Y., Shen, X., Chiu, T., Green, T., & Zhu, M.-J. (2022). Efficacy of chlorine and peroxyacetic acid to control *Listeria monocytogenes* on apples in simulated dump tank water system. *Food Microbiology, 106*, 104033. https://doi.org/10.1016/j.fm.2022.104033

Sullivan, G., & Wiedmann, M. (2020). Detection and prevalence of Listeria in U.S. produce packinghouses and fresh-cut facilities. *Journal of Food Protection, 83*, 1656–1666. https://doi.org/10.1016/j.fm.2017.07.005

Truchado, P., Elsser-Gravesen, A., Gil, M. I., & Allende, A. (2020). Post-process treatments are effective strategies to reduce *Listeria monocytogenes* on the surface of leafy greens: A pilot study. *International Journal of Food Microbiology, 313*, 108390.

Tudela, J. A., López-Gálvez, F., Allende, A., Hernández, N., Andújar, S., Marín, A., Garrido, Y., & Gil, M. I. (2019). Operational limits of sodium hypochlorite for different fresh produce wash water based on microbial inactivation and disinfection by-products (DBPs). *Food Control, 104*, 300–307. https://doi.org/10.1016/j.foodcont.2019.05.005

United Fresh Produce Association (UFPA). (2018). *Guidance on environmental monitoring and control of listeria for the fresh produce industry*. Developed by the United Fresh Food Safety & Technology Council. Available at: https://info.unitedfresh.org/hubfs/Food%20Safety%20Pubs/FINAL-UFPA-Listeria-Guidance5.pdf

Vikram, A., Tokman, J. I., Woolston, J., & Sulakvelidze, A. (2020). Phage biocontrol improves food safety by significantly reducing the level and prevalence of *Escherichia coli* O157:H7 in various foods. *Journal of Food Protection, 83*, 668–676. https://doi.org/10.4315/0362-028X.JFP-19-433

Vikram, A., Callahan, M. T., Woolston, J. W., Sharma, M., & Sulakvelidze, A. (2022). Phage biocontrol for reducing bacterial foodborne pathogens in produce and other foods. *Current Opinion in Biotechnology, 78*, 102805. https://doi.org/10.1016/j.copbio.2022.102805

Wang, H., & Ryser, E. T. (2020). Quantitative transfer and sanitizer inactivation of *Salmonella* during simulated commercial dicing and conveying of tomatoes. *Food Control, 107*, 106762. https://doi.org/10.1016/j.foodcont.2019.106762

Yu, H., Lin, Z., Lin, M. S., Neal, J. A., & Sirsat, S. A. (2022). Consumers' knowledge and handling practices associated with fresh-cut produce in the United States. *Food Security, 11*(14), 2167. https://doi.org/10.3390/foods11142167. PMID: 35885411; PMCID: PMC9318181.

Zapaśnik, A., Sokołowska, B., & Bryła, M. (2022). Role of lactic acid bacteria in food preservation and safety. *Food Security, 11*, 1283.

Zhang, X., Niu, Y. D., Nan, Y., Stanford, K., Holley, R., McAllister, T., & Narváez-Bravo, C. (2019). SalmoFresh™ effectiveness in controlling *Salmonella* on romaine lettuce, mung bean sprouts, and seeds. *International Journal of Food Microbiology, 305*, 108250. https://doi.org/10.1016/j.ijfoodmicro.2019.108250

Zhao, X., Li, L., Zhang, Q., Li, M., Hu, M., Luo, Y., Xu, X., Chen, Y., & Liu, Y. (2023). Characterization of the Clostridium perfringens phage endolysin cpp-lys and its application on lettuce. *International Journal of Food Microbiology, 405*, 110343. https://doi.org/10.1016/j.ijfoodmicro.2023.110343

Zhou, B., Luo, Y., Nou, X., Lyu, S., & Wang, Q. (2015). Inactivation dynamics of *Salmonella enterica*, *Listeria monocytogenes*, and *Escherichia coli* O157:H7 in wash water during simulated chlorine depletion and replenishment processes. *Food Microbiology, 50*, 88–96. https://doi.org/10.1016/j.fm.2015.03.004

Disinfection Technologies Applied in Indoor Hydroponic System

Jim Junhui Huang, Dan Li, and Weibiao Zhou

1 Introduction

Hydroponic cultivation technique is a soilless approach of horticultural practices for executing plant growths or crop production in water, in which the systems, equipments or facilities developed for carrying out hydroponic cultivation are called hydroponic systems (Polo-López et al., 2013). Currently for hydroponic cultivation/production systems, they are mainly classified into five categories, namely continuous drip system, deep water raft culture (DWC) system, flood and drain system, nutrient film technique (NFT) system and the wick and capillary action system (Riggio et al., 2019). As well recognized, these systems might undergo different internalization risks of human pathogens. In this scenario, the application of liquid nutrient solutions in hydroponic systems creates a dissimilar growing environment, compared with the conventional soil-based cultivation, via elevating the opportunity to potentially introduce pathogens from specific fungi (e.g., *Fusarium* spp., *Phytophthora* spp. *Pythium* sp. and *Olpidium* spp.), microalgae (e.g., *Chlorella* spp.), bacteria (e.g., *Escherichia coli* O157:H7 and *Salmonella typhimurium*) and even virus (e.g., norovirus) to easily spread throughout the whole systems and the crops by means of the rapid circulation operations of these solutions in the system (Wang & Kniel, 2016; Dong et al., 2022; McClure et al., 2023). This situation calls for efficient disinfection approaches for fully removing these pathogens for preventing from crop production loses and avoiding threatening human health by consuming these crops (Polo-López et al., 2013).

Before the discussion on various disinfection technologies applied in the indoor hydroponic system for antimicrobial purpose, it is important to look into the real

J. J. Huang · D. Li · W. Zhou (✉)
Department of Food Science and Technology, National University of Singapore, Singapore, Singapore
e-mail: weibiao@nus.edu.sg

D. Li (ed.), *Antimicrobial Strategies in the Food System: Updates, Opportunities, Challenges*, https://doi.org/10.1007/978-3-031-95056-8_9

situations in the system. Using *Salmonella typhimurium* as the example, *S. typhimurium* could present on lettuce and induce cross-contamination in the hydroponic farming environment, such presence might last for as long as 6 weeks under the absence of disinfection application (Li et al., 2022). On the other hand, longer exposure duration of plants to *Salmonella*-contaminated water might lead to remarkably higher concentrations of internalized *S. typhimurium* than those exposed for period as short as 1 day (Ge et al., 2014). This indicated that longer contact of plants to *S. typhimurium* could aggravate such contamination. Of note, strong biofilm-forming bacteria could efficiently support the colonization of *S. typhimurium* on the building material of hydroponic system, potentially, they were possible to result in high spread risk of *S. typhimurium* to the whole hydroponic farming facilities and further to crops grown. As documented, the strong biofilm formers (e.g., *Corynebacterium tuberculostearicum* C2 and *Pseudoxanthomonas mexicana* C3) could help to achieve better attachment and colonization of *Salmonella* on the coupons of polyvinyl chloride (PVC, a popular building material for hydroponic systems currently) incubated in hydroponic nutrient solution at 25 °C, by remarkably facilitating *Salmonella* growth (Tham et al., 2024).

Noteworthily, packaged leafy green grown in the indoor hydroponic system of a controlled-environment farm might undergo risk to be contaminated by *S. typhimurium*. This could be seen in a 2021 case reporting an outbreak of illness associated with *S. typhimurium* occurred at four states of U.S.A. by consuming such products, in which a total of 31 patients including 4 hospitalized were recorded. The outbreak reasons were mainly concluded by ignoring their indoor hydroponic pond water as agricultural water that might contact the greens chronically, neglecting to protect the packaged soilless growth media and packaging pallets stored outdoors, as well as indulging condensate dripping formed on chiller water pipelines to drip on the leafy vegetables and on the production conveyors (McClure et al., 2023).

As above stated, not only bacteria, but also viruses are easy to spread through the hydroponic system and recirculate in nutrient solution, and eventually reach the roots and edible tissues of cultivars. As reported, when a human norovirus (huNoV) surrogate [e.g., murine norovirus (MNV)] was inoculated in the recirculated water with about 3 log PFU/mL MNV after 7 days of growths of kale and mustard microgreens, its levels could be still remained from 2.73 to 3.54 log copies/sample during the harvest period from day 8 to 12. Additionally, MNV was found in both the roots and edible tissues of these cultivars after 2 h post-inoculation and kept stable at the first 12 h. Although the titer of this virus was detected to gradually decrease from day 8 (about 2.5 log PFU/sample) to 12 (about 1.5 log PFU/sample), its RNA concentration was still remained at a higher degree at around 4.0 to 5.5 log copies/sample (Wang & Kniel, 2016). These phenomena indicated that appropriate disinfection measures were necessary also for eliminating virus's influences on hydroponic system. Besides bacteria and viruses, microalgae are also recognized as a crucial issue to be addressed in hydroponic systems due to their disadvantages on seizing dissolved oxygen and competing nutrients in the hydroponic solutions with crops, thus impacting on the physiologies of crops and eventually decreasing crop productions (Thakulla et al., 2022).

Based on the above cases and situations, it is very essential to carry out the corresponding disinfection technologies to fully remove pathogens from indoor hydroponic systems, for maintaining the sustainable crop productivities and keeping safe and pathogen-free crops for consumption. Nowadays, the disinfection technologies can be categorized based on physical, chemical and biological means.

2 Physical Disinfection Techniques

The physical disinfection techniques are the application of physical conditions/ methods such as electromagnetic waves (e.g., ultraviolet lights and gamma rays), filtration, pulsed electric field and pH to treat hydroponic systems, or to sanitize the surfaces of the corresponding system facilities, for seeking to fully remove pathogens from the systems.

2.1 Electromagnetic Wave Disinfection Techniques

The electromagnetic waves for disinfecting hydroponic systems normally include ultraviolet lights such as ultraviolet band A (UVA, 320–400 nm), ultraviolet band B (UVB, 280–320 nm) and ultraviolet band C (UVC, 100–280 nm), along with solar radiation because it also include ultraviolet wavebands in its spectrum, as well as gamma ray.

2.1.1 UV Light Disinfection Techniques

The mechanism to use UV light for sanitizing pathogens is to carry out the non-ionizing radiation triggered by UV light to form 4 thymine dimerization (DNA kink) on the DNA chain of the pathogens, thus damage the replication of DNA in the pathogens, and eventually kill the pathogens (Huang & Tu, 2001). Practically, UV light disinfection techniques have long been adopted as a kind of efficient and clean manner for applying in hydroponic crop surface sanitization. In an early trial using effluents to cultivate hydroponic lettuce, the application of UV light (wavelength not mentioned) as a supplemental disinfection measure was found to maintain a low contaminated levels of *E. coli*, thermotolerant and total coliforms, as well as keep an absent environment of *Salmonella* spp. and helminth eggs on the leaves of lettuce, furthermore, these microorganisms could not be detected on lettuce's root system (Keller et al., 2005).

2.1.1.1 UVA

UVA has been proved that it could efficiently disinfect hydroponic nutrient solutions in plant factories. As found, the inactivation rate of *Escherichia coli* ATCC25922 by UVA-LED at 385 nm was positively related to its irradiation intensity, this could be seen that the decreased log value of this bacterium was achieved by near 6.0 under 1.2 kW/m^2 of UVA irradiation intensity for 6 min, compared with approximately 0.3, 2.6 and 4.0 under 0.3, 0.6 and 0.9 kW/m^2 for the same treatment duration, respectively (Tsunedomi et al., 2018). This indicated that there was a positive relation between UVA intensity and inactivation effect. In recent year, a UVA disinfection system equipped with online microbial monitoring function and microfluidic chip has been developed for simultaneously executing online monitoring when carrying out UVA disinfection, in which the working power of UVA sterilizer applied was 29.98-fold of UVA irradiance, while the UVA disinfection rate on nutrient solution was positively related to the UVA irradiation dose (Wang et al., 2022).

2.1.1.2 UVB

Besides UVA, the recycling of hydroponic nutrient solutions could also be achieved by UVB radiation. In a recent trial, the application of UVB-LED (280 nm) for 5 min was found to remarkably suppress the mycelia growth and degrade DNA of *Pythium aphanidermatum* which might induce damping-off, these exposed fungal mycelia was found incapable of infecting cucumber thereafter (Hashiguchi et al., 2024).

2.1.1.3 UVC

UVC light is potent in removing almost all the microorganisms in a dose-dependent manner due to its high-energy and high-frequency property that can fully destory or broken the microbial DNA structures. Currently, UVC light has been widely applied in hydroponic systems due to its effectiveness in killing microorganisms. For example, the application of UVC (254 nm) with an average intensity of 4800 μW/cm^2 (1440 mJ/cm^2 of UVC dose per treatment) weekly by 5 min for 4 weeks was found to lead to 4 log reduction of the bacteria in the water reserviors of a vertical and indoor hydroponic system, such UVC irradiation could also inhibit bacterial and microalgal biofilm growth in the system (Lee et al., 2015).

In fact, UVC exposure dose, the appropriate auxiliary manner (e.g., solution flow rate), and the tolerances of treated microorganism species to UVC, were the major factors to be concerned for disinfecting the pathogens in hydroponic system. This could be seen when using UVC (270–285 nm) LED lamps, the higher light output might lead to higher sterilization efficiencies on four microorganisms including *Clavibacter michiganensis*, *E. coli*, *Fusarium oxysporum* and *Pseudomonas cichorii* (Kim et al., 2020). In this scenario, the combination of different hydroponic nutrient solution flow rates (up to 10 mL/s) and UVC LED light outputs (300, 500 and 700 mW) showed that the maximal flow rate, no matter under what kind of output, could help to completely eliminate these above-tested pathogens (Kim et al., 2020). This revealed that the higher flow rate might facilitate the targeted pathogens in the solution to expose to UVC more frequently, thus kill these pathogens through using less UVC dose.

As above mentioned, the tolerance doses of different pathogens to the same UVC wavelength were various. As found, the UVC (253.7 nm) radiation doses ranging from 30 to 40 mJ/cm^2 could disinfect a majority of conidia of *Fusarium oxysporum*, while a lower dose ranging from 20 to 30 mJ/cm^2, can achieve the same purpose to sanitize zoospores of *Pythium aphanidermatum*, in water. However, although there was tolerance dose difference between species, it was clear that no matter for *F. oxysporum*, or for *P. aphanidermatum*, their colony-forming abilities were reduced logarithmically, with the gradual increase of UVC radiation doses in this case (Sutton et al., 2000). When such UVC was applied to hydroponic nutrient solutions, the doses, provided by a collimated-beam or a flow-through UV-radiation apparatus, needed to be enhanced by up to 254 and 213 mJ/cm^2 in nutrient solutions for pepper and tomato crops, respectively (Sutton et al., 2000). The presences of nutrients might increase the turbidity of the solutions, thus the UVC dose was required to enhance so that the similar disinfection effect could be achieved because UVC, as a short wavelength UV light, is easier to be absorbed by the environmental substances, compared with UVA and UVB.

Adequate UVC doses are critical for fully removing all microorganisms from the plants and hydroponic system, while inadequate UVC doses can only eliminate parts of them although all the species seemed to be remarkably inhibited, the survived pathogens can still infect the plants and lead to plant diseases. For example, when tomato roots were inoculated with *Pythium aphanidermatum* and exposed under UVC (253.7 nm), all the pathogens and non-target bacterial populations on the roots were recoreded significantly decreased under all the gradually increasing doses (i.e., 19, 38, 59 and 88 mJ/cm^2). However, such reductions neither led to the decrement of *Pythium* root rot, nor resulted in the increment of tomato fruit yield, although they facilitated the remarkable decrement of the non-target bacterial populations in the rhizosphere (Zhang & Tu, 2000). Noteworthily, no matter pre-harvest UVC intervention approach, or post-harvest one, can not offset such inadequate UVC dose in eliminating pathogens. As reported, pre-harvest intervention approach by exposing UVC (254 nm, 950 mJ/cm^2) to irrigation water for mung bean sprouts was found to reduce the concentration of internalized *S. typhimurium* by 1.84 log CFU/g but its level was still as high as 3.72 log CFU/g, while post-harvest intervention approach by treating irrigation water with chlorine wash and UVC (254 nm, 778 mJ/cm^2) exposure simultaneously did not efficiently remove these internalized pathogens (the lowest level by 4.19 log CFU/g) (Ge et al., 2014). These results indicated that pre- and post-harvest intervention approaches did not function in hydroponic system to fully eliminate *Salmonella* from mung bean sprouts and extra disinfection procedures are required.

Of note, UVC light not only could disinfect the targeted pathogens, but also could regulate the metabolite contents of crops simultaneously. As documented, the application of UVC (< 280 nm) to mung and groundnut beans for 15 min was found to decrease their root infecting fungi counts and simultaneously increase the total chlorophyll and carbohydrate contents of their microgreens (Siddiqui et al., 2011). The cause could be that UVC light reduced the impact of the root infecting fungi to compete nutrients with the plants, the microgreens could thus absorb adequate nutrients to support their own growths, through synthesizing more intracellular chlorophylls and carbohydrates.

2.1.1.4 Gamma Ray

Gamma ray, belonging to the member of electromagnetic spectrum, is a kind of high-energy ray with wavelength shorter than 0.01 nm, followed by energy and frequency higher than 124 keV and 3×10^{19} Hz, respectively. In the hydroponic system, the radiation dose increment of gamma ray could help to efficiently disinfect *E. coli* O157:H7/pGFP in lettuce. As a related case shown, the ratios of *E. coli* O157:H7/pGFP-positive lettuces to numbers of lettuces tested were significantly decreased from 3/8, 1/8 to 0/8, and 4/8, 1/8 to 0/8 in Hoagland's media contaminated with 10^5 and 10^7 CFU/mL of *E. coli* O157:H7/pGFP, respectively, when the radiation dose (kGy) of γ-ray was increased from 0.25, 0.50 to 0.75 (Nthenge et al., 2007).

2.1.1.5 Solar UV Radiation

Because solar spectrum contains UV lights and plenty of UVA and UVB could reach the earth's surface daily (Huang & Tu, 2001), solar UV radiation can be developed as an efficient disinfection technique in hydroponic system. By using such radiation, the required processing time for disinfecting two major plant pathogenic bacteria (e.g., *Clavibacter michiganensis* and *Pseudomonas syringae*) at 10^7 CFU/mL was found negatively related to the processing temperature in the hydroponic system, where the minimum time was 420, 45 and 15 min achieved by 50, 55, and 60 °C, respectively. Furthermore, the introduction of 2.79 MJ/m^2 of solar radiation dose including 95.48 kJ/m^2 of UV light has been recorded to kill these two bacteria after 1 h of treatment at 50 °C, this result indicated that the combination of heat (50 °C) and solar UV radiation could significantly decrease the time to disinfect both bacteria from 7 to 1 h (Farhadi et al., 2022).

2.1.2 Solar Water Disinfection Technique

Solar water disinfection, using the heat energy rather than rays from solar radiation, is a traditional technology for sanitizing hydroponic cultures in a convenient, free and clean way. As reported in a 1000 m^2 greenhouse, normally 400–1000 L of hydroponic cultures could be daily disinfected, in which the evacuated tube collector (ETC) and flat plate collector (EPC) were usually used as the major solar collectors to convert solar energy into heat for disinfection purpose. The advantage of solar water disinfection is that it can reduce the performance of chemical controls to prevent from contaminating environment, for example, it can minimize the use of fungicide under an acceptable level and cost (Farhadi, 2021).

2.2 *Filtration-Based Disinfection Techniques*

Physical filtration technology is a clean and efficient means to eliminate unwanted pathogens by means of gravity or pressure. Nowadays, filtration-based disinfection techniques applied in hydroponic system mainly include microfiltration membrane disinfection technique, slow sand filtration (SSF) technique and cross-flow filtration (CFF) disinfection technique.

2.2.1 Microfiltration Membrane Disinfection Technique

A microfiltration membrane disinfection system was developed and applied in the closed hydroponics of a greenhouse for fully removing *Pseudomonas solanasearum* (a reference microorganism) from the nutrient solution by using 0.3 μm of membrane pores. This system could achieve 57.6 m^3/day of filtration capacity and 0.2% of the total amount of permeate was reported to be discharged out of the nutrient solution through automatic reverse flow technology after the membrane cleaning step. After such operation, the filtered nutrient solution was thus recycled and ready for the next hydroponic cultivation (Ohtani et al., 2000).

2.2.2 Slow Sand Filtration (SSF) Technique

SSF technique is an inexpensive, low-enery-consumption and environmentally-friendly approach for sanitizing recirculating nutrient solutions in the enclosed hydroponic system through eliminating pathogens by means of long fine sand pathway. Employing *E. coli* elimination efficiency as an index to evaluate SSF technique, it was found that heating and aeration could be used for enhancing the disinfection efficiency of SSF, where the effluent *E. coli* concentrations from filters through both processings were more than 4.00 log of reductions, compared with the influent ones. Furthermore, the vertical distribution of the tested viable *E. coli* in such slow sand filter demonstrated that the *E. coli* concentrations were significantly decreased from 10^6 CFU/mL in the Schmutzdecke to 10^2 CFU/mL at the 10 cm position just below sand layer surface. Besides *E. coli*, this SSF technique was also found efficient in eliminating the fungal pathogens such as *Fusarium oxysporum* and *Pythium helicoides* (Mine et al., 2007).

2.2.3 Cross-Flow Filtration (CFF) Disinfection Technique

CFF, also termed tangential flow filtration (TFF), is a kind of filtration system by which the feed flows mainly pass through filter surface tangentially, rather than vertically travels into the filter. Such arrangement lets the filtrate permeate through the membrane, while leads the concentrate accumulated and circulated at the opposite side of membrane (Henry, 2018). As recorded, CFF could be used in the recirculating hydroponic system to efficiently inhibit related plant disease through filtering out the associated pathogens. A previous study on CFF has showed its capability to minimize the occurrence of root disease in pepper infected by *Pythium myriotylum*. Normally, the sizes of zoospore cysts of *P. myriotylum* were around 8–10 μm, while pore-sizes of filters smaller than 5 μm were capable of fully removing such pathogen. Thus, the suppression effect of filter on plant disease was subjected to the different filter brands used that possessed different sizes of pores, in which the suppression effect of the combined filters was better than that of the single filter, this is because the overlap of the two filters may generate smaller pores. For example, the double-filter application with 0.5 and 1.0 μm polymate brand filters could efficiently protect pepper plants cultivated in NFT system by 100% removal of zoospore cysts, while the single-filter application (Honeycomb and Polypure brands) with pore-size from 1 to 30 μm could only remove 85–95% of zoospore cysts (Schuerger & Hammer, 2009).

2.3 *Pulsed Electric Field (PEF) Disinfection Technique*

PEF was a recently developed disinfection technique for continuously sanitizing the recirculating nutrient solution in hydroponic system. This technique required a pulsed power generator for generating repetitive pulses in the amplitude between 0 and 5 kV with a 10 μs of pulse period and a 50 Hz of pulse frequency, furthermore, a parallel-plate treatment chamber was also needed to receive continuous processing. In a validation trial, the PEF efficiency was proved to inactivate 99% of *Fusarium oxysporum* in the nutrient solution, in which the levels of inorganic components were not significantly affected by such PEF operation (Chen et al., 2020).

2.4 *Ultrasound Disinfection Technique*

Appropriate ultrasonic treatment duration is a value-added and environmentally-friendly approach to efficiently sanitize pathogens grown on crops while maintaining the normal growth of hydroponic crops. As recently reported, the application of a high acoustic power density (APD) ultrasound emitted by an ultrasound probe driven by a Vibra-Cell processor at 25% power, combined with 55 °C of mild heat water for 5 min, was found to achieve a comparable decrement of *E. coli* O157:H7 and sprout yield of the sanitated radish, broccoli and kale seeds, the same as those treated by FDA-recommended 20 min, 20,000 mg/L chlorine treatment. While the ultrasonic treatment for 10 min could lead to a remarkably higher decrement of *E. coli* O157:H7 in these three seed types, compared with the chlorine treatment. However, a lower germination rate ($p < 0.05$) was also achieved (Dong et al., 2022). This indicated that although longer ultrasonic treatment could significantly sanitize the targeted pathogens, it also negatively affected crop growths. Thus, the ultrasonic treatment of 5 min was enough for disinfecting these pathogens on the crops as chlorine did, while simultaneously reducing the energy cost and the damage to the crops.

2.5 Photodynamic Inactivation (PDI) Disinfection Technique

PDI utilizes light and a dye-based photosensitizer, through the participation of oxygen, to overproduce reactive oxygen species (ROS) in the field of illumination that kills target bacteria and virus finally (Hasenleitner & Plaetzer, 2019; Pucelik & Dąbrowski, 2022). As noted, PDI is a potential mitigation strategy for using in hydroponic farming systems to improve microbial safety. It was reported that the application of rose bengal-mediated PDI on *Salmonella*-contaminated lettuce seeds could remarkably reduce *Salmonella* from 3.90 log CFU per lettuce plant to 2.77 log CFU per lettuce plant by week six (Li et al., 2022).

3 Chemical Disinfection Techniques

3.1 pH Disinfection Technique

The manipulation of pH could be an efficient approach to sanitize recirculating nutrient solution in hydroponic system, which can be achieved through decreasing the pH value in solution. Normally, the down-regulation of pH may suppress the growth of plants, this is because the availabilities of micronutrients like B, Cu, Mn and Zn might be increased, whereas that of Mo might be decreased when solution pH is reduced. Such availabilities might lead to the toxicities of B, Cu, Mn and Zn, and the deficiency of Mo to the plants, as well as result in nutrient disorders eventually. Meanwhile, lower pH could efficiently kill the pathogens and thus sanitize the nutrient solution. For example, the survival of *Clavibacter michiganensis* (a bacterium causing bacterial canker in tomato) was completely inhibited within 24 h by nutrient solution under a pH of 4.0. Additionally, the population of this pathogen tested in greenhouse NFT culture at pH 5.0 was lower than that at pH 6.5. Under this scenario, the occurrence of bacterial canker (11 out of 48 plants) of root-inoculated tomato plants by *C. michiganensis* at pH 5.0 was also found significantly lower than that (34 out of 48 plants) at pH 6.5 (Huang & Tu, 2001).

Therefore, the strategies to minimize the inhibition of nutrient solution on plants under lower pH are to down-regulate the concentrations of B, Cu, Mn and Zn, while up-regulating that of Mo beforehand, whereas such low pH could also significantly suppress the growths of pathogens in hydroponic system. This idea seems to make possibility that lowering solution pH could completely disinfect hydroponic system while maintaining normal growth of plants simultaneously. In a recent study, this strategy was applied in hydroponic basil plants. It was executed through decreasing the concentrations of B, Cu, Mn and Zn by one-half, whereas doubling that of Mo. The results showed that the basil plant Nufar grown under DWC system at pH 4.0 for fortnight did not remarkably decrease its biomass, meanwhile, this pH value could efficiently inhibit the severity of root rot infected by *Pythium aphanidermatum*, while those basil plants grown under a conventional pH 5.5 without micronutrient adjustment were badly infected by *P. aphanidermatum*, simultaneously, their growths were significantly inhibited (Gillespie et al., 2020).

3.2 Hypochlorite Disinfection Technique

Hypochlorite is capable of disinfecting virus derived from hydroponic plants. As documented, an electrolytic disinfector, designed for sanitizing irrigation water in greenhouses by containing an electrolysis reactor equipped with titanium electrodes, could be used for generating potassium hypochlorite and then chlorine. When they were injected into the nutrient solution employed for cultivating tomato per week through a sensor, the dispersal of *Pepino mosaic* virus (PepMV) in the tomato plant was found greatly inhibited. Further investigation found that the application at 0.2 mg free chlorine/L nutrient solution for 60 min, or that at 0.5 mg/L for 30 min, could guarantee the inactivation of PepMV and did not lead to phytotoxicity. Moreover, the suppression of PepMV dispersal to tomato plants could significantly decrease the amount of the unmarketable tomato fruits that were commonly discolored affected by PepMV (Bandte et al., 2016).

Besides virus, the electrolytically-generated potassium hypochlorite (KClO) could show its disinfection efficacy against the economically-essential fungal plant pathogens such as *Fusarium oxysporum* (Synder and Hans) and *Rhizoctonia solani* (Kühn) in hydroponic tomatoes. However, its capability to sanitise the nutrient solution is subjected to the application rates to pathogens, processing time and solution pH value. For example, the combination of 2 mg free chlorine/L of electrolytically-generated KClO and 30 min at pH 6.0 could fully disinfect *F. oxysporum* but show minor influence on the viability of *R. solani*. In another scenario, the application of KClO solution at 0.5 mg free chlorine/L [780 mV of oxidation-reduction potential (ORP)] into a recirculating nutrient solution weekly for 60 min could suppress the dispersal of both pathogens during a 16-week practical period (Rodriguez et al., 2018).

Actually, not only virus and fungi, but also bacteria are the targeted microorganisms that hypochlorite can demonstrate its disinfection roles. By comparison, appropriate sodium hypochlorite (NaOCl) concentration was better than those of H_2O_2 and sodium percarbonate (SPC) to remove biofilm formed by *Corynebacterium tuberculostearicum* C2 and *Pseudoxanthomonas mexicana* C3 on polyvinyl chloride (PVC) coupons in hydroponic nutrient solutions. It was found that NaOCl at 50 ppm could significantly decrease C2 and C3 counts to reach below 1 log CFU/cm^2 within 12 h of treatment, instead, no matter 3% H_2O_2 or 1% SPC could not achieve such consequence. While practicing on hydroponic farming facilities, the concentration of NaOCl was required to enhance by up to 500 ppm, for fully removing the biofilm. One of the causes to achieve such high concentration was that the existence of organic substances in hydroponic nutrient solutions might accumulate thicker biofilm during plant cultivation, another cause could be the higher persistence of the biofilm formed by natural bacteria multispecies. Luckily, such high sanitizer concentration and treatment duration (500 ppm NaOCl for 12 h) did not markedly inhibit subsequent plant growth, however, the high concentration of the by-product of chlorination (i.e., chlorate) was observed in the hydroponic solution of the sanitized systems prior to rinsing (Tham et al., 2024).

As the above case stated, appropriate concentration of chlorine is essential for maintaining good disinfection effect by chlorination and keeping good quality of cultivated plants simultaneously. In a related study, the application of 0.55 m/L of sodium hypochloride was found to achieve the highest fresh weight and vitamin C concentration in *Lactuca sativa* Mantecosa (a kind of lettuce), as well as generate the less toxicity symptomes like chlorosis, compared with those under 0 (the control), 5.5 and 11 mg/L of sodium hypochloride. Noteworthily, 11 mg/L of treatment might badly induce the necrosis and chlorosis of lettuce leaves and thus lower their qualities (Premuzic et al., 2007).

Althoug hypochlorite has displayed its effectiveness in disinfecting pathogens, the safety of hypochlorite used as the nutrient solution disinfectant and the residue issues of chlorate (ClO_3^-) and perchlorate (ClO_4^-) in hydroponic system are worthy to be concerned. In a recent study, the uses of 2.5, 5.0 and 7.5 mg/L for three times in per 14-day's interval, respectively, did not found to negatively affect the growth, gas exchange parameters, and nutrient contents of tomato plants and its fruits, furthermore, the residues of chlorate and perchlorate were not observed on tomato fruits (Lykogianni et al., 2023).

Importantly in the chlorination technique, the residue issues of chlorate, along with perchlorate in the crop-based foods such as fruits and vegetables grown in the recirculating hydroponic system should be critically noticed. This is because such chlorate and perchlorate are actually the by-products of hypochlorite that is the disinfectant commonly used for sanitizing plant production facilities including hydroponic systems. Furthermore, these compounds (e.g., chlorate and perchlorate) could result in acute toxicological impacts on human health. Moreover, the application of hypochlorite on the hydroponic systems might more or less lead to the accumulations of ClO_3^- and ClO_4^- on the fruits and vegetables. In a study trying to reveal the relation between hypochlorite treatment and chlorate/perchlorate accumulations in this scenario, the 1% of KClO was applied into a recirculating tomato production system weekly for 90 min. When the free chlorine levels reached 1 and

2 mg/L, respectively, the ClO_3^- content in the tomato was increased to 0.22 and 0.25 mg ClO_3^-/kg FW, compared with the control (0.01 mg ClO_3^-/kg FW), while the co-occurrence of ClO_3^- and ClO_4^- was absent. Interestingly, ClO_3^- was found to trigger oxidative stress in tomoto, resulting that the contents of lycopene and β-carotene in tomato were enhanced by up to 33.5 and 23.9%, respectively (Dannehl et al., 2016).

3.3 Ozonation Disinfection Technique

Ozone (O_3), a colorless and unstable gas with a pungent odor, can demonstrate strong disinfection effectiveness against virus and bacteria in water, similar to chlorination. It is generated through an oxygen molecule (O_2) to collide with an oxygen atom (O) which is dissociated from another oxygen molecule by the energy source such as thunderstorm or ozone generator (Leung, 2023). Due to such potent disinfectant property, ozone is widely applied in sanitizing water, including those nutrient and culture solutions from hydroponic system. Recently, some emerging technologies termed as Ozone microbubbles (OMB) technique and Cold plasma ozone treatment technique have been developed as the efficient disinfection methods for this purpose (Kobayashi et al., 2011; Zhou et al., 2024).

In a related study, OMB technique was found to demonstrate a very high disinfecting activity against the phytopathogens such as *Fusarium oxysporum* and *Pectobacterium carotovorum* located in the infected plant roots. This is because OMB possesses higher solubility and can maintain in the water for a long while, compared with ozone millibubbles, due to their smaller sizes. Additionally, the initial concentration of dissolved ozone in the OMB-treated hydroponic solutions could be an important factor to enhance its disinfectant activity against both phytopathogens (Kobayashi et al., 2011). These outcomes supported that OMB could be a novel disinfectant against phytopathogens in the hydroponic cultivation system. For another technique, cold plasma ozone treatment has been recently reported as an efficient technique in disinfecting recirculating hydroponic systems. An appropriate case was its capability to sanitize tomato brown rugose fruit virus (ToBRFV) in a hydroponic greenhouse. It was found that the viabilities of ToBRFV in an inoculum with 1:1000 of virus dilution ratio were completed inhibited when it was exposed to ozone for 48 min, furthermore, a shorter exposure duration (24 min) could be achieved to reach the identical disinfection effect when the dilution ratio was increased to 1:10,000 (Zhou et al., 2024).

For this technique, some phenomena are worthy of being noticed. Temperature does not significantly affect the disinfection efficiency of ozonated water on *Fusarium oxysporum* conidia from hydroponic nutrient solution. This could be seen in a previous study showing that different treatment temperatures at 15, 25 and 30 °C led to almost the identical disinfection efficiencies on the pathogen *F. oxysporum* in an inorganic soil-less nutrient medium. Additionally, higher initial ozone concentration at 1.0 ppm could rapidly wipe out *F. oxysporum* conidia by up to 4 log orders, which inactivation speed was significantly faster than those below 1.0 ppm (e.g., 0.4, 0.6 and 0.8 ppm) (Igura et al., 2004). This indicated that the disinfection efficiency of ozone could be guaranteed in a faster manner when its initial concentration was reached a relatively higher level.

Looking into the feeding mode of ozonation disinfection technique, oxygen-feed ozonation disinfection strategy was found more efficient than air-feed one in sanitizing *F. oxysporum*, the cause is that pure oxygen feed in ozone generator can supply higher ozone concentration to fully eliminate the suspended spores in the solution, compared with that provided by air-feed ozone generator. For example, it was found that oxygen-feed ozonation at 5 or 21 °C could 100% disinfect the pathogen (0% viability) through supplying ozone dosage by up to 88 mgO_3/L, while under air-feed ozonation at 5 °C, there was a linear relation between the viability of *F. oxysporum* and ozone dosage, in which the lowest estimated viability (19%) of pathogen required the treatment of ozone concentration by up to 63 mgO_3/L. When the temperature was increased to 21 °C, the viability result was shown linear in O_3 and quadratic in O_3 simultaneously, where the lowest estimated viability (37%) needed at least 29 mgO_3/L to achieve (Msayleb et al., 2022).

Certainly, there should have concerns that aqueous-state ozone residue in the nutrient solution might generate stress to the crops cultivated and subsequently affect their qualities. A recent study answered this question by showing that the hydroponic nutrient solution with dissolved ozone concentration (DOC) by up to 4.0 mg/L demonstrated no negative influence on the growth and root antioxidant of the lettuce, indicating that ozonation made to nutrient solution is practicable as a replaceable disinfection approach, for eliminating chemical residues or wastes produced by traditional disinfectants and water resources (Zheng et al., 2020).

3.4 Hydrogen Peroxide (H_2O_2) Disinfection Technique

H_2O_2 is a kind of efficient disinfectant and potent oxidizer that can kill various forms of microorganisms including microalgae, bacteria, fungi and virus, through generating destructive hydroxyl radicals to oxidize DNA, membrane lipids and other cellular components under appropriate concentrations, where the byproducts are only water and oxygen which are safe to the environment. However, it might take a longer time for H_2O_2 to display its antimicrobial activity (Freeman & Auer, 2012). H_2O_2 has been widely used in hydroponic system to eliminate different origins of microalgae, bacteria and fungi doing harm to the cultivated crops in recent years, as below reviewed.

As studied, H_2O_2 could efficiently suppress the infestation of microalgae grown in phenolic foam cells of hydroponic cultivation system for lettuce, in a 15-day's cultivation duration (Caixeta et al., 2018). In another ebb and flow hydroponic system for cultivating two cultivars of lettuce (i.e., 'Green Forest' and 'Tropicana'), many kinds of microalgal species including *Microspora tumidula*, Centric diatoms, Pennate diatoms, *Leptolyngbya*, *Tetraspora*, *Gloeocystis*, *Chlamydomonas* and

Scenedesmus were identified, in which *M. tumidula* was the dominant population. In this scenario, the applications of H_2O_2 products such as 15 mL of ZeroTol (27.1% of H_2O_2 and 2.0% of peroxyacetic acid), 60 mL of ZeroTol, 15 mL of PERpose Plus (33.0% of H_2O_2) and 3% hydrogen peroxide treatments, were reported to achieve 96.0, 99.2, 94.0 and 97.9% of decrements in the concentrations of this *M. tumidula*, respectively. Fortunately, the 3% of H_2O_2, excluding ZeroTol and PERpose Plus, did not negatively affect the growths of both lettuce cultivars (e.g., shoot fresh weight) (Hendrickson et al., 2022). This indicated that 3% of H_2O_2 can be applied to sanitize the ebb and flow hydroponic system during the cultivation for fully removing microalgae, while keeping normal growth of this crop simultaneously.

In the cases of pepper cultivars such as 'Early Jalapeno' and 'Lunchbox Red', as well as tomato ones including 'Little Sicily' and 'Geronimo' using the same H_2O_2 products, the applications of 70 mL of ZeroTol and PERpose Plus per week could significantly decrease the counts of microalgae grown in ebb and flow hydroponic systems, along with microalgal dry weights and their chlorophyll *a* contents, while promoting the growth of both crops (Thakulla et al., 2022). The causes for appropriate H_2O_2 concentration to improve the growths of pepper and tomato were that H_2O_2, the stabilized ROS, could act as the signaling molecule in photosynthesis, transpiration, translocation and respiration of plants (Ślesak et al., 2007). Additionally, H_2O_2 could regulate the activity of sucrose phosphate synthase, an essential enzyme for the generation of sucrose from triose phosphates throughout photosynthesis, as well as positively modulate the activities of metabolic and antioxidant enzymes for facilitating plant development (Uchida et al., 2002). Moreover, H_2O_2 was capable of adjusting endogenous plant hormones for elevating chlorophyll content in plants such as tomato and mung bean to further improve photosynthesis and development (Fariduddin et al., 2014; Orabi et al., 2015).

Besides microalgae, H_2O_2 could also eliminate different origins of fungi and bacteria under different concentrations in hydroponic system. In lab-scale tests, the EC_{50} values of H_2O_2 to sanitize different biofilm-forming rhizogenic *Agrobacterium*, the causative bacterium strain of hairy root disease (HRD) found during hydroponic cultivation of cucumber and tomato, can be largely changed from 18.8 to 600.0 ppm. While in pilot-scale trials, the concentration of H_2O_2 to significantly remove *Agrobacterium* population was subjected to whether it was catalase-negative strain that required 50 ppm or catalase-positive strain that needed 100 ppm (Bosmans et al., 2016). In a recent case, the application of 5% H_2O_2 could markedly decrease molds and yeast log count and the amounts of *E. coli*, coliforms and total aerobic bacteria in roselle seeds, compared with H_2O_2-free control (Phornvillay et al. 2022).

Although H_2O_2 demonstrates potent antimicrobial activity against various pathogens, its roles played in plant growths are varied. H_2O_2 could negatively influence the germination of lettuce seeds (Caixeta et al., 2018). In another scene, however, the 5% H_2O_2 could improve the germination rate of roselle seeds instead, as well as significantly enhance the levels of the total phenolic content, total flavonoid content, DPPH and ascorbic acid content of the roselle microgreens, compared with the control (Phornvillay et al. 2022). From the positive point of view, H_2O_2 is capable of executing seed decontamination treatment to enhance plant seed germination. As documented, H_2O_2 under from 3% to 10% could be used as a tool to carry out the alfalfa seeds decontamination treatments for sprouted-seeds production, in which 8% of concentration was confirmed to decrease food-borne pathogens by up to 4 log CFU/g (Holliday et al., 2001). Especially, H_2O_2 could also achieve better germination rates of seeds (e.g., those from pea and *Zizinnia elegans*) after seed decontamination treatments (Barba-Espín et al., 2012; Szopińska, 2014).

From the negative point of view, however, H_2O_2 under high concentrations could display phytotoxic effects on plants. The phytotoxicity of H_2O_2 to suppress the growth of plants and decrease their biomass is because the decomposition of H_2O_2 to form the unstable free radical oxygen molecule that could damage healthy plant cell tissue including living root cells, which subsequently affects the absorption of nutrients through roots and finally leads to the reduction of plant (e.g., lettuce) fresh weight (Eicher-Sodo et al., 2019; Lau & Mattson, 2021). Fortunately, the application of organic fertilizers on hydroponic cultivation could be a solution to prevent plant from H_2O_2 toxicity. As reported, the application of 37.5 and 75.0 mg/L of H_2O_2 using inorganic-mineral-based conventional fertilizers (21-5-20 of N-P_2O_5-K_2O) to grow lettuce could significantly inhibit its growth and even lead to the death of this plant. However, when 37.5 mg/L of H_2O_2 was applied to the same lettuce species using organic fertilizers (4-4-1 of N-P_2O_5-K_2O), there was no remarkably negative effect of this sanitizer on lettuce yield. The reason was that the presence of biofilm in the rootzone, due to the provision of organic fertilizer, might be acted as shield to react with free radical oxygen molecules generated by H_2O_2, thus reduce the damage of H_2O_2 to root cells, and eventually result in normal growth of lettuce, compared with the conventional ones that did not help to form visible biofilm on the lettuce rootzone (Lau & Mattson, 2021).

3.5 *Fungicide Disinfection Technique*

Chemical control strategies can be used in hydroponic system to treat tomato *Fusarium* wilt. *Fusarium* wilt induced by wilt pathogen *Fusarium oxysporum* Klotz (a fungus) is the most serious soil-borne disease that is easy to infect hydroponic cultivation system through dust, tools, equipments or seeds. In a chemical control trial, the introduction of fungicides into the system could be a solution to deal with this disease. It was reported that prochloraz and carbendazim were the effective fungicides in suppressing mycelial growth of *F. oxysporum*, which median effective concentrations (EC_{50}) were 0.019 and 0.235 μg/mL, respectively. After 2-week's treatment, the preventive effects of prochloraz at 0.4 μg/mL and carbendazim at 5.0 μg/mL could reach by up to 69.6 and 87.0%, respectively (Song et al., 2004). These results indicated that tomato wilt disease in the hydroponic system could be efficiently treated by appropriate concentrations of fungicides.

4 Physicochemical Disinfection Techniques

Physicochemical disinfection techniques are normally to combine two or more of physical (e.g., UV light, visible light, ultrasound and electricity) and chemical (e.g., O_3, H_2O_2, chloride and Fenton) disinfection means to carry out the sanitization procedures for killing the undesired pathogens. In hydroponic system, the advanced oxidation processes (AOPs) are the popular techniques developed recently for efficiently removing pathogens from the designated places. Normally speaking, AOPs are involved in the combined treatments of two or more of the inducers or stimulators, at least one from physical means and another from chemical side. So far, AOPs could be divided into non-electrochemical advanced oxidation processes (NEAOPs) that apply non-electrical means such as UV light, visible light and ultrasound to combine with specific chemicals (e.g., O_3, H_2O_2, chloride and Fenton) for disinfection purpose, as well as electrochemical advanced oxidation process (EAOPs) that use electrochemical way to sanitize the pathogens.

4.1 Non-electrochemical Advanced Oxidation Processes (NEAOPs)

NEAOPs are a category of cutting-edge technologies employed for disinfection of pathogens or removal of hazardous organic materials in water through oxidation via the reactions with hydroxyl radicals (HO•) *in situ* produced by inducers or stimulators including ozonation, H_2O_2, UV radiation, Fenton, ultrasonic and photocatalysis (Polo-López et al., 2013; Fan & Song, 2020). In an NEAOPs study involving in H_2O_2 and UVC in combination with microbubbles (MBs) techniques, the different combinations of 5% H_2O_2 + UVC, 5% H_2O_2 + MBs and 5% H_2O_2 + UVC + MBs were found to display their potent antimicrobial activity against total aerobic bacteria, total coliforms, *E. coli*, as well as yeast and mold log count in roselle seeds, through forming the higher HO• by up to 5.25×10^{-13} M, compared with the control. Furthermore, these NEAOPs combinations could facilitate the significant improvements of total flavonoid content and DPPH radical scavenging activity of roselle microgreens after seed decontamination treatments, as compared to the control (Phornvillay et al. 2022).

As above-mentioned, the photo-Fenton technique is one of the efficient NEAOPs for disinfecting irrigation water in hydroponic systems. In a related photo-Fenton trial applying ferrous sulphate (Fe^{2+}) and ferric nitrate (Fe^{3+}) as iron sources, the application of 5 mg/L of Fe^{3+} (89.5 μM), along with 10 mg/L of H_2O_2 (294.0 μM), followed by 2.5 kJ/L of solar UV dose for 60 min could achieve the highest disinfection rate of *Phytophthora capsici* zoospores under 250 mL of small scale. The cause for achieving by Fe^{3+}, rather than by Fe^{2+} might be owing to the cell wall property of zoospores, as well as the dissimilar role displayed by the intracellular iron speciation (Polo-López et al., 2013). In another case, the biochar supported maghemite nanoparticles (M/biochar), formed by the pyrolysis-induced *Cymbopogon flexuosus* leaves attached by ferric chloride under visible light, have been reported to sanitize *E. coli* in water in a photo-Fenton like process, in which the ROS generated in the reaction was found to hinder the oxido-reductase protein system of *E. coli* and subsequently inhibit its metabolic activity (Basu et al., 2021).

The mechanism of photo-Fenton is associated with the production of HO• through the interaction between H_2O_2 using Fe^{2+} as catalyst and UV radiation as the stimulator, in the ferrous ion recovering cycle. This could be achieved by means of the stable zero-charge H_2O_2 that can freely penetrate cellular membranes to react with intracellular ferrous ion, the produced HO• could subsequently oxidize the organelles and intracellular structures, and eventually kill the cells (Mico et al., 2010; Polo-López et al., 2013). The related reactions were as below shown:

$$Fe^{2+} + H_2O_2 \rightarrow Fe^{3+} + OH^- + HO\cdot \rightarrow Fe(OH)^{2+} + HO\cdot \quad (1)$$

$$Fe(OH)^{2+} + hv \rightarrow Fe^{2+} + HO\cdot \quad (2)$$

4.2 *Electrochemical Advanced Oxidation Processes (EAOPs)*

In hydroponic system, EAOPs normally use electrochemical approach to decompose chloride into reactive chlorine species (RCS) such as $Cl_{2(aq)}$, OCl^- and HOCl, or H_2O_2 into HO•, through generating them at the electrode surface by applying direct electro-oxidation processes, for sanitizing and cleaning pathogens in nutrient solutions, in which the electrodes employed are the essential tools to electrolyze the targeted chemicals into the desired disinfectants for sanitization purpose (Barrera-Díaz et al., 2014). Currently, the dimensionally stable anodes (DSA) and boron-doped diamond (BDD) anodes applied in EAOPs have been reported in hydroponic systems for disinfecting pathogens (Lévesque et al., 2019, 2020, 2022, 2023).

4.2.1 DSA-Based EAOP

Electrochemical disinfection based on DSA is an emerging technology for recirculating hydroponics. The mechanism involved is to let an electrochemical flow cell (EFC) with DSA electrodes carry out regenerative fashion like regenerative hypochlorination to continuously produce free chlorine [e.g., $Cl_{2(aq)}$, OCl^- and HOCl] from chloride contained in solution, these primary disinfectants formed could subsequently inactivate the targeted pathogens such as *Rhizoctonia solani* and *Fusarium oxysporum* in the recirculating fertigation solutions (Lévesque et al., 2019; Lévesque et al., 2022). As recognized in a concrete case, this technology could achieve a 4 log decrement of *F. oxysporum* in fertigation solutions and thus avoided pathogenicity in the floriculture crops like *Cyclamen persicum*. In addition, this DSA technology did not significantly affect plant morphology and physiology during the operation, moreover, its energy consumption in combination with EFC was two-fold in electricity use less than that by traditional EAOPs (Lévesque et al., 2022).

In a recent study, DSA-based regenerative *in situ* electrochemical hypochlorination (R*is*E^{Hc}) was applied to treat recirculating nutrient solutions used for culturing hydroponic lettuce. As noted, the presence of ammonium in the solutions was the cause to generate chloramines, which led to the reductions in lettuce biomass by 53 and 83%, due to the different current density applications of 2.27 and 4.55 mA/cm^2, respectively. Such phytotoxic effects induced by chloramines could be avoided by adding UV light to photodecompose chloramines, or by adopting nitrate-based fertilizer to refrain from the generation of this compound. Under these two treatments, the growth of lettuce was not significantly inhibited, even under 4.55 mA/cm^2. Expectantly after doing so, EFC treatment under 2.27 mA/cm^2 led to a 13% increment of lettuce shoot biomass (Lévesque et al., 2023).

4.2.2 BDD-Based EAOP

Electrochemical disinfection based on BDD anodes is also efficient for recirculating hydroponics. As documented, a BDD-based EAOP was applied to inactivate plant pathogen *Rhizoctonia solani* in hydroponic fertigation solutions, in which the BDD anodes and stainless-steel cathodes were involved. This electrochemical system using 9.09 mA/cm^2 in cooperation with 20 mg/L of chloride were reported to fully disinfect *R. solani*, where less than 0.17 mg/L of free chlorine that was more active to supply the electrode surface EAOP reactions, was presented (Lévesque et al., 2020).

5 Biological Disinfection Techniques

The biological disinfection techniques are the application of biological agents (e.g., bacteriophage) to deal with nutrient solutions in hydroponic systems, through letting bacteriophage infect the targeted pathogens or bacteria in the solutions and then remove them from the systems. So far, bacteriophages have been proved to be the efficient tools in disinfecting nutrient solution running in hydroponic system by means of their lytic properties and the corresponding bacteriophage disinfection technique has been developed. In a trial to deal with hairy root disease induced by rhizogenic *Agrobacterium* biovar 1, six bacteriophages named OLIVR specific to this *Agrobacterium* were isolated from the corresponding infected greenhouses. Their capabilities to infect their hosts were subsequently observed but their respective performances were found different in reducing *Agrobacterium* levels in nutrient solutions, for example, OLIVR1 among these six bacteriophages could decrease *Agrobacterium* level by up to 4 log units. However, the issue on phage resistance should be noticed. Except for the above OLIVR1, phage resistance could be found in other bacteriophages such as OLIVR4 and OLIVR5 which did not always fully reduce *Agrobacterium* level till undetectable (Fortuna et al., 2023).

6 Conclusion and future prospectives

This chapter updated and summarized the recent advances and findings on the disinfection technologies applied in indoor hydroponic systems, from the viewpoints of physical, chemical and biological disinfection approaches, these informative contents could be helpful to a wide range of indoor agricultural practitioners by providing various cost-effective manners for disinfecting hydroponic farming facilities and sustaining crop yields simultaneously. The mutual advantages of these techniques are that they are highly efficient in cleaning the hydroponic systems by maximally removing the environmental microorganisms (e.g., food-borne pathogens) attached to the surfaces of facilities. However, their drawbacks are also needed to be concerned. Although physical disinfection techniques are normally performed in a clean and environmentally friendly manner, they usually require consuming electrical energies, which might more or less enhance the total disinfection cost, furthermore, some special physical disinfection techniques (e.g., UVC and gamma ray) may be harmful to the operators. Additionally, the toxicities of chemical agents for disinfecting hydroponic systems should be paid close attention as their residuals, without appropriate cleaning and removal after disinfection procedure, may get high chance to attach to the surfaces of hydroponic facilities, which may be absorbed by the crops in the subsequent cultivation and further affect the health of consumers when they consume these crops. For biological agents (e.g., virus), attention should be focused on whether such viruses applied are also harmful or toxic to human while they are used in hydroponic systems for disinfection purpose.

As such, the future directions should be focused on developing more highly efficient, cost-saving, and low-toxic disinfection technologies to fully remove various pathogens attached to the surfaces of hydroponic systems, meanwhile, the productivities, qualities, and appearances of crops grown in the hydroponic systems should be maximally maintained by balancing the adopted disinfection techniques with the cultivars, in order to create a win-win situation to contribute indoor fresh produce in a safe and high-quality way.

References

Bandte, M., Rodriguez, M. H., Schuch, I., Schmidt, U., & Buettner, C. (2016). Plant viruses in irrigation water: Reduced dispersal of viruses using sensor-based disinfection. *Irrigation Science, 34*, 221–229.

Barba-Espín, G., Hernández, J. A., & Diaz-Vivancos, P. (2012). Role of H2O2 in pea seed germination. *Plant Signaling & Behavior, 7*(2), 193–195.

Barrera-Díaz, C., Canizares, P., Fernández, F. J., Natividad, R., & Rodrigo, M. A. (2014). Electrochemical advanced oxidation processes: an overview of the current applications to actual industrial effluents. *Journal of the Mexican Chemical Society, 58*(3), 256–275.

Basu, A., Behera, M., Maharana, R., Kumar, M., Dhal, N. K., Tamhankar, A. J., et al. (2021). To unsnarl the mechanism of disinfection of Escherichia coli via visible light assisted heterogeneous photo-Fenton reaction in presence of biochar supported maghemite nanoparticles. *Journal of Environmental Chemical Engineering, 9*(1), 104620.

Bosmans, L., Van Calenberge, B., Paeleman, A., Moerkens, R., Wittemans, L., Van Kerckhove, S., et al. (2016). Efficacy of hydrogen peroxide treatment for control of hairy root disease caused by rhizogenic agrobacteria. *Journal of Applied Microbiology, 121*(2), 519–527.

Caixeta, V., Mata, A., Curvelo, C., Tavares, W., Ferreira, L., & Pereira, A. (2018). Hydrogen peroxide for insect and algae control in a lettuce hydroponic environment. *The Journal of Agricultural Science, 10*(8), 221.

Chen, R., Zhang, D., Zhu, N., Liu, C., Novac, B., & Zhong, C. (2020). Continuous disinfection of fusarium oxysporum in nutrient solution by pulsed electric field. In *2020 IEEE international conference on high voltage engineering and application (ICHVE)* (pp. 1–4). IEEE.

Dannehl, D., Schuch, I., Gao, Y., Cordiner, S., & Schmidt, U. (2016). Effects of hypochlorite as a disinfectant for hydroponic systems on accumulations of chlorate and phytochemical compounds in tomatoes. *European Food Research and Technology, 242*, 345–353.

Dong, M., Park, H. K., Wang, Y., & Feng, H. (2022). Control Escherichia coli O157: H7 growth on sprouting brassicacae seeds with high acoustic power density (APD) ultrasound plus mild heat and calcium-oxide antimicrobial spray. *Food Control, 132*, 108482.

Eicher-Sodo, M., Gordon, R., & Zheng, Y. (2019). Characterizing the phytotoxic effects of hydrogen peroxide on common microgreen species and lettuce cultivars. *HortTechnology, 29*(3), 283–289.

Fan, X., & Song, Y. (2020). Advanced oxidation process as a postharvest decontamination technology to improve microbial safety of fresh produce. *Journal of Agricultural and Food Chemistry, 68*(46), 12916–12926.

Farhadi, R. (2021). Elimination of chemical controls in hydroponic cultures using solar energy. *Journal of Cleaner Production, 313*, 127861.

Farhadi, R., Teimourlou, R. F., & Ghosta, Y. (2022). Elimination of plant pathogenic bacteria by solar ultraviolet radiation in hydroponic systems. *The Journal of Agricultural Science, 28*(2), 342–350.

Fariduddin, Q., Khan, T. A., & Yusuf, M. (2014). Hydrogen peroxide mediated tolerance to copper stress in the presence of 28-homobrassinolide in Vigna radiata. *Acta Physiologiae Plantarum, 36*, 2767–2778.

Fortuna, K. J., Holtappels, D., Venneman, J., Baeyen, S., Vallino, M., Verwilt, P., et al. (2023). Back to the roots: Agrobacterium-specific phages show potential to disinfect nutrient solution from hydroponic greenhouses. *Applied and Environmental Microbiology, 89*(4), e00215–e00223.

Freeman, D. E., & Auer, J. A. (2012). Instrument preparation, sterilization, and antiseptics. In *Equine surgery* (pp. 98–111). WB Saunders.

Ge, C., Rymut, S., Lee, C., & Lee, J. (2014). Salmonella internalization in mung bean sprouts and pre-and postharvest intervention methods in a hydroponic system. *Journal of Food Protection, 77*(5), 752–757.

Gillespie, D. P., Kubota, C., & Miller, S. A. (2020). Effects of low pH of hydroponic nutrient solution on plant growth, nutrient uptake, and root rot disease incidence of basil (Ocimum basilicum L.). *HortScience, 55*(8), 1251–1258.

Hasenleitner, M., & Plaetzer, K. (2019). In the right light: Photodynamic inactivation of microorganisms using a LED-based illumination device tailored for the antimicrobial application. *Antibiotics, 9*(1), 13.

Hashiguchi, A., Ino, M., & Ueno, M. (2024). The reutilization of hydroponic nutrient solutions and disease control by LED irradiation. *Environmental Control in Biology, 62*(1), 11–15.

Hendrickson, T. D., Dunn, B. L., Goad, C., Hu, B., & Singh, H. (2022). Effects of hydrogen peroxide products on basil, lettuce, and algae in an ebb and flow hydroponic system. *Horticulturae, 8*(7), 569.

Henry, J. D. (2018). Cross flow filtration. In *Recent developments in separation science* (pp. 205–225). CRC Press.

Holliday, S. L., Scouten, A. J., & Beuchat, L. R. (2001). Efficacy of chemical treatments in eliminating salmonella and Escherichia coli O157: H7 on scarified and polished alfalfa seeds. *Journal of Food Protection, 64*(10), 1489–1495.

Huang, R., & Tu, J. C. (2001). Effects of nutrient solution pH on the survival and transmission of *Clavibacter michiganensis* ssp. *michiganensis* in hydroponically grown tomatoes. *Plant Pathology, 50*(4), 503–508.

Igura, N., Fujii, M., Shimoda, M., & Hayakawa, I. (2004). Research note: Inactivation efficiency of ozonated water for *fusarium oxysporum* conidia under hydroponic greenhouse conditions. *Ozone Science and Engineering, 26*(5), 517–521.

Keller, R., Perim, K., Semionato, S., Zandonade, E., Cassini, S., & Gonçalves, R. F. (2005). Hydroponic cultivation of lettuce (Lactuca sativa) using effluents from primary, secondary and tertiary+ UV treatments. *Water Science and Technology: Water Supply, 5*(1), 95–100.

Kim, B. S., Youm, S., & Kim, Y. K. (2020). Sterilization of harmful microorganisms in hydroponic cultivation using an ultraviolet LED light source. *Sensors and Materials, 32*.

Kobayashi, F., Ikeura, H., Ohsato, S., Goto, T., & Tamaki, M. (2011). Disinfection using ozone microbubbles to inactivate *Fusarium oxysporum* f. sp. *melonis* and *Pectobacterium carotovorum* subsp. *carotovorum*. *Crop Protection, 30*(11), 1514–1518.

Lau, V., & Mattson, N. (2021). Effects of hydrogen peroxide on organically fertilized hydroponic lettuce (*Lactuca sativa* L.). *Horticulturae, 7*(5), 106.

Lee, S., Ge, C., Bohrerova, Z., Grewal, P. S., & Lee, J. (2015). Enhancing plant productivity while suppressing biofilm growth in a windowfarm system using beneficial bacteria and ultraviolet irradiation. *Canadian Journal of Microbiology, 61*(7), 457–466.

Leung, F. (2023). *Ozone pollution and plant health: Understanding the impacts and solutions for sustainable agriculture*. Elsevier.

Lévesque, S., Graham, T., Bejan, D., Lawson, J., Zhang, P., & Dixon, M. (2019). Inactivation of Rhizoctonia solani in fertigation water using regenerative *in situ* electrochemical hypochlorination. *Scientific Reports, 9*(1), 14237.

Lévesque, S., Graham, T., Bejan, D., Lawson, J., Zhang, P., & Dixon, M. (2020). An electrochemical advanced oxidation process (EAOP) for the inactivation of Rhizoctonia solani in fertigation solutions. *Canadian Journal of Plant Science, 100*(4), 415–424.

Lévesque, S., Graham, T., Bejan, D., & Dixon, M. (2022). Comparative analysis of regenerative in situ electrochemical hypochlorination and conventional water disinfection technologies for growing ornamental crops with recirculating hydroponics. *Agricultural Water Management, 269*, 107673.

Lévesque, S., Graham, T., Bejan, D., Lawson, J., & Dixon, M. (2023). Prevention of phytotoxic effects of regenerative in situ electrochemical hypochlorination in recirculating hydroponic systems. *HortScience, 58*(1), 107–113.

Li, Y., Zwe, Y. H., Tham, C. A. T., Zou, Y., Li, W., & Li, D. (2022). Fate and mitigation of salmonella contaminated in lettuce (Lactuca sativa) seeds grown in a hydroponic system. *Journal of Applied Microbiology, 132*(2), 1449–1456.

Lykogianni, M., Bempelou, E., Karavidas, I., Anagnostopoulos, C., Aliferis, K. A., & Savvas, D. (2023). Impact of sodium hypochlorite applied as nutrient solution disinfectant on growth, nutritional status, yield, and consumer safety of tomato (Solanum lycopersicum L.) fruit produced in a soilless cultivation. *Horticulturae, 9*(3), 352.

McClure, M., Whitney, B., Gardenhire, I., Crosby, A., Wellman, A., Patel, K., et al. (2023). An outbreak investigation of salmonella typhimurium illnesses in the United States linked to packaged leafy greens produced at a controlled environment agriculture indoor hydroponic operation–2021. *Journal of Food Protection, 86*(5), 100079.

Mico, M. M., Chourdaki, S., Bacardit, J., & Sans, C. (2010). Comparison between ozonation and photo-fenton processes for pesticide methomyl removal in advanced greenhouses. *Ozone Science and Engineering, 32*(4), 259–264.
Mine, Y., Suematsu, M., & Sugiyama, N. (2007). Aeration and heating improve slow sand filtration as a disinfection system for closed hydroponic systems. *Environmental Control in Biology, 45*(3), 179–187.
Msayleb, N., Kanwar, R., Wu, H., & van Leeuwen, J. (2022). Effects of ozonation on the viability of fusarium oxysporum conidia in hydroponic nutrient solutions. *Ozone Science and Engineering, 44*(1), 96–104.
Nthenge, A. K., Weese, J. S., Carter, M., Wei, C. I., & Huang, T. S. (2007). Efficacy of gamma radiation and aqueous chlorine on Escherichia coli O157: H7 in hydroponically grown lettuce plants. *Journal of Food Protection, 70*(3), 748–752.
Ohtani, T., Kaneko, A., Fukuda, N., Hagiwara, S., & Sase, S. (2000). SW – soil and water: development of a membrane disinfection system for closed hydroponics in a greenhouse. *Journal of Agricultural Engineering Research, 77*(2), 227–232.
Orabi, S. A., Dawood, M. G., & Salman, S. R. (2015). Comparative study between the physiological role of hydrogen peroxide and salicylic acid in alleviating the harmful effect of low temperature on tomato plants grown under sand-ponic culture. *Science in Agriculture, 9*(1), 49–59.
Phornvillay, S., Yodsarn, S., Oonsrithong, J., Srilaong, V., & Pongprasert, N. (2022). A novel technique using advanced oxidation process (UVC/H2O2) combined with micro-nano bubbles on decontamination, seed viability, and enhancing phytonutrients of roselle microgreens. *Horticulturae, 8*(1), 53.
Polo-López, M. I., Oller, I., & Fernández-Ibáñez, P. (2013). Benefits of photo-Fenton at low concentrations for solar disinfection of distilled water. A case study: *Phytophthora capsici*. *Catalysis Today, 209*, 181–187.
Premuzic, Z., Palmucci, H. E., Tamborenea, J., & Nakama, M. (2007). Chlorine disinfection: effects on hydroponics lettuce. *Journal of Applied Horticultre, 9*(1), 62–65.
Pucelik, B., & Dąbrowski, J. M. (2022). Photodynamic inactivation (PDI) as a promising alternative to current pharmaceuticals for the treatment of resistant microorganisms. In *Advances in inorganic chemistry* (Vol. 79, pp. 65–108). Academic.
Riggio, G. M., Jones, S. L., & Gibson, K. E. (2019). Risk of human pathogen internalization in leafy vegetables during lab-scale hydroponic cultivation. *Horticulturae, 5*(1), 25.
Rodriguez, M. H., Bandte, M., Gaskin, T., Fischer, G., & Büttner, C. (2018). Efficacy of electrolytically-derived disinfectant against dispersal of *fusarium oxysporum* and *Rhizoctonia solani* in hydroponic tomatoes. *Scientia Horticulturae, 234*, 116–125.
Schuerger, A. C., & Hammer, W. (2009). Use of cross-flow membrane filtration in a recirculating hydroponic system to suppress root disease in pepper caused by *Pythium myriotylum*. *Phytopathology, 99*(5), 597–607.
Siddiqui, A., Dawar, S. H. A. H. N. A. Z., Zaki, M. J., & Hamid, N. (2011). Role of ultra violet (UV-C) radiation in the control of root infecting fungi on groundnut and mung bean. *Pakistan Journal of Botany, 43*(4), 2221–2224.
Ślesak, I., Libik, M., Karpinska, B., Karpinski, S., & Miszalski, Z. (2007). The role of hydrogen peroxide in regulation of plant metabolism and cellular signalling in response to environmental stresses. *Acta Biochimica Polonica, 54*(1), 39–50.
Song, W., Zhou, L., Yang, C., Cao, X., Zhang, L., & Liu, X. (2004). Tomato *Fusarium* wilt and its chemical control strategies in a hydroponic system. *Crop Protection, 23*(3), 243–247.
Sutton, J. C., Yu, H., Grodzinski, B., & Johnstone, M. (2000). Relationships of ultraviolet radiation dose and inactivation of pathogen propagules in water and hydroponic nutrient solutions. *Canadian Journal of Plant Pathology, 22*(3), 300–309.
Szopińska, D. (2014). Effects of hydrogen peroxide treatment on the germination, vigour and health of seeds. *Folia Horticulturae, 26*(1), 19–29.

Thakulla, D., Dunn, B. L., Goad, C., & Hu, B. (2022). Timing and rates of two products using hydrogen peroxide (H_2O_2) to control algae in ebb and flow hydroponic systems. *HortScience, 57*(1), 32–39.

Tham, C., et al. (2024). Microbial analysis and sanitization of hydroponic farming facilities in Singapore. bioRxiv, 2024–04

Tsunedomi, A., Miyawaki, K., Masamura, A., Nakahashi, M., Mawatari, K., Shimohata, T., et al. (2018). UVA-LED device to disinfect hydroponic nutrient solution. *The Journal of Medical Investigation, 65*(3.4), 171–176.

Uchida, A., Jagendorf, A. T., Hibino, T., Takabe, T., & Takabe, T. (2002). Effects of hydrogen peroxide and nitric oxide on both salt and heat stress tolerance in rice. *Plant Science, 163*(3), 515–523.

Wang, Q., & Kniel, K. E. (2016). Survival and transfer of murine norovirus within a hydroponic system during kale and mustard microgreen harvesting. *Applied and Environmental Microbiology, 82*(2), 705–713.

Wang, X., Fang, W., & Zhao, Z. (2022). Design of UVA ultraviolet disinfection system for nutrient solution residual liquid and development of microbial online monitoring system. *Sustainable Forestry, 15*(1), 173.

Zhang, W., & Tu, J. C. (2000). Effect of ultraviolet disinfection of hydroponic solutions on Pythium root rot and non-target bacteria. *European Journal of Plant Pathology, 106*, 415–421.

Zheng, L., Liu, C., & Song, W. (2020). Effect of ozonated nutrient solution on the growth and root antioxidant capacity of substrate and hydroponically cultivated lettuce (*Lactuca sativa*). *Ozone Science and Engineering, 42*(3), 286–292.

Zhou, J., Gilliard, A., & Ling, K. S. (2024). Tomato Brown rugose fruit virus is transmissible through a greenhouse hydroponic system but may be inactivated by cold plasma ozone treatment. *Horticulturae, 10*(4), 416.

Antimicrobial Strategies in Animal Husbandry

Mohamed-Yousif Mohamed, Hazim Khalifa, Ahmed Elolimy, and Ihab Habib

1 Current Status of Antimicrobial Use in Animal Husbandry

Antibiotic use has become crucial in modern intensive livestock farming practices to prevent and treat microbiological infections, primarily bacteria, that cause animal illnesses. Food animals are reported to consume approximately 63,151 tons of antibiotics in 2010; by 2030, this amount is expected to rise by 67%, whereas the human population is expected to increase by just 13% during the same period (Laxminarayan et al., 2015; Van Boeckel et al., 2015). According to another estimate, nearly 80% of all food-producing animals and birds today are given antibiotics for some or all of their lives (Lee et al., 2001). Eighty percent of all antibiotics consumed in the United States each year come from food animals (Obimakinde et al., 2017), with antimicrobial intake in cattle, poultry, and pigs predicted to be 45, 148, and 172 mg/kg on average per year worldwide (Hosain et al., 2021). Historically, wealthy nations like the USA and several European nations were the primary consumers of antibiotics in food animals (Elliott et al., 2017). Currently, antibiotic usage in livestock has been rising quickly in developing South and

M.-Y. Mohamed · H. Khalifa · I. Habib (✉)
Veterinary Public Health Research Laboratory, Department of Veterinary Medicine, College of Agriculture and Veterinary Medicine, United Arab Emirates University, Al Ain, United Arab Emirates
e-mail: i.habib@uaeu.ac.ae

A. Elolimy
Veterinary Public Health Research Laboratory, Department of Veterinary Medicine, College of Agriculture and Veterinary Medicine, United Arab Emirates University, Al Ain, United Arab Emirates

Department of Integrative Agriculture, United Arab Emirates University, Al Ain, United Arab Emirates

D. Li (ed.), *Antimicrobial Strategies in the Food System: Updates, Opportunities, Challenges*, https://doi.org/10.1007/978-3-031-95056-8_10

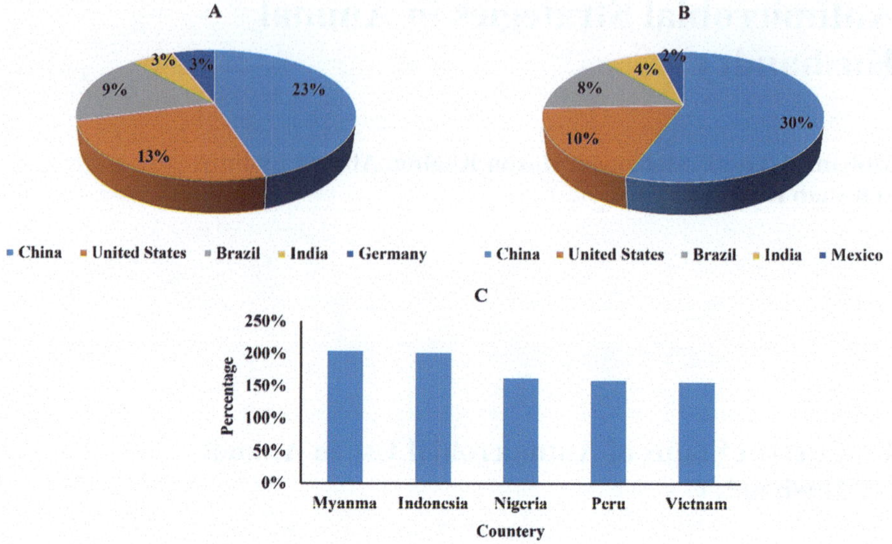

Fig. 1 Top five countries in antimicrobial consumption in livestock. (**a**) Top five countries in antimicrobial consumption in livestock in 2010. (**b**) Top five countries in antimicrobial consumption in livestock in 2030 (projected). (**c**) Top five countries with the highest anticipated percentage increases in antibiotic consumption between 2010 and 2030

Southeast Asian countries with β-lactams, tetracyclines, aminoglycosides, lincosamides, quinolones, polypeptides, amphenicols, macrolides, sulfonamides are the most commonly used antibiotics for food animals (Hosain et al., 2021). Furthermore, according to a previous estimate, the five countries with the highest percentages of global antibiotic consumption in food animal production in 2010 were China (23%), the US (13%), Brazil (9%), India (3%), and Germany (3%) (Fig. 1) (Van Boeckel et al., 2015). It is also predicted that China will account for 30% of the ranking by 2030, followed by the US (10%), Brazil (8%), India (4%), and Mexico (2%). The five countries with the highest anticipated percentage increases in antibiotic consumption by 2030 are presumably Myanmar (205%), Indonesia (202%), Nigeria (163%), Peru (160%), and Vietnam (157%), out of the 50 with the highest levels of antibiotics used in livestock in 2010.

Nevertheless, as different antimicrobials work better against different pathogens, the antimicrobials used in animal production vary by species and nation (Wu, 2018). For instance, tetracyclines (33%), penicillins (26%), and sulfonamides (11%) were the three antimicrobials most frequently marketed in food animal production in Europe in 2014 (Wu, 2018). In contrast, the top three antimicrobials sold in the USA in 2015 were tetracyclines (44%), ionophores (30%), and penicillins (6%) (FDA, 2011). According to a recent study, the UK broiler sector extensively uses the antibiotic fluoroquinolones, which were outlawed in the USA 10 years ago due to possible AMR. Cephalosporins, a family of β-lactam antibiotics produced initially from the fungus and designated for human use, are routinely utilized in the UK cattle industry, according to a recent BBC television documentary (Wu, 2018). In African countries, the circumstances are marginally different. A three-year (2010–2012) retrospective investigation on the use of antibiotics in livestock production in three states in southwest Nigeria revealed that tetracyclines, fluoroquinolones, and β-lactams/aminoglycosides were the most often used antimicrobials (33.6, 26.5, and 20.4%, respectively) (Adesokan et al., 2015). Antimicrobials such as penicillins, tetracyclines, fluoroquinolones, and aminoglycosides are commonly used by livestock keepers in various African countries, including Ghana, with the primary goal of preventing and treating illnesses in food animals (Boamah & Agyare, 2016). A recent study assessing the use of antibiotics in food animals in Rwanda, central Africa, found that up to 97% of farmers use antibiotics in their animals (Manishimwe et al., 2017), with prophylactic use being the primary justification for using these medications.

The World Health Organization has classified certain antimicrobials as critically important antimicrobials (CIA) due to their significance for human health (WHO, 2011). Furthermore, high-priority CIA (HP-CIA) is assigned to macrolides, fluoroquinolones, and third and fourth-generation cephalosporins. Sales of CIAs such as macrolides, fluoroquinolones, and third and fourth-generation cephalosporins for food-producing animals accounted for almost 10% of all sales in Europe in the 29 countries that participated in European Surveillance of Veterinary Antimicrobial Consumption (ESVAC) in 2014 (Wu, 2018). On the other hand, the UK's total use of CIAs in animals has been comparatively consistent over the last 5 years, accounting for 1.1% of all antibiotic use in 2015 (VMD, 2017). The amount of medically essential antimicrobials used in the USA went from 7687 tons in 2009 to 9702 tons in 2015. During that same period, the percentage of domestic food animal use climbed from 61% to 62%, although the use of macrolides increased by 12% (VMD, 2017). There are severe data gaps in the usage of CIA and HP-CIA in livestock animals in Asian and African nations.

Nonetheless, several independent studies revealed how extensively specific CIA and/or HP-CIA were used in livestock from Asia and Africa. According to a 2019 assessment of antimicrobial usage in livestock in Thailand, Vietnam, and Indonesia, the most often ingested types of antibiotics were sulfonamides, tetracyclines, and penicillins (Coyne et al., 2019). This result was consistent with that of the European Union, whose top three antimicrobials in 2020 were macrolides (8.8%), sulfonamides (9.9%), and penicillins (31.1% of total consumption) (EMA, 2021). Tetracyclines were also widely utilized on animals in African nations. According to an assessment, tetracyclines accounted for 32.2% of all veterinary antibiotics used by commercial chicken producers and pharmaceutical stores in Africa, whereas sulfonamides accounted for 20.8% of all antibiotics used (Mubito et al., 2014).

One of the key elements influencing antimicrobial use in animal husbandry is the common practices of antimicrobial use in various livestock systems. Antimicrobials are frequently given prophylactically in intense confinement systems, such as industrial-scale feedlots and poultry farms, to prevent disease outbreaks brought on by high stocking densities. Sometimes, feed is supplemented with subtherapeutic amounts to promote growth, increase feed efficiency, and hasten weight gain. Since antimicrobials are routinely used rather than in reaction to infections that have been detected, this approach has sparked worries regarding the development of AMR (Marshall & Levy, 2011). In contrast, pasture-based livestock systems focus on providing animals access to natural forage and open spaces. In these systems, the use of antibiotics is typically more focused and prioritizes the treatment of infections that have been diagnosed.

Nevertheless, problems occur when livestock from pasture-based systems are transferred to more intense production settings, increasing their antibiotic exposure (Graham et al., 2007). Furthermore, the absence of veterinary supervision and limited access to diagnostic services in backyard and small-scale farming operations may result in the improper administration of antibiotics. Antimicrobial misuse can be caused by a lack of resources and understanding, which increases the risk of AMR development and residues in animal products (Van Boeckel et al., 2015).

2 Drivers and Challenges of Antimicrobial Use

The use of antibiotics in animal husbandry is a complex process that is impacted by numerous variables. Antimicrobials are essential for preventing and treating livestock diseases and preserving animal health and productivity. However, there are serious problems with these medications when they are overused or misused, especially when AMR is present. Here, we examine the several elements that influence the application of antimicrobials in animal husbandry. These factors include disease dynamics, regulatory frameworks, animal welfare, economic concerns, and the larger agricultural context.

The main factor influencing the usage of antibiotics in animal husbandry is disease dynamics within animal populations. Farmers' decisions to administer these medications are influenced by the incidence and prevalence of infectious diseases, particularly in intensive farming systems where animals are frequently housed in close quarters. The objective is to minimize the impact of diseases on the general health of the herd or flock by preventing and controlling outbreaks (Albernaz-Gonçalves et al., 2022). In addition, economic factors also have a big influence on how antimicrobials are used in animal husbandry. Decisions on veterinarian care costs, possible financial losses from illness, and the worth of the animals under their care must be made by farmers. Antimicrobial use is frequently seen as an economical way to safeguard the investment in cattle by reducing financial losses brought on by slower growth rates, less efficient feed, and higher death rates (Roskam et al., 2023).

Concerns about animal welfare have also driven the use of antimicrobials to alleviate suffering caused by diseases. It is seen as a responsible practice to treat sick animals promptly to protect the animals' health under the farmer's care. In the process of making decisions, it is essential to strike a balance between ethical considerations and effective illness management (Roskam et al., 2023). This was also connected to the antimicrobial regulatory framework, which significantly influences how farmers practice their craft. Tight laws, such as those limiting specific medications and dosages, can deter improper usage, but lax enforcement can encourage misuse (WHO, 2017). Therefore, the regulatory framework is one crucial factor in determining the responsible use of antibiotics in animal husbandry. Nevertheless, farmers' understanding of diseases, how to use antibiotics, and substitutes for antibiotics affects their ability to make decisions and how much of them to use. Programs for raising awareness and educating farmers are essential in encouraging responsible use because they provide the knowledge they need to make wise decisions (Grave et al., 2010).

The use of antimicrobials in animal husbandry is a practice deeply intertwined with economic, social, and regulatory aspects. Economic factors are crucial when making judgments on the use of antibiotics in animal husbandry. Farmers constantly assess the costs and benefits of veterinary care, antibiotics, and other interventions against the possible financial losses they may incur from illness (Raboisson et al., 2020). In addition to economic considerations, the market needs and customer preferences for food goods also impact the antimicrobial usage economy. A growing trend towards antibiotic-free and responsibly produced animal products prompts farmers to consider alternative strategies.

The public's perception and consumer confidence are directly related to the social aspect of the use of antibiotics in the farming of animals. Worries about animal welfare and food safety influence social expectations for responsible consumption of antibiotics. As public awareness of the consequences of antibiotic usage grows, farmers are faced with matching their approaches to social standards. In contrast, in some rural communities, the sustainability of local economies and farmers' livelihoods are frequently linked to the usage of antimicrobials. Disease control and prevention benefit farmers as well as the people who depend on agriculture by maintaining the general stability of farming operations (Barrett et al., 2021).

The regulatory framework significantly impacts how antibiotics are used in animal husbandry. Regulations at the international and national levels seek to manage the use of antibiotics, frequently with an emphasis on halting the development of antibiotic resistance. Regulations dictate which medications, how much of them, and when to stop using them; this reflects a dedication to appropriate use. Antimicrobial rules are still difficult to establish and enforce despite regulatory attempts. Gaps in attaining optimal antimicrobial stewardship are caused by inconsistent regulatory approaches, inadequate surveillance, and disparate enforcement capacities (WHO, 2020).

Antimicrobials in animal agriculture can significantly affect animal health and food safety. One of the major concerns is the development of AMR in bacteria, as the overuse or misuse of antimicrobials in animals can lead to the emergence of resistant strains of bacteria, which may compromise the effectiveness of these drugs in treating infections in animals and humans. Furthermore, excessive antimicrobial use in animal husbandry contributes to the transmission of antimicrobial-resistant bacteria to humans, increasing the risk of zoonotic infections. Resistant bacteria can cause difficult-to-treat human infections, increasing morbidity, mortality, and healthcare costs. According to a more current estimate from the Centers for Disease

Control and Prevention, the United States has an annual economic loss of 28 billion dollars due to AMR-associated infections, with an additional 12.4 billion dollars lost to productivity (Dadgostar, 2019). The disease burden resulting from AMR is anticipated to be higher in lower- and middle-income countries, particularly in Sub-Saharan Africa. However, the absence of an effective disease reporting and surveillance system has resulted in a lack of realistic financial estimates to support this expectation. The challenges of enforcing a strict and total prohibition on the use of antibiotics in food animals, which could have detrimental effects on animal welfare, productivity, health, and food costs, exacerbate this issue. However, improvements in feed management and hygiene have been shown to lessen the detrimental impact of such prohibitions on the productivity and health of animals (Ma et al., 2021). In order to regulate the use of antibiotics in animals, the World Organization for Animal Health (OIE), the FAO, and the WHO are working together (Maron et al., 2013). Information on antibiotic intake and resistance is being gathered by international monitoring systems such as the European Antimicrobial Resistance Surveillance Network and the National Antimicrobial Resistance Monitoring System (NARMS) for Enteric Bacteria (Grundmann et al., 2011). "Combating drug resistance: no action today, no cure tomorrow" was the subject of World Health Day 2011, an initiative to encourage nations to be more aggressive in combatting bacterial resistance. Moreover, the World Health Assembly approved a worldwide action plan to address antibiotic resistance in 2015. Furthermore, firm commitments to manage antibiotic resistance were expressed at the G20 Summit in Hangzhou and the 2016 UN High-Level Meeting on Antibiotic Resistance (Qu et al., 2019).

In addition to developing resistance in microorganisms, the misuse of antibiotics in food animals has resulted in animal microbiome dysbiosis and the accumulation of drug residue in processed animal products and animal waste. Unfortunately, antimicrobial usage is still largely unregulated here, as it is in many underdeveloped nations. Consequently, inadequate protocols significantly contribute to the presence of antimicrobial residues in animal-derived food. In various nations, tetracyclines account for most antibiotic residues discovered in animal foods, with β-lactams following closely behind. This is not shocking, considering how frequently these medications are used in veterinary care across the globe. The reduced incidence of β-lactamases may be explained by their inherent hydrolysis reaction, which makes them unstable (Van et al., 2020). Since antibiotic residues can lead to several health hazards, including toxicity, teratogenicity, tumors, antimicrobial resistance, and allergic reactions, antibiotics in food animals must be strictly monitored to reduce antibiotic abuse.

Apart from the earlier ramifications of using antibiotics in animal husbandry, the overuse of antibiotics might conceal underlying health problems in animals, which can result in the emergence of chronic diseases. This may be because antibiotics may make an animal more susceptible to infection by either suppressing normal flora and making it more likely for pathogens to colonize a site (the "competitive effect") or, if given at the time of exposure to a resistant pathogen, by facilitating the infection due to a selective effect (the "selective effect"). In its Framework document, the FDA raises concern about the use of antibiotics in food animals, pointing to a possible rise in the pathogen load in the animal's digestive system and an increased risk of sickness for consumers (McEwen & Fedorka-Cray, 2002).

3 Alternatives to Conventional Antimicrobial Use

In light of the imperative to reduce environmental contamination, there arises a necessity for the exploration of alternative natural substances to supplant the conventional antimicrobial agents employed in animal industry practices.

3.1 Probiotics Alternative to Antimicrobials

Probiotics represent a viable alternative to conventional antimicrobial agents for mastitis treatment, thus offering a potential strategy to mitigate the emergence of antibiotic-resistant bacterial strains (Table 1). Urakawa et al. (2022) conducted a study demonstrating the efficacy of probiotic supplementation with *Bacillus subtilis* C-3102 in reducing the incidence of mastitis in dairy cattle with a prior history of the condition. Their intervention decreased the medication period, days of milk disposal, and somatic cell count (SCC) in milk compared to a control group lacking antimicrobial treatment. Moreover, probiotic *B. subtilis* C-3102 supplementation alleviated systemic and hepatic oxidative stress by reducing cortisol and malondialdehyde metabolites in plasma, respectively, thereby facilitating expedited recovery from mastitis in dairy cows (Urakawa et al., 2022). Additionally, probiotic supplementation enhanced innate and adaptive immune responses, as evidenced by an augmentation in $CD11c^{+}$ $CD172a^{high}$ dendritic cells and $CD4^{+}$ T cells, respectively (Urakawa et al., 2022). These findings collectively underscore the potential of probiotic *B. subtilis* C-3102 as a valuable alternative to antibiotics in controlling mastitis in dairy cattle.

Several studies have explored the potential of lactic acid bacteria (LAB) in mitigating mastitis in dairy cows. For instance, Fukuyama et al. (2020) performed a study examining the impact of LAB strains isolated from the environment of beef cattle on bovine mammary epithelial (BME) cells challenged with lipopolysaccharide (LPS) derived from *Escherichia coli* in vitro. BME cells were pre-stimulated with various LAB strains for 12, 24, and 48 h, followed by LPS challenge for 12 h (Fukuyama et al., 2020). The study's findings revealed that different LAB strains exerted suppressive effects on the inflammatory response induced by LPS in BME cells through diverse mechanisms (Fukuyama et al., 2020). Notably, *Lactobacillus acidophilus* CRL2074 was found to downregulate the expression of pro-inflammatory genes such as *IL1B*, *IL1A*, *CCL2*, *CXCL3*, and *CXCL8*, while simultaneously upregulating three negative regulators of Toll-like receptors (TLRs), including *TNFAIP3*, *SIGIRR*, and *TOLLIP* genes (Fukuyama et al., 2020). This

Table 1 Studies utilizing probiotics as alternatives to antimicrobials in animal husbandry

Probiotics	Model	Delivery	Results	References
Bacillus subtilis C-3102	Mastitis, dairy cows	Oral feeding	↓ Mastitis incidence ↓ Medication days ↓ Days of discarded milk ↓ SCC[a] ↓ Blood cortisol and TBARS[b] ↑ $CD4^{+}$ T cells and $CD11c^{+}$ $CD172a^{high}$ dendritic cells ↓ Sensitivity of peripheral granulocytes to cyclophilin	Urakawa et al. (2022)
Lactobacillus acidophilus CRL2074	Mastitis, BME[c] cells challenged with LPS[d] for 12 hours	*In vitro*	↓ Inflammatory genes (*IL1B*, *IL1A*, *CCL2*, *CXCL3*, and *CXCL8*) ↑ Negative regulators of TLRs, including TNFAIP3, SIGIRR, and TOLLIP genes	Fukuyama et al. (2020)
Lactobacillus rhamnosus CRL2084	Mastitis, BME cells challenged with LPS for 12 hours	*In vitro*	↓ Inflammatory genes (*IL1B*, *CCL2*, and *CXCL8*) ↑ *TOLLIP* gene	Fukuyama et al. (2020)
Lactobacillus casei BL23	Mastitis, BME cells infected with *S. Aureus*	*In vitro*	↓ Inflammatory genes (*TNF*, *IL1A*, *IL1B*, *IL6*, and *CXCL8*)	Souza et al. (2018)
Lactobacillus casei	Mastitis, BME cells infected with *S. Aureus* bovine strains, RF122 and Newbould305	*In vitro*	↓ Pathogenic invasion	Bouchard et al. (2013)

[a]*SCC* somatic cell count
[b]*TBARS* thiobarbituric acid reactive substances
[c]*BME* bovine mammary epithelial
[d]*LPS* lipopolysaccharide derived from *Escherichia coli*

resulted in a suppressed inflammatory state in BME cells following LPS challenge (Fukuyama et al., 2020). Furthermore, pre-stimulation of BME cells with *Lactobacillus rhamnosus* CRL2084 attenuated inflammation by downregulating *IL1B*, *CCL2*, and *CXCL8* genes while upregulating *TOLLIP* gene expression (Fukuyama et al., 2020). These findings underscore the immunomodulatory properties of LAB in mitigating inflammation associated with intramammary infections caused by key pathogens involved in bovine mastitis.

Staphylococcus aureus is a predominant pathogen implicated in mastitis among dairy cows. An in vitro investigation by Souza et al. (2018) highlighted the anti-inflammatory characteristics of *Lactobacillus casei* BL23 in BME cells infected with *S. aureus*. *L. casei* downregulated essential pro-inflammatory genes, including *TNF*, *IL1A*, *IL1B*, *IL6*, and *CXCL8*, following 8 hours of infection in *S. aureus*-stimulated

BME cells (Souza et al., 2018). In support of this study, Bouchard et al. (2013) showed earlier that *L. casei* strains reduced the invasion of BME cells by two *S. aureus* bovine strains, RF122 and Newbould305, by 60 to 80%. Importantly, this protective effect was observed without any adverse effects on the viability and morphology of BME cells (Bouchard et al., 2013). These findings elucidate the antagonistic relationship between LAB and *S. aureus* within mammary cells, potentially reducing mastitis in dairy cattle.

3.2 Plant-Based Extracts and Essential Oils as Natural Alternatives

Several studies suggested plant extracts as natural alternatives to antimicrobials, as summarized in Table 2. Oxidized β-carotene (OxBC) is a phytochemical compound naturally occurring in plants, formed due to the interaction between β-carotene and oxygen (Riley et al., 2023). OxBC exhibits alternative antimicrobial properties by imparting beneficial immune-modulating effects. For instance, (McDougall, 2021)

Table 2 Studies utilizing plant-based extracts as alternatives to antimicrobials in animal Husbandry

Treatment	Model	Delivery	Results	References
Oxidized β-carotene	Mastitis, dairy cows	Oral feeding	↓ Mastitis incidence ↓ Intramammary infection	McDougall (2021)
MPFF[a]	Mastitis, dairy cows	Intramammary infusion	↑ Cure rate	Gutiérrez-Reinoso et al. (2023)
Ricinus communis leaf ethyl acetate fraction extract	Mastitis, pathogenic bacteria	*In vitro*	↓ Pathogenic *E. coli* and *S. aureus*	Kebede and Shibeshi (2022)
Qicao Rukang powder	Mastitis, dairy cows	Oral feeding	↓ Pro-inflammatory cytokines in blood (IL-1β, IL-6, and INF-γ)	Imam et al. (2021)
Rhodomyrtus tomentosa leaves extract	Mastitis, BME[b] cells	*In vitro*	↓ Pathogenic *S. aureus*	Mordmuang et al. (2019)
Terminalia chebula extract	Mastitis, BME cells	*In vitro*	↓ Pathogenic *S. aureus, E. coli, Pseudomonas aeruginosa*, and *Bacillus megaterium*	Kher et al. (2019)
Avocado seeds extract	Mastitis, BME cells	*In vitro*	↓ Pathogens internalization by 60–80% ↓ Cytokine TNF-α proinflammatory	Báez-Magaña et al. (2019)

[a]*MPFF* micronized purified flavonoid fraction
[b]*BME* bovine mammary epithelial

demonstrated that administering 300 mg/cow/day of OxBC increased the bacteriological cure rate, suppressed the prevalence of intramammary infection, and reduced the incidence of clinical mastitis in pasture-fed cows with subclinical mastitis during the early lactation period. The micronized purified flavonoid fraction (MPFF) is a biocompatible active polyphenolic compound derived from flavonoid glycosides, demonstrating various antimicrobial, anti-inflammatory, and phlebotonic properties. (Gutiérrez-Reinoso et al., 2023) showed that intramammary infusion of MPFF enhanced the cure rate on day three post-partum in mastitis-diagnosed dairy cows infected with *Staphylococcus* spp. during late lactation. Nor-3 demethoxyisoguaiacin, an antibacterial compound isolated from the *Larrea tridentate* plant, holds promise as an alternative treatment against multidrug-resistant bacteria associated with mastitis in dairy cows (Morales-Ubaldo et al., 2022). Furthermore, (Kebede & Shibeshi, 2022) reported the bactericidal activities of the *Ricinus communis* leaf ethyl acetate fraction extract against pathogenic bacteria involved in mastitis, such as *E. coli* and *S. aureus*.

Qicao Rukang powder, a Chinese herbal compound, has demonstrated antimicrobial efficacy against various pathogens, including *Aeromonas* spp., *Pseudomonas* spp., *Acinetobacter* spp., and *Enterococcus* spp., commonly associated with subclinical mastitis in dairy cows (Imam et al., 2021). Additionally, Qicao Rukang powder supplementation to dairy cows has been shown to reduce pro-inflammatory cytokines in the blood serum, including IL-1β, IL-6, and INF-γ (Imam et al., 2021). Consequently, treated cows exhibited increased milk fat, protein, and total solids (Imam et al., 2021). These findings underscore the promising antibacterial and anti-inflammatory properties of Qicao Rukang powder, suggesting its potential as an alternative to conventional treatments for subclinical mastitis in dairy cows.

An *in vitro* study explored the effects of the ethanolic extract from *Rhodomyrtus tomentosa* leaves (RTL) on BME cells revealed RTL antibacterial activities against *S. aureus*, a key pathogen associated with bovine mastitis (Mordmuang et al., 2019). Additionally, RTL extract decreased the number of intracellular small-colony variants within BME cells (Mordmuang et al., 2019). Another *in vitro* study conducted by Kher et al. (2019) on the antibacterial activity of the ethyl acetate extract derived from the *Terminalia chebula* plant revealed significant antimicrobial effects against a spectrum of pathogens, including *S. aureus*, *E. coli*, *Pseudomonas aeruginosa*, and *Bacillus megaterium*. These bacterial strains were isolated from crossbred dairy cows with subclinical mastitis in India (Kher et al., 2019). The Mexican avocado (*Persea americana* var. drymifolia) has been identified as a source of molecules possessing anti-inflammatory and immunomodulatory properties (Báez-Magaña et al., 2019). A lipid-rich extract obtained from avocado seeds was found to reduce bacterial internalization by 60–80% within BME cells (Báez-Magaña et al., 2019). Additionally, it exhibited a suppressive effect on TNF-α proinflammatory cytokine production (Báez-Magaña et al., 2019); ultimately, these interventions hold the potential to mitigate inflammation and facilitate expedited recovery of infected BME cells.

Plant essential oils (EOs) have been extensively investigated for their antibacterial properties (Table 3). Konputtar et al. (2022) assessed the efficacy of

Table 3 Studies utilizing essential Oils (EOs) as alternatives to antimicrobials in animal husbandry

Treatment	Model	Delivery	Results	References
Lemongrass EOs	Mastitis, pathogenic *S. aureus,* and *Staphylococcus chromogenes*	*In vitro*	↑ Minimum inhibitory concentration ↓ Minimum bactericidal concentration	Konputtar et al. (2022)
Thymus vulgaris, *Origanum vulgare*, *Syzygium aromaticum*, and *Coriandrum sativum* EOs	Mastitis, pathogenic *S. aureus*	*In vitro*	↓ Pathogenic biofilm formation	Albuquerque et al. (2023)
Thymol, carvacrol, and oregano EOs	Mastitis, milk samples	*In vitro*	↓ Pathogenic *Serratia* spp. and *Proteus* spp	Tomanić et al. (2022)
Cinnamon EOs	Endometritis, pathogenic *E. coli,* and *Trueperella pyogenes*	*In vitro*	↓ Minimum inhibitory concentration ↓ Minimum bactericidal concentration	Paiano et al. (2023)
Coriander, thyme, clove, cinnamon cassia, peppermint, lemongrass, and winter savory EOs	Bovine respiratory disease, dairy calves	*In vitro*	↓ Pathogenic *Pasteurella multocida*	Bismarck et al. (2022)

EOs derived from five plant sources—lavender, lemongrass, peppermint, eucalyptus, and thyme—against isolates of *S. aureus* and *Staphylococcus chromogenes* obtained from clinical cases of mastitis in dairy cows. The study revealed that lemongrass EOs exhibited the most potent antibacterial activity, as evidenced by the highest minimum inhibitory concentration and minimum bactericidal concentration among the tested EOs (Konputtar et al., 2022). Thus, lemongrass EO is a promising alternative for combating *Staphylococcus* species associated with mastitis in dairy cows. In another study, Albuquerque et al. (2023) examined the antimicrobial potential of 20 EOs against strains of *S. aureus*, a prominent bacterium in clinical bovine mastitis. The findings highlighted the solid antimicrobial activity of EOs derived from *Thymus vulgaris*, *Origanum vulgare*, *Syzygium aromaticum*, and *Coriandrum sativum*, which effectively reduced biofilm formation by more than 80% in *S. aureus* (Albuquerque et al., 2023). Furthermore, a recent study by Tomanić et al. (2022) explored the therapeutic efficacy of three EOs against mastitis pathogens belonging to bovine *Serratia* spp. and *Proteus* spp., using milk samples collected from dairy cows with mastitis. The findings demonstrated the antibacterial activity of thymol, carvacrol, and oregano EOs against the targeted pathogens (Tomanić et al., 2022).

Clinical endometritis poses significant economic burdens on dairy farms, primarily due to its adverse effects on fertility and milk production, along with the added expenses associated with medications used for treatment. While antibiotic therapy remains the primary approach for managing uterine diseases, its widespread use in dairy herds can contribute to the emergence of bacterial resistance, leading to treatment failures. In addressing this concern, Paiano et al. (2023) explored the antibacterial activity of EOs derived from thyme, oregano, and cinnamon against *E. coli* and *Trueperella pyogenes* isolated from dairy cows infected with clinical endometritis. Paiano et al. (2023) indicated that cinnamon exhibited the lowest minimum inhibitory concentration and minimum bactericidal concentration against the targeted pathogens (Paiano et al., 2023). Consequently, cinnamon has emerged as a promising alternative for treating clinical endometritis in dairy cows. Another *in vitro* study conducted by Bismarck et al. (2022) proposed EOs supplementation as a potential alternative to antibiotic treatment for bovine respiratory disease. The research revealed that a blend of EOs derived from coriander, thyme, clove, cinnamon cassia, peppermint, lemongrass, and winter savory effectively inhibited *Pasteurella multocida*, a pathogen isolated from the respiratory tract of dairy calves (Bismarck et al., 2022).

3.3 Innovative Strategies for Disease Prevention and Control

Extracellular Vesicles (EVs) are nano-sized structures encapsulated by double lipid membranes, primarily functioning as messengers between distant cells and exerting significant influence on immune response modulation and inflammation regulation in target cells (Kalluri & LeBleu, 2020). Notably, milk and colostrum are rich sources of EVs, presenting promising potential as natural alternatives in combating microbial infections (Santoro et al., 2023). In a recent study, Mecocci et al. (2022) demonstrated that pre-treatment of the IPEC-J2 cell line with colostrum-derived EVs (colosEVs) obtained from Piedmontese cows, followed by infection with enteropathogenic *E. coli*/enteroaggregative *E. coli* (EPEC/NTEC) strains, reduced adherence of these pathogens to cell surfaces. This observation underscores the role of colosEVs in modulating interactions between hosts and pathogens. Furthermore, (Santoro et al., 2023) reported that colosEVs downregulated the expression of Toll-like receptor 5 (*TLR5*), cyclic GMP-AMP synthase (*CGAS*), and stimulator of interferon genes (*STING*) genes in cells pre-treated with colosEVs. These findings signify the immunomodulatory and anti-inflammatory effects of colosEVs in an *in vitro* model of neonatal calf diarrhea.

The Pm11 antimicrobial peptide, derived nonribosomally from pleurocidin, an alpha-helical cationic peptide family found in *Pleuronectes americanus* (Zhang et al., 2016), was investigated by Popitool et al. (2022) in an in vitro study focusing on its effects on bovine mastitis. The study demonstrated that Pm11 reduced the cell counts of mastitis-associated pathogenic bacteria, including *S. agalactiae* SCM1084 and *S. uberis* SCM1310, from 10^5 to 0 CFU/mL within 1 hour (Popitool et al., 2022). Furthermore, Pm11 exhibited efficacy against *E. coli* SCM1249 and *S. aureus* CM967, reducing their cell counts from 10^5 to 0 CFU/mL within 4 h (Popitool et al., 2022). These findings underscore the *in vitro* antibacterial activity of the Pm11 antimicrobial peptide against common bovine mastitis pathogens.

Phage therapy has recently regained attention as a potential treatment option. (Guo et al., 2021) conducted a study to evaluate the efficacy of three lytic phages, namely SYGD1 and SYGMH1 belonging to Myoviridae, and SYGE1 belonging to Autographiviridae. These phages were isolated from sewage samples collected from a dairy farm and were assessed for their activity against *E. coli* (Guo et al., 2021). The results unmasked that cocktails comprising these three phages reduced *E. coli* counts, SCC, and inflammatory factors. Furthermore, these phages alleviated the symptoms of mastitis in cattle and demonstrated comparable efficacy to antibiotic treatment (Guo et al., 2021). Thus, phages may represent a promising alternative therapy for mastitis caused by drug-resistant strains of *E. coli* in bovine mastitis.

4 Technological Innovations in Disease Management

Recent developments in technology, such as data-driven techniques, precision farming, the development of early detection diagnosis, and the monitoring and management of animal health, have changed how disease is managed in animal industry and provided more accurate and efficient methods for tracking and addressing health issues in cattle.

4.1 Role of Technology in Monitoring and Managing Animal Health

Technology is essential for controlling and tracking animal health since it offers up-to-date information and insights into animal management. The Internet of Things, which uses sensors and gadgets to gather information on animal behavior, environmental factors, and health metrics, is one of the most critical technologies in this area (Zeineldin et al., 2021). Wearable technology, such as smart-collars and ear tags with sensors, can monitor vital signs and activity levels and identify early indicators of disease in cattle (Alipio & Villena, 2023). By warning producers about possible health problems, such as variations in body temperature, breathing rate, or eating behavior, these systems allow for the proactive implementation of management techniques. Technology not only enables real-time monitoring but also data-driven decision-making for the control of animal health. Large datasets can be analyzed by advanced analytics and artificial intelligence algorithms to find patterns, trends, and correlations that might point to future health issues or disease outbreaks (De Vries et al., 2023). By utilizing these on-farm insights, farmers can make well-informed decisions about disease prevention, treatment, and general herd management techniques.

4.2 Precision Farming and Data-Driven Approaches

Precision livestock farming (PLF) is a method that uses technology to maximize animal productivity while lowering input requirements and environmental effects (García et al., 2020). PLF uses various technologies, including robotics, sensors, and data analytics, to monitor and manage each animal individually (Aquilani et al., 2022). The capacity to adjust feed and nutrition management based on the needs of individual animals is one of the main advantages of precision farming. Feed intake monitoring systems, for instance, can keep track of how much each animal eats and modify feeding schedules based on the health status to guarantee the best possible nutrition and health. Furthermore, veterinarians can employ focused treatments to stop disease outbreaks and use fewer antibiotics thanks to precision farming's data-driven methods (Morrone et al., 2022). Farmers can detect risk factors and take proactive steps to reduce them by examining historical data on disease incidence, environmental factors, and animal health metrics. Predictive modeling algorithms can predict disease outbreaks using environmental factors and animal health data (Wisnieski et al., 2021). This technology enables veterinarians to take preventive measures to stop the spread of the disease.

4.3 Advances in Diagnostics for Early Disease Detection

Improvements in diagnostic technologies have made it possible to detect diseases in animals earlier and treat them promptly by allowing for better early illness detection (Tommasoni et al., 2023). More sensitive and quick diagnostic procedures like polymerase chain reaction (PCR) and next-generation sequencing (NGS) replace traditional methods like bacterial culture and serological testing. Early diagnosis of infectious disorders is made possible by the quick and precise identification of pathogens in clinical samples by PCR-based techniques. These assays shorten diagnosis times and enable prompt treatment by directly detecting infections from clinical samples like blood, tissue, or excrement. Furthermore, by enabling high-throughput sequencing of whole microbial genomes, NGS technologies have transformed epidemiology and microbial genomics (Harris et al., 2023). This method helps establish focused control measures by offering insightful information about diseases' genetic diversity, evolution, and transmission dynamics.

Real-time monitoring, data-driven decision-making, precision farming, and sophisticated diagnostics for early illness identification have all been made possible by technology advancements, which have revolutionized disease management in animal husbandry. Farmers may maximize animal health and wellbeing, reduce disease risk, and increase productivity and sustainability in livestock production by utilizing technology.

5 Biosecurity Measures for Disease Prevention

In an era marked by increasing globalization, rapid urbanization, and intensified agricultural practices, the threat of infectious diseases looms large over human and animal populations alike. Biosecurity measures, encompassing a range of protocols and practices designed to minimize the risk of disease transmission, have emerged as indispensable tools in safeguarding public health, ensuring food safety, and preserving the integrity of ecosystems (Youssef et al., 2021). Whether in the context of livestock production, healthcare facilities, or natural ecosystems, effective biosecurity strategies play a pivotal role in mitigating the spread of pathogens, curbing the emergence of antimicrobial resistance, and averting the devastation wrought by epidemics and pandemics (Maye & Chan, 2020).

The concept of biosecurity transcends traditional boundaries, encompassing a holistic approach that integrates surveillance, risk assessment, containment, and response mechanisms. From farms and aquaculture facilities to hospitals and border checkpoints, implementing biosecurity measures demands interdisciplinary collaboration, informed decision-making, and proactive engagement with stakeholders at all levels of society (Stoneham et al., 2021; Hao et al., 2022). As emerging infectious diseases pose formidable challenges to global health security, prioritizing biosecurity as a cornerstone of disease prevention and control has never been more urgent.

This chapter explores the interconnected topics of biosecurity, antimicrobial stewardship, and integrated disease prevention strategies. By examining key components such as biosecurity infrastructure, hygiene protocols, vaccination strategies, and risk communication frameworks, we aim to elucidate the multifaceted nature of biosecurity and its pivotal role in safeguarding public health while enhancing resilience against infectious threats. Through a comprehensive understanding of biosecurity principles and their application in real-world contexts, stakeholders can better navigate the complex landscape of infectious disease management and build more resilient systems capable of withstanding the challenges of the twenty-first century.

5.1 Importance of Biosecurity in Reducing the Need for Antimicrobials

Biosecurity plays a crucial role in reducing the need for antimicrobials by preventing the introduction and spread of infectious diseases (Subasinghe et al., 2023). Effective biosecurity measures, such as proper sanitation, hygiene protocols, and restricted access, significantly reduce the risk of disease introduction into animal populations or healthcare settings. By preventing disease outbreaks, biosecurity measures minimize the need for antimicrobial treatment, as sick animals or patients are less likely to require antibiotics (Paramitadevi et al., 2023). Additionally, biosecurity practices, including quarantine procedures, isolation of sick individuals, and control of vectors, help limit the spread of pathogens within and between populations, interrupting disease transmission chains. This reduction in disease transmission ultimately decreases the demand for antimicrobials to treat infected individuals. Moreover, a healthy environment created through effective biosecurity measures supports optimal immune function in animals and humans. By reducing stressors and minimizing exposure to pathogens, biosecurity helps maintain robust immune responses, reducing susceptibility to infections and lessening the need for antimicrobial interventions (Subasinghe et al., 2023).

Furthermore, overuse and misuse of antimicrobials contribute to developing antimicrobial resistance (AMR), a major global health threat. By minimizing the occurrence of infectious diseases, biosecurity measures help preserve the efficacy of antimicrobial agents by reducing selective pressure on microbial populations. This ensures that antimicrobials effectively treat severe infections in humans and animals (Mohamed et al., 2022). Economically, implementing biosecurity protocols can lead to significant cost savings by reducing the need for antimicrobial treatments, veterinary interventions, and productivity losses associated with disease outbreaks. Moreover, preventing diseases through biosecurity measures can help maintain market access for agricultural products and reduce the economic impact of antimicrobial resistance on healthcare systems (Hoskins et al., 2023).

Biosecurity plays a pivotal role in reducing the need for antimicrobials by preventing disease transmission, preserving immune function, and minimizing the emergence and spread of antimicrobial resistance. By prioritizing biosecurity practices across various sectors, we can promote sustainable and responsible antimicrobial use while safeguarding public health, animal welfare, and environmental integrity.

5.2 *Best Practices for Farm-Level Biosecurity*

Best practices for farm-level biosecurity are essential to prevent the introduction and spread of diseases among livestock populations (Table 4). Restricted access entails limiting access to the farm premises to authorized personnel and visitors. Implementing controlled entry points and ensuring proper signage indicating biosecurity protocols are critical components of this measure (Maye & Chan, 2020). Quarantine and testing involve isolating newly acquired animals before introducing them to the main herd or flock. Conducting appropriate testing and health screenings helps identify any potential diseases before integration (Narmandakh & Sakurai, 2021).

Sanitation and hygiene are paramount for maintaining high standards of cleanliness throughout the farm, including animal housing, equipment, and facilities. Regularly cleaning and disinfecting surfaces, equipment, and vehicles minimizes the risk of disease transmission (Paramitadevi et al., 2023). Separation of species involves keeping different livestock species separate to prevent cross-species

Table 4 Best practices for farm-level biosecurity

Best practice	Description	References
Restricted access	Limit authorized personnel and visitors' access to the farm premises. Implement controlled entry points and proper signage	Maye and Chan (2020)
Quarantine and testing	Isolate newly acquired animals before introducing them to the main herd or flock. Conduct appropriate testing and health screenings	Narmandakh and Sakurai (2021)
Sanitation and hygiene	Maintain high standards of cleanliness throughout the farm. Regularly clean and disinfect surfaces, equipment, and vehicles	Paramitadevi et al. (2023)
Separation of species	Keep different livestock species separate to prevent cross-species transmission. Use separate equipment, feed, and water sources	Narmandakh and (Sakurai 2021)
Biosecurity zones	Divide the farm into areas based on disease risk. Implement stricter measures in high-risk areas such as entry points and housing	Narmandakh and Sakurai (2021)
Vector control measures	Implement strategies to control pests and vectors that can transmit diseases. Use traps, barriers, and appropriate insecticides	Narmandakh and Sakurai (2021)
Personnel training	Provide comprehensive training to farm workers on biosecurity protocols and practices. Ensure understanding and adherence to procedures	Delabbio (2006)
Health monitoring	Establish a regular health monitoring program for livestock. Identify and isolate sick animals promptly to prevent the spread of illness	Perez et al. (2023)
Vaccination programs	Implement tailored vaccination programs for specific disease risks. Consult with a veterinarian for appropriate vaccination schedules	Jama et al. (2022)
Record-keeping	Maintain accurate records of animal movements, health status, vaccinations, and biosecurity-related activities	Sanderson and Gnad (2002)
Emergency preparedness	Develop and regularly update an emergency response plan for disease outbreaks or health emergencies. Ensure personnel are familiar with it	Narmandakh and Sakurai (2021)

transmission of diseases. Using separate equipment, feed, and water sources for different groups of animals is essential for this practice. Biosecurity zones divide the farm into areas based on disease risk and animal health status. Implementing stricter biosecurity measures in high-risk areas such as entry points and animal housing facilities is crucial. Vector control measures involve implementing strategies to control pests and vectors that can transmit diseases, such as rodents, flies, and mosquitoes. Using traps, barriers, and appropriate insecticides helps minimize their presence on the farm (Narmandakh & Sakurai, 2021).

Personnel training is vital for providing comprehensive training to farm workers on biosecurity protocols and practices. Ensuring that all personnel understand their role in maintaining biosecurity and adhere to proper procedures is essential for effective implementation. Health monitoring entails establishing a regular health monitoring program for livestock, including routine check-ups and disease surveillance. Promptly identifying and isolating sick animals helps prevent the spread of illness (Perez et al., 2023). Vaccination programs involve implementing vaccination programs tailored to the specific disease risks in the region. Consulting with a veterinarian to develop a vaccination schedule appropriate for the livestock species and production system is essential. Record-keeping is crucial for maintaining accurate records of animal movements, health status, vaccinations, and biosecurity-related activities. These records provide valuable information for disease management and traceability (Sanderson & Gnad, 2002).

Emergency preparedness involves developing and regularly updating an emergency response plan to address disease outbreaks or other health emergencies. Ensuring that all personnel are familiar with the plan and know their roles and responsibilities is essential. By implementing these best practices for farm-level biosecurity, producers can minimize the risk of disease introduction and spread, protect animal health and welfare, and safeguard the sustainability of their operations.

5.3 Integrated Approaches to Disease Prevention

Integrated approaches to disease prevention involve combining strategies and interventions across multiple sectors to address the complex nature of disease transmission and emergence. These approaches recognize the interconnectedness of human, animal, and environmental health and aim to mitigate disease risks holistically (Ali & Katz, 2015). One Health is a collaborative approach acknowledging the interconnection between human health, animal health, and the environment. Integrated disease prevention efforts leverage expertise from multiple disciplines, including medicine, veterinary science, ecology, and environmental health, to address health challenges at the human-animal-environment interface (Habib et al., 2021).

Surveillance and early warning systems involve integrated disease surveillance systems that monitor potential human and animal health threats in real time. By collecting and analyzing data on disease occurrence, outbreaks, and trends, these systems enable early detection and rapid response to emerging infectious diseases, zoonotic infections, and other health threats (Rodríguez-Prieto et al., 2015). Risk assessment and management frameworks integrate various methodologies to evaluate disease outbreaks' likelihood and potential impact across diverse sectors and populations. By identifying high-risk regions, vulnerable demographics, and critical control points, these assessments guide targeted interventions and resource allocation for effectively mitigating disease risks (Robertson, 2020).

Vaccination and immunization programs are fundamental to disease prevention efforts, providing immunity against various infectious diseases in humans and animals. Integrated vaccination strategies consider the needs of both human and animal populations, aiming to achieve herd immunity and reduce disease transmission within communities (Hutubessy et al., 2023). Vector control and environmental management involve integrated vector control measures targeting vectors such as mosquitoes, ticks, and flies that transmit infectious diseases to humans and animals. These efforts reduce vector breeding sites and minimize disease transmission by implementing environmental management strategies, such as habitat modification and water sanitation (WHO, 2020).

Health education and behavior change communication encompass integrated health education programs promoting awareness, knowledge, and behavioral change among communities to prevent disease transmission. By fostering informed decision-making, promoting hygiene practices, and addressing cultural beliefs and practices, these programs empower individuals and communities to adopt preventive measures. Capacity building and training prioritize integrated approaches to strengthen healthcare systems, veterinary services, and public health infrastructure. These efforts improve disease detection, surveillance, and response capabilities by enhancing the skills and expertise of healthcare workers, veterinarians, and community health workers (Nicholls et al., 2020; Mohamed & Habib, 2023). By advocating for multisectoral collaboration, resource mobilization, and policy coherence, stakeholders can create an enabling environment for effective disease prevention and control (Janakiram & Dye, 2020).

By integrating these approaches into comprehensive disease prevention strategies, stakeholders can address the complex and interconnected nature of health challenges, promote sustainable development, and safeguard the health and well-being of populations worldwide.

6 Case Studies and Successful Implementations

Antimicrobial resistance (AMR) poses a significant threat to both animal and human health, prompting the need for effective antimicrobial strategies in animal husbandry (Mohamed & Habib, 2023). Several case studies and successful implementations highlight the importance of proactive measures in combating AMR while maintaining animal welfare and sustainable food production.

Denmark's Antibiotic Reduction Program: Denmark is often cited as a success story in reducing antimicrobial use in animal production. In the 1990s, Denmark implemented a national strategy to curb antimicrobial use in agriculture, including strict regulations on antibiotic prescriptions, surveillance programs, and education initiatives for farmers. Through these efforts, Denmark successfully reduced antibiotic consumption in animal husbandry while maintaining high animal health and productivity (Levy, 2014).

The Netherlands' "One Health" Approach: The Netherlands adopted a comprehensive "One Health" approach to address antimicrobial resistance, recognizing the interconnectedness of human, animal, and environmental health. This approach involves collaboration between human and veterinary healthcare sectors and environmental stakeholders to promote responsible antimicrobial use in animal husbandry. As a result, the Netherlands has seen significant reductions in antimicrobial consumption in livestock production, accompanied by improved animal health and food safety (van Herten et al., 2020).

Norwegian Salmon Farming: The Norwegian aquaculture industry has implemented stringent antimicrobial stewardship practices in salmon farming. By focusing on disease prevention measures such as improved hygiene, vaccination programs, and optimized husbandry practices, Norwegian salmon producers have minimized the need for antimicrobial treatments. As a result, Norway has maintained a low prevalence of antimicrobial resistance in its aquaculture sector while ensuring sustainable growth in salmon production (Lillehaug et al., 2018).

Organic Livestock Production: Organic livestock production systems often prioritize preventive healthcare practices and limit the use of antimicrobials. Case studies from organic farms demonstrate successful implementation of holistic management approaches, including pasture-based systems, diversified crop-livestock integration, and reduced stocking densities. These practices promote animal health and resilience, reducing the reliance on antimicrobial treatments while meeting consumer demand for sustainably produced food (Kijlstra & Eijck, 2006).

Global Antibiotic Reduction Initiatives: Various international initiatives and collaborations aim to address antimicrobial resistance in animal husbandry globally. Organizations such as the World Organization for Animal Health (OIE) and the Food and Agriculture Organization of the United Nations (FAO) work with countries to develop guidelines, promote best practices, and build capacity for responsible antimicrobial use in livestock production. Successful implementation of these initiatives relies on multi-stakeholder engagement, policy support, and investment in research and infrastructure (Badau, 2021; Wernli et al., 2022).

Case studies and successful implementations in antimicrobial strategies in animal husbandry demonstrate the effectiveness of proactive measures in reducing antimicrobial use while ensuring animal health, welfare, and sustainable food production. By adopting a One Health approach, fostering collaboration, and promoting best practices, countries and stakeholders can mitigate the threat of antimicrobial resistance and safeguard the future of animal agriculture. Table 5 presents a summary of the antimicrobial reduction strategies under discussion.

6.1 Successful Initiatives in Reducing Antimicrobial Use

Implementing antibiotic stewardship programs in healthcare settings, along with public awareness campaigns, regulatory measures, alternative therapies, diagnostic tools, education and training, multidisciplinary collaboration, and global action plans, has proven highly effective in combating antimicrobial resistance. Antibiotic stewardship programs involve preauthorization requirements, appropriate antibiotic use guidelines, and regular monitoring of prescribing practices. Public awareness campaigns stress prudent antibiotic use, while regulatory measures restrict antibiotic use in agriculture and veterinary medicine. Alternative therapies like probiotics and phage therapy offer alternatives to traditional antibiotics, and rapid diagnostic tests

Table 5 Summary of antimicrobial reduction strategies

Country/ program	Approach and initiatives	Outcomes and achievements	References
Denmark: Antibiotic reduction program	Strict regulations on antibiotic prescriptions Surveillance programs Education initiatives	Reduced overall antibiotic consumption in animal husbandry Maintained high levels of animal health and productivity	Levy (2014)
Netherlands: The Netherlands' "one health" approach	Comprehensive "one health" approach Collaboration between human, animal, and environmental sectors	Significant reductions in antimicrobial consumption in livestock production Improvements in animal health and food safety	van Herten et al. (van Herten et al., 2020)
Norwegian: Norwegian Salmon farming	Stringent antimicrobial stewardship practices Focus on disease prevention measures	Low prevalence of antimicrobial resistance in the aquaculture sector Sustainable growth in salmon production	Lillehaug et al. (2018)
Organic livestock production	Prioritize preventive healthcare practices Limit the use of antimicrobials	Successful implementation of holistic management approaches Promotion of animal health and resilience Meeting consumer demand for sustainably produced food	Kijlstra and Eijck (2006)
Global antibiotic reduction initiatives	International collaborations and initiatives Guidelines development Capacity building	Address antimicrobial resistance on a global scale Promotion of best practices in livestock production Multi-stakeholder engagement and policy support	Badau (2021); Wernli et al. (2022)

help reduce unnecessary prescriptions. Education and training empower healthcare providers to make evidence-based decisions, while multidisciplinary collaboration ensures comprehensive interventions. Global action plans provide strategic frameworks for addressing antimicrobial resistance comprehensively, emphasizing collaboration, surveillance, research, and innovation at national and international levels (Pollack & Srinivasan, 2014; Satterfield et al., 2020; Vasala et al., 2020; Silva et al., 2020; Munkholm & Rubin, 2020; More, 2020; Godman et al., 2021; Habib & Mohamed, 2022). Table 6 provides a concise overview of the various strategies, along with their descriptions and references.

6.2 *Showcasing Farms or Regions with Effective Strategies*

Showcasing farms or regions with practical strategies in antimicrobial stewardship provides valuable insights into successful approaches to combating antimicrobial

Table 6 Overview of various strategies for successful initiatives in reducing antimicrobial use

Strategy	Description	References
Preauthorization requirements	Programs involving preauthorization requirements, guidelines for appropriate antibiotic use, and regular monitoring of antibiotic prescribing practices.	Pollack and Srinivasan (2014)
Public awareness campaigns	Campaigns aimed at educating healthcare providers and the general public about the dangers of antibiotic overuse, emphasizing appropriate antibiotic use.	Godman et al. (2021)
Regulatory measures	Regulatory measures implemented by governments and health agencies to restrict antibiotic use in agriculture and veterinary medicine.	More (2020)
Alternative therapies	Promotion of alternative therapies like probiotics, phage therapy, and bacteriophage-derived enzymes as alternatives to traditional antibiotic treatments.	Silva et al. (2020)
Diagnostic tools	Development and adoption of rapid diagnostic tests to identify the cause of infection and distinguish between bacterial and viral infections, reducing unnecessary antibiotic prescriptions.	Vasala et al. (2020)
Education and training	Continuing education and training programs for healthcare providers on antimicrobial stewardship principles and best practices to reduce inappropriate antibiotic prescribing.	Satterfield et al. (2020)
Multidisciplinary collaboration	Collaboration among healthcare providers, policymakers, veterinarians, pharmacists, researchers, and other stakeholders to address factors contributing to antimicrobial resistance.	Habib and Mohamed et al. (2022)
Global action plans	Global initiatives providing strategic frameworks for countries to address antimicrobial resistance comprehensively, emphasizing collaboration, surveillance, research, and innovation.	Munkholm and Rubin (2020)

resistance in animal husbandry (Table 7). Examples include Danish Poultry Producers, whose proactive implementation of antimicrobial stewardship practices has significantly reduced antibiotic reliance while maintaining high standards of animal welfare and food safety (Emborg et al., 2001); Norwegian Salmon Aquaculture, recognized for sustainable practices and responsible antimicrobial use, maintaining low levels of antimicrobial resistance through preventive measures like robust vaccination programs and strict hygiene protocols (Lillehaug et al., 2018); Organic Dairy Farms worldwide, demonstrating sustainable approaches through holistic management practices and minimizing antibiotic use while prioritizing animal welfare and environmental stewardship (Morgans et al., 2021); Dutch Pig Farms, achieving substantial reductions in antimicrobial consumption through strict regulations, targeted interventions, and innovative management practices (Bergevoet, 2019); and Swedish Broiler Production, demonstrating significant reductions in antimicrobial consumption through stringent regulations, adherence to strict antibiotic prescribing guidelines, and investment in research and innovation (Björkman et al., 2021). These examples provide valuable lessons for producers and

Table 7 Examples of effective antimicrobial stewardship in animal husbandry

Farm or region	Description	References
Danish poultry producers	Denmark's poultry industry has implemented antimicrobial stewardship practices, including enhanced biosecurity protocols, routine health monitoring, and alternative feed additives to promote gut health and immunity. Danish poultry farms have significantly reduced antimicrobial use while maintaining high animal welfare and food safety standards.	Emborg et al. (2001)
Norwegian Salmon aquaculture	Norway's salmon aquaculture industry prioritizes preventive measures such as robust vaccination programs, strict hygiene protocols, and optimized feeding practices to minimize antimicrobial treatments. These efforts have led to low levels of antimicrobial resistance in salmon farming, ensuring the industry's long-term viability.	Lillehaug et al. (2018)
Organic dairy farms	Organic dairy farms worldwide emphasize holistic management practices, including pasture-based systems, outdoor grazing access, and low-stress handling techniques to promote animal health and reduce disease incidence. These farms prioritize preventive healthcare measures and minimize antibiotic use, demonstrating sustainable dairy production.	Morgans et al. (2021)
Dutch pig farms	The Netherlands' pig farming industry has made significant strides in reducing antimicrobial use through strict regulations, targeted interventions, and industry-wide collaboration. Innovative management practices, improved biosecurity measures, and selective breeding for disease resistance have led to substantial antimicrobial consumption reductions.	Bergevoet (2019)
Swedish broiler production	Sweden's broiler production industry implements stringent regulations and industry initiatives to reduce antimicrobial use. Adherence to strict antibiotic prescribing guidelines, robust biosecurity measures, and investment in research for alternative disease management strategies have resulted in significant antimicrobial consumption reductions.	Björkman et al. (2021)

stakeholders, promoting collaboration, sharing best practices, and supporting innovation to combat antimicrobial resistance and ensure the sustainability of animal production systems.

6.3 *Extracting Lessons Learned and Best Practices*

Extracting lessons learned and best practices from farms or regions with practical strategies in antimicrobial stewardship offers valuable insights for promoting responsible antimicrobial use in animal husbandry. Key lessons and best practices include adopting a holistic approach, emphasizing preventive measures such as robust biosecurity protocols, vaccination programs, and optimized management practices to minimize the need for antimicrobial treatments and promote animal health and welfare (Asiimwe et al., 2021) (Table 8). Collaboration among stakeholders, including farmers, veterinarians, researchers, policymakers, and industry associations, is crucial for promoting best practices in antimicrobial stewardship. Successful farms and regions actively share knowledge, participate in collaborative networks, and leverage expertise from multiple disciplines to develop and implement effective strategies (Mohamed et al., 2019). Data-driven decision-making is prioritized, with regular tracking of antimicrobial usage, disease prevalence, and treatment outcomes enabling farms to identify trends, evaluate interventions' effectiveness, and make evidence-based adjustments to their practices (Singh et al., 2024). Continuous improvement is a hallmark, with regular evaluation of antimicrobial use practices, impact assessment on animal health and productivity, and identification of optimization opportunities (Morgans et al., 2021). Education and training are crucial in promoting responsible antimicrobial use among stakeholders.

Successful farms and regions invest in training programs, workshops, and educational resources to increase awareness about antimicrobial resistance, promote best disease prevention and management practicesand empower stakeholders to make informed decisions regarding antimicrobial use (Levy, 2014). Policy support and regulatory frameworks are essential for creating an enabling environment for antimicrobial stewardship. Successful farms and regions benefit from clear guidelines, regulations, and incentives that promote responsible antimicrobial use, encourage compliance with best practices, and support innovation and capacity building (More, 2020). Consumer engagement and transparency are increasingly important, with successful farms and regions openly communicating with consumers about their antimicrobial use practices, demonstrating commitment to animal welfare and food safety, and providing transparent information about their production methods (Regan et al., 2023). By extracting lessons learned and best practices, stakeholders can identify actionable insights, replicate successful approaches, and drive continuous improvement in antimicrobial use practices across the animal agriculture sector.

7 Challenges and Future Outlook

Antimicrobial reduction strategies in animal husbandry are critical for addressing concerns related to AMR, ensuring animal welfare, and safeguarding public health. However, the successful implementation of such strategies faces persistent obstacles that warrant careful consideration. The financial burden of alternative treatments is one enduring barrier to antimicrobial reduction approaches. Some farmers, especially those with low means, may be discouraged from old antimicrobial-dependent practices due to the more significant initial expenses associated with implementing biosecurity measures, improving animal management practices, and adopting vaccination programs (Queenan et al., 2016). Additionally, more enduring issues are caused by farmers' and veterinary professionals' lack of knowledge and training. Antimicrobial reduction measures in animal husbandry are ineffective because of a

Table 8 Key lessons and best practices in antimicrobial stewardship

Key lessons and best practices	Description	References
Holistic approach	Successful farms and regions adopt a holistic approach to antimicrobial stewardship, emphasizing preventive measures such as robust biosecurity protocols, vaccination programs, and optimized management practices. These farms minimize the need for antimicrobial treatments and promote animal health and welfare by addressing underlying factors contributing to disease incidence.	Asiimwe et al. (2021)
Collaboration and knowledge sharing	Collaboration among stakeholders, including farmers, veterinarians, researchers, policymakers, and industry associations, is crucial for promoting best practices in antimicrobial stewardship. Successful farms and regions actively share knowledge, participate in collaborative networks, and leverage expertise from multiple disciplines to develop and implement effective strategies.	Mohamed et al. (2019)
Data-driven decision making	Farms and regions prioritize data collection, monitoring, and analysis to inform decision-making in antimicrobial stewardship. By regularly tracking antimicrobial usage, disease prevalence, and treatment outcomes, these farms identify trends, evaluate intervention effectiveness, and adjust evidence-based management practices.	Singh et al. (2024)
Continuous improvement	Continuous improvement is fundamental to successful antimicrobial stewardship programs. Farms and regions regularly evaluate antimicrobial use practices, assess their impact on animal health and productivity, and identify optimization opportunities. By fostering a culture of continuous learning and innovation, these farms stay abreast of emerging challenges and opportunities in antimicrobial stewardship.	Morgans et al. (2021)
Education and training	Education and training are crucial for promoting responsible antimicrobial use among stakeholders. Successful farms and regions invest in training programs, workshops, and educational resources to increase awareness about antimicrobial resistance, promote best disease prevention and management practices, and empower stakeholders to make informed decisions regarding antimicrobial use.	Levy (2014)
Policy support and regulatory frameworks	Policy support and regulatory frameworks create an enabling environment for antimicrobial stewardship. Successful farms and regions benefit from clear guidelines, regulations, and incentives that promote responsible antimicrobial use, encourage compliance with best practices, and support innovation and capacity building.	More (2020)
Consumer engagement and transparency	Consumer engagement and transparency are vital considerations in antimicrobial stewardship. Successful farms and regions communicate openly with consumers	Regan et al. (2023)

(continued)

Table 8 (continued)

Key lessons and best practices	Description	References
	about antimicrobial use practices, demonstrate commitment to animal welfare and food safety, and provide transparent information about their production methods. By building trust and credibility, these farms contribute to promoting sustainable and responsible agriculture.	

lack of training on disease prevention and management strategies and awareness of alternative options.

Nevertheless, stakeholders are uncertain about antimicrobial reduction approaches because of the lack of established norms and legislation. Uncertainty regarding compliance and the effectiveness of alternative tactics may cause farmers to use inconsistent practices due to the absence of defined regulatory frameworks and benchmarks (OIE, 2016). Additionally, resistance to change within the agricultural community and the interconnected nature of the global food supply chain are persistent obstacles (O'Neill, 2015). Farmers who are used to using antimicrobials in traditional ways may be reluctant to adopt new methods because of perceived dangers, unknowns, or worries about possible harm to the production and health of their animals. In addition, some farmers could be reluctant to use antimicrobial reduction strategies because they worry about being competitive in the global market, where laws and regulations can change (O'Neill, 2015).

The future of antimicrobial strategies in animal husbandry is marked by a paradigm shift towards sustainable and holistic approaches to address emerging challenges. Increasing concerns over antibiotic resistance have prompted reevaluating conventional practices and exploring alternatives such as probiotics, prebiotics, and phytogenics. Probiotics, consisting of beneficial microorganisms, have gained attention for their potential to enhance animal health and productivity while reducing the need for antibiotics (Leistikow & Bal, 2020). On the other hand, prebiotics support the growth of beneficial bacteria in the gut, promoting a balanced microbial community. Additionally, phytogenics, derived from plants, exhibit antimicrobial properties and contribute to improved animal performance (Mnisi et al., 2022). These alternatives align with the global trend toward reducing antibiotics in animal agriculture and emphasize the importance of preventive strategies to maintain animal health. This approach enables farmers to make informed decisions, leading to targeted antimicrobial use and better overall livestock health management. Furthermore, there is a growing emphasis on vaccination as a preventive measure, with ongoing research focusing on developing effective vaccines against prevalent diseases in livestock.

Addressing emerging issues in antimicrobial strategies also involves focusing on precision farming and digital technologies. Precision farming integrates data-driven decision-making to optimize animal health and production. Implementation of sensors, monitoring devices, and data analytics allows for real-time assessment of animal well-being, enabling timely intervention and reducing the need for broad-spectrum antibiotics (Yamin et al., 2023). Furthermore, advancements in genetic selection and breeding techniques contribute to developing animals with enhanced disease resistance, reducing the reliance on antimicrobials. As the industry navigates these future trends, a comprehensive overview is crucial for stakeholders to stay informed and adapt to sustainable practices. This overview encompasses a wide range of disciplines, from microbiology and veterinary science to data analytics and genetics, reflecting the interdisciplinary nature of the evolving landscape of antimicrobial strategies in animal husbandry.

Cross-species transmission of antimicrobial-resistant bacteria is a significant concern, necessitating a holistic One Health approach that considers the interconnectedness of human, animal, and environmental health (Khalifa et al., 2020). The globalization of supply chains further complicates the scenario, demanding international cooperation and coordinated efforts to address AMR globally. Consumer awareness and demands for sustainable and responsibly produced food products also influence farming practices, pressuring the industry to balance economic viability and responsible antimicrobial use.

In conclusion, navigating the future of antimicrobial strategies in animal husbandry requires a multidisciplinary approach. Adapting to emerging trends and addressing challenges demands collaboration between researchers, farmers, policymakers, and consumers. By embracing innovative solutions and staying informed about the evolving landscape, the agricultural sector can contribute to a sustainable and resilient animal and human health future.

References

Adesokan, H. K., Akanbi, I. O., Akanbi, I. M., & Obaweda, R. A. (2015). Pattern of antimicrobial usage in livestock animals in South-Western Nigeria: The need for alternative plans. *The Onderstepoort Journal of Veterinary Research, 82*. https://doi.org/10.4102/ojvr.v82i1.816

Albernaz-Gonçalves, R., Olmos Antillón, G., & Hötzel, M. J. (2022). Linking animal welfare and antibiotic use in pig farming-a review. *Animals, 12*, 216. https://doi.org/10.3390/ani12020216

Albuquerque, K. R. S., Purgato, G. A., Píccolo, M. S., Rodrigues, F. F., Pizziolo, V. R., Muñoz, G. D., Rossi, C. C., & Diaz, M. A. N. (2023). Formulations of essential oils obtained from plants traditionally used as condiments or traditional medicine active against *Staphylococcus aureus* isolated from dairy cows with mastitis. *Letters in Applied Microbiology, 76*. https://doi.org/10.1093/lambio/ovad034

Ali, A., & Katz, D. L. (2015). Disease prevention and health promotion: how integrative medicine fits. *American Journal of Preventive Medicine, 49*, S230–S240. https://doi.org/10.1016/j.amepre.2015.07.019

Alipio, M., & Villena, M. L. (2023). Intelligent wearable devices and biosensors for monitoring cattle health conditions: A review and classification. *Smart Health, 27*, 100369. https://doi.org/10.1016/j.smhl.2022.100369

Aquilani, C., Confessore, A., Bozzi, R., Sirtori, F., & Pugliese, C. (2022). Review: precision livestock farming technologies in pasture-based livestock systems. *Animal, 16*, 100429. https://doi.org/10.1016/j.animal.2021.100429

Asiimwe, B. B., Kiiru, J., Mshana, S. E., Neema, S., Keenan, K., Kesby, M., Mwanga, J. R., Sloan, D. J., Mmbaga, B. T., Smith, V. A., Gillespie, S. H., Lynch, A. G., Sandeman, A., Stelling, J., Elliott, A., Aanensen, D. M., Kibiki, G. E., Sabiiti, W., & Holden, M. T. G. (2021). Protocol for an interdisciplinary cross-sectional study investigating the social, biological and community-level drivers of antimicrobial resistance (AMR): holistic approach to unravel antibacterial resistance in East Africa (HATUA). *BMJ Open, 11*, e041418. https://doi.org/10.1136/bmjopen-2020-041418

Badau, E. (2021). A one health perspective on the issue of the antibiotic resistance. *Parasite, 28*, 16. https://doi.org/10.1051/parasite/2021006

Báez-Magaña, M., Ochoa-Zarzosa, A., Alva-Murillo, N., Salgado-Garciglia, R., & López-Meza, J. E. (2019). Lipid-rich extract from Mexican avocado seed (persea Americana var. Drymifolia) reduces *Staphylococcus aureus* internalization and regulates innate immune response in bovine mammary epithelial cells. *Journal of Immunology Research, 2019*, 1–10. https://doi.org/10.1155/2019/7083491

Barrett, J. R., Innes, G. K., Johnson, K. A., Lhermie, G., Ivanek, R., Greiner Safi, A., & Lansing, D. (2021). Consumer perceptions of antimicrobial use in animal husbandry: A scoping review. *PLoS One, 16*, e0261010. https://doi.org/10.1371/journal.pone.0261010

Bergevoet, R. H. M. (2019). Economics of antibiotic usage on Dutch farms: The impact of antibiotic reduction on economic results of pig and broiler farms in The Netherlands. In *Wageningen economic research* (pp. 19–26).

Bismarck, D., Becker, J., Müller, E., Becher, V., Nau, L., & Mayer, P. (2022). Screening of antimicrobial activity of essential oils against bovine respiratory pathogens – Focusing on *Pasteurella multocida. Planta Medica, 88*, 274–281. https://doi.org/10.1055/a-1726-9291

Björkman, I., Röing, M., Sternberg Lewerin, S., Stålsby Lundborg, C., & Eriksen, J. (2021). Animal production with restrictive use of antibiotics to contain antimicrobial resistance in Sweden-a qualitative study. *Frontiers in Veterinary Science, 7*. https://doi.org/10.3389/fvets.2020.619030

Boamah, V. E., & Agyare, C. (2016). Antibiotic practices and factors influencing the use of antibiotics in selected poultry farms in Ghana. *Journal of Antimicrobial Agents, 2*. https://doi.org/10.4172/2472-1212.1000120

Bouchard, D. S., Rault, L., Berkova, N., Le Loir, Y., & Even, S. (2013). Inhibition of *Staphylococcus aureus* invasion into bovine mammary epithelial cells by contact with live *Lactobacillus casei. Applied and Environmental Microbiology, 79*, 877–885. https://doi.org/10.1128/AEM.03323-12

Coyne, L., Arief, R., Benigno, C., Giang, V. N., Huong, L. Q., Jeamsripong, S., Kalpravidh, W., McGrane, J., Padungtod, P., Patrick, I., Schoonman, L., Setyawan, E., Harja Sukarno, A., Srisamran, J., Ngoc, P. T., & Rushton, J. (2019). Characterizing antimicrobial use in the livestock sector in three south east Asian countries (Indonesia, Thailand, and Vietnam). *Antibiotics, 8*, 33. https://doi.org/10.3390/antibiotics8010033

Dadgostar, P. (2019). Antimicrobial resistance: implications and costs. *Infection and Drug Resistance, 12*, 3903–3910. https://doi.org/10.2147/IDR.S234610

De Vries, A., Bliznyuk, N., & Pinedo, P. (2023). Invited review: examples and opportunities for artificial intelligence (AI) in dairy farms. *Applied Animal Science, 39*, 14–22. https://doi.org/10.15232/aas.2022-02345

Delabbio, J. (2006). How farm workers learn to use and practice biosecurity. *Journal of Extension, 44*, 6.

Elliott, K. A., Kenny, C., & Madan, J. (2017). *A global treaty to reduce antimicrobial use in livestock*. Center for Global Development.

Emborg, H., Ersbøll, A., & Wegener, H. (2001). The effect of discontinuing the use of antimicrobial growth promoters on the productivity in the Danish broiler production. *Preventive Veterinary Medicine, 50*, 53–70.

European Medicines Agency (EMA). (2021). European surveillance of veterinary antimicrobial consumption. Sales of veterinary antimicrobial agents in 31 European countries in 2019 and

2020. Available at: https://www.ema.europa.eu/en/documents/report/sales-veterinary-antimicrobial-agents-31-european-countries-2019-2020-trends-2010-2020-eleventh_en.pdf

Food and Drug Administration (FDA). (2011). *Summary report on antimicrobials sold or distributed for use in food-producing animals*. FDA.

Fukuyama, K., Islam, M. A., Takagi, M., Ikeda-Ohtsubo, W., Kurata, S., Aso, H., Vignolo, G., Villena, J., & Kitazawa, H. (2020). Evaluation of the immunomodulatory ability of lactic acid bacteria isolated from feedlot cattle against mastitis using a bovine mammary epithelial cells in vitro assay. *Pathogens, 9*, 410. https://doi.org/10.3390/pathogens9050410

García, R., Aguilar, J., Toro, M., Pinto, A., & Rodríguez, P. (2020). A systematic literature review on the use of machine learning in precision livestock farming. *Computers and Electronics in Agriculture, 179*, 105826. https://doi.org/10.1016/j.compag.2020.105826

Godman, B., Egwuenu, A., Haque, M., Malande, O. O., Schellack, N., Kumar, S., Saleem, Z., Sneddon, J., Hoxha, I., Islam, S., Mwita, J., et al. (2021). Strategies to improve antimicrobial utilization with a special focus on developing countries. *Lifestyles, 11*, 528. https://doi.org/10.3390/life11060528

Graham, J. P., Boland, J. J., & Silbergeld, E. (2007). Growth promoting antibiotics in food animal production: an economic analysis. *Public Health Reports, 122*, 79–87. https://doi.org/10.1177/003335490712200111

Grave, K., Torren-Edo, J., & Mackay, D. (2010). Comparison of the sales of veterinary antibacterial agents between 10 European countries. *The Journal of Antimicrobial Chemotherapy, 65*, 2037–2040. https://doi.org/10.1093/jac/dkq247

Grundmann, H., Klugman, K. P., Walsh, T., Ramon-Pardo, P., Sigauque, B., Khan, W., Laxminarayan, R., Heddini, A., & Stelling, J. (2011). A framework for global surveillance of antibiotic resistance. *Drug Resist, 14*, 79–87. https://doi.org/10.1016/j.drup.2011.02.007

Guo, M., Gao, Y., Xue, Y., Liu, Y., Zeng, X., Cheng, Y., Ma, J., Wang, H., Sun, J., Wang, Z., & Yan, Y. (2021). Bacteriophage cocktails protect dairy cows against mastitis caused by drug resistant *Escherichia coli* infection. *Frontiers in Cellular and Infection Microbiology, 11*. https://doi.org/10.3389/fcimb.2021.690377

Gutiérrez-Reinoso, M. A., Uquilla, J. B., Guamaní, J. L., Caiza, Á. E., Carrera, R. P., & Garcia-Herreros, M. (2023). Intramammary infusion of micronised purified flavonoid fraction (MPFF) in mastitis-diagnosed dairy cows naturally infected by *Staphylococcus* spp. in the late lactation. *Veterinary Sciences, 10*, 335. https://doi.org/10.3390/vetsci10050335

Habib, I., & Mohamed, M. Y. I. (2022). Foodborne infections in the Middle East. In *Food safety in the Middle East* (pp. 71–107). Elsevier. https://doi.org/10.1016/B978-0-12-822417-5.00005-2

Habib, I., Mohamed, M. Y. I., & Khan, M. (2021). Current state of *Salmonella*, *Campylobacter* and *Listeria* in the food chain across the Arab countries: A descriptive review. *Food Security, 10*(10), 2369. https://doi.org/10.3390/foods10102369

Hao, R., Liu, Y., Shen, W., Zhao, R., Jiang, B., Song, H., Yan, M., & Ma, H. (2022). Surveillance of emerging infectious diseases for biosecurity. *Science China. Life Sciences, 65*, 1504–1516. https://doi.org/10.1007/s11427-021-2071-x

Harris, M., Fasolino, T., Ivankovic, D., Davis, N. J., & Brownlee, N. (2023). Genetic factors that contribute to antibiotic resistance through intrinsic and acquired bacterial genes in urinary tract infections. *Microorganisms, 11*, 1407. https://doi.org/10.3390/microorganisms11061407

Hosain, M. Z., Kabir, S. M. L., & Kamal, M. D. M. (2021). Antimicrobial uses for livestock production in developing countries. *Veterinary World, 14*, 210–221. https://doi.org/10.14202/vetworld.2021.210-221

Hoskins, J. L., Rempoulakis, P., Stevens, M. M., & Dominiak, B. C. (2023). Biosecurity and management strategies for economically important exotic tephritid fruit fly species in Australia. *Insects, 14*, 801. https://doi.org/10.3390/insects14100801

Hutubessy, R., Lauer, J. A., Giersing, B., Sim, S. Y., Jit, M., Kaslow, D., & Botwright, S. (2023). The full value of vaccine assessments (FVVA): a framework for assessing and communicating the value of vaccines for investment and introduction decision-making. *BMC Medicine, 21*, 229. https://doi.org/10.1186/s12916-023-02929-0

Imam, B. H., Oladejo, A. O., Wu, X., Yang, J., Ma, X., Shen, W., Wei, J., Yan, Z., & Ding, X. (2021). Anti-inflammatory and antibacterial potential of qicao rukang powder in bovine subclinical mastitis. *Evidence-Based Complementary and Alternative Medicine, 2021*, 1–10. https://doi.org/10.1155/2021/2148186

Jama, A., Appelqvist, E., Kulane, A., Karregård, S., Rubin, J., Nejat, S., Habersaat, K. B., Jackson, C., Butler, R., Lindstrand, A., & Godoy-Ramirez, K. (2022). Design and implementation of tailored intervention to increase vaccine acceptance in a Somali community in Stockholm, Sweden – Based on the tailoring immunization Programmes approach. *Public Health Practice, 4*, 100305. https://doi.org/10.1016/j.puhip.2022.100305

Janakiram, C., & Dye, B. A. (2020). A public health approach for prevention of periodontal disease. *Periodontology 2000, 84*, 202–214. https://doi.org/10.1111/prd.12337

Kalluri, R., & LeBleu, V. S. (2020). The biology, function, and biomedical applications of exosomes. *Science, 1979*, 367. https://doi.org/10.1126/science.aau6977

Kebede, B., & Shibeshi, W. (2022). In vitro antibacterial and antifungal activities of extracts and fractions of leaves of Ricinus communis Linn against selected pathogens. *Veterinary Medicine and Science, 8*, 1802–1815. https://doi.org/10.1002/vms3.772

Khalifa, H. O., Oreiby, A. F., Abd El-Hafeez, A. A., Okanda, T., Haque, A., Anwar, K. S., Tanaka, M., Miyako, K., Tsuji, S., Kato, Y., & Matsumoto, T. (2020). First report of multidrug-resistant carbapenemase-producing bacteria coharboring mcr-9 associated with respiratory disease complex in pets: potential of animal-human transmission. *Antimicrobial Agents and Chemotherapy, 65*. https://doi.org/10.1128/AAC.01890-20

Kher, M. N., Sheth, N. R., & Bhatt, V. D. (2019). In vitro antibacterial evaluation of terminalia chebula as an alternative of antibiotics against bovine subclinical mastitis. *Animal Biotechnology, 30*, 151–158. https://doi.org/10.1080/10495398.2018.1451752

Kijlstra, A., & Eijck, I. A. J. M. (2006). Animal health in organic livestock production systems: a review. *Wageningen Journal of Life Sciences, 54*, 77–94. https://doi.org/10.1016/S1573-5214(06)80005-9

Konputtar, A., Aengwanich, W., Saraphol, B., & Yossapol, M. (2022). Antimicrobial activity of polyphenols extracted from Thai medical plants on extended-spectrum beta-lactamase-producing *Escherichia coli* isolates from healthy dairy cows. *Polish Journal of Veterinary Sciences, 25*, 501–510. https://doi.org/10.24425/pjvs.2022.142038

Laxminarayan, R., et al. (2015) The economic costs of withdrawing antimicrobial growth promoters from the livestock sector. OECD Food, Agric and Fisheries Papers 78. https://doi.org/10.1787/5js64kst5wvl-en

Lee, M. H., Lee, H. J., & Ryu, P. D. (2001). Public health risks: chemical and antibiotic residues – Review. *Asian-Australasian Journal of Animal Sciences, 14*, 402–413. https://doi.org/10.5713/ajas.2001.402

Leistikow, I., & Bal, R. A. (2020). Resilience and regulation, an odd couple? Consequences of safety-II on governmental regulation of healthcare quality. *BMJ Quality and Safety, 29*, 1–2. https://doi.org/10.1136/bmjqs-2019-010610

Levy, S. (2014). Reduced antibiotic use in livestock: how Denmark tackled resistance. *Environmental Health Perspectives, 122*, A160–A165. https://doi.org/10.1289/ehp.122-A160

Lillehaug, A., Børnes, C., & Grave, K. (2018). A pharmaco-epidemiological study of antibacterial treatments and bacterial diseases in Norwegian aquaculture from 2011 to 2016. *Diseases of Aquatic Organisms, 128*, 117–125. https://doi.org/10.3354/dao03219

Ma, F., Xu, S., Tang, Z., Li, Z., & Zhang, L. (2021). Use of antimicrobials in food animals and impact of transmission of antimicrobial resistance on humans. *Biosafety and Health, 3*, 32–38. https://doi.org/10.1016/j.bsheal.2020.09.004

Manishimwe, R., Nishimwe, K., & Ojok, L. (2017). Assessment of antibiotic use in farm animals in Rwanda. *Tropical Animal Health and Production, 49*, 1101–1106. https://doi.org/10.1007/s11250-017-1290-z

Maron, D. F., Smith, T. J., & Nachman, K. E. (2013). Restrictions on antimicrobial use in food animal production: an international regulatory and economic survey. *Globalization and Health, 9*, 48. https://doi.org/10.1186/1744-8603-9-48

Marshall, B. M., & Levy, S. B. (2011). Food animals and antimicrobials: impacts on human health. *Clinical Microbiology Reviews, 24*, 718–733. https://doi.org/10.1128/CMR.00002-11

Maye, D., & Chan, K. W. (2020). On-farm biosecurity in livestock production: farmer behaviour, cultural identities and practices of care. *Emerging Topics in Life Sciences, 4*, 521–530. https://doi.org/10.1042/ETLS20200063

McDougall, S. (2021). Evaluation of fully oxidised β-carotene as a feed ingredient to reduce bacterial infection and somatic cell counts in pasture-fed cows with subclinical mastitis. *New Zealand Veterinary Journal, 69*, 285–293. https://doi.org/10.1080/00480169.2021.1924091

McEwen, S. A., & Fedorka-Cray, P. J. (2002). Antimicrobial use and resistance in animals. *Clinical Infectious Diseases, 34*, S93–S106. https://doi.org/10.1086/340246

Mecocci, S., De Paolis, L., Zoccola, R., Fruscione, F., De Ciucis, C. G., Chiaradia, E., Moccia, V., Tognoloni, A., Pascucci, L., Zoppi, S., Zappulli, V., Chillemi, G., Goria, M., Cappelli, K., & Razzuoli, E. (2022). Antimicrobial and immunomodulatory potential of cow colostrum extracellular vesicles (ColosEVs) in an intestinal in vitro model. *Biomedicine, 10*, 3264. https://doi.org/10.3390/biomedicines10123264

Mnisi, C. M., Mlambo, V., Gila, A., Matabane, A. N., Mthiyane, D. M. N., Kumanda, C., Manyeula, F., & Gajana, C. S. (2022). Antioxidant and antimicrobial properties of selected phytogenics for sustainable poultry production. *Applied Sciences, 13*, 99. https://doi.org/10.3390/app13010099

Mohamed, M.-Y. I., & Habib, I. (2023). Pathogenic *E. coli* in the food chain across the Arab countries: A descriptive review. *Food Security, 12*, 3726. https://doi.org/10.3390/foods12203726

Mohamed, M.-Y. I., Abu, J., Abdul-Aziz, S., Zakaria, Z., Rashid, A., & Awad, E. A. (2019). Occurrence of antibiotic resistant *C. jejuni* and *E. coli* in wild birds, chickens, environment and humans from Orang Asli Villages in Sungai Siput, Perak, Malaysia. *American Journal of Animal and Veterinary Sciences, 14*, 158–169. https://doi.org/10.3844/ajavsp.2019.158.169

Mohamed, M.-Y. I., Abu, J., Zakaria, Z., Khan, A. R., Abdul Aziz, S., Bitrus, A. A., & Habib, I. (2022). Multi-drug resistant pathogenic *Escherichia coli* isolated from wild birds, chicken, and the environment in Malaysia. *Antibiotics, 11*, 1275. https://doi.org/10.3390/antibiotics11101275

Morales-Ubaldo, A. L., Gonzalez-Cortazar, M., Zaragoza-Bastida, A., Meza-Nieto, M. A., Valladares-Carranza, B., et al. (2022). Nor 3′-Demethoxyisoguaiacin from larrea tridentata is a potential alternative against multidrug resistant bacteria associated with bovine mastitis. *Molecules, 27*, 3620. https://doi.org/10.3390/molecules27113620

Mordmuang, A., Brouillette, E., Voravuthikunchai, S. P., & Malouin, F. (2019). Evaluation of a Rhodomyrtus tomentosa ethanolic extract for its therapeutic potential on *Staphylococcus aureus* infections using in vitro and in vivo models of mastitis. *Veterinary Research, 50*, 49. https://doi.org/10.1186/s13567-019-0664-9

More, S. J. (2020). European perspectives on efforts to reduce antimicrobial usage in food animal production. *Irish Veterinary Journal, 73*, 2. https://doi.org/10.1186/s13620-019-0154-4

Morgans, L. C., Bolt, S., Bruno-McClung, E., van Dijk, L., Escobar, M. P., Buller, H. J., Main, D. C. J., & Reyher, K. K. (2021). A participatory, farmer-led approach to changing practices around antimicrobial use on UK farms. *Journal of Dairy Science, 104*, 2212–2230. https://doi.org/10.3168/jds.2020-18874

Morrone, S., Dimauro, C., Gambella, F., & Cappai, M. G. (2022). Industry 4.0 and precision livestock farming (PLF): An up to date overview across animal productions. *Sensors, 22*, 4319. https://doi.org/10.3390/s22124319

Mubito, E., et al. (2014). Antimicrobial use in the poultry industry in Dar-Es-Salaam, Tanzania and public health implications. *American Journal of Research Communication, 2*, 51–63.

Munkholm, L., & Rubin, O. (2020). The global governance of antimicrobial resistance: a cross-country study of alignment between the global action plan and national action plans. *Globalization and Health, 16*, 109. https://doi.org/10.1186/s12992-020-00639-3

Narmandakh, D., & Sakurai, T. (2021). The impact of quarantine against foot-and-mouth disease in Mongolia on pastoralists' farming performance and welfare. *Japanese Journal of Agricultural Economics, 23*, 137–142.

Nicholls, H. L., John, C. R., Watson, D. S., Munroe, P. B., Barnes, M. R., & Cabrera, C. P. (2020). Reaching the end-game for gwas: Machine learning approaches for the prioritization of complex disease loci. *Frontiers in Genetics, 11*. https://doi.org/10.3389/fgene.2020.00350

O'Neill, J. (2015). Antimicrobials in agriculture and the environment: Reducing unnecessary use and waste. The review on antimicrobial resistance. Available at: https://amr-review.org/sites/default/files/Antimicrobials%20in%20agriculture%20and%20the%20environment%20-%20Reducing%20unnecessary%20use%20and%20waste.pdf

Obimakinde, S., Fatoki, O., Opeolu, B., & Olatunji, O. (2017). Veterinary pharmaceuticals in aqueous systems and associated effects: an update. *Environmental Science and Pollution Research, 24*, 3274–3297. https://doi.org/10.1007/s11356-016-7757-z

Paiano, R. B., de Sousa, R. L. M., Bonilla, J., Moreno, L. Z., de Souza, E. D. F., Baruselli, P. S., & Moreno, A. M. (2023). In vitro effects of cinnamon, oregano, and thyme essential oils against *Escherichia coli* and *Trueperella pyogenes* isolated from dairy cows with clinical endometritis. *Theriogenology, 196*, 106–111. https://doi.org/10.1016/j.theriogenology.2022.11.010

Paramitadevi, Y. V., Priadi, C. R., Rahmatika, I., Rukmana, A., & Moersidik, S. S. (2023). Integration of water, sanitation, and hygiene program with biosecurity: A one health approach to reduce the prevalence and exposure of antibiotic-resistant bacteria in the livestock community. *International Journal of One Health, 9*, 181–193.

Perez, M. M., Cabrera, E. M., & Giordano, J. O. (2023). Effects of targeted clinical examination based on alerts from automated health monitoring systems on herd health and performance of lactating dairy cows. *Journal of Dairy Science, 106*, 9474–9493. https://doi.org/10.3168/jds.2023-23477

Pollack, L. A., & Srinivasan, A. (2014). Core elements of hospital antibiotic stewardship programs from the centers for disease control and prevention. *Clinical Infectious Diseases, 59*, S97–S100. https://doi.org/10.1093/cid/ciu542

Popitool, K., Wataradee, S., Wichai, T., Noitang, S., Ajariyakhajorn, K., Charoenrat, T., Boonyaratanakornkit, V., & Sooksai, S. (2022). Potential of Pm11 antimicrobial peptide against bovine mastitis pathogens. *American Journal of Veterinary Research, 1–6*, 1–6. https://doi.org/10.2460/ajvr.22.06.0096

Qu, J., Huang, Y., & Lv, X. (2019). Crisis of antimicrobial resistance in China: Now and the future. *Frontiers in Microbiology, 10*. https://doi.org/10.3389/fmicb.2019.02240

Queenan, K., Häsler, B., & Rushton, J. (2016). A one health approach to antimicrobial resistance surveillance: is there a business case for it? *International Journal of Antimicrobial Agents, 48*, 422–427. https://doi.org/10.1016/j.ijantimicag.2016.06.014

Raboisson, D., Ferchiou, A., Sans, P., Lhermie, G., & Dervillé, M. (2020). The economics of antimicrobial resistance in veterinary medicine: Optimizing societal benefits through mesoeconomic approaches from public and private perspectives. *One Health, 10*, 100145. https://doi.org/10.1016/j.onehlt.2020.100145

Regan, Á., Sweeney, S., McKernan, C., Benson, T., & Dean, M. (2023). Consumer perception and understanding of the risks of antibiotic use and antimicrobial resistance in farming. *Agriculture and Human Values, 40*, 989–1001.

Riley, W. W., Nickerson, J. G., Mogg, T. J., & Burton, G. W. (2023). Oxidized β-carotene is a novel phytochemical immune modulator that supports animal health and performance for antibiotic-free production. *Animals, 13*, 289. https://doi.org/10.3390/ani13020289

Robertson, I. D. (2020). Disease control, prevention and on-farm biosecurity: the role of veterinary epidemiology. *Engineering, 6*, 20–25. https://doi.org/10.1016/j.eng.2019.10.004

Rodríguez-Prieto, V., Vicente-Rubiano, M., Sánchez-Matamoros, A., Rubio-Guerri, C., Melero, M., Martínez-López, B., Martínez-Avilés, M., Hoinville, L., Vergne, T., Comin, A., Schauer, B., Dorea, F., Pfeiffer, D. U., & Sánchez-Vizcaíno, J. M. (2015). Systematic review of surveillance systems and methods for early detection of exotic, new and re-emerging diseases in animal populations. *Epidemiology and Infection, 143*, 2018–2042. https://doi.org/10.1017/S095026881400212X

Roskam, J. L., Oude Lansink, A. G. J. M., & Saatkamp, H. W. (2023). The economic value of antimicrobial use in livestock production. *Antibiotics, 12*, 1537. https://doi.org/10.3390/antibiotics12101537

Sanderson, M. W., & Gnad, D. P. (2002). Biosecurity for reproductive diseases. *Veterinary Clinics of North America. Food Animal Practice, 18*, 79–98. https://doi.org/10.1016/S0749-0720(02)00003-8

Santoro, J., Mukhopadhya, A., Oliver, C., Brodkorb, A., Giblin, L., & O'Driscoll, L. (2023). An investigation of extracellular vesicles in bovine colostrum, first milk and milk over the lactation curve. *Food Chemistry, 401*, 134029. https://doi.org/10.1016/j.foodchem.2022.134029

Satterfield, J., Miesner, A. R., & Percival, K. M. (2020). The role of education in antimicrobial stewardship. *The Journal of Hospital Infection, 105*, 130–141. https://doi.org/10.1016/j.jhin.2020.03.028

Silva, D. R., et al. (2020). Probiotics as an alternative antimicrobial therapy: Current reality and future directions. *Journal of Functional Foods, 73*, 104080. https://doi.org/10.1016/j.jff.2020.104080

Singh, H., Neha, K., Kumar, R., Kaushik, P., Singh, A. K., & Singh, G. (2024). Role of infrastructure and operation in disease prevalence in dairy farms: groundwork for disease prevention-based antibiotic stewardship. *Preventive Veterinary Medicine*, 106158. https://doi.org/10.1016/j.prevetmed.2024.106158

Souza, R. F. S., Rault, L., Seyffert, N., Azevedo, V., Le Loir, Y., & Even, S. (2018). Lactobacillus casei BL23 modulates the innate immune response in *Staphylococcus aureus*-stimulated bovine mammary epithelial cells. *Beneficial Microbes, 9*, 985–995. https://doi.org/10.3920/BM2018.0010

Stoneham, G., Hester, S. M., Li, J. S., Zhou, R., & Chaudhry, A. (2021). The boundary of the market for biosecurity risk. *Risk Analysis, 41*, 1447–1462. https://doi.org/10.1111/risa.13620

Subasinghe, R., Alday-Sanz, V., Bondad-Reantaso, M. G., Jie, H., Shinn, A. P., & Sorgeloos, P. (2023). Biosecurity: Reducing the burden of disease. *Journal of the World Aquaculture Society, 54*, 397–426. https://doi.org/10.1111/jwas.12966

Tomanić, D., Božin, B., Kladar, N., Stanojević, J., Čabarkapa, I., Stilinović, N., Apić, J., Božić, D. D., & Kovačević, Z. (2022). Environmental bovine mastitis pathogens: prevalence, antimicrobial susceptibility, and sensitivity to thymus vulgaris l., thymus serpyllum l. and origanum vulgare l. essential oils. *Antibiotics, 11*, 1077. https://doi.org/10.3390/antibiotics11081077

Tommasoni, C., Fiore, E., Lisuzzo, A., & Gianesella, M. (2023). Mastitis in dairy cattle: on-farm diagnostics and future perspectives. *Animals, 13*, 2538. https://doi.org/10.3390/ani13152538

Urakawa, M., Zhuang, T., Sato, H., Takanashi, S., Yoshimura, K., Endo, Y., Katsura, T., Umino, T., Tanaka, K., Watanabe, H., Kobayashi, H., Takada, N., Kozutsumi, T., Kumagai, H., Asano, T., Sazawa, K., Ashida, N., Zhao, G., Rose, M. T., Kitazawa, H., Shirakawa, H., Watanabe, K., Nochi, T., Nakamura, T., & Aso, H. (2022). Prevention of mastitis in multiparous dairy cows with a previous history of mastitis by oral feeding with probiotic Bacillus subtilis. *Journal of Animal Science, 93*, e13764. https://doi.org/10.1111/asj.13764

Van Boeckel, T. P., Brower, C., Gilbert, M., Grenfell, B. T., Levin, S. A., Robinson, T. P., Teillant, A., & Laxminarayan, R. (2015). Global trends in antimicrobial use in food animals. *Proceedings of the National Academy of Sciences, 112*, 5649–5654. https://doi.org/10.1073/pnas.1503141112

van Herten, J., Buikstra, S., Bovenkerk, B., & Stassen, E. (2020). Ethical decision-making in zoonotic disease control: how do one health strategies function in The Netherlands? *Journal of*

Agricultural and Environmental Ethics, 33, 239–259. https://doi.org/10.1007/s10806-020-09828-x

Van, T. T. H., Yidana, Z., Smooker, P. M., & Coloe, P. J. (2020). Antibiotic use in food animals worldwide, with a focus on Africa: Pluses and minuses. *Journal of Global Antimicrobial Resistance, 20*, 170–177. https://doi.org/10.1016/j.jgar.2019.07.031

Vasala, A., Hytönen, V. P., & Laitinen, O. H. (2020). Modern tools for rapid diagnostics of antimicrobial resistance. *Frontiers in Cellular and Infection Microbiology, 10*. https://doi.org/10.3389/fcimb.2020.00308

VMD. (2017) UK veterinary antibiotic resistance and sales surveillance report (UK-VARSS 2016). Available at: https://apps.who.int/iris/bitstream/handle/10665/255747/9789241512411-eng.pdf;jsessionid=2AEDBDB0305FF6555D5E84960B6727C9?sequence=1

Wernli, D., Harbarth, S., Levrat, N., & Pittet, D. (2022). A 'whole of United Nations approach' to tackle antimicrobial resistance? A mapping of the mandate and activities of international organisations. *BMJ Global Health, 7*, e008181. https://doi.org/10.1136/bmjgh-2021-008181

Wisnieski, L., Amrine, D. E., & Renter, D. G. (2021). Predictive modeling of bovine respiratory disease outcomes in feedlot cattle: A narrative review. *Livestock Science, 251*, 104666. https://doi.org/10.1016/j.livsci.2021.104666

World Health Organization (WHO). (2011) WHO list of critically important antimicrobials (CIA), 3rd revision. Available at: http://www.who.int/foodsafety/publications/antimicrobials-third/en/

World Health Organization (WHO). (2017). Global action plan on antimicrobial resistance. Available at: https://www.amcra.be/swfiles/files/WHO%20actieplan_90.pdf

World Health Organization (WHO). (2020). Multisectoral approach to the prevention and control of vector-borne diseases: A conceptual framework

World Organisation for Animal Health (OIE). (2016). OIE strategy on antimicrobial resistance and the prudent use of antimicrobials. Available at: https://www.woah.org/fileadmin/Home/eng/Media_Center/docs/pdf/PortailAMR/EN_OIE-AMRstrategy.pdf

Wu, Z. (2018). Antimicrobial use in food animal production: situation analysis and contributing factors. *Frontiers of Agricultural Science and Engineering, 5*, 301. https://doi.org/10.15302/J-FASE-2018207

Yamin, D., Uskoković, V., Wakil, A., Goni, M., Shamsuddin, S., Mustafa, F., Alfouzan, W., Alissa, M., Alshengeti, A., Almaghrabi, R., Fares, M., Garout, M., Al Kaabi, N., Alshehri, A., Ali, H., Rabaan, A., Aldubisi, F., Yean, C., & Yusof, N. (2023). Current and future technologies for the detection of antibiotic-resistant bacteria. *Diagnostics, 13*, 3246. https://doi.org/10.3390/diagnostics13203246

Youssef, D. M., Wieland, B., Knight, G. M., Lines, J., & Naylor, N. R. (2021). The effectiveness of biosecurity interventions in reducing the transmission of bacteria from livestock to humans at the farm level: A systematic literature review. *Zoonoses and Public Health, 68*, 549–562. https://doi.org/10.1111/zph.12807

Zeineldin, M., Elolimy, A. A., Reddy, P. R. K., Abdelmegeid, M., Mellado, M., Elghandour, M. M. M. Y., & Salem, A. Z. M. (2021). *Sustainable agriculture reviews* (p. 54). Springer International Publishing. https://doi.org/10.1007/978-3-030-76529-3

Zhang, M., Wei, W., Sun, Y., Jiang, X., Ying, X., Tao, R., & Ni, L. (2016). Pleurocidin congeners demonstrate activity against *Streptococcus* and low toxicity on gingival fibroblasts. *Archives of Oral Biology, 70*, 79–87. https://doi.org/10.1016/j.archoralbio.2016.06.008

Antimicrobial Strategies in Food Processing and Post-processing

Thi Anh Ngoc Tong, Thi Luyen Cao, and Bao Nguyen Pham

1 Introduction

The need to produce high-quality, pathogen-free food depends on pre- and post-harvest technology. Antimicrobial strategies in food processing and post-processing are critical to enhance food safety, prevent microbial contamination, and extend shelf life while maintaining quality. In addition to ensuring food safety and hygiene according to standards, the application of traditional processing technologies, emerging technologies, natural and chemical antimicrobials was also discussed.

2 Food Microbial Contamination

2.1 Introduction to Microbial Contamination

Microorganisms play various beneficial roles in the production of food, with many species being used to create specific products like fermented foods. However, certain species can present serious industrial, economic, and medical challenges when they are contaminated with food products. The microbial contamination can occur during harvesting, food processing, and distribution (Malhotra et al., 2015). The primary

T. A. N. Tong (✉)
Can Tho University, Can Tho, Vietnam
e-mail: ttangoc@ctu.edu.vn

T. L. Cao
An Giang University, An Giang, Vietnam

Vietnam National University, Ho Chi Minh city, Vietnam

B. N. Pham
Tra Vinh University, Tra Vinh, Vietnam

D. Li (ed.), *Antimicrobial Strategies in the Food System: Updates, Opportunities, Challenges*, https://doi.org/10.1007/978-3-031-95056-8_11

microbiological contaminants in food include bacteria, microfungi, parasites, and viruses. These microorganisms can spoil food, causing undesirable changes that may lead to illness. Some also produce toxins, resulting in food poisoning or intoxication (Fu et al., 2016; Chatterjee & Abraham, 2018). The inadvertent colonization of pathogenic microorganisms is referred to as microbiological contamination. These pathogens can proliferate within equipment utilized in the food industry, facilitating their entry during both food production and processing stages, ultimately leading to food spoilage. Microbial contamination has emerged as a critical concern for food industries globally. In addition to affecting food processing, contamination can also compromise food packaging. Despite stringent hygienic measures being implemented, effectively mitigating contamination remains a challenge.

Foodborne illnesses caused by microbial contamination present a significant public health concern due to the high risk posed by various biological hazards. These hazards contribute to personal suffering, preventable fatalities, and avoidable economic strain (Parra et al., 2014; Tesson et al., 2020). Respiratory infections and diarrheal diseases are the most common illnesses, with the greatest negative impact on the growth of infants and young children. Diarrheal diseases are traditionally linked to water quality and sanitation issues (Kamala & Kumar, 2018). The high rate of illnesses caused by foodborne microorganisms underscores the importance of microbiological quality as the top food safety concern (Prüβ, 2020; Tropea, 2022). The biggest challenge in protecting people from foodborne diseases is to gather and apply knowledge about the sources and transmission routes of pathogens in food. This content outlines the key aspects of microbial contamination in food.

2.2 Common Microorganisms in Food Contamination

Microorganisms, including bacteria, molds, and yeasts, are prevalent in the environment and can colonize food products, making microbiology crucial to food safety, production, and preservation. While beneficial microbes are used in fermentation processes for various food items, microbial contamination can occur at any stage of food production, leading to visible alterations in food, including changes in color, the development of powdery mycelial growth, degradation of organoleptic properties, and the emergence of effervescence on food surfaces, thereby compromising food quality and safety. Microbial contamination is a critical concern in food systems, requiring vigilant control measures to protect public health. Advances in detection technologies, as well as increased awareness and regulatory oversight, have made significant progress in reducing contamination risks. Controlling microbial contamination is vital for public health, and advancements in detection technologies, along with regulatory oversight, have reduced risks. Ongoing education, research, and adherence to food safety practices remain essential in preventing future outbreaks and ensuring the safety of the global food supply.

Microbial contamination in food refers to the unintended presence of harmful microorganisms. These microorganisms can thrive in various food environments, leading to spoilage, foodborne diseases, and significant economic losses. Food can be contaminated by a wide range of microorganisms, which can be classified into four main categories such as bacteria, virus, parasites, fungi and mold (Tropea, 2022).

2.2.1 Bacteria Involved in Food Contamination

Numerous studies on bacterial contamination in food have revealed that prokaryotic cells can thrive and colonize within food, utilizing the nutrients available for their growth. While bacteria are commonly used in various food production processes, many species can still cause significant foodborne illnesses, often resulting in symptoms such as diarrhea, vomiting, and fever. Bacteria involved in food contamination included food spoilage and pathogenic bacteria.

2.2.1.1 Food Spoilage Bacteria

Food spoilage bacteria are distinguishable by the degradation they cause in food, manifesting as alterations in texture, appearance, and the emission of undesirable odors and flavors. These microorganisms flourish in environments rich in moisture, warmth, and nutrients, targeting various food types such as dairy, meat, seafood, fruits, and vegetables (Kaczmarek et al., 2019). As these bacteria multiply, they decompose food components, leading to spoilage. Lactic acid bacteria (e.g., *Lactobacillus*) are prominent in fermented products, imparting a sour flavor and aroma as well as gas-forming. *Lipolytic* bacteria (e.g., *Pseudomonas*) accelerate rancidity by breaking down lipids. While slime-forming bacteria (e.g., *Streptococcus*) further deteriorate food by generating gases and producing slimy textures (Lorenzo et al., 2018). Though food affected by spoilage bacteria may be unappetizing, it is typically not harmful unless consumed in large quantities, which may result in mild gastrointestinal discomfort.

2.2.1.2 Pathogenic Bacteria

Pathogenic bacteria, unlike spoilage bacteria, are unseen hazards that can induce severe illness even in minimal amounts. These pathogens cannot be detected through sensory evaluation and pose serious health risks, especially to susceptible populations such as young children, pregnant women, the elderly, and immunocompromised individuals (Lorenzo et al., 2018). Notable pathogenic bacteria include *Escherichia coli* (*E. coli*), with strains like *E. coli* O157:H7 known to produce toxins that can cause serious gastrointestinal distress and renal impairment. Inadequate food handling, particularly with raw meats, is a common source of contamination. Another example is *Staphylococcus aureus (S. aureus)*, a bacterium commonly found on human skin and in the nasal passages. It produces a heat-stable toxin that can lead to rapid-onset food poisoning (Durand et al., 2018).

A comprehensive overview of the bacteria involved in foodborne infections, including the bacterial species, contamination sources, affected food products, and their characteristics are presented in Table 1.

2.2.2 Viruses Involved in Food Contamination

Viruses play a significant role in foodborne illnesses, and while they are smaller and simpler than bacteria, they can cause serious health issues when transmitted through contaminated food or water. Unlike bacteria, viruses do not multiply in food, but they can survive on food surfaces and in liquids for extended periods, especially in unsanitary conditions. These pathogens are often spread through improper handling and poor sanitation (Table 2).

2.2.3 Parasites

Beside viruses, major groups of foodborne parasites cause parasitic diseases. Numerous parasites, including various helminths and protozoa, can be transmitted through food. In developing countries, parasitic infections are commonly contracted by consuming undercooked fish, crabs, mollusks, meat, raw aquatic plants like watercress, or raw vegetables contaminated with human or animal feces (Dorny et al., 2009). The symptoms of foodborne parasitic infections vary depending on the parasite involved. Helminth infections can lead to symptoms such as abdominal pain, diarrhea, muscle pain, cough, skin lesions, malnutrition, weight loss, and neurological issues, depending on the type and severity of the infection. Protozoa like *Cryptosporidium* spp., *Giardia intestinalis*, and *Cyclospora cayetanensis* primarily cause diarrhea and other gastrointestinal symptoms (Lorenzo et al., 2018; Ortega, 2019).

Cryptosporidium, an enteric protozoan parasite belonging to the order Eucoccidiorida, caused 30–50% of the deaths among children under the age of five and has been ranked as the second leading cause of diarrhea and death in this age group of children after rotavirus (Ryan & Hijjawi, 2015). Among 27 determined species, *Cryptosporidium hominis* and *Cryptosporidium parvum* play important roles in most of infections in human being (Ryan & Hijjawi, 2015) (Fig. 1).

2.2.4 Yeast and Molds Involved in Food Contamination

The diversity of yeast and fungi as food spoilers has been explored in different food products. Fungi and molds play a significant role in food contamination, often leading to food spoilage and the production of harmful mycotoxins, which pose health risks to humans and animals (Matthews et al., 2024). These organisms thrive

Table 1 Common bacteria involved in food contamination

Taxonomy	Contaminated source/products affected	Characteristics	References
Escherichia coli (*E. coli*)	Feces, contaminated water/soil, dairy sewage, meat, vegetables	It causes severe foodborne illness; certain strains of E. coli, especially *E. coli* O157:H7, are highly pathogenic, causing symptoms ranging from mild diarrhea to severe kidney damage.	Bravo et al. (2014), Bintsis (2017), Ekici and Dümen (2019)
Salmonella	Eggs, poultry, unpasteurized milk, juice products	It can survive for extended periods of time in an acidic environment, causing severe food poisoning, leading to symptoms like diarrhea, fever, and abdominal cramps.	Bintsis (2017), Kaczmarek et al. (2019) Ehuwa et al. (2021)
Listeria monocytogenes (L. monocytogenes)	Refrigerated foods, sausages, dairy, deli meats	Thrives in cold environments, affects ready-to-eat foods, poses a serious threat to pregnant women, newborns, and people with weakened immune systems.	Thakur et al. (2018), Meloni et al. (2014), Gómez et al. (2014)
Staphylococcus aureus	Salads, pastries, deli meats, ice cream, bakery, fresh juice	The third leading cause of confirmed foodborne illness globally. The illness occurs due to the consumption of preformed enterotoxins present in improperly stored foods, leading to rapid onset of food poisoning symptoms like vomiting and nausea.	Kaczmarek et al. (2019), Solano et al. (2013), Fetsch et al. (2014), Kaczmarek et al. (2019)
Clostridia	Soils, canned or vacuum-packaged foods; intestinal tract of animals and humans	Anaerobic, generate butyric acid and contribute to food spoilage by breaking down proteins through anaerobic proteolysis. Spore-forming, *Clostridium botulinum (C. botulinum)* releases toxin, which causes botulism, a life-threatening illness.	Dürre (2021), Lorenzo et al. (2018), Garvey (2023), Kaczmarek et al. (2019)
Vibrio parahaemolyticus (V. parahaemolyticus)	Seafood, raw or undercooked fish and shellfish	Certain species of *Vibrio* can cause gastrointestinal illness and, in severe cases, bloodstream infections.	Guin et al. (2019), Bravo et al. (2014), Ndraha et al. (2022)
	Various food types		

(continued)

Table 1 (continued)

Taxonomy	Contaminated source/products affected	Characteristics	References
Shiga toxin-producing bacteria		Caused by unsanitary conditions/ environments, particularly after natural disasters	Frank et al. (2011), Gómez et al. (2014)
Campylobacter	Raw/undercooked poultry	This genus causes gastroenteritis with symptoms like diarrhea, fever, and abdominal pain.	Ngulukun (2017), Lorenzo et al. (2018)
Pseudomonas spp.	Dairy, meat, poultry, produce	Thrives in adverse conditions, involved in spoilage, causing slime formation, off-color, rancidity in spoiled food.	Bloomfield et al. (2024), Chang et al. (2024)
Lactic Acid Bacteria (LAB)	Dairy products (yogurt, cheese), juice, meat	Through fermentation, these bacteria produce lactic acid, along with acetyl methylcarbinol and diacetyl, which contribute to spoilage, sour smell.	Matthews et al. (2024), Kaczmarek et al. (2019)
Streptococcus	Fermented dairy products	Some species of Streptococcus genus can lead to undesirable fermentation and off-flavors.	Hernández-Cortez et al. (2017), Lorenzo et al. (2018)

Table 2 Viruses Involved in Food Contamination

Taxonomy	Contaminated source/ products affected	Characteristics	References
Norovirus	Ready-to-eat foods, shellfish, contaminated water, salads	Highly contagious, spreads through contaminated food, water, and surfaces, resistant to heat and disinfectants. Causing gastroenteritis: vomiting, diarrhea, stomach cramps, nausea.	CDC (2024), Lorenzo et al. (2018)
Hepatitis A Virus (HAV)	Shellfish, fruits, vegetables, and infected individuals	HAV infects the liver, transmitted through fecal-contaminated food or water, long incubation period (2–6 weeks), causing liver inflammation, jaundice, fatigue, nausea, and abdominal pain.	Lorenzo et al. (2018, CDC (2024)
Rotavirus	Contaminated water, raw vegetables, and fruits	Affects young children and infants, causes severe dehydration due to diarrhea vomiting, fever, abdominal pain.	Lorenzo et al. (2018, CDC (2024)
Astrovirus	Contaminated water, shellfish, and food prepared by infected handlers	Contributing to foodborne gastroenteritis symptoms, including mild diarrhea, nausea, and vomiting, particularly in young children and the elderly.	Lorenzo et al. (2018), Bosch et al. (2014)
Sapovirus	Contaminated water, shellfish, and ready-to-eat foods	Causing gastroenteritis: diarrhea, vomiting, stomach pain, fever transmitted through fecal-oral route	Lorenzo et al. (2018)

in moist, warm environments and can affect a variety of foods, including grains, nuts, fruits, vegetables, and dairy products (Kamala & Kumar, 2018). Certain species of fungi and molds can produce severe toxins, which are life-threatening to humans when ingested, especially concerning in staple foods like grains and nuts (Ekwomadu & Mwanza, 2023).

2.2.4.1 Yeast

Yeast has a wide application in food industry, particularly fermented products. Whilst, the spoilage yeasts of foods and beverages accounts for noteworthy economic losses. Pitt (2009) observed that only around 10 yeast species are primarily responsible for the spoilage of processed and packaged foods. These species include *Dekkera bruxellensis, Zygosaccharomyces bisporus, Schizosaccharomyces pombe, Issatchenkia orientalis, Debaryomyces hansenii, Candida holmii, Pichia membranifaciens, Zygosaccharomyces bailii, Saccharomyces cerevisiae, Zygosaccharomyces rouxii,* and *Kloeckera apiculata* (Table 3).

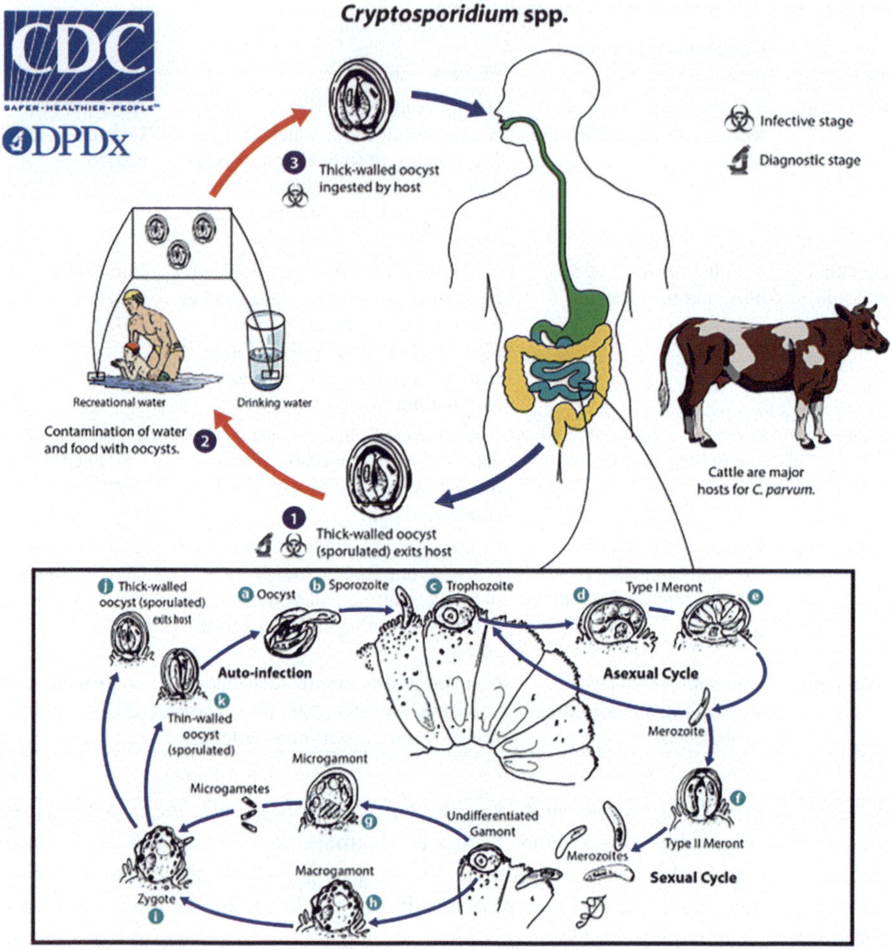

Fig. 1 Life cycle of Cryptosporidium (Adapted from Ortega, 2019)

2.2.4.2 Molds

Molds, classified as multicellular filamentous fungi, are recognized by their fuzzy or cottony appearance, indicative of food spoilage. Their identification can often be made visually, based on texture (dry, powdery, velvety, or gelatinous) and pigmentation (black, red, gray, etc.) (Chatterjee & Abraham, 2018; Matthews et al., 2024). Molds develop through branching filaments called hyphae, which form a mycelium colony. Unlike bacteria and yeasts, molds thrive in lower moisture environments and grow optimally between 25 °C and 30 °C. They require free oxygen for surface growth and prefer acidic environments (pH 2–8.5) (Chatterjee & Abraham, 2018; Lorenzo et al., 2018). Molds significantly contribute to food spoilage, posing substantial risks to food quality and safety. Their presence can lead to the

Table 3 Degradation caused by Yeast activity in food products (Lorenzo et al., 2018)

	Fresh fruits, veggies	Beverages	Canned foods	Mayonnaise/dressing	Butter/cream/cheese	Yoghurts	Bread	Meats
Surface growth	•		•		•		•	•
Discoloration	•		•	•	•		•	•
Gas forming		•	•	•		•		
Cloudiness		•	•	•				
Sediments		•						
Films		•	•	•		•		
Off-flavors	•	•	•	•	•	•	•	•
Texture changes	•		•	•	•	•		•

Table 4 Common molds in food deterioration

Taxonomy	Characteristics	References
Aspergillus	*Aspergillus* molds are highly prevalent and play a significant role in food spoilage. This genus thrives in environments with high sugar and salt concentrations, making it common in various low-moisture foods.	Lorenzo et al. (2018), Perrone and Gallo (2017)
Rhizopus	Contributing to the spoilage of vegetables, fruits, and bread. The most common species include *Rhizopus arrhizus* and *Rhizopus stolonifer.*	Lorenzo et al. (2018)
Penicillium	With a significant global presence and considerable economic influence on human life. In nature, its primary role is the breakdown of organic matter, where certain species lead to severe decay as pre- and postharvest pathogens on food crops and produce various mycotoxins.	Lorenzo et al. (2018), Perrone and Gallo (2017), Otero et al. (2020)
Alternaaria	Spoils crops like potatoes and tomatoes and produces toxins. Common species include *Alternaria citri* and *Alternaria brassicae.*	Lorenzo et al. (2018)
Bothrytis	Significant pathogens affect various economically crucial crops, including grapevines, tomatoes, bulb flowers, and ornamental plants.	Lorenzo et al. (2018)
Fusarium	*Fusarium* spp. are major contributors to food safety issues, especially in cereal crops. This genus is economically vital plant pathogens and opportunistic human pathogens.	Ekwomadu and Mwanza (2023)

deterioration of various food products, resulting in unpleasant odors, off-flavors, and undesirable textures (Adams & Moss, 2000). Moreover, certain mold species produce mycotoxins, which can have serious health implications for consumers, including acute toxicity and long-term carcinogenic effects. The economic impact of mold-related spoilage is also considerable, as it leads to increased food waste and financial losses for producers and retailers (Kamala & Kumar, 2018) (Table 4).

Mycotoxins are secondary metabolitc toxin, produced by fungi, commonly contaminating foods like fruits and vegetables (Khan et al., 2024). The major mycotoxins and their effects include:

- Aflatoxin: Produced mainly by Aspergillus, it affects DNA, RNA, and proteins, causing mutagenic and carcinogenic effects. Common aflatoxins are AFB_1 (aflatoxin B_1), AFB_2, AFG_1, AFG_2, and AFM_1/AFM_2
- Ochratoxin: Formed by species like *Penicillium* and *Aspergillus,* this toxin contaminates beans, cereals, dried fruits, eggs, and milk. It poses significant health risks.
- Fumonisins: Primarily produced by *Fusarium* species, these carcinogenic toxins contaminate corn-based foods. Consumption can lead to severe conditions like porcine lung edema and brain necrosis (Table 5).

Table 5 Molds and their toxin (Khan et al., 2024; Patial et al., 2018)

Molds	Mycotoxin
Aspergillus flavus; A. parasiticus	Aflatoxins
A. flavus	Cyclopiazonic acid
Penicillium viridicatum	Citreoviridin
A. ochraceus; A. carbonarius; Penicillium verrucosum	Ochratoxin A
P.citrinum; P. expansum	Citrinin
Fusarium sporotrichioides; F. poae	T-2 toxin
F. sporotrichioides; F. poae	Diacetoxyscirpenol
F.culmorum; F. graminearum	Deoxynivalenol (DON)
F.culmorum; F. graminearum	Zearalenone
F. verticillioides; F. proliferatum	Fumonisins
Alternaria alternata	Tenuazonic acid
Claviceps purpurea	Ergot alkaloids
Penicillium expansum, Aspergillus spp.	Patulin
Penicillium islandicum	Rugulosin
Aspergillus oryzae	Maltoryzine
Aspergillus flavus; Aspergillus versicolor	Sterigmatocystin

2.3 *Sources of Microbial Contamination*

To safeguard consumers from the harmful effects of food microbiological contamination, it is essential to enhance understanding of the sources and transmission pathways of pathogens (Pakdel et al., 2023). Addressing food safety involves not only recognizing how microorganisms enter the food supply chain but also understanding the conditions that favor their growth and identifying effective control measures (Marriott et al., 2018). This section outlines key aspects of microbiological contamination, including mechanisms, common contaminants, and associated foodborne diseases.

2.3.1 Pathways of Microbial Contamination in Food

Microorganisms can enter food through a variety of pathways, which can occur at multiple stages in the food production and handling process (Fig. 2):

Water Contamination
Contaminated water is one of the most common vectors for microbial contamination. Water used in irrigation, food processing, or preparation can introduce bacteria, viruses, or parasites into the food chain (Marriott et al., 2018). Associated contaminants including pathogens like *E. coli*, *Salmonella, Norovirus*, and *Giardia* can enter the food supply if the water is not properly treated. These microbial contaminants can directly affect raw fruits, vegetables, seafood, and foods prepared with contaminated water are at higher risk (Fu et al., 2016).

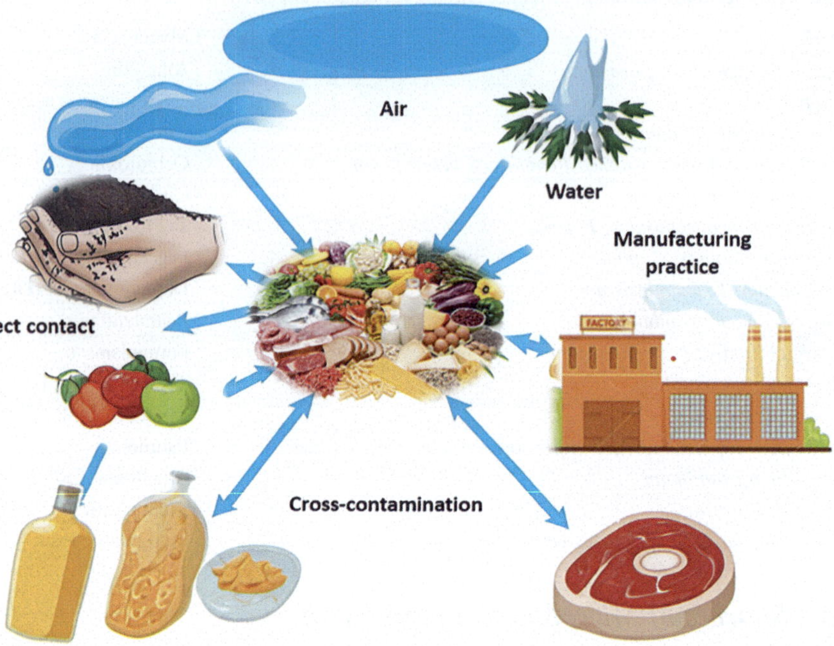

Fig. 2 Exposure pathways

Airborne Contamination
Airborne microbes, including bacteria, mold spores, and viruses, can settle on food surfaces during processing, packaging, and storage. Poor ventilation, unsanitary environments, and lack of proper air filtration increase this risk (Fu et al., 2016). Associated contaminants are fungal spores (e.g., *Aspergillus* spp.), and bacteria like *L. monocytogenes* and *S. aureus* can become airborne and contaminate foods. Ready-to-eat foods, processed meats, and cheeses can be contaminated if exposed to contaminated air during production or storage (Nowshad et al., 2021).

Soil Contamination
Soil harbors a wide variety of microorganisms, including bacteria, fungi, and parasites. These pathogens can contaminate fruits, vegetables, and other crops through direct contact with soil, or via the use of animal manure or contaminated irrigation water (O'Connor et al., 2022). In soil the prominent contaminants are *C. botulinum* (which causes botulism), *L. monocytogenes, Salmonella* spp., and mycotoxin-producing fungi like *Fusarium spp.* are common soil-borne pathogens. Therefore, root vegetables, leafy greens, and crops grown in soil-rich environments are particularly susceptible to contamination (Alegbeleye et al., 2018).

Cross-Contamination
Microbial contamination can occur when microorganisms are transferred from one surface or object to another. This often happens in food preparation environments where contaminated utensils, cutting boards, or hands come into contact with food (Nerín et al., 2016; Thi et al., 2021). Raw meats, poultry, and seafood are common sources of pathogens that can be transferred to other foods, particularly ready-to-eat items. *Salmonella, Campylobacter, E. coli, and Norovirus* are common contaminants that can spread through cross-contamination (Nowshad et al., 2021). Foods at risk are any food can be affected by cross-contamination, but foods consumed raw (like salads and fruits) are at greater risk due to the absence of a cooking step that could kill the pathogens (Kaczmarek et al., 2019).

2.4 Consequences of Microbial Contamination

The effects of microbial contamination in food systems can be severe, including:

Foodborne Illnesses
Consumption of contaminated food can lead to outbreaks of foodborne illnesses. Especially, microbial food-borne illnesses represent a significant public health concern due to the high likelihood of contamination by various biological hazards, leading to personal suffering, preventable fatalities, and an unnecessary economic burden. Consumption of contaminated food can lead to outbreaks of foodborne illnesses. Especially, microbial food-borne illnesses represent a significant public health concern due to the high likelihood of contamination by various biological hazards, leading to personal suffering, preventable fatalities, and an unnecessary economic burden. The consumption of contaminated food caused 7.69% of people in the world, equivalent to 600 million consumers suffer from foodborne illness annually. Among 56 million death annually, foodborne illness caused 420,000 death, accounted for 7.5% total annual death worldwide (Lee & Yoon, 2021).

Common symptoms associated with foodborne pathogens include nausea, vomiting, abdominal pain, diarrhea, fever, and loss of appetite. However, microbial contamination of food can be controlled or prevented through the implementation of good manufacturing practices, rigorous raw material inspections, and strict adherence to the cold chain at both the industrial and retail levels (Prüβ, 2020).

Public Health Concerns
Populations at higher risk, including infants, the elderly, pregnant women, and those with weakened immune systems, face a greater likelihood of severe consequences from foodborne infections. This heightened vulnerability emphasizes the public health concerns surrounding microbial contamination of food.

Widespread illness: Each year, millions of people worldwide suffer from foodborne diseases. Common foodborne pathogens, such as *Salmonella*, *E. coli*, *Listeria*, and *Norovirus*, cause gastrointestinal issues including nausea, vomiting, diarrhea, and abdominal pain. Severe infections can lead to complications such as kidney failure, neurological disorders, and even death, especially in vulnerable populations like children, the elderly, and immunocompromised individuals (Tropea, 2022).

Antibiotic resistance: Some foodborne pathogens have developed resistance to antibiotics due to the overuse of antimicrobials in food animal production. This makes treatment of foodborne infections more challenging and increases the risk of outbreaks of resistant bacteria (Ngoc et al., 2022), further complicating public health efforts to control microbial food-borne illnesses (Okaiyeto et al., 2024).

Economic Impact

Microbial contamination of food imposes significant global economic burdens across sectors. Healthcare systems face high costs due to hospitalizations, medical consultations, and public health responses (Tropea, 2022). Foodborne illnesses lead to lost productivity, especially in sectors like food processing and hospitality, adding to the overall costs. The food industry suffers from expensive product recalls, supply chain disruptions, and reputational damage, with some businesses forced to close. International trade is alsosuffers, as contaminated food products may face import bans, impacting agriculture economies (Prüβ, 2020).

3 Antimicrobial Technologies of Food Processing

Along with advancements in food science and technology, especially in the digital era, people are increasingly accessing scientific knowledge in this field in an easy and multidimensional way. Consequently, there is a growing demand for fresh products with minimal thermal processing during production and with maximum retention of nutritional quality and sensory properties. This poses a challenge that how to preserve food long enough to ensure it remains safe for transport, market distribution and consumption after processing. The conventional technologies have been employed to prevent microbial contamination, ensure food safety, and extend shelf life. In addition; the application of green, safe, and eco-friendly methods for food production gains a growing concern for not only enhancing food safety and quality but also reducing environmental impacts, conserving resources, and supporting a more sustainable food industry. In response to these demands, non-thermal processing methods are continually applied in food processing to eliminate pathogenic and spoilage microorganisms with maximal retention of the nutritional content and sensory value of food (Jadhav et al., 2021; Ucar et al., 2021).

3.1 Physical Technologies

3.1.1 Antibacterial Technology by Reducing Water Activity in Food

Water is essential for microorganisms for transporting nutrients and metabolites, removal of cellular wastes, and maintenance of cell shape. In food products, water is present as free and bound water. Bound water is not available for biological functions. Only free water is important for microbial growth (Suryadi & Mun'im, 2021) If free water in the environment is reduced by either removing water or adding solutes (sugar, flour, salts, etc.) or hydrophilic colloids (proteins, carbohydrates, lipids, etc.), the free water from the cells (hypotonic) flows outside (hypertonic environment) until equilibrium is established. Water loss will cause osmotic shock and plasmolysis in cells, and the cells do not grow (Erkmen & Bozoglu, 2016).

Water activity (a_w) is defined as the ratio of the vapor pressure of water in a food product to the vapor pressure of pure water at the same temperature, and is used as a criterion for the availability of water for microbial growth, enzymatic and non-enzymatic browning reactions, and lipid oxidation (Nakagawa & Oyama, 2019).

$$a_w = \frac{p}{p_0} = \frac{ERH}{100} \tag{1}$$

P = partial vapor pressure of food moisture at temperature T
P_0 = saturation vapor pressure of pure water at T
ERH = equilibrium relative humidity at T.

The relationship between water activity and microbial activity is complex as different microorganisms have varying thresholds for growth and survival at different a_w levels. Some microorganisms have very effective mechanisms to protect their cells up to a certain level of reduction in a_w. Microorganisms show relative resistance against osmotic stress (reduced a_w) by transporting solutes from the environment (such as K^+ ions, amino acids, and polyols). Differences in a_w between cells and the environment activate the osmoregulatory capacities of cells. Tolerance to low a_w differs among microorganisms due to differences in osmoregulation capacities to maintain homeostasis (Erkmen & Bozoglu, 2016).

The relationship between water content and water activity of food products is indicated by the sorption isotherm (Fig. 3).

At a low water content ($<50\%$), even minor changes in this parameter lead to major changes in water activity. For that reason, the sorption isotherm of a food with lower water content is shown with an expanded ordinate in Fig. 3b, as compared with Fig. 3a.

Figure 3b shows that the desorption isotherm, indicating the course of a drying process, lies slightly above the adsorption isotherm on the storage of moisture-sensitive food. As a rule, the position of the hysteresis loop changes when adsorption and desorption are repeated with the same sample. The effect of water activity on processes that can influence food quality is presented in Fig. 4. Decreased water activity retards the growth of microorganisms, slows enzyme catalyzed reactions (particularly involving hydrolases) and, lastly, retards non-enzymatic browning. In contrast, the rate of lipid autoxidation increases in dried food systems.

One of the options for decreasing water activity through drying, salting, or sugar addition and thus improving the shelf life of food. Adding to salt, glycerol, sorbitol

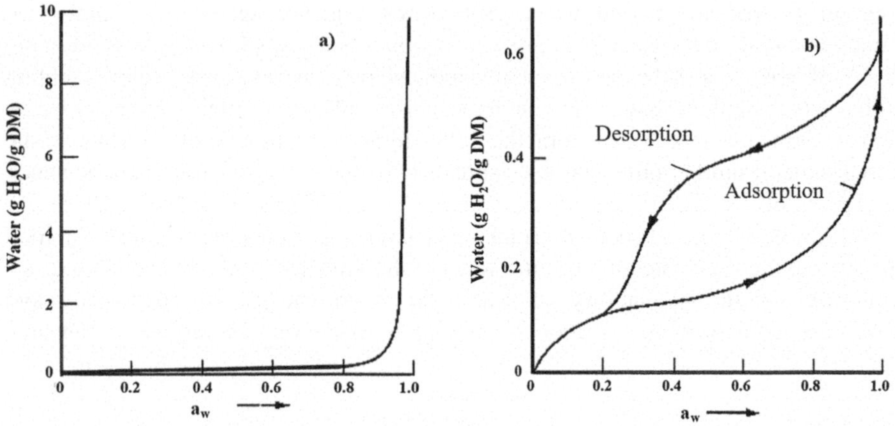

Fig. 3 Moisture sorption isotherm, (**a**) Food with high moisture content; (**b**) Food with low moisture content (*DM* Dry Matter). (Fellows, 2022)

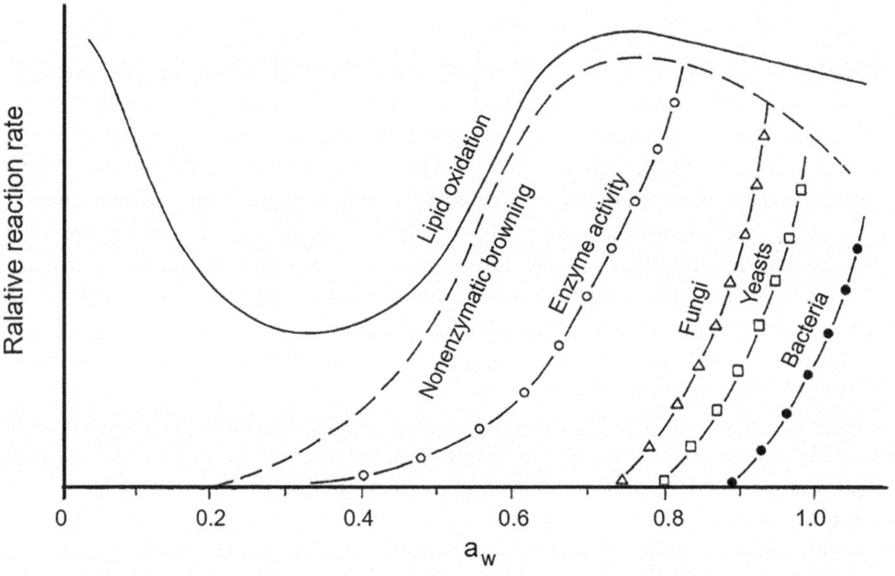

Fig. 4 Food stability as a function of water activity (Belitz & Grosch, 2013)

and sucrose with high water binding capacities as humectants have potentially reduced the water activity (Manjunatha et al., 2023). However, they are also sweeteners and would be objectionable from a consumer standpoint in many foods in the concentrations required to regulate water activity (Belitz & Grosch, 2013).

Water activity also affects the metabolic activity of microorganisms. In the fermented products, the organic acids produced during fermentation not only

Table 6 Effect of a_w on microbial activity (Erkmen & Bozoglu, 2016)

a_w	Effects on microorganisms	Examples of foods
1.00–0.95	Some yeasts, Gram- negative rods, bacterial spores	Fresh foods, foods containing 40% sucrose or 7% salt (canned foods, cheeses, several sausages, bread, etc.)
0.95–0.91	Most cocci, bacilli, lactobacilli, some molds	Foods containing 50% sucrose or 15% salt (mayonnaise, hard cheeses, jams, etc.)
0.94	Growth and toxin production by *C. botulinum*	Sausages
0.91–0.88	Most yeasts	Foods containing 65% sucrose or 15% salt (fruit jams, margarine, fruit juice concentrates, some hard cheeses, etc.)
0.86	Growth of *S. aureus*	Flour, cakes, rice, beans
0.86–0.80	Most molds	Foods containing 15–20% water (fruit cake, sweetened condensed milk, salted meat, salami, etc.)
0.80–0.75	Halophilic bacteria	Foods with 26% salt or very high sugar content (salted fish, molasses, jam, etc.)
0.75–0.65	Xerophilic molds	Foods containing less than 10% water (dates, figs, rolled oats, nuts, etc.)
0.68	Limit for fungi	Dry milk, peanut butter
0.65–0.60	Osmophilic yeasts	Confectionary products, dried fruits containing 15–20% water, honey
<0.60	No microbial growth	Dried milk, instant coffee, dried egg, spices, crackers, flour, cereals, and so on

lower the pH but also reduce the water activity, creating unfavorable conditions for microbial growth (Dejene et al., 2021; Lee & Lee, 2024). Furthermore, the use of natural preservatives such as essential oils and bioactive compounds, can also be affected by the water activity of food products (Schnur et al., 2021). The antimicrobial activity of chitosan, for example, is influenced by the water activity of the food products, with higher efficacy observed at lower water activity levels (Jung et al., 2010). The importance of considering water activity levels when applying food preservation strategies that utilize natural antimicrobial agents. The selection of techniques for reducing water activity was dependent on the type of food products and the desired quality attributes.

The Table 6 shows that microbial growth occurs in the a_w range of 0.7 to 1.0. Bacteria are required at $a_w > 0.9$, while yeasts can grow at $a_w > 0.8$. Significantly more tolerant of low a_w values are molds, whose growth is limited in the a_w range of 0.75 to 0.9 (Muhlbauer & Muller, 2020). The minimum water activity for microbial growth and reactions is shown in Fig. 4.

Effect of water activity on microbial, enzymic and chemical changes to foods are shown in the Table 7.

In Table 8 the interaction of a_w with temperature, pH, oxygen and carbon dioxide, or chemical preservatives, has an important effect on the inhibition of microbial growth. When any one of the other environmental conditions is suboptimal for a given microorganism, the effect of reduced a_w is enhanced. This permits the

Table 7 Moisture content and water activity of foods (Fellows, 2022)

Food	Moisture content (%)	a_w	Minimum a_w required for the growth of microorganisms		Degree of protection required
			a_w	Microorganism	Package to prevent moisture loss
Ice (°C)	100	1.00[a]			
Fresh meat (e.g. beef, chicken) and fish	69–73	0.99			
Raw vegetables (e.g. broccoli, carrots, cauliflower, peppers)	92–96	0.99			
Raw fruits (apples, oranges, grapes)	84–92	0.98			
Fresh root crops (e.g. potatoes, cassava)	70–80		0.97	Most Gram-negative bacteria	
Cooked meats	62	0.91–0.98	0.95	*Pseudomonas* spp., *Bacillus* spp., *Clostridium perfringens* inhibited	
			0.93	Toxin production by *S. aureus*	
Bread	35–40	0.96			
Ice (−10 °C)	100	0.91[a]	0.90	Most Gram-positivebacteria, *Salmonella* spp., *V. parahaemolyticus*, *C. botulinum*, *Lactobacillus* spp. Inhibited	
Sausages, syrups	50–60	0.87–0.91	0.88	Most yeasts	
Jams, marmalades, jellies	30–35	0.86	0.83–0.87	Aflatoxin production by *Aspergillus flavus*	
Rice, beans, peas	13–16	0.80–0.87	0.86	*S. aureus* inhibited	*Minimum protection or no packaging required*
Salami	60	0.82	0.82	Growth of *A. flavus*	
Ice (−20 °C)	100	0.82[a]			
Fruit cake/Christmas pudding	25–28	0.80	0.80	Lower limit for most enzyme activity	

Beef jerky	23	€80	0.80	Most molds	
Marzipan	15	€75	0.75	Halophilic bacteria	
Wheat flour	14	€72			
Peanut butter	2	€70			
Ice (−50 °C)	100	€62[a]			
Dried fruits (e.g. raisins)	15–20	€60–0.65	0.62–0.60	Xerophilic molds osdosmophilic yeasts	
Dried spices	6–12	€35–0.58			
Macaroni	10	€45			
Cocoa powder, egg powder	5	€40			Package to prevent moisture uptake
Boiled sweets	3	€30			
Cookies/biscuits, chocolate	5–6	€20	0.25	Maximum heat resistance of bacterial spores	
Dried vegetables	5	€20			
Dried milk	3	€11			
Potato crisps (chips in the United States)	1.5	€08			

[a]Vapour pressure of ice divided by vapour pressure of water

Table 8 Interaction of a_w, pH and temperature to preserve foods (Fellows, 2022)

Food	pH	a_w	Shelf life	Storage conditions
Fresh meat	>4.6	>0.95	Days	Preserve by chilling
Fresh vegetables			Weeks	'Stable' while respiring, shelf life may be extended by modified atmosphere packaging
Bread			Days	Low a_w in crust, preservative (e.g. calcium propionate) may be added to inhibit mold growth
Milk			Days	Preserve by chilling
Cured meat		0.90	Weeks	Preserve using salt, sodium nitrate/nitrite and chilling
Fermented sausage		<0.90	Months	Preserve using salt, low pH due to lactic acid and low a_w
Fruit cake			Weeks	Preserve by low a_w and heat during baking
Dried milk			Months	Preserve by low a_w and packaging
Pickled vegetables	<4.6	0.90	Months	Preserve using salt, low pH due to lactic acid and packaging
Yoghurt		<0.90	Weeks	Preserve using low pH due to lactic acid, optional added

combination of several mild control mechanisms that result in the preservation of food without any substantial loss of nutritional value or sensory characteristics (Table 8).

3.1.1.1 Techniques for Reducing Water Activity by Thermal Processes

The evaporation or concentration process is defined as the physical removal of water in a product. Evaporation, or concentration by boiling, is the partial removal of water from liquid foods by boiling off water vapour. It increases the solids content of a food and hence preserves it by a reduction in water activity. During evaporation, sensible heat is transferred from steam to the food, to raise the temperature to its boiling point. Latent heat of vaporisation is then supplied by the steam to form bubbles of vapour, which leave the surface of the boiling liquid. The rate of evaporation is determined by both the rate of heat transfer into the food and the rate of mass transfer of vapour from the food. The selection of appropriate evaporation processes such as conventional evaporation, vacuum evaporation, rotary evaporation, and microwave concentration is essential to optimize both the quality and safety of food products (Elik et al., 2016; da Costa Kassacula et al., 2023).

Evaporation is also used to pre-concentrate foods (for example fruit juice, milk and coffee) before drying, freezing or sterilisation and hence to reduce their weight and volume. This saves energy in subsequent operations and reduces storage, transport and distribution costs. There is also greater convenience for the consumer (for example fruit drinks for dilution, concentrated soups, tomato or garlic pastes, sugar) or for the manufacturer (for example liquid pectin, fruit concentrates for use in ice cream or baked goods) (Fellows, 2022).

The drying process removes free water from food, resulting in lower $a_{w,}$ limiting chemical reactions, microbial and enzymatic activities which increases the stability and shelf life (de Oliveira et al., 2020; Zhang et al., 2022). Also, drying reduces significantly the volume, saving the costs of packaging, preservation and transportation (Song et al., 2022). Various drying techniques, including hot air drying, microwave drying, and vacuum drying, have been employed to reduce water activity (Song et al., 2022; Chen et al., 2022a). Additionally, the advanced pretreatments of ultrasonic and microwave can enhance drying rates and improve the quality of final dried products (Mothibe et al., 2014).

Drying is a complex heat and mass transfer process influenced by factors like food composition, moisture content, shape, thickness, airflow, and humidity (Ahmed, 2018). Depending on the temperature and time of exposure, some microbial cells can die or sublethally injured during drying; depending on the storage conditions, microbial cells can also die rapidly at the initial stage and then at a slow rate during storage. Spores generally survive and remain viable during storage in dried food. Most of the microorganisms surviving in drying are thermoduric, but Gram-negative bacteria may also survive, and can cause food spoilage (Erkmen & Bozoglu, 2016).

3.1.1.2 Techniques for Reducing Water Activity by Osmotic Dehydration

Osmotic dehydration is the process of removing water by immersion of water-containing cells into a concentrated aqueous solution (Ciurzyńska et al., 2018). The driving force for water removal is the concentration gradient between the solid and the intracellular fluid. The solute is unable to diffuse through the membrane into the cells. This mass transport process involves simultaneous water and solute movement, with water removal occurring mainly through diffusion and capillary flow, while solute uptake or leaching occurs by diffusion or active transport. Osmotic dehydration from cells can be achieved by adding solutes into foods. Solutes, such as NaCl, sucrose, glucose, and inverted sugar, are used to adjust a_w of foods (Dermesonlouoglou et al., 2019). These solutes do not freely enter the microbial cells. The osmotic dehydration in reducing water activity is influenced by several factors, including concentration gradient, temperature, and duration of treatment (Dermesonlouoglou et al., 2019; Zhao et al., 2014).

The osmotically dehydrated food product can be further processed (generally by air, freeze, or vacuum drying methods) to obtain a shelf-stable product or the dehydration process can be used as a pretreatment for canning, freezing, and minimum processing (Tirado et al., 2017; Jiménez et al., 2020). The use of the osmotic dehydration process in the food industry has several advantages: (i) quality improvement in terms of color, flavor, and texture, (ii) energy efficiency, (iii) reduction in packaging and distribution costs, (iv) elimination of need for chemical treatment, and (v) product stability and retention of nutrients during storage (Erkmen & Bozoglu, 2016).

3.1.2 Antimicrobial Technology by Thermal Process

3.1.2.1 Effect of Heat on Destruction of Microorganisms

Heat denatures proteins, which destroys enzyme activity and enzyme-controlled metabolism in microorganisms and food cells. The rate of destruction of many microorganisms is a first-order reaction: that is when food is heated to a temperature that is high enough to inactivate contaminating microorganisms, the same percentage die in a given time interval regardless of the numbers present initially. This is known as the 'logarithmic order of death' and is described by a 'death rate curve' (Fig. 5).

The time needed to destroy 90% of the microorganisms (to reduce their numbers by a factor of 10) is referred to as the 'decimal reduction time' or *D* value. The *D* value is a critical parameter in designing pasteurization processes (Peñalver-Soto et al., 2021). *D* values differ for different microbial species (Table 9) and a higher *D* value indicates greater heat resistance.

There are two important implications arising from the decimal reduction time:

1. The higher the number of microorganisms present in a raw material, the longer it takes to reduce the numbers to a specified level. In commercial operation the number of microorganisms varies in each batch of raw materials, but it is difficult to recalculate process times for each batch of food. A specific temperature-time

Fig. 5 Death rate curve

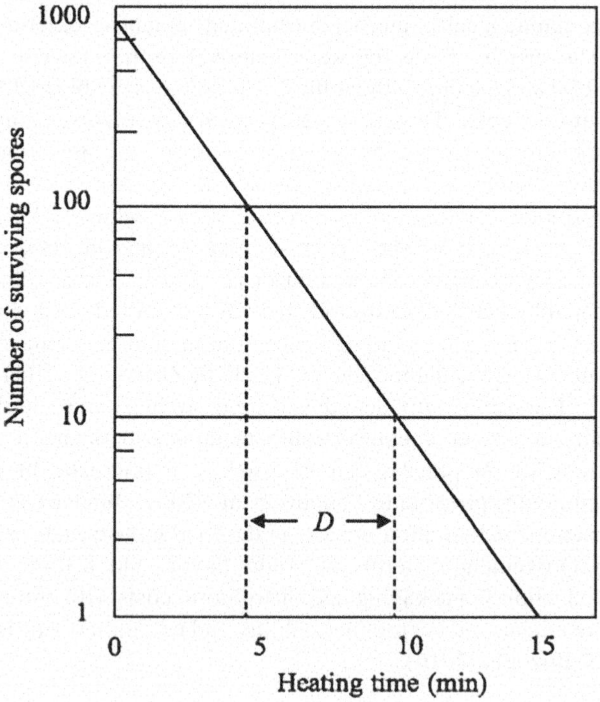

Table 9 Heat resistance of selected pathogens (Fellows, 2022)

Microorganism	*D* Value (min)	z Value	Temperature (°C)	Substrate/typical food
Vegetative cells				
Aeromonas hydrophila	2.2–6.6	5.2–7.7	48	Milk
Bacillus stearothermophilus	3.0–4.0	9–10	–	Vegetables, milk
B. subtilis	0.3–0.76	4.1–7.2	–	Milk products
B. cereus	3.8	36	–	Milk
Campylobacter jejuni	0.62–2.25	–	55–56	Beef/lamb/chicken
C. jejuni	0.74–1.0	–	55	Skim milk
Clostridium sporogenes	0.7–1.5	8.8–11.1	–	Meats
Cl. thermosaccharolyticum	3.0–4.0	7.2–10.0	–	Vegetables
E. coli O111:B4	5.5–6.6	–	55	Skim/whole milk
E. coli O157:H7	4.1–6.4	–	57.2	Ground beef
E. coli O157:H8	0.26–0.47	5.3	62.8	Ground beef
L. monocytogenes	0.22–0.58	5.5	63.3	Milk
L. monocytogenes	1.6–16.7	–	60	Meat products
S. aureus	6	–	60	Meat macerate
S. aureus	3	–	60	Pasta
S. aureus	0.9	9.5	60	Milk
Salmonella senftenberg	276–480	18.9	70–71	Milk chocolate
S. senftenberg	0.56–1.11	4.4–5.6	65.5	Various foods
S. typhimurium	396–1050	17.7	70–71	Milk chocolate
S. typhimurium	2.13–2.67		57	Ground beef
V. cholerae	0.35–2.65	17–21	60	Crab/oyster
V. parahaemolyticus	0.02–2.5	5.6–12.4	55	Clam/crab
V. parahaemolyticus	10–16	5.6–12.4	48	Fish homogenate
Yersinia enterocolitica	0.067–0.51	4–5.78	60	Milk
Spores				
B. subtilis	30.2	9.16	88	0.1% NaCl
B. cereus	1.5–36.2	6.7–10.1	95	Various
C. botulinum 62A	0.61–2.48	7.5–11.6	110	Vegetable products
C. botulinum B	0.49–12.42	7.4–10.8	110	Vegetable products
C. botulinum E	6.8–13	9.78	74	Seafood
C. perfringens	6.6	–	104.4	Beef gravy
C. thermosaccharolyticum	3.3	10.2	124	Molasses

combination is therefore selected to process every batch of a particular product, and adequate preparation procedures are used to ensure that the raw material has a satisfactory and uniform microbiological quality.

2. Because microbial destruction takes place logarithmically, it is theoretically possible to destroy all cells only after heating for an infinite time. Processing therefore aims to reduce the number of surviving microorganisms by a predetermined amount. This gives rise to the concept of ‘commercial sterility’

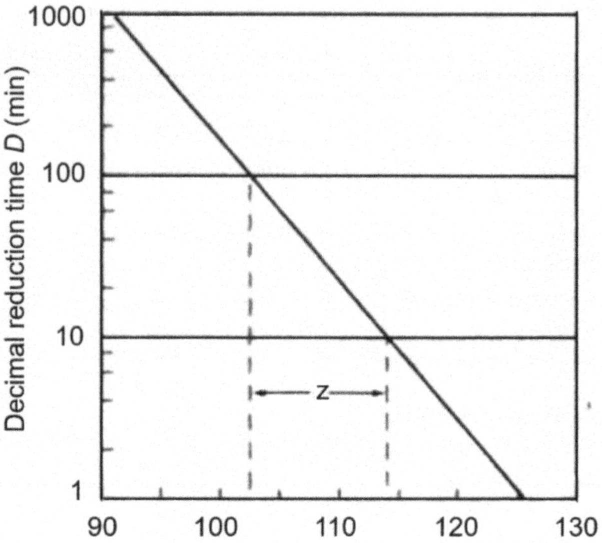

Fig. 6 Thermal death time (TDT) curve. Microbial destruction is faster at higher temperatures (e.g. 100 min at 102.5 °C has the same lethal effect as 10 min at 113 °C)

The destruction of microorganisms is temperature dependent; cells die more rapidly at higher temperatures. By collating *D* values at different temperatures, a semilogarithmic thermal death time (TDT) curve is constructed (Fig. 5).

The slope of the TDT curve is termed the *z* value and is defined as the number of degrees Celsius required to bring about a tenfold change in decimal reduction time (10.5 °C in Fig. 5). The *D* and *z* values are used to characterise the heat resistance of a microorganism and its temperature dependence, respectively (Fig. 6).

There are a large number of factors that determine the heat resistance of microorganisms, but general statements of the effect of a given variable on heat resistance are not always possible. The following factors are known to be important:

1. Type of microorganism: Different species and strains show wide variation in their heat resistance (Table 9). Spores are much more heat resistant than vegetative cells.
2. Incubation conditions during cell growth or spore formation. These include
   (a) temperature (spores produced at higher temperatures are more resistant than those produced at lower temperatures),
   (b) age of the culture (the stage of growth of vegetative cells affects their heat resistance),
   (c) culture medium used (e.g. mineral salts and fatty acids influence the heat resistance of spores).
3. Conditions during heat treatment. The important conditions are as follows:
   (a) pH of the food (pathogenic and spoilage bacteria are more heat resistant near to neutrality; yeasts and fungi can tolerate more acidic conditions but are less heat resistant than bacterial spores),
   (b) water activity of the food influences the heat resistance of vegetative cells; in addition, moist heat is more effective than dry.

3.1.2.2 Pasteurisation

Pasteurisation is a relatively mild heat treatment in which liquid foods are mostly heated to below 100 °C, a termed pasteurisation to indicate the mild heat treatment employed. In low-acid foods (pH > 4.5), such as milk or liquid egg, it is used to minimise public health hazards from pathogenic microorganisms and to extend the shelf life of foods for several days or weeks. In acidic foods, such as fruit juices (pH < 4.5), it is used to extend the shelf life by several weeks by the destruction of spoilage microorganisms (mainly yeasts or molds) and/or enzyme inactivation. In both types of food, only minor changes are caused to the sensory characteristics or nutritive value. The pasteurization process is influenced by factors such as the food matrix, the initial microbial load, and the desired shelf life (Chiozzi et al., 2022).

Theory

The sensible heat required to raise the temperature of a liquid during pasteurisation is found using:

$$Q = mc^{*}(\theta_{\mathrm{A}} - \theta_{\mathrm{B}}) \tag{2}$$

where Q (W) = specific rate of heat transfer, m (kg s^{-1}) = mass flow rate, c (kJ kg^{-1} °C^{-1}) = specific heat capacity and $(\theta_{\mathrm{A}} - \theta_{\mathrm{B}})$ (°C) = temperature change.

The extent of the heat treatment required to stabilise a food is determined by the D value of the most heat-resistant enzyme or micro-organism which may be present. As flavours, colours and vitamins are also characterised by D values, pasteurisation conditions can be optimised for retention of nutritional and sensory quality by the use of high-temperature short-time (HTST) conditions. For example, in milk processing the lower-temperature longer-time process operating at 63 °C for 30 min (the holder process) causes greater changes to flavour and a slightly greater loss of vitamins than HTST processing at 71.8 °C for 15 s. Higher temperatures and shorter times (for example 88 °C for 1 s, 94 °C for 0.1 s or 100 °C for 0.01 s for milk) are described as higher-heat shorter-time processing or 'flash pasteurisation'.

Pathogens are unable to grow in acidic foods (pH < 4.5) and pasteurisation is based on the inactivation of heat-resistant enzymes or acid-tolerant spoilage micro-organisms such as lactic acid bacteria, yeasts and molds. For example, in fruit juices, pectinesterase is more heat resistant than yeasts and Gram-positive nonspore-forming bacteria, and processing is designed to inactivate this enzyme to enable the required shelf life to be obtained. Juice pasteurisation is also intended to inactivate other enzymes, including polyphenoloxidase (tyrosinase or catechol oxidase), peroxidase and polygalacturonase, which produce undesirable changes in sensory quality and nutritive value of products. Polyphenoloxidase causes browning and degradation of pigments and other polyphenols, leading to discolouration and loss of antioxidant activity. Peroxidase is involved in plant metabolic processes, including catabolism of auxins, lignification of cell walls and browning reactions that catalyse oxidation processes. Pectinesterase and polygalacturonase cause the breakdown of pectin and other cell wall materials, resulting in products that have reduced viscosity and undesirable organoleptic properties (Marszałek et al., 2017). High-temperature-short-time (HTST) pasteurisation reduces pectinesterase by 95.3% and polyphenoloxidase by 90.9% in apple juice (Aguilar-Rosas et al., 2013) and completely inactivates polyphenoloxidase, peroxidase and pectinesterase in apricot nectar (Fellows, 2022).

3.1.2.3 Sterilisation

Heat sterilisation is a unit operation in which foods are heated at a sufficiently high temperature and for a sufficiently long time to destroy vegetative microbial cells and spores and to inactivate enzymes. As a result, packaged sterilised foods have a shelf life over 6 months at ambient temperatures. The foods are also precooked and require minimum heating before consumption, thereby increasing their convenience. However, the severe heat treatment during the older process of in-container sterilisation (canning or bottling) may produce substantial changes in nutritional and organoleptic qualities of foods (Ramesh et al., 2016; Tareke et al., 2002).

Developments in processing technology aim to reduce the damage to nutrients and sensory components by either reducing the time of processing in containers using flexible pouches, or by processing foods at higher temperatures for shorter times before packaging, using 'aseptic' or 'ultrahigh-temperature' (UHT) processing. UHT processing typically involves heating foods to 130 °C–150 °C for a few seconds and then filling the product into presterilised containers.

Theory

The aims of in-container sterilisation are

- to destroy pathogenic microorganisms to ensure food safety;
- to inactivate enzymes and kill spoilage microorganisms to ensure the required shelf life;
- to produce the required sensory properties by adequate cooking of the product, without substantial changes to the nutritional quality.

The *z* values of vitamins and the chemicals that contribute to sensory characteristics are four to seven times higher than those of microorganisms (Table 10). Therefore, for every 10 °C rise in processing temperature, there is approximately a doubling of cooking effect, whereas microbial inactivation is increased 10 times. This has given rise to the concept of a 'cook-value' (or *C* value), which is needed to achieve the required change in sensory characteristics (e.g. altering the texture of canned meats or adequate cooking of canned vegetables). The processing time required to achieve a *C* value is longer than that needed for sterilisation.

The length of time required to sterilise a food is influenced by the

- heating conditions;
- heat resistance of microorganisms likely to be present in the food and their numbers;

Table 10 z values for microorganisms compared to heat-vulnerable components of foods (Fellows, 2022)

Component	z Value (°C)
Bacterial spores	7–12
Microbial cells	4–8
Enzymes	3–50
Vitamins	25–30
Proteins	15–37
Sensory factors	
Overall	25–47
Texture-softening	25–47
Pigments	24–50

- types and activity of enzymes;
- pH of the food;
- size and shape of the container;
- physical state of the food.

To determine the process time for a given food, it is necessary to have information about both the heat resistance of microorganisms, particularly heat-resistant spores or enzymes that are likely to be present, and the rate of heat penetration into the food.

Heat Resistance of Microorganisms

The factors that influence heat resistance of microorganisms or enzymes and their characterisation by *D*- and z values are described in Table 5. In low-acid foods (pH > 4.5), the heat resistant, spore-forming microorganism, *C. botulinum*, is the most dangerous pathogen likely to be present. Under anaerobic conditions inside a sealed container, it can grow to produce a powerful exotoxin, botulin, which is sufficiently potent to be 65% fatal to humans if untreated. There are seven forms of botulinum toxin, types A-G, four of which (types A, B, E and rarely F) cause human botulism (WHO, 2023). *C. botulinum* is ubiquitous in soil and it is therefore likely to be found in small numbers on any raw material that has contact with soil, or be transferred by equipment or operators to other foods. Because of the extreme hazard from botulin, the destruction of this microorganism is therefore a minimum requirement of heat sterilisation, such as canned vegetables and meats with pH > 4.5 (Lazárková et al., 2023). Normally, foods receive more than this minimum treatment as other more heat-resistant spoilage bacteria may also be present. *C. botulinum* cannot grow in more acidic foods (pH 4.5–3.7), and other microorganisms (e.g. yeasts and fungi), or heat-resistant enzymes, are more important causes of food spoilage, and these are used to establish processing times and temperatures. In acidic foods (pH < 3.7), enzyme inactivation is the main reason for processing and heating conditions are less severe (sometimes referred to as 'pasteurisation').

Thermal destruction of microorganisms has long been assumed to take place logarithmically at high temperatures, although different thermal destruction kinetics have since been found. A logarithmic death rate means that theoretically a sterile product cannot be produced with certainty no matter how long the process time. However, the probability of survival of a single microorganism can be predicted using details of the heat resistance of the particular microbial strain and the temperature and time of heating. This gives rise to a concept known as 'commercial sterility'. For example, a process that reduces cell numbers by 12 decimal reductions (a 12*D* process) applied to a raw material that contains 1000 spores (Świder et al., 2023) per container would reduce microbial numbers to 10^{-9} per container, or the probability of one microbial spore surviving in one billion (Kamarudin et al., 2022) containers processed. Commercial sterility means in practice that heat processing inactivates substantially all vegetative cells and spores, which if present would be capable of growing in the food under defined storage conditions.

A thermal process that produces commercial sterility in low-acid canned foods may be defined as: 'that process by which all *C. botulinum* spores and all other pathogenic bacteria have been destroyed, as well as more heat-resistant microorganisms which, if present, could produce spoilage under normal conditions of nonrefrigerated storage and distribution (Tucker & Featherstone, 2021).

However, if foods that contain more heat-resistant spoilage microorganisms are given a 12*D* process, this would result in overprocessing and excessive loss of quality. In practice, a 2*D*-8*D* process is used to give the most economical level of food spoilage consistent with adequate food quality and safety. Because of the comparatively lower heat resistance of *C. botulinum,* the probability of survival remains similar to that obtained in a 12*D* process. The spoilage probability can be expressed as follows:

$$\frac{1}{n} = \frac{n_0}{10^{\left(\frac{F}{D}\right)}}$$

where n is the number of containers of processed product, n_0 is the initial number of spoilage microorganisms per container, F is thermal death time (TDT) required (the process time) and *D* is decimal reduction time.

The equation can be used in several ways to calculate

1. the number of containers that can be processed before there is a probability of one spoiled container;
2. an acceptable number of containers that can be processed before one contains a spoilage microorganism;
3. the process time needed to achieve an acceptable level of spoilage;
4. the level of spoilage that could be expected from a process that has a known F value

For the safety and quality reasons, it must be kept at a low-level of microbial loads on raw materials by hygienic handling and proper processing conditions and methods must guarantee uniform heat distribution across all containers. Any failure in these procedures would increase the initial numbers of cells and hence increase the incidence of spoilage after processing (Rahman, 2020).

3.1.3 Antimicrobial Technology by the Removal of Thermal

3.1.3.1 Effect of Heat on Microbial Growth

Temperature can influence enzymatic reactions and microbial growth in food systems. Foodborne pathogens and spoilage microorganisms are able to grow in a range of temperatures. The following categories of microorganisms have been established based on their growth temperature ranges: psychrophiles (subzero –20 °C), psychrotrophs (0–20 °C), mesophiles (20–45 °C), and thermophiles (55–65 °C) (Jay et al., 2008). Psychrotrophic microorganisms are considered cold-tolerant and ubiquitous in natural food environments. Growth temperatures for yeasts and molds demonstrates a broad range of 10–35 °C, with some species growing at temperatures below and above these thresholds (Demirci et al., 2020).

Psychrotrophs are most commonly observed within the *Pseudomonas* and *Enterococcus genera*, growing well at refrigerated temperatures and causing spoilage of refrigerated products such as meats, poultry and vegetables (Jay et al., 2008). With the refrigeration of food products, a reduction in the growth and types of microorganisms occurs once temperatures begin to decrease. For instance, at approximately 12 °C, growth of strict anaerobes is stalled along with *C. perfringens* and proteolytic strains of *C. botulinum* (type A and some of types B and F). At approximately 3 °C, nonproteolytic strains of *C. botulinum* (type E and some of types B and F) are unable to grow. Although 3 °C was once thought to prevent the growth of all pathogens, *L. monocytogenes* and *Yersinia enterocolitica* have shown capable of growth at temperatures below 1 °C. Storage and preservation of foods through freezing at a temperature of approximately 18 °C prevents microbial growth but cannot be considered a kill step alone and commonly used in combination with other inactivation treatments (Demirci et al., 2020).

Thermophiles are frequently associated with canning/retort products, belonging to the genera such as *Bacillus*, *Clostridium*, and *Geobacillus* (Jay et al., 2008). Heat pasteurization techniques are commonly used in industrial processing, targeting the inactivation of spoilage microorganisms and vegetative, non spore forming pathogens. Commercial sterilization (canning/retort) is used to provide a treatment that kills all microorganisms. However, *C. botulinum* spores may still be present though unable to germinate in the low pH or low a_w environment of the product (Montville et al., 2012).

The above classification applies to foodborne microorganisms. There are other microorganisms (bacterial and archaeal species) which can grow optimally at temperatures between 80 °C and 110 °C, called hyperthemophiles. They have been isolated from all types of hot terrestrial and marine environments, including natural and man-made environments. Their enzymes have unique structural and functional properties which provide high thermostability and optimal activity at temperatures above 70 °C (Demirci et al., 2020).

3.1.3.2 Cold Storage

Fresh meats, fish, dairy products and some types of fruits and vegetables, together with chilled processed foods, are stored at temperatures <5 °C to meet safety, quality and legal requirements. The rate of biochemical reactions in fresh foods caused by naturally occurring enzymes changes logarithmically with temperature. Chilling therefore reduces the rate of enzymic changes and retards respiration and senescence in fresh foods. The factors that control the shelf life of fresh crops in chill storage include

- type of crop and variety or cultivar;
- degree of maturity;
- part of the crop selected (the fastest growing parts have the highest metabolic rates and the shortest storage lives);
- condition of the food at harvest (e.g. the presence of mechanical damage or microbial contamination);
- temperature at harvest;
- relative humidity (RH) of the storage atmosphere;
- gas composition of storage atmosphere.

For fresh fruits and vegetables, the optimum conditions is essential to preserve quality. At these optimum conditions to achieve the required storage life and production of/sensitivity to ethylene, cold storage causes little or no reduction in the eating quality or nutritional value. However, excessive storage times, incorrect temperatures and mechanical damage to crops can cause significant changes, including enzymes browning, vitamin C losses, and wilting and weight loss due to transpiration.

Control over the temperature of storage is the most important factor that affects the rate of deterioration of fish and seafoods. Typically, the quality of white fish remains acceptable for 10–20 days if it is kept at 1–2 °C with ice, whereas iced fatty fish such as herring remain acceptable for a shorter time (4–6 days). Seafoods (e.g. shrimps, lobsters, crabs, mussels, clams, squid and scallops) are stored at or near 0 °C until they are sold or processed.

Temperature of food products represents the internal energy of the food, specifically the energy of chemical molecules making up the foods. The molecular energy positively correlates to their agitation and vibration. Therefore, the molecular agitation and vibration is slower at lower temperature, resulting in the decreased rates of reactions in foods preserved at low temperature.

The rate of most reactions occurring in foods decreases 2–5 times when storage temperature decreases for 10 °C. The reduction in reaction rates according to the decrease in temperature at the interval of 10 °C is described as an exponential reduction and expressed as Q10 concept. However, kinetics of reactions involving by enzymes and microorganisms are complicated, not only explained by Q10 equation (Augusto et al., 2018).

$$Q_{10} = \frac{reaction\ rate\ at\ T = Ti + 10^{\circ}C}{reaction\ rate\ at\ T = Ti}$$

3.1.3.3 Freezing

Freezing is a unit operation that is intended to preserve foods without causing significant changes to their sensory qualities or nutritional value. It involves a reduction in the temperature of food to below its freezing point, which is achieved commercially using either mechanical refrigeration or cryogens. It causes a proportion of the water in the food to change state to form ice crystals. The immobilisation of water as ice and the resulting concentration of dissolved solutes in unfrozen water lower the water activity (a_w) of the food. Preservation is achieved by a combination of low temperatures that reduce biochemical, enzymic and microbial activity, reduced water activity and, for vegetables, pretreatment by blanching. When correct freezing, storage and thawing procedures are followed, there are only small changes to the nutritional value and sensory quality of foods. However, slow freezing, temperature abuse during frozen storage and thawing can each damage the quality of foods.

The main groups of commercially frozen foods are as follows:

baked goods (e.g. bread, cakes, fruit and meat pies);
fish fillets and seafoods (e.g. cod, plaice, shrimps and crab meat), including fish fingers, fish cakes or prepared dishes with an accompanying sauce;
vegetables (e.g. peas, green beans, sweetcorn, spinach, Brussels sprouts and potatoes);
fruits (e.g. strawberries, raspberries and blackcurrants) either whole or pureed, or as juice concentrates;
meats as carcasses, boxed joints or cubes and meat products (e.g. sausages, beefburgers and reformed steaks);
prepared foods (e.g. pizzas, desserts, ice cream, ready meals and cook-freeze dishes).

Theory

There are different stages involved in lowering the temperature of a food below its freezing point. First, sensible heat is removed and, in fresh foods, heat produced by respiration is also removed. This is termed the 'heat load' and is important in determining the correct size of freezing equipment for a particular production rate. Next, latent heat is removed when water changes state to form ice. Most foods contain a large proportion of water (Table 11), which has a high specific heat (4.182 kJ kg^{-1} K^{-1}) and a high latent heat of crystallisation (334 kJ kg^{-1}). A substantial amount of energy is therefore needed to remove sensible and latent heat to form ice crystals. The latent heat of other components of the food (e.g. fats) must also be removed before they can solidify, but in most foods, they are present in smaller amounts and require removal of a relatively small amount of heat for crystallisation to take place. Energy for freezing is supplied as electrical energy, which is used to compress refrigerants in mechanical freezing equipment or to compress cryogens.

Table 11 Water contents and freezing points of selected foods (Fellows, 2022)

Food	Water content (%)	Freezing point (°C)
Fruits	87–95	−0.9 to −2.7
Milk	87	−0.5
Vegetables	78–92	−0.8 to −2.8
Egg[s]	74	−0.5
Fish	65–81	−0.6 to −2.0
Meats	55–70	−1.7 to −2.2

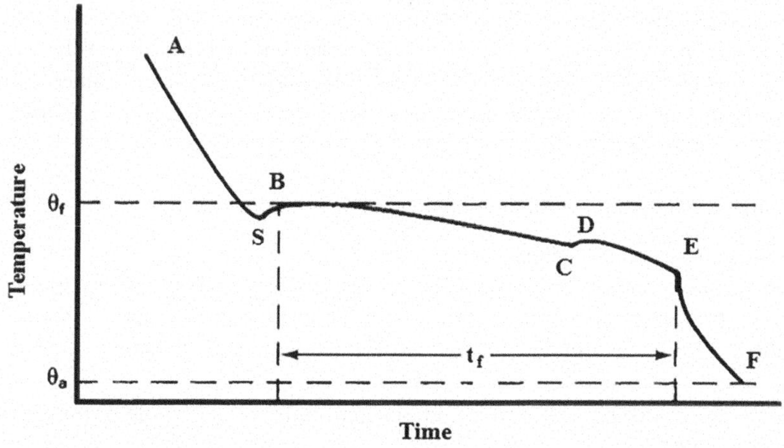

Fig. 7 Time–temperature relationship during freezing (Fellows, 2022)

If the temperature is monitored at the thermal centre of a food (the point that cools most slowly) as heat is removed, a characteristic curve is obtained (Fig. 7). The six components of the curve are as follows:

- AS: The food is cooled to below its freezing point θf which, except for pure water, is always below 0 °C (Table 11). At point S, the water remains liquid, although the temperature is below the freezing point. This phenomenon is known as 'supercooling', which may be as much as 10 °C below the freezing point, and is the period in which nucleation begins. The length of the supercooling period depends on the type of food and the rate at which heat is removed.
- SB: The temperature rises rapidly to the freezing point as ice crystals begin to form and latent heat of crystallisation is released.
- BC: Heat is removed from the food at the same rate as before, but it is latent heat being removed as ice forms and the temperature therefore remains almost constant at the freezing point. The freezing point is gradually depressed by the increase in solute concentration in the unfrozen liquor and, as more ice is formed, the temperature falls slightly. Once stable nuclei have formed, they continue to grow and it is during this stage that the major part of the ice is formed (Fig. 8).

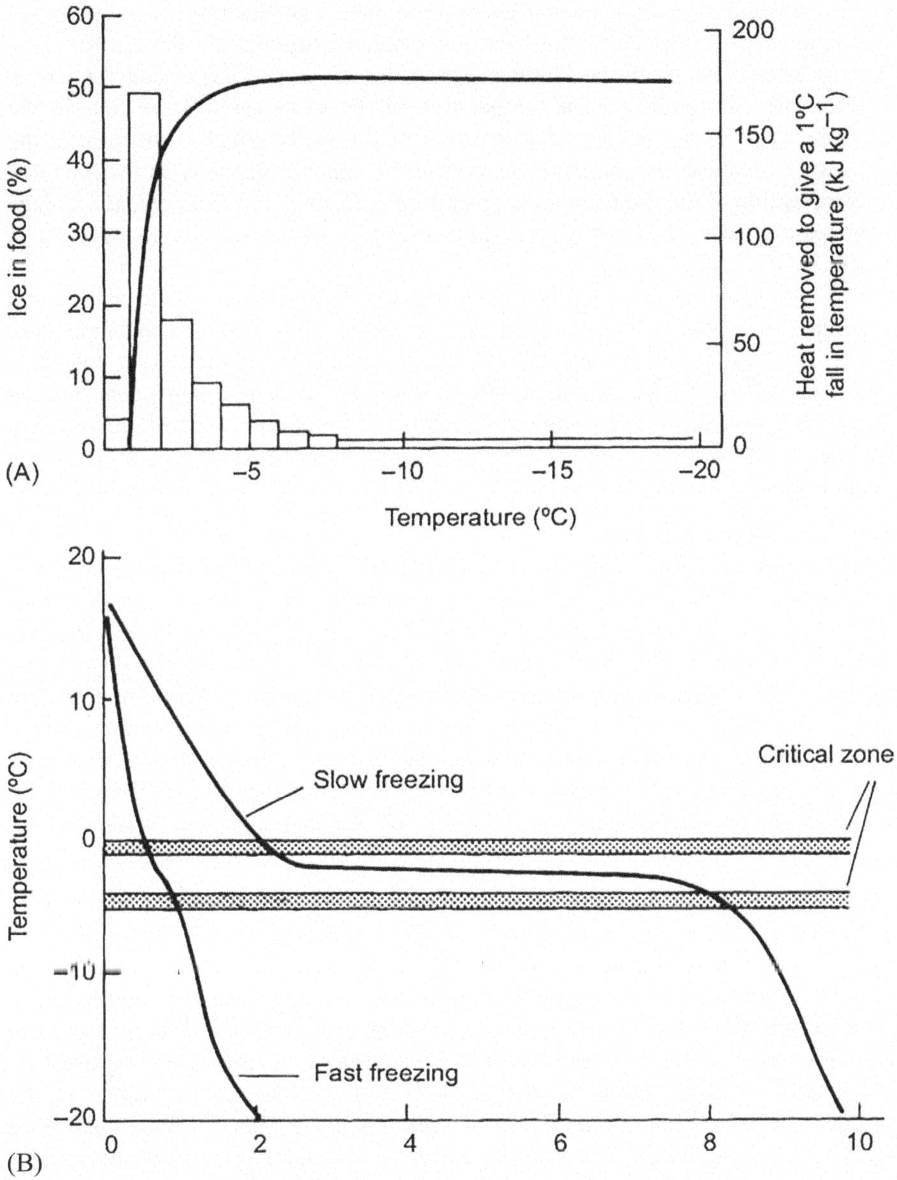

Fig. 8 Freezing: (**A**) ice formation at different freezing temperatures and (**B**) temperature changes of food through the critical zone (Fellows, 2022)

- CD: One of the solutes becomes supersaturated and crystallises out. The latent heat of crystallisation is released and the temperature rises to the 'eutectic' temperature for that solute

- DE: Crystallisation of water and solutes continues. The total time (tf) taken for ice crystal growth (known as the 'freezing plateau') depends on the rate of mass transfer of water from the liquid phase to the ice crystal nuclei and the rate at which heat is removed. The temperature of the ice-water mixture falls to the temperature of the freezer. A proportion of the water remains unfrozen at the temperatures used in commercial freezing; the amount depends on the type and composition of the food and the temperature of storage. For example, at a storage temperature of −20 °C the percentage of frozen water is 88% in lamb, 91% in fish and 93% in egg albumin
- EF: If freezing is continued below commercial temperatures, ice formation and solute concentration continue until no more water can be frozen. The temperature falls as sensible heat is removed from the ice. The temperature (θa) at point F is known as the 'glass transition temperature' of the amorphous concentrated solution. When a critical, solute-dependent concentration is reached, the physical state of the unfrozen liquid changes from a viscoelastic liquid to a brittle, amorphous solid glass.

Effects on Microorganisms

Freezing prolongs the shelf life of products by slowing microbial growth and enzyme activity. In general, the lower the temperature of frozen storage, the lower is the rate of microbiological and biochemical changes. However, freezing and frozen storage have a variable effect on microorganisms and do not inactivate enzymes. The negative effect of freezing on microorganisms is due to temperature shock, concentration of extracellular solutes, toxicity of intracellular solutes, dehydration and ice formation. Quick freezing and thawing result in higher microbial survival rates than those found for slow freezing and thawing. Relatively high storage temperatures (between −4 °C and − 10 °C) have a greater lethal effect on microorganisms than do lower temperatures (between −15 °C and − 30 °C).

Different types of microorganism also vary in their resistance to low temperatures: vegetative cells of yeasts, molds and Gram-negative bacteria (e.g. Coliforms and *Salmonella* spp.) are most easily destroyed; Gram-positive bacteria (e.g. *S. aureus* and *Enterococci)* and mold spores are more resistant, and bacterial spores (especially *Bacillus* spp. and *Clostridium* spp. such as *C. botulinum)* are virtually unaffected by low temperatures. Other factors that affect the microbiological quality of frozen foods are the physical and chemical characteristics of the product, the prefreezing microbiological quality and the extent of handling during and after freezing. If enzymes are not inactivated, the disruption of cell membranes by ice crystals allows them to react to a greater extent with concentrated solutes. The main changes to frozen foods during storage include degradation of pigments, loss of vitamins, residual enzyme activity, and oxidation of lipids (Fellows, 2000).

3.2 *Antibacterial Technology by Fermentation Process*

3.2.1 Principles of Fermentation Process

Fermented foods are among the oldest processed foods. Today they continue to form major sectors of the food processing industry, including baked products, alcoholic drinks, yoghurt, cheese and soy products among many others. During food fermentations, the controlled action of selected micro-organisms is used to alter the texture of foods, preserve foods by production of acids or alcohol, or to produce subtle flavours and aromas which increase nutritional value of foods (Asagbara et al., 2019).

Food fermentations can be grouped into those in which the main products are lactic acid, acetic acid, or alcohol and carbon dioxide. These products help control the growth of microorganisms, thus extending shelf life and ensuring safety (Anal, 2019). Many food fermentations involve complex mixtures of microorganisms or sequences of microbial populations that develop as changes take place in the pH, redox potential or substrate availability.

The main advantages of fermentation process are as follows (Fellows, 2022; Hasan et al., 2014):

- the use of mild conditions of pH and temperature, which maintain the nutritional properties and sensory characteristics of the food;
- enrichment of foods with vitamins, proteins, essential amino acids or fatty acids, or detoxification by removal of antinutritional factors (e.g., phytate, trypsin inhibitor) or toxins (e.g., aflatoxins);
- the production of foods that have diverse flavours, aromas or textures, many of which cannot be achieved by other methods;
- low-energy consumption due to the mild operating conditions;
- acceptable investment and low operating costs.

3.2.2 Bacteriocins and Antimicrobial Ingredients

The production of bacteriocins-antibacterial peptides secreted by lactic acid bacteria-further contributes to the antimicrobial efficacy of fermented products (Qureshi et al., 2017). Bacteriocins are biologically active (or bioactive) peptides that are produced by proteases secreted by some stains of lactic acid bacteria and fungi, which inhibit the growth of spoilage or pathogenic bacteria (Behera et al., 2018). An example is nisin, produced by *Lactococcus lactis* in European cheeses, which prevents growth of *Clostridium tyrobutyricum* and thus prevents off-flavour development and 'blowing' of Swiss-type cheese during ripening. Nisin is also effective against *L. monocytogenes* and, although it has been added to cultures in the past, its production by *L. lactis* is a cheaper and more effective method of removing this pathogen from cheese. *L. lactis* has also been used to inhibit the growth of *C. botulinum* in processed cheese and other dairy products, processed meats, fish (Suganthi et al., 2012) vegetables, soups, sauces and beer (Fellows, 2022).

The sequence of lactic acid bacteria in a fermentation is determined mainly by their acid tolerance. For example, in milk, *Streptococcus liquifaciens*, *Lactococcus* (formerly *Streptococcus*) lactis or the closely related *Streptococcus cremoris* are inhibited when the lactic acid content reaches 0.7–1.0%. They are then outgrown by more acid-tolerant species including *Lactobacillus casei* (1.5–2.0% acid) and *Lactobacillus bulgaricus* (2.5–3.0% acid). Similarly, in vegetable fermentations, *Lactobacillus* spp. are stronger acid producers than *Streptococcus* spp. Among the four main groups of lactic acid bacteria, *Streptococcus spp.* and *Pediococcus* spp. groups are homolactic fermenters, *Leuconostoc* spp. group is heterolactic fermenters and *Lactobacillus* spp. group exhibit strain-dependent metabolic pathways.

In food fermentations, particularly those that involve low-acid substrates (for example milk and meat), a starter culture is added to rapidly generate large numbers of the desired micro-organism, and thus reduce fermentation times and inhibit growth of pathogens and spoilage bacteria. In other fermentations, the natural flora is sufficient to reduce the pH rapidly and to prevent the growth of undesirable micro-organisms. Lactic acid bacteria, including *Leuconostoc spp.*, *Lactobacillus* spp. and *Pediococcus* spp. produce a range of bacteriocins. An example is *Pediococcus acidilactici*, which when used in fermented meat, has the potential to inhibit spoilage bacteria and thus reduce the need for nitrate addition (Fellows, 2000).

3.3 Antimicrobial Technology by Food Additive

3.3.1 Definition of Food Additive

Food additive means any substance not normally consumed as a food by itself and not normally used as a typical ingredient of the food, whether or not it has nutritive value, the intentional addition of which to food for a technological (including organoleptic) purpose in the manufacture, processing, preparation, treatment, packing, packaging, transport or holding of such food results, or may be reasonably expected to result, (directly or indirectly) in it or its byproducts becoming a component of or otherwise affecting the characteristics of such foods. The term does not include "contaminants" or substances added to food for maintaining or improving nutritional qualities (Alimentarius, 2023).

3.3.2 Role of Food Additive

Food additives may be classified by one of six primary functions they serve: preservation, improvement in nutritional value, addition or replacement of color, addition or replacement of flavor, improvement in texture, or processing aids (Branen & Haggerty, 2002).

3.3.3 Mechanism of Action of Food Additives for Microorganisms

The mechanism of action of food additives on microorganisms varies significantly, influencing microbial behavior, growth, and metabolic processes. These additives can either inhibit microbial growth or alter microbial metabolism, leading to changes in food safety and quality. The following sections outline key mechanisms through which food additives interact with microorganisms.
Antibiotic Tolerance Modulation

- Benzaldehyde (BZH): This food additive has been shown to induce fluoroquinolone tolerance in *E. coli* by inhibiting flagella formation and reducing proton motive force (PMF), which helps expel antibiotics from bacterial cells (Xiao et al., 2024).
- Impact on Respiration: BZH also suppresses bacterial respiration and reactive oxygen species (ROS) production, contributing to its tolerance effects (Xiao et al., 2024).

Biogenic Amine Reduction

- Plant Additives: Certain plant-derived additives can significantly reduce biogenic amine (BA) formation in fermented foods by exerting antimicrobial effects and inhibiting microbial decarboxylases (Świder et al., 2023; Abedini et al., 2023).
- Natural and Synthetic Additives: Both types of additives have been identified to either increase or decrease BA production, depending on their nature and the food matrix involved (Abedini et al., 2023).

Gut Microbiota Interaction

- Alteration of Microbial Composition: Food additives can influence the abundance and composition of gut microbiota, potentially affecting digestion, metabolism, and overall health (Zhou et al., 2023). Long-term exposure to these additives may lead to significant changes in microbial communities, raising concerns about their health implications.

While food additives can enhance food safety and quality, their effects on microorganisms can lead to unintended consequences, such as the development of antibiotic tolerance or alterations in gut microbiota composition. This duality highlights the need for careful consideration in the use of food additives.

3.3.4 MaxLevel of Food Additive Used Antimicrobials

FDA defines antimicrobial agents as "substances used to preserve food by preventing the growth of microorganisms and subsequent spoilage, including fungistats, mold and rope inhibitors." The European Union definition of "preservatives" is the same as the US definition of antimicrobials, i.e., those compounds that "extend the shelf-life of foods by protecting against loss of quality caused by microorganisms or protecting against the growth of illness-causing microorganisms".

Food antimicrobials are added to extend the shelf life and maintain the quality of the food products due to the inhibition of spoilage microorganisms. Food antimicrobials have been used exclusively to control the growth of specific foodborne pathogens and spoilage bacteria. Organic acids against pathogens on beef carcasses, nisin and lysozyme against *C. botulinum* in pasteurized process cheese, and lactate and diacetate against *L. monocytogenes* in processed meats. Thus, food antimicrobials are being increasingly applied for the inhibition or inactivation of microorganisms in food products. The selection of the appropriate antimicrobial agent depends on several key factors, including the specifications of the antimicrobial activity, the chemical properties of the antimicrobial, the physicochemical properties and composition of the food product in question, and the type of storage or processing and storage system used. The selection of antibacterial substances and malevel of them can be referred to the following Tables 12 and 13.

Table 12 Traditional or regulatory-approved food antimicrobials (Taylor & Doores, 2020)

Compound(s)	Microbial target	Primary food applications
Acetic acid, acetates, diacetates, dehydroacetic acid	Yeasts, bacteria	Baked goods, condiments, confections, dairy products, fats/oils, meats, sauces
Benzoic acid, benzoates	Yeasts, molds	Beverages, fruit products, margarine
Dimethyl dicarbonate	Yeasts	Beverages
Lactic acid, lactates	Bacteria	Meats, fermented foods
Lactoferrin	Bacteria	Meats
Lauric arginate	Bacteria	Meats
Lysozyme	*C. botulinum*, other bacteria	Cheese, casings for frankfurters, cooked meat, and poultry products
Natamycin	Molds	Cheese
Nisin	*C. botulinum*, other bacteria	Cheese, casings for frankfurters, cooked meat, and poultry products
Nitrite, nitrate	*C. botulinum*	Cured meats
Parabens (alkyl esters (propyl, methyl, heptyl) of *p*-hydroxybenzoic acid)	Yeasts, molds, bacteria (Gram-positive)	Beverages, baked goods, syrups, dry sausage
Propionic acid, propionates	Molds	Bakery products, dairy products
Sorbic acid, sorbates	Yeasts, molds, bacteria	Most foods, beverages, wines

Table 13 Regulatory-approved food antimicrobial designations in codex Alimentarius (Organization WH, 2019)

INS	Food additive	Functional class	Maximum level depending on the type of food
260	Acetic acid, glacial	Acidity regulator, Preservative	GMP
210	Benzoic acid	Preservative	800–1000 mg/kg
211	Sodium benzoate	Preservative	800–1000 mg/kg
212	Potassium benzoate	Preservative	800–1000 mg/kg
213	Calcium benzoate	Preservative	800–1000 mg/kg
928	Benzoyl peroxide	Bleaching agent, Flour treatment agent, Preservative	75–100 mg/kg
263	Calcium acetate	Acidity regulator, Preservative, Stabilizer	GMP
282	Calcium propionate	Preservative	3000-GMP mg/kg
290	Carbon dioxide	Carbonating agent, Foaming agent, Packaging gas, Preservative, Propellant	GMP
242	Dimethyl dicarbonate	Preservative	200–250 mg/kg
385	Calcium disodium ethylenediaminetetraacetate	Antioxidant, Colour retention agent, Preservative, Sequestrant	25–1000 mg/kg
386	Disodium ethylenediaminetetraacetate	Antioxidant, Colour retention agent, Preservative, Sequestrant, Stabilizer	25–800 mg/kg
239	Hexamethylene tetramine	Preservative	25 mg/kg
214	Ethyl para-hydroxybenzoate	Preservative	36–1500 mg/kg
218	Methyl para-hydroxybenzoate	Preservative	36–1500 mg/kg
384	Isopropyl citrates	Antioxidant, Preservative, Sequestrant	100–200 mg/kg
234	Lauric arginate ethyl ester	Preservative	50–315 mg/kg
1105	Lysozyme	Preservative	GMP-500 mg/kg
235	Natamycin	Preservative	6–10 mg/kg
231	ortho-Phenylphenol	Preservative	12 mg/kg
232	Sodium ortho-phenylphenol	Preservative	12 mg/kg
339 (iii)	Trisodium phosphate	Acidity regulator, Emulsifier, Humectant, Preservative, Sequestrant, Stabilizer, Thickener	2200 mg/kg
280	Propionic acid	Preservative	GMP-3000 mg/kg
262 (i)	Sodium acetate	Acidity regulator, Preservative, Sequestrant	GMP-6000 mg/kg
262 (ii)	Sodium diacetate	Acidity regulator, Preservative, Sequestrant	500–4000 mg/kg
281	Sodium propionate	Preservative	GMP-3000 mg/kg
200	Sorbic acid	Antioxidant, Bleaching agent, Flour treatment agent, Preservative	500 mg/kg

(continued)

Table 13 (continued)

INS	Food additive	Functional class	Maximum level depending on the type of food
201	Sodium sorbate	Antioxidant, Bleaching agent, Flour treatment agent, Preservative	200–10,000 mg/kg
202	Potassium sorbate	Antioxidant, Preservative	200–10,000 mg/kg
203	Calcium sorbate	Antioxidant, Bleaching agent, Flour treatment agent, Preservative	200–10,000 mg/kg
220	Sulfur dioxide	Antioxidant, Bleaching agent, Flour treatment agent, Preservative	15–1000 mg/kg
221	Sodium sulfite	Antioxidant, Preservative	15–1000 mg/kg
222	Sodium hydrogen sulfite	Antioxidant, Sequestrant	15–1000 mg/kg
223	Sodium metabisulfite	Antioxidant, Bleaching agent, Flour treatment agent, Preservative	15–1000 mg/kg
224	Potassium metabisulfite	Antioxidant, Bleaching agent, Flour treatment agent, Preservative	15–1000 mg/kg
225	Potassium sulfite	Antioxidant, Preservative	15–1000 mg/kg
	Sodium thiosulfate	Antioxidant, Bleaching agent, Flour treatment agent, Preservative	15–1000 mg/kg

3.4 *Antimicrobial Packaging*

Antimicrobial packaging was developed to protect food from microbial contamination, extend the shelf life of food and help to reduce food waste. Five common methods have been used to develop efficient antimicrobial packaging systems (Limbo & Khaneghah, 2015) as follows:

(i) the incorporation of sachets or pads that contain volatile antimicrobial compounds;
(ii) the incorporation of volatile and nonvolatile antimicrobial compounds directly into the structure of the polymers;
(iii) the use of a coating or adsorbing antimicrobial compounds onto the surfaces of the polymers in contact with a food;
(iv) the immobilization of antimicrobial agents within polymers using methods ion or covalent linkages;
(v) the application of polymers, such as chitosan, that can inherently act as antimicrobial compounds (Fig. 9).

Some studies on the application of packaging in combination with antimicrobials are summarized in the following Table 14.

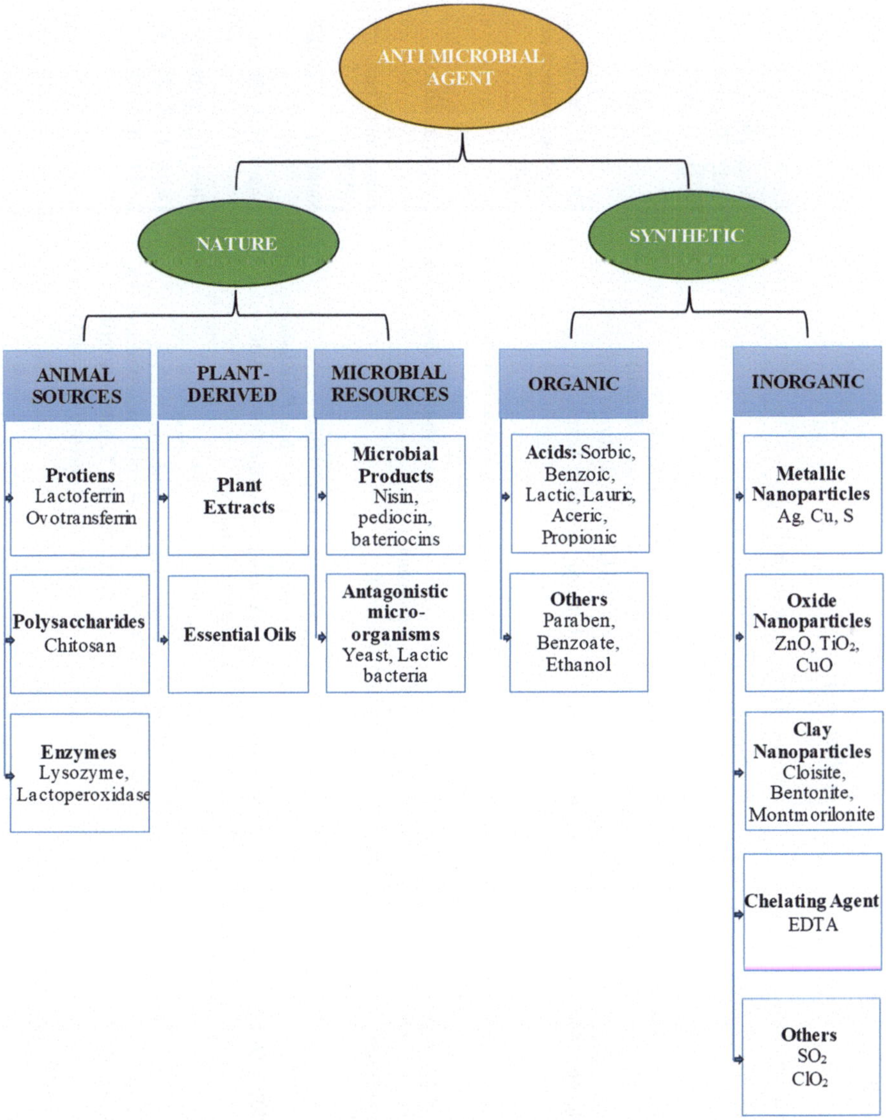

Fig. 9 Grouping of antimicrobial agents (Apdaped from Kamarudin et al., 2022)

Table 14 Studies on antimicrobial packaging

Film matrix		Application in product preservation	Target organisms	References
Solid substrate	Functional additives			
CMC	Chi + gellan/LAC NPs	Strawberry	ABC	Duarte and Picone (2022), Tavassoli et al. (2021)
GEL+CNF	(i) LAC (ii)QUE	NA	(i) LAC (*S. aureus* < *E. coli*) (ii) QUE (*S. aureus* > *E. coli*) (iii) LAC/ QUE (*S. aureus* < *E. coli*)	
NaAlg	CuS nanocomposite	NA	*L. monocytogenes; E. coli*	Roy and Rhi (2020), Saravanakumar et al. (2020)
NaAlg	CNCs, CuO	Fresh-cut Pepper	*E. coli, S. typhi, S. aureus, Candida albicans, Trichoderma* spp.	
NaAlg and GEL	MMT and clove essential oil	Cheese	*E. coli, L. monocytogenes* and *Penicillium expansum*	Ding et al. (2020)
NaAlg	MMT nanocomposite infused with star anise oil	Cherry tomato	*E. coli*	Zhang and Chen (2023)
Chi + NaAlg	(i) LAC (ii)QUE	Pork meat	↓ Growth of TVC, PBC, *Enterobacteriaceae,* and *Pseudomonas* spp.	Montone et al. (2023)
CMC	Chi + TPP/ LAC NPs	Strawberry	↓ Aerobic mesophiles/lowest MIC values against *S. aureus*	Duarte et al. (2023)
CMCS- GEL	ZnO	NA	*S. aureus* < *E. coli*	Chen et al. (2022b)
BC	LAC	Fresh sausages	*S. aureus* < *E. coli*	Padrao et al. (2016)
GEL	(i) LAC/free (ii) LYS/free	Salmon fillet	LAC/ LYS (↓ Aerobic mesophiles of 0.8, psychrotrophs of 1.5, enterobacteriaceae of 0.3, lactic acid bacteria of 0.3, and pseudomonas of 0.3 log)	Rollini et al. (2016)

PS	(i) LAC/free (ii) LYS/free	Minced pork meat	(i) LAC (*L. innocua* < *E. coli*) (ii) LYS (*L. innocua* > *E. coli*) (iii) LAC/ LYS (*L. innocua* > *E. coli*)	Montone et al. (2023)
CS	Cinnamic acid and ferulic acid	Chicken breast	Inhibition of *E. coli* and *L. innocua*	Ordoñez et al. (2021)
CS & GEL	Curcumin	Sausages	Reduced the growth of *Lysteria* species	Tosati et al. (2018)
CS	Thymol nanoparticles	Blueberries and cherry tomatoes	*E. coli, S. typhi, S. aureus and L. innocua.*	Medina et al. (2019)
Agar, alginate, collagen	AgNPs and grapefruit seed extract	Fresh potatoes	*L. monocytogenes; E. coli*	Wang and Rhim (2015)
Sweet potato starch	Thyme essential oil	fresh spinach leaves	*S. Typhi and E. coli*	Issa et al. (2017)
Alginate	*Lacticaseibacillus casei* and *Bifidobacterium animalis* subsp. *lactis*	NA	*L. innocua CIP8011 and E. coli O157:H7 FP605/03*	Bambace et al. (2021)
Alginate	Propolis extract	*Colossoma macropomum* fillet	*E. coli, Pseudomonas aeruginosa, S. aureus, L. monocytogenes*	Cruz et al. (2021)
Alginate, chitosan and carboxymethyl cellulose	*Bifidobacterium lactis*, *Lactiiplantibacillus acidophilus*, and *Lactiplantibacillus casei*	Soft cheese	*S. aureus, Salmonella Typhimurium, L. monocytogenes, E. coli, B. cereus Aspergillus niger Aspergillus flavus*	El-Sayed et al. (2021)
Alginate	Cinnamon essential oil and citric acid	Sliced cooked ham	*L. monocytogenes, E. coli*	Lee et al. (2023)

Notes: *ABC* aerobic bacteria count, *BC* bacterial cellulose, *Chi* Chitosan, *CMCS* Carboxymethyl chitosan, *CMC* carboxymethyl cellulose, *CNB* chitosan nanofiber, *CNCs* cellulose nanocrystals, *CS* Cassava starch, *GEL* Gellatin, *LAC* Lactoferrin, *LYS* lysozyme, *MMT* Montmorillonite, *NPs* nanoparticles, *NA* not analyzed, Na-Alg; sodium alginate, *PS* potato starch, *PBC* psychotropic bacteria count, *QUE* quercetin, *TVC* total viable count

3.5 Emerging Technologies

3.5.1 High Hydrostatic Pressure Processing (HPP)

HPP has found a growing acceptance in the food industry for producing high-quality foods such as meat products and then seafood and fish products. HPP uses high pressure to remove pathogens without significant heat, thus retaining its nutritional and sensory properties (Koutchma, 2014).

HPP is a novel technique that emerged in food production within the last decade and more recently for pharmaceutical and medical applications. Key advantages of HPP such as instant and homogenous transmittance of pressure in the treated product and the possibility of processing at ambient temperatures represent an attractive powerful tool to implement mild postpackaging processing for a broad range of foods. Theoretically pressure-temperature operational conditions for HPP treatment of foods may range from 300 to 1000 MPa at −22 °C to 121 °C, depending on the process and the nature of the food resulting in the processes such as sterilization, pasteurization, or shelf-life extension with added value. The volume reduction of water and water-based foods at 600 MPa and room temperature is approximately 15%. Fluids undergo adiabatic heating during compression and cooling during decompression. Adiabatic heat of compression is the instantaneous volumetric temperature change in materials during compression or decompression is often called the heat of compression. As follows from Table 15, significant differences in compression heating behavior have been observed in foods of different composition.

Compressibility is an intrinsic property of the material and is defined by the balance between attractive and repulsive potentials. Compression of a liquid decreases the average intermolecular distance and tends to reduce rotational and translational motion. Food material is considered to contain molecules that occupy space in excess of that needed for close packing. This excess is called "free volume" and it is this volume that is reduced in initial compression. At elevated pressures, when the free volume has largely disappeared, reduction in van der Waals dimensions may occur and the compressibility is greatly diminished.

The governing principles of HPP processing are based on the assumption that foods under high pressure in a vessel follow the isostatic rule regardless of their size

Table 15 Compression heating (°C/100 MPa) factors of selected food substances determined at initial product temperature of 25 °C (Koutchma, 2014)

Food substances	Temperature increase (°C/100 MPa)
Orange juice, tomato salsa, skim milk, salmon fish	2.6–3.0
Carbohydrates	2.6–3.6
Proteins	2.7–3.3
Mayonnaise	5.0–7.2
Extracted beef fat	6.2–9.1
Olive oil	6.3–8.7

or shape. According to this rule, pressure is instantaneously and uniformly transmitted throughout a sample, whether the sample is in direct contact with the pressure medium or hermetically sealed in a flexible package. Therefore, in contrast to thermal processing, the time necessary for HPP processing should be independent of the sample size.

The effect of HPP processing on food chemistry and microbiology is governed by the Le Chatelier's principle. High pressure stimulates some phenomena (e.g., phase transitions, chemical reactions, and changes in molecular configuration) that are accompanied by a decrease in volume but opposes reactions that involve an increase in volume. The effects of pressure on protein stabilization are also governed by this principle, i.e., the negative changes in volume that occur with an increase in pressure cause an equilibrium shift toward bond formation. Alongside this, the breaking of ionic bonds is also enhanced by HPP, as this leads to a volume decrease due to the electrostriction of water. Moreover, hydrogen bonds are stabilized by high pressure, as their formation involves a volume decrease. Pressure does not generally affect covalent bonds. Consequently, HPP can disrupt large molecules and microbial cell structures, such as enzymes, proteins, lipids, and cell membranes, and leave small molecules such as vitamins and flavor components unaffected.

Packaging is a vital factor for any preservation method to be successful since it protects the food products from recontamination during processing and adverse environmental conditions during storage. For HPP, food is packaged in the form of pouches before the application of high pressure, and the food product and the package are both exposed to the processing conditions, so the entire pack remains a "secure unit" until the consumer opens it. Prepackaging helps to prevent contamination from the pressure medium and improves HPP efficiency. The severe processing conditions used in high-pressure applications may cause changes in the structure of packaging materials and therefore the barrier and mechanical properties of the package may be altered. When high-pressure and -temperature combinations are used in order to achieve sterilization effect, prepackaged food products need to be preheated to a targeted temperature before high pressure is applied (Koutchma, 2014).

High-Pressure Pasteurization

Pasteurization treatment typically employs pressures in the range of 600 MPa at or near ambient temperatures (high-pressure low-temperature process, HPLT) for a specific holding time. Inactivation of pathogenic vegetative bacteria is the primary objective and product requires refrigeration storage.

Pressure-assisted thermal processing (PATP) or pressure-assisted thermal sterilization (PATS) sterilization treatment use elevated pressures (500,900 MPa) combined with a several minutes heat (90–121 °C) exposure at pressure to sterilize low acid foods (high-pressure-high-temperature process, HPHT). Inactivation of bacterial spores is the primary objective and product normally stored at ambient temperature (Koutchma, 2014).

Process Parameters

Cycle time

The total time for loading, closing the vessel, compression, holding, and decompression and unloading is commonly referred to as the "cycle time." Commercial processing times can range from a pulse of a few seconds to over 20 min.

Pressure

Process pressure (P) refers to the holding pressure during the sample treatment. It has been recommended to control and record pressure at ±0.5% (electronic) or ± 1.0% (dial display) level of accuracy.

Temperature

Initial temperature of pressure-transmitting fluid and the process vessel must be documented, as they are integral conditions for microbial inactivation. The reference thermocouple sensor should be located at the cold point or equivalent zone within the pressure vessel and calibrated to an accuracy of ±0.5 °C.

It is important to consider that all compressible substances change temperature during physical compression and this is an unavoidable thermodynamic effect. In general, the temperature of both the product and the pressure-transmitting fluid may rise by 20–40 °C during HPP treatment. Solid metallic materials do not experience significant compression heating. Therefore, the temperature increase may vary in foods with relatively complex compositions (Koutchma, 2014).

Product parameters

HPP process variables may be further complicated by the presence of variability of food composition (e.g., fat content and its distribution), pH, and water activity of the food.

pH: Inactivation of pathogenic microorganisms in food in pH < 4.6 is a requirement for HPP pasteurization at pressures above 200 MPa and at temperatures less than 45 °C. Reduced pH is generally synergistic with pressure in eliminating microorganisms (Koutchma, 2014). Studying the effect of water activity on the HPP inactivation of *L. monocytogenes*. Processing parameters used were: pressure (400–600 MPa), a_w (0.94–0.98), and holding time (180–420 s). The relationship between the listed variables was described through a polynomial equation, the three variables influenced the microorganism inactivation, where a_w showed the highest influence. Processing pressure and holding time showed a positive influence on the inactivation of the microorganism. On the other hand, the pathogen inactivation decreased with the reduction of the a_w in the emulsified meat product, showing a clear baroprotection of the microorganism by the lower a_w (Ferreira et al., 2025).

3.5.2 Pulsed Electric Fields (PEF)

PEF technology effectively inactivates microorganisms while retaining food quality, making it suitable for juices and other sensitive products (Barbosa-Cánovas et al., 2022).

Phospholipids are one of the components of cell membranes, having amphipathic properties with polar (hydrophilic) heads and non-polar (hydrophobic) tails. These molecules are arranged in a bilayer, in which the non-polar tails face inwards and the polar heads face outwards interacting with the surrounding aqueous environment.

When treated with electrical pulses, electrophoresis is initiated by water molecules into the lipid bilayer of the membrane, leading to the reorientation of lipids with polar head groups towards these water molecules. The theory of the electrophoretic mechanism is mainly based on the formation of hydrophilic pores, however recent studies have shown that other factors may contribute to increased membrane permeability, such as chemical changes to lipids and changes in membrane protein function (Kotnik et al., 2019). The application of short electric pulses of high voltages for short duration to biological cells and tissues to enhance membrane permeability, is becoming a pertinent method in different applications including applications in food technological, medicinal, and biotechnological and environmental processes. However, depending on the research field, those processes/techniques appear in the scientific literature with different terms, namely electroporation, electropermeabilization, electropulsation or PEF treatment, and are characterized by subtle differences (Kotnik et al., 2019; Jaeger et al., 2009).

The application of an electric field to a biological cell can lead to three different outcomes depending on (i) the electric field strength, (ii) the duration of exposure, and (iii) the membrane recovery rate (Mahnič-Kalamiza & Miklavčič, 2022). In case the electric field strength and the duration of exposure are insufficient to achieve pore formation, electroporation is unachievable, and there is no effect on the cell's permeability and viability (Mahnič-Kalamiza & Miklavčič, 2022). On the condition that the electric field strength exceeds the critical threshold value of the potential difference across a membrane (transmembrane potential), typically between 0.5 and 1.5 V, repulsion between the charge-carrying molecules induces pore formation in the cell membrane (Barba et al., 2015). Depending on the selected parameters of PEF treatment, reversible or irreversible electroporation can occur, with the intensity of treatment determining whether the electroporation can cause temporary or permanent injuries for microbial cells (Jaeger et al., 2009; Lytras et al., 2024).

3.5.3 Dielectric, Ohmic and Infrared Heating

Conventional thermal processing can produce large temperature gradients and uneven temperature distributions within a product. In contrast the five electromagnetic heating techniques [infrared (IR), microwave, radio frequency, ohmic and magnetic induction heating] heat foods, particularly solid foods, faster and more efficiently than thermal processing.

The advantages of dielectric and ohmic heating compared to conventional heating can be summarised as:

- Rapid volumetric heating throughout the food without localised overheating or hot surfaces, which results in minimum heat damage and no surface browning.
- Heat transfer is not limited by boundary films and energy conversion efficiencies are high (60% to 100%), compared to heat from boilers via heat exchangers, which has an energy conversion efficiency of 25–30%.
- Equipment is small, compact and suited to automatic control.
- There is no risk of contaminating foods by products of combustion (Table 16).

Effects on Foods and Microorganisms of Microwave Heating The effects of electromagnetic energy on the destruction of vegetative microbial cells and changes to food components are similar to those found using conventional methods of heating. However, the more rapid microwave and RF radio frequency heating results in shorter processing times and hence fewer changes to nutritional and sensory properties (Fig. 10). As in conventional heating, heat-sensitive vitamins (e.g. ascorbic acid) undergo losses.

Effects on Foods and Microorganisms of Ohmic Heating Ohmic heating is a high temperature short time (HTST) process and therefore has similar benefits to other methods of rapid heating that destroy microorganisms before there are adverse effects on nutrients or organoleptic qualities (see Sect. 3.1.2.1 for details of *D*- and *z* values). It also causes similar changes in foods to those found in conventional heating, such as starch gelatinisation, melting of fats and coagulation of proteins. In addition to the lethal thermal effect on microorganisms, there are also nonthermal lethal effects caused by the electric field. These are (i) electroporation (formation of pores in cell membranes due to the increased electrical potential across the membrane, resulting in leakage); (ii) membrane rupture caused by the voltage drop across the membrane and (iii) cell lysis, causing disruption to internal components of the cell. These effects also increase the diffusion of materials from solid food particles into the carrier liquid. Losses of materials from food cells only alter the nutritional value if the liquid is not consumed, in, for example, ohmic blanching (Fellows, 2022).

Table 16 Applications of dielectric, ohmic and infrared heating (Fellows, 2022)

Method of heating	Applications
Microwave	Cooking, thawing, melting, finish-drying, freeze drying, tempering, pasteurisation, sterilisation, rendering, frying, blanching
Radio frequency	Drying, baking, cooking (e.g. processed meat)
Ohmic	UHT sterilisation, pasteurisation
Infrared	Drying, baking/roasting, frying, blanching, thawing, freeze drying, cooking, surface pasteurisation

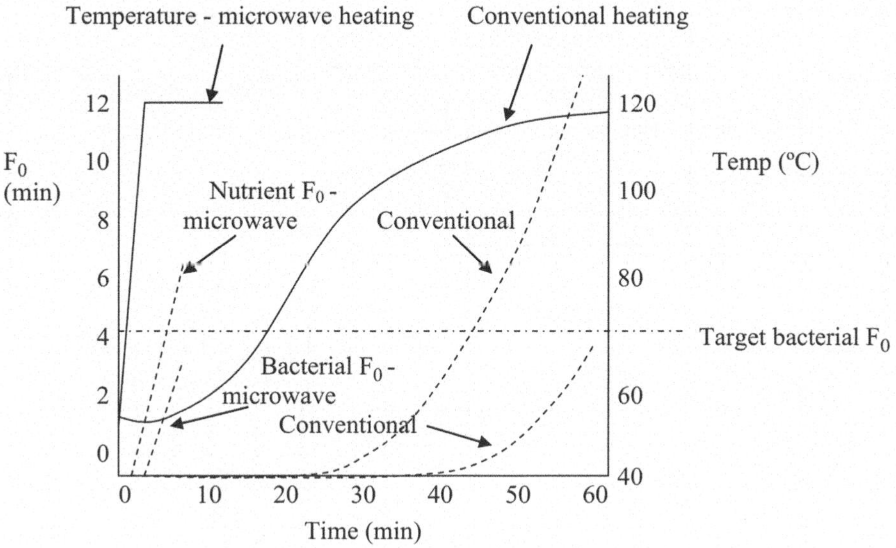

Fig. 10 Quality parameters for microwave and conventional heating (F_o represents accumulated lethality) (Fellows, 2022; Datta & Davidson, 2000)

Effects on Foods and Microorganisms of Infrared Heating The rapid IR surface heating changes the flavour and colour of foods due to the Maillard reactions and protein denaturation and also seals moisture and flavour or aroma compounds into the interior of the food. These changes are similar to those that occur during baking. IR heating also has similar effects on microorganisms to those described during baking. Details of the interaction of IR radiation with food components and the effects on quality are also given by Fellows (2022).

3.5.4 Pulsed Light Technology

Pulsed light technology (PLT) has been approved by the U.S. Food and Drug Administration (FDA) since 1996 as a technology applied in food processing and handling, as well as in microbial decontaminating on food surfaces (Wang et al., 2023).

The application of PLT in food industry has increased over the past 20 years as a non-thermal antimicrobial technology for packed and unpacked foods in forms of solid or liquid states (Dhar & Chakraborty, 2023; Chen et al., 2021). PLT produces locally and mild heat in treated food but minimally causes adverse effects on nutrition and sensory quality of foods and effectively inactivate microorganisms (Obileke et al., 2022). Therefore, it can be applied widely in food industry to overcome disadvantages of thermal processes.

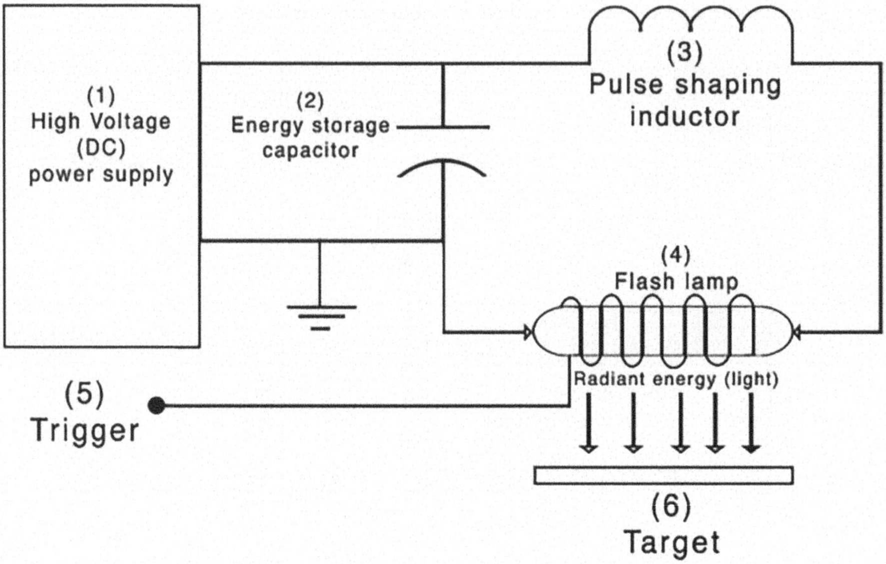

Fig. 11 Schematic diagram of a typical pulsed light system. (Adapted from Ganessingh et al., 2018)

PLT involves the generation of intense-power and short-duration pulses of a broad-spectrum light at the wavelength of 200–1100 nm (John & Ramaswamy, 2018; Mandal et al., 2020). PLT has also been referred to high-intensity pulsed UV light, pulsed UV light, high-intensity and broad-spectrum UV light, intense light pulsed and pulsed white light (Rowan, 2019). A typical PL system consists of a high-voltage power generator, an energy storage capacitor, a pulse shaping inductor, gas-discharge flash lamp and a trigger that initiates discharging the electrical energy to the flash lamp (Fig. 11) (Mahendran et al., 2019).

PLT inactivates wide range of microorganisms such as bacteria, fungi, yeasts, bacterial spores, and viruses through photochemical, photothermal and photophysical mechanisms (Wang et al., 2023; Zenklusen et al., 2018; Cai et al., 2024). Photochemical pathways cause DNA or RNA damage due to UV exposure and involve reactive oxygen species. Photothermal and photophysical effects lead to localized heating, membrane disruption, vacuole expansion, and protein leakage (Dhar & Chakraborty, 2023). The antimicrobial mechanisms of PLT can be evaluated through the determination of protein leakage, analysis of cellulose reactive oxygen species (ROS), measurement of adenosine triphosphate (ATP) concentration, and observation methods of cell morphology such as confocal laser scanning microscopy (CLSM) and scanning electron microscope (SEM) observation (Cai et al., 2024).

The antimicrobial effectiveness of PLT depends on many factors such as microbial species and strain, food ingredients, thickness of solid foods, transparency of liquid foods, pulse duration, and energy supply (Wang et al., 2023; John &

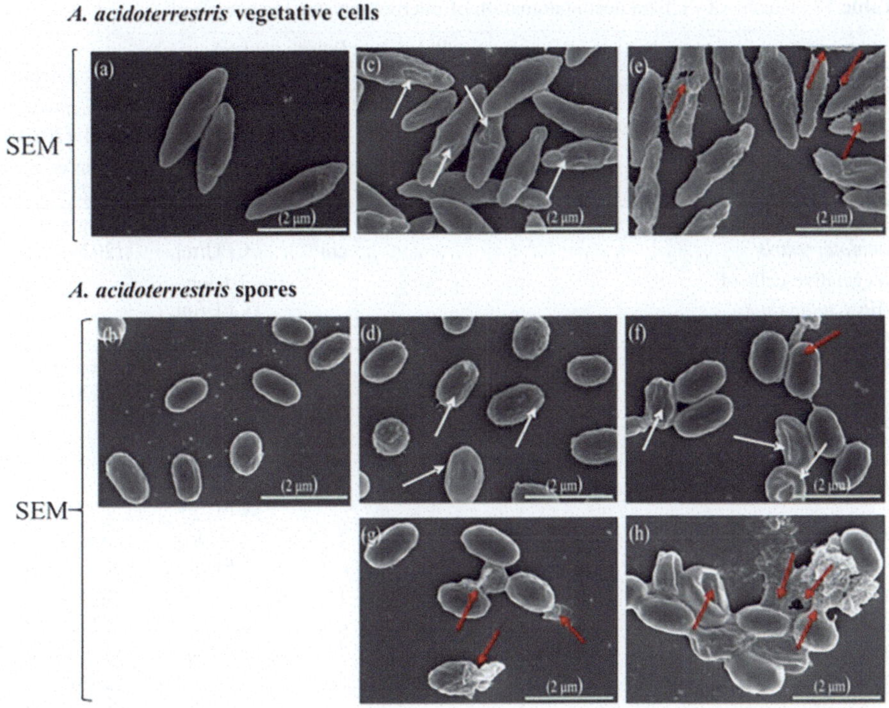

Fig. 12 Scanning electron microscopy images of *A. acidoterrestris* vegetative cells and spores after PL treatment. Images of (**a–h**) represents the cells or spores treated at different fluences of 0, 13.50, 27.00, 40.50, 54.00 J/cm^2, respectively. The white arrows indicate sites of deformation on vegetative cells and spores. The red arrows indicate sites of damage on vegetative cells and spores. (Adapted form Cai et al., 2024)

Ramaswamy, 2018; Ding et al., 2024). It is reported that the resistance of microorganisms to PLT increases in the order of vegetative bacteria, bacterial endospores or spores, and pigmented conidiospores (John & Ramaswamy, 2018). The PLT punctures bacterial membranes of both vegetative cells and spores (Fig. 12). However, bacterial spores show much stronger resistance to PLT than vegetative cells (Cai et al., 2024). Dhar and Chakraborty (2023) applied PLT to bael juice for inactivation of pathogenic, or spoilage microorganisms and enzymes and observed that *Saccharomyces cerevisiae* is more sensitive to PLT than *Escherichia coli.* Table 17 shows some application of PLT to some foods for inactivation of microorganisms.

3.5.5 Non-thermal Plasma Technology

Non-thermal plasma (NTP) is an emerging technology has been increasingly applied in food industry recent years for various technical purpose such as inactivating microbial growth and enzyme activity, maintaining and enhancing sensory

Table 17 Effects of PLT on decontamination of microorganisms in some foods

Target microorganisms	Kinds of food	Pulse duration	Pulse energy	Log reduction	References
Listeria innocua	Ready-to-eat dry-cured ham	Not reported	8.4 J/ cm^2	1–2 log CFU/cm^2	Fernández et al. (2020)
B. cereus spore	Mesquite flour	28 s	7.13 J/ cm^2	1.69 log CFU/g	Chen et al. (2021)
Spores of *Alicyclobacillus acidoterrestris*	Apple juice	3 s	12 J/ cm^2	3.5 log CFU/mL	Ding et al. (2024)
Vegetative cells of *Alicyclobacillus acidoterrestris*				2.1 log CFU/mL	
Cronobacter sakazakii	Wheat flour	60 s	17.46 J/ cm^2	4.15 log CFU/g)	Chen et al. (2020)
Enterococcus faecium				3.31 log CFU/g)	
B. cereus				1.53 log CFU/g)	
Salmonella Typhimurium	Pecan halves	40 s	1.27 J/ cm^2	3.43 log CFU/pecan	Gyawali et al. (2024)
Hepatitis A virus *Murine norovirus-1*	Cranberries	8736 μs	11.52 J/ cm^2	> 3.5 log cycles	Kim et al. (2024)
Salmonella	Cherry tomato	30 s	31.5 J/ cm^2	2.3 log	Leng et al. (2020)

properties of food products, sterilizing surface, and decontaminating surfaces of machinery system in food processing (Ucar et al., 2021). The application of NTP to inactivation of microbes in food industry has been recently studied in term of its antimicrobial efficacy, nutritional and sensory retaining, and its environmental benefits (Pankaj et al., 2018; Niedźwiedź et al., 2019; Lisboa et al., 2024).

The principle of NTP generation is the application of energy to a flowing gas, resulting in the ionization of gas molecules and the formation of reactive species such as ions, electrons, radicals, and UV photons (Fig. 13) (Lisboa et al., 2024). The generated reactive species probably damage different targets of microorganism cells including cell membrane, DNA, and enzymes, leading leakage of intracellular substances and the inhibition on cell growth. Antimicrobial mechanism of NTP against Gram-positive bacteria probably differ from that against Gram-negative ones. It is proposed that the growth of Gram-positive bacteria is inhibited due to the damage on DNA and enzymes, whereas the decreased viability of Gram-negative is possibly attributed to the cell membrane damage (Calvo et al., 2020).

NTP exhibits the microbial decontamination on wide range of microorganisms such as vegetative Gram-negative and Gram-positive bacteria, yeasts, molds, viruses, and endospores (Ucar et al., 2021). NTP technology involving ionized gas treatment at ambient temperature effectively eliminate pathogenic bacteria on food surfaces of various ready-to-eat foods, particularly on cut apple, where bacterial

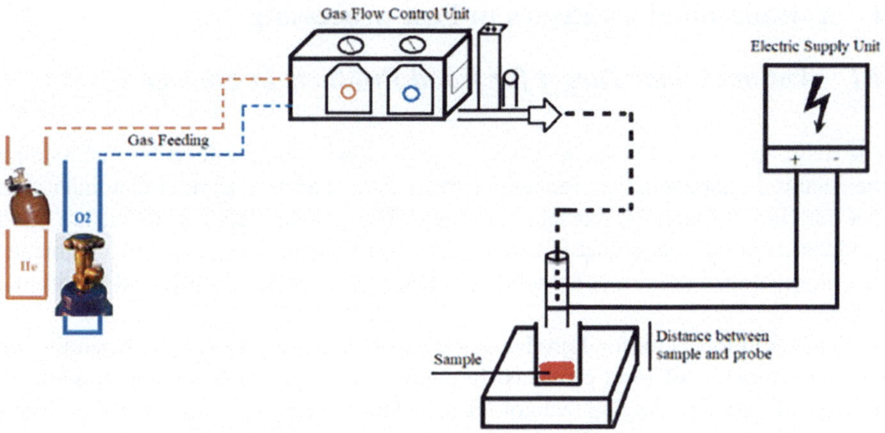

Fig. 13 Model of dielectric barrier discharge (DBD) plasma system. (Adapted from Ucar et al., 2021)

populations of two tested *Salmonella* strains and *E. coli* O157 were reduced by 1.3–1.8 log units after 15 min of exposure (Calvo et al., 2020). In addition, NTP can be applied indirectly for packed food to inactivate microorganisms, especially pathogenic microorganisms such as SARS-CoV-2, contaminating on the external surface of food packaging. The indirect treatment of packed food becomes promising strategy to prevent the spread out of microorganisms causing infectious diseases in food processing and distribution chain (Capelli et al., 2021; Sani et al., 2023). Tarabová, Tampieri (2021) has applied NTP in processing of fresh-pressed apple juice to inactivate *E. coli* ATCC 25922 and *Saccharomyces cerevisiae* S228C. After treatment, a greater (6 log) reduction in contamination was observed for the bacteria *E. coli* compared to the yeast *S. cerevisiae*, whereas there is no affect on physical properties of the juice, such as pH, conductivity, color, Brix degree, and transmittance. In addition, chemical ingredients, such as sugar, organic acids, polyphenols are unaffected when they together treated in juice. NTP causes a temporary increase in nitrite and hydrogen peroxide concentration, however the concentration of these two substances decreases within 24 h after treatment.

NTP treatment does not release harmful waste to environment therefore it is approved as eco-friendly process (Ucar et al., 2021). In addition, NTP decontaminates food surfaces at ambient temperature in short time, therefore it retains nutritional value and sensory properties of NTP-treated food, and allows food still fresh after microbial decontamination. However, NTP treatment accelerates lipid oxidation and increases thiobarbituric acid reactive substance value in high-fat foods due to the generation of reactive oxygen species such as zone, singlet oxygen, and peroxide (Ganesan et al., 2021; Wang et al., 2021). NTP treatment also causes discoloration in red meat such as fresh beef due to the accumulation of metmyoglobin after NTP treatment (Wang et al., 2021). Further studies are needed on approaches for minimizing adverse effects of NTP treatment on red meat and high-fat products.

4 Antimicrobial Strategies in Post-processing

4.1 Antimicrobial Target for Food Products in Storage

The main objective of post-processing is to preserve food products by maintaining their natural characteristics, remaining fresh, safe, and free from contaminants, or deterioration throughout the storage stage. The primary goal is to ensure food products against contamination sources by microorganisms, harmful chemicals, and physical or foreign matters while ensuring their safety, quality, and nutritional value.

The requirements for food products in the storage stage vary greatly depending on the characteristics of food products. Implementing appropriate storage conditions, utilizing effective packaging technologies, and monitoring environmental conditions are crucial strategies to reach the food quality, food safety and food integrity.

– Storage in cool and dry storage conditions

Dry products including grains and dried fruits, require cool and dry storage conditions, with low humidity to inhibit microbial growth (Chayjan & Esna-Ashari, 2010). The vacuum packaging or the use of patches can further enhance the stability of the dry products in storage (Parvanov & Dinkov, 2017). Moreover, monitoring the temperature and humidity of stored grains is crucial in controlling mycotoxin production (Ibrahim et al., 2016).

Packaged and processed foods often have extended by storing in a cool, dry place away from direct sunlight to prevent the degradation of microoragnisms (Jamróz & Kopel, 2020). In addition, the active packaging incorporated with antimicrobial agents and/or oxygen scavengers can further enhance the safety and quality of processed foods during storage (Jamróz & Kopel, 2020; Xing et al., 2021).

– Storage in refrigerated storage conditions

Dairy products, such as milk, cheese, and yogurt, are stored in refrigerated storage conditions at or below 4 °C to inhibit the growth of spoilage microorganisms and pathogens (Min et al., 2017). Additionally, the materials of packaging play a crucial role in inhibiting the growth of microbial spoilage (Otero et al., 2017).

Meat and seafood products are highly perishable and require stringent storage conditions to prevent spoilage and foodborne illnesses. The ideal storage temperature for fresh meat is typically between 0 °C and 4 °C, while seafood should be stored at temperatures close to 0 °C to minimize microbial growth (Otero et al., 2017; Kuts et al., 2022).

Fruits and vegetables are stored in refrigerated and controlled relative humidity to prevent respiration rates, water loss and maintain quality (Uchechukwu-Agua et al., 2015). Additionally, the application of natural preservatives, such as essential oils, can help extend the shelf life of fruits and vegetables by reducing microbial contamination (Denkova-Kostova et al., 2021).

– Storage in the frozen storage conditions

The required temperature for frozen food products is typically at or below −18 °C (0 °F) to inhibit microbial growth and enzymatic activity (Zhang et al., 2016). The quality degradation can occur due to textural and flavor changes and lipid oxidation over time (Zhao et al., 2022; Rachman et al., 2021). Temperature fluctuations, packaging, and duration of storage are involved in the quality of frozen food products in storage. The integration of smart technologies, such as Internet of Things (IoT) systems, into cold storage facilities for monitoring temperature and humidity can significantly enhance food quality management (Mohammed et al., 2022). The implementation of energy management practices by IoT-enabled sensing technology especially in the food manufacturing industry can help to reduce energy consumption and carbon emissions while maintaining product quality (Jagtap et al., 2022). On the other hand, temperature buffering and energy storage during food storage can be achieved by different formulations and applications of phase change materials (PCMs). PCM is used in various temperature ranges in the food industry to save energy and improve energy efficiency by capturing wasted heat or cold energy and maintaining a stable temperature (Hu et al., 2023).

– Requirements for food storage facilities

Food storage facility and layout are designed to allow for easy cleaning and disinfection, maintenance, and adequate drainage systems to prevent water accumulation (Oladipo-Adekeye & Tabit, 2021). Regular cleaning and sanitation of food storage areas are paramount for preventing microbial contamination. Facilities must implement a routine cleaning schedule that includes the sanitization the facility of shelves, floors, and walls (Martins et al., 2019). Regular inspections and maintenance for temperature accuracy and functionality of refrigeration units are necessary to maintain optimal storage conditions (Akabanda et al., 2017). Comprehensive training programs should focus on educating staff about the importance of personal hygiene, proper food handling, and the significance of maintaining clean storage environments (Duret et al., 2019). Regular audits and microbiological sampling can help identify risk areas for improvement in hygiene practices within food storage facilities (Mendagudali et al., 2016).

4.2 Antimicrobial Target for Food Products in Transportation and Distribution

Antimicrobial strategies during post-processing are essential to enhance product safety and shelf life. Depending on the requirements of the specific product and consumer preferences, the following strategies may be considered to minimize the adverse effects of microorganisms on the product during post-processing.

Transportation vehicles such as trucks, ships, or vans used for food products must be kept clean, well-maintained, and designed for easy cleaning to prevent contamination. Specialized transportation vehicles in direct contact with food during transportation must be made from materials that do not contaminate the food, are easy to clean, and must protect the food from external environmental factors such as dust and insects. The transportation vehicles used during transportation should be cleaned and sanitized on a regular basis. Loading temperature and sanitation conditions of the vehicle should be controlled. The best strategy to prevent growth and development of pathogens is to keep food products at recommended transport conditions depending on specific types of products, particularly temperature. Sufficient equipment or associations must be equipped to control temperature, humidity, ventilation, and other technical parameters implemented according to the manufacturer's specifications during transportation. Workers directly handling food during transportation must be in good health, free from infectious diseases, and maintain good personal hygiene practices.

Food business distributors should provide training on good hygienic practices, including those for minimizing cross-contamination. The WHO Five Keys to Safer Food Manual contains useful information for food handlers to minimize the transmission of foodborne illness, including resistant infections (WHO. FA, 2023).

5 Alternative Antimicrobial and Food Preservation Strategies

5.1 Antimicrobial Agents from Natural Substances as Safe and Prioritized Alternatives to Synthetic Food Additives

Chemically antimicrobial additives have been applied widely in food industry to eliminate and inhibit the growth of microorganisms that cause food spoilage and poisoning, resulting in extending the shelf-life of food products, ensuring microbial food safety, reducing food waste, and improving cost-effectiveness in food business (Batiha et al., 2021; Pisoschi et al., 2018). Among chemical additives used for minimization of the microbial contamination, sorbic acid and its salts, benzoic acid and its salts, propionic acid and its salts, sulfites, nitrite, and nitrates have been used extremely in various food products for inhibiting the growth of a broad spectrum of microorganisms such as bacteria, yeasts, molds, and fungi (Pisoschi et al., 2018; Wu et al., 2022; Lalani et al., 2024). However, the presence of synthetic additive residues in food products as a result of improper use can cause allergic reactions such as asthma, nettle rash, intestinal pains, nausea, and diarrhea, as well as antibiotic resistance, potentially leading to serious health issues (Lisboa et al., 2024; Pisoschi et al., 2018; Wu et al., 2022; Chipley, 2020; Serwecińska, 2020). In serious cases, the consumption of synthetic antimicrobials can cause seizures and anaphylaxis, leading in death (Pisoschi et al., 2018). In addition, the improper use of chemical antimicrobials may detrimentally impact organoleptic properties of food products and stimulate the resistance of microorganism to common chemical antimicrobials (Batiha et al., 2021; Steels et al., 2000; Warth, 1985). These detrimental effects of synthetical antimicrobials point out the need for effective and safe antimicrobial additives that are naturally derived (Batiha et al., 2021; Gyawali et al., 2015). Natural antimicrobials or bio-antimicrobials may originate from plants, animals and microorganisms (Batiha et al., 2021).

5.1.1 Plant-Originated Components as Antimicrobial Agents

Plant-originated components such as polyphenols, flavonoid compounds, and essential oils, exhibiting antimicrobial and antioxidant activities have been added to food as natural pre-servatives and incorporated to packaging materials for preparation of active packaging (Lisboa et al., 2024; Redondo-Blanco et al., 2020).

Polyphenol and Flavonoid Compounds

Polyphenol compounds are secondary metabolites in plants, characterized by one or more hydroxyl groups attached to aromatic rings (Oulahal & Degraeve, 2022). These compounds are found in varying concentrations across different plant species, some high-polyphenol foods include spices, dried herbs, cocoa products, berries, certain seeds and nuts, and vegetables like olives and globe artichokes (Lisboa et al., 2024). Additionally, by-products and waste from phenolic-rich edible plants are considered cost-effective and abundant sources of antimicrobial phenolics, presenting significant potential for various applications (Batiha et al., 2021). The antibacterial properties stem from their ability to disrupt bacterial cell structures, interfere with enzymes, and generate oxidative stress (Lisboa et al., 2024; Chen et al., 2023a). Table 18 below are examples of common polyphenol compounds, which are utilized in antibacterial activity.

Mechanism of Antibacterial Action

Cell wall disruption: Polyphenols are able to damage the structural integrity of bacteria's cell wall. This disruption weakens the mechanical strength of the cells by making them less resistant to osmotic and other stresses. It is explained through the interactions involving their hydroxyl (–OH) groups, which enable hydrogen bonding with microbial cell membrane. Depending on their hydrophobic nature, polyphenols may either accumulate on the cell surface, penetrate the membrane, or cross into the cytoplasm (Chen et al., 2023a). Inside the cell, polyphenol compounds can interact with cellular components or disrupt intracellular pH. Gram-negative bacteria, with their hydrophilic outer membranes, are typically less sensitive to hydrophobic components compared to Gram-positive bacteria (Oulahal & Degraeve, 2022). In addition, polyphenols may interfere with the function of membrane transport proteins. However, the antibaterial effectiveness of plant-originated extract can rely on the interactions between various polyphenolic compounds (Ferdes, 2018).

The disruption of bacterial cell membrane structure and function: The bacterial cell membrane hosts numerous enzyme systems essential for physiological processes, so its damage can directly or indirectly disrupt metabolism, leading to

Table 18 Example of potential plant-originated antimicrobial compounds

Potential plant-originated components		Structure	Objesctive microorganisms	Extracted source	References
Phenolic	Gallic acid $C_7H_6O_5$	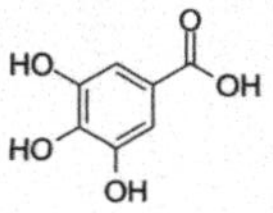	*E. coli;* *Salmonella;* *Pseudomonas aeruginosa;* *S. aureus;* *L. monocytogenes;* *B. cereus*	Fruits: Berries, pomegranates, grapes; Veggies: beans, carrots, tomatoes; Herbs and spices: cloves, cinamon, tea; Nuts and seeds: walnuts and hazelnuts	Pacheco-Ordaz et al., (2018), Oulahal and Degraeve (2022) Skroza et al. (2019)
	Salicylic acid $C_7H_6O_3$	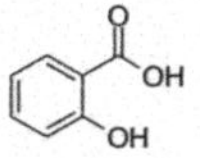	*S. aureus;* *Enterococcus faecalis* *E. coli; B. cereus* *P. aeruginosa*	Willow Bark (*Salix* spp.), Wintergreen, Spiraea plant Herbs and spices: thyme, oregano and curry leaves Low concentration in olive, nuts, potato, broccoli and cucumber	Adamczak et al. (2019) Oulahal and Degraeve (2022)
	Caffeic acid $C_9H_8O_4$	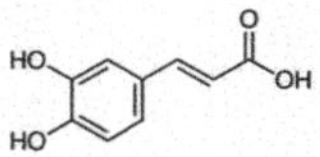	*E. coli;* *S. aureus;* *B. cereus;* *S. Infantis;* *Salmonella typhimurium* *Lactobacillus rhamnosus*	Fruits and veggies: berries, apples, citrus fruits, tomatoes and beets; Grains and seeds; Beverages: coffee, tea; Herbs and spices: basil, thyme and oregano	Skroza et al. (2019) Chen et al. (2023a)
Flavonols	Quercetin $C_{15}H_{10}O_7$	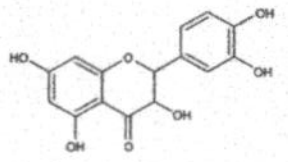	*E. coli* *S. aureus* *S. typhimurium* *Lactobacillus rhamnosus*	Fruits: apples, citrus fruits, berries, grapes; Vegetable: onions, capers, kale, broccoli, spinach; Herbs and spices: parsley, cilantro, dill, caperberries; Other sources: green tea, buckwheat	Abdulhussain Kareem and Razavi (2020) Chen et al. (2023a)
	Rutin $C_{27}H_{30}O_{16}$		*S. aureus;* *E. faecalis;*		Adamczak et al. (2019)

		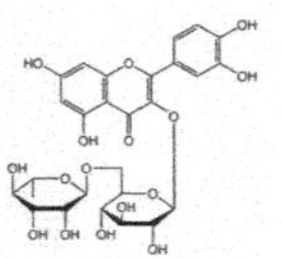	*E. coli* *P. aeruginosa*	Fruits: citrus fruits, apples, berries, figs; Veggies: asparagus, tomatoes; Other sources: green tea, black tea, buckwheat	
Flavanols	Catechin $C_{15}H_{14}O_6$	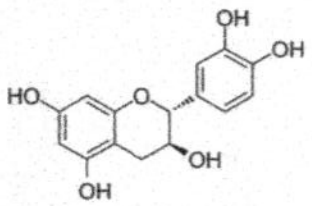	*E. coli;* *S. aureus;* *B. cereus;* *S. Infantis* *S. typhimurium* *L. rhamnosus*	Fruits and veggies: grapes (red and dark varieties), berries, peas; Dark chocolate; Herbs: Ginkgo biloba, cinnamon; Beverages: green and black tea, red wine	Skroza et al., (2019), Chen et al. (2023a)
	Epicatechin $C_{15}H_{14}O_6$	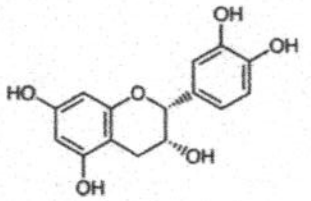			
Anthocyanin		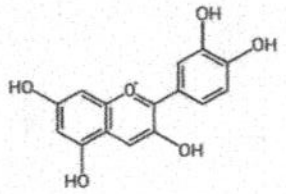	*S. aureus* *Listeria* spp.	Being found in plants, primarily contributing to the red, purple, and blue colors of vegetables, fruits, flowers.	Chen et al. (2023a)
Tannins	Tannic acid $C_{76}H_{52}O_{46}$	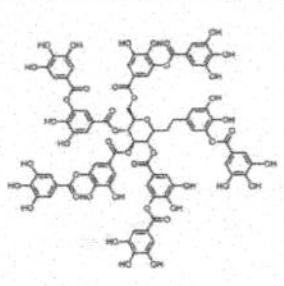	*S. aureus*	Fruits: apple (seed and peel), persimmons (unripe fruit), pomegrantes; Black and green tea; Cinnamon bark; Oak bark and leaves; Skin and husk od Nuts and seeds	Oulahal and Degraeve (2022), Chen et al. (2023a)
Stilbenoids	Resveratrol $C_{14}H_{12}O_3$	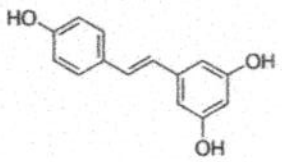	*E. coli;* *S. aureus;* *B. cereus;* *S. Infantis;* *E. faecalis*	Fruits: grapes, berries; Other sources: peanuts, soybeans, vines, pines, cocoa, Japanese knotweed	Oulahal and Degraeve (2022)

bacterial death. Polyphenols are known to increase membrane permeability, causing leakage of small molecules such as potassium ions and ATP, as well as larger molecules like nucleic acids and proteins. Particularly, membrane fluidity, vital for bacterial adaptation to environmental changes, can be compromised by polyphenols (Chen et al., 2023a). For instance, gallic acid and ferulic acid can penetrate the cell membrane in their undissociated forms via passive diffusion, triggering localized hyper-acidification. This leads to the formation of ruptures or pores, resulting in cell death. Additionally, polyphenols are able to interact with bacterial enzymes and structural proteins, denaturing them and disrupting essential cellular functions or interfere bacterial membranes by integrating into lipid bilayers, causing increased permeability and leakage of cellular contents (Oulahal & Degraeve, 2022). For instance, catechins, combining their gallate moiety and resorcinol ring, interact with bacterial membrane lipid bilayers, limiting phospholipid movement and causing colonization or structural compromise of the cell membrane (Nie et al., 2021).

Oxidative Stress Induction: Polyphenol compounds can generate reactive oxygen species (ROS), leading to oxidative damage in bacterial cells, including DNA, proteins, and lipids that damages bacterial cell walls and releases intracellular components. Reactive oxygen species (ROS), generated through nonspecific attacks, make it challenging for bacteria to develop resistance, leading to their efficient elimination (Chen et al., 2023a). Additionally, polyphenols can act as prooxidants in the presence of redox-active metals like iron and copper, participating in Fenton reactions that amplify ROS production. This mechanism underscores the role of polyphenols in generating oxidative stress, which effectively inhibits bacterial survival and proliferation (Chen et al., 2023a) (Fig. 14).

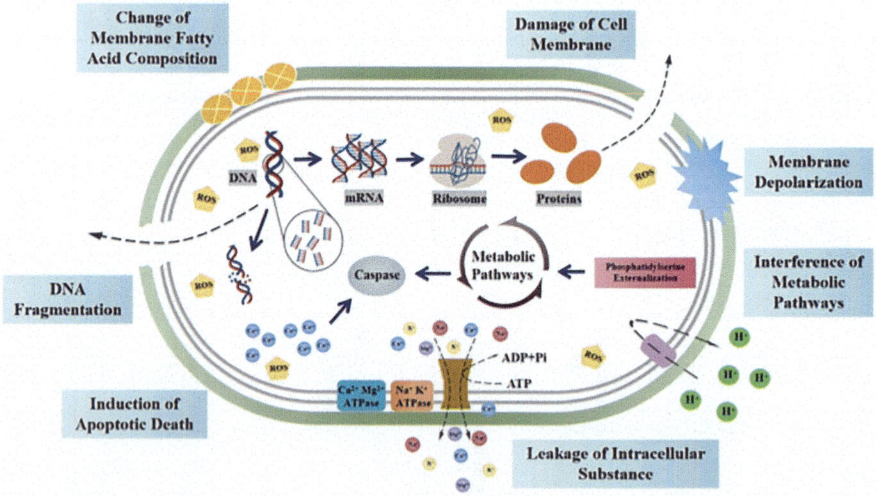

Fig. 14 The demonstration of main actions of plant-originated compounds in bacteria cells (Adapted from Chen et al., 2023a)

5.1.2 Animal-Originated Antimicrobials

Lysozyme

Mammals, poultry, insects, fish, eggs, viruses, and fungi (Abdulhussain Kareem & Razavi, 2020; Davidson et al., 2015). Lysozyme from avian eggs and mammalian milk has been applied in food industry as a safe antimicrobial additive (Khorshidian et al., 2022; Syngai & Ahmed, 2019). Antimicrobial activity of lysozyme is ascribed to decomposition of the bacterial cell walls by hydrolysis of the β-1,4-glycosidic bond between N-acetyl-D-glucosamine and N-acetylmuramic acid molecules in peptidoglycan layer of the cell bacterial walls (Khorshidian et al., 2022; Syngai & Ahmed, 2019). Lysozyme exhibits antibacterial activity against Gram-positive bacteria more effectively than that against Gram-negative bacteria. This is attributed to the possibility of lysozyme reaching to peptidoglycan in the cell walls of Gram-positive bacteria more easily than reaching to peptidoglycan in the cell walls of Gram-negative bacteria. The β-1,4-glycosidic bond in the cell wall of Gram-negative bacteria are obstructed by the lipopolysaccharide layer, causing the limited contact between lysozyme and reaction cites (Batiha et al., 2021).

Various technological and environmental factors, such as pH, temperature, salt concentration, and osmotic strength can affect the stability and antimicrobial activity of lysozyme in food and this varies based on the lysozyme types (Khorshidian et al., 2022) (Fig. 15).

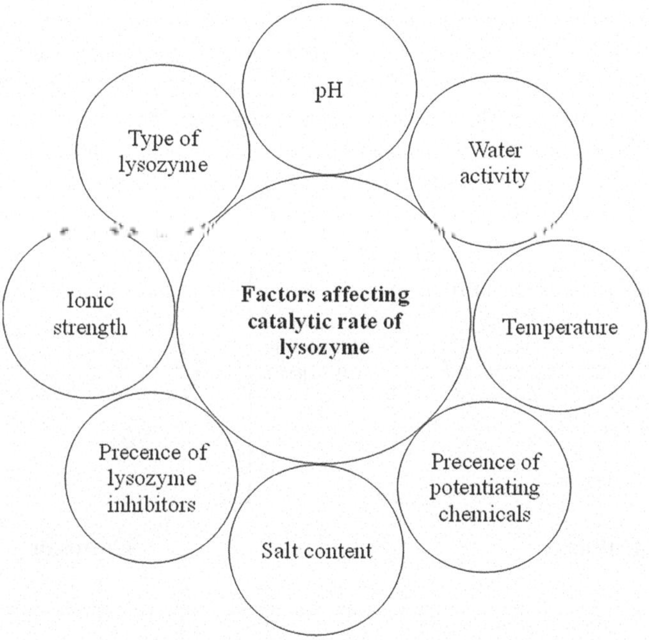

Fig. 15 Factors affecting the catalytic rate of lysozyme (Gautam et al., 2024)

Various advanced research techniques have been proposed to enhance lysozyme activity against Gram-negative bacteria, aiming to increase its application in the food industry. Lysozymes have been addition to cheeses, beverages, sausages, frankfurters, and other products for preservative purpose (Khorshidian et al., 2022; Nawaz et al., 2022). It has also been incorporated in packaging materials such as cellulose acetate, zein, cellulose, and pullulan for the fabrication of antimicrobial packaging (Silva et al., 2018; Barbiroli et al., 2012). The superior characteristics of lysozyme compared to synthetic antimicrobials are its stability across a wide range of pH levels (pH 2–10) and temperatures (4–95 °C) and enhancement of human immune system. For these reasons, lysozyme is widely used in the food industry (Khorshidian et al., 2022; Batiha et al., 2021).

Chitosan

Chitosan is a cationic hetero-polysaccharide comprising of D-glucosamine and N-acetyl-D-glucosamine molecules linked by β-1,4-glycosidic linkages (Fig. 16) (Chandrasekaran et al., 2020). Chitosan can be obtained from the deacetylation of chitin in highly basic solution by chemical methods (Bahndral et al., 2024). It can also be microbially synthesized from chitin using *Streptomyces griseoincarnatus* RB7AG (Behera et al., 2024). Chitin can be produced by demineralizing and deproteinizing of crustacean shells (Souza et al., 2020). Chitin can be also obtained from cell walls of mushroom and fungi, such as *Aspergillus niger*, *Mucor rouxii*, and *Penicillium notatum* (Bahndral et al., 2024). The degree of deacetylation of chitosan positively correlated with electrostatic interaction-induced repulsion between chitosan chains, viscosity, and antimicrobial efficacy (Chen et al., 2023b). In addition, molecular weight of chitosan is reported to impact the antimicrobial property of chitosan independently and more strongly than degree of N-deacetylation (Sekiguchi et al., 1993). Chitosan is non-dissolvable in water but dissolves well in weak acidic solutions at the pH values below 6.3. Acidic chitosan solution can be used as based material for nontoxic, biodegradable, and antimicrobial food packaging. The improvement of physical, chemical, and biological

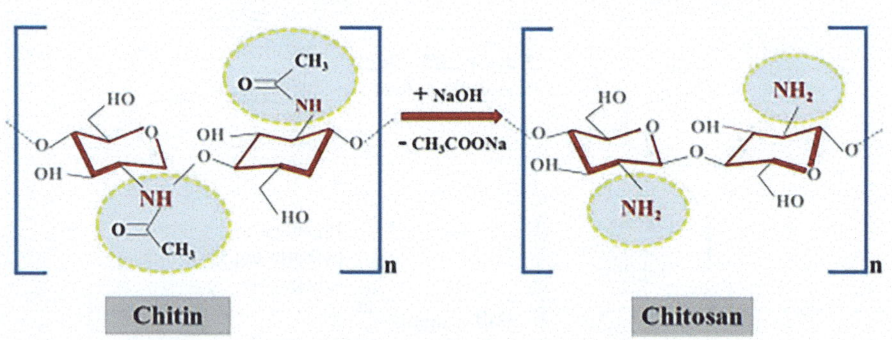

Fig. 16 Production of chitosan from deacetylation of chitin. (Adapted from Bahndral et al., 2024)

properties of chitosan have been investigated using diversified methods such as nanotechnology and modifying chemical structure of chitosan to obtained various derivatives (Chen et al., 2023b; Sahariah & Másson, 2017). Although chitosan is approved as GRAS, the toxicological properties of modified chitosan for antimicrobial purpose in food industry have been concerned and investigated, mainly on the migration of its derivatives in food matrix (Ahmad et al., 2023; Enescu et al., 2020). The migration tests have shown that the migration rate of chitosan derivatives in food are still acceptable, however comprehensive researches need to be investigated to determine the in vivo toxicology of its derivatives to ensure chemical safety for consumers (Chen et al., 2023b). Besides molecular weight and degree of deacetylation, the antimicrobial efficacy of chitosan and its derivatives depends on other factors such as pH, source of chitosan, food ingredients, the interaction with other food ingredients, chitosan concentration, and microbial species (Hosseinnejad & Jafari, 2016).

Chitosan possesses antibacterial and antifungal properties and has been extensively studied as a potential natural antimicrobial agent in the pharmaceutical, cosmetic, and food industries. Chitosan exhibits activity against some strains of fungi (Al-Zahrani et al., 2021), yeast (Savard et al., 2002), and bacteria (Li & Zhuang, 2020).

Antimicrobial mechanisms of chitosan are explained by the interaction between chitosan molecules and bacterial membranes, which is induced by the electrostatic attraction between cationic groups (NH^{3+}) in chitosan chains with negatively charged groups on the membrane surface of microbial cells (Hosseinnejad & Jafari, 2016; Li et al., 2010). In addition, the hydrolysis of peptidoglycans in the microbial wall may occur, resulting in the disruption of the membrane permeability, the change in osmosis, and the leakage of cellular ions and other substances. Consequently, the growth of microorganisms is inhibited (Hosseinnejad & Jafari, 2016; Goy et al., 2009). The morphology of the damaged cell membranes of *E. coli* can be observed via TEM image for elucidation of the antibacterial activity of chitosan (Hosseinnejad & Jafari, 2016) (Fig. 17).

Another hypothesis is the formation of chitosan – microbial DNA complex, which inhibit the syntheses of mRNA and protein. Chitosan is proposed to penetrate through microorganism membranes and contact to nucleic acids (Hosseinnejad & Jafari, 2016). It is also believed that chitosan can chelate minerals and nutrition compounds, causing the lack of essential substances for the growth of microorganism (Goy et al., 2009). In addition, the antimicrobial of chitosan may be due to the formation of chitosan membrane surrounding microbial cells, obstructing the transportation of nutrients and oxygen for the growth of microbes (Hosseinnejad & Jafari, 2016).

At present, the common methods to improve the antibacterial activity of chitosan are chemical modification of chitosan and the combination of chitosan with other antibacterial materials such as metal, metal oxide (Li & Zhuang, 2020), polyphenols,

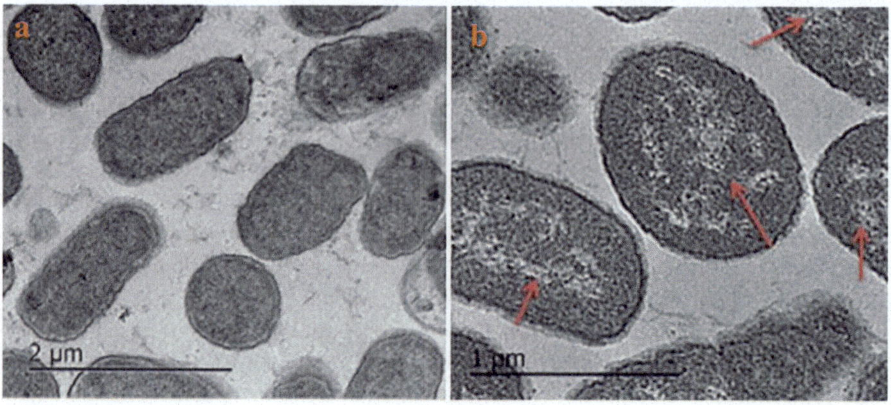

Fig. 17 TEM images of E. coli: (**a**) untreated cells; and (**b**) cells treated with chitosan (Hosseinnejad & Jafari, 2016)

and Metal-polyphenol coordination networks (Li et al., 2010). Chitosan-zinc, chitosan-Au, chitosan-Cu, chitosan-sil have improved the antibacterial properties of chitosan (Lin et al., 2022).

5.1.3 Microbial-Originated Antimicrobials

Microbial preservatives commonly utilize beneficial bacteria, these include inhibitory compounds, well-noted ones as organic acids (like lactic acid), bacteriocins, hydrogen peroxide. These substances reduce the pH level and create a condition that is undesirable for spoilage organisms and pathogens, effectively inactivating their growth (Lisboa et al., 2024).

Organic Acids

Organic acids play an essential role in food preservation due to their ability to inhibit harmful microorganisms and extend shelf life. Lactic acid, acetic acid, citric acid, and propionic acid are natural products of fermentation and are simply added to foods (Batiha et al., 2021). Organic acids can directly affect spoilage microorganisms and pathogenic bacteria by lowering the pH of the environment, which hinders their growth. With the growing demand for safe, natural foods with fewer additives, organic acids are predicted to become a key preservation solution in the future food industry (Agriopoulou et al., 2020; Ibrahim et al., 2021). This preservation method is environmentally friendly and is widely preferred in modern food industry practices.

Bacteriocins

Bacteriocins, natural antimicrobial peptides produced by lactic acid bacteria (LAB) and some other bacterial species, are highly effective for food preservation due to their ability to inhibit or kill pathogenic and spoilage bacteria (Johnson et al., 2018). These compounds offer substantial benefits, including extending food shelf life, enhancing food safety, reducing the reliance on synthetic preservatives or additives, and minimizing the need for excessive heat treatment (Mihaylova-Garnizova et al., 2024). Similar with ogranic acids, bacteriocins are regarded as a safe, environmentally friendly, and natural preservation method that aligns with the growing consumer demand for maintaining biological safety, eco-friendliness and healthy food products (Bhattacharya et al., 2022).

Hydrogen Peroxide
Hydrogen peroxide (H_2O_2) eliminates spoilage microorganisms and pathogens based on strong oxidative properties, which disrupt the cellular structures of microbes, effectively inhibiting their growth on food surfaces and within processing systems (Finnegan et al., 2010; Linley et al., 2012). Its decomposition characteristic, combined with its antimicrobial capability help maintain the quality and safety of the products that has made hydrogen peroxide increasingly popular in natural and eco-friendly food processing practices (Lisboa et al., 2024) (Table 19).

Examples of practical applications of microbial-originated antimicrobials

1. Meat Preservation: Lactic acid and acetic acid are commonly used to treat the surface of meat products, preventing the growth of pathogenic bacteria such as *E. coli* and *Salmonella* (Yu et al., 2021; Rossi et al., 2023).
2. Fruits and Vegetables: Citric acid is often used to preserve fruits and vegetables to inhibit mold and spoilage bacteria. Spraying or soaking fruit in a citric acid solution not only retains natural color but also helps extend the storage life (Zhang et al., 2023).
3. Dairy Products: Propionic acid is widely used in cheese production to prevent spoilage bacteria and mold growth, thereby extending the product's shelf life without impacting flavor (Bücher et al., 2021).
4. Grains and Bread: Propionic and acetic acids are also added during the production of bread and cereal products to prevent mold and spoilage bacteria growth. This extends shelf life while keeping products mold-free and safe for consumption (Bücher et al., 2021).
5. Dairy Products: Nisin, a widely used bacteriocin, is added to cheese to inhibit the growth of *C. botulinum* and *L. monocytogenes*, thereby extending shelf life and increasing safety (Hassan et al., 2021).
6. Meat and Seafood: Plantaricin isolated from *L. plantarum* or Pediocin isolated from *Pediococcus* spp., which present the ability to inhibit *Listeria* spp., *S. aureus and E. coli* are commonly used in preserving deli meats and seafood. It helps control pathogenic bacteria, enhancing food safety in these products (Batiha et al., 2021; Abdulhussain Kareem & Razavi, 2020).

Table 19 Mechanisms of different microbial-originated antimicrobials

Mechanisms	References
Organic acids kill or inhibit bacteria through the following mechanisms	
pH Reduction: When added to food, organic acids lower the environmental pH, creating a hostile setting for harmful microorganisms to grow and thrive. Many types of bacteria cannot survive in highly acidic environments, thereby extending the food's shelf life. Osmotic Pressure: Organic acids, especially weak acids, dissolve to form non-ionized molecules. These molecules can penetrate bacterial cell membranes and then dissociate within the cell, causing an increase in intracellular H^+ concentration, disrupting pH balance, depleting energy, and ultimately leading to cell death. Enzyme Inhibition: Some organic acids inhibit the function of essential enzymes within bacterial cells, affecting their metabolism, reducing their growth rate, or killing the bacteria. Inhibition of Transport and Nutrient Absorption: Organic acids can also disrupt nutrient absorption in bacteria, thereby limiting their growth and development.	Ben Braïek and Smaoui (2021), Batiha et al. (2021), Ji et al. (2023), Zhang et al. (2023)
Bacteriocins inhibit or kill bacteria through several mechanisms, depending on their type and structure. The main mechanisms include	
Membrane Permeabilization: Bacteriocins, like nisin, often create pores in bacterial cell membranes, leading to membrane destabilization and leakage of essential intracellular components such as ions, amino acids, or macromolecules, ultimately resulting in cell death. Degradation of Cellular Components: Some bacteriocins act on intracellular bacterial components like DNA or RNA, disrupting or inhibiting replication and metabolism. Inhibition of Cell Wall Synthesis: Certain bacteriocins interfere with bacterial cell wall synthesis, disrupting cell structure and causing cell lysis. Inhibition of Transport Systems: Some bacteriocins inhibit bacterial transport systems and ionic regulation, making bacteria unable to maintain intracellular balance.	Preciado et al. (2016), Vishweshwaraiah et al. (2021), Mihaylova-Garnizova et al. (2024), Talib et al. (2024)
Hydrogen peroxide eradicates bacteria, fungi, and pathogens through the following mechanisms	
Membrane Oxidation: Hydrogen peroxide acts as a strong oxidizing agent that, upon contact with the microbial cell membrane, breaks down lipid and protein structures. This weakens the cell membrane, leading to leakage of intracellular contents and ultimately causing cell death. Free Radical Generation: Hydrogen peroxide can generate free radicals (such as hydroxyl radicals, OH·) when interacting with metal ions, which damage the DNA, RNA, and proteins within bacterial cells. These highly reactive radicals can rapidly degrade essential cellular components in microorganisms. Enzyme Inhibition: Hydrogen peroxide can inhibit key enzymes vital for microbial survival, particularly those involved in metabolism and cellular defense against oxidative stress. When these enzymes are inhibited, bacterial growth and development are significantly hampered.	Linley et al. (2012), Finnegan et al. (2010), Talib et al. (2024), Lisboa et al. (2024)

7. Hurdle Technology: Bacteriocins are used in combination preservation systems (hurdle technology) alongside other methods like pH control, temperature management, or oxygen restriction. In such systems, bacteriocins act as an additional protective barrier, reducing the risk of contamination and spoilage (Batiha et al., 2021).

5.2 *Biodegradable and Antimicrobial Food Packaging*

There have been increased concerns on the consumption of healthy and safety food with a minimally thermal treatment to keep the nutrients and fresh-liked sensory properties. However, the mild treatments on food may result in the short shelf life of food products, consequently leading the increase in food waste and losses due to food spoilage and deterioration (Upadhyay et al., 2024). This fosters the development of innovative methods that preserve food nutrients, maintain sensory properties, extend the shelf life, and ensure chemical and microbial safety of preserved food (Fellows, 2022). Besides development of eco-friendly and non-thermal processes applied in food industry, the application of active food packaging in food preservation has been increasingly investigated and evaluated across various aspects, such as effectiveness in inhibiting and killing microorganisms, the side effects of them on food chemical composition and sensory properties, benignity to human health, application cost, and environmental friendliness (Lisboa et al., 2024; Ahmed et al., 2022; Honarvar et al., 2016) (Table 20).

6 Conclusion and Future Trends

This chapter discusses antimicrobial technologies and strategies in food processing and post-processing, highlighting the importance of selecting appropriate methods to minimize microbial contamination, improve food safety, and extend shelf life. At the present time, the application of new technologies for antimicrobial treatment and food preservation during food production and post processing are continuously being developed and updated. The negative impacts from climate change, sea level rise, environmental pollution and resources degradation present significant challenges for food growers and producers. The use of alternative technologies that effectively inactivate pathogens while reducing energy costs and carbon emissions is essential. Environmentally friendly solutions are necessary to ensure the safety and quality of food products during handling and storage. As the food industry continues to evolve, these strategies will play a crucial role in promoting sustainability and efficiency in food production and preservation.

Table 20 Biodegradable and antimicrobial food packaging

Based materials	Antimicrobial agents	Concentration of antimicrobial agents	Target microorganisms	References
	Nano particles			
Tragacanth Hydroxypropyl methylcellulose Beeswax	Silver nanoparticles	2–8%	*B. cereus* *S. aureus* *Streptococcus pneumoniae* *L. monocytogenes* *E. coli* *Salmonella* Typhimurium *Pseudomonas aeruginosa* *Klebsiella pneumoniae*	Bahrami et al. (2019)
Starch	ZnO-chitosan nanoparticles	1.0–5.0%	*E. coli* *S. aureus*	Hu et al. (2019)
Zein Sodium alginate	Titanium dioxide (TiO2) nanoparticles	0.5%	*S. aureus* *E. coli*	Amjadi et al. (2020)
	Esstiential oils			
Bombacaceae gum	Cinnamon leaf essential oil	0.75–1.25%	*L. monocytogenes* *S. aureus* *E. coli* O157:H7 *L. monocytogenes* *Salmonella* Typhimurium	Cao and Song (2020)
Guar gum	Orange oil/Curcumin	Orange oil: 1–2% Curcumin: 0.1%	*E. coli* *Bacillus subtilis*	Aydogdu et al. (2020)
Sago starch Guar gum	Carvacrol Citral	Carvacrol: 0.75% Citral: 1.0%	*B. cereus* MTCC 430 *E. coli MTCC 443*	Dhumal et al. (2019)
Chitosan Gum arabic	Cinnamon essential oil Clove essential oil	5–10%	*S. aureus ATCC 25923* *E. coli ATCC 25922*	Xu et al. (2019)
Fish gelatin	Carvacrol	0.6%	*E. coli O157:H7 ATCC 32158* *S. aureus ATCC 25923*	Neira et al. (2019)

	Crude plant extracts			
Gelatin Chitosan	Cinnamon, guarana, rosemary, boldo-do-chile extracts			Bonilla and Sobral (2016)
Carrageenan	Grapefruit seed extract	0.6–13.3 μg/mL	*Escherichia coli* *Staphylococcus aureus* *B. cereus*	Kanmani and Rhim (2014)
	Other substances			
Chitosan and gelatin	Lauroyl arginate ethyl	0.1%	*Listeria monocytogenes (UNIMORE 19115),* *E. coli (UNIMORE 40522),* *Salmonella Typhimurium (UNIMORE 14028)* *Campylobacter jejuni* (UNIMORE 33250)	Haghighi et al. (2019)
Sodium alginate	Lysozyme	4.285 g/L	*Kocuria rhizophila* (ATCC 9341) *Listeria innocua* (ATCC 33090)	Amara et al. (2016)

References

Abdulhussain Kareem, R., & Razavi, S. H. (2020). Plantaricin bacteriocins: As safe alternative antimicrobial peptides in food preservation – A review. *Journal of Food Safety, 40*(1), e12735.

Abedini, A., Sadighara, P., Sani, M. A., & McClements, D. J. (2023). The impact of synthetic and natural additives on biogenic amine production in food products. *Food Bioscience, 56*, 103295.

Adamczak, A., Ożarowski, M., & Karpiński, T. M. (2019). Antibacterial activity of some flavonoids and organic acids widely distributed in plants. *Journal of Clinical Medicine, 9*(1), 109.

Adams, M., & Moss, M. (2000). *Non-bacterial agents of foodborne illness.*

Agriopoulou, S., Stamatelopoulou, E., Sachadyn-Król, M., & Varzakas, T. (2020). Lactic acid bacteria as antibacterial agents to extend the shelf life of fresh and minimally processed fruits and vegetables: Quality and safety aspects. *Microorganisms, 8*(6), 952.

Aguilar-Rosas, S., Ballinas-Casarrubias, M., Elias-Ogaz, L., Martin-Belloso, O., & Ortega-Rivas, E. (2013). Enzyme activity and colour changes in apple juice pasteurised thermally and by pulsed electric fields. *Acta Alimentaria, 42*(1), 45–54.

Ahmad, A., Qurashi, A., & Sheehan, D. (2023). Nano packaging–Progress and future perspectives for food safety, and sustainability. *Food Packaging Shelf Life, 35*, 100997.

Ahmed, J. (2018). Drying of vegetables: principles and dryer design. In *Handbook of vegetables vegetable processing* (pp. 381–405).

Ahmed, M. W., Haque, M. A., Mohibbullah, M., Khan, M. S. I., Islam, M. A., Mondal, M. H. T., et al. (2022). A review on active packaging for quality and safety of foods: Current trends, applications, prospects and challenges. *Food Packaging and Shelf Life, 33*, 100913.

Akabanda, F., Hlortsi, E. H., & Owusu-Kwarteng, J. (2017). Food safety knowledge, attitudes and practices of institutional food-handlers in Ghana. *BMC Public Health, 17*, 1–9.

Alegbeleye, O. O., Singleton, I., & Sant'Ana AS. (2018). Sources and contamination routes of microbial pathogens to fresh produce during field cultivation: A review. Food microbiology: Fundamentals. *Frontiers in Microbiology, 73*, 177–208.

Alimentarius, C. (2023). *Codex alimentarius commission procedural manual.*

Al-Zahrani, S. S., Bora, R. S., & Al-Garni, S. M. J. B. (2021). Antimicrobial activity of chitosan nanoparticles. *Biotechnology Biotechnological Equipment, 35*(1), 1874–1880.

Amara, C. B., Eghbal, N., Oulahal, N., Degraeve, P., & Gharsallaoui, A. (2016). Properties of lysozyme/sodium alginate complexes for the development of antimicrobial films. *Food Research International, 89*, 272–280.

Amjadi, S., Almasi, H., Ghorbani, M., & Ramazani, S. (2020). Preparation and characterization of TiO2NPs and betanin loaded zein/sodium alginate nanofibers. *Food Packaging and Shelf Life, 24*, 100504.

Anal, A. K. (2019). Quality ingredients and safety concerns for traditional fermented foods and beverages from Asia: A review. *Fermentation, 5*(1), 8.

Asagbara, A., Onawola, O., & Okeagu, M. (2019). Identification of critical control points in some Nigerian fermented foods and seasonings. *World Journal of Innovative Research, 6*(4), 55–60.

Augusto, P. E., Soares, B. M., & Castanha, N. (2018). Conventional technologies of food preservation. In *Innovative technologies for food preservation* (pp. 3–23). Elsevier.

Aydogdu, A., Radke, C. J., Bezci, S., & Kirtil, E. (2020). Characterization of curcumin incorporated guar gum/orange oil antimicrobial emulsion films. *International Journal of Biological Macromolecules, 148*, 110–120.

Bahndral, A., Shams, R., & Choudhary, P. (2024). Plant-based chitosan for the development of biodegradable packaging materials. In *Carbohydrate polymer technologies and applications* (p. 100598).

Bahrami, A., Mokarram, R. R., Khiabani, M. S., Ghanbarzadeh, B., & Salehi, R. (2019). Physico-mechanical and antimicrobial properties of tragacanth/hydroxypropyl methylcellulose/beeswax edible films reinforced with silver nanoparticles. *International Journal of Biological Macromolecules, 129*, 1103–1112.

Bambace, M. F., Alvarez, M. V., & Moreira, M. R. (2021). Ready-to-eat blueberries as fruit-based alternative to deliver probiotic microorganisms and prebiotic compounds. *LWT, 142*, 111009.

Barba, F. J., Parniakov, O., Pereira, S. A., Wiktor, A., Grimi, N., Boussetta, N., et al. (2015). Current applications and new opportunities for the use of pulsed electric fields in food science and industry. *Food Research International, 77*, 773–798.

Barbiroli, A., Bonomi, F., Capretti, G., Iametti, S., Manzoni, M., Piergiovanni, L., et al. (2012). Antimicrobial activity of lysozyme and lactoferrin incorporated in cellulose-based food packaging. *Food Control, 26*(2), 387–392.

Barbosa-Cánovas, G. V., Donsì, F., Yildiz, S., Candoğan, K., Pokhrel, P. R., & Guadarrama-Lezama, A. Y. (2022). Nonthermal processing technologies for stabilization and enhancement of bioactive compounds in foods. *Food Engineering Reviews*, 1–37.

Batiha, G. E.-S., Hussein, D. E., Algammal, A. M., George, T. T., Jeandet, P., Al-Snafi, A. E., et al. (2021). Application of natural antimicrobials in food preservation: Recent views. *Food Control, 126*, 108066.

Behera, S. S., Ray, R. C., & Zdolec, N. (2018). *Lactobacillus plantarum* with functional properties: an approach to increase safety and shelf-life of fermented foods. *BioMed Research International, 2018*(1), 9361614.

Behera, S. S., Nivedita, S., Behera, H. T., Mojumdar, A., Sahoo, B. N., & Ray, L. (2024). Bio synthesis of chitosan from shrimp shell chitin using Streptomyces griseoincarnatus RB7AG: Characterization and application as coating material to enhance shelf life of tomatoes (var. Ruby pusa) in post-harvest storgae condition. *Bioactive Carbohydrates and Dietary Fibre, 32*, 100455.

Belitz, I. H.-D., & Grosch, I. W. (2013). *Food chemistry*. Springer.

Ben Braïek, O., & Smaoui, S. (2021). Chemistry, safety, and challenges of the use of organic acids and their derivative salts in meat preservation. *Journal of Food Quality, 2021*(1), 6653190.

Bhattacharya, D., Nanda, P. K., Pateiro, M., Lorenzo, J. M., Dhar, P., & Das, A. K. (2022). Lactic acid bacteria and bacteriocins: Novel biotechnological approach for biopreservation of meat and meat products. *Microorganisms, 10*(10), 2058.

Bintsis, T. (2017). Foodborne pathogens. *AIMS Microbiology, 3*(3), 529.

Bloomfield, S. J., Palau, R., Holden, E. R., Webber, M. A., & Mather, A. E. (2024). Genomic characterization of *Pseudomonas* spp. on food: implications for spoilage, antimicrobial resistance and human infection. *BMC Microbiology, 24*(1), 20.

Bonilla, J., & Sobral, P. J. (2016). Investigation of the physicochemical, antimicrobial and antioxidant properties of gelatin-chitosan edible film mixed with plant ethanolic extracts. *Food Bioscience, 16*, 17–25.

Bosch, A., Pintó, R. M., & Guix, S. (2014). Human astroviruses. *Clinical Microbiology Reviews, 27*(1), 1048–1071.

Branen, A. L., & Haggerty, R. J. F. A. (2002). Introduction to food additives. *Food additives, 2*, 1–9.

Bravo, D., de Alba, M., & Medina, M. (2014). Combined treatments of high-pressure with the lactoperoxidase system or lactoferrin on the inactivation of *Listeria monocytogenes*, *Salmonella* Enteritidis and *Escherichia coli* O157: H7 in beef carpaccio. *Food Microbiology, 41*, 27–32.

Bücher, C., Burtscher, J., & Domig, K. J. (2021). Propionic acid bacteria in the food industry: An update on essential traits and detection methods. *Comprehensive Reviews in Food Science and Food Safety, 20*(5), 4299–4323.

Cai, R., Ma, Y., Wang, Z., Yuan, Y., Guo, H., Sheng, Q., et al. (2024). Inactivation activity and mechanism of pulsed light against *Alicyclobacillus acidoterrestris* vegetative cells and spores in concentrated apple juice. *International Journal of Food Microbiology, 413*, 110576.

Calvo, T., Prieto, M., Alvarez-Ordóñez, A., & López, M. (2020). Effect of non-thermal atmospheric plasma on food-borne bacterial pathogens on ready-to eat foods: morphological and physicochemical changes occurring on the cellular envelopes. *Foods, 9*(12), 1865.

Cao, T. L., & Song, K. B. (2020). Development of bioactive Bombacaceae gum films containing cinnamon leaf essential oil and their application in packaging of fresh salmon fillets. *LWT, 131*, 109647.

Capelli, F., Tappi, S., Gritti, T., de Aguiar Saldanha Pinheiro, A. C., Laurita, R., Tylewicz, U., et al. (2021). Decontamination of food packages from SARS-CoV-2 RNA with a cold plasma-assisted system. *Applied Sciences, 11*(9), 4177.

CDC. (2024). *Norovirus*. https://www.cdc.gov/norovirus/index.html

Chandrasekaran, M., Kim, K. D., & Chun, S. C. (2020). Antibacterial activity of chitosan nanoparticles: A review. *Processes, 8*(9), 1173.

Chang, G., Li, Q., Wang, T., Zhang, B., Wu, W., Lv, C., et al. (2024). Characterization of *Pseudomonas* spp. contamination and in situ spoilage potential in pasteurized milk production process. *Food Research International, 188*, 114463.

Chatterjee, A., & Abraham, J. (2018). Microbial contamination, prevention, and early detection in food industry. In *Microbial contamination and food degradation* (pp. 21–47). Elsevier.

Chayjan, R., & Esna-Ashari, M. (2010). Modeling of heat and entropy sorption of maize (cv. Sc704): neural network method. *Research in Agricultural Engineering, 56*(2), 69–76.

Chen, D., Wiertzema, J. R., Peng, P., Cheng, Y., Wang, Y., Liu, J., et al. (2020). Catalytic intense pulse light inactivation of Cronobacter sakazakii and other pathogens in non-fat dry milk and wheat flour. *Food Chemistry, 332*, 127420.

Chen, D., Mosher, W., Wiertzema, J., Peng, P., Min, M., Cheng, Y., et al. (2021). Effects of intense pulsed light and gamma irradiation on *Bacillus cereus* spores in mesquite pod flour. *Food Chemistry, 344*, 128675.

Chen, X., Liu, Y., Zhang, R., Zhu, H., Li, F., Yang, D., et al. (2022a). Radio frequency drying behavior in porous media: A case study of potato cube with computer modeling. *Foods, 11*(20), 3279.

Chen, J., Luo, L., Cen, C., Liu, Y., Li, H., & Wang, Y. (2022b). The nano antibacterial composite film carboxymethyl chitosan/gelatin/nano ZnO improves the mechanical strength of food packaging. *International Journal of Biological Macromolecules, 220*, 462–471.

Chen, X., Lan, W., & Xie, J. (2023a). Natural phenolic compounds: Antimicrobial properties, antimicrobial mechanisms, and potential utilization in the preservation of aquatic products. *Food Chemistry*, 138198.

Chen, Y., Liu, Y., Dong, Q., Xu, C., Deng, S., Kang, Y., et al. (2023b). Application of functionalized chitosan in food: A review. *International Journal of Biological Macromolecules, 235*, 123716.

Chiozzi, V., Agriopoulou, S., & Varzakas, T. (2022). Advances, applications, and comparison of thermal (pasteurization, sterilization, and aseptic packaging) against non-thermal (ultrasounds, UV radiation, ozonation, high hydrostatic pressure) technologies in food processing. *Applied Sciences, 12*(4), 2202.

Chipley, J. R. (2020). *Sodium benzoate and benzoic acid. Antimicrobials in food* (pp. 41–88). CRC Press.

Ciurzyńska, A., Cichowska, J., Kowalska, H., Czajkowska, K., & Lenart, A. (2018). Osmotic dehydration of Braeburn variety apples in the production of sustainable food products. *International Agrophysics, 32*(1).

Cruz, A. I. C., da Cruz Costa, M., Mafra, J. F., Ferreira, M. A., Miranda, F. M., Costa, J. A., et al. (2021). A sodium alginate bilayer coating incorporated with green propolis extract as a powerful tool to extend *Colossoma macropomum* fillet shelf-life. *Food Chemistry, 355*, 129610.

da Costa Kassacula, S. D. F., Carlos, M., & Barros, A. A. C. (2023). Production of concentrated red Pitaya (*Hylocereus Undatus*) juice by vacuum rvaporation. *Research, Society Development, 12*(5), e19612541652-e.

Datta, A. K., & Davidson, P. M. (2000). Microwave and radio frequency processing. *Journal of Food Science, 65*, 32–41.

Davidson, P., Cekmer, H. B., Monu, E., & Techathuvanan, C. (2015). JHonaffs, quality. In *The use of natural antimicrobials in food: An overview* (pp. 1–27).

de Oliveira, F., Lima, C. A., Lopes, A. M., Marques, D. A. V., Druzian, J. I., Pessoa Junior, A., et al. (2020). Microbial colorants production in stirred-tank bioreactor and their incorporation in an alternative food packaging biomaterial. *Journal of Fungi, 6*(4), 264.

Dejene, F., Regasa Dadi, B., & Tadesse, D. (2021). In vitro antagonistic effect of lactic acid bacteria isolated from fermented beverage and finfish on pathogenic and foodborne pathogenic microorganism in Ethiopia. *International Journal of Microbiology, 2021*(1), 5370556.

Demirci, A., Feng, H., & Krishnamurthy, K. (2020). *Food safety engineering*. Springer.

Denkova-Kostova, R. S., Goranov, B. G., Teneva, D. G., Tomova, T. G., Denkova, Z. R., Shopska, V., et al. (2021). Bio-preservation of chocolate mousse with free and immobilized cells of *Lactobacillus plantarum* D2 and lemon (*Citrus lemon* L.) or grapefruit (*Citrus paradisi* L.) zest essential oils. *Acta Scientiarum Polonorum Technologia Alimentaria, 20*(1), 5–16.

Dermesonlouoglou, E. K., Pantelaiaki, K., Andreou, V., Katsaros, G. J., & Taoukis, P. S. (2019). Osmotic pretreatment for the production of novel dehydrated tomatoes and cucumbers. *Journal of Food Processing Preservation, 43*(7), e13968.

Dhar, R., & Chakraborty, S. (2023). Pasteurization of bael fruit (*Aegle marmelos*) juice using high-intensity pulsed light treatment. *Food Control, 152*, 109826.

Dhumal, C. V., Ahmed, J., Bandara, N., & Sarkar, P. (2019). Improvement of antimicrobial activity of sago starch/guar gum bi-phasic edible films by incorporating carvacrol and citral. *Food Packaging and Shelf Life, 21*, 100380.

Ding, J., Huang, D., Wang, W., Lu, Y., Dong, W., Zong, L., et al. (2020). Significantly improve the water and chemicals resistance of alginate-based nanocomposite films by a simple in-situ surface coating approach. *International Journal of Biological Macromolecules, 156*, 1297–1307.

Ding, H., Wang, Z., Yuan, Y., & Yue, T. (2024). Inactivation of *Alicyclobacillus acidoterrestris* in apple juice using pulsed light. *Food Control, 163*, 110439.

Dorny, P., Praet, N., Deckers, N., & Gabriël, S. (2009). Emerging food-borne parasites. *Veterinary Parasitology, 163*(3), 196–206.

Duarte, L. G., & Picone, C. S. (2022). Antimicrobial activity of lactoferrin-chitosan-gellan nanoparticles and their influence on strawberry preservation. *Food Research International, 159*, 111586.

Duarte, L. G., Ferreira, N. C., Fiocco, A. C. T., & Picone, C. S. (2023). Lactoferrin-chitosan-TPP nanoparticles: Antibacterial action and extension of strawberry shelf-life. *Food Bioprocess Technology, 16*(1), 135–148.

Durand, G., Javerliat, F., Bes, M., Veyrieras, J.-B., Guigon, G., Mugnier, N., et al. (2018). Routine whole-genome sequencing for outbreak investigations of *Staphylococcus aureus* in a national reference center. *Frontiers in Microbiology, 9*, 511.

Duret, S., Hoang, H. M., Derens-Bertheau, E., Delahaye, A., Laguerre, O., & Guillier, L. (2019). Combining quantitative risk assessment of human health, food waste, and energy consumption: The next step in the development of the food cold chain? *Risk Analysis, 39*(4), 906–925.

Dürre, P. (2021). The genus Clostridium. In *Practical handbook of microbiology* (pp. 325–338). CRC Press.

Ehuwa, O., Jaiswal, A. K., & Jaiswal, S. (2021). *Salmonella*, food safety and food handling practices. *Foods, 10*(5), 907.

Ekici, G., & Dümen, E. (2019). *Escherichia coli* and food safety. In *The universe of Escherichia coli*. IntechOpen.

Ekwomadu, T. I., & Mwanza, M. (2023). *Fusarium* fungi pathogens, identification, adverse effects, disease management, and global food security: a review of the latest research. *Agriculture, 13*(9), 1810.

Elik, A., Yanık, D. K., Maskan, M., & Göğüş, F. (2016). Influence of three different concentration techniques on evaporation rate, color and phenolics content of blueberry juice. *Journal of Food Science Technology, 53*, 2389–2395.

El-Sayed, H. S., El-Sayed, S. M., Mabrouk, A. M., Nawwar, G. A., & Youssef, A. M. (2021). Development of eco-friendly probiotic edible coatings based on chitosan, alginate and

carboxymethyl cellulose for improving the shelf life of UF soft cheese. *Journal of Polymers and the Environment, 29*, 1941–1953.
Enescu, D., Dehelean, A., Gonçalves, C., Cerqueira, M. A., Magdas, D. A., Fucinos, P., et al. (2020). Evaluation of the specific migration according to EU standards of titanium from Chitosan/Metal complexes films containing TiO2 particles into different food simulants. A comparative study of the nano-sized vs micro-sized particles. *Food Packaging and Shelf Life, 26*, 100579.
Erkmen, O., & Bozoglu, T. F. (2016). Food preservation by reducing water activity. (Vol. 2, pp. 44–58).
Fellows, P. (2000). *Food processing technology: Principles and practice*. Woodhead publishing
Fellows, P. J. (2022). *Food processing technology: principles and practice*. Woodhead Publishing.
Ferdes, M. (2018). Antimicrobial compounds from plants. In *Fighting antimicrobial resistance* (pp. 243–271). IAPC-OBP.
Fernández, M., Hospital, X. F., Cabellos, C., & Hierro, E. (2020). Effect of pulsed light treatment on Listeria inactivation, sensory quality and oxidation in two varieties of Spanish dry-cured ham. *Food Chemistry, 316*, 126294.
Ferreira, N. B. M., Rodrigues, M. I., & Cristianini, M. (2025). Modeling the protective effect of water activity of a meat emulsion matrix model on *Listeria monocytogenes* inactivation by high pressure processing. *Food Control, 167*, 110751.
Fetsch, A., Contzen, M., Hartelt, K., Kleiser, A., Maassen, S., Rau, J., et al. (2014). *Staphylococcus aureus* food-poisoning outbreak associated with the consumption of ice-cream. *International Journal of Food Microbiology, 187*, 1–6.
Finnegan, M., Linley, E., Denyer, S. P., McDonnell, G., Simons, C., & Maillard, J.-Y. (2010). Mode of action of hydrogen peroxide and other oxidizing agents: differences between liquid and gas forms. *Journal of Antimicrobial Chemotherapy, 65*(10), 2108–2115.
Frank, C., Werber, D., Cramer, J. P., Askar, M., Faber, M., van der Heiden, M., et al. (2011). Epidemic profile of Shiga-toxin–producing *Escherichia coli* O104: H4 outbreak in Germany. *New England Journal of Medicine, 365*(19), 1771–1780.
Fu, L., Valentino, H., & Wang, Y. (2016). Bacterial contamination in food production. In *Antimicrobial food packaging* (pp. 35–43). Elsevier.
Ganesan, A. R., Tiwari, U., Ezhilarasi, P., & Rajauria, G. (2021). Application of cold plasma on food matrices: A review on current and future prospects. *Journal of Food Processing and Preservation, 45*(1), e15070.
Ganessingh, V., Sahibdeen, R., & Maharaj, R. (2018). An evaluation of the impact of novel processing technologies on the phytochemical composition of fruits and vegetables. In *Phytochemicals-Source of Antioxidants and Role in Disease Prevention*. IntechOpen.
Garvey, M. (2023). Foodborne clostridioides species: pathogenicity, virulence and biocontrol options. *Microorganisms, 11*(10), 2483.
Gautam, P. B., Sharma, R., Atbhaiya, Y., Gandhi, K., & Mann, B. (2024). Thermosonication: A technique to inactivate the plasmin system in milk. *International Dairy Journal*, 105997.
Gómez, D., Azón, E., Marco, N., Carramiñana, J. J., Rota, C., Ariño, A., et al. (2014). Antimicrobial resistance of *Listeria monocytogenes* and *Listeria innocua* from meat products and meat-processing environment. *Food Microbiology, 42*, 61–65.
Goy, R. C., Britto, D., & Assis, O. B. (2009). A review of the antimicrobial activity of chitosan. *Polímeros, 19*, 241–247.
Guin, S., Saravanan, M., Chowdhury, G., Pazhani, G. P., Ramamurthy, T., & Das, S. C. (2019). Pathogenic *Vibrio parahaemolyticus* indiarrhoeal patients, fish and aquatic environments and their potential for inter-source transmission. *Heliyon, 5*(5).
Gyawali, R., Hayek, S., & Ibrahim, S. A. (2015). Plant extracts as antimicrobials in food products: Mechanisms of action, extraction methods, and applications. In *Handbook of natural antimicrobials for food safety and quality* (Vol. 49, pp. 49–62).

Gyawali, R., Degala, H. L., Biswal, A. K., Bardsley, C. A., & Mahapatra, A. K. (2024). Effects of intense pulsed light on inactivation of *Salmonella* Typhimurium and quality characteristics of pecan halves. *LWT*, 116344.

Haghighi, H., De Leo, R., Bedin, E., Pfeifer, F., Siesler, H. W., & Pulvirenti, A. (2019). Comparative analysis of blend and bilayer films based on chitosan and gelatin enriched with LAE (lauroyl arginate ethyl) with antimicrobial activity for food packaging applications. *Food Packaging and Shelf Life, 19*, 31–39.

Hasan, M., Sultan, M., & Mar-E-Um, M. (2014). Significance of fermented food in nutrition and food science. *Journal of Scientific Research, 6*(2), 373–386.

Hassan, H., St-Gelais, D., Gomaa, A., & Fliss, I. (2021). Impact of nisin and nisin-producing *Lactococcus lactis* ssp. lactis on *Clostridium tyrobutyricum* and bacterial ecosystem of cheese matrices. *Foods, 10(4):898.*

Hernández-Cortez, C., Palma-Martínez, I., Gonzalez-Avila, L. U., Guerrero-Mandujano, A., Solís, R. C., & Castro-Escarpulli, G. (2017). Food poisoning caused by bacteria (food toxins). In *Poisoning: From specific toxic agents to novel rapid and simplified techniques for analysis* (p. 33). InTechOpen.

Honarvar, Z., Hadian, Z., & Mashayekh, M. (2016). Nanocomposites in food packaging applications and their risk assessment for health. *Electronic Physician, 8*(6), 2531.

Hosseinnejad, M., & Jafari, S. M. (2016). Evaluation of different factors affecting antimicrobial properties of chitosan. *International Journal of Biological Macromolecules, 85*, 467–475.

Hu, X., Jia, X., Zhi, C., Jin, Z., & Miao, M. (2019). Improving the properties of starch-based antimicrobial composite films using ZnO-chitosan nanoparticles. *Carbohydrate Polymers, 210*, 204–209.

Hu, C., Sha, L., Huang, C., Luo, W., Li, B., Huang, H., et al. (2023). Phase change materials in food: Phase change temperature, environmental friendliness, and systematization. *Trends in Food Science & Technology, 104167.*

Ibrahim, F., Jalal, H., Khan, A., Asghar, M., Iqbal, J., Ahmed, A., et al. (2016). Prevalence of aflatoxigenic *Aspergillus* in food and feed samples from Karachi. *Journal of Infection and Molecular Biology, 4*, 1–8.

Ibrahim, S. A., Ayivi, R. D., Zimmerman, T., Siddiqui, S. A., Altemimi, A. B., Fidan, H., et al. (2021). Lactic acid bacteria as antimicrobial agents: Food safety and microbial food spoilage prevention. *Foods, 10*(12), 3131.

Issa, A., Ibrahim, S. A., & Tahergorabi, R. (2017). Impact of sweet potato starch-based nanocomposite films activated with thyme essential oil on the shelf-life of baby spinach leaves. *Foods, 6*(6), 43.

Jadhav, H. B., Annapure, U. S., & Deshmukh, R. R. (2021). Non-thermal technologies for food processing. *Frontiers in Nutrition, 8*, 657090.

Jaeger, H., Schulz, A., Karapetkov, N., & Knorr, D. (2009). Protective effect of milk constituents and sublethal injuries limiting process effectiveness during PEF inactivation of *Lb. rhamnosus. International Journal of Food Microbiology, 134*(1–2), 154–161.

Jagtap, S., Rahimifard, S., & Duong, L. N. (2022). Real-time data collection to improve energy efficiency: A case study of food manufacturer. *Journal of Food Processing and Preservation, 46*(8), e14338.

Jamróz, E., & Kopel, P. (2020). Polysaccharide and protein films with antimicrobial/antioxidant activity in the food industry: A review. *Polymers, 12*(6), 1289.

Jay, J. M., Loessner, M. J., & Golden, D. A. (2008). *Modern food microbiology*. Springer.

Ji, Q.-Y., Wang, W., Yan, H., Qu, H., Liu, Y., Qian, Y., et al. (2023). The effect of different organic acids and their combination on the cell barrier and biofilm of *Escherichia coli. Foods, 12*(16), 3011.

Jiménez, N., Bassama, J., Soto, M., Dornier, M., Pérez, A. M., Vaillant, F., et al. (2020). Coupling osmotic dehydration with heat treatment for green papaya impregnated with blackberry juice solution. *International Journal of Food Science Technology, 55*(6), 2551–2561.

John, D., & Ramaswamy, H. S. (2018). Pulsed light technology to enhance food safety and quality: A mini-review. *Current Opinion in Food Science, 23*, 70–79.

Johnson, E. M., Jung, D. Y.-G., Jin, D. Y.-Y., Jayabalan, D. R., Yang, D. S. H., & Suh, J. W. (2018). Bacteriocins as food preservatives: Challenges and emerging horizons. *Critical Reviews in Food Science and Nutrition, 58*(16), 2743–2767.

Jung, E. J., Youn, D. K., Lee, S. H., No, H. K., Ha, J. G., & Prinyawiwatkul, W. (2010). Antibacterial activity of chitosans with different degrees of deacetylation and viscosities. *International Journal of Food Science Technology, 45*(4), 676–682.

Kaczmarek, M., Avery, S. V., & Singleton, I. (2019). Microbes associated with fresh produce: Sources, types and methods to reduce spoilage and contamination. In *Advances in applied microbiology* (Vol. 107, pp. 29–82). Elsevier.

Kamala, K., & Kumar, V. P. (2018). Food products and food contamination. In *Microbial contamination and food degradation* (pp. 1–19). Elsevier.

Kamarudin, S. H., Rayung, M., Abu, F., Ahmad, S. B., Fadil, F., Karim, A. A., et al. (2022). A review on antimicrobial packaging from biodegradable polymer composites. *Polymers, 14*(1), 174.

Kanmani, P., & Rhim, J.-W. (2014). Development and characterization of carrageenan/grapefruit seed extract composite films for active packaging. *International Journal of Biological Macromolecules, 68*, 258–266.

Khan, R., Anwar, F., & Ghazali, F. M. (2024). A comprehensive review of mycotoxins: Toxicology, detection, and effective mitigation approaches. *Heliyon.*

Khorshidian, N., Khanniri, E., Koushki, M. R., Sohrabvandi, S., & Yousefi, M. (2022). An overview of antimicrobial activity of lysozyme and its functionality in cheese. *Frontiers in Nutrition, 9*, 833618.

Kim, H.-J., Jubinville, E., Goulet-Beaulieu, V., & Jean, J. (2024). Inactivation of murine norovirus and hepatitis A virus on various frozen fruits using pulsed light. *International Journal of Food Microbiology, 424*, 110851.

Kotnik, T., Rems, L., Tarek, M., & Miklavčič, D. (2019). Membrane electroporation and electropermeabilization: mechanisms and models. *Annual Review of Biophysics, 48*(1), 63–91.

Koutchma, T. (2014). *Adapting high hydrostatic pressure (HPP) for food processing operations.* Academic.

Kuts, O., Verbytskyi, S., Bokova, S., Kozachenko, O., & Patsera, N. (2022). Food security and food safety issues of the long-term storage of meat and milk products. *Animal Husbandry Products Production and Processing, 1*, 162–167.

Lalani, A. R., Rastegar-Pouyani, N., Askari, A., Tavajohi, S., Akbari, S., & Jafarzadeh, E. (2024). Food additives, benefits, and side effects: A review article. *Journal of Chemical Health Risks, 14*(1).

Lazárková, Z., Kratochvílová, A., Salek, R. N., Polášek, Z., Šiška, L., Pětová, M., et al. (2023). Influence of heat treatment on the chemical, physical, microbiological and sensorial properties of pork liver pâté as affected by fat content. *Foods, 12*(12), 2423.

Lee, E.-B., & Lee, K. (2024). Woodfordia fruticosa fermented with lactic acid bacteria impact on foodborne pathogens adhesion and cytokine production in HT-29 cells. *Frontiers in Microbiology, 15*, 1346909.

Lee, H., & Yoon, Y. (2021). Etiological agents implicated in foodborne illness world wide. *Food Science of Animal Resources, 41*(1), 1.

Lee, M., Rüegg, N., & Yildirim, S. (2023). Evaluation of the antimicrobial activity of sodium alginate films integrated with cinnamon essential oil and citric acid on sliced cooked ham. *Packaging Technology and Science, 36*(8), 647–656.

Leng, J., Mukhopadhyay, S., Sokorai, K., Ukuku, D. O., Fan, X., Olanya, M., et al. (2020). Inactivation of *Salmonella* in cherry tomato stem scars and quality preservation by pulsed light treatment and antimicrobial wash. *Food Control, 110*, 107005.

Li, J., & Zhuang, S. (2020). Antibacterial activity of chitosan and its derivatives and their interaction mechanism with bacteria: Current state and perspectives. *European Polymer Journal, 138*, 109984.

Li, X.-f., Feng, X.-q., Yang, S., Fu, G.-q., Wang, T.-p., & Su, Z.-x. (2010). Chitosan kills *Escherichia coli* through damage to be of cell membrane mechanism. *Carbohydrate Polymers, 79*(3), 493–499.

Limbo, S., & Khaneghah, A. M. (2015). Active packaging of foods and its combination with electron beam processing. In *Electron beam pasteurization and complementary food processing technologies* (pp. 195–217). Elsevier.

Lin, X., Zhang, H., Li, S., Huang, L., Zhang, R., Zhang, L., et al. (2022). Polyphenol-driving assembly for constructing chitin-polyphenol-metal hydrogel as wound dressing. *Carbohydrate Polymers, 290*, 119444.

Linley, E., Denyer, S. P., McDonnell, G., Simons, C., & Maillard, J.-Y. (2012). Use of hydrogen peroxide as a biocide: new consideration of its mechanisms of biocidal action. *Journal of Antimicrobial Chemotherapy, 67*(7), 1589–1596.

Lisboa, H. M., Pasquali, M. B., dos Anjos, A. I., Sarinho, A. M., de Melo, E. D., Andrade, R., et al. (2024). Innovative and sustainable food preservation techniques: Enhancing food quality, safety, and environmental sustainability. *Sustainability, 16*(18), 8223.

Lorenzo, J. M., Munekata, P. E., Dominguez, R., Pateiro, M., Saraiva, J. A., & Franco, D. (2018). Main groups of microorganisms of relevance for food safety and stability: General aspects and overall description. In *Innovative technologies for food preservation* (pp. 53–107). Elsevier.

Lytras, F., Psakis, G., Gatt, R., Cebrián, G., Raso, J., & Valdramidis, V. (2024). Exploring the efficacy of pulsed electric fields (PEF) in microbial inactivation during food processing: A deep dive into the microbial cellular and molecular mechanisms. *Innovative Food Science & Emerging Technologies, 95*, 103732.

Mahendran, R., Ramanan, K. R., Barba, F. J., Lorenzo, J. M., López-Fernández, O., Munekata, P. E., et al. (2019). Recent advances in the application of pulsed light processing for improving food safety and increasing shelf life. *Trends in Food Science & Technology, 88*, 67–79.

Mahnič-Kalamiza, S., & Miklavčič, D. (2022). The phenomenon of electroporation. In *Pulsed electric fields technology for the food industry: Fundamentals and applications* (pp. 107–141).

Malhotra, B., Keshwani, A., & Kharkwal, H. (2015). Antimicrobial food packaging: Potential and pitfalls. *Frontiers in Microbiology, 6*, 611.

Mandal, R., Mohammadi, X., Wiktor, A., Singh, A., & Pratap, S. A. (2020). Applications of pulsed light decontamination technology in food processing: An overview. *Applied Sciences, 10*(10), 3606.

Manjunatha, B., Supraja, N., Yedoti, V., & Vijaykumar, B. (2023). Bacteriological, preparation and optimization for lactose hydrolyzed sweet concentrate whey using protein hydrolysate in ice creams. *Current Journal of Applied Science, 42*(37), 1–12.

Marriott, N. G., Schilling, M. W., Gravani, R. B., Marriott, N. G., Schilling, M. W., & Gravani, R. B. (2018). Food contamination sources. In *Principles of food sanitation* (pp. 83–91).

Marszałek, K., Krzyżanowska, J., Woźniak, Ł., & Skąpska, S. (2017). Kinetic modelling of polyphenol oxidase, peroxidase, pectin esterase, polygalacturonase, degradation of the main pigments and polyphenols in beetroot juice during high pressure carbon dioxide treatment. *LWT-Food Science Technology, 85*, 412–417.

Martins, W. S., de Campos Leite, A. B., Martins, R. L., da Silva, J. O., & de Carvalho, B. S. (2019). Assessment of frozen seafood good storage practices in the 21st supply deposit of the Brazilian Army. *Brazilian Journal of Veterinary Research and Animal Science, 56*(2), e151385-e.

Matthews, K. R., Kniel, K. E., & Critzer, F. J. (2024). *Food microbiology: an introduction.* Wiley.

Medina, E., Caro, N., Abugoch, L., Gamboa, A., Díaz-Dosque, M., & Tapia, C. (2019). Chitosan thymol nanoparticles improve the antimicrobial effect and the water vapour barrier of chitosan-quinoa protein films. *Journal of Food Engineering, 240*, 191–198.

Meloni, D., Consolati, S. G., Mazza, R., Mureddu, A., Fois, F., Piras, F., et al. (2014). Presence and molecular characterization of the major serovars of *Listeria monocytogenes* in ten Sardinian fermented sausage processing plants. *Meat Science, 97*(4), 443–450.

Mendagudali, R. R., Akka, K. D., Swati, I. A., Shedole, D. T., & Bendigeri, N. (2016). Knowledge, attitude, and practices of food safety among women of Khaza bazar, the urban field practice area of KBN Institute of Medical Sciences, Kalaburagi, Karnataka. *International Journal of Medical Science and Public Health, 5*(3), 516–520.

Mihaylova-Garnizova, R., Davidova, S., Hodzhev, Y., & Satchanska, G. (2024). Antimicrobial peptides derived from bacteria: Classification, sources, and mechanism of action against multidrug-resistant bacteria. *International Journal of Molecular Sciences, 25*(19), 10788.

Min, M., Bunt, C. R., Mason, S. L., Bennett, G. N., & Hussain, M. A. (2017). Effect of non-dairy food matrices on the survival of probiotic bacteria during storage. *Microorganisms, 5*(3), 43.

Mohammed, M., Riad, K., & Alqahtani, N. (2022). Design of a smart IoT-based control system for remotely managing cold storage facilities. *Sensors, 22*(13), 4680.

Montone, A. M. I., Malvano, F., Taiano, R., Capparelli, R., Capuano, F., & Albanese, D. (2023). Alginate coating charged by hydroxyapatite complexes with lactoferrin and quercetin enhances the pork meat shelf life. *Foods, 12*(3), 553.

Montville, T. J., Matthews, K. R., & Kniel, K. E. (2012). *Food microbiology: An introduction.* ASM Press.

Mothibe, K. J., Zhang, M., Mujumdar, A. S., Wang, Y. C., & Cheng, X. (2014). Effects of ultrasound and microwave pretreatments of apple before spouted bed drying on rate of dehydration and physical properties. *Drying Technology, 32*(15), 1848–1856.

Muhlbauer, W., & Muller, J. (2020). *Drying atlas: Drying kinetics and quality of agricultural products.* Woodhead Publishing.

Nakagawa, H., & Oyama, T. (2019). Molecular basis of water activity in glycerol–water mixtures. *Frontiers in Chemistry, 7*, 731.

Nawaz, N., Wen, S., Wang, F., Nawaz, S., Raza, J., Iftikhar, M., et al. (2022). Lysozyme and its application as antibacterial agent in food industry. *Molecules (Basel, Switzerland), 27*(19), 6305.

Ndraha, N., Huang, L., Wu, V. C., & Hsiao, H.-I. (2022). *Vibrio parahaemolyticus* in seafood: Recent progress in understanding influential factors at harvest and food-safety intervention approaches. *Current Opinion in Food Science, 48*, 100927.

Neira, L. M., Martucci, J. F., Stejskal, N., & Ruseckaite, R. A. (2019). Time-dependent evolution of properties of fish gelatin edible films enriched with carvacrol during storage. *Food Hydrocolloids, 94*, 304–310.

Nerín, C., Aznar, M., & Carrizo, D. (2016). Food contamination during food process. *Trends in Food Science Technology, 48*, 63–68.

Ngoc, T. T. A., Tu, N. C., Trang, P. N., Ha, N. C., & Miyamoto, T. (2022). *Pangasius Hypophthalmus* Viscera as a potential vector of bacterial cross-contamination and resistance of *Escherichia coli* to antibiotics. *Current Research in Nutrition Food Science Journal, 10*(2), 711–719.

Ngulukun, S. S. (2017). Taxonomy and physiological characteristics of *Campylobacter* spp. In *Campylobacter* (pp. 41–60). Elsevier.

Nie R-Z, Dang M-Z, Ge Z-Z, Huo Y-Q, Yu B, Tang S-W. Influence of the gallate moiety on the interactions between green tea polyphenols and lipid membranes elucidated by molecular dynamics simulations. Biophysical Chemistry 2021;274:106592-.

Niedźwiedź, I., Waśko, A., Pawłat, J., & Polak-Berecka, M. (2019). The state of research on antimicrobial activity of cold plasma. *Polish Journal of Microbiology, 68*(2), 153–164.

Nowshad, F., Mustari, N., & Khan, M. S. (2021). Overview of microbial contamination of foods and associated risk factors. In *Techniques to measure food safety quality: microbial, chemical, sensory* (pp. 11–29).

O'Connor, J., Mickan, B. S., Siddique, K. H., Rinklebe, J., Kirkham, M., & Bolan, N. S. (2022). Physical, chemical, and microbial contaminants in food waste management for soil application: A review. *Environmental Pollution, 300*, 118860.
Obileke, K., Onyeaka, H., Miri, T., Nwabor OF, Hart, A., Al-Sharify, Z. T., et al. (2022). Recent advances in radio frequency, pulsed light, and cold plasma technologies for food safety. *Journal of Food Process Engineering, 45*(10), e14138.
Okaiyeto, S. A., Sutar, P. P., Chen, C., Ni, J.-B., Wang, J., Mujumdar, A. S., et al. (2024). Antibiotic resistant bacteria in food systems: Current status, resistance mechanisms, and mitigation strategies. *Agriculture Communications*, 100027.
Oladipo-Adekeye, O. T., & Tabit, F. T. (2021). The food safety knowledge of street food vendors and the sanitary compliance of their vending facilities, Johannesburg, South Africa. *Journal of Food Safety, 41*(4), e12908.
Ordoñez, R., Atarés, L., & Chiralt, A. (2021). Physicochemical and antimicrobial properties of cassava starch films with ferulic or cinnamic acid. *LWT, 144*, 111242.
Organization WH. (2019). *Evaluation of certain food additives: Eighty-sixth report of the joint FAO/WHO expert committee on food additives*. World Health Organization.
Ortega, Y. R. (2019). Protozoan parasites. In *Food microbiology: Fundamentals frontiers in microbiology* (pp. 667–691). Wiley.
Otero, L., Pérez-Mateos, M., & López-Caballero, M. (2017). Hyperbaric cold storage versus conventional refrigeration for extending the shelf-life of hake loins. *Innovative Food Science & Emerging Technologies, 41*, 19–25.
Otero, N., Vidal, S., Ares, F., Coto, I., Gontad, F., Delgado, T., et al. (Ed.). (2020). Manufacturing of molds by multiphoton polymerization for micro-replication of optically enhanced surfaces. In *Laser 3D manufacturing VII*. SPIE.
Oulahal, N., & Degraeve, P. (2022). Phenolic-rich plant extracts with antimicrobial activity: An alternative to food preservatives and biocides? *Frontiers in Microbiology, 12*, 753518.
Pacheco-Ordaz, R., Wall-Medrano, A., Goñi, M. G., Ramos-Clamont-Montfort, G., Ayala-Zavala, J. F., & González-Aguilar, G. (2018). Effect of phenolic compounds on the growth of selected probiotic and pathogenic bacteria. *Letters in Applied Microbiology, 66*(1), 25–31.
Padrao, J., Gonçalves, S., Silva, J. P., Sencadas, V., Lanceros-Méndez, S., Pinheiro, A. C., et al. (2016). Bacterial cellulose-lactoferrin as an antimicrobial edible packaging. *Food Hydrocolloids, 58*, 126–140.
Pakdel, M., Olsen, A., & Bar, E. M. S. (2023). A review of food contaminants and their pathways within food processing facilities using open food processing equipment. *Journal of Food Protection, 86*(12), 100184.
Pankaj, S. K., Wan, Z., & Keener, K. M. (2018). Effects of cold plasma on food quality: A review. *Foods, 7*(1), 4.
Parra, P. A., Kim, H., Shapiro, M. A., Gravani, R. B., & Bradley, S. D. (2014). Home food safety knowledge, risk perception, and practices among Mexican-Americans. *Food Control, 37*, 115–125.
Parvanov, P., & Dinkov, D. (2017). Proposals for desiccated bee pollen production and labeling. *Journal of Nutrition and Human Health, 1*, 13–16.
Patial, V., Asrani, R. K., & Thakur, M. (2018). Food-borne mycotoxicoses: Pathologies and public health impact. In *Foodborne diseases* (pp. 239–274). Elsevier.
Peñalver-Soto, J. L., Garre, A., Aznar, A., Fernández, P. S., & Egea, J. A. (2021). Dynamics of microbial inactivation and acrylamide production in high-temperature heat treatments. *Foods, 10*(11), 2535.
Perrone, G., & Gallo, A. (2017). Aspergillus species and their associated mycotoxins. In *Mycotoxigenic fungi: Methods protocols* (pp. 33–49).
Pisoschi, A. M., Pop, A., Georgescu, C., Turcuş, V., Olah, N. K., & Mathe, E. (2018). An overview of natural antimicrobials role in food. *European Journal of Medicinal Chemistry, 143*, 922–935.
Pitt, J. (2009). *Fungi and food spoilage*. Springer.

Preciado, G., Michel, M., Villarreal-Morales, S., Flores-Gallegos, A., Aguirre-Joya, J., Morlett-Chávez, J., et al. (2016). Bacteriocins and its use for multidrug-resistant bacteria control. *Antibiotic Resistance*, 329–349.

Prüβ, B. M. (2020). Microbes in our food, an ongoing problem with new solutions. *Antibiotics, 9*(9), 584.

Qureshi, A. S., Zhang, J., da Costa, S. L., & Bao, J. (2017). Antibacterial peptide secreted by Pediococcus acidilactici enables efficient cellulosic open L-lactic acid fermentation. *ACS Sustainable Chemistry Engineering, 5*(10), 9254–9262.

Rachman, A. B., Akazawa, T., & Ogawa, M. (2021). Olive leaf water extract protects chicken breast sausages against quality deterioration induced by frozen storage. *The Journal of Poultry Science, 58*(3), 200–209.

Rahman, M. S. (2020). *Handbook of food preservation*. CRC Press.

Ramesh, T., Nayak, B., Amirbahman, A., Tripp, C. P., & Mukhopadhyay, S. (2016). Application of ultraviolet light assisted titanium dioxide photocatalysis for food safety: A review. *Innovative Food Science Emerging Technologies, 38*, 105–115.

Redondo-Blanco, S., Fernández, J., López-Ibáñez, S., Miguélez, E. M., Villar, C. J., & Lombó, F. (2020). Plant phytochemicals in food preservation: Antifungal bioactivity: A review. *Journal of Food Protection, 83*(1), 163–171.

Rollini, M., Nielsen, T., Musatti, A., Limbo, S., Piergiovanni, L., Hernandez Munoz, P., et al. (2016). Antimicrobial performance of two different packaging materials on the microbiological quality of fresh salmon. *Coatings, 6*(1), 6.

Rossi, G. A. M., Link, D. T., Bertolini, A. B., Tobias, F. L., & Mioni, M. S. R. (2023). A descriptive review of the use of organic acids and peracetic acid as a decontaminating strategy for meat. *eFood, 4*(4), e104.

Rowan, N. J. (2019). Pulsed light as an emerging technology to cause disruption for food and adjacent industries–Quo vadis? *Trends in Food Science & Technology, 88*, 316–332.

Roy, S., & Rhim, J.-W. (2020). Effect of CuS reinforcement on the mechanical, water vapor barrier, UV-light barrier, and antibacterial properties of alginate-based composite films. *International Journal of Biological Macromolecules, 164*, 37–44.

Ryan, U., & Hijjawi, N. (2015). New developments in *Cryptosporidium* research. *International Journal for Parasitology, 45*(6), 367–373.

Sahariah, P., & Másson, M. (2017). Antimicrobial chitosan and chitosan derivatives: A review of the structure–activity relationship. *Biomacromolecules, 18*(11), 3846–3868.

Sani, I. K., Aminoleslami, L., Mirtalebi, S. S., Sani, M. A., Mansouri, E., Eghbaljoo, H., et al. (2023). Cold plasma technology: Applications in improving edible films and food packaging. *Food Packaging and Shelf Life, 37*, 101087.

Saravanakumar, K., Sathiyaseelan, A., Mariadoss, A. V. A., Xiaowen, H., & Wang, M.-H. (2020). Physical and bioactivities of biopolymeric films incorporated with cellulose, sodium alginate and copper oxide nanoparticles for food packaging application. *International Journal of Biological Macromolecules, 153*, 207–214.

Savard, T., Beaulieu, C., Boucher, I., & Champagne, C. P. (2002). Antimicrobial action of hydrolyzed chitosan against spoilage yeasts and lactic acid bacteria of fermented vegetables. *Journal of Food Protection, 65*(5), 828–833.

Schnur, S. E., Amachawadi, R. G., Baca, G., Sexton-Bowser, S., Rhodes, D. H., Smolensky, D., et al. (2021). Antimicrobial activity of sorghum phenolic extract on bovine foodborne and mastitis-causing pathogens. *Antibiotics, 10*(5), 594.

Sekiguchi, S., Miura, Y., Kaneko, H., Nishimura, S., Nishi, N., Iwase, M., et al. (1993). Molecular weight dependency of antimicrobial activity by chitosan oligomers. In *Food hydrocolloids: Structures, properties, functions* (pp. 71–76). Springer.

Serwecińska, L. (2020). Antimicrobials and antibiotic-resistant bacteria: A risk to the environment and to public health. *Water, 12*(12), 3313.

Silva, N. H., Vilela, C., Almeida, A., Marrucho, I. M., & Freire, C. S. (2018). Pullulan-based nanocomposite films for functional food packaging: Exploiting lysozyme nanofibers as antibacterial and antioxidant reinforcing additives. *Food Hydrocolloids, 77*, 921–930.

Skroza, D., Šimat, V., Smole Možina, S., Katalinić, V., Boban, N., & Generalić Mekinić, I. (2019). Interactions of resveratrol with other phenolics and activity against food-borne pathogens. *Food Science & Nutrition, 7*(7), 2312–2318.

Solano, R., Lafuente, S., Sabate, S., Tortajada, C., de Olalla, P. G., Hernando, A. V., et al. (2013). Enterotoxin production by *Staphylococcus aureus*: An outbreak at a Barcelona sports club in July 2011. *Food Control, 33*(1), 114–118.

Song, S., Huang, X., Liu, Y., & Zhang, Q. (2022). Drying characteristics and quality analysis of hot air-assisted radio frequency and hot-air drying of jujube (Zizyphus jujube Miller cv. Jinsixiaozao). *Engenharia Agrícola, 42*(1), e20210112.

Souza, V. G., Pires, J. R., Rodrigues, C., Coelhoso, I. M., & Fernando, A. L. (2020). Chitosan composites in packaging industry – Current trends and future challenges. *Polymers, 12*(2), 417.

Steels, H., James, S., Roberts, I., & Stratford, M. (2000). Sorbic acid resistance: The inoculum effect. *Yeast, 16*(13), 1173–1183.

Suganthi, V., Selvarajan, E., Subathradevi, C., & Mohanasrinivasan, V. (2012). Lantibiotic nisin: Natural preservative from Lactococcus lactis. *International Research Journal of Pharmacy, 3*, 13–19.

Suryadi, H., & Mun'im, A. (2021). Effect of gamma irradiation on the caffeoylquinic acid derivatives content, antioxidant activity, and microbial contamination of Pluchea indica leaves. *Heliyon, 7*(8).

Świder, O., Roszko, M. Ł., & Wójcicki, M. (2023). The inhibitory effects of plant additives on biogenic amine formation in fermented foods–A review. *Critical Reviews in Food Science Nutrition*, 1–26.

Syngai, G. G., & Ahmed, G. (2019). Lysozyme: A natural antimicrobial enzyme of interest in food applications. In *Enzymes in food biotechnology* (pp. 169–179). Elsevier.

Talib, A., Samad, A., Hossain, M. J., Muazzam, A., Anwar, B., Atique, R., et al. (2024). Modern trends and techniques for food preservation. *Food and Life,*(1), 19-32.

Tarabová, B., Tampieri, F., Maran, E., Marotta, E., Ostrihoňová, A., Krewing, M., et al. (2021). Chemical and antimicrobial effects of air non-thermal plasma processing of fresh apple juice with focus on safety aspects. *Foods, 10*(9), 2055.

Tareke, E., Rydberg, P., Karlsson, P., Eriksson, S., & Törnqvist, M. (2002). Analysis of acrylamide, a carcinogen formed in heated foodstuffs. *Journal of Agricultural Food Chemistry, 50*(17), 4998–5006.

Tavassoli, M., Sani, M. A., Khezerlou, A., Ehsani, A., & McClements, D. J. (2021). Multifunctional nanocomposite active packaging materials: Immobilization of quercetin, lactoferrin, and chitosan nanofiber particles in gelatin films. *Food Hydrocolloids, 118*, 106747.

Taylor, T. M., & Doores, S. X. (2020). Organic acids. In *Antimicrobials in food* (pp. 133–190). CRC Press.

Tesson, V., Federighi, M., Cummins, E., de Oliveira, M. J., Guillou, S., & Boué, G. (2020). A systematic review of beef meat quantitative microbial risk assessment models. *International Journal of Environmental Research Public Health, 17*(3), 688.

Thakur, M., Asrani, R. K., & Patial, V. (2018). *Listeria monocytogenes*: A food-borne pathogen. In *Foodborne diseases* (pp. 157–192). Elsevier.

Thi, A. N. T., Kittirath, P., Abiola, S. D., & Ha, N. C. (2021). Evaluation of street food safety and hygiene practices of food vendors in Can Tho city of Vietnam. *Current Research in Nutrition Food Science Journal, 9*(1), 158–171.

Tirado, D., González-Morelo, K. P. M., Ahumada, O., & Correa, D. (2017). Osmotic dehydration and hot-air drying of pineapple (*Ananas comosus*). International Journal of. *Engineering and Technology, 9*(6), 4193–4202.

Tosati, J. V., de Oliveira, E. F., Oliveira, J. V., Nitin, N., & Monteiro, A. R. (2018). Light-activated antimicrobial activity of turmeric residue edible coatings against cross-contamination of Listeria innocua on sausages. *Food Control, 84*, 177–185.

Tropea, A. (2022). Microbial contamination and public health: An overview. *International Journal of Environmental Research Public Health, 19*(12), 7441.

Tucker, G., & Featherstone, S. (2021). *Essentials of thermal processing*. Wiley.

Ucar, Y., Ceylan, Z., Durmus, M., Tomar, O., & Cetinkaya, T. (2021). Application of cold plasma technology in the food industry and its combination with other emerging technologies. *Trends in Food Science Technology, 114*, 355–371.

Uchechukwu-Agua, A. D., Caleb, O. J., & Opara, U. L. (2015). Postharvest handling and storage of fresh cassava root and products: A review. *Food and Bioprocess Technology, 8*, 729–748.

Upadhyay, P., Zubair, M., Roopesh, M., & Ullah, A. (2024). An overview of advanced antimicrobial food packaging: emphasizing antimicrobial agents and polymer-based films. *Polymers, 16*(14), 2007.

Vishweshwaraiah, Y. L., Acharya, A., Hegde, V., & Prakash, B. (2021). Rational design of hyperstable antibacterial peptides for food preservation. *NPJ Science of Food, 5*(1), 26.

Wang, L.-F., & Rhim, J.-W. (2015). Preparation and application of agar/alginate/collagen ternary blend functional food packaging films. *International Journal of Biological Macromolecules, 80*, 460–468.

Wang, X., Wang, Z., Zhuang, H., Nasiru, M. M., Yuan, Y., Zhang, J., et al. (2021). Changes in color, myoglobin, and lipid oxidation in beef patties treated by dielectric barrier discharge cold plasma during storage. *Meat Science, 176*, 108456.

Wang, Y., Zheng, Y., Shang, J., Wu, D., Zhou, A., Cai, M., et al. (2023). Pulsed light reduces postharvest losses of Chinese bayberries by affecting fungal microbiota during cold storage. *Food Control, 146*, 109524.

Warth, A. D. (1985). Resistance of yeast species to benzoic and sorbic acids and to sulfur dioxide. *Journal of Food Protection, 48*(7), 564–569.

WHO. FA. (2023). *Foodborne antimicrobial resistance – Compendium of Codex standards (First revision)*. Codex Alimentarius Commission.

Wu, L., Zhang, C., Long, Y., Chen, Q., Zhang, W., & Liu, G. (2022). Food additives: From functions to analytical methods. *Critical Reviews in Food Science and Nutrition, 62*(30), 8497–8517.

Xiao, X., Ma, C., Zhang, H., Liu, W., Huang, Y., Meng, C., et al. (2024). The food additive benzaldehyde confers a broad antibiotic tolerance by modulating bacterial metabolism and inhibiting the formation of bacterial flagella. *International Journal of Molecular Sciences, 25*(16), 8843.

Xing, H., Li, J., Zhang, T., Pan, F., & Liu, J. (2021). Preparation of porous cross-linked CS/PVA freshness indicator film and its recognition property of carbon dioxide. *Journal of Food Processing and Preservation, 45*(9), e15280.

Xu, T., Gao, C., Feng, X., Huang, M., Yang, Y., Shen, X., et al. (2019). Cinnamon and clove essential oils to improve physical, thermal and antimicrobial properties of chitosan-gum arabic polyelectrolyte complexed films. *Carbohydrate Polymers, 217*, 116–125.

Yu, H. H., Chin, Y.-W., & Paik, H.-D. (2021). Application of natural preservatives for meat and meat products against food-borne pathogens and spoilage bacteria: A review. *Foods, 10*(10), 2418.

Zenklusen, M. H., Coronel, M. B., Castro, M. Á., Alzamora, S. M., & González, H. H. L. (2018). Inactivation of *Aspergillus carbonarius* and *Aspergillus flavus* in malting barley by pulsed light and impact on germination capacity and microstructure. *Innovative Food Science & Emerging Technologies, 45*, 161–168.

Zhang, M., & Chen, H. (2023). Development and characterization of starch-sodium alginate-montmorillonite biodegradable antibacterial films. *International Journal of Biological Macromolecules, 233*, 123462.

Zhang, X., Wang, H., Li, M., Wu, N., & Xu, X. (2016). Near-freezing temperature storage (– 2 C) for extension of shelf life of chilled yellow-feather broiler meat: A special breed in Asia. *Journal of Food Processing and Preservation, 40*(2), 340–347.

Zhang, Q., Wan, F., Zang, Z., Jiang, C., Xu, Y., & Huang, X. (2022). Effect of ultrasonic far-infrared synergistic drying on the characteristics and qualities of wolfberry (*Lycium barbarum* L.). *Ultrasonics Sonochemistry, 89*, 106134.

Zhang, W., Jiang, Y., & Zhang, Z. (2023). The role of different natural organic acids in postharvest fruit quality management and its mechanism. *Food Frontiers, 4*(3), 1127–1143.

Zhao, J. H., Hu, R., Xiao, H. W., Yang, Y., Liu, F., Gan, Z. L., et al. (2014). Osmotic dehydration pretreatment for improving the quality attributes of frozen mango: effects of different osmotic solutes and concentrations on the samples. *Journal of Food Science Technology, 49*(4), 960–968.

Zhao, B., Fu, S., Li, H., Li, H., Liu, C., & Chen, Z. (2022). Effect of storage conditions on the quality of frozen steamed bread. *International Journal of Food Science & Technology, 57*(1), 695–704.

Zhou, X., Qiao, K., Wu, H., & Zhang, Y. (2023). The impact of food additives on the abundance and composition of gut microbiota. *Molecules, 28*(2), 631.

Sanitization of Wet and Dry Surface Biofilms in Food Processing Environments

Zejia Lin and Dan Li

1 Introduction

The enduring existence of pathogenic biofilms in food processing environments poses a significant threat to global food safety, contributing to millions of foodborne illnesses per year (WHO, 2024). Although traditional hydrated biofilms, known as wet surface biofilms (WSB), have been thoroughly studied, recent findings highlight a distinctive biofilm phenotype—dry surface biofilms (DSB)—that grows in low-water-activity (a_w) environments, which is a common characteristic for food facilities processing low-moisture foods (LMFs) such as spices, powdered dairy, and cereals (Alonso et al., 2023). In contrast to WSB, which is developed on visibly moist surfaces, DSB arises on seemingly "dry" surfaces via a microscale fluid film of 1.5 μm in thickness, facilitating bacterial survival despite stringent sanitation measures (Rhodeland et al., 2020). This persistence arises not just from diminished water availability but also from a fundamental reorganization of biofilm structure and function: DSB demonstrates extracellular polymeric substances (EPS) matrices that are 50–80% denser, a rise in cell wall thickness of 15–20%, and a spore-like metabolic dormancy, together providing 10–100 times better durability against heat, desiccation, and chlorine-based sanitizers in comparison to WSB (Almatroudi et al., 2018; Parvin et al., 2023; Lin et al., 2024).

Microbial contamination in LMFs has caused serious public health incidents. In 2020, an outbreak of *Salmonella* infections was reported in peanut butter products, affecting 14 states in the U.S (CDC, 2022). Similarly, *Cronobacter sakazakii*

Z. Lin · D. Li (✉)
Department of Food Science & Technology, National University of Singapore, Singapore, Singapore
e-mail: dan.li@nus.edu.sg

D. Li (ed.), *Antimicrobial Strategies in the Food System: Updates, Opportunities, Challenges*, https://doi.org/10.1007/978-3-031-95056-8_12

contamination in infant formula production lines led to a global recall, after lethal neonatal infections in 2022 (Mousavi et al., 2023). These cases underscore microbial threat in LMFs. Moreover, the microorganisms identified in these cases need further examination for association with and development of DSB. However, the limited knowledge regarding methodologies for studying DSB and as well as effective sanitizing methods used in the food industry makes it difficult to control DSB for the food industry. Thus, the purpose of this chapter is to better understand the novel DSB morphology in contrast to the WSB. Furthermore, the variations in adhesion and transfer ability, and biocidal resistance between WSB and DSB will be explored. Novel sanitization techniques appropriate for low-moisture situations during food processing will also be discussed. Finally, emerging DSB research will be examined, focusing on methodologies for examining DSB biofilms in the food industry, as well as an innovative omics study to uncover cellular interactions inside DSB.

2 Morphological Distinctions Between Wet and Dry Surface Biofilms

The structural development of biofilms between wet and dry conditions demonstrates significant variation in microbial survival strategies. In moist conditions, WSB demonstrates a markedly diverse dispersion of cell clusters, whereas DSB that form under desiccation stress are uniformly distributed on the surface, as seen in *Salmonella* biofilms via scanning electron microscopy (SEM) (Fig. 1) according to Lin et al. (2024). Despite the limited study on the properties of DSB, the lack of fluid may serve as an explanation for their distinctive structure. Unlike the WSB, where bacteria may easily congregate for nutrition exchange or collaboration (Nadell et al., 2016), bacterial mobility in DSB is limited, necessitating that bacteria endure in situ.

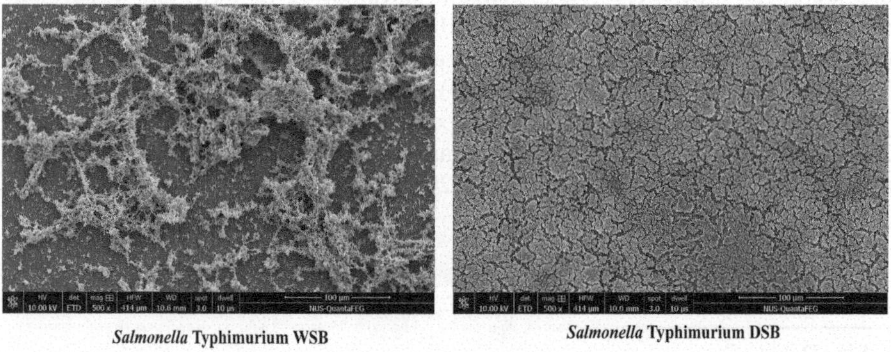

Fig. 1 SEM images of *S.* Typhimurium WSB and DSB with 100 μm scale bar. (Images are obtained from Lin et al., 2024)

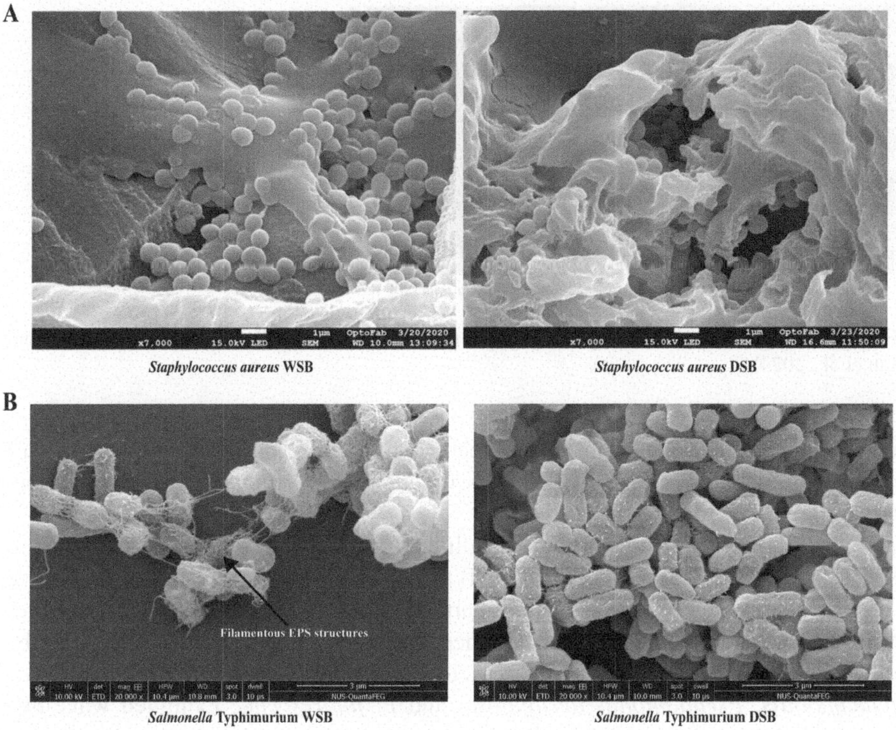

Fig. 2 SEM images of *S*. aureus WSB and DSB, with 1 μm scale bar (**a**) (Images are obtained from Parvin et al., 2023). SEM images of *S*. Typhimurium WSB and DSB with 3 μm scale bar (**b**) (Images are obtained from Lin et al., 2024)

Moreover, WSB (e.g., *Staphylococcus aureus* and *S*. Typhimurium) exhibit a low-density filamentous structure of EPS, where these loose networks might provide partial exposure of microbial cells at the matrix surface, enhancing nutrient absorption and movement (Fig. 2a). Conversely, DSB features its dense EPS structure, whereby the polysaccharide-rich matrix completely envelops the cells, covering individual cocci or rods (Fig. 2b) (Parvin et al., 2023; Lin et al., 2024). This high-density barrier not only impedes antimicrobial penetration but also reduces water loss, a vital adaptation for life on abiotic surfaces under dry environments (Ziege et al., 2021). The structural stratification of biofilms also varies between wet and dry conditions. In Lin et al. (2024) work, confocal laser scanning microscopy (CLSM) cross-sectional images reveal that *Salmonella* WSB develops thin (about 5 μm), homogenous layers with vertically arranged cells, allowing rapid metabolite exchange. By contrast, *Salmonella* DSB creates layered structures like a "sandwich," thick (~15 μm) (Fig. 3). The outer layer is largely composed of nonviable cells, which provide a protective barrier against disinfectants. Live cells are contained in the center of this bacterial barrier, resulting in a compartmentalized survival strategy

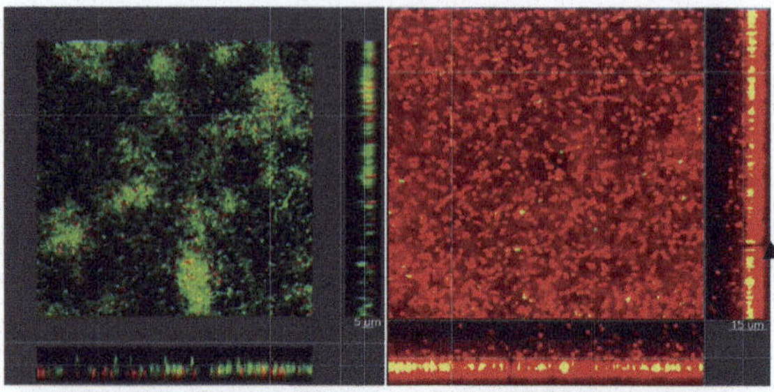

Fig. 3 CLSM section views images of *S.* Typhimurium WSB and DSB. (Images are obtained from Lin et al., 2024)

not observed in WSB. This spatial partitioning to resource constraints in dry conditions, whereby exterior cell layers function as sacrificial elements to protect the community.

In addition to biofilm structure, cellular morphology experiences significant species-specific alterations between WSB and DSB. On the one hand, Gram-positive *S. aureus* increases its cell sizes and rearranges into tetrad or cubical forms rather than their regular grape-like shapes under discussion (Parvin et al., 2023). On the other hand, Gram-negative rods such as *Escherichia coli*, *Sphingomonas paucimobilis*, *Pseudomonas aeruginosa*, and *P. fluorescence* exhibited wave-like forms, forming larger waves as they became increasingly filamentous (De Goffau et al., 2009; Marra et al., 2024). Transmission electron microscopy (TEM) images indicated that *S. aureus* in DSB shows a 20% increase in cell wall thickness (66.52 ± 5.24 nm vs. 55.69 ± 5.86 nm) with enhanced peptidoglycan production, compared to WSB (Fig. 4a) (Parvin et al., 2023). Gram-negative *Salmonella* in WSB revealed a distinct plasma membrane and outer membrane, separated by the periplasm, while the DSB cells were encased in thick capsules (Fig. 4b) (Lin et al., 2024). Many investigations have observed the increased production of capsular polysaccharides in microbes under desiccation stress, since these capsular may help to retain water within the cell to avoid desiccation (Ophir & Gutnick, 1994; Stukalov et al., 2008; Whitfield, 2006). These structural fortifications significantly differ from WSB, whose thinner cell walls and permeable membranes highlight growth efficiency rather than environmental durability.

In the food processing environment specifically, these morphological variations could have functional effects. Conventional chemical sanitizers might be less effective against the compacted EPS matrix and layered structure of DSB as disinfectants could not completely penetrate the protective outer barrier and enable the inner surviving subpopulations to rapidly recolonize surfaces. Although WSB's homogenous cell distribution and loose EPS structure help to increase antimicrobial penetration, they yet carry risks of rapid regrowth in wet environments. Therefore, this issue emphasizes the requirement of hydration state-specific antimicrobial

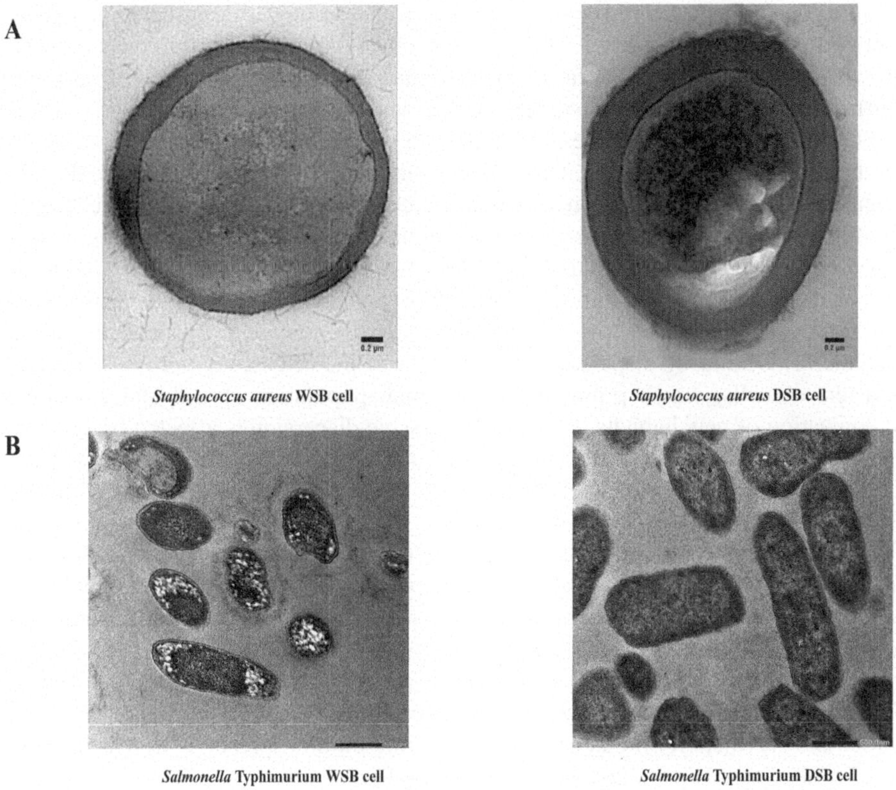

Fig. 4 TEM images of *S. aureus* WSB and DSB, with 0.2 μm scale bar (**a**) Images are obtained from Parvin et al., 2023). TEM images of *S.* Typhimurium WSB and DSB with 600 nm scale bar (**b**) Images are obtained from Lin et al., 2024)

strategies as DSB's resistance results not only from metabolic inactivity but also from a complex protective morphology created by evolutionary adaptability to desiccation.

3 Adhesion and Transfer Ability of WSB and DSB

Biofilm adhesion and transfer are critical for microbial persistence in food industrial environments. Kretschmer et al. (2021) investigated the effect of biofilm wettability and stiffness on these behaviors, demonstrating that wet and dry biofilms differ significantly in their adhesion abilities and material transfer. WSB, due to its hydrophilic and flexible nature, adheres strongly to food surfaces, increasing contact and adhesive strength. Because of its viscosity and increased surface wettability, WSB is more likely to suffer cohesive failure, which occurs when only parts of the biofilm detach and the rest stay on the surface, resulting in less material transfer between interacting surfaces. DSB, on the other hand, is hydrophobic and stiff, displaying less adhesion because their rigidity inhibits their ability to adapt to surfaces. This results in less surface contact and decreases adhesive forces. Thus, DSB mostly undergoes adhesive failure, in which the whole biofilm detaches without leaving much behind. This causes more material transfer during the detachment process. These adhesion and transfer characteristics are strongly related to biofilm dispersion mechanisms, which are divided into active and passive processes (Wille & Coenye, 2020). Active dispersion is a biologically controlled process in which microbial cells in a biofilm actively depart the community in response to environmental cues such as food availability, oxygen content, or quorum-sensing signals. This kind of dispersion often results in the release of portions or small clusters of cells, which promote the microbial population spread and allow the colonization of new habitats. Conversely, passive dispersion is affected by external physical forces, such as surface interactions, mechanical disturbance, or fluid shear. Thus, from the finding of Kretschmer et al. (2021), the active dispersal is mostly likely to play a vital role during the adhesion and transfer of WSB, as the early biofilm community has reached maturity, necessitating the active release of bacteria to seek nutrients in a new habitat. Passive dispersion, on the other hand, might be the main force for DSB adherence and transmission, since complete biofilm structures are more likely to detach whole than disintegrate. Overall, this research found that wettability has a greater impact on biofilm adherence and transfer than stiffness. Thus, understanding the interaction between biofilm physical features and dispersion processes is critical for predicting and controlling biofilm spread, as well as improving biofilm management tactics in food production.

Though it is clear that the moisture level of food contact surfaces plays a crucial role in bacterial adherence or transfer, different bacteria might exhibit varying behaviors in this regard. Table 1 summarizes the adhesion and transfer studies on different bacterial biofilms on wet and dry surfaces. Specifically, Rodriguez et al. (2007) employed a universal quantification method to assess the impact of hydration level on *Listeria monocytogenes* biofilms that were grown on stainless steel (SS), as well as the biofilm transfer to bologna and salami, at a pressure of 45 kPa for 30 s. In general, the transfer of *Listeria* from SS to both foods increased substantially as biofilms became drier. Upon transfer, the efficiency of transfer for bologna and hard salami increased from 0.2 to 0.51 and from 2 to 3.8, respectively, when the relative humidity (RH) of the biofilm dried from 94% to 38%. The findings indicated that the interactions between cells and surfaces within biofilms weakened as they were desiccated, increasing bacterial transfer. Interestingly, it has been reported that mechanical removal such as wiping could easily remove WSB, while barely removing DSB, while also indicating DSB might possess a stronger ability to adhesion and transfer compared to WSB (Duggan et al., 2024). In contrast, Sattar et al. (2001) also designed a quantitative protocol for evaluating the transfer of *S. aureus* from fabrics to surfaces of finger pads and discovered that the *S. aureus* transmission was significantly greater on wet surfaces compared to dry surfaces. While numerous research has examined bacterial adherence and transmission on wet and dry surfaces

Table 1 Adhesion and transfer studies of WSB and DSB

Microorganism	Biofilm type	Adhesion and transfer	References
4 strains of *L. monocytogenes* (LM10, LM21, LM27, LM29)	WSB & DSB	SS to bologna and salami	Rodriguez et al. (2007)
Bacillus subtilis 3610	WSB & DSB	Adhere to aluminum surface	Kretschmer et al. (2021)
STEC *E. coli* O103:H2	WSB, DSB	SS and polyurethane to beef	Nan et al. (2022)
L. monocytogenes N53-1	DSB	SS to salmon	Hansen and Vogel (2011)
14 strains of STEC *E. coli*	DSB	SS and polystyrene to lettuce	Adator et al. (2018)
S. Typhimurium ATCC SL1344	DSB	Agar plates to coupon	Duggan et al. (2024)
4 strains of *Salmonella enterica* isolated from beef	WSB	SS and polyvinylchloride to beef products	Wang et al. (2022a)
2 strains of *Lactobacillus* and 1 strain of *L. monocytogenes*	WSB	Adhere to glass surface	Jara et al. (2020)
2 strains of *L. monocytogenes* (ST9, ST87)	WSB	Glass to cantaloupe	Wang et al. (2022b)

(Adator et al., 2018; Hansen & Vogel, 2011), conflicting observations persist due to the lack of standardized methodologies and the influence of bacterial species, making cross-study comparisons difficult. Specifically, there are several gaps in our knowledge of biofilm adherence and transmission between WSB and DSB. First, the absence of standardized models that could be used for systematic comparison between the WSB and DSB makes it difficult to draw consistent findings from several investigations. Second, further studies of adhesion and transfer behaviors between Gram-positive and Gram-negative DSB are needed to identify potential differences in their persistence and propagation. Third, real-time analytical approaches for linking desiccation-induced EPS breakage to bacterial cell release kinetics are unclear. Addressing these gaps will help to distinguish between universal biofilm activities and strain-specific responses, allowing the development of prediction models for biofilm cross-contamination in diverse food processing environments.

4 Sanitization Tolerance Between WSB and DSB

Pathogenic biofilms are reported to show biocidal resistance upon sanitization in food processing environments, with a rising concern that DSB might exhibit elevated resistance to cleaning and disinfection compared to WSB (Alonso et al., 2023). Thus, the following discussion aims to review the sanitization tolerance in WSB vs DSB.

The EPS matrix serves as a fundamental protective barrier in all biofilms; however, its properties vary between WSB and DSB. In WSB, the EPS acts as a hydrated gel that obstructs sanitizer invasion and protects cells. In DSB, cells produce more EPS/capsular material. A crystal violet experiment showed that *Salmonella* DSB contains much more biomass than WSB, suggesting that bacteria might promote EPS production upon desiccation stress (Lin et al., 2024; Weaver et al., 2015). Furthermore, the EPS matrix within DSB is often formed as thick capsules around individual cells rather than as filamentous networks (Lin et al., 2024). This capsular EPS has hygroscopic characteristics that facilitate water retention and exhibit low permeability, thus reducing water loss and preventing contact with biocides (Whitfield, 2006). Overall, the EPS matrix in DSB functions as a "protective coating" for individual cells during desiccation and exposure to sanitizers, which differs from WSB, where EPS offers protection to the whole community.

WSB cells tend to remain metabolically active (growth and division) under a nutrient-rich, wet environment, making them vulnerable to biocides that target active cellular functions (e.g. membrane function, respiration) (Mazaheri et al., 2021; Bridier et al., 2015). On the other hand, DSB cells would enter a dormant state during desiccation, as indicated by a single-cell transcriptome investigation of *Salmonella* DSB, demonstrating that most cells become metabolically inactive in DSB (Lin et al., 2024). Because inactive cells do not actively absorb chemicals or produce reactive metabolites, they may be more resistant to many disinfectants (particularly alcohols and oxidative agents) (Edwards et al., 2016). In summary, DSB behaves like a community of stressed, semi-dormant cells, while WSB includes a greater percentage of actively metabolizing cells—a critical differential that influences biocide effectiveness.

Lastly, biocidal resistance may also be significantly influenced by cellular adaptations. Bacteria in DSB can develop unique physiological and morphological modifications. For instance, SEM images revealed significant variations in the morphology of *S.* Typhimurium cells in DSB after disinfectant treatment, as compared to water control. The disinfected cells in DSB had a condensed and plump appearance, along with the presence of perforations, which were speculated as transmembrane tunnels (Jawale & Lee, 2014). Besides, an outstanding characteristic of the *S.* Typhimurium DSB was the existence of tubular appendages, often linking with neighboring cells. A few of these appendages were also visible in the water control group, but there were significant increases in such structures after disinfectant treatment. While the reason for the augmented appendages remains uncertain, it may be theorized that the modified cells enhance the production of these structures as a defensive reaction, in order to stabilize the biofilm.

Numerous research has investigated the biocidal resistance between WSB and DSB, consistently demonstrating that DSB is much more challenging to kill or remove with biocides. Lin et al. (2024) cultivated *S.* Typhimurium as WSB and DSB on SS and then applied commonly used sanitizers to them. The results were notable: a 1-min exposure to 70% isopropyl alcohol resulted in a 4.3 log CFU/surface reduction in the WSB, but the DSB exhibited only around a 1.5 log reduction. Likewise, 200 μg/mL sodium hypochlorite resulted in a 3.5 log reduction in the WSB but about a 2.8 log reduction in the DSB. In both cases, the DSB exhibited a greater viable cell count after treatment, suggesting promoted biocidal tolerance. Other studies have observed similar results with DSB. Nkemngong et al. (2020) discovered in disinfection effectiveness testing that quaternary ammonium compound (QAC) disinfectants effective against WSB often failed to eliminate DSB. This implies that sanitizers that are effective in wet-cleaned areas of a food facility may not be as effective in low-moisture zones, where DSBs develop. What's worse, organic residues may enhance the protection of DSB cells (Lin et al., 2023), thus DSB might need extended contact durations or innovative biocidal chemicals for successful management.

In addition to chemical disinfection, DSB is reported to be more tolerant to thermal treatment than WSB. Almatroudi et al. (2018). found that *S. aureus* DSB maintained its ability to grow in culture even after being exposed to the severe dry-heat treatment of 100 °C for 60 min. After undergoing autoclaving, the samples showed that around 70% of the bacteria in DSB remained viable, demonstrated by live/dead staining via CLSM. However, WSB was only able to recover after being subjected to dry-heat treatment of 100 °C for up to 10 min, and cells in WSB did not recover after being autoclaved, even though around 50% of the cells were still alive, as confirmed by live/dead staining. Lastly, a study discovered that *S.* Typhimurium ATCC14028 DSB could survive longer than WSB under ambient temperature (Lee et al., 2020). *S.* Typhimurium WSB and DSD were developed on SS, rubber, and plastic, respectively, and stored at ambient temperature for 35 days. After that, enumeration was performed and the reduction of DSB on SS, plastic, rubber, and eggshell was 3.3, 3.2, and 3.5 log CFU/coupon, while the reduction of WSB was 4.4, 3.5, and 4.5 log CFU/, respectively.

Despite the rising recognition of DSB as a severe food safety risk, substantial information gaps still remain. For example, there are no established protocols for evaluating disinfectants against DSB. Regulatory organizations, such as the US Environmental Protection Agency (EPA), presently lack a standard efficacy test method for DSB claims, which implies that sanitizers are not regularly examined for DSB disinfection. This is a significant discrepancy since normal WSB testing may overestimate a product's actual performance in LMF facilities. Future research should test disinfectant methods against DSB and advocate guidelines that address DSB issues.

5 Novel Sanitization Methods against DSB

In low-moisture conditions, DSBs might provide a great challenge, since bacteria can survive in a dry biofilm form with decreased metabolic activity but maintain pathogenicity activities (Lin et al., 2024). Moreover, sanitization in the LMF industry often calls for avoiding water (to prevent product spoiling or microbiological propagation), therefore limiting the use of conventional water-based disinfectants

(Alonso et al., 2023). These challenges have inspired studies on novel sanitizing techniques targeting DSB. Below, we summarize several novel techniques including natural bioactive compounds, nanotechnologies, antimicrobial peptides, biological enzymes and phages, and advanced physical treatments (Table 2)—highlighting their mechanisms of action, advantages and limitations, and appropriate application situations.

Plant-derived antimicrobials are emerging as "green" biofilm control treatments. Many herbal extracts, essential oils, and purified phytochemicals can disturb biofilm formation or integrity by utilizing multi-targeted processes (Pruteanu et al., 2020). Essential oils (EOs), for example, are rich in aromatic compounds that can penetrate biofilms and interfere with bacterial communication and adhesion (Li et al., 2022). Some EOs disrupt the biofilm matrix and suppress quorum-sensing signals at sub-lethal doses without always destroying the cells, which helps to avoid initial bacterial attachment and buildup (Shamim et al., 2023). Oregano oil, for instance, has demonstrated disruptive effects on *Staphylococcus* and *E. coli* biofilms, reducing biofilm biomass and even preventing biofilm formation (Gherasim et al., 2021). Some natural substances could inhibit the regulation of virulence. For instance, ajoene produced from garlic reduces regulatory RNAs and rhamnolipid synthesis in *P. aeruginosa*, therefore restricting biofilm development (Harjai et al., 2010). One of the advantages of such natural agents is their generally low human toxicity and multiple modes of action, making it harder for bacteria to develop resistance (Kerekes et al., 2019). In addition, many are also food-grade, ideal for use in the food industry where residuals have to be safe. They can also synergize with traditional biocides or antibiotics, e.g., tea tree oil combined with ciprofloxacin yielded ~70% biofilm biomass reduction in *P. aeruginosa*, far more than either alone (Rosato et al., 2020). Overall, natural antimicrobials are appropriate for surfaces in food processing or healthcare when non-toxic, residue-free treatment is required. They may be utilized as surface sprays, as edible films or coatings, or to supplement cleaning solutions. Certain plant components, such as flavonoids, could be given in dry conditions via alcohol or vapor.

Nanotechnology provides novel tools to improve biofilm sanitization. Metal nanoparticles (NPs) such as silver and zinc oxide are robust antimicrobials with special biofilm-penetrating properties (Cacciatore et al., 2021). Their mechanisms include the gradual release of metal ions that disrupt microbial enzymes and membranes, and direct physical interactions with cells producing reactive oxygen species. Furthermore, the nanoscale size also allows particles to penetrate the biofilm matrix more easily and contact embedded bacteria (Méndez-Vilas, 2011). NPs have broad-spectrum activity and can kill dormant biofilm cells that might resist antibiotics. They could also be built for triggered release or targeted distribution. For example, encapsulating a natural oil such as thymol into polymeric nanoparticles promoted its antibiofilm efficacy against *S. aureus* compared to thymol alone (Zhang et al., 2014). Polymeric or lipid-based nanocarriers (nanospheres, liposomes, nanoemulsions) could solubilize and preserve natural antimicrobials, releasing them gradually for sustained action (Blanco-Padilla et al., 2014). This might help to sustain residual sanitizing action on surfaces. However, a major concern is safety

Table 2 Emerging sanitization to biofilm disinfection: mechanisms, advantages, limitations and applications

Sanitization method	Mechanism	Advantages	Limitations	Applications
Essential oils	Disrupt bacterial membranes, inhibit quorum sensing	Natural, low toxicity, effective against diverse biofilms	Odor issues, may need high concentrations for efficacy	Food processing, surface coatings
Phytochemicals	Block biofilm synthesis, inhibit bacterial signaling	Multi-targeted, can prevent biofilm formation	Variable potency, and stability issues, require optimization	Food safety, biofilm prevention
Metal nanoparticles	Release metal ions, generate reactive oxygen species	Broad-spectrum, penetrates biofilms, prolonged effect	Potential toxicity, regulatory concerns, cost	Food packaging
Nano-encapsulated antimicrobials	Controlled release of antimicrobial agents, better penetration	Enhances efficacy and stability of natural compounds	High cost, formulation complexity	Encapsulation for food and pharmaceutical use
Antimicrobial peptides	Membrane disruption, targeting persister cells	Targets biofilm-resistant cells, reduce resistance risk	Stability issues, potential cytotoxicity, costly production	Food surface treatments, high-risk areas
Enzymatic cleaners	Degrade biofilm matrix, facilitating bacterial removal	Non-toxic, environmentally friendly, effective against mixed-species biofilms	Requires specific conditions, may need secondary biocides	Food industry cleaning, industrial sanitation
Bacteriophage	Lysis of bacterial cells, biofilm matrix degradation	Highly specific, effective against biofilms without disrupting microbiota	Host specificity, potential for bacterial resistance	Food safety, hospital sanitation, water treatment
Ultraviolet	DNA damage, ROS generation, biofilm inactivation	Non-chemical, continuous disinfection is possible	Limited penetration requires direct exposure	Food surfaces, industrial equipment
Cold plasma	Ionized gas generating oxidative stress, biofilm penetration	Dry method, effective against persistent biofilms	Expensive, requires special equipment	Healthcare, food industry, packaging sterilization
Ozone and chlorine dioxide gas	Strong oxidative agents breaking down biofilms	No residue	Toxic at high levels, requires controlled application	Food production, storage, industrial cleaning
Steam heating	Heat-induced biofilm removal and microbial death	Chemical-free, reaches crevices, highly effective	Not suitable for heat-sensitive surfaces	Food industry (e.g., dairy, breweries)

and regulatory approval. Certain metal NPs can be toxic to human cells or the environment, which currently limits their use on food-contact surfaces (Brandelli, 2020). Moreover, the cost and scalability of nano-formulations can be challenges. Nonetheless, nano-formulated disinfectants are promising in the food industry and medical device coatings where improved penetration and long-lasting impact are required. For instance, nano-coatings with embedded antimicrobial agents can create self-disinfecting surfaces. In food processing, edible nanoscale coatings or sprays containing essential oils or enzymes could prevent biofilm buildup while being safe for foods (Cacciatore et al., 2021). Thus, future research is required to continue to balance the powerful antimicrobial performance of nanomaterials with safety by using biocompatible carriers and minimizing free metal release.

Antimicrobial peptides (AMPs) are small, often cationic peptides produced by organisms as part of innate defense, and they are emerging as antibiofilm agents. Many AMPs can insert into bacterial membranes to form pores, leading to rapid cell death even in stationary-phase or biofilm cells. Their amphipathic characteristics could also help them penetrate biofilms or induce immune responses (Luo & Song, 2021). One such example is melittin, a peptide derived from bee venom that has shown the capacity to pass biofilm defenses and eradicate latent persister cells within biofilms (Azmy et al., 2024). AMPs have broad-spectrum efficacy even against multidrug-resistant bacteria, and their mechanism (membrane disruption) is less susceptible to conventional resistance mechanisms. They can act on both Gram-positive and Gram-negative biofilms and may even disperse biofilms by triggering stress responses (Tavares et al., 2020). Nonetheless, the limitation of AMPs is their instability (susceptibility to proteolysis) and the high cost of production. Certain AMPs may also exhibit hemolytic or cytotoxic properties at elevated doses, raising safety issues for direct application on food or medical surfaces (Greco et al., 2020). Also, their salt sensitivity and short half-life can reduce efficacy in real-world conditions (Han et al., 2016). Despite this, AMPs still hold promises in food sanitation, they might be applied if immobilized on surfaces or used in combination with enzymes to facilitate penetration. Future research is demanded to formulate AMPs in gels, coatings, or encapsulated forms to protect them and target their delivery to biofilms.

Enzymatic treatments target the biofilm's structural matrix instead of the bacteria directly. Enzyme-based cleaners apply enzymes, such as proteases, DNases, polysaccharide depolymerases, or amylases that can break down the EPS that holds the biofilm together (Coughlan et al., 2016). These enzymes loosen and remove the biofilm from surfaces by breaking down proteins, polysaccharides, or eDNA in the matrix. Once the biofilm structure is damaged, conventional disinfectants can more easily kill or wash away the cells. A tailored approach is often needed: the choice of enzymes (or enzyme cocktail) depends on the biofilm's composition, and combining enzymes that target different components results in improved removal, particularly in mixed-species biofilms (Walker et al., 2007). Enzymatic cleaners can greatly improve cleaning efficacy without strong chemicals, and they work under moderate conditions. They are non-toxic and biodegradable, making them suitable for the food industry (e.g., enzyme-based detergents in Clean-in-Place systems). Successful cases include commercial proteases removing protein-rich biofilms in dairy plants and breweries (Coughlan et al., 2016). However, enzymes can be inhibited by substances in the soil/biofilm. One instance is that proteases were less effective when milk residues were present, as the milk neutralized the enzyme action (Augustin et al., 2004). Enzymatic treatments generally do not outright kill bacteria, so a secondary biocide or thorough rinse is required. Besides that, enzymes also have optimal pH and temperature ranges and may be costly.

Bacteriophages (phages), viruses that infect bacteria, have re-emerged as highly specific biofilm control agents. Phages can penetrate biofilms and lyse the inside cells, therefore lowering the bacterial burden. Importantly, many lytic phages produce depolymerase or other enzymes during infection that degrade the biofilm matrix, promoting biofilm dispersal (Liu et al., 2023). This dual action—attacking both the bacteria and the matrix—is a unique feature of the phage treatment. Phages have great specificity for their host species, which is an advantage in targeting pathogens without harming beneficial flora. For example, phage P100 (targeting *L. monocytogenes*) is a USDA-approved product for food safety; it induces cell lysis and enzymatic EPS disintegration and has achieved significant reduction of *Listeria* biofilms on stainless steel (Soni & Nannapaneni, 2010). Another research indicated that biofilm removal might be much improved by a modified *E. coli* phage carrying a biofilm-degrading enzyme (dispersin B) (Lu & Collins, 2007). Phages can self-replicate at the site of infection as long as their host bacteria are present, providing a sustained effect. They are natural and generally safe for humans and can be applied in food environments. Whereas the host specificity of bacteriophages means a given phage target only a narrow range of bacteria—a biofilm with multiple species may require a cocktail of phages. Aside from that, bacteria may acquire phage resistance, which might reduce their virulence or make them more vulnerable to other stresses (Liu et al., 2023). Phage treatments are suitable for food processing facilities (to target persistent strains like *Listeria* or *Salmonella* in niches and drains), and enzyme and phage approaches can be combined: e.g., using enzymes to open the matrix and then phages to kill the bacteria, or using phage-derived enzymes (lysins) as stand-alone agents (Azeredo et al., 2021). These biological strategies are highly targeted and environmentally friendly, aligning with the push for sustainable biofilm control.

Several non-traditional physical disinfection methods have emerged to tackle biofilms, many of which are particularly useful on dry surfaces or where chemical use must be minimized. Ultraviolet (UV) irradiation is being refined in forms like pulsed UV light, which delivers high-intensity light in flashes. UV-C light damages microbial DNA and can inactivate biofilm cells given sufficient exposure. Pulsed UV has enhanced penetration and can disrupt biofilms more efficiently than continuous low-intensity UV (Alvarez-Ordóñez et al., 2019). Cold plasma technology is an emerging dry disinfection method that uses ionized gas (often air or noble gases) to generate a cocktail of reactive species (oxygen, nitrogen radicals, etc.). These reactive species can penetrate and oxidize biofilm components, leading to cell death and matrix disruption (Zhu et al., 2020). Cold plasmas are effective at inactivating a broad range of microorganisms on surfaces, including cracks and crevices, without heat damage. Electron beam (e-beam) irradiation is another high-tech approach, where high-energy electrons are used to sterilize surfaces. E-beams break DNA strands and can kill bacteria within biofilms on packaging or equipment surfaces (Fan et al., 2023). Compared to traditional gamma irradiation, e-beams can be turned on/off instantly and leave no residual radioactivity, making them suitable for food packaging or medical device sterilization.

Beyond these, gaseous chemical treatments are valuable for dry or hard-to-reach biofilms. Ozone (O_3) and gaseous chlorine dioxide (ClO_2) can permeate the environment and oxidize microbial cells and biofilm matrices on contact. They are applied in enclosed spaces (e.g., empty food storage room) to decontaminate surfaces without liquid residue. Ozone, for example, is a powerful oxidant that could kill bacteria and break down the EPS matrix, then transform it into oxygen without leaving any harmful residue (Panebianco et al., 2022). ClO_2 gas is often used to disinfect food processing equipment and ventilation systems due to its low corrosiveness and high efficacy at low concentrations (Jiang et al., 2024). Another physical approach is steam heating, which uses pressured hot steam to physically remove biofilms while killing bacteria in the meantime (Prestes et al., 2024). Steam is chemical-free and can access narrow channels on equipment; it's often used in organic food processing or winery/brewery settings to sanitize without chemicals.

6 Emerging Research on DSB and Future Aspects

6.1 DSB Model Standardization

DSB represents an emerging field in biofilm research, especially within the food industry, while the majority of current studies focus on the medical field (Alonso et al., 2023). Consequently, there is a lack of standardized methodologies for producing DSB in laboratory environments, particularly in contrast to the well-established protocol for WSB formation. Existing biofilm models (e.g., EPA standard procedures) mostly apply to moist environments and are restricted to certain species (Nkemngong et al., 2020), hence leaving a gap in understanding how to reliably grow and study biofilms under low-moisture conditions. Researchers have developed numerous *in vitro* models for both WSB and DSB, each with distinct methodologies. Below are three recent representative studies that exhibit novel methodologies in DSB model creation, emphasizing their methodological distinctions along with respective benefits and limitations.

Amaeze et al. (2020) developed a high-throughput, low-cost DSB model by adapting the CDC's dry biofilm model (DBM) from Almatroudi et al. (2015) into a 24-well microtiter plate format. In their system, *S. aureus* biofilms were cultivated under repeated cycles of partial drying and nutrient exposure. Their protocol has advantages in terms of scalability and cost-efficiency; running several replicates or conditions in parallel is feasible by using standard plates and a simple humidity chamber. This is particularly useful for screening interventions (e.g. disinfectants) against DSB. A limitation is that the protocol is relatively lengthy (up to 12 days of incubation), and it was developed with a specific organism (*S. aureus*); different species or mixed communities might require further optimization. Furthermore, while the microplate design aims to simulate a dry surface, it may not adequately represent the reality of bigger-scale industrial surfaces (such as equipment) in food processing facilities.

Chaggar et al. (2024) adapt to a current standardized WSB method for DSB development. Using a CDC biofilm reactor, they initially grew *S.* Typhimurium, *L. monocytogenes*, and *P. aeruginosa* WSB (mono- or multi-species) on glass coupons, following the EPA MLB SOP MB-19 protocol. This approach produced mature WSB on the coupons by first batch growth in a low-nutrient medium followed by a 24-h continuous flow phase. To develop DSB, the coupons with established wet biofilms were taken from the reactor and air-dried at room temperature for 24, 48, or 72 h. By comparing bacterial survival at each desiccation time point to the initial WSB (0 h), the study showed that pathogens can persist in dried biofilms, though with gradual viability loss over time. This wet-to-dry model has many benefits including its application to mixed-species biofilms, which better mimics the variety in food processing settings, and its use of a standardized, repeatable WSB culture approach, therefore guaranteeing uniform biofilm growth before desiccation. It also directly integrates with existing disinfectant efficacy testing frameworks (since EPA-standard biofilms were the starting point), which could facilitate adoption. A major drawback, nevertheless, is that the DSB is produced by the fast desiccation of a WSB, which could not exactly reflect the structural and physiological evolution of biofilms developed in progressively drier circumstances. The procedure limits throughput and accessibility by requiring specialist equipment—a CDC reactor and coupon handling tool. Furthermore, their drying procedure did not actively regulate ambient humidity, hence repeatability of moisture levels might differ across laboratories or climatic environments. Despite these limitations, this study filled an important gap by extending disinfection test models to include dry biofilms, highlighting the resilience of pathogens like *Salmonella* and *Listeria* in low-moisture states.

Lin et al. (2024) emphasized a single-species *Salmonella* DSB model and offered a side-by-side comparison with its WSB equivalent. Their methodology was modified for foodborne *S*. Typhimurium from Amaeze et al. (2020). Using diluted (5%) TSB medium, biofilms were developed in static 60-mm petri plates. For the WSB, the culture was maintained wet for 7 days; after an initial 48-h growth, the medium was replenished, and the biofilm continued to remain wet (with one media refresh at mid-point) until day 7. In contrast, the DSB experienced sequential hydration and dehydration cycles: after the first 48 h of growth, the liquid was drained, and the dish was placed in a desiccator with 49% RH for 48 h. The biofilm was then briefly rehydrated with fresh TSB for 6 h, followed by a second drying phase at 49% RH for 66 h. This yielded a 168-h (7-day) old DSB that could be analyzed in parallel with the 7-day WSB. The controlled RH drying phases avoid total desiccation shock, thereby enabling the biofilm to adapt to drying over time. The benefit of Lin et al. (2024) method is that it closely simulates how biofilms could naturally grow on occasionally wet surfaces in low-moisture food facilities. By using microscopy, this method allowed for thorough studies of morphological variations between WSB and DSB, therefore clarifying the distinct resilience of DSB-grown cells. Moreover, the setup is relatively simple (Petri dishes in a desiccator), which makes it easily accessible for microbiology laboratories without flow reactors. However, its focus on a single organism in a controlled environment limits it in terms of capturing multi-species interactions or the impact of actual food residues on biofilm growth. This approach also needs cautious humidity control to be repeatable and is also time-intensive—a full week per cycle. Still, this model offers a useful framework for researching DSB in the context of food safety, and its effective application of a standardized cycle, with designated durations and RH, helps to support more general efforts in DSB model standardizing.

In conclusion, these three studies show various strategies for developing WSB and DSB in the lab. Amaeze et al. (2020) highlighted high-throughput static models for DSB (originally targeting healthcare surfaces) and showed that changing environmental conditions (temperature, humidity, agitation) might greatly affect biofilm production and tolerance. Chaggar et al. (2024) combine dynamic growth systems (biofilm reactor) with a post-growth drying phase, successfully integrating the dependability of WSB protocols with the realities of desiccation in food processing environments. Lin et al. (2024) propose a strategy, combining static culture with cyclic drying to replicate natural wet-dry variations and directly comparing the end-states of WSB and DSB under controlled settings. Despite their differences, all three methods contribute to an expanding toolkit for DSB research, by guiding toward standardized procedures capable of adaptation across labs.

6.2 DSB Cellular Interaction

Using single-cell analysis, researchers have investigated the functional heterogeneity inside *Salmonella* DSBs for the first time, revealing an intricate structure of interactions among cells in the biofilm community (Lin et al., 2024). This work found eight distinct groups of *Salmonella* cells in DSBs, each with a particular function based on their gene expression pattern. In short, more than half of the cells dropped into two categories with extremely low expression activities, suggesting they were in a dormant state to survive dry circumstances (Laskowska & Kuczyńska-Wiśnik, 2020). Two others triggered an SOS-associated DNA repair pathway to repair damage caused by dryness (Mérida-Floriano et al., 2021), while one group had high levels of catalase activities, an enzyme that helps protect against oxidative stress (Farrow et al., 2018). Notably, several subgroups continued to express genes associated with virulence, suggesting they may still pose a risk even under adverse conditions. Another small group showed high expression levels of CspA protein, which is associated with persistent cell development, enabling some bacteria to remain significantly resistant to stress (Kim & Wood, 2010). These findings suggest that DSB communities are not uniform; instead, they contain a mix of inactive, stress-resistant, and potentially virulent cells that together improve the biofilm's ability to survive.

This study also revealed differences that were not visible in earlier research using bulk transcriptomic analysis. Earlier studies found that *Salmonella* under dry conditions slowed down their metabolic activities and pathogenicity expression (Abdelhamid & Yousef, 2020). Single-cell research, however, found some cells remaining active and with virulence gene expression maintained. The coexisting of both dormant and active groups within DSB likely confers collective resilience: cells mounting antioxidant defenses and DNA repair can mitigate local damage and preserve viability, while those with enhanced adhesion or membrane integrity help maintain biofilm structure under stress (Roy Chowdhury et al., 2022). Such diversity suggests that subgroups might interact synergistically, in which case the function of one subgroup serves the larger population. However, it remains unclear how these subpopulations are spatially organized or whether they communicate through chemical signals in a low-moisture matrix. Furthermore, real-world DSBs often contain mixed microbial species, and the effects of inter-species interactions in dry biofilms on community heterogeneity and resilience are still mostly unknown. Future studies should combine spatial transcriptomic methods and multi-species models to clarify the interactions and regulatory mechanisms underlying DSB community resilience.

7 Conclusion

Foodborne pathogens form biofilms in various situations within the food industry, leading to widespread food contamination and incidences of food poisoning globally. Considering the financial losses experienced by food manufacturers and the increasing need for food safety, food firms, and researchers must intensify their efforts in developing effective techniques to manage biofilm in the food industry. Managing biofilms in the food industry is a challenging task due to the potential for different types of biofilms to develop under diverse processing circumstances. For example, a wet environment may lead to the growth of WSB, whereas a dry environment might result in the development of DSB. Extensive studies have thoroughly examined the features of classical WSB, while DSB has just lately started to get attention. DSB differed from WSB in terms of its morphology, adhesion behavior, and antimicrobial resistance. However, it is crucial to conduct further research on DSB since our understanding of this form of biofilms is still restricted compared to WSB. Moreover, it is worth noting that outbreaks of LMF linked with DSB continue to recur intermittently.

Several novel sanitization approaches applied in dry processing environments have been reviewed, but the necessity to enhance the efficacy of sanitization remains a necessary reality. One potential approach is the omics application of biofilm study to systematically elucidate the molecular mechanisms of signal interaction, cell differentiation, and cellular response to a variety of sanitization treatments. This method has the potential to identify any susceptible molecules for a targeted antibiofilm strategy by providing valuable insights on critically altered genes, proteins, or metabolites during biofilm formation or disinfection.

References

Abdelhamid, A. G., & Yousef, A. E. (2020). Collateral adaptive responses induced by desiccation stress in Salmonella enterica. *LWT, 133*, 110089.

Adator, E. H., Cheng, M., Holley, R., McAllister, T., & Narvaez-Bravo, C. (2018). Ability of Shiga toxigenic Escherichia coli to survive within dry-surface biofilms and transfer to fresh lettuce. *International Journal of Food Microbiology, 269*, 52–59.

Almatroudi, A., Hu, H., Deva, A., Gosbell, I. B., Jacombs, A., Jensen, S. O., Whiteley, G., Glasbey, T., & Vickery, K. (2015). A new dry-surface biofilm model: An essential tool for efficacy testing of hospital surface decontamination procedures. *Journal of Microbiological Methods, 117*, 171–176. https://doi.org/10.1016/j.mimet.2015.08.003

Almatroudi, A., Tahir, S., Hu, H., Chowdhury, D., Gosbell, I. B., Jensen, S. O., Whiteley, G. S., Deva, A. K., Glasbey, T., & Vickery, K. (2018). Staphylococcus aureus dry-surface biofilms are more resistant to heat treatment than traditional hydrated biofilms. *Journal of Hospital Infection, 98*, 161–167.

Alonso, V. P. P., Gonçalves, M. P. M., de Brito, F. A. E., Barboza, G. R., de Oliveira Rocha, L., & Silva, N. C. C. (2023). Dry surface biofilms in the food processing industry: An overview on surface characteristics, adhesion and biofilm formation, detection of biofilms, and dry sanitization methods. *Comprehensive Reviews in Food Science and Food Safety, 22*, 688–713.

Alvarez-Ordóñez, A., Coughlan, L. M., Briandet, R., & Cotter, P. D. (2019). Biofilms in food processing environments: Challenges and opportunities. *Annual Review of Food Science and Technology, 10*, 173–195. https://doi.org/10.1146/annurev-food-032818-121805

Amaeze, N., Akinbobola, A., Chukwuemeka, V., Abalkhaila, A., Ramage, G., Kean, R., Staines, H., Williams, C., & Mackay, W. (2020). Development of a high throughput and low cost model for the study of semi-dry biofilms. *Biofouling, 36*, 403–415.

Augustin, M., Ali-Vehmas, T., & Atroshi, F. (2004). Assessment of enzymatic cleaning agents and disinfectants against bacterial biofilms. *Journal of Pharmacy and Pharmaceutical Science, 7*, 55–64.

Azeredo, J., García, P., & Drulis-Kawa, Z. (2021). Targeting biofilms using phages and their enzymes. *Current Opinion in Biotechnology, 68*, 251–261.

Azmy, H. A., Sofy, A., Aboseidah, A., El-Morsi, E., Hmed, A. A., & Elmorshedy, H. A. (2024). Combating multidrug resistance: The potential of antimicrobial peptides and biofilm challenges. *Int J IJISRT, 236*, 307–335.

Blanco-Padilla, A., Soto, K. M., Hernández Iturriaga, M., & Mendoza, S. (2014). Food antimicrobials nanocarriers. *The Scientific World Journal, 2014*, 837215. https://doi.org/10.1155/2014/837215

Brandelli, A. (2020). The interaction of nanostructured antimicrobials with biological systems: Cellular uptake, trafficking and potential toxicity. *Food Science and Human Wellness, 9*, 8–20. https://doi.org/10.1016/j.fshw.2019.12.003

Bridier, A., Sanchez-Vizuete, P., Guilbaud, M., Piard, J.-C., Naitali, M., & Briandet, R. (2015). Biofilm-associated persistence of food-borne pathogens. *Food Microbiology, 45*, 167–178.

Cacciatore, F. A., Brandelli, A., Malheiros, P., & d. S. (2021). Combining natural antimicrobials and nanotechnology for disinfecting food surfaces and control microbial biofilm formation. *Critical Reviews in Food Science and Nutrition, 61*, 3771–3782. https://doi.org/10.1080/10408398.2020.1806782

CDC. (2022). https://www.fda.gov/food/outbreaks-foodborne-illness/outbreak-investigation-salmonella-peanut-butter-may-2022#:~:text=May%2020%2C%202022,of%20illnesses%20in%20this%20outbreak

Chaggar, G. K., Bryant, D. B., Chen, R., Fajardo, D., Jules-Culver, Z. A., Drolia, R., & Oliver, H. F. (2024). Development of Salmonella enterica serovar Typhimurium, Listeria monocytogenes, and Pseudomonas aeruginosa multi-species in vitro dry surface biofilm models: Insights into resilience and persistence in low-moisture environments. *Food Control, 166*, 110703. https://doi.org/10.1016/j.foodcont.2024.110703

Coughlan, L. M., Cotter, P. D., Hill, C., & Alvarez-Ordóñez, A. (2016). New weapons to fight old enemies: Novel strategies for the (bio)control of bacterial biofilms in the food industry. *Frontiers in Microbiology, 7*. https://doi.org/10.3389/fmicb.2016.01641

De Goffau, M. C., Yang, X., Van Dijl, J. M., & Harmsen, H. J. (2009). Bacterial pleomorphism and competition in a relative humidity gradient. *Environmental Microbiology, 11*, 809–822.

Duggan, K., Shepherd, M., & Maillard, J.-Y. (2024). Susceptibility of salmonella enterica Typhimurium dry surface biofilms to disinfection. *Journal of Food Safety, 44*, e13117. https://doi.org/10.1111/jfs.13117

Edwards, A. N., Karim, S. T., Pascual, R. A., Jowhar, L. M., Anderson, S. E., & McBride, S. M. (2016). Chemical and stress resistances of Clostridium difficile spores and vegetative cells. *Frontiers in Microbiology, 7*. https://doi.org/10.3389/fmicb.2016.01698

Fan, L., Li, L., Shang, F., Xie, Y., Duan, Z., Cheng, Q., & Zhang, Y. (2023). Study on antibacterial mechanism of electron beam radiation on Aspergillus flavus. *Food Bioscience, 51*, 102197.

Farrow, J. M., III, Wells, G., & Pesci, E. C. (2018). Desiccation tolerance in Acinetobacter baumannii is mediated by the two-component response regulator BfmR. *PLoS One, 13*, e0205638.

Gherasim, O., Popescu, R. C., Grumezescu, V., Mogoşanu, G. D., Mogoantă, L., Iordache, F., Holban, A. M., Vasile, B. Ş., Bîrcă, A. C., Oprea, O.-C., et al. (2021). MAPLE coatings embedded with essential oil-conjugated magnetite for anti-biofilm applications. *Materials, 14*, 1612.

Greco, I., Molchanova, N., Holmedal, E., Jenssen, H., Hummel, B. D., Watts, J. L., Håkansson, J., Hansen, P. R., & Svenson, J. (2020). Correlation between hemolytic activity, cytotoxicity and systemic in vivo toxicity of synthetic antimicrobial peptides. *Scientific Reports, 10*, 13206.

Han, H. M., Gopal, R., & Park, Y. (2016). Design and membrane-disruption mechanism of charge-enriched AMPs exhibiting cell selectivity, high-salt resistance, and anti-biofilm properties. *Amino Acids, 48*, 505–522.

Hansen, L. T., & Vogel, B. F. (2011). Desiccation of adhering and biofilm listeria monocytogenes on stainless steel: Survival and transfer to salmon products. *International Journal of Food Microbiology, 146*, 88–93.

Harjai, K., Kumar, R., & Singh, S. (2010). Garlic blocks quorum sensing and attenuates the virulence of Pseudomonas aeruginosa. *FEMS Immunology & Medical Microbiology, 58*, 161–168. https://doi.org/10.1111/j.1574-695X.2009.00614.x

Jara, J., Pérez-Ramos, A., del Solar, G., Rodríguez, J. M., Fernández, L., & Orgaz, B. (2020). Role of Lactobacillus biofilms in Listeria monocytogenes adhesion to glass surfaces. *International Journal of Food Microbiology, 334*, 108804. https://doi.org/10.1016/j.ijfoodmicro.2020.108804

Jawale, C. V., & Lee, J. H. (2014). Salmonella enterica serovar enteritidis ghosts carrying the Escherichia coli heat-labile enterotoxin B subunit are capable of inducing enhanced protective immune responses. *Clinical and Vaccine Immunology, 21*, 799–807.

Jiang, Y., Qiao, Y., Jin, R., Jia, M., Liu, J., He, Z., & Liu, Z. (2024). Application of chlorine dioxide and its disinfection mechanism. *Archives of Microbiology, 206*, 400.

Kerekes, E. B., Vidács, A., Takó, M., Petkovits, T., Vágvölgyi, C., Horváth, G., Balázs, V. L., & Krisch, J. (2019). Anti-biofilm effect of selected essential oils and main components on mono- and polymicrobic bacterial cultures. *Microorganisms, 7*, 345.

Kim, Y., & Wood, T. K. (2010). Toxins Hha and CspD and small RNA regulator Hfq are involved in persister cell formation through MqsR in Escherichia coli. *Biochemical and Biophysical Research Communications, 391*, 209–213.

Kretschmer, M., Schüßler, C. A., & Lieleg, O. (2021). Biofilm adhesion to surfaces is modulated by biofilm wettability and stiffness. *Advanced Materials Interfaces, 8*, 2001658.

Laskowska, E., & Kuczyńska-Wiśnik, D. (2020). New insight into the mechanisms protecting bacteria during desiccation. *Current Genetics, 66*, 313–318.

Lee, K.-H., Lee, J.-Y., Roy, P. K., Mizan, M. F. R., Hossain, M. I., Park, S. H., & Ha, S.-D. (2020). Viability of Salmonella Typhimurium biofilms on major food-contact surfaces and eggshell treated during 35 days with and without water storage at room temperature. *Poultry Science, 99*, 4558–4565.

Li, Y.-X., Erhunmwunsee, F., Liu, M., Yang, K., Zheng, W., & Tian, J. (2022). Antimicrobial mechanisms of spice essential oils and application in food industry. *Food Chemistry, 382*, 132312.

Lin, Z., Wang, G., Zhang, K., Jiang, S., Li, S., & Yang, H. (2023). Metabolomics investigation of global responses of Cronobacter sakazakii against common sanitizing in infant formula processing environments. *Food Research International, 172*, 113162.

Lin, Z., Liang, Z., He, S., Chin Fion Wei, L., Huang, D., Hong, Y., Wang, X., & Li, D. (2024). Salmonella dry surface biofilm: Morphology, single-cell landscape, and sanitization. *Applied and Environmental Microbiology, 90*, e01623–e01624. https://doi.org/10.1128/aem.01623-24

Liu, X., Yao, H., Zhao, X., & Ge, C. (2023). Biofilm formation and control of foodborne pathogenic bacteria. *Molecules, 28*, 2432.

Lu, T. K., & Collins, J. J. (2007). Dispersing biofilms with engineered enzymatic bacteriophage. *Proceedings of the National Academy of Sciences, 104*, 11197–11202.

Luo, Y., & Song, Y. (2021). Mechanism of antimicrobial peptides: Antimicrobial, anti-inflammatory and antibiofilm activities. *International Journal of Molecular Sciences, 22*, 11401.

Marra, D., Orillo, E., Toscano, G., Petala, M., Karapantsios, T. D., & Caserta, S. (2024). The role of air relative humidity on the wettability of Pseudomonas fluorescens AR11 biofilms. *Colloids and Surfaces B: Biointerfaces, 237*, 113831.

Mazaheri, T., Cervantes-Huamán, B. R. H., Bermúdez-Capdevila, M., Ripolles-Avila, C., & Rodríguez-Jerez, J. J. (2021). Listeria monocytogenes biofilms in the food industry: Is the current hygiene program sufficient to combat the persistence of the pathogen? *Microorganisms, 9*, 181.

Méndez-Vilas, A. (2011). Nanoparticles and their potential application as antimicrobials. In *Science against microbial pathogens: Communicating current research and technological advances*. FORMATEX.

Mérida-Floriano, A., Rowe, W. P. M., & Casadesús, J. (2021). Genome-wide identification and expression analysis of SOS response genes in salmonella enterica Serovar Typhimurium. *Cells, 10*, 943.

Mousavi, Z. E., Hunt, K., Koolman, L., Butler, F., & Fanning, S. (2023). Cronobacter species in the built food production environment: A review on persistence, pathogenicity, regulation and detection methods. *Microorganisms, 11*, 1379.

Nadell, C. D., Drescher, K., & Foster, K. R. (2016). Spatial structure, cooperation and competition in biofilms. *Nature Reviews Microbiology, 14*, 589–600.

Nan, Y., Rodas-Gonzalez, A., Stanford, K., Nadon, C., Yang, X., McAllister, T., & Narváez-Bravo, C. (2022). Formation and transfer of multi-species biofilms Containing E. coli O103: H2 on food contact surfaces to beef. *Frontiers in Microbiology, 13*, 863778.

Nkemngong, C. A., Voorn, M. G., Li, X., Teska, P. J., & Oliver, H. F. (2020). A rapid model for developing dry surface biofilms of Staphylococcus aureus and Pseudomonas aeruginosa for in vitro disinfectant efficacy testing. *Antimicrobial Resistance & Infection Control, 9*, 1–9.

Ophir, T., & Gutnick, D. L. (1994). A role for exopolysaccharides in the protection of microorganisms from desiccation. *Applied and Environmental Microbiology, 60*, 740–745. https://doi.org/10.1128/aem.60.2.740-745.1994

Panebianco, F., Rubiola, S., & Di Ciccio, P. A. (2022). The use of ozone as an eco-friendly strategy against microbial biofilm in dairy manufacturing plants: A review. *Microorganisms, 10*, 162.

Parvin, F., Rahman, M. A., Deva, A. K., Vickery, K., & Hu, H. (2023). Staphylococcus aureus cell wall phenotypic changes associated with biofilm maturation and water availability: A key contributing factor for chlorine resistance. *International Journal of Molecular Sciences, 24*, 4983.

Prestes, F. S., Yotsuyanagi, S. E., Alonso, V. P., & Nascimento, M. S. (2024). Dry sanitization in the food industry: A review. *Current Opinion in Food Science, 57*, 101166.

Pruteanu, M., Hernández Lobato, J. I., Stach, T., & Hengge, R. (2020). Common plant flavonoids prevent the assembly of amyloid curli fibres and can interfere with bacterial biofilm formation. *Environmental Microbiology, 22*, 5280–5299.

Rhodeland, B., Hoeger, K., & Ursell, T. (2020). Bacterial surface motility is modulated by colony-scale flow and granular jamming. *Journal of the Royal Society Interface, 17*, 20200147.

Rodriguez, A., Autio, W. R., & Mclandsborough, L. A. (2007). Effect of biofilm dryness on the transfer of listeria monocytogenes biofilms grown on stainless steel to bologna and hard salami. *Journal of Food Protection, 70*, 2480–2484.

Rosato, A., Sblano, S., Salvagno, L., Carocci, A., Clodoveo, M. L., Corbo, F., & Fracchiolla, G. (2020). Anti-biofilm inhibitory synergistic effects of combinations of essential oils and antibiotics. *Antibiotics, 9*, 637.

Roy Chowdhury, A., Sah, S., Varshney, U., & Chakravortty, D. (2022). Salmonella Typhimurium outer membrane protein A (OmpA) renders protection from nitrosative stress of macrophages by maintaining the stability of bacterial outer membrane. *PLoS Pathogens, 18*, e1010708. https://doi.org/10.1371/journal.ppat.1010708

Sattar, S., Springthorpe, S., Mani, S., Gallant, M., Nair, R., Scott, E., & Kain, J. (2001). Transfer of bacteria from fabrics to hands and other fabrics: Development and application of a quantitative method using Staphylococcus aureus as a model. *Journal of Applied Microbiology, 90*, 962–970.

Shamim, A., Ali, A., Iqbal, Z., Mirza, M. A., Aqil, M., Kawish, S. M., Siddiqui, A., Kumar, V., Naseef, P. P., Alshadidi, A. A. F., et al. (2023). Natural medicine a promising candidate in combating microbial biofilm. *Antibiotics, 12*, 299.

Soni, K. A., & Nannapaneni, R. (2010). Removal of Listeria monocytogenes biofilms with bacteriophage P100. *Journal of Food Protection, 73*, 1519–1524.

Stukalov, O., Korenevsky, A., Beveridge, T. J., & Dutcher, J. R. (2008). Use of atomic force microscopy and transmission electron microscopy for correlative studies of bacterial capsules. *Applied and Environmental Microbiology, 74*, 5457–5465.

Tavares, T. D., Antunes, J. C., Padrão, J., Ribeiro, A. I., Zille, A., Amorim, M. T. P., Ferreira, F., & Felgueiras, H. P. (2020). Activity of specialized biomolecules against gram-positive and gram-negative bacteria. *Antibiotics, 9*, 314.

Walker, S. L., Fourgialakis, M., Cerezo, B., & Livens, S. (2007). Removal of microbial biofilms from dispense equipment: The effect of enzymatic pre-digestion and detergent treatment. *Journal of the Institute of Brewing, 113*, 61–66.

Wang, R., King, D. A., & Kalchayanand, N. (2022a). Evaluation of salmonella biofilm cell transfer from common food contact surfaces to beef products. *Journal of Food Protection, 85*, 632–638. https://doi.org/10.4315/JFP-21-334

Wang, Z., Ma, Y., Li, Z., Wang, Y., Liu, Y., & Dong, Q. (2022b). Characterization of Listeria monocytogenes biofilm formation kinetics and biofilm transfer to cantaloupe surfaces. *Food Research International, 161*, 111839. https://doi.org/10.1016/j.foodres.2022.111839

Weaver, L., Webber, J., Hickson, A., Abraham, P., & Close, M. (2015). Biofilm resilience to desiccation in groundwater aquifers: A laboratory and field study. *Science of the Total Environment, 514*, 281–289.

Whitfield, C. (2006). Biosynthesis and assembly of capsular polysaccharides in Escherichia coli. *Annual Review of Biochemistry, 75*, 39–68.

WHO. (2024). https://www.who.int/data/gho/data/themes/who-estimates-of-the-global-burden-of-foodborne-diseases#:~:text=Each%20year%2C%20an%20estimated%20600,from%20foodborne%20diseases%20every%20year

Wille, J., & Coenye, T. (2020). Biofilm dispersion: The key to biofilm eradication or opening Pandora's box? *Biofilms, 2*, 100027. https://doi.org/10.1016/j.bioflm.2020.100027

Zhang, Y., Niu, Y., Luo, Y., Ge, M., Yang, T., Yu, L., & Wang, Q. (2014). Fabrication, characterization and antimicrobial activities of thymol-loaded zein nanoparticles stabilized by sodium caseinate–chitosan hydrochloride double layers. *Food Chemistry, 142*, 269–275. https://doi.org/10.1016/j.foodchem.2013.07.058

Zhu, Y., Li, C., Cui, H., & Lin, L. (2020). Feasibility of cold plasma for the control of biofilms in food industry. *Trends in Food Science & Technology, 99*, 142–151.

Ziege, R., Tsirigoni, A.-M., Large, B., Serra, D. O., Blank, K. G., Hengge, R., Fratzl, P., & Bidan, C. M. (2021). Adaptation of Escherichia coli biofilm growth, morphology, and mechanical properties to substrate water content. *ACS Biomaterials Science & Engineering, 7*, 5315–5325.

Antimicrobial Packaging Systems

Maryam Zabihzadeh Khajavi, Anton Nikiforov, Nathalie De Geyter, Frank Devlieghere, and Peter Ragaert

1 Introduction

Packaging is a pivotal step in food processing, dedicated to preserving the food quality and ensuring the safety throughout storage, transportation, and use. At every stage, from production to market display and the hands of the final consumer, food encounters various elements and potential contamination risks (Cruz et al., 2018; Teck Kim et al., 2014). The primary objective of conventional packaging is to effectively maintain the quality and safety of food. It strives to meet both industrial demands and consumer preferences while minimizing environmental impact. This includes addressing concerns related to contaminants, gas composition, microbial spoilage, and potential physical or mechanical damage (Han, 2014; Han et al., 2018; Majid et al., 2018).

Over time, increased consumer awareness regarding the potential harmful effects of chemical-based preservatives intentionally added to food products for microbial spoilage reduction has significantly shifted consumer preferences. Simultaneously,

M. Zabihzadeh Khajavi
Faculty of Bioscience Engineering, Department of Food Technology, Safety and Health, Research Unit Food Microbiology and Food Preservation (FMFP), Ghent University (UGent), Ghent, Belgium

Faculty of Engineering and Architecture, Department of Applied Physics, Research Unit Plasma Technology (RUPT), Ghent University (UGent), Ghent, Belgium

A. Nikiforov · N. De Geyter
Faculty of Engineering and Architecture, Department of Applied Physics, Research Unit Plasma Technology (RUPT), Ghent University (UGent), Ghent, Belgium

F. Devlieghere · P. Ragaert (✉)
Faculty of Bioscience Engineering, Department of Food Technology, Safety and Health, Research Unit Food Microbiology and Food Preservation (FMFP), Ghent University (UGent), Ghent, Belgium
e-mail: Peter.Ragaert@UGent.be

D. Li (ed.), *Antimicrobial Strategies in the Food System: Updates, Opportunities, Challenges*, https://doi.org/10.1007/978-3-031-95056-8_13

increased awareness of healthy dietary habits has resulted in an increased demand for minimally processed and fresh-like foods. The removal of chemical-based preservatives from food formulations, coupled with a focus on producing fresher foods, poses a significant challenge for industries. The vulnerability of these products to microbial contamination results in a substantial economic loss due to increased spoilage and subsequent food wastage (Karanth et al., 2023; Mesías et al., 2021). Various microorganisms, including bacteria, molds, and yeasts, have the potential to cause spoilage of packaged food. This susceptibility relies on the features of the food and predominant environmental parameters such as temperature, and amount of oxygen and carbon dioxide in the packaging. These parameters should be considered when designing packaging materials. To address the increasing demand for safety, the elimination of microbial contamination, and prolonging shelf life, there is a requirement for novel engineered food packaging materials. These materials should offer enhanced functionality compared to conventional packaging, providing more effective solutions to meet the challenges posed by different types of microorganisms and environmental conditions (Sun et al., 2022; Lee et al., 2015).

One of the promising strategies entails the development of packaging systems with inherent antimicrobial activity. Within these systems, antimicrobial agents are intricately integrated to exert control or inhibition over microbial growth within the packaged food during the full storage period. The overarching objective is to extend the shelf life of food while concurrently upholding its standards of quality and safety (Corrales et al., 2014; Khaneghah et al., 2018). Antimicrobial agents exhibit specific inhibitory activities against particular microorganisms, necessitating their selection based on the prevalent microorganisms associated with spoilage in the final product. It is imperative to meticulously ascertain the requisite levels of antimicrobial agents for optimal inhibition. Additionally, the method employed for the introduction of antimicrobial agents must not compromise the inherent functionality of the packaging (Kapetanakou & Skandamis, 2016; Sung et al., 2013).

Over the years, an array of antimicrobial packaging systems has been developed with the overarching objectives of enhancing safety and prolonging shelf life. These systems, however, come with their distinct advantages and disadvantages. This chapter provides a comprehensive overview of various antimicrobial packaging systems employed in food packaging, with a meticulous examination of their respective merits, drawbacks, and overall efficacy.

2 Active Antimicrobial Packaging

The term ‘active packaging’ typically denotes a transition in the defensive function of a passive packaging system to a more proactive role. It functions as a medium for interaction among the packaging, product, and the environment. This interaction encompasses a range of physical, chemical, and biological activities designed to modify the inherent environment of the packaged food. The ultimate aim is to enhance the sustainability, safety, quality, and shelf life of the food. Technically, active packaging entails the intentional integration of structures that facilitate the uptake or emission of certain compounds into or from the enclosed environment or the packaged food (Yam & Lee, 2012; Yildirim et al., 2018).

In a comprehensive classification, active packaging systems can be delineated into distinct categories (Lee et al., 2015; Wyrwa & Barska, 2017):

1. Absorption Mechanisms (Scavengers): Comprising compounds proficient in absorbing undesirable gases or other components from the internal environment of the package, including oxygen, moisture, ethylene, carbon dioxide, and odors/flavors.
2. Gas Emission Systems (Emitters): Encompassing substances designed to discharge gases or volatile components with advantageous properties into the headspace of the food packaging.
3. Structure-Based Delivery: Characterized by components meticulously designed to release beneficial components, such as antioxidants and antibacterial agents, directly into the packed food through intimate contact.

Within these categories, the antimicrobial-based active packaging system stands out as a particularly promising technique. Antimicrobial packaging has emerged as a commendable alternative technique aimed at diminishing, limiting or preventing the growth of spoilage and pathogenic microorganisms in food items (Jideani & Vogt, 2016). The antimicrobial activity inherent in such packaging systems typically relies on the extension of lag phases, a decrease in live cell counts of targeted microbes or a decrease in growth rate, resulting in the reduction and prevention of microbial growth. The engineering of these antimicrobial packaging systems is centered around fulfilling three primary objectives: maintaining quality, ensuring safety, and extending shelf life (Bastarrachea et al., 2011).

3 Antimicrobial Compounds in Food Packaging

Numerous studies have been conducted to meet the increasing need for active antimicrobial packaging materials. The research explores suitable antimicrobial agents for integration into food packaging systems, categorizing them into natural and synthetic types (Duda-Chodak et al., 2023). This section introduces various antimicrobial agents, highlighting their key features and mechanisms. The application of these agents will be discussed later in relation to different types of antimicrobial food packaging.

3.1 Natural Antimicrobial Agents

Natural antimicrobial agents are crucial resources for antimicrobial packaging, being both natural and health-friendly, with higher consumer acceptance compared to synthetic counterparts. Despite their numerous benefits, they have limitations, including susceptibility to high temperatures that can alter their structure and reduce functionality. They may also affect taste or odor, interact with food components leading to reduced efficiency, and face challenges with higher costs and limited availability. (Zhu, 2021; Zhou et al., 2018; Punia Bangar et al., 2021). This restricts their widespread application in developing antimicrobial packaging materials.

Essential Oils Essential oils (EOs) are diverse blends of volatile organic compounds derived from plants, including ketones, aldehydes, phenols, esters, ethers, oxides, terpenes, monoterpene and sesquiterpene alcohols, and more. Their antimicrobial efficiency is mainly due to aldehydes, phenols, and oxygenated terpenoids. EOs, being hydrophobic, disrupt microbial membranes and mitochondria by interacting with lipids, increasing permeability, and resulting in the release of ions and cellular contents, ultimately causing cellular death with a loss of vital components. Additionally, EOs inhibit virulence factors linked to cell walls and can attach to particular regulatory gene products in microorganisms, thereby disrupting their translation. EOs containing hydrocarbon monoterpenes can facilitate the absorption of other antimicrobial agents by interacting with active sites located on the cell wall of microorganisms. Therefore, combining EOs with other antimicrobial agents enhances their effectiveness. EOs exhibit inhibitory properties against bacteria, molds, and yeast (Solórzano-Santos & Miranda-Novales, 2012; Bassolé & Juliani, 2012).

Polyphenolic Compounds Polyphenolic compounds, characterized by many hydroxyl groups, exert an inhibitory impact by reacting with the bacterial cell wall, damaging cell membrane structure, and causing leakage of cellular contents. Acting as antioxidants, polyphenols hinder the development and activity of reactive oxygen species (ROS) while scavenging free radicals. As a result, they decrease the redox potential of the surrounding environment or growth medium. This dual role - acting as antioxidants and exhibiting bactericidal or bacteriostatic effects - makes polyphenols effective against undesirable microorganisms. Key mechanisms by which polyphenols exert antibacterial action include reacting with proteins, prevention the synthesis of nucleic acid, induction of DNA damage, preventing the formation of cell wall or interacting with it, modifying the functions of cytoplasmic membrane, depleting the energy metabolism, preventing the development of biofilm, and deprivation of substrates and metals (Zhu, 2021; Makarewicz et al., 2021).

Bacteriocins Antimicrobial peptides (AMPs) are short chains of amino acids, varying in length from five to over a hundred, that serve as the host's defense mechanism and are produced by bacteria. Examples of bacteriocins, a type of AMP, such as, enterocin, nisin, leucocin, bifidin, lactocin, pediocin, and lactococcin A. Bacteriocins, characterized by a positive charge and hydrophobic components, efficiently bind to microbial cell membranes by engaging in electrostatic interactions with phosphate groups (negatively charged). Subsequently, the hydrophobic segments insert into the membrane, creating pores (Santos et al., 2018; Kumar et al., 2018; Lei et al., 2019).

Organic Acids Organic acids can be found in form of dissociated and undissociated. The antimicrobial efficiency of an organic acid is influenced by its pKa value, representing the pH at which an acid is equally dissociated and undissociated. Organic acids enter cells in a charged form through the cell membrane. Upon entry, in a higher pH environment, they separate into charged particles—protons and anions. Importantly, these charged particles cannot exit through the cell membrane, disrupting cell function. Both organic acids and their salts demonstrate antimicrobial activity against bacteria and fungi (Guan & Liu, 2020; Hauser et al., 2016).

Enzymes Enzymes are being employed more frequently as antimicrobial agents in packaging materials and directly in food. Their efficacy lies in inhibiting bacterial biofilm formation, achieved through direct attacks on microorganisms or catalyzing reactions that lead to the production of antibacterial products. Three primary categories of enzymes exhibit antimicrobial and antibiofilm properties are: proteolytic enzymes, polysaccharide-degrading enzymes, and oxidative enzymes (Sharma et al., 2022a).

Enzymes act against a biofilm through various mechanisms, such as degrading structures that facilitate cell adhesion or surface attachment, inhibiting biofilm formation, detaching an established biofilm, or enhancing susceptibility to antimicrobials. Lysozyme, dispersin B, proteinase K, and DNAse I are notable enzymes with significant anti-biofilm activity, with lysozyme being the most frequently utilized in food packaging materials. Lysozyme is effective against Gram-positive bacteria with a peptidoglycan-rich membrane (90%), targeting the β-1-4 glycosidic bond in peptidoglycan. However, it is not as efficint against Gram-negative bacteria due to the protective layer of lipopolysaccharides (LPS) on their cell walls (Ferraboschi et al., 2021).

Enzymes like DNAse I, proteinase K, and dispersin B can dissolve established biofilms, with their effectiveness depending on the biofilm type. Dispersin B, for example, can break down the polysaccharide poly-N-acetylglucosamine found in the developed biofilm by certain Gram-positive bacteria (Galie et al., 2018).

Bacteriophages Exploring the use of bacteriophages in packaging development is a fascinating but somewhat debated concept. Bacteriophages, being viruses, necessitate a host for replication in the environment, requiring direct contact with food is applied as coatings. The bacteriophage life cycle involves host identification, infection, replication, bacterial lysis, and the subsequent release of new phage particles into the surrounding environment. Notably, several commercial bacteriophage formulations have received approval from regulatory authorities such as the European Food Safety Authority (EFSA) and the US Food and Drug Administration (FDA), being designated as Generally Recognized as Safe (GRAS) (Poaska & Sokoowska, 2019; Moye et al., 2018).

Allyl Isothiocyanate Allyl isothiocyanate (AIT) is found in cruciferous plants, including mustard, wasabi, and horseradish. It is notable for its substantial antimicrobial activity in vapor form against various spoilage and pathogenic microorganisms, even at low levels. Therefore, AIT can be utilized for developing antimicrobial packaging that allows for a slow release, helping to extend the shelf life of foods. AITs disrupt cell membranes, resulting in the release of cellular metabolites (Lin et al., 2000).

Chitosan Chitosan, a biobased and biodegradable polymer, possesses inherent antimicrobial properties, making it ideal for direct use in packaging. It can also function as an antimicrobial coating on primary packaging materials or be integrated in various forms like fibers, powders, and nanoparticles within the packaging matrix. Chitosan is a deacylated derivative of the chitin polymer, which itself is structured with β-(1,4)-linked N-acetyl-β-D-glucosamine. Research indicates that chitosan exhibits both antibacterial and antifungal activities. The polycationic nature of chitosan has a crucial role in its antimicrobial efficacy (Huq et al., 2022).

In acidic conditions, the positively charged C-2 position of chitosan facilitates interaction with negatively charged membranes of microbial cells. This interaction results in the release of proteins and intracellular components, ultimately culminating in cell death. Furthermore, upon entering cell nuclei, chitosan binds to DNA, disrupting mRNA synthesis and halting amino acid production (Yan et al., 2021).

Chitosan's antifungal activity is attributed to its inhibition of spore germination, causing alterations in gene expression and modifying the activity of enzymes involved in hyphae growth. Additionally, it increases intracellular ROS levels. Moreover, chitosan alters the cell membrane permeability of baker's yeast, preventing the uptake of glucose, a crucial nutrient for yeast growth (Lopez-Moya et al., 2019).

Vanillin Vanillin is classified as an aromatic hydrocarbon and is renowned for its flavor and fragrance. However, it also exhibits inhibitory properties against bacteria, molds, and yeast. Its primary mode of action is as a membrane-active compound, causing the dissipation of ion gradients and inhibiting respiration, with the extent of impact being species-specific (Fitzgerald et al., 2004). Furthermore, vanillin holds the GRAS status, allowing its safe use in developing antimicrobial packaging materials.

3.2 Synthetic Antibacterial Agents

This category of antimicrobial agents is distinguished by its high thermal stability, making it an ideal choice for engineering antimicrobial packaging materials, especially those involving thermal processes in their manufacture, such as the hot-melt extrusion process during packaging conversion. Their capacity to endure high temperatures makes them an advantageous option for ensuring the effectiveness and integrity of antimicrobial features during the production of packaging materials (Huang et al., 2019).

Metal-Based Nanoparticles Metals and metal oxides are utilized in form of nanoparticles (NPs) because of their small dimentions, which aids in penetrating bacterial cells. While the specific mechanism of how metal NPs and metal oxide NPs exert toxicity against bacteria remains unclear, it is hypothesized to involve the production of free radicals and ROS. ROS induce oxidative stress, influence the expression of gene, and damage crucial cellular products like DNA and proteins, affecting mitochondrial function and leading to disruption of the cellular membrane and wall, ultimately resulting in bacterial cell death. Additionally, direct contact with the bacterial surface is thought to cause cell wall damage and disrupt electron transport. The antibacterial effectiveness of metal NPs is influenced by the release of metal ions. Studies have shown that Gram-positive bacteria exhibit greater resistance to NPs compared to Gram-negative bacteria. This resistance is linked to the higher presence of peptidoglycan layers and teichoic acids in Gram-positive bacteria. Metal NPs can interact with multiple units of amino acids and carbohydrates within the peptidoglycan, and they are more efficiently captured by the negatively charged peptidoglycan in Gram-positive bacteria (Sharma et al., 2020). Among the metal NPs known for their antimicrobial properties, the most frequently used in food packaging include copper (Cu), silver (Ag), selenium (Se), and gold (Au) (El Gohary et al., 2023; Paidari & Ibrahim, 2021; Zamindar et al., 2020; Marrez et al., 2019). Additionally, in the category of metal oxide NPs, zinc oxide (ZnO), titanium oxide (TiO_2), and silica (SiO_2) are the most frequently employed (Venkatesan & Rajeswari, 2019; Phuinthiang et al., 2023; Kim et al., 2022).

Quaternary Ammonium Salts Quaternary ammonium salts (QAS) find extensive use in various industries due to the straightforward development of the desired material through functionalization of their heterocyclic nitrogen atom. The positively charged head of QAS attracts the negatively charged bacterial cell membrane. This attraction damages the bacteria, causing leakage and breakdown, ultimately resulting in bacterial death. Apart from electrostatic attraction, QAS also have amphiphilic properties, allowing them to act like detergents. This versatility makes QAS effective against a broad spectrum of bacteria, fungi, and viruses (Min et al., 2020; Qiu et al., 2023).

Parabens Parabens, including methyl-, ethyl-, propyl-, and butylparaben, are esters of para-hydroxybenzoic acid commonly used as antimicrobial agents. These compounds exhibit a wide range of activity, particularly against yeast and molds. They also maintain stability over a broad range of pH levels and storage conditions, making them suitable for antimicrobial packaging in various food categories. Their antimicrobial action works by disrupting microbial cell membranes and inhibiting enzyme functions and cellular processes, ultimately leading to cell death (Davidson, 2020).

Sorbates Sorbates are salts of sorbic acid, with potassium sorbate and sodium sorbate being the most commonly used forms. These compounds are primarily effective against molds and yeasts, and to a lesser extent, some bacteria. Their antimicrobial activity involves two main mechanisms: first, lowering the pH to inhibit microbial enzyme activity, and second, disrupting microbial cell membranes, which interferes with their function and prevents growth (Uz & Altnkaya, 2011; Kuplennik et al., 2015).

Sodium Propionate Sodium propionate is chemically synthesized from sodium hydroxide and propionic acid, a natural short-chain fatty acid found in various foods. This compound is particularly effective against molds, even at low concentrations, and somewhat effective against certain bacteria and yeasts. It works best in acidic environments, where it disrupts microbial cell membranes and metabolic processes. However, it is crucial to highlight that when used at high concentrations, sodium propionate may impart a slightly bitter taste to food products, potentially affecting their sensory qualities (Thanakkasaranee et al., 2018).

Sodium Benzoate Sodium benzoate is formed by the reaction of benzoic acid with sodium hydroxide. It is effective against a broad range of microorganisms, including bacteria, molds, and yeasts, with a greater impact on molds and yeasts. Sodium benzoate works primarily by lowering the pH of its environment, which inhibits microbial growth by disrupting enzyme function and cellular processes. It is most effective in acidic conditions (pH $\leq$ 7) and less effective in non-acidic environments, which can limit its use in certain food products (Shahmohammadi et al., 2016; Birck et al., 2016). Additionally, some individuals may experience allergic reactions or sensitivities to sodium benzoate (Piper & Piper, 2017).

N-Halamine N-halamines are a class of compounds with nitrogen-halogen covalent bonds. They are synthesized by introducing halogens (chlorine, bromine, or iodine) into nitrogen-containing substrates like amides, imides, or amines. N-halamines are effective against a broad spectrum of bacteria, fungi, and viruses. Their antimicrobial activity comes from halogen atoms, such as chlorine, which can be released as hypohalous acids upon contact with microbes. This process disrupts microbial cell walls and causes cell death. Additionally, N-halamines can regenerate their antimicrobial properties through rehalogenation, making them particularly useful in applications like antimicrobial packaging for long-lasting protection (Liu et al., 2020; Bastarrachea et al., 2014; Bastarrachea & Goddard, 2015).

Ethanol Ethanol works as an antimicrobial agent by denaturing proteins and disrupting microbial cell membranes, resulting in the inactivation or death of microorganisms like bacteria, viruses, and fungi. Its broad-spectrum effectiveness makes it a versatile antimicrobial. Ethanol can be incorporated into antimicrobial sachets to release ethanol within food packaging systems, providing ongoing protection against microbial contamination (Kapetanakou et al., 2015).

Chlorine Dioxide (ClO_2) Chlorine dioxide is a yellow-green gas at room temperature and is highly soluble in water. It is a potent oxidizing agent, often more efficient at lower concentrations compared to chlorine. Chlorine dioxide kills microorganisms by disrupting their cell membranes and interfering with protein synthesis and enzyme activity. It has a broad-spectrum efficacy, targeting a broad range of pathogens including bacteria, viruses, fungi, and protozoa. This makes it particularly useful against foodborne pathogens and spoilage organisms. Chlorine dioxide can be integrated into packaging materials or used in sachets that slowly release the gas, providing ongoing antimicrobial activity over time (Ray et al., 2013; Singh et al., 2021).

4 Types of Antimicrobial Packaging Systems

The predominant classification of food packaging systems encompasses either a packaging material/food system or a packaging material/headspace/food system. Consequently, these components serve as viable mediums for the integration of antimicrobials, with the primary objective of mitigating or eliminating the proliferation of pathogenic and food spoilage microorganisms on food surfaces (Han, 2003).

Antimicrobial packaging systems can be categorized into two types: migrating and nonmigrating. This classification is based on the antimicrobial agent and how it interacts with the packaging material and food matrix. In migrating systems, the active agent is released into the package headspace and from there absorbed by the food product or it is directly migrating onto the food surface from the packaging material. Conversely, nonmigrating systems keep the active agent immobilized within the material. While effective antimicrobial activity in migrating systems does not necessarily hinge on direct contact between the packaging material and the food product, such contact is essential for the efficacy of antimicrobial activity in nonmigrating systems (Fadiji et al., 2023).

Figure 1 presents an array of antimicrobial packaging systems. In A:1 and A:2, headspace systems are depicted, where the initial integration of a volatile antimicrobial agent into the matrix layer facilitates its subsequent release into the headspace, establishing contact with the food product. A:2 highlights a sophisticated headspace system featuring a control layer designed to precisely regulate the permeability of the

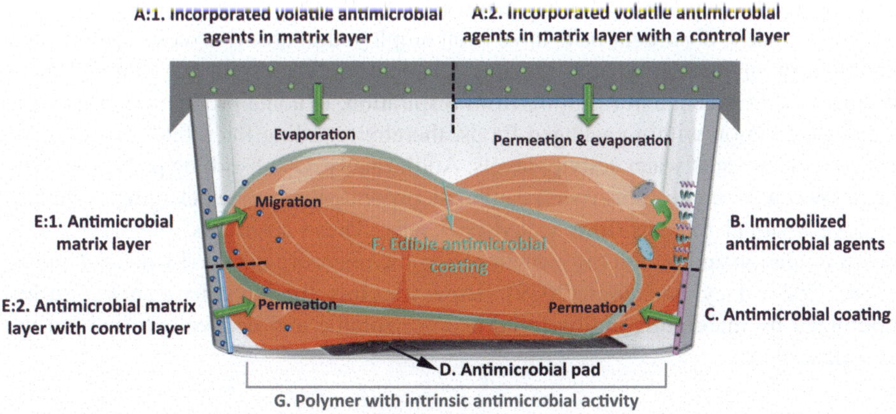

Fig. 1 Various antimicrobial systems applicable to food packaging

volatile antimicrobial agent, ensuring the maintenance of a specific headspace concentration. System B entails the immobilization of antimicrobial agents on the surface of the packaging material, constituting a nonmigratory system necessitating direct contact with the food for microorganism inactivation. In system C, a coating imbued with antibacterial properties is meticulously applied to the packaging material. Sachets or pads, capable of generating or containing antibacterial agents, can be strategically positioned within the packaging box, as exemplified in D. For systems E:1 and E:2, antimicrobial agents are intricately embedded into the matrix of the packaging materials, permitting controlled release through diffusion between the layers. To modulate the release rate, the antimicrobial matrix layer can be judiciously laminated with a control layer (E:2). Alternatively, an edible antibacterial coating can be discerningly applied on the surface of the food itself, as exemplified in F (Han, 2003; Brody et al., 2001). Additionally, the packaging materials can be made from a polymer with intrinsic antimicrobial properties, as demonstrated in system G.

In the next sections, a comprehensive introduction to illustrate the different antimicrobial systems will be provided, presenting the research findings to elucidate their efficiency. This examination will meticulously outline the advantages and disadvantages inherent to these systems.

4.1 Emitting Sachets and Absorbent Pads

Antimicrobial-emitting sachets and absorbent pads are inserted into the packaging, creating an antimicrobial active packaging system as shown in Fig. 1 (D type). This system plays a crucial function in food preservation, particularly on a commercial scale. The emitting sachets are infused with antimicrobial components effective against foodborne microorganisms. Upon release, these volatile antimicrobial agents permeate the food packaging system, effectively reducing or preventing the growth of spoilage and pathogenic microorganisms.

Traditionally, absorbent pads have been employed to absorb excessive moisture, particularly in meat products, serving as a purge. Their primary purpose is to mitigate moisture content resulting from respiration, exudate production, and water vapor permeation within packaged foods, thereby retarding microbial growth associated with elevated water activity levels. Additionally, O_2-absorbing pads or sachets (also known as oxygen absorbers) can be incorporated into the packaging system to remove residual oxygen from sealed packages. These absorbers effectively slow down oxidation and inhibit the growth of aerobic microorganisms (Esquivel-Chávez et al., 2021; Dey & Neogi, 2019). However, this functionality can be further optimized by integrating antibacterial agents (volatile and non-volatile) within the absorbent pads.

4.1.1 Antimicrobial Emitting Sachets

Two methods are employed in the production of antimicrobial-releasing sachets: one involves generating antimicrobial compounds in situ and releasing them, while the other entails carrying and releasing pre-loaded antimicrobials. The process for antimicrobial-generating sachets consists of two steps: introducing the system that produces antimicrobial compounds inside the sachet and sealing it. On the other hand, antimicrobial-carrying sachets follow a three-step procedure: adding the antimicrobial agent to the carrier, placing the carrier into the sachet, and sealing it.

Carrier-emitting sachets operate on the principle of controlled release of antimicrobial agents into a sealed packaging system. This release occurs through a permeable sachet material, which must endure handling and transportation to prevent the release of internal components. Placed within an enclosed system alongside food items, the sachet is shielded from the external environment by the main packaging material with specific barrier properties tailored to each application, aiming to preserve the modified gas composition within the packaging headspace.

4.1.1.1 Carrier Materials

Support materials, commonly referred to as carriers, has a pivotal role in facilitating and regulating the release of active compounds within packaging headspaces. The choice of material for incorporating volatile compounds significantly impacts the loading and releasing capacities of antimicrobial sachets. In commercial antimicrobial ethanol emitters sachets, food-grade ethanol is released from a carrier like SiO_2 powder contained in sachets fabricayed of paper and ethylene-vinyl acetate (EVA) (Pasqualone, 2019). Carrier materials extend beyond powders to include natural or synthetic polymers, resin, beads, gels, and porous minerals. As an example, β-cyclodextrin (β-CD), a natural polymer, has been successfully utilized for carrying garlic EO to develop an antibacterial emitting sachet aimed at controlling microbial growth in fresh cut tomatoes (Ayala Zavala & Gonzalez Aguilar, 2010). Additionally, in certain studies, porous high-density polyethylene (HDPE) resin has proven effective when integrated with various essential oils. This innovative approach has been employed to create antibacterial sachets for the packaging of sliced bread and papaya (Espitia et al., 2011; Passarinho et al., 2014).

4.1.1.2 Sachet Materials

To ensure optimal diffusion of active agent within the sachet without limiting factors, sachet materials need to be porous or semipermeable. Various materials, including semipermeable or perforated polymer films, porous nonwoven fabrics, and papers, have been employed to overwrap carriers (Ozdemir & Floros, 2004; Hempel et al., 2013; Pereira de Abreu et al., 2012).

The release rate of the active agents to the packaging is influenced by various sachet material properties such as permeability, pore dimensions, and thickness (Otoni et al., 2016). To prevent leakage of the carrier material and subsequent uncontrolled release of the active agents, a sealable layer must be presented in the sachet material. While heat sealing is common for sachet materials, cold seal coatings may be employed when the active agent or the sachet material is sensitive to high temperature. This limitation is often overcome by using adhesive layers or aluminum tape (Han et al., 2014).

4.1.2 Absorbent Pads

Fresh meat, fish, and cut fruits and vegetables often produce exudates during storage, which can negatively impact both the sensory qualities and microbial integrity of packaged food. The liquid released, rich in micro- and macronutrients, fosters an environment suitable for the growth of pathogens and spoilage microorganisms. To counteract this, antimicrobial absorbent pads are widely employed to mitigate the risk of microbial growth and preserve the sensory attributes of packaged foods. Typically positioned at the bottom of plastic trays, containers, or even within bags for food display, these pads effectively absorb the exuded liquid throughout the storage period while concurrently releasing loaded antimicrobial agents into the food. To effectively meet these criteria, absorbent pads must adhere to specific requirements, including (Otoni et al., 2016):

1. Rapid Absorption: The absorbent pad should swiftly absorb released liquid, regardless of the packaging system's orientation (horizontal or inclined).
2. Liquid Retention: The absorbent pads must securely retain the released liquid within their structure, even under the weight of the food product throughout the entire storage period.
3. Mechanical Strength: The absorbent pads need to exhibit sufficient mechanical resistance to facilitate high-speed automation during the filling process, preventing rupture under the weight of the packed food.
4. Visual Preservation: The absorbent pads should help maintain the visual appeal of the packed food. Additionally, the release of antibacterial agents should not impact the sensory characteristics of the packed food.
5. Cost Considerations: The application of antibacterial absorbent pads should be economically viable, ensuring that it does not substantially increase costs for the consumer.

Commonly, absorbent pads consist of upper and lower layers along with an internal layer housing the absorbent material and antibacterial agents. The upper and lower layers can be constructed from impermeable thermoplastic material or coated cellulosic material to prevent direct contact with the packed food. Perforations are introduced on the surface of both upper and lower layers, or at least on one layer, to facilitate the absorption of released liquids.

Despite the widespread use of absorbent pads, they may have drawbacks related to food preservation. One issue arises when the capability of absorbent pad is limited under a compression load exerted by the packed food. Under such conditions, an part of released liquid tends to accumulate on the upper surface of the absorbent pad, impeding successful absorption. Another limitation concerns the perforations on the outer layers of the absorbent pad. Proper spacing is crucial to avoid film tearing, which could release the absorbent material and potentially lead to safety concerns (Otoni et al., 2016).

4.1.3 Antimicrobial Agents in Antimicrobial Emitting Sachets and Absorbent Pads

Recent advancements in food preservation have significantly diversified the range of antimicrobial compounds integrated into the development of antimicrobial sachets and absorbent pads (Ahmed et al., 2022). In the domain of antimicrobial sachets, the selection of antimicrobial agents primarily focuses on volatile compounds. EOs, AITC, ClO_2, and ethanol are widely integrated into sachets designed for antimicrobial packaging. The distinctive advantage of volatile antimicrobials lies in their capacity to operate in gaseous phases, eliminating the need for direct contact between the food and packaging material. This allows the antimicrobial properties to reach virtually any food matrix placed in the packaging system (Pereira de Abreu et al., 2012). Moreover, the vapor phase of essential oils has demonstrated superior antimicrobial efficacy compared to methods involving direct contact solutions (Velázquez-Nuñez et al., 2013).

Despite these advantages, challenges persist with volatile antimicrobial agents. Issues such as stability during the storage time, sensory perception, and low heat resistance stability impose limitations on production approaches. Table 1 provides an overview of the volatile antimicrobial agents integrated into emitting sachets for food packaging applications.

Conversely, in the development of antimicrobial absorbent pads, both non-volatile and volatile antimicrobial agents can be employed. EOs, NPs, and bacteriophages, in particular, find extensive use in this context. Studies outlining the progress in antimicrobial pads for food packaging purposes are demonstrated in Table 2.

4.1.4 Drawbacks and Constraints

Despite the numerous advantages of using emitting sachets or absorbent pads, particularly the absence of requirements for adapting the packaging material production line with this technique, they still have a few disadvantages. The placement of sachets or pads in the packaging is commonly done manually, leading to increased packaging time and decreased productivity due to an additional operational step (Pereira de Abreu et al., 2012). Additionally, many antimicrobial agents, especially

Table 1 Summary of selected studies on the application of antimicrobial emitting sachets in food packaging systems

Antimicrobial agent	Carrier	Food/simulant	Target microorganism	Antimicrobial activity	References
Garlic EO	β-CD	Fresh-cut tomato	Total aerobic mesophilic bacteria (TAMB) Total coliforms Fecal coliforms Total yeast and mold count (TYMC) *Escherichia coli* (*E. coli*)	The microbial growth in fresh-cut tomatoes was significantly impacted by garlic EO. Despite an increase in TAMB and TYMC during storage garlic EO showed a beneficial effect, slowing down the microbial growth compared to the control sample.	Ayala Zavala and Gonzalez Aguilar (2010)
Cinnamon EO Oregano EO Lemongrass EO	High-absorption polymeric resin	Papaya	TAMB *Alternaria alternata* (*A. alternata*) *Fusarium semitectum* (*F. semitectum*) *Lasiodiplodia theobromae* (*L. theobromae*) *Rhizopus stolonifera* (*R. stolonifera*)	Cinnamon, oregano, and lemongrass sachets significantly reduced TAMB and fungal growth during storage, with cinnamon EO showing the greatest reduction	Espitia et al. (2012)
Oregano EO	High-absorption polymeric resin	Sliced bread	TYMC *E. coli* *Salmonella Enteritidis* (*S. Enteritidis*)	Oregano EO-containing sachets exhibited in vitro antimicrobial effects against the tested bacteria and reduced the growth of yeasts and molds on sliced bread.	Passarinho et al. (2014)
Rosemary EO Thyme EO	Microcellular foam starch	Mozzarella cheese	*Listeria monocytogenes* (*L. monocytogenes*) Lactic acid bacteria (LAB) Total aerobic bacteria (TAB)	Rosemary and thyme oil volatiles released from the sachet reduced *L. monocytogenes* growth by 2.5 log colony forming units (CFU) g^{-1} on day	Han et al. (2014)

				9 at 10 °C. The oils also inhibited LAB and TAB, resulting in a 1.2 and 1.4 log CFU g^{-1} reduction respectively, after 15 days at 10 °C.	
Basil EO *Pimenta dioica* EO	*β*-CD	Culture media	*Staphylococcus aureus* (*S. aureus*) *E. coli* *L. monocytogenes* *Pseudomonas aeruginosa* (*P. aeruginosa*) *S.* Enteritidis *Botryotinia nivea* (*B. nivea*)	*Pimenta dioica* EO demonstrated greater antimicrobial activity against the studied microorganisms in comparison to basil EO.	Marques et al. (2019)
Oregano EO	Polyvinyl alcohol (PVA)	Fresh-cut iceberg lettuce	*Dickeya chrysanthemi* (*D. chrysanthemi*) TYMC TAMB	Oregano EO prevented *D. chrysanthemi* growth by 3.9 log CFU/5 pieces over a period of 5 days at 20 °C. It also significantly reduced TYMC and TAMB by 2.1 log CFU/5 pieces and 1.5 log CFU/5 pieces, respectively.	Chang et al. (2017)
Peppermint EO	Starch-based microcellular foam	Strawberry	TYMC TAMB	Packaged fruits with sachets had lower microbial loads. The TAMB in the treated samples was 1.7 Log CFU g^{-1}, in comparison with 3.7 Log CFU g^{-1} in control samples. Additionally, the TYMC in treated fruits was 1.3 Log CFU g^{-1}, while control fruits had 1.57 Log CFU g^{-1}.	Amiri et al. (2022)

(continued)

Table 1 (continued)

Antimicrobial agent	Carrier	Food/simulant	Target microorganism	Antimicrobial activity	References
Geraniol EO	Paper coated with ethylene vinyl alcohol (EVOH)	Sliced bread/ Culture media	*S. aureus* *E. coli* Total plate count (TPC) TYMC	The antimicrobial sachet was more efficient against *S. aureus* compared to *E. coli* in culture media. Bread stored with the sachet showed a shelf life extension of over 3 weeks, due to lower levels of TPC and TYMC compared to the blank.	Petchwattana et al. (2021)
Eugenol EO Citral EO	Porous starch	Sliced bread	*Penicillium roqueforti* (*P. roqueforti*) *Aspergillus niger* (*A. niger*)	Eugenol was significantly more destructive to the cell membrane of *A. niger* than citral.	Ju et al. (2020)
Eugenol Carvacrol *Trans*-anethole *Trans*-cinnamaldehyde α-Pinene	Polylactic acid (PLA) pellets	Organic wild rocket	*Pectobacterium carotovorum* (*P. carotovorum*)*Pseudomonas fluorescens* (*P. fluorescens*) *E. coli* *S. enteritidis*	In the packaging experiment, eugenol and trans-anethole masked the off-odor of wild rocket, but there was no clear impact on the microbial growth.	Wieczyska et al. (2016)
Lemongrass EO	β-CD	Shrimp	*E. coli* *S. enteritidis*	The inclusion complex had antimicrobial activity against *E. coli* and *S. aureus*, prolonging the shelf life of shrimp up to 12 h at room temperature (28 ± 2 °C) by inhibiting bacterial growth on the shrimp's surface.	Yunilawati et al. (2023)
Thyme EO	Modified starch/Agave fructans	Mango	*Colletotrichum gloeosporioides* (*C. gloeosporioides*)	In mango, 0.10 g of sachets lessened the frequency and intensity of *C. gloeosporioides*	Esquivel-Chávez et al. (2021)

				to 75% and 77%, while 0.20 g reduced them to 65% and 41%, respectively, compared to the blank.	
Plai EO	Palm wood powder	Broccoli	*S. aureus* *E. coli* *L. monocytogenes* *Salmonella* Typhimurium TYMC TAB	The sachet inhibited microorganism growth in broccoli for at least 20 days. It demonstrated complete bacteria inhibition for at least two uses, with an 80% reduction after three uses. Additionally, it controlled TAB and TYMC growth in broccoli, meeting microbiological criteria for up to 14 days.	Matan and Kongchoosi (2023)
Clove and/or cinnamon EO	Polyvinylpyrrolidone-carboxymethyl cellulose-bacterial cellulose-guar gum (PVP-CMC-BC-GG) based hydrogel films	Gouda cheese	*S. aureus* *E. coli* *Bacillus cereus* (*B. cereus*) *Klebsiella.* spp. *Candida albicans* (*C. albicans*) *Aspergillus.* spp.	The use of sachets prolonged the shelf life of cheese by over 10–12 days compared to conventional polyethylene (PE) packaging. Cinnamon EO exhibited superior inhibition properties compared to clove oil and was effective at a lower concentration.	Bandyopadhyay et al. (2020)
Clove EO	Paraffin wax	Peach	*Fusarium sporotrichioides* (*F. sporotrichioides*)	The sachet achieved a maximum 96.31% reduction in in vitro spore germination of pathogenic fungi. In vivo, it significantly reduced lesion diameter, disease severity, and extended the shelf life of peaches by over 2 weeks against *F. sporotrichioides* at ambient temperatures (7–38.5 °C).	Irshad et al. (2023)

(continued)

Table 1 (continued)

Antimicrobial agent	Carrier	Food/simulant	Target microorganism	Antimicrobial activity	References
Oregano EO	Starch/Agave fructans	Avocado	*C. gloeosporioides* *Colletotrichum acutatum* (*C. acutatum*) *Diaporthe passiflorae* (*D. passiflorae*) *Neoscytalidium hyalinum* (*N. hyalinum*)	Oregano oil sachets completely prevented the growth of various fungi in vitro at 30 °C for 12 days. On avocados inoculated with *C. gloeosporioides*, the sachets decreased anthracnose infection without affecting the color or firmness of the fruit, relative to the control	Colín-Chávez et al. (2022)
Litsea cubeba EO	Bamboo	Dried fish	*Aspergillus flavus* (*A. flavus*)	Results indicated that the sachet, reused up to three times, completely inhibited *A. flavus* on dried fish, extending shelf life to 14 days relative to the control sample lasting three days.	Suhem et al. (2022)
Peppermint EO	Paraffin-coated activated carbon	Brown rice	*A. flavus*	Sachet displayed high antifungal activity against *A. flavus* on brown rice surfaces. Additionally, the shelf life of brown rice was extended from 10 to a minimum of 60 days in tropical climates	Chaemsanit et al. (2019)
Michelia alba EO	HDPE Polypropylene (PP)/PE Polyamide (PA)/PE PLA Cellophane	Culture media / and brown rice	*A. flavus*	Antifungal volatiles from HDPE and PP/PE sachets with 300 μL of EO completely inhibited *A. flavus* growth on culture media and brown rice for at least 20 days under accelerated	Matan et al. (2023)

				conditions (25 °C and 80% RH). Treated brown rice showed no spore germination or deformed hyphae, unlike the control with swelling spores and regular structure.	
Eugenol EO Carvacrol EO *Trans*-anethole EO	PLA pellets	Organic ready-to-eat iceberg lettuce	Coliforms	Eugenol, carvacrol, and *trans*-anethole sachets exhibited antimicrobial activities against coliforms, with reductions in log CFU g^{-1} of −1.38, −0.91, and −0.93, respectively, compared to the blank.	Wieczyska and Cavoski (2018)
Clove EO	Chitosan	Dry cake	*S. aureus* *E. coli*	When chitosan was used with clove EO in a ratio of more than 1:1, it fully stopped the growth of *E. coli* and *S. aureus*. The dry cakes, containing active capsules, showed decreased bacterial growth until the tenth day of storage, in comparison to the control..	Sharma et al. (2023)
Garlic and holy basil EO	Chitosan-coated pectin beads	Chilled stir-fried pork with basil leaves on rice	*S. aureus* *E. coli* *L. monocytogenes* *P. fluorescens* *B. cereus* *Clostridium perfringens* (*C. perfringens*)	Beads, at 2.20 and 4.40 g, efficiently suppressed the growth of all studied bacteria.	Sriwattana et al. (2023)

(continued)

Table 1 (continued)

Antimicrobial agent	Carrier	Food/simulant	Target microorganism	Antimicrobial activity	References
AIT	Calcium alginate beads	Spinach leaf	*E. coli* TYMC	*E. coli* and TYMC were similarly sensitive to AIT vapor. On spinach leaves with an initial count of 5.6 log CFU/leaf, *E. coli* reduced by 1.6–2.6 log CFU/leaf at 4 °C and 2.1–5.7 log CFU/leaf at 25 °C during 5 days. Reductions were markedly higher at 85% relative humidity (RH) than at 43% RH. Control leaves showed reductions not exceeding 1.2 log CFU/leaf in any testing conditions.	Seo et al. (2012)
AIT	HDPE-based resin	Peanut	*A. flavus*	A tenfold decrease in *A. flavus* survival occurred within 1 week, decreasing by 4.81 log cycles after 60 days. No traces of AITC were detected in the grains throughout storage, and volatile AITC in headspaces declined, with 92.4% reduced within the first 15 days and none detected after 30 days.	Otoni et al. (2014)
AIT	PE pouch containing mustard flour plus water/Filter paper containing AITC that sandwiched between EVOH (external side) and an adhesive layer of low density polyethylene (LDPE) (inner side)	Mozzarella cheese	*Penicillium digitatum* (*P. digitatum*) *Aspergillus parasiticus* (*A. parasiticus*)	AIT-treated samples displayed no fungal growth for 60 days. Aflatoxin B1 was only present in the control in trays.	Tracz et al. (2018)

AIT	Filter paper containing AITC that sandwiched between EVOH (external side) and an adhesive layer of LDPE (inner side)	Chilled pizza	*Penicillium nordicum* (*P. nordicum*)	The lowest AITC dose in vitro (0.125 μL/L) reduced mycelial growth by 66%, while higher doses suppressed visual growth for 7 days. In pizza, all tested concentrations substantially delayed fungal growth in a dose-dependent way. AITC stickers have the capability to prolong the shelf life of chilled pizza.	Lopes et al. (2018)
AIT	Filter paper soaked with AITC	Mung bean sprout and fresh-cut iceberg lettuce	TYMC Total mesophilic aerobic bacteria *Pseudomonas* spp. Enterobacteria LAB	Results indicate that AITC (0.81–1.41 μg/mL) has powerful antimicrobial characteristics against the native microflora of iceberg lettuce and mung bean sprouts. However, its application significantly affected the quality of both products. Respiration activity decreased, but discolorations and texture changes were noticeable even at minimal concentrations (0.11–0.16 μg/mL), which did not exhibit antimicrobial properties.	Kramer et al. (2018)
ClO_2	ClO_2 gas generating dry powder in zip lock paper/PE pouches	Raw chicken breast	*Salmonella enterica* subspecies enterica serovar Typhimurium (*S. Typhimurium*) *L. monocytogenes*	The greatest decrease in modified atmosphere packaging (MAP) plus ClO_2 (compared with MAP alone) was 0.68 log CFU g^{-1} for *S. Typhimurium* and 1.87 log CFU g^{-1} for *L. monocytogenes*.	Shin et al. (2011)

(continued)

Table 1 (continued)

Antimicrobial agent	Carrier	Food/simulant	Target microorganism	Antimicrobial activity	References
ClO_2	Sodium chlorite ($NaClO_2$) precursor and a ferric chloride acid ($FeCl_3$) activator packed in gas-permeable sachets	Black pepper, cumin, and sesame seed	*Salmonella enterica* (*S. enterica*)	*S. enterica* numbers on treated spices decreased by up to 2.50 log CFU g^{-1} over 30 days of storage, relative to the blank sample.	Golden et al. (2019)
ClO_2	$NaClO_2$ precursor and a $FeCl_3$ activator packed in gas-permeable sachets	Strawberry and blueberry	*S. enterica* *E. coli* *L. monocytogenes*	After 3 h, gaseous ClO_2 treatment reduced *E. coli*, *L. monocytogenes*, and *S. enterica* levels on strawberries by 2.0 ± 0.83, 3.5 ± 0.18, and 2.9 ± 0.20 log CFU g^{-1}, respectively. Similarly, on blueberries, the reduction was 2.3, 2.1 ± 0.28, and 2.2 ± 0.63 log CFU g^{-1} for *E. coli*, *L. monocytogenes*, and *S. enterica*, respectively.	Luu et al. (2021)
ClO_2	$NaClO_2$ packet in HDPE fibers sachet	Tomato	*S. Typhimurium*	The generated gaseous ClO_2 significantly decreased surface-inoculated *S. Typhimurium* by 4 log CFU/fruit to a level below detection (<5 CFU/fruit) within 24 h. It had no impact on the surface color and firmness of the fruits during the entire 144-h storage period at 22 °C.	Zhou et al. (2018)
ClO_2	ClO_2 slurry packed in semi-permeable plastic film	Cherry/grape tomato	*A. alternata* *E. coli* *S. enterica*	The sachet application decreased *E. coli* and *A. alternata* by 2.9–4.7 and 1.6 to 4.0 log CFU g^{-1}, respectively, within 14 days of storage at 20 °C.	Sun et al. (2017a)

ClO_2	ClO_2 slurry packed in semi-permeable plastic film	Grape tomato	*A. alternata* *E. coli*	The ClO_2 pouch demonstrated significant antimicrobialproperties, reducing *E. coli* by 3.08 log CFU g^{-1} and *Alternaria alternata* by 2.85 log CFU g^{-1} after 14 days of storage.	Sun et al. (2017b)
ClO_2	Extruded EVA with citric acid (CA), sprayed with a $NaClO_2$ solution, and heat-pressed afterwards	Culture media	*Salmonella enterica* subspecies enterica serovar Montevideo (*S. enterica*)	Total inactivation of *S. enterica* (2.13 log CFU/plate) was achieved with 15% CA labels in 2 h; 7.5% CA labels partially inactivated (1.03 log CFU/plate) in up to 6 h.	Saade et al. (2017)
ClO_2	ClO_2 aqueous solution was encapsulated in gelatin by the double emulsion method and then incorporated into a PVA matrix	Blueberry	*E. coli* *L. monocytogenes* *Kytococcus sedentarius* (*K. sedentarius*) TYMC	The inhibition rates for *L. monocytogenes* and *E. coli* were 93.69% and 95.55%, respectively, with successful inhibition of *K. sedentarius*. ClO_2 self-releasing films applied to blueberry preservation slowed quality degradation and prevented mold contamination during a 14-day storage period. Blueberries stored in the antibacterial film group showed increased anthocyanin accumulation.	Su et al. (2023)
Ethanol	Distillery ethanol/brandy added in cotton/cellulose absorbent cloths	Pomegranate aril	LAB TYMC	Vapors from both volatile antimicrobials significantly inhibited LAB and TYMC at all storage temperatures. For	Kapetanakou et al. (2015)

(continued)

Table 1 (continued)

Antimicrobial agent	Carrier	Food/simulant	Target microorganism	Antimicrobial activity	References
				example, at 4 °C on day 23, when TVC on the controls reached 6.9 log CFU g^{-1}, the counts on arils treated with distillery ethanol or brandy were as follows: 4.9 log CFU g^{-1} (1 mL ethanol) > 3.9 log CFU g^{-1} (1 mL brandy) > 2.2 log CFU g^{-1} (2 mL ethanol) > 1.2 log CFU g^{-1} (2 mL brandy)	
Ethanol	Alcoholic beverages in cotton/ cellulose absorbent cloths	Pork meat	TPC *Pseudomonas* spp. *Brochothrix thermosphacta* (*B. thermosphacta*) LAB TYMC Enterobacteriaceae	All applied alcoholic beverages were effective against LAB, *Pseudomonas* spp., and *B. thermosphacta*.	Kapetanakou et al. (2014)
Ethanol	Commercial ethanol emitter	Chinese steamed bread	TPC TYMC	On day 2 of storage, the TPC was reduced by 3.7 and 4.4 CFU g^{-1} by ethanol (1% v/v) and ethanol (3% v/v), respectively. TYMC were inhibited throughout the storage period.	Sheng et al. (2015)

Table 2 Summary of selected studies on the application of antimicrobial absorbent pads in food packaging systems

Antimicrobial agent	Antimicrobial concentration	Absorbent material	Food matrix	Target microorganism	Antimicrobial activity	Reference
Bacteriophages (BFSE16, BFSE18, PaDTA1, PaDTA9, PaDTA10, and PaDTA11)	The pads were treated with 20%, 50%, and 80% volumes of a bacteriophage solution with a concentration of 10^9 plaque forming units (PFU) mL^{-1}, proportional to the mass of the pads.	Commercial absorbent pad	Culture media	*S. Typhimurium*	Three phage concentrations in pads reduced log cycles by 4.36, 3.66, and 0.87 at 15 °C, with an average of 0.55 at 10 °C. Higher concentrations had a more significant impact, showcasing increased infection capacity. Viable phages remained on the pad during the 48-hour treatment.	Gouvêa et al. (2016)
Rosemary and oregano EOs	1% or 5% v/v in the polymer solution	Electrospun fiber pad of cellulose acetate	Culture media	*S. aureus* *E. coli* *C. albicans*	Pad with 5% v/v oregano oil demonstrated the most effective antimicrobial and antibiofilm properties, particularly against *E. coli* and *C. albicans*.	Liakos et al. (2017)
Carvacrol, thymol, linalool, and *a*- and *β*-pinene EOs	Two emulsions named Etnic (60% thyme, 30% oregano, cumin) and Mediterranean (60% thyme, 40% oregano) were prepared and a volume of 400 mL/m^2 of emulsion was sprayed onto the pads	Cellulosic pads	Minced beef meat	*Pseudomonas putida* (*P. putida*) *Pseudomonas fragi* (*P. fragi*) *P. fluorescens* *Enterococcus faecalis* (*E. faecalis*) *Lactococcus lactis* (*L. lactis*) *S. enterica* *Campylobacter jejuni* (*C. jejuni*) *S. aureus*	Mediterranean and Etnic emulsion pads induced a consistent 0.3–0.8 Log10 CFU g^{-1} decrease in psychrophilic microbiota in minced meat and hamburger stored for 12 and 15 days.	Agrimonti et al. (2019)

(continued)

Table 2 (continued)

Antimicrobial agent	Antimicrobial concentration	Absorbent material	Food matrix	Target microorganism	Antimicrobial activity	Reference
Cinnamon EO	313.07 μL/g in artificial nano-porous alumina silicates (Permutite)	Permutite composite containing Ag^+, Zn^{2+} and Ag^+/Zn^{2+} loaded with EO were incorporated into the ethyl cellulose pad	Chinese bayberry	*A. niger* *Penicillium* sp. (P. sp.)	The Ag^+/Zn^{2+}-permutite/cinnamon EO antimicrobial pads showed significantly less decay compared to the control group.	Niu et al. (2018a)
Lysozyme	Nano-hybrid loading 1.0, 2.5, 5.0 wt %	Electrospun fiber pad of PA11 containing halloysite nanotubes (HNTs) filled with lysozyme	Chicken slice	*Pseudomonas* spp. (*P.* spp.)	The initial bacterial count was low but increased by four orders of magnitude for the PA11 membrane pad and two orders for the pad with a composite filled with 5 wt % HNTs–lysozyme after 6 days. Over time, the bacterial count consistently rose, always lower for the composite material. Notably, the PA11+ 5 wt % HNTs–lysozyme sample reached a plateau at 13 days, one order of magnitude lower than the PA11 (control) sample.	Bugatti et al. (2018)
N-halamine	Soaked 2 g of absorbent pad cellulose in 10 mL of 0.01%, 0.05%, and 0.1% *N*-halamine solutions in 95% ethanol for 1 min.	A highly absorbent fluff pulp layer, separated from food products by a non-permeable top layer	Raw beef	Table *P.* spp. LAB *Enterobacteriaceae*	The pads significantly lowered microbial loads in meat samples in comparison to the control on day 11, reducing by an average	Ren et al. (2018)

		of PE film perforated with one-way valves			of 1 log CFU g^{-1}. They effectively extended the shelf life of refrigerated raw beef.	
Pinosylvin (PS)	Pads with 0.08–0.8 mg/cm^2 of pinosylvin	Perforated cellulose/PP absorbent pads	Chicken meat	*Campylobacter* spp. (*C.* spp.) LAB TPC *P.* spp.	At 4 °C, pads with 0.4 mg PS/cm^2 showed anti- *C.* spp. activity in chicken fillets and exudates. Despite no impact on *P.* spp., these coated pads reduced LAB, and TPC.	Silva et al. (2018)
Ag NPs	Mixed 2 g of cellulose from absorbent pads with 200 mL of freshly prepared 1% w/v silver nitrate ($AgNO_3$) stock solutions for 1 h under gentle magnetic stirring	A single PE layer	Fresh-cut melon or kiwi Poultry and beef meat	TPC LAB TYMC	Ag-loaded absorbers reduced microorganism and TYMC by up to 99.9% in kiwi fruit and melon juices. In poultry and beef meat drips, microbial loads remained 90% below controls.	Lloret et al. (2012)
ZnO, copper oxide (CuO), and Ag NPs	Immersed 100 mg of cotton pads into a 50 mL NPs suspension containing 10 mg of each NP (ZnO, CuO, and Ag)	Cotton pads	Culture media	*S. aureus* *E. coli* *L. monocytogenes* *Staphylococcus epi dermis* (*S. epidermidis*)	ZnO and Ag NPs were more efficient against Gram-negative (*E. coli*) than Gram-positive bacteria (*L. monocytogenes*, *S. epidermis*, and *S. aureus*). However, CuO NPs showed similar antibacterial activity against both types.	Shankar and Rhim (2017)

(continued)

Table 2 (continued)

Antimicrobial agent	Antimicrobial concentration	Absorbent material	Food matrix	Target microorganism	Antimicrobial activity	Reference
Ag NPs	Mixed 2 g of cellulose from absorbent pads with 200 mL of double-distilled water containing 1% and 2.5% $AgNO_3$ for 1 h	Cellulose fiber pad	Beef meat	TAB LAB *P.* spp. *Enterobacteriaceae*	The pad reduced major microbial groups by an average of 1 log CFU g^{-1} throughout storage. Ag ions significantly lowered TAB and *P.* spp., while LAB showed less sensitivity. *Enterobacteriaceae* levels remained below detection in the presence of silver.	Fernandez et al. (2010)
Ag NPs	–	PLA fiber	Culture media	*E. coli*	Pad reduced *E. coli* by approximately 4 log.	Durdag et al. (2014)

EOs, have a strong or unpleasant odor that can adversely impact consumers' sensory perception when opening the packaging. Most reports on applications of antimicrobial emitting sachets do not take into account the sensory impact (Moshe Dvir et al., 2019).

In the case of applying NPs as antibacterial agents, consumer acceptance poses a challenge. Research indicates that consumers remain resistant to accepting novel approaches based on nanotechnology for food purposes (Henchion et al., 2019; Kalia & Parshad, 2015). Furthermore, Loose sachets and pads might be confused with food or one of its ingredients, like salt or dressing, posing a potential risk (Dainelli et al., 2008). To mitigate the risk of unintentional ingestion, they must be clearly labeled with a "do not eat" sign.

Some studies have indicated that consumers are reluctant to use sachets due to concerns about possible accidental ingestion, the potential for disintegration and contamination of food, and a reluctance to perceiving anything other than the actual food (Aday & Yener, 2015). Lastly, the constraints also include unsuitability for liquid foods, since the contact with the sachet material could result in spillage of their contents (Rooney, 2005). The application of emitting sachets or absorbent pads in food packaging offers significant benefits, such as reducing food waste and extending shelf life, both of which positively impact sustainability. However, these components can hinder the recyclability of the main packaging system, as they are often made from different materials that are not recyclable. If they cannot be easily separated, they can complicate the recycling process. Therefore, to enhance sustainability, further research is needed to develop biodegradable or recyclable alternatives and to assess their impact on the entire packaging system (Priyanka et al., 2023; Alves et al., 2023).

4.2 Antimicrobial Coated Films on the Packaging Material

In antimicrobial food packaging, antimicrobial agents can either be integrated directly into polymers to create antimicrobial packaging materials or applied as coatings on polymer surfaces, as shown in Fig. 1c. The former method often involves thermal processing, such as film extrusion or compression molding. Consequently, it is best suited for heat-stable antimicrobial agents, and a relatively higher content of these agents is usually necessary to achieve optimal effectiveness. While this direct incorporation negatively alters bulk properties like mechanical or barrier characteristics in most cases, it can also be advantageous by enhancing specific features of the polymer. For instance, the inclusion of nanoparticles into the polymer matrix is recognized as an effective approach for reinforcing mechanical strength and enhancing the barrier properties (Zabihzadeh Khajavi et al., 2020).

On the flip side, the development of antimicrobial coatings employs nonthermal technology, providing the flexibility to choose from a wider range of antimicrobial agents. This approach preserves the main characteristics of the polymeric substrate. As a nonthermal technique, surface antimicrobial coating presents itself as a viable alternative to traditional thermal film-making processes (Sung et al., 2013). Various methodologies have been explored for engineering antimicrobial coatings on polymer surfaces, yielding promising results at the laboratory scale. However, transitioning these approaches from lab-scale concepts to mass production poses a significant challenge (Tiwari, 2017). In the subsequent sections, both laboratory and potentially scalable coating methods for antimicrobial-coated films will be introduced and explored.

4.2.1 Advanced Laboratory Coating Techniques

Various methods have been employed for the development of antimicrobial coatings at the laboratory level. These techniques typically involve straightforward setups; however, some of these approaches can be scaled up by integrating them into a continuous manufacturing system. Figure 2 illustrates different laboratory-scale coating manufacturing techniques, which will be elaborated on in the following sections.

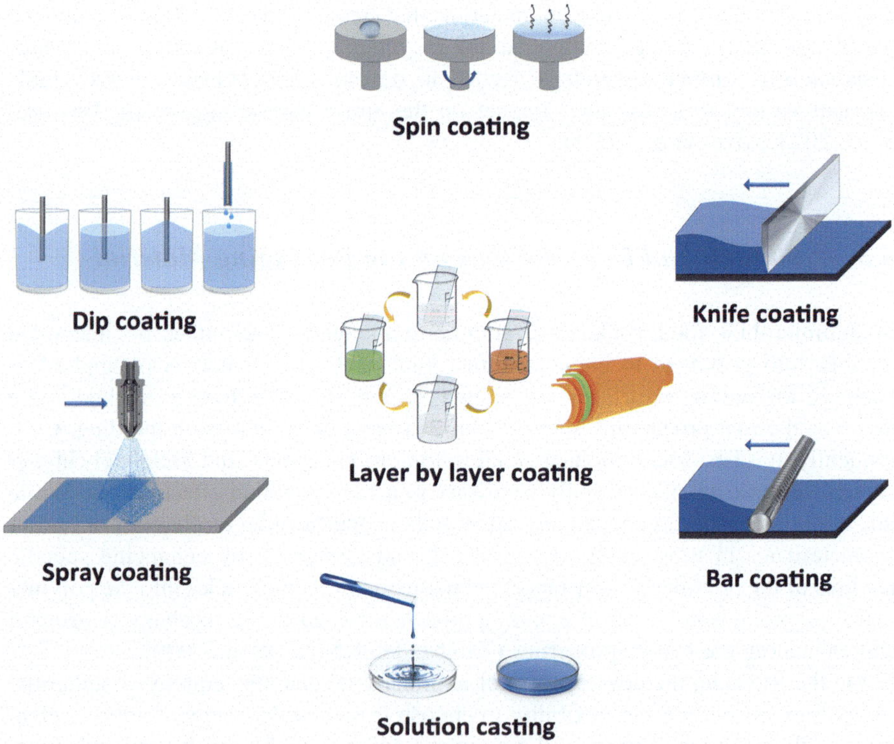

Fig. 2 Laboratory methods for creating antimicrobial coatings on bulk packaging materials

Knife Coating Knife coating, also referred to as the spread or blade coating method, is a straightforward approach that deposits coating material onto a polymeric substrate utilizing tools like a doctor knife, a blade, or a spreader. In this method, the thickness of the coating is adjusted by the gap size between the knife and the substrate, the movement speed of the knife or the substrate, as well as the rheological features of the coating solution (Shim, 2013). However, this method can only deposit coatings with a thickness above 10 μm, and thinner coatings are not feasible. Generally, this technique is considered an economically simple method for manufacturing approximately thick and uniform coatings on flexible substrates, with a medium level of solution waste. The coating deposition speed in this approach is fast, reaching up to 10 m/min, but the required drying time for the deposited coating without applying a heating process is relatively slow and can last for hours. Because of its ease of use and affordability, and high yield, this technique can be scaled up for industrial applications (Gooch, 2010).

Bar Coating Bar coating, also referred to as the rod or Mayer bar coating method, bears resemblance to knife coating. In this method, an excessive amount of coating material is initially placed on the substrate and is subsequently spread across using a bar or rod entwined with wire. The thickness of the coating in this technique can be adjusted by the wire diameter and the rheological characteristics of the coating solution, with the coating thickness limited to values higher than 1 μm. Similar to knife coating, the drying step in this method is slow, with a medium level of solution waste. However, in bar coating, the deposition speed is slower and constrained to 2–15 mm/min. Despite the method's low cost and simplicity, scalability for industrial levels is limited, primarily due to its low deposition yield (McKeen, 2017).

Spray Coating In this method, the coating solution or suspension is atomized into droplets using a spray jet, and then transferred through the nozzle towards the substrate via a carrier gas to form the coating. This approach is commonly employed on both lab and industrial scales. It offers several advantages, including low solution waste, fast deposition process ((in seconds), and a rapid drying process (10–30 min). The morphology and thickness of the coating can be adjusted by different operating parameters, like the applied spray jet, nozzle configuration, pressure of the carrier gas, distance between the nozzle and the substrate, deposition time, and the features of the coating solution or suspension, such as viscosity, particle size, and the volatility of the solvent. Generally, coatings ranging from nanometers (25 nm) to several hundreds of micrometers can be deposited using this method (Moridi et al., 2014; Pawlowski, 2009).

Dip Coating Dip coating is extensively utilized in small-scale research for creating thin coatings. This method comprises five steps:

1. Immersion: The substrate is immersed in a coating solution at a fixed speed.
2. Dwelling: The immersed substrate remains in contact with the coating solution for a specific required time.
3. Withdrawal: The substrate is removed from the coating solution.
4. Drying: Excess coating material is removed from the substrate, and the solvent evaporates, leading to the formation of a dry, thin coating on the substrate.

The coating thickness can be adjusted by the removal speed and the characteristics of the coating solution, ranging from 20 nm to 50 μm. However, this technique has limitations for scaling up to industrial applications, such as a low deposition speed (0.6–3 m/min), a slow drying process (orders of hours), and high solution waste (Dastan et al., 2016).

Solution Casting In this approach, the coating solution is poured onto the surface of the desired substrate. The thickness of the coating can be adjusted by both the characteristics of the coating solution and the applied volume of the solution on the surface area of the substrate, ranging from 50 to 1000 μm. This technique boasts a high coating deposition speed and low solution waste. However, the drying step of the deposition could last for hours, inhibiting its scalability for commercial usage (Mathew & Oksman, 2014).

Spin Coating In the spin coating approach, the coating solution is applied to the centre of a substrate and evenly distributed by spinning at a high velocity. This method is limited to small, flat substrates, making it suitable only for laboratory-level applications. Uniform coatings with thickness varying from micrometers to nanometers can be achieved through this technique, depending on the features of the coating solution, spin speed, and time. However, it is important to note that this method comes with high solution waste and a low coating deposition speed (Mouhamad et al., 2014).

Layer by Layer Coating Layer by Layer (LbL) coating presents an economical and straightforward technique for manufacturing multilayer coatings on a substrate. In this method, multiple alternating layers can be applied on the substrate through various interactions such as electrostatic, hydrophobic, charge-surface, cation-dipole interactions, or hydrogen bonds, or a combination of these forces. One of the most common approaches for developing LbL coatings through electrostatic interactions involves the alternating immersion of a charged substrate into oppositely charged coating solutions, with washing steps in between and an optional drying process, contributing to the formation of a continuous layering of positively and negatively charged materials. Various techniques, such as dip, spin, or spray-assisted deposition methods, can be applied to form the coating layers. The coating thickness in this method can be varied from nano to microscale, depending on the characteristics of the coating solution (such as temperature, ionic strength, concentration, and pH), the number of deposited layers, and the approach applied to assist coating formation. The deposition speed, as well as the drying step of the deposition, is relatively slow, depending on the number of layers, and can last minutes to hours. In addition, the solution waste in this technique is high, and these mentioned disadvantages limit the scalability of this method (Ariga & Hill, 2007; Richardson et al., 2016).

4.2.2 Scalable Coating Techniques

The production of commercial antimicrobial coatings faces a key limitation—the absence of a continuous manufacturing method suitable for large-scale coating. Previous research in this domain has predominantly concentrated on innovating antimicrobial coatings through the previously mentioned lab-scale techniques, followed by the characterization of the physicochemical properties of the resulting coatings. However, some of these methods prove unsuitable for continuous mass production (Widsten et al., 2017).

In Fig. 3, a few scalable techniques that hold promise for continuous production are summerized. These methods will be introduced and elaborated upon in the subsequent sections.

Electrospinning/Spraying Electrospinning and electrospraying represent electrohydrodynamic methodologies applying electric potential to produce ultrafine fibers or particles within the micro- to nano-scale range. This technique, characterized by its simplicity and efficiency, emerges as a suitable choice for fabricating

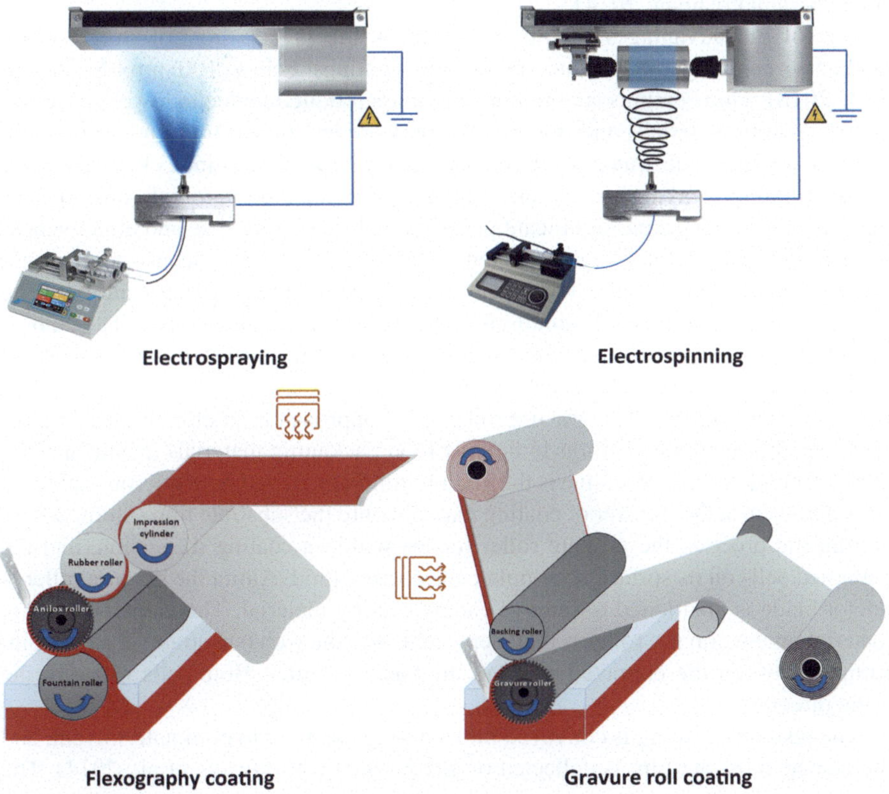

Fig. 3 Scalable methods for creating antimicrobial coatings on bulk packaging materials

antimicrobial coatings on both laboratory and industrial scales (Alghoraibi & Alomari, 2018). As indicated in Fig. 3, the apparatus comprises three main components: a spinneret or emitter equipped with a pump, a grounded collector, and a high-voltage power supply interfacing between these constituents. In this technique, a polymer solution containing antimicrobial agents is pumped through the spinneret or emitter, leading to the formation of a stable Taylor cone. This cone is stabilized by liquid surface tension, electrostatic force, and gravity. In **the electrospray technique**, where the polymer solution has low viscosity, the electrostatic stretch over the surface tension in the formed cone is relatively low, resulting in the formation of small, highly charged liquid droplets that radially disperse due to Coulomb repulsion. On the other hand, in **electrospinning**, where the polymer solution has higher viscosity, the electrostatic force reshapes the droplet into a pointed form, creating a charged fibrous jet. The selected substrate can be placed on the collector so that the electrospun fibers or electrosprayed particles can be continuously deposited on it. The diameter of the polymer fiber or particles coated on the substrate can be regulated by factors such as applied voltage, speed of polymer solution pumping, emitter geometry, distance between emitter and collector, collector rotational speed, temperature, and properties of the polymer solution (Bhushani & Anandharamakrishnan, 2014).

One of the advantages of this technique is the precise control over coating thickness, enabling the production of thin coatings below 100 nm (Munteanu et al., 2018). Thin coatings are preferable in mass production due to lower consumption of coating materials, and importantly, they are less potent to influence the bulk features of the substrate. Additionally, antimicrobial coatings fabricated by electrospinning/spraying act as an optimal carrier for antimicrobials due to their high specific surface area, significant porosity, and small pore size, and effectiveness even at low antimicrobial agent content. Moreover, a coaxial electrospinning and electrospraying setup can be developed by implementing two emitters, allowing the deposition of two immiscible solutions on the substrate simultaneously. The coaxial method improves processability and can be expanded from a laboratory scale to an industrial level (Jaworek, 2007).

Gravure Roll Coating The gravure roll coating approach is an effective method for applying antimicrobial coatings to flexible food packaging materials (Bautista et al., 2017). Central to this technique is the gravure roller, engraved with micron-patterned cells that control the transfer of coating material onto the substrate film (Shim, 2019). During the process, the gravure roller rotates within a coating dispersion, and the patterned cells on its surface accumulate the coating fluid. Along the gravure roller, a doctor blade is positioned to remove excess coating material. The gathered coating material on the gravure roller is then deposited onto the web (substrate) at a nip point formed between the engraved roll and the backing roller. Both rolls rotate in the same direction.

The resulting coating is conveyed into a drying chamber to eliminate solvent, and the coated substrate film is collected on the product roll (Hwang et al., 2014). The thickness of the coating is precisely adjusted by the speed of the backing roller, the web speed, and the speed and diameter of the gravure roller. Remarkably, this method allows for the production of extremely thin coatings, with thicknesses as low as 20 nm. The deposition process in this approach is rapid, and the waste of coating material is minimal, turning it into a highly appropriate method for commercial purposes. In the field of food packaging, gravure roll coating has been thoroughly studied for depositing barrier coatings on packaging materials, as well as for printing. In the packaging materials printing system, an additional roll for supplying antimicrobial coating material can be integrated. This allows the system to simultaneously print and apply an antimicrobial coating to the packaging material (Park et al., 2016; Sharma et al., 2022b).

Flexography Coating Flexography is a widely used method for printing and adding antimicrobial coatings to flexible packaging films. Similar to gravure roll coating, it requires a low viscosity coating material. As can be seen in Fig. 3, a fountain roller dips into a tray of coating material. As it turns, the roller collects the material and transfers it to the anilox roller, which has tiny cells engraved on it to hold the coating fluid. A doctor blade on top removes excess fluid, and the remaining is sent to the rubber roller. The substrate moves between the rubber roller and the impression cylinder. Pressure from the impression cylinder spreads the coating fluid onto the film surface. The coated film then goes through a drying unit to make the end product (Hewson et al., 2006). The size and design of the anilox roller cells are crucial for a precise coating transfer to the rubber roller. This method efficiently produces a very thin, smooth coating on the film in a short time period (Ayers, 2018).

4.2.3 Immobilized Antimicrobial Agents on the Surface of Packaging Material

While multiple reviews often categorize the immobilization of antimicrobial agents onto the surface of a substrate as a distinct approach for developing antimicrobial materials (Khaneghah et al., 2018; Sung et al., 2013), it can also be viewed as a coating technique. In this method, antimicrobial agents are applied to the surface of a substrate, resembling a coating or coverage, as indicated in Fig. 1 (B type) (Fu & Dudley, 2021). Surface immobilization occurs through covalent or noncovalent bonds between antimicrobial agents and the substrate. This approach creates a nonmigrating packaging system, necessitating direct contact with the food matrix. Compared to other methods in antimicrobial packaging engineering, the risk of consumer ingestion is significantly lower due to limited migration.

To immobilize antimicrobial agents onto the substrate's surface, compatible functional groups are essential for both the agent and substrate. Antimicrobial agents such as bacteriocins, chitosan, lactoferrin, lysozyme, organic acids, and polyamines are suitable options because of their functional groups (Khaneghah et al., 2018). However, common polymeric materials in the food packaging industry often have inert surfaces with low surface energy levels, leading to inadequate bonding capacity with antimicrobial agents. Consequently, a surface activation step is usually necessary before immobilizing antimicrobial agents. Physical methods, like corona treatment, ultraviolet light, gamma irradiation, and surface plasma treatment, are typically favored over chemical approches because of their environmental sustainability and potential for industrial scalability. Once activated, the polymer surface contains reactive groups, such as carbonyl, hydroxyl, carboxylic acid, or nitrogen-containing groups, allowing selected antimicrobials to be immobilized onto the activated polymer surface through covalent or noncovalent means (Goddard & Hotchkiss, 2007). Noncovalent immobilization arises from electrostatic interactions between the activated substrate and antimicrobial agents with opposite charges. On the other hand, covalent immobilization involves the direct formation of covalent bonds between antimicrobials and the activated polymer surface. Occasionally, the formation of covalent attachments or conjugations may require a spacer or crosslinker. Common spacers for this purpose include macromolecules like dextran, chitosan, polyethylene glycol (PEG), ethylenediamine, and polyethyleneimine (PEI), capable of facilitating a shared covalent bond between the substrate and the antimicrobial. They can also facilitate covalent bond formation between the activated film and the antimicrobial without becoming part of that bond. It is important to note that covalent bonding of antimicrobial agents can impact the conformation of certain antimicrobials, such as peptides, potentially resulting in reduced antimicrobial activity (Goddard & Hotchkiss, 2007). It has been shown that bacteriophages can be covalently immobilized on polymer surfaces to serve as antimicrobial agents. In this patent, three bacteriophages—Fluorescens (target host: *P. fluorescens*), ϕPFS (target host: *P. fluorescens*), and ϕLAX (target host: Leuconostoc amnigenum)—were used at concentrations of 3×10^9, 2×10^9, and 1.1×10^{10}, respectively. Each bacteriophage (2 mL) was sprayed onto a 20 cm × 20 cm corona-discharged PP film. After drying, the films were used as packaging for baby spinach. The results showed a significant extension in shelf life, and the packaging demonstrated antimicrobial activity against bacteria, including *Campylobacter*, *Salmonella*, *Pseudomonas*, *Pectobacterium*, and *Citrobacter* (Clark & Mattey, 2020).

4.2.4 Drawbacks and Constraints

As mentioned earlier, the majority of approaches for producing antimicrobial coatings are nonthermal or involve minimal thermal processes, primarily aimed at accelerating the drying phase. This is often brief due to the thin nature of the coatings. Consequently, a wide range of antimicrobial agents, including those with low thermal stability, can be integrated into the antimicrobial coatings. Tables 3 provide a summary of recent research on antimicrobial coatings developed for food applications. Despite the various advantages of antimicrobial coatings, the commercial availability of such coatings for food packaging applications is limited. Several factors contribute to this limitation in mass production. Firstly, compliance with regulatory requirements, such as migration of antimicrobial agents, food contact notification, and safety evaluations, poses challenges for engineered antimicrobial coatings seeking approval for use as food packaging material. For substances not

Table 3 Antimicrobial agents used in antimicrobial coatings for food packaging applications

Antimicrobial agents	Substrate	Antimicrobial concentration	Coating method	Food/ simulant matrix	Target microorganism	Antimicrobial activity	Reference
Clove EO	Linear low density polyethylene (LLDPE)	11 mg/cm^2 on LLDPE film	Knife coating	Minced chicken/ Tryptic soy broth (TSB)	*L. monocytogenes* *S. Typhimurium*	No bacteria survived on chicken after the fifth day at 4 °C. In TSB, a 4 to 5 log decrease occurred after day 1, and by day 7, bacterial survival was not observed	Mulla et al. (2017)
Molecule (MW 500–10,000) with structures from guanidine, glucosamine, or amino acids	Conventional polymers used for food packaging	0.2 mg/m^2–300 mg/m^2	Conventional coating method	Fish	TPC	Extending the shelf life of fish during cold storage.	Nakayama et al. (2018)
TiO_2 NPs	LDPE	0.05, 0.08, 0.11% (w/v) in coating solution	Bar coating	Lettuce	*E. coli*	TiO_2 NPs-coated films exhibited greater effectiveness against *E. coli* when exposed to UV light compared to those exposed to fluorescent light.	Othman et al. (2014)

(continued)

Table 3 (continued)

Antimicrobial agents	Substrate	Antimicrobial concentration	Coating method	Food/ simulant matrix	Target microorganism	Antimicrobial activity	Reference
Lauroyl arginate ethyl (LAE)	Polyethylene terephthalate (PET)	5, 10, 20% (w/w) of LAE in coating solution	Bar coating	TSB plus 0.6% yeast extract	*E. coli*	A minimum LAE concentration of 10% is required for total inhibition, achieving a reduction of more than 6 logs.	Apicella et al. (2019)
Ag NPs	LDPE	3% (w/w) of Ag precursor in coating solution	Spray coating	Mueller Hinton agar (MHA)/ Chicken breast fillet	*P. fluorescens* *S. aureus* *E. coli* Total coliforms LAB TPC	Extending the shelf life of packaged chicken from 6.5 to 9 days at 4 °C.	Azlin-Hasim et al. (2018)
Chitosan, ZnO NPs	LDPE	0.08 mg/cm^2 on LDPE film	Spray coating	Nutrient broth (NB)	*S. enterica* *E. coli* *S. aureus*	Complete inactivation of the studied pathogens in NB after 24 h of incubation at 37 °C.	Al-Naamani et al. (2016)
Carvacrol EO	LDPE	–	LBL	Chicken/ Phosphate buffered saline (PBS)	*Aeromonas hydrophila* (*A. hydrophila*) TAB	45% decrease in TAB for chicken stored at 4 °C for 24 h; 85% decrease in *A. hydrophila* viability in PSB after 24 h at 30 °C.	Tas et al. (2019)

Ag NPs	LDPE	0.5, 2, 5% (w/w) $AgNO_3$ in coating solution	LBL dip coating	MHA	*P. fluorescens* *S. aureus*	Ag NPs-coated LDPE film generated inhibitory zones ranging from 66 to 396 mm^2, dependent on AgNPs levels and the number of coatings.	Azlin-Hasim et al. (2016)
Chitosan, black cumin EO	PE	1% (w/v) of chitosan and black cumin oil in coating solution	LBL dip coating	Chicken/ Nutrient agar (NA)	*S. aureus* *E. coli* TAMB Psychrotrophic bacteria count (PBC)	Suppressed TAMB and PBC during 5-day storage at 4 °C for chicken; more efficient against *E. coli* compared to *S. aureus* on NA after 24 h at 37 °C.	Takma and Korel (2019)
Natamycin	PLA	0.18 to 0.25 mg/ dm^{2} on PLA film	Sol-gel dip coating	Semi-soft cheese	Indigenous mold on cheese	Natamycin-coated films effectively inhibit mold growth on cheese throughout a 30-day period at 4 °C.	Lantano et al. (2014)
ε-Poly(lysine) (ε-PL)	Bilayer cellulose/ chitosan	0.5% and 1% (w/v) of ε-PL in coating solution (sol)	Sol-gel dip coating	Tryptic soy agar (TSA)/ TSB/ Chicken breast meat	*L. monocytogenes*	Coatings with 1% ε-PL showed a 1.5 log decrease throughout a 12-day storage at 4 °C.	Lantano et al. (2014)

(continued)

Table 3 (continued)

Antimicrobial agents	Substrate	Antimicrobial concentration	Coating method	Food/ simulant matrix	Target microorganism	Antimicrobial activity	Reference
Zataria Multiflora Boiss. EO (ZEO)	Bilayer PP/zein	10% (w/w): ratio of ZEO to zein powder	Solution casting	Cow's milk/ TSB	*L. monocytogenes*, *E. coli*	Around a 1 log decrease on day 6 of milk storage at 4 °C; 2.99 and 3.08 log decrease of *L. monocytogenes* and *E. coli* in TSB after 6 days at 4 °C.	Kashiri et al. (2017)
ZEO	Bilayer PP/Carboxymethyl cellulose (CMC)	1% to 4% (v/v) of ZEO in CMC film-forming dispersion	Solution casting	MHA	*S. aureus* *E. coli*, *S. Typhimurium* *B. cereus* *Pseudomonas aeruginosa* (*P. aeruginosa*)	PP/CMC coated with 4% ZEO exhibited inhibitory zone diameters exceeding 20 mm for all tested bacteria.	Honarvar et al. (2017)
Chitosan	PE	2.3% (w/w) in coating solution	Electrospraying	Selective culture media	*Listeria* spp. (*L.* spp.) *E. coli* *Salmonella* spp. (*S.* spp.)	Achieved over 90% bacterial inhibition after 48 h of incubation.	Stoleru et al. (2016a)
Chitosan	PVA	Weight ratio of Chitosan: PVA = 25:75	Electrospraying	Culture media	*E. coli* *S. aureus*	Achieved 99% bacterial inhibition after a 24 h incubation at 37 °C.	Liu et al. (2018)

Chitosan, clove EO	PLA	Chitosan and clove oil: both 1.5% w/w in coating solution	Electrospinning	Soybean casein digest lecithin polysorbate medium (SCDLP)	*L. monocytogenes* *E. coli* *S. Typhimurium*	Attained 100% inhibition of *L. monocytogenes* and *S. Typhimurium*, along with 91% inhibition of *E. coli* after a 48 h incubation.	Munteanu et al. (2018)
Eugenol EO	Bilayer poly (3-hydroxybutyrate) (PHB) / poly (3-hydroxybutyrate-*co*-3 hydroxyvalerate) (PHBV)	Weight ratio of eugenol: PHBV = 15:85	Electrospinning	TSB	*E. coli* *S. aureus*	Achieved a reduction of over 3 logs after a 24 h incubation at 24 °C.	Figueroa-Lopez et al. (2020)
Oregano EO Citral EO	Bilayer PP/EVOH	7.5% (w/w) in EVOH coating solution	Gravure roll coating	Salad	TAB PBC TYMC Total enterobacteria	Oregano and citral EO-coated films demonstrated reductions of 1.38 log and 2.13 log, respectively, against enterobacteria, 1.08 log and 1.23 log against TAB, and approximately 2 log against TYMC and PBC for salad following 1 day storage at 4 °C.	Muriel-Galet et al. (2013)

(continued)

Table 3 (continued)

Antimicrobial agents	Substrate	Antimicrobial concentration	Coating method	Food/ simulant matrix	Target microorganism	Antimicrobial activity	Reference
Nisin	LLDPE	12.97 IU/cm^2 on LLDPE	Gravure roll coating	TSA	*Listeria innocua* (*L. innocua*) *Micrococcus luteus* (*M. luteus*)	Efficiently suppressing the growth of *L. innocua* and *M. luteus* on TSA.	Perna (2016)
Mango seed extract, tannic acid, feijoa skin extract	Cryovac® BLUS 50	20% (w/w) of plant extract in coating solution	Flexography coating	MHA or TSA plus 0.6% yeast extract	16 common spoil-age and pathogenic bacteria of fish and meat	Tannic acid was found to be the most effective phenolic plant extract examined, resulting in the largest inhibitory zone of 8.3 mm.	Widsten et al. (2017)
Nisin	PET/LLDPE	–	Flexography coating	Turkey deli meat	TAB Total coliforms *S. aureus*	Nisin-coated film achieved a 2 log decrease in TAB on day 5 of turkey storage at 5 °C, although it was not effective against total coliforms.	Ayers (2018)
Lactoferrin	PLA	136 μg/mL in coating solution	Immobilization	Specific culture media	*L. monocytogenes* *E. coli* *S. Typhimurium*	All tested bacteria were completely inhibited after a 48 h incubation period.	Stoleru et al. (2016b)

N-halamine Cationic primary amine groups (polyethyleneimine (PEI)) used as precursor)	PP/Polypropylene-*graft*-maleic anhydride (PP-g-MA)	120.2 ± 3.9 nmol/cm^2 of N-halamine, and 334.2 ± 33.0 nmol/cm^2 of primary amine groups on PP/PP-g-MA film	Spin coating assisted immobilization	TSB with horse serum	*E. coli*	99.9% reduction in contact with cationic primary amine groups, >99.99% decrease after post-chlorination of N-halamine, both after 2 h incubation at 32 °C. With horse serum (50,000 ppm): 90% decrease as prepared, 75–80% decrease as N-halamine.	Bastarrachea and Goddard (2016)
Ag NPs	Chitosan/PET	5% (w/w) in coating solution	Immobilization	Soft white cheese	TAMB Total coliforms TMYC	All studied microorganisms were undetectable on day 7 during cheese storage at 40 °C.	Singh et al. (2017)
ε-PL	PP/PP-g-MA	~3.75 mg/cm^2 on PP/PP-g-MA film	Spin coating assisted immobilization	Apple juice, Milk	*E. coli* *L. innocua*	Approximately 90% reduction in *E. coli* in apple juice after 15 min at 25 °C with UV light. 99.99–99.999% lower *L. innocua* load than control in inoculated milk after 21 days at 4 °C.	Bastarrachea (2019)

(continued)

Table 3 (continued)

Antimicrobial agents	Substrate	Antimicrobial concentration	Coating method	Food/ simulant matrix	Target microorganism	Antimicrobial activity	Reference
Chitosan	PP	2.2, 3.8% (w/w) chitosan on PP graft film	Immobilization	MHA	*E. coli* *S. aureus* *Ralstonia solanacearum* (*R. solanacearum*) *Clavibacter michiganensis* (*C. michiganensis*)	Inhibition zone >40 mm for all bacteria after 48 h at 37 °C.	Abudonia et al. (2018)

included in Regulation (EU 10/2011), their application in antimicrobial coatings requires a case-by-case assessment of migration and toxicological characteristics, with a stringent maximum migration limit of 0.01 mg/kg food (Dainelli et al., 2008).

Secondly, the difference between laboratory development and industrial-scale continuous production creates hurdles in commercializing these products. Properties like solid content and viscosity, often overlooked throughout laboratory development, become crucial factors for effective industrial coating. Issues such as slow processing speed and variable coating quality arise when transitioning these formulations to industrial-scale production (Perna, 2016). Uniformity and evenness of the coating on the substrate are essential in mass production, as a lack of uniformity leads to irregular distribution of antimicrobial agents, resulting in inadequate concentration on certain parts of the packaged food surface (Fu et al., 2016).

Thirdly, the vulnerability of antimicrobials at the surface of a film to degradation because of potential interactions with food components poses challenges. For instance, nisin and essential oils may degrade in the presence of lipids, proteins, or enzymes (Castro-Rosas et al., 2017). Antimicrobials incorporated into a coating, excluding nonmigratory immobilized antimicrobials, are more prone to migration into food due to their improved proximity, further complicating mass production challenges (Jokar & Abdul Rahman, 2014). Notably, due to the thin nature of these coatings, only a limited amount of antimicrobial agents can be incorporated. This can be problematic when higher concentrations are required for effective performance, particularly in contact with food. Some antimicrobial agents may become trapped by food components, reducing their activity, which necessitates a higher content of antimicrobial agents to achieve the desired efficiency (Wang et al., 2020; Bahmid & Siddiqui, 2024).

4.3 Antimicrobial Polymers as Packaging Material

Polymers, owing to their inherent characteristics, are widely and effectively utilized as materials for food packaging. Enhancing their functionality with antimicrobial properties has garnered increased attention from both academic and industrial viewpoints. Polymers can serve as a matrix to incorporate antimicrobial agents (as indicated in Fig. 1a, e), typically achieved through processes like thermal extrusion or compression molding. Consequently, only antimicrobial agents with heat resistance can be effectively integrated into the polymer matrix. The physicochemical properties of the polymer, like hydrophilicity or molecular weight, significantly affect the antimicrobial property. This influence extends to aspects like the homogeneous dispersion of antibacterial substances and the release rate into the contact food (Huang et al., 2019). Alternatively, polymers with intrinsic antimicrobial activity, whether natural or synthetic, can be employed for developing antimicrobial packaging materials, as shown in Fig. 1g. This approach circumvents the challenges associated with the diffusion of antimicrobial agents through the polymeric matrix and their migration into the packed food. Generally, the development of antimicrobial polymers involves thermal processes for polymerization (Haktaniyan & Bradley, 2022). Another method entails the chemical modification of polymers that inherently lack antimicrobial activity (Sedlarik, 2013). In the following sections, various categories of polymeric materials with antimicrobial activity will be introduced and explored.

4.3.1 Polymers with Antimicrobial Activity

4.3.1.1 Natural Polymers

The exploration of naturally occurring polymers with inherent antimicrobial properties has been a captivating research area for years, primarily due to their widespread abundance in nature. Several noteworthy natural antimicrobial polymers, such as chitin, pectin, agarose, carrageen, and ε-poly-l-lysine, have been identified. Numerous studies have focused on the application of these polymers in the field of food packaging, along with various techniques aimed at enhancing their somewhat limited antimicrobial activity. Typically, these polymers are utilized to create antimicrobial coatings on the primary packaging material. Their direct use as the primary packaging material is constrained by their suboptimal mechanical, barrier, and thermal stability characteristics. Nevertheless, current research efforts are directed towards the development of biobased or biodegradable packaging materials through the partial integration of these natural polymers into the primary matrix. These polymers can be introduced in diverse forms, such as polymers, nanofibers, and even nanoparticles (Yadav et al., 2017; Li et al., 2021; Aga et al., 2021).

4.3.1.2 Synthetic Polymers

Following the discovery of robust antimicrobial activity in certain synthetic polymers, extensive investigations have been directed towards identifying the most effective structural elements that contribute to optimal antimicrobial properties. The objective has been to either create new polymers with antimicrobial structural elements (direct polymerization from antimicrobial units) or introduce these elements into the existing polymer structures (copolymerization of polymeric monomers and antibacterial units) (Muñoz-Bonilla & Fernández-García, 2012).

In this section, synthetic polymers exhibiting antimicrobial activity will be categorized based on their chemical structures, as this is the primary characteristic used to delineate divisions and analyze their capabilities.

Polymers with Quaternary Nitrogen Atoms Polymers containing quaternary ammonium groups stand out as extensively studied polymeric biocides. These polymers bear a positive charge, enabling them to engage in destructive interactions with the negatively charged cytoplasmic membranes or bacterial cell walls, primarily composed of phosphatidylethanolamine (70%). The positive charge enhances their effectiveness in interacting with Gram-positive bacterial cells, thanks to the loosely packed polyglycane outer layer, facilitating extensive penetration of the polymer chain into the cell to engage with the cytoplasmic membrane. On the other hand, Gram-negative bacterial cells exhibit a higher degree of protection against the adverse effects of polymeric biocides. This protection arises from the presence of a supplementary membrane composed of a phospholipid bilayer structure, providing enhanced defense for the inner cytoplasmic membrane (Grigoras, 2021).

Guanidine Containing Polymers Polyguanidines and polybiguanides emerge as exceptional antimicrobial polymers, boasting remarkable features such as excellent water solubility, outstanding biocidal effectiveness, non-toxicity, and a broad antimicrobial spectrum. The primary antimicrobial mechanism of these polymer groups involves electrostatic interactions with bacterial cell membranes, resulting in the release of cellular contents. Similar to polymers with quaternary nitrogen atoms, this group of antimicrobial polymers demonstrates higher effectiveness against Gram-positive bacteria than against Gram-negative bacteria (Leong et al., 2020). Typically, synthesis involves the reaction of chlorcyan, cyanamid, or dicyanamid (for polybiguanides) with a diamine, or through the polycondensation of a guanidinium salt with a diamine. Research indicates that a mean molecular weight of 800 Da for guanidine-based oligomers is essential for effective antimicrobial activity (Tan et al., 2022).

Polymers Mimicking Natural Peptides AMPs, also referred to as host defense peptides, play a vital role in the innate immune response and are generated by all living organisms. These peptides exhibit broad and potent antimicrobial activity against bacteria, viruses, and fungi. Typically, AMPs carry a positive charge and have both hydrophobic and hydrophilic regions, enabling them to dissolve in water and traverse lipid-rich membranes. Despite their numerous advantages, such as their broad spectrum of activity, the widespread application of AMPs has been impeded primarily by the high manufacturing costs and their susceptibility to proteolysis (Erdem Büyükkiraz & Kesmen, 2022). Consequently, in recent years, there has been an increasing interest in the development of synthetic antimicrobial peptides. Synthetic antimicrobial polymers that mimic natural peptides can be categorized into various groups, including synthetic peptides featuring natural peptide frameworks (oligomers of L-α-amino acids), structures not found in nature (such as D-α-amino acids, β-peptides, or peptoids), arylamide backbones, phenylene ethynylene backbone polymers, and polynorbornene derivatives (Salas-Ambrosio et al., 2020; Rani et al., 2021).

Halogen Polymers Halogenated polymers encompass polymers containing halogen elements, which can be incorporated into polymer molecules via substitution or addition reactions. Among these, fluorine-containing polymers stand out as particularly attractive due to their distinctive characteristics. These polymers demonstrate water and oil repellency, which is due to the low polarizability and high electronegativity of fluorine atoms. Additionally, they boast excellent resistance to chemicals, heat, aging, and weathering, as well as a low dielectric constant and extremely low surface energy. The antimicrobial properties of fluorine-containing polymers are associated with their surface activity and significant hydrophobicity (Chaschin et al., 2023). Another significant category within halogen polymers is polymeric N-halamines. N-halamines are compounds that contain one or more nitrogen–halogen covalent bonds, created through the chlorination of imide, amide, or amine groups. Polymers derived from N-halamines are considered promising antimicrobial candidates due to their ability to achieve fast and comprehensive microbial eradication across a broad spectrum, without causing environmental concerns. Furthermore, the probability of microorganisms developing resistance is highly unlikely. In N-halamines, halogen atoms covalently bonded to nitrogen atoms gradually release active halogen species (oxidative halogens like Cl^+ or Br^+) into the environment (Tang et al., 2020).

Polymers Containing Phospho and Sulfo Derivatives The antimicrobial mechanism of polymers containing quaternary phosphonium or sulfonium compounds closely resembles that of those with quaternary ammonium, effectively eliminating bacteria by disrupting the membrane and cell wall. Generally, polycationic biocides with phosphonium exhibit superior efficacy compared to polymers with quaternary ammonium salt. Moreover, the antimicrobial activity increases with a higher number of phosphonium units in the polymer (Sommers et al., 2022). Some studies have also suggested that the alkyl chain length can impact the efficiency of antimicrobial activity (Nonaka et al., 2003).

Additionally, polyanions, including polymers with phosphonate and sulfonate groups, exhibit antimicrobial activity. Typically, under physiological conditions, cell membranes carry an excess of negative charge, making it challenging for polyanions to interact with them. However, in a mildly acidic environment, phospholipids become protonated, enabling polyacids like poly(styrene sulfonic acid) to interact with the membrane. These interactions involve not only electrostatic binding but also hydrogen and hydrophobic bindings (Huang et al., 2020a).

Presently, there is considerable interest in macromolecular units combining cationic and anionic moieties, such as zwitterionic structures like phosphorylcholine, phosphobetaine, sulfobetaine, and carboxybetaine. While polycations can disrupt bacterial membranes, they exhibit toxicity and a tendency to bind proteins. In contrast, zwitterionic polymers inhibit protein adsorption, as well as bacterial adhesion (Ma et al., 2020).

Phenol and Benzoic Acid Derivative Polymers Phenols serve as biocidal agents targeting bacterial membranes. They interact with the cell surface, leading to cell death by disrupting the cell membrane and releasing cellular contents. Additionally, phenols induce intracellular coagulation of cytoplasmic components, which further contributes to cell death or inhibits cell growth (Manuja et al., 2013). Park et al. developed antimicrobial polymers using vinyl monomers that contained phenol and benzoic acid pendant groups. Their research involved assessing antimicrobial activities through the halo zone test against *S. aureus*, *P. aeruginosa* bacteria, *Aspergillus fumigatus* (*A. fumigatus*), and *Penicillium pinophilum* (*P. pinophilum*) fungi. Results indicated that polymerization notably reduced the antimicrobial activity of the monomers (Park et al., 2001).

Benzaldehyde derivatives are well-known environmentally friendly antimicrobial agents with broad-spectrum inhibitory effects. They are utilized as bactericides, fungicides, and algaecides (Kenawy et al., 2006). For instance, an antifouling polymer comprising *p*-acryloyloxybenzaldehyde and copolymers with methyl methacrylate (MMA) were successfully synthesized, exhibiting antimicrobial efficacy against *Bacillus megaterium (B. megaterium)* and *P. aeruginosa* bacteria via the paper disc method. It was observed that the homopolymer demonstrated slightly greater antibacterial efficacy compared to the copolymers (Iemma et al., 2010).

4.3.2 Organometallic Coordination Polymers

Unlike traditional organic polymeric antibacterial materials, organometallic coordination polymers are a unique category of antibacterial polymers created from inorganic metal ions and polymer ligands. They possess exceptional efficacy against bacteria owing to their distinctive spatial structure and components, with antibacterial mechanisms often resembling those of metal ions. Organometallic coordination polymers are typically synthesized through self-assembly processes. Metal ions form bonds with the polymer through p-bonds to carbon, coordinate bonds to components with free electron pairs, or p/p-bonds to other components. Among the different metal ions used for coordination, Ag, Ni, Zn, Cu, and Au are the most frequently selected for their excellent antibacterial properties and coordination abilities. In these polymers, a polymer with a lone pair of electrons can function as a ligand, with nitrogen and oxygen often serving as binding sites (e.g., polypyridine or polycarboxylates). In addition, the antibacterial property can be finely tuned by manipulating various parameters, such as the metal ion type, dimensions and spatial arrangement, bond length, and the extent of electron delocalization (Yan et al., 2016; Morgan et al., 2018).

Beyond its antibacterial properties, the ability of coordinating ligand bonds between metal ion and polymer to dynamically form or cleave, imparts unique functionalities to coordination polymers. The balance of coordination between the central ions and the polymer ligand can be disrupted by changes in the environment's pH, leading to a stimulus-responsive antibacterial pattern through the controlled release of metal ions. As an illustration, Huang et al. reported on a polysaccharide complex involving Ag ions. In their research, the Ag ion was integrated into carboxymethylated agarose, establishing a coordinate bond with the carboxyl group. The release of the Ag ion to deactivate bacteria occurred when the pH transitioned to an acidic state (Huang et al., 2020b). This dynamic response highlights the potential for coordination polymers to exhibit tailored and responsive antibacterial functions.

4.3.3 Chemical Modification of Polymers to Achieve Antimicrobial Activity

Integrating antibacterial properties into inactive polymers is a common approach, in addition to the previously discussed polymerization strategies. In contrast to polymerization, the modification method is not limited by spatial hindrance or active shielding. While some antibacterial modification approaches may slightly degrade the fundamental characteristics of the original polymer, such as mechanical properties, they often lead to additional enhancements in certain properties, like wettability or thermostability. Typically, the modification process is more straightforward and more cost-effective compared to polymerization (Luo et al., 2021).

This section introduces processes that involve modifying polymers through chemical bonds with antimicrobial structures. In contrast, in immobilized antimicrobial agent systems, the antimicrobial agents are applied to the polymer surface (in the final form of food packaging materials) through surface activation rather than being integrated into the polymer structure. However, in the chemical modification method, the antimicrobial agents are chemically incorporated into the polymer structure, which can then be used to create food packaging materials. Antibacterial agents chemically bonded on the surface of the polymer exhibit notable features, including stability, safety, and nonleaching. The modification approach is mainly carried out in polymers with active groups, such as chitosan, cellulose, and PVA. The antibacterial agents used for modification vary from small molecules to polymers, and these will be further discussed in the following sections.

Covalent Incorporation of Low Molecular Weight Antimicrobial Compounds The covalent integration of low molecular weight antimicrobial compounds into the polymer is regarded as a direct and adaptable method, primarily achieved through a chemical extrusion process. A technique has been developed for antibacterial modification of chitosan, introducing quaternary ammonium compounds, using the amino group as the reactive site. The modified chitosan demonstrated improved antibacterial activity, leveraging the combined effect of two different inactivation mechanisms (Yin et al., 2020). In another study, Kougia et al. demonstrated the effectiveness of combining the low concentrations of the hydrophilic monomer acrylic acid (AA) with the covalently bonded bacteriostatic group vinyl benzyl dimethylhexadecylammonium chloride (VBCHAM) in the copolymer P(AA-co-VBCHAM88). This led to a significant bacteriostatic effect against *P. aeruginosa* and *E. faecalis*, resulting in a 6-log reduction in some instances. (Kougia et al., 2015). Furthermore, Orlando *et al.* suggested functionalizing cellulose with glycidyl trimethylammonium chloride (GTMAC) and glycidyl hexadecyl ether for use in antibacterial wound dressings. These active agents were employed to covalently modify the hydroxyl groups of glucose via a heterogeneous reaction under basic aqueous conditions. The resulting product exhibited a 53% reduction in *S. aureus* and a 43% reduction in *E. coli* within the first 24 h (Alvarado et al., 2019).

Coupling of Antimicrobial Peptides Covalent attachments between AMPs and functional polymers can be accomplished using various reactions, where the coupling method is particularly important. It is essential that AMPs and polymers possess functional groups capable of reacting with each other. Popular methods for conjugation include amide bond formation and click chemistry (Sun et al., 2018). Muszanska et al. conducted a study where they activated PF127-COOH using trifluoroacetic acid NHS ester and subsequently reacted it with the AMP (ILPWRWPWWPWRR-NH_2) to create AMP-functionalized antiadhesive polymer brush coatings. In vitro experimental results showed that the coatings performed exceptionally well in decreasing biofilm formation and improving tissue integration (Muszanska et al., 2014). In another study, Liu and co-workers performed covalent conjugation by attaching polymaleic anhydride (PMA) to the N-termini of the monomeric peptides RWRW-NH_2 and RRWW-NH_2, while protecting all side chains to avoid cross-linking. This led to an increased antimicrobial activity against *E. coli* and *B. subtilis*, along with a simultaneous rise in hemolytic activity (Liu et al., 2006).

Grafting Other Antimicrobial Polymers Another approach for incorporating antimicrobial properties into the base polymer involves grafting antimicrobial natural or synthetic polymers onto them. For instance, Ojogbo et al. demonstrated the grafting of polyhexamethylene guanidine hydrochloride (PHGH) onto starch, which was subsequently added to the PLA matrix. The results indicated that polymer surfaces exhibited high antimicrobial potency upon contact with *E. coli* and *B. subtilis* (Ojogbo et al., 2020). In another study, poly[2-(tertbutylamino)ethyl methacrylate] (PtBAEMA) was grafted onto PP, both with and without the addition of a compatibilizer agent (PP-g-MAH), through reactive extrusion, and then processed as fibers. The antimicrobial properties of obtained PP was analyzed using a cell counting method for *E. coli* microbes at a concentration of 10^8 CFU/mL. Results showed that Bacteria in the non-compatibilized blend died more quickly than those in the compatibilized material, suggesting that a reduced amount of PtBAEMA is present at the fiber surface (Thomassin et al., 2007).

4.3.4 Polymers with Antimicrobial Organic/Inorganic Compounds

The most straightforward non-chemical bond modification entails the physical blending of polymer materials with antibacterial agents. It is crucial to choose compatible antibacterial agents and combine them with a processable method that leverages their respective strengths, taking into account their benefits and drawbacks. Commonly added antibacterial agents include both inorganic and organic agents with low molecular weight, as well as antimicrobial polymers. However, a challenge associated with the direct blend process is the potential uncontrollable release mode. The uncontrollable leaching of antibacterial agents can lead to environmental contamination and a decline in antibacterial properties (Luo et al., 2021). This issue can be resolved by applying a thin layer coating on top of the antimicrobial packaging to regulate the release of the antimicrobial agents, as shown in Fig. 1. E:2. Table 4 summarizes studies illustrating advancements in antimicrobial polymers for food packaging applications.

Addition of Low Molecular Weight Antimicrobial Compounds The most common low molecular weight inorganic antimicrobial compounds for addition include metal or metal oxide particles, while for organic compounds, various antimicrobial agents have been introduced into polymers such as triclosan and vancomycin (Khaneghah et al., 2018; Videira-Quintela et al., 2021). For instance, Shuai et al. performed a study on the physical antibacterial modification of PLA and polyglycolic acid (PGA) by incorporating Ag. They synthesized Ag NPs in situ in mesoporous bioactive glass, which were subsequently integrated into the polymer though melt extrusion to confer antibacterial properties (Shuai et al., 2019). To enhance the dispersion of NPs within a polymer matrix, a patented method involves adding a colloidal suspension with a metal content, preferably 1–3% by weight (using $AgNO_3$ in this study), into an appropriate extruder through a degassing module. This process allows most or all of the carrier liquid to evaporate, ensuring a even distribution of the metal throughout the polymer layer and preventing bubble formation. When this metal-containing layer absorbs water under ambient conditions, antimicrobial metal ions are generated. These ions then diffuse through the moisture in the layer, activating their antimicrobial properties within the resulting food packaging (Schiffmann & Schlösser, 2011).

In another study by Hao et al., a film designed to release capsaicin based on pH changes was developed. Capsaicin was encapsulated in chitosan and subsequently incorporated into a polymer blend made up of polydopamine (PDA) and alginate, resulting in the formation of the film. The pH-sensitive nature of chitosan facilitated the conformational transformation of the structure, enabling the intelligent release of capsaicin and prolonging its shelf life by delaying excessive release (Hao et al., 2019). Furthermore, cellulose nanofiber (CNF) was modified with rosin and utilized as a reinforcing filler in a PLA matrix in another study. The obtained film was further coated with chitosan to form a two-layer composite film suitable for antimicrobial food packaging. This composite film showed outstanding antimicrobial effectiveness against *E. coli* and *Bacillus subtilis* (*B. Subtilis*) (Niu et al., 2018b).

Blends with Antimicrobial Polymers As previously noted, polymers can be combined with synthetic or natural antimicrobial polymers to achieve antimicrobial properties. Additionally, in the case of immiscible polymer blends, various compatibilizers can be employed to enhance the interface and overall compatibility. For example, a study by Haghighi et al. explored the development of an antimicrobial blended composite using microbial-derived extracellular polymeric substances and chitosan for fruit packaging. The findings demonstrated effective inhibition against *E. coli*, establishing its efficiency as a food packaging material (Muthulakshmi et al., 2021). In a separate investigation, the correlation between the PVA/Polyvinylpyrrolidone (PVP)/PEG/chitosan ratio in composite blends and antimicrobial activity against *S. aureus* was examined. Through statistical analysis and the derived equation, it was determined that the most economically optimized

Table 4 Antimicrobial agents used in antimicrobial polymers for food packaging applications

Antimicrobial agents	Polymer matrix	Antimicrobial concentration	Production method	Food/simulant matrix	Target microorganism	Antimicrobial activity	References
Organically modified montmo rillonite (OMMT) and Ag-loaded montmorillonite (Ag-MMT)	PE/Thermoplastic starch (TPS)	0.5, 1.25, 2.5 wt%	Melt extrusion, followed by blown-film extrusion	Culture media	*E. coli* *L. monocytogenes*	The composite films displayed substantial antibacterial effectiveness against *L. monocytogenes* and *E. coli*, particularly with higher concentrations of incorporated nanoclays, with a stronger impact on *L. monocytogenes*.	Hong et al. (2022)
Ag NPs containing silica nanotube carrier	PE, PP, PS, polyester, vinyl chloride, polyvinylidene chloride, PET, and phenol resins	1–2 wt%	Melt extrusion	Culture media	*E. coli* *S. aureus*	The antibacterial activity for *E. coli* was 6.1 log, indicating an antibacterial efficacy of 99.99% or higher. In comparison, the antibacterial activity for *S. aureus* was 3.5 log, corresponding to an efficacy of over 99.90%	Park (2010)
Ag, ZnO NPs	LDPE	0.25, 1, 1.5, 5 wt%	Melt extrusion, followed by blown-film extrusion	Orange juice	*Lactobacillus plantarum* (*L. plantarum*)	Ag and ZnO NPs in orange juice nanocomposite packages reduced *L. plantarum* numbers. Ag NPs exhibited stronger antimicrobial effects compared to ZnO NPs.	Emamifar et al. (2011)

(continued)

Table 4 (continued)

Antimicrobial agents	Polymer matrix	Antimicrobial concentration	Production method	Food/ simulant matrix	Target microorganism	Antimicrobial activity	References
Cloisite 30B, trimethylhexadecyl ammonium and cetylpyridinium chloride monohydrate modified MMT	Polycaprolactone (PCL)	3 wt%	Melt extrusion, followed by blown-film extrusion	Culture media	*E. coli* *S. aureus*	Adding 3 wt% organoclay to the PCL matrix significantly reduced bacterial count, achieving up to 96% growth inhibition. It demonstrated greater effectiveness against *S. aureus* compared to *E. coli* in all samples.	Hadj-Hamou et al. (2017)
Nisin, chitosan, potassium sorbate (PS) or Ag substituted zeolite (AgZeo)	LDPE	2 wt%	Melt extrusion, followed by blown-film extrusion	Chicken drumstick	TAMB Total coliforms TMYC	Samples packed in bags with chitosan, nisin, AgZeo, and PS showed lower TAMB (1.03, 0.98, 0.51, and 0.17 log, respectively) compared to control bags. Additionally, active bags resulted in lower microbial counts and reduced thiobarbituric acid reactive substance (TBARS) values compared to the control bags.	Soysal et al. (2015)
Nisin or natamycin	Sodium caseinate	0.1 w/w% nisin, 0.5% w/w% natamycin	Melt extrusion, followed by blown-film extrusion, or	Culture media	*A. niger* *Kocuria rhizophila (K. rhizophila)*	After 24 h, *K. rhizophila* growth inhibition zones around nisin-incorporated films for solution casting	Colak et al. (2016)

			heat-press. Direct solution casting.			(control), extrusion blowing, and heat pressing were 15.5 ± 0.9 mm, 9.8 ± 0.2 mm, and 8.6 ± 1.0 mm. Heat-pressing did not further inactivate nisin, indicating limited inactivation during extrusion-blowing. The smaller diameter observed around films containing nisin pellets, compared to those directly cast, indicated that nisin inactivation primarily occurred during melt extrusion. Natamycin films prevented *A. niger* growth. Unlike nisin films, heat-pressed films exhibited larger inhibition zones compared to blown films, suggesting partial natamycin inactivation occurred during the extrusion-blowing process.	

(continued)

Table 4 (continued)

Antimicrobial agents	Polymer matrix	Antimicrobial concentration	Production method	Food/simulant matrix	Target microorganism	Antimicrobial activity	References
Pomegranate peel extract (PPE)	PE in monolayer form and in a two-layer format, covered either by PP or PET.	1.5 w/w%	Melt extrusion, followed by blown-film extrusion	Pork	TPC	By day 6, the TPC of pork samples in PPE-PE, PPE-PE/PET, PPE-PE/PP, and PE/PET films were 5.78, 5.49, 5.58, and 5.72 log CFU g^{-1}, respectively. In comparison, the TVC of pork samples in PE film was 6.54 log CFU g^{-1}. This suggests that PPE-PE/PET and PPE-PE/PP films exhibited effective antibacterial properties.	Hu et al. (2017)
Sorbic acid (SA)	PP	2, 4 and 6 w/w %	Melt extrusion, followed by blown-film extrusion	Culture media	*E. coli* *S. aureus* *A. niger*	PP-SA films demonstrated antibacterial activity against *E. coli* and *S. aureus*, along with stronger antifungal effects on *A. niger*.	Fasihnia et al. (2018)
Vermiculite (VER) modified by cation exchange with hexadecyltrimethylammonium ($HDTMA^+$) bromide	PE	7 wt%	Melt extrusion, followed by heat-press.	Culture media	*E. faecalis* *E. coli* *S. aureus*	The most sensitive Gram-positive bacteria stopped growing after 24 h, with a consistent minimum inhibitory concentration (MIC) of 0.014% (w/v). Gram-	Hundáková et al. (2015)

						negative bacteria showed growth inhibition at 0.041–10% (w/v) depending on the $HDTMA^+$ content in the samples.	
Cellulose and lignin	PLA Glycidyl methacrylate (GMA) grafted PLA (GMA-g-PLA)	1 and 3 wt%	Melt extrusion, followed by blown-film extrusion	Culture media	*Pseudomonas syringae* (*P. syringae*)	The PLA nanocomposite containing 3 wt% lignin and 1 wt% cellulose showed antibacterial activity compared to PLA within the first few hours. The GMA-g-PLA nanocomposite containing 3 wt% lignin formulation displayed the highest and prolonged antibacterial activity.	Yang et al. (2016)
Native and carboxymethylated plantain flour	Polystyrene (FS)	flour/PS blends at a 50: 50 ratio	Reactive extrusion with chromium octanoate ($Cr(Oct)_3$) as a catalyst (Cat), followed by thermomolding.	Culture media	*L. monocytogenes* *E. coli*	*L. monocytogenes* exhibited "extremely sensitive" reactions to Cat-containing films. Surprisingly, films without Cat also displayed antimicrobial activity against *L. monocytogenes*, suggesting the possible production of compounds like acrylamide, hydroxymethylfurfural, and furan via Maillard reactions in these non-Cat films.	Herniou Julien and Gutiérrez (2021)

(continued)

Table 4 (continued)

Antimicrobial agents	Polymer matrix	Antimicrobial concentration	Production method	Food/simulant matrix	Target microorganism	Antimicrobial activity	References
Sodium propionate (SP)	PP	0.5, 1, 2, 3, and 5 wt%.	Melt extrusion	Bread	*E. coli*, *S. aureus* TMYC	These composite films demonstrated increased antimicrobial activity against both Gram-negative and Gram-positive microorganisms. The interaction of SP with moisture from the bread changed the pH, leading to the breakdown of fungal cellular structures and a reduction in bacterial growth rate.	Thanakkasaranee et al. (2018)
Chitosan	PLA/Tributyl O-acetylcitrate (TBoAC)	1 wt%	Melt extrusion	Culture media	*E. coli, S. aureus*	Incorporating 1 wt% chitosan into PLA/TBoAC resulted in log reductions of 2.73 and 5.54 against *S. aureus* and *E. coli*, respectively.	Darie-Ni et al. (2021)
CuO and ZnO NPs	LDPE	2 wt%	Melt extrusion	Culture media	*B. Subtilis* *Enterobacter aerogenes* (*E. aerogenes*)	CuO-containing nanocomposites inhibited the growth of both bacteria, showing a stronger antibacterial effect compared to ZnO-containing nanocomposites, likely due to their smaller size.	Esmailzadeh et al. (2021)

Ag, ZnO, TiO_2 NPs	PLA	0.5, 1.0, and 2.5 wt%	Melt extrusion, followed by blown-film extrusion	Culture media	*E. coli* *S. aureus*	The antibacterial activity of PLA-NPs films revealed that Ag, ZnO, and TiO_2 NPs contribute to their antimicrobial effect, with PLA-Ag films demonstrating the strongest activity. However, increasing the contents of TiO_2 and ZnO to 2.5 wt% reduced their antibacterial activity.	Pušnik Črešnar et al. (2021)
Kraft lignin (KL) and tannin	PLA	0.5, 1.0, and 2.5 w/w%	Melt extrusion	Culture media	*E. coli*	After 24 h, PLA-tannin with 2.5% tannin achieved 88.5% antimicrobial activity, while lower tannin content (1.0% and 0.5%) resulted in 58% and 20% bacterial activities, respectively. For PLA-KL, a high KL content (2.5%) showed no antimicrobial activity, but a low content of 0.5% exhibited nearly 82% antibacterial activity.	Črešnar et al. (2022)

(continued)

Table 4 (continued)

Antimicrobial agents	Polymer matrix	Antimicrobial concentration	Production method	Food/ simulant matrix	Target microorganism	Antimicrobial activity	References
ZnO NPs	LDPE	2 and 4 wt%	Melt extrusion, followed by heat-press.	Culture media	*B. Subtilis* *E. aerogenes*	Nanocomposites with 4 wt% ZnO had greater antibacterial impact than those with 2 wt%. *B. subtilis* were more susceptible to ZnO films compared to *E. aerogenes*.	Esmailzadeh et al. (2016)
Ag NPs	TPS	0.006 and 0.015 wt%	Melt extrusion, followed by compression molding.	Culture media	*E. coli*	The addition of 0.015 wt% Ag NPs to TPS effectively delayed the growth of *E. coli*.	Ceballos et al. (2021)

polymer blend the concentrations of PVA, PVP, PEG, and chitosan should be 0.002, 0.002, 0.002, and 0.001 mg/mL, respectively, to achieve an impressive inhibition percentage of 99.162% (Doan & Tran, 2023). In another study, the impact of incorporating chitosan into plasticized PLA on antimicrobial activity was explored. The developed blends exhibited strong antifungal activity against *Aspergillus brasiliensis* (*A. brasiliensis*), *Penicillium corylophilum* (*P. corylophilum*), and *Fusarium graminearum* (*F. graminearum*), along with strong antimicrobial properties against *S. aureus* and *E. coli* (Râpă et al., 2016).

4.3.5 Drawbacks and Constraints

The development of antimicrobial polymers has witnessed significant scientific progress in recent decades, showing rapid advancements. Synthetic antimicrobial polymers offer advantages like prolonged antimicrobial activity and a non-leaching mechanism. However, their high production costs present a challenge, and more economical polymerization processes are sought (Luo et al., 2021). Natural antimicrobial polymers are cost-effective but often exhibit inferior mechanical and barrier properties in comparison to traditional synthetic polymers, limiting their widespread use. They are commonly blended with conventional polymers to enhance performance (Bustamante-Torres et al., 2022). The prevalent method for producing antimicrobial composite polymers involves adding antimicrobial components to the polymer matrix. However, concerns arise regarding the safety of antimicrobial agents migrating into packed foods, particularly in case of nanomaterials. Nanoparticles exhibit different properties from their macro-sized counterparts, making toxicity assessments challenging. Some nanomaterials may pose potential health risks, although evidence linking this to nanomaterial applications in the food industry is inconclusive. Risk assessment of nanocomposite packaging materials is a major challenge, prompting concerns from both consumers and regulatory bodies. Current regulations in the European Union limit nanomaterial migration from food packaging nanocomposites to 10 mg dm^2. While most migration values stay below this limit, ongoing technical evaluations prompt periodic updates to regulations based on new information related to food contact nanocomposite plastic materials (Huang et al., 2015; EFSA, 2012; Ilyas et al., 2021).

While there have been significant developments in engineering antimicrobial polymers, there is still ample room for further progress. The development of antibacterial polymer materials requires consideration of various parameters, including safety, performance enhancement (e.g., durable, broad-spectrum, quick antibacterial action, non-toxicity, and multiple antibacterial modes of action), and industrial feasibility. Processing procedures should preserve the initial functions, structures, and antibacterial characteristics of the polymers. Additionally, factors like recyclability and natural degradation, especially in the case of biodegradable polymers, need attention. The presence of antimicrobial agents can significantly influence these features, and their impact should be carefully considered during the design and development of antimicrobial polymers (Hernández-García et al., 2021).

4.4 Antimicrobial Edible Coating on Food

Edible antimicrobial packaging is becoming a practical solution for preserving food quality and extending shelf life, working by postponing microbial spoilage and delivering efficient moisture and gas barrier capabilities. Technological advancements in this field have yielded positive leads to the improvement of longevity of food. The global edible packaging market, valued at US$ 1004.00 million in 2022, is projected to expand at a compound annual growth rate (CAGR) of 6.29% from 2023 to 2030 (Nair et al., 2023). Antimicrobial edible coatings are primarily developed by incorporating active agents like plant extracts, EOs, organic acids, natural antimicrobial polymers, and nanomaterials into edible packaging. Edible packaging can be made from proteins (such as gelatin, whey protein, casein, and soy protein), polysaccharides (including starch, chitosan, pectin, and cellulose derivatives), lipids (like beeswax, carnauba wax, and shellac), or a combination of these materials. Beyond antimicrobial benefits, these coatings promise to improve the mechanical, physical, barrier, and other features of edible materials, improving overall food product quality (Hamed et al., 2022). Edible packaging is commonly applied in two forms: preformed films wrapped around food products and direct application of the edible coating onto the food items as shown in Fig. 1f. In this section, various methods for forming edible films and depositing them directly onto food products will be introduced and discussed.

4.4.1 Methods of Edible Film Formation

Edible film, often utilized as a solid matrix wrapping around food products, serves as primary packaging without impacting sensory or nutritional attributes. In such cases, it is designed to be flavorless, colorless, and inconspicuous. Additionally, edible films can be employed as pouches or sachets, especially in energy drinks and meal replacement shakes. These films, when introduced to cold or hot liquids, dissolve, releasing their contents. This not only provides customers with convenience and portion control but also contributes to reducing solid waste. Edible films can be developed through two distinct methods: wet processes (solvent casting) and dry processes (extrusion) (Suhag et al., 2020).

Solution Casting The casting method, previously described for coating deposition and indicated in Fig. 2, is also applicable to the production of edible films, commonly employed at laboratory and pilot scales. The drying process in casting is a crucial step for enhancing intra-molecular relationships between polymer chains and achieving a suitable microstructure. Quick-drying methods in film casting can negatively impact physical and structural properties. Air dryers such as hot air ovens, tray dryers, microwaves, and vacuum dryers are typically used for film casting to aid in solvent removal and film detachment (Velaga et al., 2018). A consistent, defect-free edible film is desired, often achieved through the use of plasticizers for improved mechanical strength and uniform microstructure (Sanyang et al., 2015). The casting method offers advantages such as cost-effectiveness, simplicity, and homogenous particle packaging. However, limitations include the restriction to simple forms (sheets and tubes), potential solvent trapping, protein denaturation, and long drying times unsuitable for commercial production. Converting film fabrication from lab-scale to production scale poses challenges because of various variables impacting quality, such as heating, speed, and temperature combinations. Achieving consistent development for commercial scales remains a significant challenge (Jensen et al., 2015; Karki et al., 2016).

Extrusion The extrusion method, extensively utilized in commercial-scale polymer film production, is also integral to the fabrication of edible films and packaging materials. In this technique, components are introduced into the extruder barrel from a hopper, where they undergo gradual melting facilitated by the mechanical energy from turning screws and heating elements. Subsequently, the molten polymer is shaped by a die and undergoes cooling to achieve hardening, as indicated in Fig. 4. Recognized as a dry process, extrusion thrives with minimal water or solvent content, although plasticizers may be incorporated to enhance film flexibility. Critical parameters governing the extrusion process include moisture levels, screw rotation speed, barrel heating temperature, die size, pressure levels, and energy input (Gaspar-Cunha et al., 2022).

Notably, the advantages of extrusion encompass its reduced processing time, minimal energy consumption, and enhanced mechanical and optical characteristics such as elongation and transparency. These attributes render it a cost-efficient and high-performance process within the food sector's commercial production landscape. Furthermore, the lack of solvents, the convenience of handling high-viscosity polymers and enhanced control over feed residence time and mixing, especially during the integration of antimicrobial agents, underscore the method's efficacy. Moreover, extrusion offers superior versatility in shaping edible films and generates a broader array of forms compared to solvent casting. Nevertheless, drawbacks include the process's limitation to heat-resistant and low-moisture materials, which may restrict the utilization of specific polymers. Moreover, the initial investment in specialized machinery and subsequent maintenance costs pose additional considerations in adopting extrusion for edible film production (Nastaj & Wilczyski, 2021).

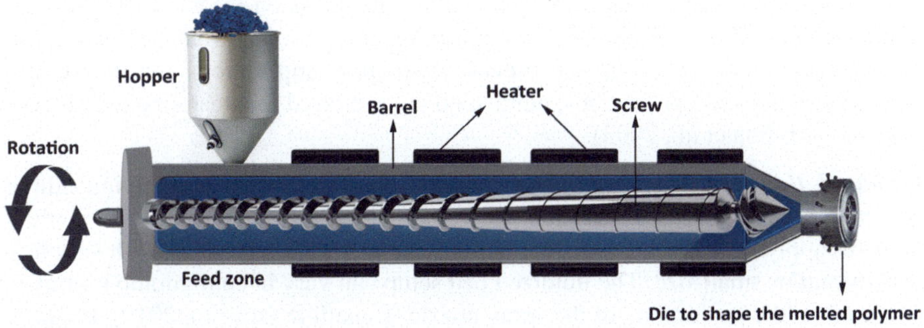

Fig. 4 Melt extrusion process for forming films

4.4.2 Methods of Applying Edible Coatings on Food Surfaces

Various techniques can be utilized to deposit edible coatings onto food surfaces, including dipping, spraying, panning, and fluidized-bed methods. The choice of deposition method relies on several factors, including the food's characteristics such as shape, size, ripening types (including climacteric and non-climacteric in fruits), surface attributes, and the desired coating thickness (Suhag et al., 2020).

Dipping Dipping stands as the most prevalent method for coating food products, involving the dipping of a food sample into the coating dispersion, as previously detailed in the coating section and indicated in Fig. 2. This approach is extensively employed for whole fruits and vegetables. Typically, fruits and vegetables go through a submersion period of 5–30 s in the edible coating formulation, ensuring uniformity across their often intricate surfaces. While the dipping method proves effective for most fruits, challenges arise when coating cut surfaces, where the hydrophilic nature clashes with the hydrophobic coating material (Skurtys, 2010). To address this, LBL coating assembly stands as a potential solution. Despite its straightforwardness and cost-effectiveness, the dipping approach is primarily utilized in laboratory settings. However, it may present challenges, including coating dilution, potential accumulation of waste or impurities, and microbial development in the dipping vat. Additionally, drawbacks involve the dilution and potential detachment of the natural wax coating on fruits and vegetables after dipping, impacting its functionality (Soares et al., 2016).

Spraying Spraying emerges as the predominant method for coating food products, as detailed in the coating section and indicated in Fig. 2. In this approach, the top surface of the food product is initially coated and dried, followed by a separate application to cover the bottom surface. The spray-coating technique is known for its versatility, capable of forming thin or thick films on food surfaces, offering precise control over thickness through careful viscosity adjustments in the coating solutions and the design of the spray nozzle. This technology ensures a consistent coating with uniform thickness. Moreover, spraying systems maintain the integrity of the layer, enable temperature regulation of the coating solution and support automatic continuous production. However, it is worth noting that highly viscous solutions may pose challenges in spray applications, often making dipping methods more suitable. This adaptation could result in an increased coating thickness on the food item surface (Andrade et al., 2012). While bilayer applications may be disadvantageous because of the requirement for a four-step process (two spray applications and two drying phases) in industrial applications, emulsified are preferred to streamline the process (Bosquez-Molina et al., 2003).

Fluidized Bed The fluidized-bed technique is widely used for applying edible coatings in both the food industry and research settings. Typically, this method involves applying a thin layer of coating material to dry food particles with very low density and/or small size. The fluidized bed setup can vary in configuration, primarily based on the positioning of the spray nozzle (Benelli & Oliveira, 2019). Figure 5

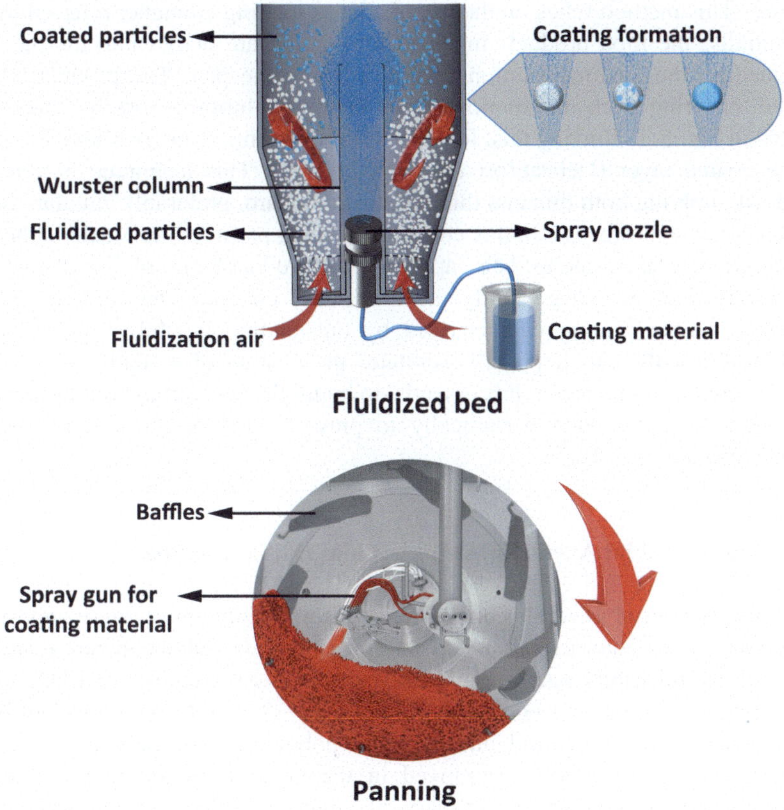

Fig. 5 Fluidized bed (bottom spray system) and panning methods for applying edible coatings on food particles

illustrates a bottom spray system. In this process, food particles are initially loaded into the chamber. An airstream is then blown upwards through the bottom plate, which is equipped with meshes, imparting sufficient velocity to the particles. This results in a circular flow of particles within the chamber, as depicted in the figure. Once the particles are fluidized, a coating solution or suspension is sprayed onto the surface of the fluidized powder through multiple nozzles to form a shell-like structure (Tric et al., 2023).

The size of the food particles in this method is crucial, ideally around 100 μm. Particles that are either larger cannot be stably fluidized, or smaller ones tend to form excessive agglomerations (Bertuzzi & Slavutsky, 2016). This technique offers shorter processing times, ensures complete coverage, and minimizes the development of clusters of coated items. However, it is more expensive compared to other coating methods and necessitates a greater quantity of coating solution because of losses in the Wurster column while spraying. Moreover, this method is only suitable for specific food products that meet the size criteria and are not prone to damage from fluidization (Bozkurt et al., 2023).

Panning This method relies on the use of a large rotating container referred to as a pan. Initially, the food products intended for coating are placed into the pan, and subsequently, the coating solution is sprayed onto the pan. The product is then tumbled, ensuring even distribution of the coating solution across its surface, as depicted in Fig. 5. Following this, forced air at room temperature or above is used to dry the coated layer (Debeaufort & Voilley, 2009). This technique is versatile, capable of applying both thin and thick coatings on hard, preferably circular-shaped particles in a batch process. In this context, extruded particles are ideally suited for the panning technique due to their capacity to be produced in round or oval shapes and various sizes, which generally facilitates easy coating. The method yields a transparent edible coating characterized by outstanding flexibility and a glossy finish. Additionally, this approach facilitates the coating of a significant quantity of food products in one batch. It is essential to highlight that, throughout the process, water must be evaporated periodically to prevent food products from sticking together (Priya et al., 2023).

4.4.3 Antimicrobial Agents in Antimicrobial Edible Coatings

As the materials used in edible coatings are consumed, obtaining safety approval is imperative before commercial use. Therefore, both the edible material and the integrated antimicrobial agents must be recognized as safe by the FDA before application for commercial purposes (Paidari et al., 2021). Generally, plant-based extracts such as EOs, natural antimicrobial polymers, especially chitosan, and nanoparticles are extensively employed in the creation of antimicrobial edible coatings. Table 5 provides a summary of studies highlighting progress in the development of antimicrobial edible coatings for food applications.

4.4.4 Drawbacks and Constraints

Despite the numerous advantages offered by antimicrobial edible coatings, such as extending shelf life, preventing moisture loss, and providing gas barrier properties, there are certain drawbacks associated with their use. In some instances, applied coatings can lead to undesirable changes in texture, diminishing consumer acceptance of the products. They may also alter the ripening process of fruits in an undesirable manner. Additionally, sensory-related alterations may be observed when incorporating antimicrobial compounds (Panchal et al., 2022).

Moreover, some components or materials utilized in coating preparation could provoke allergic reactions upon consumption. For example, the application of certified natural plant extracts and EOs has been known to cause allergic reactions and, in some cases, toxic effects varying with the dosage. Consequently, it is crucial to implement proper measures to regularly evaluate the toxicity and allergenic properties of extracts and EOs utilized in edible coating production (Peerzada et al., 2023).

Table 5 Antimicrobial agents used in antimicrobial edible coatings for food packaging applications

Antimicrobial agents	Matrix	Antimicrobial concentration	Coating method	Food/ simulant matrix	Target microorganism	Antimicrobial activity	References
Oregano or clove EOs	Whey protein isolate (WPI)	10 and 20 g kg^{-1}	Dipping	Chicken breast fillet	TAMB *Enterobacteriaceae* PBC LAB *P*. spp	Films with 20 g kg^{-1} of oregano EO doubled the storage time of chicken breast to 13 days, while ensuring microbiological groups stayed within recommended limits for distribution and consumption.	Fernández-Pan et al. (2014)
ZnO NPs	Chitosan (100%Q), alginate (100%A) and blends of 50%Q–50% A; 90%Q–10%A and 90%A–10%Q	1% v/v	Dipping	Guava	–	The study found that applying ZnO nano-structured coatings to either 100% Q or 90% Q–10% A, can extend guava shelf life to 20 days due to antimicrobial activity, in contrast to uncoated fruits lasting only 7 days.	Arroyo et al. (2020)
Blackberry pomace powder (BPP)	CMC incorporated with various contents of pectin and bacterial cellulose fibrils (BCF)	1% w/v	Casting	NB	*E. coli*	The blackberry pomace powder exhibited a moderate antimicrobial effect, with the fastest release observed when the pectin concentration was highest and the BCF content was lowest.	Isopencu et al. (2021)

(continued)

Table 5 (continued)

Antimicrobial agents	Matrix	Antimicrobial concentration	Coating method	Food/ simulant matrix	Target microorganism	Antimicrobial activity	References
Beeswax and chitosan	Glucomannan	–	Dipping	Salak	*S. aureus* *E. coli*	The glucomannan–beeswax–chitosan coating reduced *S. aureus* and *E. coli* growth by 10% compared to the control (glucomannan-coated salak fruit). Peeled salak fruit with this coating remained mold-free for 3 days at room temperature.	Meindrawan et al. (2020)
ZnO NPs	Chitosan/gum arabic	0.1 to 1 w/w%	Dipping	Avocado	–	The edible coating prolonged the shelf life of avocados to 7 days under ambient conditions, surpassing the 4-day deterioration observed in uncoated avocados.	Le et al. (2021)
Plant extracts from licorice, eucalyptus, sage, and thyme	Alginate	2 w/v%	Casting	Culture media	*E. coli* *P. aeruginosa* *B. cereus* *L. monocytogenes* *S. aureus*	The coatings successfully inhibited or reduced the growth of all bacteria except with the sage extract coating. Films with licorice and eucalyptus inhibited Gram-positive bacteria, while none of them inhibited Gram-negative bacteria.	Lopes et al. (2023)

Citrus sinensis EO	Alginate	–	Dipping	Tomato	*Salmonella typhi (S. typhi)* *L. monocytogenes*	The edible coating on tomatoes significantly reduced the total mesophilic bacteria, including *S. typhi* and *L. monocytogenes*, in comparison to uncoated tomatoes.	Das et al. (2020)
ZnO NPs	Chitosan/gum arabic	0.1, 0.3, 0.5, and 1.0 w/w %	Dipping	Banana	*S. aureus* *E. coli,* *B. subtilis.*	ZnO NPs demonstrated strong antibacterial properties. The coating notably enhanced the quality and extended the shelf life of bananas stored at 35 °C and 54% relative humidity.	La et al. (2021)
Lemongrass, thyme, and oregano EOs and PPE and grape pomace extract (GPE)	Sodium alginate and chitosan	1.5 and 3 v/v %	Dipping	Pomegranate	*Botrytis cinerea* (B. cinerea) *P.* spp	Coatings with sodium alginate and chitosan, containing lemongrass EO (15 or 30 µL/mL), fully inhibited spore germination and minimized decay in pomegranates.	Kawhena et al. (2021)
Citral	Sodium alginate	0.1, 0.5 and 1 v/v%	Dipping	Fresh cut pineapple	*S. enterica* *L. monocytogenes*	A 0.5% citral coating on pineapple caused a notable reduction (2.52 and 2.23 log CFU g^{-1}) in *S. enterica* and *L. monocytogenes*. A higher citral concentration, however, negatively affected texture and sensory acceptance of the pineapple samples.	Prakash et al. (2020)

(continued)

Table 5 (continued)

Antimicrobial agents	Matrix	Antimicrobial concentration	Coating method	Food/ simulant matrix	Target microorganism	Antimicrobial activity	References
Trans-resveratrol (R)-encapsulated NPs (RNPs)	Chitosan	–	Dipping	Strawberry	*S. enteritidis* *L. monocytogenes* *E. coli* *S. aureus.*	Encapsulated R showed significant antimicrobial activity against all strains. Combining RNPs with low-molecular-weight CS in the coating resulted in excellent antimicrobial effectiveness against foodborne pathogens. Notably, the chitosan coating with RNP demonstrated significantly greater antimicrobial activity against *E. coli* compared to chitosan or RNPs alone, indicating a synergistic effect.	Hu et al. (2023)
Citrus paradise EO	*Lallemantia iberica* seed mucilage (LISM)	0, 0.5, 1, 1.5, and 2 v/v%	Dipping	Lamb	TBC PBC *E. coli* *S. aureus*	Incorporating EO into the edible coating effectively inhibited microbial growth, including TBC, PBC, *E. coli*, *S. aureus*, and fungi.	Zanganeh et al. (2021)
Immobilized bacteriocin-producing LAB	Alginate, maltodextrin	–	Dipping	Fresh cheese	*L. monocytogenes* TAMB Enterobacteriaceae	Coating with immobilized Lactococcus cells on cheeses significantly	Silva et al. (2022)

						reduced *L. monocytogenes* surface contamination, and prevented TAMB growth by the sixth and eighth day respectively at 4 °C, and inhibited the migration of *L. monocytogenes* into the cheeses.	
Orange peel EO	Carnauba wax	0, 1, 2, 3, 4, and 5 wt%	Dipping	Strawberry	*Penicillium expansum* (*P. expansum*) TPC	The coating with orange EO protected the strawberries against mold. The highest TPC (7.14 log CFU g^{-1}) was obtained for the blank sample after 20 days, while the minimum TPC (3.12 log CFU g^{-1}) was found on the coating containing 5 wt % EO. The Optimum EO concentration, considering sensory effects, was reported to be 3 wt % EO.	Saeed et al. (2021)
Grapefruit seed extract (GSE) in form of NPs	Alginate in form of NPs	1% alginate +1% GSE in dipping solution	Dipping	Shrimp	TBC	The uncoated shrimps surpassed microbiological limits, reaching 7.87 log CFU g^{-1} by day 4 of storage, while the NP-based coated shrimps remained below the limit even on day 8 of storage.	Baek et al. (2021)

(continued)

Table 5 (continued)

Antimicrobial agents	Matrix	Antimicrobial concentration	Coating method	Food/ simulant matrix	Target microorganism	Antimicrobial activity	References
Chitosan-tripolyphosphate-thymol	Chitosan or a mixture of quinoa protein-chitosan	4.4 ± 0.1 mg/mL	Molding	Culture media	*S. Typhimurium* *E. coli* *E. aerogenes* *P. aeruginosa* *L. innocua* *S. aureus*	Films containing NPs showed higher antimicrobial activity compared to control films. Over time, these films swelled upon contact with inoculated agar, allowing antimicrobial NPs to desorb and spread radially, increasing their contact area with the bacteria compared to the control.	Villaneuva et al. (2019)

Certain coating methods can incur significant costs, ultimately contributing to price increases for commodities. It is crucial to point out that antimicrobial edible coatings are applicable to a restricted range of food products, specifically solid foods with distinct characteristics. Careful consideration and monitoring are essential to balance the benefits and potential drawbacks associated with their application (Akhtar et al., 2015).

5 Engineered Antimicrobial Surfaces Using Plasma Technology

Numerous research efforts and advancements have been made in engineering packaging materials with excellent antimicrobial properties, particularly in the field of food packaging, over the past decades. Despite the significant contributions and remarkable achievements of these studies, it should be emphasized that most of these methods rely on wet chemistry or thermal processes, which are not environmentally friendly. Consequently, the industry is actively seeking more sustainable green technologies that can benefit this field (Gupta et al., 2022; Zhang et al., 2022). Most research on developing sustainable antimicrobial food packaging focuses on combining biodegradable materials with natural antimicrobial agents to reduce environmental impact. However, from a technological standpoint, the extraction of natural antimicrobial agents and biopolymers, as well as the final production of antimicrobial materials, primarily rely on wet chemistry. Less attention has been given to more sustainable technologies for processing these materials into their final applications (Al-Tayyar et al., 2020; Zhao et al., 2023; Vidal et al., 2023). One promising technology in this regard is plasma surface engineering, which has the potential to act as a substitute for conventional processes currently used for developing antimicrobial packaging due to its unique advantages (Nikiforov et al., 2022). Plasma, often considered the fourth state of matter, it is a gaseous mixture containing electrons, ions, radicals, and neutral particles, which is macroscopically neutral. Plasma is generally formed as a result of gas ionization by providing electrical energy. The highly reactive species present in plasma can be directly controlled by regulating the electric power input (Li et al., 2018). As a result, plasma can be employed as an adjustable reactive physicochemical environment to establish diverse plasma-based surface engineering approaches for surface activation, coating deposition, and surface nano-structuring to achieve antimicrobial properties (Nikiforov et al., 2022; Ma et al., 2022). One attractive characteristic of plasma-based surface engineering is that it is limited to the surface layer, affecting only a few nanometers and leaving the bulk properties of the treated materials intact. Furthermore, this technology is solvent-free, energy-efficient, reliable, and scalable for industrial applications (Pillai & Thomas, 2023). Additionally, plasma-based surface engineering is a versatile method that can be applied to various materials, including polymers, metals, ceramics, and composites (Ma et al., 2021). Non-equilibrium plasmas are particularly significant in plasma-based surface engineering research due to their low temperature. Unlike equilibrium plasmas, these low-temperature plasmas operate far from local thermal equilibrium (LTE), exhibiting substantial variations in heavy particles and electron temperatures. Non-equilibrium plasmas demonstrate high chemical reactivity at low gas temperatures, even at room temperature, due to processes driven by electrons while the heavy particles remain at a low temperature. This characteristic makes it possible to utilize these plasmas for treating heat-sensitive polymeric materials used in food packaging applications. Such plasma acts as a chemical reactor with a high concentration of relatively low-energy active species in the gas phase, suitable for surface etching and plasma polymerization (Fridman et al., 2023; Boulos et al., 2023). In addition, non-equilibrium atmospheric pressure plasmas are significantly important in this field because they eliminate the need for costly, space-restricted vacuum equipment (Fanelli & Fracassi, 2017).

5.1 *Surface Chemical Etching*

Surface etching is a modification process used to remove specific areas or layers from a material's surface to create desirable nano-scale patterns, as indicated in Fig. 6 (Puliyalil & Cvelbar, 2016). It is well-known that certain nanostructured surfaces can exhibit antimicrobial activity, inspired by the natural antimicrobial properties observed on cicada wings due to their nanostructured patterns (Pogodin et al., 2013). Typically, two mechanisms contribute to the antimicrobial characteristics of these nanopatterned surfaces: rupturing bacterial cell membranes, as shown in Fig. 6, and reducing the available surface area for bacterial contact. The promising feature of nano-scale patterned antimicrobial surfaces is that they do not pose concerns about chemical-induced cytotoxicity or the development of drug-resistant bacteria. This is because the killing mechanisms are entirely mechanical, inducing stress on the bacterial cells (Linklater et al., 2017). In general, low pressure plasma is typically used for surface etching to create well-structured nanopatterns resulting from the elevated energy levels of ions and reactive species. In contrast, atmospheric pressure plasma provides a high density of active species, but with lower energy.

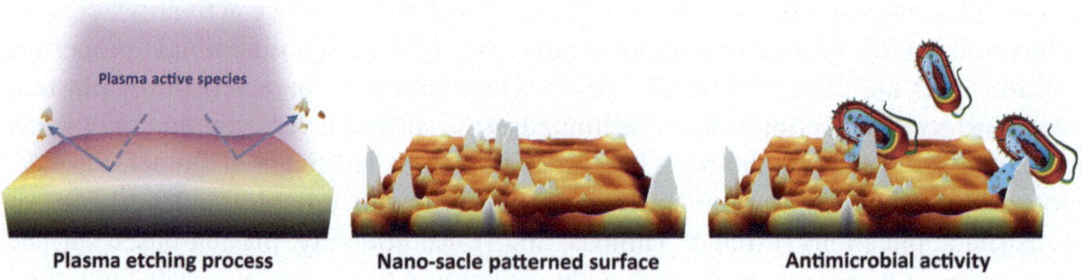

Fig. 6 Surface chemical etching for developing nano-scale patterned surfaces and the mechanism of antimicrobial activity

Consequently, surface etching using atmospheric pressure plasma is still in its early stages, and the amount of research conducted in this area is limited compared to low pressure plasma.

Dimitrakellis et al. performed nanotexturing on ZnO/poly(methyl methacrylate) (PMMA) nanocomposite coatings using atmospheric pressure plasma etching. The researchers employed a dielectric barrier discharge (DBD) plasma device working within a gas mixture of helium and oxygen (O_2). This process allowed them to selectively remove the PMMA matrix from the surface, revealing the ZnO particles and forming a hierarchical topography (Dimitrakellis et al., 2019). In a similar study, Dimitrakellis et al. used the same plasma device to etch ZnO/PMMA coatings, increasing the exposure of ZnO particles on the surface by removing the organic PMMA phase. The findings indicated increased antimicrobial effectiveness against *E. coli* after plasma etching (Dimitrakellis et al., 2021). In another study by Rezaei et al., an atmospheric pressure oxygen DBD device was used for the surface modification of PMMA. The authors reported that increasing the plasma treatment time resulted in rougher surface areas due to enhanced plasma etching. Additionally, the antimicrobial properties of the surface against *E. coli* was intensified after plasma treatment due to the increased roughness from plasma etching (Rezaei et al., 2016).

5.2 *Surface Activation, Plasma-Induced Grafting and Plasma Curing*

In addition to using plasma-based approaches to regulate surface topography and introduce nanostructures, these methods can also modify the surface chemistry, which is beneficial for engineering antimicrobial surfaces. When packaging materials are subjected to a plasma environment—comprising excited atomic, molecular, ionic, and free-radical species—multiple processes can occur, like plasma activation or plasma-induced graft polymerization, as shown in Fig. 7. The specific outcome depends on the type of plasma used and the applied operational parameters. In the plasma activation process, a gas discharge is used to directly treat the material surface, incorporating desirable functional groups to modify characteristics such as surface adhesion and wettability, as indicated in Fig. 7a. Gases like N_2, O_2, or NH_3 are commonly used to introduce oxygen or nitrogen functional groups, thereby enhancing surface hydrophilicity (Lamichhane et al., 2023; Ma et al., 2023). Additionally, inert gases like argon and helium are often employed for surface activation to form free radicals, facilitating the incorporation of oxygen functional groups after exposure to the ambient environment (Bitar et al., 2018). Plasma activation can also be part of a two-stage process combined with additional polymerization methods, referred to as plasma-induced graft polymerization. In this method, plasma activation creates free radicals and functional groups on the material surface, which then form covalent bonds with selected monomers under specific conditions, resulting in the formation of a top-layer coating, as shown in Fig. 7b (Kyzio & Kyzio, 2018).

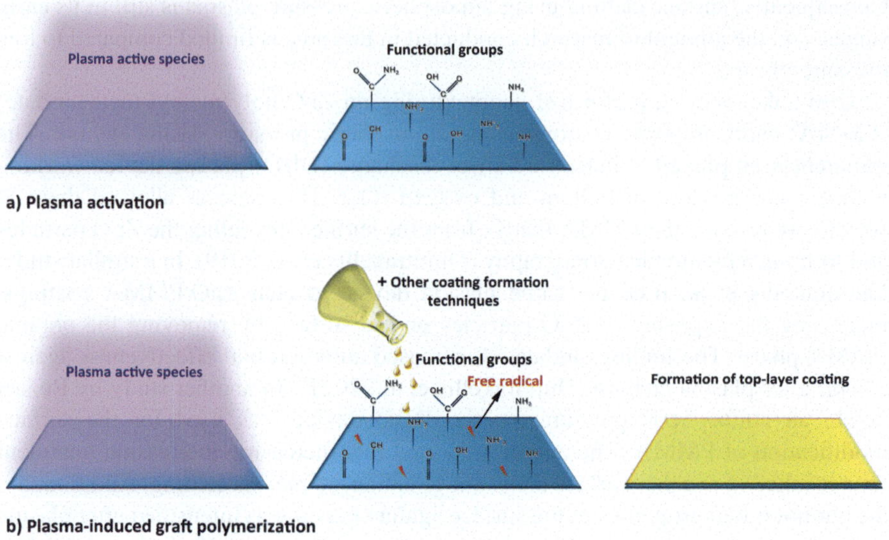

Fig. 7 Plasma activation and plasma-induced graft polymerization for engineering antimicrobial surfaces

Studies show that plasma activation can enhance the antimicrobial loading on material surfaces and control the release of these agents by modifying surface properties. For example, β-tricalcium phosphate ceramics treated with an atmospheric pressure helium plasma jet were subsequently loaded with the antibiotic doxycycline hyclate and compared to ceramics loaded without plasma activation. The results demonstrated that plasma treatment increased the surface O/C ratio and generated surface charges, thereby altering the surface chemistry without affecting the morphology. These surface modifications led to higher antibiotic loading levels and a more precise control over antibiotic release (Canal et al., 2016).

In another similar study, an atmospheric pressure corona discharge using air as discharge gas was employed to activate the surface of PP surgical meshes. The results indicated that plasma surface activation functionalized the surface with polar oxygen groups, resulting in higher surface wettability and increased surface roughness. These modifications created more bonding sites for attaching ampicillin molecules, leading to approximately three times higher ampicillin loading in the mesh compared to untreated meshes (Labay et al., 2015). Moradi et al. used an air atmospheric pressure plasma jet to activate the surface of LDPE, enhancing its ability to load potato peel extract. This treatment aimed to create an antimicrobial coating to prolong the shelf life of chicken thighs. The plasma-treated LDPE film, loaded with 20% potato peel extract, showed a higher release rate of phenolic compounds and greater antimicrobial activity compared to untreated samples. This improvement increased the shelf life of raw chicken thighs while stored in cold conditions by up to 6 days. The enhanced effectiveness is linked to the formation of binding sites on the LDPE surface due to plasma activation, which improved the loading efficiency of the extract (Moradi et al., 2023). Kongboonkird et al. studied the effect of argon DBD plasma surface activation on the antimicrobial activity of a chitosan film. The findings showed that 60 s of plasma treatment resulted in the formation of oxygen-containing groups on the surface and etching, resulting in enhanced surface roughness. The existence of these oxygen-containing groups, along with the increased surface roughness, enhanced the antimicrobial activity compared to untreated chitosan (Kongboonkird et al., 2023). In another study, an air atmospheric pressure plasma jet was employed for the surface activation of polyimide (PI) composite nonwoven fabrics to enhance their antimicrobial properties. After plasma activation, the antimicrobial rate improved by 24.9% against *S. aureus* and 50.6% against *E. coli* compared to untreated samples. The enhanced antimicrobial activity was ascribed to the generation of more nitrogen-functional groups on the surface, such as NH_3^+ and NH_4^+, as well as the presence of free, locally reactive oxygen and nitrogen species, including O, O_3, H_2O_2, NO, NO_2, and their derivatives, known for their antimicrobial characteristics (Lu et al., 2024).

Plasma activation can also be used to induce the on-demand delivery of loaded antimicrobial agents. Gaur et al. developed a PVA hydrogel matrix incorporated with the antibiotic gentamicin. They employed an argon atmospheric plasma jet to activate the hydrogel, leading to the efficient release and targeted delivery of gentamicin, which effectively eradicated bacteria both in the planktonic state and within a biofilm. They also investigated the efficiency of plasma-activated delivery for other antimicrobial agents, including silver and cetrimide, which were also successful (Gaur et al., 2023).

Kliewer et al. conducted air atmospheric pressure plasma surface activation on PE foils to induce graft polymerization of quaternary ammonium compounds (QAC) based on vinylbenzyl trimethylammonium chloride (VBTAC) monomers. The plasma treatment facilitated subsequent radical polymerization in highly concentrated aqueous monomer solutions, resulting in the formation of a homogeneous polymer layer with a thickness of 300 nm and a remarkable charge density. The material obtained in this study demonstrated robust antimicrobial activity against *P. aeruginosa*, *E. coli*, *S. aureus*, and *S. epidermidis* bacteria (Kliewer et al., 2020). In another study, air DBD plasma operating at atmospheric pressure was employed to generate oxygen-containing functional groups on the surface of wiping materials composed of PET and cellulose. This treatment facilitated surface interaction with antimicrobial alkyldimethylbenzylammonium chloride (ADBAC) agents (Song et al., 2019).

Plasma activation can also be used for cross-linking active agents on the surface and grafting them to the substrate, a process known as plasma curing. In plasma curing, the substrate is immersed in a solution of the selected active agents, dried, and then treated with plasma, as indicated in Fig. 8. This approach is particularly effective for active agents that contain double bonds in their structure, such as vinyl groups. Tsou et al. compared the efficiency of conventional plasma-induced grafting of quaternary methyl diallyl ammonium salt on the surface of PP nonwoven with plasma curing after immersing the material in a solution of quaternary methyl diallyl ammonium salt and subsequent drying. Both processes aimed to achieve

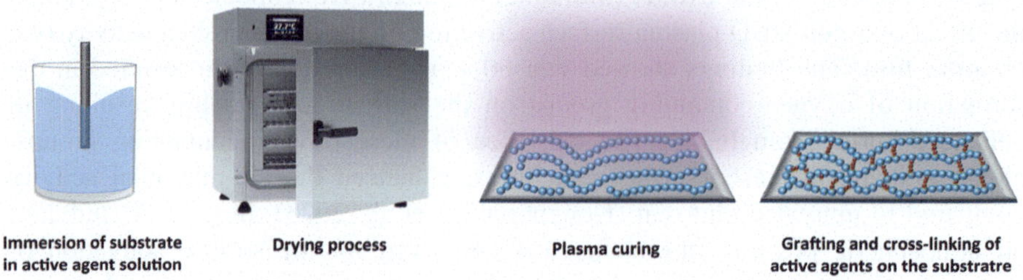

Fig. 8 Plasma curing process for grafting and cross-linking the active agents on the surface of a substrate

antibacterial and hydrophilic surface properties and employed the same plasma setup operating in either oxygen or argon. The results indicated that plasma curing with an oxygen discharge was more effective, incorporating higher levels of peroxides or carboxylic acid groups onto the PP nonwoven and enhancing hydrophilic properties. The PP fabric obtained through plasma curing exhibited higher antibacterial properties compared to the fabric treated with the traditional plasma grafting process (Tsou et al., 2018). In the plasma curing process, the presence of a cross-linker can enhance the stability and durability of the coating. Malshe et al. employed an air atmospheric pressure plasma jet to impregnate diallyldimethylammonium chloride (DADMAC) onto a 50–50 nylon-cotton fabric with pentaerythritol tetraacrylate (PETA) as a cross-linking agent. Initially, the fabrics were treated with a 4 wt% monomer solution containing 5 mol% cross-linker. The plasma-cured fabric demonstrated a strong antimicrobial effect against *Klebsiella pneumoniae* (*K. pneumoniae*) and *S. aureus* (Malshe et al., 2012).

5.3 Plasma Polymerization

Plasma-based techniques can be directly implemented to deposit thin layer coatings on selected substrate materials. To achieve this, appropriate precursors should be introduced into the discharge plasma area as a form of a vapor or liquid, as shown in Fig. 9. Plasma polymerization involves complex reactions that differ significantly from those in conventional polymerization techniques. During plasma polymerization, precursor molecules collide with highly energetic electrons and other plasma active species present in the discharge plasma, causing the precursor molecules to break into electrons, ions, radicals, and excited molecules. These particles randomly recombine and deposit on the substrate (Coad et al., 2022). Unlike conventional polymerization methods, this technique involves continuous fragmentation and random recombination of precursor fragments, leading to the formation of highly cross-linked polymers. The coating characteristics can be adjusted through varying the plasma operational parameters, and the nature of the precursor plays a critical

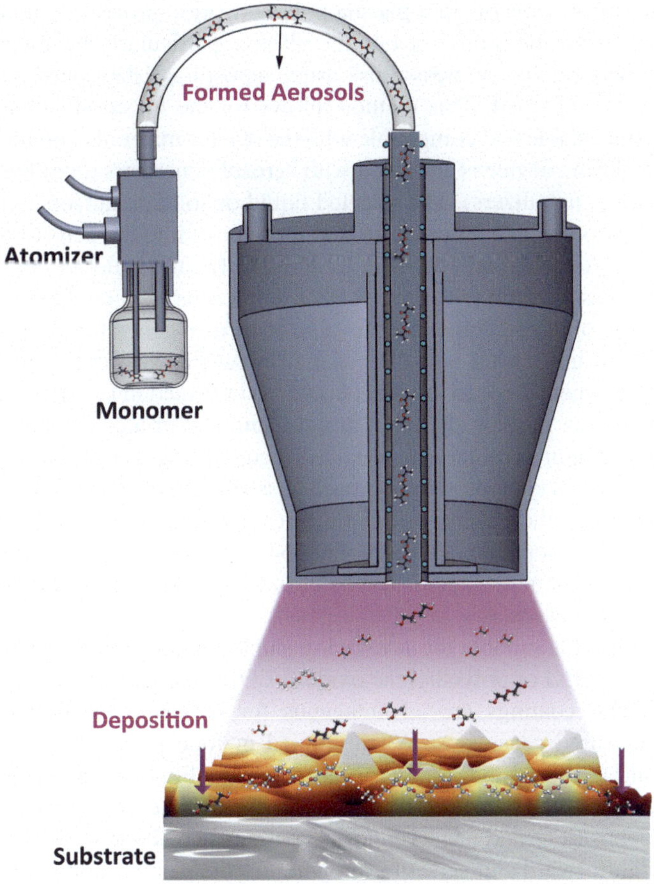

Fig. 9 A schematic of an AAPD process

role. Consequently, optimization of the plasma polymerization process is necessary for each specific precursor (Asandulesa et al., 2010; Hegemann et al., 2019). Plasma polymers typically possess a highly cross-linked structure with significant branching, contributing to their chemical and thermal stability. This method can produce ultra-thin coatings that do not affect the substrate's bulk properties. Additionally, it allows precise control over deposition growth and final thickness. By adjusting plasma operating conditions such as input power, frequency, and the distance between the plasma and the substrate, the properties of the plasma polymerized coating can be finely tuned. Plasma polymers also generally exhibit excellent adhesion to various substrates, including metals, ceramics, and other polymers, due to the strong bonds formed by the energetic species in the plasma with the substrate surface (Biederman, 2004; Carneiro de Oliveira et al., 2021).

One of the main concerns in a plasma polymerization process is the appropriate injection method for the precursor into the plasma, particularly for precursors with low vapor pressure. To address this issue, aerosol-assisted plasma deposition (AAPD) can be employed. This method allows for the injection of any precursor that is adequately stable and compatible with the reactor materials into the plasma. In this case, the plasma device is equipped with aerosol generators (commonly referred to as atomizers or nebulizers). The selected liquid or solid precursors, which can be used pure or dissolved in a solvent, are first introduced into the aerosol generator to form aerosols. Aerosols are defined as a mixture of liquid or solid particles suspended in a gas medium with droplet sizes below tens of micrometers, depending on the properties of the precursor and the type of aerosol generator used. The formed aerosols are then introduced into the plasma. In AAPD, two main approaches are typically used to inject the feed into the plasma, as indicated in Fig. 10. In approach I, the monomer (either pure or dispersed in a solvent) (Ia) or a dispersion of an active agent (such as an antimicrobial agent) in the precursor (Ib) is atomized in the plasma using a single aerosol generator. In approach II, a solution or suspension of the active agent is atomized in the plasma using one aerosol generator, while the precursor is injected through a separate auxiliary line or a second aerosol generator. It is important to note that active agents can be utilized to form depositions even without the presence of a precursor.

The approach Ia is suitable for developing single-component depositions, such as various forms of PEG or polyethylene oxide (PEO) coatings that feature antifouling properties. These coatings are advantageous for preventing the attachment and growth of undesired organisms. When applied in food packaging materials, they can significantly enhance the safety, shelf life, and quality of packaged food. On the other hand, approaches Ib and II are more appropriate for engineering multi-component depositions and nanocomposite coatings in a single step. These methods are both straightforward and versatile, allowing for the incorporation of both organic and inorganic active agents into the plasma polymer coating. Method II offers better control over the relative abundance of active agents in the final coating compared to method Ib. In method Ib, where atomization is conducted through a single atomizer, adjusting the concentration in the deposited coating necessitates altering the

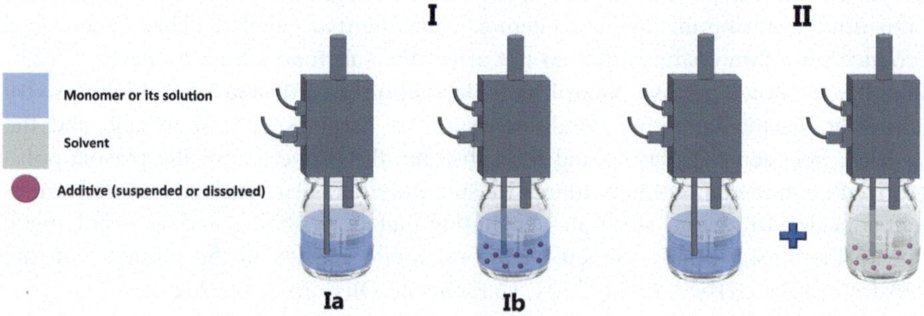

Fig. 10 Scheme of possible coupling of atomization sources to atmospheric pressure plasma

composition of the dispersion in the liquid feed. Conversely, in method II, the two coating components are introduced through separate sources, enabling continuous adjustment of their relative abundance in the plasma feed by varying the flow rates in the aerosol generator or the auxiliary line. Examples of antimicrobial coating deposition using AAPD will be provided below.

Da Ponte et al. developed antifouling PEO-like coatings employing an aerosol-assisted atmospheric pressure DBD plasma operated with helium and fed with aerosols of a tetraethylene glycol dimethyl ether (TEGDME) precursor. The author achieved antifouling coatings by adjusting multiple plasma parameters such as precursor flow ratio, frequency, and plasma input power. Through these adjustments, they successfully created PEO-like coatings that chemically resembled conventional PEO coatings (Da Ponte et al., 2012).

Nisol et al. developed a PEG-like coating with antifouling properties using an aerosol-assisted atmospheric pressure DBD plasma polymerization process with TEGDME as precursor. It should be noted that in this work, the same precursor was used to develop PEG polymer, whereas in the previous study, it was used to produce PEO. PEG and PEO are essentially the same polymer, but they are distinguished by their molecular weight: PEG typically refers to the polymer when its molecular weight is below 20,000 g/mol, and it is referred to as PEO when the molecular weight exceeds this value. The researchers evaluated the impact of plasma input power on the antifouling characteristics and found that a high plasma input power caused increased fragmentation of the monomer, leading to a loss of C-O-C functional groups in the PEG polymer and decreased protein repellence. Conversely, at lower plasma input power (< 50 W), the resulting coating chemically resembled the conventional PEG polymer and exhibited higher protein repellence (Nisol et al., 2010).

Liguori et al. developed plasma-polymerized polyacrylic acid (PAA)/Ag NPs nanocomposite coatings using an aerosol-assisted atmospheric pressure plasma jet operating in Ar gas. In this process, AA and Ag NPs dispersed in ethanol were employed as precursors and individually introduced into the plasma zone using two atomizers. The nanocomposite coating containing 5 wt% Ag NPs demonstrated excellent antimicrobial activity against *E. coli* (Liguori et al., 2016).

Fanelli et al. reported the deposition of an organic-inorganic nanocomposite coating using aerosol-assisted atmospheric pressure DBD operated with helium. In this system, a dispersion of oleate-capped ZnO NPs in n-octane was introduced into the plasma via an atomizer. The resulting coatings exhibited both the chemical properties of the oleate-capped ZnO NPs and a PE-like matrix initiated by the plasma polymerization of n-octane. The deposition featured a micro/nanostructured surface morphology, with the PE-like coating immobilizing quasi-spherical NPs agglomerates within the matrix (Fanelli et al., 2014).

In another study, an aerosol-assisted atmospheric pressure jet was employed to deposit siloxane/TiO_2 NPs nanocomposite coatings. In this process, a dispersion of NPs in tetramethyl orthosilicate (TMOS) precursor was injected into the plasma through an atomizer. Various alcohols, including methanol, octanol, and pentanol, were added to the dispersion to enhance NP stability in the precursor liquid and to prevent NP agglomeration. The influence of these different alcohols on the features of the deposited coatings was evaluated. The results indicated that the coatings obtained from TMOS/TiO_2, with or without methanol, were approximately identical. This was primarily due to the evaporation of methanol in the plasma, given its lower boiling point (65 °C) compared to the maximum temperature of the plasma measured at the jet exit (77 °C). However, the addition of octanol and pentanol to the TMOS/TiO_2 mixture resulted in the deposition of much thicker coatings. This was due to the incorporation of the alcohol precursors into the siloxane deposition, as confirmed by the observation of peaks related to the functional groups of the alcohols in the Fourier-transform infrared spectroscopy (FTIR) spectra of the coatings. This is attributable to the higher boiling points of octanol (196 °C) and pentanol (136 °C) (Dembele et al., 2011).

The AAPD process can also be employed to embed antimicrobial bioactive components in the polymer matrix, with the active component release controlled by adjusting the cross-linking degree of the polymer through plasma parameter modifications. These antimicrobial bioactive components generally have complex and sensitive structures, which must be preserved to maintain their bioactivity. This preservation is challenging in a plasma environment. However, the AAPD process offers a solution. When using an additional solvent, it forms a protective shell around the active agents, shielding them from the energetic active species in the plasma (Heyse et al., 2011). This capability will be demonstrated hereafter in some interesting examples.

Hsiao et al. employed an aerosol-assisted atmospheric pressure DBD plasma system to incorporate lysozyme into a PE polymer matrix. The authors used ethylene as precursor, introduced through an auxiliary line to the plasma zone, and for lysozyme injection, they employed an aerosol generator with two different solvents: deionized water and PBS. The results indicated that higher deposition rates were achieved with lysozyme-PBS aerosols compared to lysozyme-deionized water aerosols. However, the lysozyme showed lower viability in the PBS solution. According to this study, water proved to be a better solvent for preserving the structure of lysozyme in the plasma (Hsiao et al., 2016).

An aerosol-assisted atmospheric pressure pulsed DBD plasma was used to deposit a catechol-bearing coating known for its antimicrobial effects. A solution of dopamine acrylamide in 2-hydroxyethylmethacrylate was injected into the plasma zone through an atomizer. By varying the pulsing of the electrical discharge, the catechol content in the deposited coating was successfully controlled (Jalaber et al., 2019).

In another study, D-limonene, a cyclic monoterpene found primarily in citrus fruit peels and some other plants, was used to produce an antibacterial nano-thin coating using an aerosol-assisted atmospheric pressure jet employing Ar as discharge gas. Surface chemistry analysis indicated partial preservation of the D-limonene structure. However, oxygen-containing groups (C–O, C=O, and O–C=O), which are not originally present in the D-limonene structure, were also detected on the surface. The formation of these oxygen-containing groups is likely due to free radical reactions occurring in ambient air during plasma polymerization. The resulting coating demonstrated strong antibacterial activity against *E. coli* (Masood et al., 2023).

The AAPD process can also be employed for the encapsulation of antibacterial agents. In this regard, Lo Porto et al. prepared a vancomycin-containing nanocapsules coating using a one-step aerosol-assisted atmospheric pressure DBD plasma. In this approach, ethylene was used as precursor and introduced into the plasma directly through an auxiliary line, while a dispersion of vancomycin in water was separately atomized into the plasma. Once ethylene molecules and the aerosol droplets that contain water and vancomycin enter the plasma zone, the partial vaporization of aerosol droplets and the dissociation of water occurs, along with the fragmentation and activation of the ethylene precursor. This results in the polymerization of the precursor both on the substrate and on the surface of the aerosol droplets. As polymerization proceeds, the aerosol droplets, whether coated or uncoated, adsorb onto the surface of the substrate. With continued ethylene polymerization on the substrate, the droplets containing vancomycin become practically covered and embedded within the developing hydrocarbon-like matrix. The deposited coating indicated strong antibacterial activity against *S. aureus* (Lo Porto et al., 2018; Porto et al., 2017). A comparable study was conducted using a similar plasma polymerization process to develop gentamicin-containing nanocapsules coatings. The findings indicated that the size and quantity of the nanocapsules were influenced by the plasma input power and deposition time. The antibacterial activity of gentamicin was retained in the plasma-deposited coatings, showing strong effectiveness against both *S. aureus* and *P. aeruginosa*. Additionally, preliminary cytocompatibility tests showed that eukaryotic cells tolerated the release of gentamicin from the coatings well (Palumbo et al., 2018). A similar plasma polymerization process was also used to embed lysozyme in an organic matrix with ethylene as precursor. The findings indicated that the chemical structure of the embedded lysozyme was slightly modified by the plasma process. However, its activity remained close to that of pure, untreated lysozyme, as confirmed by tests against *Micrococcus lysodeikticus* (*M. lysodeikticus)* based on ATCC 4698 (Palumbo et al., 2015).

Research has reported the development of nanocapsules through AAPD, particularly when the film-forming precursor and the active agent are injected separately into the plasma (Wang et al., 2021). According to literature, the mechanism of nanocapsule formation involves two main steps based on the behavior of the aerosol droplets formed from active agent solutions or dispersions during the evaporation process in the plasma:

1. The solvent in the aerosol droplets evaporates at a constant rate (K), reducing the droplet diameter.
2. The active agent molecules in the droplet begin to aggregate, forming solid aggregates with a morphology dependent on the Peclet number (P_e):

$$P_e = K/D$$

where P_e is the mass diffusion coefficient of the active agent (solute) in the solvent. When solute diffusion is faster than solvent evaporation (low P_e value), the solute molecules can diffuse to the center of the droplet, rearranging and packing there. Conversely, when solvent evaporation is faster than solute diffusion (high P_e value), the solute molecules mainly remain at the droplet surface due to their slow diffusion. This leads to the formation of a solid crust, preventing further reduction of the droplet diameter. After crust formation, the solvent can still evaporate through molecular diffusion through the crust (Mezhericher et al., 2007; Sadafi et al., 2014). In the plasma zone, aerosol droplets become negatively charged due to interactions with electrons and are surrounded by precursor molecules or formed fragments. Initially, the solvent starts to evaporate, followed by the formation of solid aggregates depending on the P_e value. At high P_e, a crust forms due to solute aggregates on the surface, fixing the droplet diameter. At low P_e, packed particles are formed. The polymerization of the precursor can occur on the solid aggregates, which are later deposited on the substrate and further coated by continuous plasma polymerization of the precursor. For high P_e values, due to the crust formation and preserved droplet diameter, the deposited particles will exhibit a spherical shape, as shown in Fig. 11 (Palumbo et al., 2018; Zhang et al., 2023).

This mechanism further clarifies the results of research conducted by Lo Porto et al., who studied the effect of aerosol solution composition, including the concentration and nature of the solute, on nanocapsule formation during an AAPD process. The researchers indicated that for a specific solute with a high P_e value, increasing the concentration in the dispersion resulted in a more spherical shape in the obtained deposition due to facilitated crust formation. Conversely, in the absence of solute in the aerosol droplets (only solvent), a flat deposition was achieved due to the lack of solid aggregates. Furthermore, increasing the molecular weight (MW) of the solute facilitated crust formation on the droplet surface due to the solute's low diffusion, leading to the formation of larger nanocapsules. In contrast, a low MW solute, which diffuses quickly and packs at the center of the droplet, formed smaller nanocapsules. This also explains the deposition of flat coatings for lipid solutes, as they do not form solid aggregates. For solutes with surfactant properties (but not lipids), more and larger nanocapsules are expected due to their tendency to accumulate on the droplet surface and accelerate crust formation. The authors compared the morphology of coatings obtained by an aerosol-assisted atmospheric pressure DBD plasma using sodium dodecyl sulfate (SDS) and pentahydrate copper sulfate ($CuSO_4{\cdot}5H_2O$)

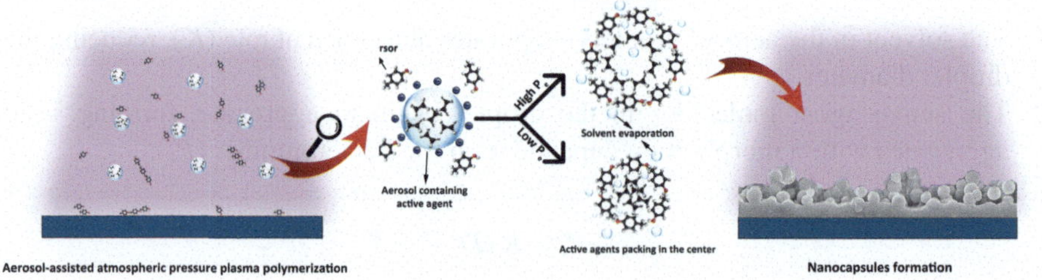

Fig. 11 Formation of nanocapsules during aerosol-assisted atmospheric pressure plasma polymerization

dispersions in water, with ethylene as precursor. Both solutes were applied at similar concentrations and had relatively similar MW. However, the coating obtained with SDS possessed more and larger nanocapsules compared to the coating obtained with $CuSO_4{\cdot}5H_2O$, due to the surfactant nature of SDS. The negatively charged aerosol droplets in the plasma can also influence nanocapsule formation. In this case, polar solutes tend to accumulate at the interface, enhancing crust formation. In contrast, non-polar solutes can freely diffuse to the center of the droplet during solvent evaporation (Palumbo et al., 2020; Lo Porto et al., 2019).

5.4 Drawbacks and Constraints

The application of plasma technology to engineer antimicrobial surfaces, particularly for antimicrobial food packaging, is a relatively new concept compared to previously discussed systems. Although significant antimicrobial efficiency has been achieved against targeted bacteria, to our knowledge, there are no reports on the use of antimicrobial plasma-engineered materials in direct contact with real food. Previous research shows that the antimicrobial efficiency can differ significantly when tested on real food compared to results obtained on agar. In many of these studies, materials that demonstrated remarkable antimicrobial activity on agar showed little or no effectiveness on real food. This discrepancy is due to the interaction between active antimicrobial agents and food components, which limits their efficiency and hinders practical application (Wang et al., 2020; Bahmid & Siddiqui, 2024). Hence, it is important to continue exploring the efficiency of plasma-engineered antimicrobial materials with real food and investigate their interactions with food components more profoundly.

References

Abudonia, K. S., Saad, G. R., Naguib, H. F., Eweis, M., Zahran, D., & Elsabee, M. Z. (2018). Surface modification of polypropylene film by grafting with vinyl monomers for the attachment of chitosan. *Journal of Polymer Research, 25*, 1–11.

Aday, M. S., & Yener, U. (2015). Assessing consumers' adoption of active and intelligent packaging. *British Food Journal, 117*(1), 157–177.

Aga, M. B., Dar, A. H., Nayik, G. A., Panesar, P. S., Allai, F., Khan, S. A., et al. (2021). Recent insights into carrageenan-based bio-nanocomposite polymers in food applications: A review. *International Journal of Biological Macromolecules, 192*, 197–209.

Agrimonti, C., White, J. C., Tonetti, S., & Marmiroli, N. (2019). Antimicrobial activity of cellulosic pads amended with emulsions of essential oils of oregano, thyme and cinnamon against microorganisms in minced beef meat. *International Journal of Food Microbiology, 305*, 108246.

Ahmed, S., Sameen, D. E., Lu, R., Li, R., Dai, J., Qin, W., et al. (2022). Research progress on antimicrobial materials for food packaging. *Critical Reviews in Food Science and Nutrition, 62*(11), 3088–3102.

Akhtar, J., Omre, P. K., & Ahmad Azad, Z. R. A. (2015). Edible coating for preservation of perishable foods: A review. *Journal of Ready to Eat Food, 2*(3), 81–88.

Alghoraibi, I., & Alomari, S. (2018). Different methods for nanofiber design and fabrication. In *Handbook of nanofibers* (pp. 1–46). Springer.

Al-Naamani, L., Dobretsov, S., & Dutta, J. (2016). Chitosan-zinc oxide nanoparticle composite coating for active food packaging applications. *Innovative Food Science and Emerging Technologies, 38*, 231–237.

Al-Tayyar, N. A., Youssef, A. M., & Al-Hindi, R. (2020). Antimicrobial food packaging based on sustainable Bio-based materials for reducing foodborne pathogens: A review. *Food Chemistry, 310*, 125915.

Alvarado, D. R., Argyropoulos, D. S., Scholle, F., Peddinti, B. S. T., & Ghiladi, R. A. (2019). A facile strategy for photoactive nanocellulose-based antimicrobial materials. *Green Chemistry, 21*(12), 3424–3435.

Alves, J., Gaspar, P. D., Lima, T. M., & Silva, P. D. (2023). What is the role of active packaging in the future of food sustainability? A systematic review. *Journal of the Science of Food and Agriculture, 103*(3), 1004–1020.

Amiri, A., Sourestani, M. M., Mortazavi, S. M. H., Kiasat, A. R., & Ramezani, Z. (2022). Fabrication of the antimicrobial sachet by encapsulation of peppermint essential oil in active packaging of strawberry fruit. *Journal of Food Processing & Preservation, 46*(12), e17181.

Andrade, R. D., Skurtys, O., & Osorio, F. A. (2012). Atomizing spray systems for application of edible coatings. *Comprehensive Reviews in Food Science and Food Safety, 11*(3), 323–337.

Apicella, A., Scarfato, P., Di Maio, L., & Incarnato, L. (2019). Sustainable active PET films by functionalization with antimicrobial bio-coatings. *Frontiers in Materials, 6*, 243.

Ariga, K., & Hill, J. (2007). Layer-by-layer assembly as a versatile bottom-up nanofabrication technique for exploratory research and realistic application. *Phys Chem Chem Phys* [Internet], *9*(19), 2319–2340. https://pubs.rsc.org/en/content/articlehtml/2007/cp/b700410a. Cited 25 Mar 2019.

Arroyo, B. J., Bezerra, A. C., Oliveira, L. L., Arroyo, S. J., de Melo, E. A., & Santos, A. M. P. (2020). Antimicrobial active edible coating of alginate and chitosan add ZnO nanoparticles applied in guavas (Psidium guajava L.). *Food Chemistry, 309*, 125566.

Asandulesa, M., Topala, I., Pohoata, V., & Dumitrascu, N. (2010). Influence of operational parameters on plasma polymerization process at atmospheric pressure. *Journal of Applied Physics, 108*(9).

Ayala Zavala, J. F., & Gonzalez Aguilar, G. A. (2010). Optimizing the use of garlic oil as antimicrobial agent on fresh cut tomato through a controlled release system. *Journal of Food Science, 75*(7), M398–M405.

Ayers, E. A. (2018). *Evaluation of a flexographically applied antimicrobial coating for ready-to-eat turkey deli meat packaging*. Clemson University.

Azlin-Hasim, S., Cruz-Romero, M. C., Cummins, E., Kerry, J. P., & Morris, M. A. (2016). The potential use of a layer-by-layer strategy to develop LDPE antimicrobial films coated with silver nanoparticles for packaging applications. *Journal of Colloid and Interface Science, 461*, 239–248.

Azlin-Hasim, S., Cruz-Romero, M. C., Morris, M. A., Cummins, E., & Kerry, J. P. (2018). Spray coating application for the development of nanocoated antimicrobial low-density polyethylene films to increase the shelf life of chicken breast fillets. *Food Science and Technology International, 24*(8), 688–698.

Baek, J. H., Lee, S. Y., & Oh, S. W. (2021). Enhancing safety and quality of shrimp by nanoparticles of sodium alginate-based edible coating containing grapefruit seed extract. *International Journal of Biological Macromolecules, 189*, 84–90.

Bahmid, N. A., & Siddiqui, S. A. (2024). Factors influencing release, absorption, and concentration of natural compounds in antimicrobial packaging. *Food Review International, 40*(6), 1554–1580.

Bandyopadhyay, S., Saha, N., Zandraa, O., Pummerová, M., & Sáha, P. (2020). Essential oil based PVP-CMC-BC-GG functional hydrogel sachet for "cheese": Its shelf life confirmed with anthocyanin (Isolated from red cabbage) bio stickers. *Food, 9*(3), 307.

Bassolé, I. H. N., & Juliani, H. R. (2012). Essential oils in combination and their antimicrobial properties. *Molecules, 17*(4), 3989–4006.

Bastarrachea, L. J. (2019). Antimicrobial polypropylene with -poly (lysine): Effectiveness under UV-A light and food storage applications. *LWT, 102*, 276–283.

Bastarrachea, L. J., & Goddard, J. M. (2015). Antimicrobial coatings with dual cationic and N-halamine character: characterization and biocidal efficacy. *Journal of Agricultural and Food Chemistry, 63*(16), 4243–4251.

Bastarrachea, L. J., & Goddard, J. M. (2016). Self-healing antimicrobial polymer coating with efficacy in the presence of organic matter. *Applied Surface Science, 378*, 479–488.

Bastarrachea, L., Dhawan, S., & Sablani, S. S. (2011, June 15). Engineering properties of polymeric-based antimicrobial films for food packaging. *Food Engineering Reviews, 3*(2), 79–93.

Bastarrachea, L. J., McLandsborough, L. A., Peleg, M., & Goddard, J. M. (2014). Antimicrobial n halamine modified polyethylene: Characterization, biocidal efficacy, regeneration, and stability. *Journal of Food Science, 79*(5), E887–E897.

Bautista, L., Molina, L., Niembro, S., García, J. M., López, J., & Vílchez, A. (2017). Coatings and inks for food packaging including nanomaterials. In *Emerging nanotechnologies in food science* (pp. 149–173). Elsevier.

Benelli, L., & Oliveira, W. P. (2019). Fluidized bed coating of inert cores with a lipid-based system loaded with a polyphenol-rich Rosmarinus officinalis extract. *Food and Bioproducts Processing, 114*, 216–226.

Bertuzzi, M. A., & Slavutsky, A. M. (2016). Standard and new processing techniques used in the preparation of films and coatings at the lab level and scale-up. In *Edible films and coatings* (pp. 21–42). CRC Press.

Bhushani, J. A., & Anandharamakrishnan, C. (2014). Electrospinning and electrospraying techniques: Potential food based applications. *Trends in Food Science and Technology, 38*(1), 21–33.

Biederman, H. (2004). *Plasma polymer films* (pp. 15–16). Imperial College Press.

Birck, C., Degoutin, S., Maton, M., Neut, C., Bria, M., Moreau, M., et al. (2016). Antimicrobial citric acid/poly (vinyl alcohol) crosslinked films: Effect of cyclodextrin and sodium benzoate on the antimicrobial activity. *LWT- Food Science and Technology, 68*, 27–35.

Bitar, R., Cools, P., De Geyter, N., & Morent, R. (2018). Acrylic acid plasma polymerization for biomedical use. *Applied Surface Science, 448*, 168–185.

Bosquez-Molina, E., Guerrero-Legarreta, I., & Vernon-Carter, E. J. (2003). Moisture barrier properties and morphology of mesquite gum-candelilla wax based edible emulsion coatings. *Food Research International, 36*(9–10), 885–893.

Boulos, M. I., Fauchais, P. L., & Pfender, E. (2023). Thermodynamic properties of non-equilibrium plasmas. In *Handbook of thermal plasmas* (pp. 385–426). Springer.

Bozkurt, S., Altay, Ö., Alemdar, F., Türker, M., Koç, M., & Kaymak-Ertekin, F. (2023). The effect of different materials on the coating of baker's yeast (Saccharomyces cerevisiae) with the fluidized bed method. *Journal of Coatings Technology and Research*, 1–16.

Brody, A. L., Strupinsky, E. P., & Kline, L. R. (2001). *Active packaging for food applications*. CRC Press.

Bugatti, V., Vertuccio, L., Viscusi, G., & Gorrasi, G. (2018). Antimicrobial membranes of bio-based PA 11 and HNTs filled with lysozyme obtained by an electrospinning process. *Nanomaterials, 8*(3), 139.

Bustamante-Torres, M., Arcentales-Vera, B., Estrella-Nuñez, J., Yánez-Vega, H., & Bucio, E. (2022). Antimicrobial activity of composites-based on biopolymers. *Macromolecules, 2*(3), 258–283.

Canal, C., Modic, M., Cvelbar, U., & Ginebra, M. P. (2016). Regulating the antibiotic drug release from -tricalcium phosphate ceramics by atmospheric plasma surface engineering. *Biomaterials Science, 4*(10), 1454–1461.

Carneiro de Oliveira, J., Airoudj, A., Kunemann, P., Bally-Le Gall, F., & Roucoules, V. (2021). Mechanical properties of plasma polymer films: A review. *SN Applied Sciences, 3*(6), 656.

Castro-Rosas, J., Ferreira-Grosso, C. R., Gómez-Aldapa, C. A., Rangel-Vargas, E., Rodríguez-Marín, M. L., Guzmán-Ortiz, F. A., et al. (2017). Recent advances in microencapsulation of natural sources of antimicrobial compounds used in food-A review. *Food Research International, 102*, 575–587.

Ceballos, R. L., Von Bilderling, C., Guz, L., Bernal, C., & Famá, L. (2021). Effect of greenly synthetized silver nanoparticles on the properties of active starch films obtained by extrusion and compression molding. *Carbohydrate Polymers, 261*, 117871.

Chaemsanit, S., Sukmas, S., Matan, N., & Matan, N. (2019). Controlled release of peppermint oil from paraffin coated activated carbon contained in sachets to inhibit mold growth during long term storage of brown rice. *Journal of Food Science, 84*(4), 832–841.

Chang, Y., Choi, I., Cho, A. R., & Han, J. (2017). Reduction of Dickeya chrysanthemi on fresh-cut iceberg lettuce using antimicrobial sachet containing microencapsulated oregano essential oil. *LWT- Food Science and Technology, 82*, 361–368.

Chaschin, I. S., Perepelkin, E. I., & Dyachenko, V. I. (2023). Study of the antimicrobial action of chitosan coating doped with fluorinated quinolines. *Polymer Science, Series B*, 1–12.

Clark, J. R., & Mattey, M. (2020). Anti-bacterial packaging. U.S. Patent Application 16/650,703. *Google Patents.*

Coad, B. R., Favia, P., Vasilev, K., & Griesser, H. J. (2022). Plasma polymerization for biomedical applications: A review. *Plasma Processes and Polymers, 19*(11), 2200121.

Colak, B. Y., Peynichou, P., Galland, S., Oulahal, N., Prochazka, F., & Degraeve, P. (2016). Antimicrobial activity of nisin and natamycin incorporated sodium caseinate extrusion blown films: A comparative study with heat pressed/solution cast films. *Journal of Food Science, 81*(5), E1141–E1150.

Colín-Chávez, C., Virgen-Ortiz, J., Miranda-Ackerman, M. A., Hernández-Cristóbal, O., Martínez-Téllez, M. Á., Esquivel-Chávez, F., & Gallegos-Santoyo, N. (2022). Induction of defense mechanisms in avocado using Mexican oregano oil-based antifungal sachet. *Future Foods, 6*, 100171.

Corrales, M., Fernández, A., & Han, J. H. (2014). Antimicrobial packaging systems. In *Innovations in food packaging* (2nd ed., pp. 133–170). Academic.

Črešnar, K. P., Zamboulis, A., Bikiaris, D. N., Aulova, A., & Zemlji, L. F. (2022). Kraft lignin/tannin as a potential accelerator of antioxidant and antibacterial properties in an active thermoplastic polyester-based multifunctional material. *Polymers (Basel), 14*(8), 1532.

Cruz, R. M. S. S., Alves, V., Khmelinskii, I., & Vieira, M. C. (2018). New food packaging systems. In *Food packaging and preservation* (pp. 63–85). Academic.

Da Ponte, G., Sardella, E., Fanelli, F., d'Agostino, R., Gristina, R., & Favia, P. (2012). Plasma deposition of PEO like coatings with aerosol assisted dielectric barrier discharges. *Plasma Processes and Polymers, 9*(11–12), 1176–1183.

Dainelli, D., Gontard, N., Spyropoulos, D., Zondervan-van den Beuken, E., & Tobback, P. (2008). Active and intelligent food packaging: Legal aspects and safety concerns. *Trends in Food Science and Technology, 19*, S103–S112.

Darie-Ni, R. N., Râpă, M., Sivertsvik, M., Rosnes, J. T., Popa, E. E., Dumitriu, R. P., et al. (2021). PLA-based materials containing bio-plasticizers and chitosan modified with rosehip seed oil for ecological packaging. *Polymers (Basel), 13*(10), 1610.

Das, S., Vishakha, K., Banerjee, S., Mondal, S., & Ganguli, A. (2020). Sodium alginate-based edible coating containing nanoemulsion of Citrus sinensis essential oil eradicates planktonic and sessile cells of food-borne pathogens and increased quality attributes of tomatoes. *International Journal of Biological Macromolecules, 162*, 1770–1779.

Dastan, D., Panahi, S. L., & Chaure, N. B. (2016). Characterization of Titania thin films grown by dip-coating technique. *Journal of Materials Science: Materials in Electronics, 27*, 12291–12296.

Davidson, P. M. (2020). Parabens. In *Antimicrobials in food* (pp. 405–420). CRC Press.

Debeaufort, F., & Voilley, A. (2009). Lipid-based edible films and coatings. In *Edible films and coatings for food applications* (pp. 135–168). Springer.

Dembele, A., Rahman, M., Reid, I., Twomey, B., MacElroy, J. M., & Dowling, D. P. (2011). Deposition of hybrid organic-inorganic composite coatings using an atmospheric plasma jet system. *Journal of Nanoscience and Nanotechnology, 11*(10), 8730–8737.

Dey, A., & Neogi, S. (2019). Oxygen scavengers for food packaging applications: A review. *Trends in Food Science and Technology, 90*, 26–34.

Dimitrakellis, P., Patsidis, A. C., Smyrnakis, A., Psarras, G. C., & Gogolides, E. (2019). Atmospheric plasma nanotexturing of organic-inorganic nanocomposite coatings for multifunctional surface fabrication. *ACS Applied Nano Materials, 2*(5), 2969–2978.

Dimitrakellis, P., Kaprou, G. D., Papavieros, G., Mastellos, D. C., Constantoudis, V., Tserepi, A., et al. (2021). Enhanced antibacterial activity of ZnO-PMMA nanocomposites by selective plasma etching in atmospheric pressure. *Micro and Nano Engineering, 13*, 100098.

Doan, L., & Tran, K. (2023). Relationship between the polymer blend using chitosan, polyethylene glycol, polyvinyl alcohol, polyvinylpyrrolidone, and antimicrobial activities against staphylococcus aureus. *Pharmaceutics, 15*(10), 2453.

Duda-Chodak, A., Tarko, T., & Petka-Poniatowska, K. (2023). Antimicrobial compounds in food packaging. *International Journal of Molecular Sciences, 24*(3), 2457.

Durdag, K., Pendleton, B., Hamlyn, R., Gunn, V., & Etchells, M. (2014). Biodegradable polymer non-woven absorbent pad with absorbency and antimicrobial chemistry. *Google Patents.*

EFSA. (2012). EFSA Panel on food contact materials flavourings and processing aids (CEF) enzymes. Scientific opinion on the safety evaluation of the substance, titanium nitride, nanoparticles, for use in food contact materials. *EFSA Journal, 10*(3), 2641.

El Gohary, H. G., Asnag, G. M., Tarabiah, A. E., Qahtan, T. F., Abdelrazek, E. M., Cevik, E., et al. (2023). Modification and development of optical, thermal, dielectric properties and antibacterial activity of PVA/SA blend by Ag/Se nanofillers: Nanocomposites for energy storage devices and food packaging applications. *Polymer Testing, 129*, 108258.

Emamifar, A., Kadivar, M., Shahedi, M., & Soleimanian-Zad, S. (2011). Effect of nanocomposite packaging containing Ag and ZnO on inactivation of Lactobacillus plantarum in orange juice. *Food Control, 22*(3–4), 408–413.

Erdem Büyükkiraz, M., & Kesmen, Z. (2022). Antimicrobial peptides (AMPs): A promising class of antimicrobial compounds. *Journal of Applied Microbiology, 132*(3), 1573–1596.

Esmailzadeh, H., Sangpour, P., Shahraz, F., Hejazi, J., & Khaksar, R. (2016). Effect of nanocomposite packaging containing ZnO on growth of Bacillus subtilis and Enterobacter aerogenes. Materials Science and Engineering: C [Internet], *58*, 1058–1063. https://www.scopus.com/inward/record.uri?eid=2-s2.0-84942474547&doi=10.1016%2Fj.msec.2015.09.078&partnerID=40&md5=4aa4f9cae1931684aa02c4e0479e8cea

Esmailzadeh, H., Sangpour, P., Shahraz, F., Eskandari, A., Hejazi, J., & Khaksar, R. (2021). CuO/LDPE nanocomposite for active food packaging application: A comparative study of its antibacterial activities with ZnO/LDPE nanocomposite. *Polymer Bulletin, 78*, 1671–1682.

Espitia, P. J. P., Soares, N. D. F. F., Botti, L. C. M., Da Silva, W. A., De Melo, N. R., & Pereira, O. L. (2011). Active sachet: Development and evaluation for the conservation of Hawaiian papaya quality. *Italian Journal of Food Science, 23*, 107.

Espitia, P. J. P., Soares, N. D. F. F., Botti, L. C. M., de Melo, N. R., Pereira, O. L., & da Silva, W. A. (2012). Assessment of the efficiency of essential oils in the preservation of postharvest papaya in an antimicrobial packaging system. *Brazilian Journal of Food Technology, 15*, 333–342.

Esquivel-Chávez, F., Colín-Chávez, C., Virgen-Ortiz, J. J., Martínez-Téllez, M. Á., Avena-Bustillos, R. D. J., Peña-Madrigal, G., et al. (2021). Control of mango decay using antifungal sachets containing of thyme oil/modified starch/agave fructans microcapsules. *Future Foods, 3*, 100008.

Fadiji, T., Rashvand, M., Daramola, M. O., & Iwarere, S. A. (2023, February 15). A review on antimicrobial packaging for extending the shelf life of food. *Process, 11*, 590 [Internet], *11*(2), 590. https://www.mdpi.com/2227-9717/11/2/590/htm. Cited 11 Jan 2024.

Fanelli, F., & Fracassi, F. (2017). Atmospheric pressure non-equilibrium plasma jet technology: General features, specificities and applications in surface processing of materials. *Surface and Coatings Technology, 322*, 174–201.

Fanelli, F., Mastrangelo, A. M., & Fracassi, F. (2014, January 28). Aerosol-assisted atmospheric cold plasma deposition and characterization of superhydrophobic organic-inorganic nanocomposite thin films. *Langmuir, 30*(3), 857–865.

Fasihnia, S. H., Peighambardoust, S. H., Peighambardoust, S. J., & Oromiehie, A. (2018). Development of novel active polypropylene based packaging films containing different concentrations of sorbic acid. *Food Packaging and Shelf Life, 18*, 87–94.

Fernandez, A., Picouet, P., & Lloret, E. (2010). Reduction of the spoilage-related microflora in absorbent pads by silver nanotechnology during modified atmosphere packaging of beef meat. *Journal of Food Protection, 73*(12), 2263–2269.

Fernández-Pan, I., Carrión-Granda, X., & Maté, J. I. J. I. (2014). Antimicrobial efficiency of edible coatings on the preservation of chicken breast fillets. *Food Control, 36*(1), 69–75.

Ferraboschi, P., Ciceri, S., & Grisenti, P. (2021). Applications of lysozyme, an innate immune defense factor, as an alternative antibiotic. *Antibiotics, 10*(12), 1534.

Figueroa-Lopez, K. J., Cabedo, L., Lagaron, J. M., & Torres-Giner, S. (2020). Development of electrospun poly (3-hydroxybutyrate-co-3-hydroxyvalerate) monolayers containing eugenol and their application in multilayer antimicrobial food packaging. *Frontiers in Nutrition, 7*, 140.

Fitzgerald, D. J., Stratford, M., Gasson, M. J., Ueckert, J., Bos, A., & Narbad, A. (2004). Mode of antimicrobial action of vanillin against Escherichia coli, Lactobacillus plantarum and Listeria innocua. *Journal of Applied Microbiology, 97*(1), 104–113.

Fridman, A. A., Keidar, M., & Choi, E. H. (2023). Introduction to nonthermal atmospheric pressure plasma: Physical and chemical basis. In *Plasma biosciences and medicine* (pp. 1–34). Springer.

Fu, Y., & Dudley, E. G. (2021). Antimicrobial coated films as food packaging: A review. *Comprehensive Reviews in Food Science and Food Safety, 20*(4), 3404–3437.

Fu, Y., Sarkar, P., Bhunia, A. K., & Yao, Y. (2016 Nov). Delivery systems of antimicrobial compounds to food. *Trends in Food Science and Technology, 1*(57), 165–177.

Galie, S., García-Gutiérrez, C., Miguélez, E. M., Villar, C. J., & Lombó, F. (2018). Biofilms in the food industry: Health aspects and control methods. *Frontiers in Microbiology, 9*, 898.

Gaspar-Cunha, A., Covas, J., & Sikora, J. (2022). Optimization of polymer processing: A review (part I-Extrusion). *Materials (Basel), 15*(1), 384.

Gaur, N., Patenall, B. L., Ghimire, B., Thet, N. T., Gardiner, J. E., Le Doare, K. E., et al. (2023). Cold atmospheric plasma-activated composite hydrogel for an enhanced and on-demand delivery of antimicrobials. *ACS Applied Materials & Interfaces, 15*(16), 19989–19996.

Goddard, J. M., & Hotchkiss, J. H. (2007). Polymer surface modification for the attachment of bioactive compounds. *Progress in Polymer Science, 32*(7), 698–725.

Golden, C. E., Berrang, M. E., Kerr, W. L., & Harrison, M. A. (2019). Slow-release chlorine dioxide gas treatment as a means to reduce Salmonella contamination on spices. *Innovative Food Science and Emerging Technologies, 52*, 256–261.

Gooch, J. W. (2010). *Encyclopedic dictionary of polymers* (Vol. 1). Springer.

Gouvêa, D. M., Mendonça, R. C. S., Lopez, M. E. S., & Batalha, L. S. (2016). Absorbent food pads containing bacteriophages for potential antimicrobial use in refrigerated food products. *LWT-Food Science and Technology, 67*, 159–166.

Grigoras, A. G. (2021). Natural and synthetic polymeric antimicrobials with quaternary ammonium moieties: A review. *Environmental Chemistry Letters, 19*(4), 3009–3022.

Guan, N., & Liu, L. (2020). Microbial response to acid stress: Mechanisms and applications. *Applied Microbiology and Biotechnology, 104*(1), 51–65.

Gupta, R. K., Guha, P., & Srivastav, P. P. (2022). Natural polymers in bio-degradable/edible film: A review on environmental concerns, cold plasma technology and nanotechnology application on food packaging-A recent trends. *Food Chemistry Advances, 1*, 100135.

Hadj-Hamou, A. S., Metref, F., & Yahiaoui, F. (2017). Thermal stability and decomposition kinetic studies of antimicrobial PCL/nanoclay packaging films. *Polymer Bulletin, 74*, 3833–3853.

Haktaniyan, M., & Bradley, M. (2022). Polymers showing intrinsic antimicrobial activity. *Chemical Society Reviews, 51*(20), 8584–8611.

Hamed, I., Jakobsen, A. N., & Lerfall, J. (2022). Sustainable edible packaging systems based on active compounds from food processing byproducts: A review. *Comprehensive Reviews in Food Science and Food Safety, 21*(1), 198–226.

Han, J. H. (2003). Antimicrobial food packaging. In R. Ahvenainen (Ed.), *Novel food packaging techniques* (pp. 50–70). CRC Press.

Han, J. H. (2014). A review of food packaging technologies and innovations. In *Innovations in food packaging* (2nd ed., pp. 3–12). Academic.

Han, J. H., Patel, D., Kim, J. E., & Min, S. C. (2014). Retardation of Listeria monocytogenes growth in mozzarella cheese using antimicrobial sachets containing rosemary oil and thyme oil. *Journal of Food Science, 79*(11), E2272–E2278.

Han, J. W., Ruiz-Garcia, L., Qian, J. P., & Yang, X. T.. (2018, July 1). Food packaging: A comprehensive review and future trends. Comprehensive Reviews in Food Science and Food Safety [Internet], *17*(4):860–877. https://onlinelibrary.wiley.com/doi/full/10.1111/1541-4337.12343. Cited 8 Jan 2024.

Hao, X., Wang, W., Yang, Z., Yue, L., Sun, H., Wang, H., et al. (2019). pH responsive antifouling and antibacterial multilayer films with self-healing performance. *Chemical Engineering Journal, 356*, 130–141.

Hauser, C., Thielmann, J., & Muranyi, P. (2016). Organic acids: Usage and potential in antimicrobial packaging. In *Antimicrobial food packaging* (pp. 563–580). Elsevier.

Hegemann, D., Nisol, B., Gaiser, S., Watson, S., & Wertheimer, M. R. (2019). Energy conversion efficiency in low-and atmospheric-pressure plasma polymerization processes with hydrocarbons. *Physical Chemistry Chemical Physics, 21*(17), 8698–8708.

Hempel, A. W., Papkovsky, D. B., & Kerry, J. P. (2013). Use of optical oxygen sensors in non-destructively determining the levels of oxygen present in combined vacuum and modified atmosphere packaged pre-cooked convenience-style foods and the use of ethanol emitters to extend product shelf-life. *Food, 2*(4), 507–520.

Henchion, M., McCarthy, M., Dillon, E. J., Greehy, G., & McCarthy, S. N. (2019). Big issues for a small technology: Consumer trade-offs in acceptance of nanotechnology in food. *Innovative Food Science and Emerging Technologies, 58*, 102210.

Hernández-García, E., Vargas, M., González-Martínez, C., & Chiralt, A. (2021). Biodegradable antimicrobial films for food packaging: Effect of antimicrobials on degradation. *Food, 10*(6), 1256.

Herniou Julien, C., & Gutiérrez, T. J. (2021). Reactive extrusion processed shape memory food packaging films made from native and carboxymethylated plantain flour/polystyrene blends. *Starch Stärke, 73*(11–12), 2100053.

Hewson, R. W., Kapur, N., & Gaskell, P. H. (2006). A theoretical and experimental investigation of tri-helical gravure roll coating. *Chemical Engineering Science, 61*(16), 5487–5499.

Heyse, P., Van Hoeck, A., Roeffaers, M. B. J., Raffin, J. P., Steinbüchel, A., Stöveken, T., et al. (2011). Exploration of atmospheric pressure plasma nanofilm technology for straightforward bio active coating deposition: Enzymes, plasmas and polymers, an elegant synergy. *Plasma Processes and Polymers, 8*(10), 965–974.

Honarvar, Z., Farhoodi, M., Khani, M. R., Mohammadi, A., Shokri, B., Ferdowsi, R., et al. (2017). Application of cold plasma to develop carboxymethyl cellulose-coated polypropylene films containing essential oil. *Carbohydrate Polymers, 176*, 1–10.

Hong, S. I:, Wang, L. F., & Rhim, J. W. (2022). Preparation and characterization of nanoclays-incorporated polyethylene/thermoplastic starch composite films with antimicrobial activity. *Food Packaging and Shelf Life, 31*, 100784.
Hsiao, C. P., Wu, C. C., Liu, Y. H., Yang, Y. W., Cheng, Y. C., Palumbo, F., et al. (2016, December 1). Aerosol-assisted plasma deposition of biocomposite coatings: Investigation of processing conditions on coating properties. *IEEE Transactions on Plasma Science, 44*(12), 3091–3098.
Hu, S., Wang, H., Han, W., Ma, Y., Shao, Z., & Li, L. (2017). Development of double layer active films containing pomegranate peel extract for the application of pork packaging. *Journal of Food Process Engineering, 40*(2), e12388.
Hu, Q., Zhou, F., Ly, N. K., Ordyna, J., Peterson, T., Fan, Z., et al. (2023). Development of multifunctional nanoencapsulated trans-resveratrol/chitosan nutraceutical edible coating for strawberry preservation. *ACS Nano, 17*(9), 8586–8597.
Huang, J. Y., Li, X., & Zhou, W. (2015). Safety assessment of nanocomposite for food packaging application. *Trends in Food Science and Technology, 45*(2), 187–199.
Huang, T., Qian, Y., Wei, J., & Zhou, C. (2019). Polymeric antimicrobial food packaging and its applications. *Polymers (Basel), 11*(3), 560.
Huang, K., Ashby, R., Fan, X., Moreau, R. A., Strahan, G. D., Nuñez, A., et al. (2020a). Phenolic fatty acid-based epoxy curing agent for antimicrobial epoxy polymers. *Progress in Organic Coatings, 141*, 105536.
Huang, W. C., Ying, R., Wang, W., Guo, Y., He, Y., Mo, X., et al. (2020b). A macroporous hydrogel dressing with enhanced antibacterial and anti inflammatory capabilities for accelerated wound healing. *Advanced Functional Materials, 30*(21), 2000644.
Hundáková, M., Tokarský, J., Valášková, M., Slobodian, P., Pazdziora, E., & Kimmer, D. (2015). Structure and antibacterial properties of polyethylene/organo-vermiculite composites. *Solid State Sciences, 48*, 197–204.
Huq, T., Khan, A., Brown, D., Dhayagude, N., He, Z., & Ni, Y. (2022). Sources, production and commercial applications of fungal chitosan: A review. *Journal of Bioresources and Bioproducts, 7*(2), 85–98.
Hwang, J. K., Bae, S., & Kim, D. S. (2014). A development and evaluation of micro-gravure coater for printed electronics. *Japanese Journal of Applied Physics, 53*(5S3), 05HC12.
Iemma, F., Puoci, F., Curcio, M., Parisi, O. I., Cirillo, G., Spizzirri, U. G., et al. (2010). Ferulic acid as a comonomer in the synthesis of a novel polymeric chain with biological properties. *Journal of Applied Polymer Science, 115*(2), 784–789.
Ilyas, R. A., Sapuan, S. M., Megashah, L. N., Ibrahim, R., Atikah, M. S. N., Ainun, Z. M. A., et al. (2021). Regulations for food packaging materials. In *Bio-based Packaging: Material, environmental and economic aspects* (pp. 467–494). Wiley.
Irshad, G., Naz, F., & Hafiz, A. A. (2023). Suppression of Fusarium sporotrichioides initiating peach fruit rot with microencapsulated antifungal sachet of clove essential oil tablets. *European Journal of Plant Pathology*, 1–15.
Isopencu, G. O., Stoica-Guzun, A., Busuioc, C., Stroescu, M., & Deleanu, I. M. (2021). Development of antioxidant and antimicrobial edible coatings incorporating bacterial cellulose, pectin, and blackberry pomace. *Carbohydrate Polymer Technologies and Applications, 2*, 100057.
Jalaber, V., Del Frari, D., De Winter, J., Mehennaoui, K., Planchon, S., Choquet, P., et al. (2019). Atmospheric aerosol assisted pulsed plasma polymerization: An environmentally friendly technique for tunable catechol-bearing thin films. *Frontiers in Chemistry, 7*, 183.
Jaworek, A. (2007). Electrospray droplet sources for thin film deposition. *Journal of Materials Science, 42*, 266–297.
Jensen, A., Lim, L. T., Barbut, S., & Marcone, M. (2015). Development and characterization of soy protein films incorporated with cellulose fibers using a hot surface casting technique. *LWT-Food Science and Technology, 60*(1), 162–170.
Jideani, V. A., & Vogt, K.. (2016, June 10). Antimicrobial packaging for extending the shelf life of bread – A review. Critical Reviews in Food Science and Nutrition [Internet], *56*(8), 1313–1324. https://www.tandfonline.com/doi/abs/10.1080/10408398.2013.768198. Cited 9 Jan 2024.

Jokar, M., & Abdul Rahman, R. (2014). Study of silver ion migration from melt-blended and layered-deposited silver polyethylene nanocomposite into food simulants and apple juice. Food Additives and Contaminants: Part A [Internet], *31*(4), 734–742. https://www.scopus.com/inward/record.uri?eid=2-s2.0-84899547699&doi=10.1080%2F19440049.2013.878812&partnerID=40&md5=2dcc8edc5beb0fd448f334634374d3fb

Ju, J., Xie, Y., Yu, H., Guo, Y., Cheng, Y., Qian, H., et al. (2020). A novel method to prolong bread shelf life: Sachets containing essential oils components. *LWT, 131*, 109744.

Kalia, A., & Parshad, V. R. (2015). Novel trends to revolutionize preservation and packaging of fruits/fruit products: Microbiological and nanotechnological perspectives. *Critical Reviews in Food Science and Nutrition, 55*(2), 159–182.

Kapetanakou, A. E., & Skandamis, P. N. (2016). Applications of active packaging for increasing microbial stability in foods: Natural volatile antimicrobial compounds. *Current Opinion in Food Science, 12*(12), 1.

Kapetanakou, A. E., Agathaggelou, E. I., & Skandamis, P. N. (2014). Storage of pork meat under modified atmospheres containing vapors from commercial alcoholic beverages. *International Journal of Food Microbiology, 178*, 65–75.

Kapetanakou, A. E., Stragkas, I. G., & Skandamis, P. N. (2015). Developing an antimicrobial packaging of ready-to-eat pomegranate arils based on vapors of brandy or distillery ethanol. *Food Research International, 69*, 141–150.

Karanth, S., Feng, S., Patra, D., & Pradhan, A. K. (2023). Linking microbial contamination to food spoilage and food waste: The role of smart packaging, spoilage risk assessments, and date labeling. *Frontiers in Microbiology, 14*, 1198124.

Karki, S., Kim, H., Na, S. J., Shin, D., Jo, K., & Lee, J. (2016). Thin films as an emerging platform for drug delivery. *Asian Journal of Pharmaceutical Sciences, 11*(5), 559–574.

Kashiri, M., Cerisuelo, J. P., Domínguez, I., López-Carballo, G., Muriel-Gallet, V., Gavara, R., et al. (2017). Zein films and coatings as carriers and release systems of Zataria multiflora Boiss. Essential oil for antimicrobial food packaging. *Food Hydrocolloids, 70*, 260–268.

Kawhena, T. G., Opara, U. L., & Fawole, O. A. (2021). A comparative study of antimicrobial and antioxidant activities of plant essential oils and extracts as candidate ingredients for edible coatings to control decay in 'Wonderful'pomegranate. *Molecules, 26*(11), 3367.

Kenawy, E. R., Imam Abdel Hay, F., Abou El Magd, A., & Mahmoud, Y. (2006). Synthesis and antimicrobial activity of some polymers derived from modified amino polyacrylamide by reacting it with benzoate esters and benzaldehyde derivatives. *Journal of Applied Polymer Science, 99*(5), 2428–2437.

Khaneghah, A. M., Hashemi, S. M. B., & Limbo, S. (2018). Antimicrobial agents and packaging systems in antimicrobial active food packaging: An overview of approaches and interactions. *Food and Bioproducts Processing, 111*, 1–19.

Kim, I., Viswanathan, K., Kasi, G., Thanakkasaranee, S., Sadeghi, K., & Seo, J. (2022). ZnO nanostructures in active antibacterial food packaging: Preparation methods, antimicrobial mechanisms, safety issues, future prospects, and challenges. *Food Review International, 38*(4), 537–565.

Kliewer, S., Wicha, S. G., Bröker, A., Naundorf, T., Catmadim, T., Oellingrath, E. K., et al. (2020). Contact-active antibacterial polyethylene foils via atmospheric air plasma induced polymerisation of quaternary ammonium salts. *Colloids and Surfaces. B, Biointerfaces, 186*, 110679.

Kongboonkird, M., Chuesiang, P., Ryu, V., & Siripatrawan, U. (2023). Effects of argon dielectric barrier discharge (DBD) cold plasma treatment on mechanical, barrier and antimicrobial properties of chitosan film. *International Journal of Food Science and Technology, 58*(2), 995–1006.

Kougia, E., Tselepi, M., Vasilopoulos, G., Lainioti, G. C., Koromilas, N. D., Druvari, D., et al. (2015). Evaluation of antimicrobial efficiency of new polymers comprised by covalently attached and/or electrostatically bound bacteriostatic species, based on quaternary ammonium compounds. *Molecules, 20*(12), 21313–21327.

Kramer, B., Wunderlich, J., & Muranyi, P. (2018). Impact of volatile allyl isothiocyanate on fresh produce. *Food Packaging and Shelf Life, 16*, 220–224.

Kumar, P., Kizhakkedathu, J. N., & Straus, S. K. (2018). Antimicrobial peptides: diversity, mechanism of action and strategies to improve the activity and biocompatibility in vivo. *Biomolecules, 8*(1), 4.

Kuplennik, N., Tchoudakov, R., Zelas, Z. B. B., Sadovski, A., Fishman, A., & Narkis, M. (2015). Antimicrobial packaging based on linear low-density polyethylene compounded with potassium sorbate. *LWT- Food Science and Technology, 62*(1), 278–286.

Kyzio, A., & Kyzio, K. (2018). Surface functionalization with biopolymers via plasma-assisted surface grafting and plasma-induced graft polymerization-materials for biomedical applications. In *Biopolymer grafting* (pp. 115–151). Elsevier.

La, D. D., Nguyen-Tri, P., Le, K. H., Nguyen, P. T. M., Nguyen, M. D. B., Vo, A. T. K., et al. (2021). Effects of antibacterial ZnO nanoparticles on the performance of a chitosan/gum arabic edible coating for post-harvest banana preservation. *Progress in Organic Coatings, 151*, 106057.

Labay, C., Canal, J. M., Modic, M., Cvelbar, U., Quiles, M., Armengol, M., et al. (2015). Antibiotic-loaded polypropylene surgical meshes with suitable biological behaviour by plasma functionalization and polymerization. *Biomaterials, 71*, 132–144.

Lamichhane, P., Acharya, T. R., Park, J., Amsalu, K. A., Park, B., & Choi, E. H. (2023). Surface activation of thin polyvinyl alcohol films by atmospheric pressure plasma jet: Influence of electron temperature. *Plasma Processes and Polymers, 20*(11), e2300102.

Lantano, C., Alfieri, I., Cavazza, A., Corradini, C., Lorenzi, A., Zucchetto, N., et al. (2014). Natamycin based sol-gel antimicrobial coatings on polylactic acid films for food packaging. *Food Chemistry, 165*, 342–347.

Le, K. H., Nguyen, M. D. B., Dai Tran, L., Thi, H. P. N., Van Tran, C., Van Tran, K., et al. (2021). A novel antimicrobial ZnO nanoparticles-added polysaccharide edible coating for the preservation of postharvest avocado under ambient conditions. *Progress in Organic Coatings, 158*, 106339.

Lee, S. Y. J., Lee, S. Y. J., Choi, D. S., & Hur, S. J. (2015, November 1). Current topics in active and intelligent food packaging for preservation of fresh foods. *95*(14), 2799–2810. https://onlinelibrary.wiley.com/doi/full/10.1002/jsfa.7218. Cited 9 Jan 2024.

Lei, J., Sun, L., Huang, S., Zhu, C., Li, P., He, J., et al. (2019). The antimicrobial peptides and their potential clinical applications. *American Journal of Translational Research, 11*(7), 3919.

Leong, J., Yang, C., Tan, J., Tan, B. Q., Hor, S., Hedrick, J. L., et al. (2020). Combination of guanidinium and quaternary ammonium polymers with distinctive antimicrobial mechanisms achieving a synergistic antimicrobial effect. *Biomaterials Science, 8*(24), 6920–6929.

Li, H. P., Ostrikov, K. K., & Sun, W. (2018). The energy tree: Non-equilibrium energy transfer in collision-dominated plasmas. *Physics Reports, 770*, 1–45.

Li, J., Tian, X., Hua, T., Fu, J., Koo, M., Chan, W., et al. (2021). Chitosan natural polymer material for improving antibacterial properties of textiles. *ACS Applied Bio Materials, 4*(5), 4014–4038.

Liakos, I. L., Holban, A. M., Carzino, R., Lauciello, S., & Grumezescu, A. M. (2017). Electrospun fiber pads of cellulose acetate and essential oils with antimicrobial activity. *Nanomaterials, 7*(4), 84.

Liguori, A., Traldi, E., Toccaceli, E., Laurita, R., Pollicino, A., Focarete, M. L., et al. (2016). Co deposition of plasma polymerized polyacrylic acid and silver nanoparticles for the production of nanocomposite coatings using a non equilibrium atmospheric pressure plasma jet. *Plasma Processes and Polymers, 13*(6), 623–632.

Lin, C. M., Preston, J. F., III, & Wei, C. I. (2000). Antibacterial mechanism of allyl isothiocyanate. *Journal of Food Protection, 63*(6), 727–734.

Linklater, D. P., Juodkazis, S., & Ivanova, E. P. (2017). Nanofabrication of mechano-bactericidal surfaces. *Nanoscale, 9*(43), 16564–16585.

Liu, Z., Deshazer, H., Rice, A. J., Chen, K., Zhou, C., & Kallenbach, N. R. (2006). Multivalent antimicrobial peptides from a reactive polymer scaffold. *Journal of Medicinal Chemistry, 49*(12), 3436–3439.

Liu, Y., Wang, S., & Lan, W. (2018). Fabrication of antibacterial chitosan-PVA blended film using electrospray technique for food packaging applications. International Journal of Biological Macromolecules [Internet], *107*(PartA), 848–854. https://www.scopus.com/inward/record.uri?eid=2-s2.0-85029689245&doi=10.1016%2Fj.ijbiomac.2017.09.044&partnerID=40&md5=9e76a8a9f8387700f786136eb9c830b2

Liu, M., Wang, F., Liang, M., Si, Y., Yu, J., & Ding, B. (2020). In situ green synthesis of rechargeable antibacterial N-halamine grafted poly (vinyl alcohol) nanofibrous membranes for food packaging applications. *Composites Communications, 17*, 147–153.

Lloret, E., Picouet, P., & Fernández, A. (2012). Matrix effects on the antimicrobial capacity of silver based nanocomposite absorbing materials. *LWT- Food Science and Technology, 49*(2), 333–338.

Lo Porto, C., Palumbo, F., Buxadera Palomero, J., Canal, C., Jelinek, P., Zajickova, L., et al. (2018). On the plasma deposition of vancomycin containing nano capsules for drug delivery applications. *Plasma Processes and Polymers, 15*(5), 1700232.

Lo Porto, C., Palumbo, F., Fracassi, F., Barucca, G., & Favia, P. (2019). On the formation of nanocapsules in aerosol assisted atmospheric pressure plasma. *Plasma Processes and Polymers, 16*(11), 1900116.

Lopes, L. F., Meca, G., Bocate, K. C. P., Nazareth, T. M., Bordin, K., & Luciano, F. B. (2018). Development of food packaging system containing allyl isothiocyanate against Penicillium nordicum in chilled pizza: Preliminary study. *Journal of Food Processing & Preservation, 42*(2), e13436.

Lopes, A. I., Melo, A., Caleja, C., Pereira, E., Finimundy, T. C., Afonso, T. B., et al. (2023). Evaluation of antimicrobial and antioxidant activities of alginate edible coatings incorporated with plant extracts. *Coatings, 13*(9), 1487.

Lopez-Moya, F., Suarez-Fernandez, M., & Lopez-Llorca, L. V. (2019). Molecular mechanisms of chitosan interactions with fungi and plants. *International Journal of Molecular Sciences, 20*(2), 332.

Lu, H., Lin, M., Li, T., Zhang, H., Feng, L., & Zhang, C. (2024). Polyimide (PI) composite spunlace nonwovens for hygiene material with excellent comfort and antimicrobial properties. *PRO, 12*(2), 354.

Luo, H., Yin, X. Q., Tan, P. F., Gu, Z. P., Liu, Z. M., & Tan, L. (2021). Polymeric antibacterial materials: Design, platforms and applications. *Journal of Materials Chemistry B, 9*(12), 2802–2815.

Lum, P., Chhetri, V. S., Janes, M. E., King, J. M., & Adhikari, A. (2021). Efficacy of gaseous chlorine dioxide in reducing Salmonella enterica, E. coli O157: H7, and Listeria monocytogenes on strawberries and blueberries. *LWT, 141*, 110906.

Ma, J., Lin, W., Xu, L., Liu, S., Xue, W., & Chen, S. (2020). Resistance to long-term bacterial biofilm formation based on hydrolysis-induced Zwitterion material with biodegradable and self-healing properties. *Langmuir, 36*(12), 3251–3259.

Ma, C., Nikiforov, A., De Geyter, N., Dai, X., Morent, R., & Ostrikov, K. K. (2021). Future antiviral polymers by plasma processing. *Progress in Polymer Science, 118*, 101410.

Ma, C., Nikiforov, A., De Geyter, N., Morent, R., & Ostrikov, K. K. (2022). Plasma for biomedical decontamination: From plasma-engineered to plasma-active antimicrobial surfaces. *Current Opinion in Chemical Engineering, 36*, 100764.

Ma, C., Nikiforov, A., Hegemann, D., De Geyter, N., Morent, R., & Ostrikov, K. (2023). Plasma-controlled surface wettability: Recent advances and future applications. *International Materials Review, 68*(1), 82–119.

Majid, I., Ahmad Nayik, G., Mohammad Dar, S., & Nanda, V. (2018). Novel food packaging technologies: Innovations and future prospective. *Journal of the Saudi Society of Agricultural Sciences, 17*(4), 454–462.

Makarewicz, M., Drod, I., Tarko, T., & Duda-Chodak, A. (2021). The interactions between polyphenols and microorganisms, especially gut microbiota. *Antioxidants, 10*(2), 188.
Malshe, P., Mazloumpour, M., El-Shafei, A., & Hauser, P. (2012). Functional military textile: plasma-induced graft polymerization of DADMAC for antimicrobial treatment on nylon-cotton blend fabric. *Plasma Chemistry and Plasma Processing, 32*, 833–843.
Manuja, R., Sachdeva, S., Jain, A., & Chaudhary, J. (2013). A comprehensive review on biological activities of p-hydroxy benzoic acid and its derivatives. *International Journal of Pharmaceutical Sciences Review and Research, 22*(2), 109–115.
Marques, C. S., Carvalho, S. G., Bertoli, L. D., Villanova, J. C. O., Pinheiro, P. F., Dos Santos, D. C. M., et al. (2019). β-Cyclodextrin inclusion complexes with essential oils: Obtention, characterization, antimicrobial activity and potential application for food preservative sachets. *Food Research International, 119*, 499–509.
Marrez, D. A., Abdelhamid, A. E., & Darwesh, O. M. (2019). Eco-friendly cellulose acetate green synthesized silver nano-composite as antibacterial packaging system for food safety. *Food Packaging and Shelf Life, 20*, 100302.
Masood, A., Ahmed, N., Razip Wee, M. F. M., Patra, A., Mahmoudi, E., & Siow, K. S. (2023). Atmospheric pressure plasma polymerisation of D-Limonene and its antimicrobial activity. *Polymers (Basel), 15*(2), 307.
Matan, N., & Kongchoosi, N. (2023). Prolonging broccoli shelf life through controlled release of plai essential oil using palm wood sachet as a novel absorbent material with reusable properties. *Journal of Food Science*.
Matan, N., Limbo, S., Piergiovanni, L., & Songsamoe, S. (2023). Controlled release of Michelia Alba oil vapour from plastic sachets to control the growth of Aspergillus flavus on brown rice and its possible mode of action. *Food Control, 146*, 109504.
Mathew, A. P., & Oksman, K. (2014). Processing of bionanocomposites: solution casting. In *Handbook of green materials: 2 Bionanocomposites: Processing, characterization and properties* (pp. 35–52). World Scientific.
McKeen, L. W. (2017). Production of films, containers, and membranes. In *Permeability prop plast elastomers* (pp. 41–60).
Meindrawan, B., Ofe, O., Susanto, C. S., Ayman, A., Mangindaan, D., & Kasih, T. P. (2020). Glucomannan-Beeswax-Chitosan antimicrobial edible coating to maintain the storage quality of Salak fruit (Salacca Zalacca). In *Macromolecular symposia* (p. 1900164). Wiley Online Library.
Mesías, F. J., Martín, A., & Hernández, A. (2021). Consumers' growing appetite for natural foods: Perceptions towards the use of natural preservatives in fresh fruit. *Food Research International, 150*, 110749.
Mezhericher, M., Levy, A., & Borde, I. (2007). Theoretical drying model of single droplets containing insoluble or dissolved solids. *Drying Technology, 25*(6), 1025–1032.
Min, T., Zhu, Z., Sun, X., Yuan, Z., Zha, J., & Wen, Y. (2020). Highly efficient antifogging and antibacterial food packaging film fabricated by novel quaternary ammonium chitosan composite. *Food Chemistry, 308*, 125682.
Moradi, D., Ramezan, Y., Eskandari, S., Mirsaeedghazi, H., & Dakheli, M. J. (2023). Plasma-treated LDPE film incorporated with onion and potato peel extract-A food packaging for shelf life extension on chicken thigh. *Food Packaging and Shelf Life, 35*, 101012.
Morgan, S. M., El Sonbati, A. Z., & El Mogazy, M. A. (2018). Polymer complexes. LXX. Synthesis, spectroscopic studies, thermal properties and antimicrobial activity of metal (II) polymer complexes. *Applied Organometallic Chemistry, 32*(4), e4264.
Moridi, A., Hassani-Gangaraj, S. M., Guagliano, M., & Dao, M. (2014). Cold spray coating: review of material systems and future perspectives. *Surface Engineering, 30*(6), 369–395.
Moshe Dvir, I., Weizman, O., Lewitus, D., Weintraub, S., Ophir, A., & Dotan, A. (2019). Antimicrobial active packaging combining essential oils mixture: Migration and odor control study. *Polymers for Advanced Technologies, 30*(10), 2558–2566.
Mouhamad, Y., Mokarian-Tabari, P., Clarke, N., Jones, R. A. L., & Geoghegan, M. (2014). Dynamics of polymer film formation during spin coating. *Journal of Applied Physics, 116*(12).

Moye, Z. D., Woolston, J., & Sulakvelidze, A. (2018). Bacteriophage applications for food production and processing. *Viruses, 10*(4), 205.

Mulla, M., Ahmed, J., Al-Attar, H., Castro-Aguirre, E., Arfat, Y. A., & Auras, R. (2017). Antimicrobial efficacy of clove essential oil infused into chemically modified LLDPE film for chicken meat packaging. *Food Control, 73*, 663–671.

Muñoz-Bonilla, A., & Fernández-García, M. (2012). Polymeric materials with antimicrobial activity. *Progress in Polymer Science, 37*(2), 281–339.

Munteanu, B. S., Sacarescu, L., Vasiliu, A. L., Hitruc, G. E., Pricope, G. M., Sivertsvik, M., et al. (2018). Antioxidant/antibacterial electrospun nanocoatings applied onto PLA films. *Materials (Basel), 11*(10), 1973.

Muriel-Galet, V., Cerisuelo, J. P., López-Carballo, G., Aucejo, S., Gavara, R., & Hernndez-Muoz, P. (2013). Evaluation of EVOH-coated PP films with oregano essential oil and citral to improve the shelf-life of packaged salad. *Food Control, 30*(1), 137–143.

Muszanska, A. K., Rochford, E. T. J., Gruszka, A., Bastian, A. A., Busscher, H. J., Norde, W., et al. (2014). Antiadhesive polymer brush coating functionalized with antimicrobial and RGD peptides to reduce biofilm formation and enhance tissue integration. *Biomacromolecules, 15*(6), 2019–2026.

Muthulakshmi, L., Annaraj, J., Ramakrishna, S., Ranjan, S., Dasgupta, N., Mavinkere Rangappa, S., et al. (2021). A sustainable solution for enhanced food packaging via a science based composite blend of natural sourced chitosan and microbial extracellular polymeric substances. *Journal of Food Processing & Preservation, 45*(1), e15031.

Nair, S. S., Trafiaek, J., & Kolanowski, W. (2023). Edible packaging: A technological update for the sustainable future of the food industry. *Applied Sciences, 13*(14), 8234.

Nakayama, T., Tanaka, K., Nagai, N., Mori, N., Shimizu, M., & Ito, Y. (2018). *Antibacterial material and freshness keeping material*. WIPO (PCT), No. 2019088199.

Nastaj, A., & Wilczyski, K. (2021). Optimization and Scale-Up for Polymer Extrusion. *Polymers (Basel), 13*(10), 1547.

Nikiforov, A., Ma, C., Choukourov, A., & Palumbo, F. (2022). Plasma technology in antimicrobial surface engineering. *Journal of Applied Physics, 131*(1).

Nisol, B., Poleunis, C., Bertrand, P., & Reniers, F. (2010). Poly (ethylene glycol) films deposited by atmospheric pressure plasma liquid deposition and atmospheric pressure plasma-enhanced chemical vapour deposition: Process, chemical composition analysis and biocompatibility. *Plasma Processes and Polymers, 7*(8), 715–725.

Niu, B., Yan, Z., Shao, P., Kang, J., & Chen, H. (2018a). Encapsulation of cinnamon essential oil for active food packaging film with synergistic antimicrobial activity. *Nanomaterials, 8*(8), 598.

Niu, X., Liu, Y., Song, Y., Han, J., & Pan, H. (2018b). Rosin modified cellulose nanofiber as a reinforcing and co-antimicrobial agents in polylactic acid/chitosan composite film for food packaging. *Carbohydrate Polymers, 183*, 102–109.

Nonaka, T., Hua, L., Ogata, T., & Kurihara, S. (2003). Synthesis of water soluble thermosensitive polymers having phosphonium groups from methacryloyloxyethyl trialkyl phosphonium chlorides–N- isopropylacrylamide copolymers and their functions. *Journal of Applied Polymer Science, 87*(3), 386–393.

Ojogbo, E., Ward, V., & Mekonnen, T. H. (2020). Functionalized starch microparticles for contact-active antimicrobial polymer surfaces. *Carbohydrate Polymers, 229*, 115422.

Othman, S. H., Abd Salam, N. R., Zainal, N., Kadir Basha, R., & Talib, R. A. (2014). Antimicrobial activity of TiO 2 nanoparticle-coated film for potential food packaging applications. *International Journal of Photoenergy, 2014*.

Otoni, C. G., Soares, N. D. F. F., da Silva, W. A., Medeiros, E. A. A., & Baffa Junior, J. C. (2014). Use of allyl isothiocyanate containing sachets to reduce Aspergillus flavus sporulation in peanuts. *Packaging Technology and Science, 27*(7), 549–558.

Otoni, C. G., Espitia, P. J. P., Avena-Bustillos, R. J., & McHugh, T. H. (2016). Trends in antimicrobial food packaging systems: Emitting sachets and absorbent pads. *Food Research International, 83*, 60–73.

Ozdemir, M., & Floros, J. D. (2004). Active food packaging technologies. *Critical Reviews in Food Science and Nutrition, 44*(3), 185–193.

Paidari, S., & Ibrahim, S. A. (2021). Potential application of gold nanoparticles in food packaging: a mini review. *Gold Bulletin, 54*, 31–36.

Paidari, S., Zamindar, N., Tahergorabi, R., Kargar, M., Ezzati, S., Shirani, N., et al. (2021). Edible coating and films as promising packaging: A mini review. *Journal of Food Measurement and Characterization, 15*(5), 4205–4214.

Palumbo, F., Camporeale, G., Yang, Y. W., Wu, J. S., Sardella, E., Dilecce, G., et al. (2015). Direct plasma deposition of lysozyme embedded bio composite thin films. *Plasma Processes and Polymers, 12*(11), 1302–1310.

Palumbo, F., Treglia, A., Lo Porto, C., Fracassi, F., Baruzzi, F., Frache, G., et al. (2018). Plasma-deposited nanocapsules containing coatings for drug delivery applications. *ACS Applied Materials & Interfaces, 10*(41), 35516–35525.

Palumbo, F., Lo Porto, C., Fracassi, F., & Favia, P. (2020). Recent advancements in the use of aerosol-assisted atmospheric pressure plasma deposition. *Coatings, 10*(5), 440.

Panchal, N., Das, K., Prabhakar, P. K., & Ghanghas, N. (2022). Edible films and coatings for fruits and vegetables: Composition, functions, and regulatory aspects. In *Edible Food Packaging Applications, Innovations and Sustainability* (pp. 191–216). Springer.

Park, C. K. (2010). *Antimicrobial food packaging film*. South Korea Patent No. 101004027.

Park, E. S., Moon, W. S., Song, M. J., Kim, M. N., Chung, K. H., & Yoon, J. S. (2001). Antimicrobial activity of phenol and benzoic acid derivatives. *International Biodeterioration & Biodegradation, 47*(4), 209–214.

Park, J., Shin, K., & Lee, C. (2016). Roll-to-roll coating technology and its applications: A review. *International Journal of Precision Engineering and Manufacturing, 17*, 537–550.

Pasqualone, A. (2019). Bread packaging: Features and functions. In *Flour and breads and their fortification in health and disease prevention* (pp. 211–222). Elsevier.

Passarinho, A. T. P., Dias, N. F., Camilloto, G. P., Cruz, R. S., Otoni, C. G., Moraes, A. R. F., et al. (2014). Sliced bread preservation through oregano essential oil containing sachet. *Journal of Food Process Engineering, 37*(1), 53–62.

Pawlowski, L. (2009). Suspension and solution thermal spray coatings. *Surface and Coatings Technology, 203*(19), 2807–2829.

Peerzada, G. J., Sinclair, B. J., Perinbarajan, G. K., Dutta, R., Shekhawat, R., Saikia, N., et al. (2023). An overview on smart and active edible coatings: Safety and regulations. *European Food Research and Technology*, 1–18.

Pereira de Abreu, D. A., Cruz, J. M., & Paseiro Losada, P. (2012). Active and intelligent packaging for the food industry. *Food Review International, 28*(2), 146–187.

Perna, M. C. (2016). *Formulation and characterization of an antimicrobial coating containing nisin for large scale food package converting processes*.

Petchwattana, N., Naknaen, P., Cha-Aim, K., Suksri, C., & Sanetuntikul, J. (2021). Controlled release antimicrobial sachet prepared from poly (butylene succinate)/geraniol and ethylene vinyl alcohol coated paper for bread shelf-life extension application. *International Journal of Biological Macromolecules, 189*, 251–261.

Phuinthiang, P., Channei, D., Ratananikom, K., Nakaruk, A., & Khanitchaidecha, W. (2023). Antibacterial properties of TiO2 nano coating on food packaging surfaces against Escherichia coli and Salmonella typhimurium. *Surface Engineering, 39*(4), 433–444.

Pillai, R. R., & Thomas, V. (2023). Plasma surface engineering of natural and sustainable polymeric derivatives and their potential applications. *Polymers (Basel), 15*(2), 400.

Piper, J. D., & Piper, P. W. (2017). Benzoate and sorbate salts: a systematic review of the potential hazards of these invaluable preservatives and the expanding spectrum of clinical uses for sodium benzoate. *Comprehensive Reviews in Food Science and Food Safety, 16*(5), 868–880.

Poaska, M., & Sokoowska, B. (2019). Bacteriophages_a new hope or a huge problem in the food industry. *AIMS Microbiology, 5*(4), 324.

Pogodin, S., Hasan, J., Baulin, V. A., Webb, H. K., Truong, V. K., Boshkovikj, V., et al. (2013). Biophysical model of bacterial cell interactions with nanopatterned cicada wing surfaces. *Biophysical Journal, 104*(4), 835–840.

Porto, C. L. O., Palumbo, F., Palazzo, G., & Favia, P. (2017). Direct plasma synthesis of nanocapsules loaded with antibiotics. *Polymer Chemistry, 8*(11), 1746–1749.

Prakash, A., Baskaran, R., & Vadivel, V. (2020). Citral nanoemulsion incorporated edible coating to extend the shelf life of fresh cut pineapples. *LWT, 118*, 108851.

Priya, K., Thirunavookarasu, N., & Chidanand, D. V. (2023). Recent advances in edible coating of food products and its legislations: A review. *Journal of Agriculture and Food Research*, 100623.

Priyanka, S., Namasivyam, R. S., Bharani, A., & John, A. (2023). Biocompatible green technology principles for the fabrication of food packaging material with noteworthy mechanical and antimicrobial properties–A sustainable developmental goal towards the effective, safe food preservation strategy. *Chemosphere, 336*, 139240.

Puliyalil, H., & Cvelbar, U. (2016). Selective plasma etching of polymeric substrates for advanced applications. *Nanomaterials, 6*(6), 108.

Punia Bangar, S., Chaudhary, V., Thakur, N., Kajla, P., Kumar, M., & Trif, M. (2021). Natural antimicrobials as additives for edible food packaging applications: A review. *Food, 10*(10), 2282.

Pušnik Črešnar, K., Aulova, A., Bikiaris, D. N., Lambropoulou, D., Kuzmi, K., & Fras Zemlji, L. (2021). Incorporation of metal-based nanoadditives into the pla matrix: Effect of surface properties on antibacterial activity and mechanical performance of pla nanoadditive films. *Molecules, 26*(14), 4161.

Qiu, Y. L., Li, Y., Zhang, G. L., Hao, H., Hou, H. M., & Bi, J. (2023). Quaternary-ammonium chitosan, a promising packaging material in the food industry. *Carbohydrate Polymers, 323*, 121384.

Rani, G., Kuroda, K., & Vemparala, S. (2021). Towards designing globular antimicrobial peptide mimics: role of polar functional groups in biomimetic ternary antimicrobial polymers. *Soft Matter, 17*(8), 2090–2103.

Râpă, M., Mitelu, A. C., Tnase, E. E., Grosu, E., Popescu, P., Popa, M. E., et al. (2016). Influence of chitosan on mechanical, thermal, barrier and antimicrobial properties of PLA-biocomposites for food packaging. *Composites. Part B, Engineering, 102*, 112–121.

Ray, S., Jin, T., Fan, X., Liu, L., & Yam, K. L. (2013). Development of chlorine dioxide releasing film and its application in decontaminating fresh produce. *Journal of Food Science, 78*(2), M276–M284.

Ren, T., Qiao, M., Huang, T. S., Weese, J., & Ren, X. (2018). Efficacy of N-halamine compound on reduction of microorganisms in absorbent food pads of raw beef. *Food Control, 84*, 255–262.

Rezaei, F., Shokri, B., & Sharifian, M. (2016). Atmospheric-pressure DBD plasma-assisted surface modification of polymethyl methacrylate: A study on cell growth/proliferation and antibacterial properties. *Applied Surface Science, 360*, 641–651.

Richardson, J. J., Cui, J., Bjornmalm, M., Braunger, J. A., Ejima, H., & Caruso, F. (2016). Innovation in layer-by-layer assembly. *Chemical Reviews, 116*(23), 14828–14867.

Rooney, M. L. (2005). Introduction to active food packaging technologies. In *Innovations in food packaging* [Internet] (pp. 63–79). Food Science Australia/CSIRO. https://www.scopus.com/inward/record.uri?eid=2-s2.0-84882541487&doi=10.1016%2FB978-012311632-1%2F50037-1&partnerID=40&md5=022cdcd8ed631548a46f6736e6f79f7b

Saade, C., Annous, B. A., Gualtieri, A. J., Schaich, K. M., Liu, L., & Yam, K. L. (2017). System feasibility: Designing a chlorine dioxide self-generating package label to improve fresh produce safety part I: Extrusion approach. *Innovative Food Science and Emerging Technologies, 43*, 102–111.

Sadafi, M. H., Jahn, I., Stilgoe, A. B., & Hooman, K. (2014). Theoretical and experimental studies on a solid containing water droplet. *International Journal of Heat and Mass Transfer, 78*, 25–33.

Saeed, M., Azam, M., Saeed, F., Arshad, U., Afzaal, M., Bader Ul Ain, H., et al. (2021). Development of antifungal edible coating for strawberry using fruit waste. *Journal of Food Processing & Preservation, 45*(11), e15956.
Salas-Ambrosio, P., Tronnet, A., Verhaeghe, P., & Bonduelle, C. (2020). Synthetic polypeptide polymers as simplified analogues of antimicrobial peptides. *Biomacromolecules, 22*(1), 57–75.
Santos, J. C. P., Sousa, R. C. S., Otoni, C. G., Moraes, A. R. F., Souza, V. G. L., Medeiros, E. A. A., et al. (2018). Nisin and other antimicrobial peptides: Production, mechanisms of action, and application in active food packaging. *Innovative Food Science and Emerging Technologies, 48*, 179–194.
Sanyang, M. L., Sapuan, S. M., Jawaid, M., Ishak, M. R., & Sahari, J. (2015). Effect of plasticizer type and concentration on tensile, thermal and barrier properties of biodegradable films based on sugar palm (Arenga pinnata) starch. *Polymers (Basel), 7*(6), 1106–1124.
Schiffmann, J., & Schlösser, H. (2011). Antimicrobial food packaging. U.S. Patent No. 7,915,325. *Google Patents*.
Sedlarik, V. (2013). Antimicrobial modifications of polymers. In *Biodegradation science* (pp. 187–204). IntechOpen.
Seo, H. S., Bang, J., Kim, H., Beuchat, L. R., Cho, S. Y., & Ryu, J. H. (2012). Development of an antimicrobial sachet containing encapsulated allyl isothiocyanate to inactivate Escherichia coli O157: H7 on spinach leaves. *International Journal of Food Microbiology, 159*(2), 136–143.
Shahmohammadi, M., Javadi, M., & Nassiri-Asl, M. (2016). An overview on the effects of sodium benzoate as a preservative in food products. *Biotechnology and Health Sciences, 3*(3), 7–11.
Shankar, S., & Rhim, J. W. (2017). Facile approach for large-scale production of metal and metal oxide nanoparticles and preparation of antibacterial cotton pads. *Carbohydrate Polymers, 163*, 137–145.
Sharma, R., Jafari, S. M., & Sharma, S. (2020). Antimicrobial bio-nanocomposites and their potential applications in food packaging. *Food Control, 112*, 107086.
Sharma, V. K., Sharma, M., Usmani, Z., Pandey, A., Singh, B. N., Tabatabaei, M., et al. (2022a). Tailored enzymes as next-generation food-packaging tools. *Trends in Biotechnology*.
Sharma, B., Singh, S., Pandey, A., Dutt, D., & Kulshreshtha, A. (2022b). Sustainable and green manufacturing of gravure printing cylinder for flexible packaging printing application. *Scientific Reports, 12*(1), 16266.
Sharma, H., Ahuja, A., Sharma, B., Kulshreshtha, A., Kadam, A., & Dutt, D. (2023). Vapor phase antimicrobial active packaging application of chitosan capsules containing clove essential oil for the preservation of dry cakes. *Food and Bioprocess Technology*, 1–11.
Sheng, Q., Guo, X. N., & Zhu, K. X. (2015). The effect of active packaging on microbial stability and quality of Chinese steamed bread. *Packaging Technology and Science, 28*(9), 775–787.
Shim, E. (2013). Bonding requirements in coating and laminating of textiles. In *Joining textiles* (Elsevier ed., pp. 309–351).
Shim, E. (2019). Coating and laminating processes and techniques for textiles. In *Smart textile coatings and laminates* (pp. 11–45). Elsevier.
Shin, J., Harte, B., Selke, S., & Lee, Y. (2011). Use of a controlled chlorine dioxide (ClO2) release system in combination with modified atmosphere packaging (MAP) to control the growth of pathogens. *Journal of Food Quality, 34*(3), 220–228.
Shuai, C., Xu, Y., Feng, P., Wang, G., Xiong, S., & Peng, S. (2019). Antibacterial polymer scaffold based on mesoporous bioactive glass loaded with in situ grown silver. *Chemical Engineering Journal, 374*, 304–315.
Silva, F., Domingues, F. C., & Nerín, C. (2018). Control microbial growth on fresh chicken meat using pinosylvin inclusion complexes based packaging absorbent pads. *LWT, 89*, 148–154.
Silva, S. P. M., Ribeiro, S. C., Teixeira, J. A., & Silva, C. C. G. (2022). Application of an alginate-based edible coating with bacteriocin-producing Lactococcus strains in fresh cheese preservation. *LWT, 153*, 112486.

Singh, A., Kumari, K., & Kundu, P. P. (2017). Extrusion and evaluation of chitosan assisted AgNPs immobilized film derived from waste polyethylene terephthalate for food packaging applications. *Journal of Packaging Technology and Reseach, 1*, 165–180.
Singh, S., Maji, P. K., Lee, Y. S., & Gaikwad, K. K. (2021). Applications of gaseous chlorine dioxide for antimicrobial food packaging: A review. *Environmental Chemistry Letters, 19*, 253–270.
Skurtys, O. (2010). *Food hydrocolloid edible films and coatings*. Nova Sciences Publishers.
Soares, N. M., Fernandes, T. A., & Vicente, A. A. (2016). Effect of variables on the thickness of an edible coating applied on frozen fish-Establishment of the concept of safe dipping time. *Journal of Food Engineering, 171*, 111–118.
Solórzano-Santos, F., & Miranda-Novales, M. G. (2012). Essential oils from aromatic herbs as antimicrobial agents. *Current Opinion in Biotechnology, 23*(2), 136–141.
Sommers, K. J., Michaud, M. E., Hogue, C. E., Scharnow, A. M., Amoo, L. E., Petersen, A. A., et al. (2022). Quaternary phosphonium compounds: An examination of non-nitrogenous cationic amphiphiles that evade disinfectant resistance. *ACS Infectious Diseases, 8*(2), 387–397.
Song, X., Cvelbar, U., Strazar, P., Vossebein, L., & Zille, A. (2019). Antimicrobial efficiency and surface interactions of quaternary ammonium compound absorbed on dielectric barrier discharge (DBD) plasma treated fiber-based wiping materials. *ACS Applied Materials & Interfaces, 12*(1), 298–311.
Soysal, Ç., Bozkurt, H., Dirican, E., Güçlü, M., Bozhüyük, E. D., Uslu, A. E., et al. (2015). Effect of antimicrobial packaging on physicochemical and microbial quality of chicken drumsticks. *Food Control, 54*, 294–299.
Sriwattana, S., Torpol, K., Prinyawiwatkul, W., & Sangsuwan, J. (2023). Efficacy of chitosan pectin beads encapsulated with combined garlic and holy basil essential oils on shelf life extension of ready to eat food. *International Journal of Food Science and Technology, 58*(2), 921–928.
Stoleru, E., Munteanu, S. B., Dumitriu, R. P., Coroaba, A., Drobot, M., Zemljic, L. F., et al. (2016a). Polyethylene materials with multifunctional surface properties by electrospraying chitosan/vitamin E formulation destined to biomedical and food packaging applications. *Iranian Polymer Journal, 25*, 295–307.
Stoleru, E., Zaharescu, T., Hitruc, E. G., Vesel, A., Ioanid, E. G., Coroaba, A., et al. (2016b). Lactoferrin-immobilized surfaces onto functionalized PLA assisted by the gamma-rays and nitrogen plasma to create materials with multifunctional properties. *ACS Applied Materials & Interfaces, 8*(46), 31902–31915.
Su, H., Chen, Z., Zhao, Y., An, J., Huang, H., Liu, R., et al. (2023). Polyvinyl alcohol film with chlorine dioxide microcapsules can be used for blueberry preservation by slow-release of chlorine dioxide gas. *Frontiers in Nutrition, 10*, 1177950.
Suhag, R., Kumar, N., Petkoska, A. T., & Upadhyay, A. (2020). Film formation and deposition methods of edible coating on food products: A review. *Food Research International, 136*, 109582.
Suhem, K., Songsamoe, S., & Matan, N. (2022). Effects of bamboo sachets containing Litsea cubeba oil on the prevention of mold for extending the shelf life of dried fish, its reusability, and action mechanisms. *LWT, 154*, 112796.
Sun, X., Zhou, B., Luo, Y., Ference, C., Baldwin, E., Harrison, K., et al. (2017a). Effect of controlled-release chlorine dioxide on the quality and safety of cherry/grape tomatoes. *Food Control, 82*, 26–30.
Sun, X., Baldwin, E., Plotto, A., Narciso, J., Ference, C., Ritenour, M., et al. (2017b). Controlled-release of chlorine dioxide in a perforated packaging system to extend the storage life and improve the safety of grape tomatoes. *Journal of Visualized Experiments: JoVE, 122*, e55400.
Sun, H., Hong, Y., Xi, Y., Zou, Y., Gao, J., & Du, J. (2018). Synthesis, self-assembly, and biomedical applications of antimicrobial peptide-polymer conjugates. *Biomacromolecules, 19*(6), 1701–1720.

Sun, X., Wang, J., Dong, M., Zhang, H., Li, L., & Wang, L. (2022). Food spoilage, bioactive food fresh-keeping films and functional edible coatings: Research status, existing problems and development trend. *Trends in Food Science and Technology, 119*, 122–132.

Sung, S. Y. Y., Sin, L. T., Tee, T. T. T., Bee, S. T. T., Rahmat, A. R., Rahman, W. A. W. A., et al. (2013). Antimicrobial agents for food packaging applications. *Trends in Food Science and Technology, 33*(2), 110–123.

Takma, D. K., & Korel, F. (2019). Active packaging films as a carrier of black cumin essential oil: Development and effect on quality and shelf-life of chicken breast meat. *Food Packaging and Shelf Life, 19*, 210–217.

Tan, J., Zhao, Y., Hedrick, J. L., & Yang, Y. Y. (2022). Effects of hydrophobicity on antimicrobial activity, selectivity, and functional mechanism of guanidinium functionalized polymers. *Advanced Healthcare Materials, 11*(7), 2100482.

Tang, X., Xu, H., Shi, Y., Wu, M., Tian, H., & Liang, J. (2020). Porous antimicrobial starch particles containing N-halamine functional groups. *Carbohydrate Polymers, 229*, 115546.

Tas, B. A., Sehit, E., Tas, C. E., Unal, S., Cebeci, F. C., Menceloglu, Y. Z., et al. (2019). Carvacrol loaded halloysite coatings for antimicrobial food packaging applications. *Food Packaging and Shelf Life, 20*, 100300.

Teck Kim, Y., Min, B., & Won, K. K. (2014). General characteristics of packaging materials for food system. In *Innovations in food packaging* (2nd ed., pp. 13–35). Academic.

Thanakkasaranee, S., Kim, D., & Seo, J. (2018). Preparation and characterization of polypropylene/ sodium propionate (PP/SP) composite films for bread packaging application. *Packaging Technology and Science, 31*(4), 221–231.

Thomassin, J. M., Lenoir, S., Riga, J., Jérôme, R., & Detrembleur, C. (2007). Grafting of poly [2-(tert-butylamino) ethyl methacrylate] onto polypropylene by reactive blending and antibacterial activity of the copolymer. *Biomacromolecules, 8*(4), 1171–1177.

Tiwari, A. (2017). *Handbook of antimicrobial coatings*. Elsevier.

Tracz, B. L., Bordin, K., Bocate, K. C. P., Hara, R. V., Luz, C., Macedo, R. E. F., et al. (2018). Devices containing allyl isothiocyanate against the growth of spoilage and mycotoxigenic fungi in mozzarella cheese. *Journal of Food Processing & Preservation, 42*(11), e13779.

Tric, B., Tritean, N., Constantinescu-Aruxandei, D., & Oancea, F. (2023). Optimization of the mung bean seed coating with alginate in a bottom-sprayed wurster fluidized bed coater. *Coatings, 13*(3), 562.

Tsou, C. H., Yao, W. H., Hung, W. S., Suen, M. C., De Guzman, M., Chen, J., et al. (2018). Innovative plasma process of grafting methyl diallyl ammonium salt onto polypropylene to impart antibacterial and hydrophilic surface properties. *Industrial and Engineering Chemistry Research, 57*(7), 2537–2545.

Uz, M., & Altnkaya, S. A. (2011). Development of mono and multilayer antimicrobial food packaging materials for controlled release of potassium sorbate. *LWT- Food Science and Technology, 44*(10), 2302–2309.

Velaga, S. P., Nikjoo, D., & Vuddanda, P. R. (2018). Experimental studies and modeling of the drying kinetics of multicomponent polymer films. *AAPS PharmSciTech, 19*, 425–435.

Velázquez-Nuñez, M. J., Avila-Sosa, R., Palou, E., & López-Malo, A. (2013). Antifungal activity of orange (Citrus sinensis var. Valencia) peel essential oil applied by direct addition or vapor contact. *Food Control, 31*(1), 1–4.

Venkatesan, R., & Rajeswari, N. (2019). Preparation, mechanical and antimicrobial properties of SiO 2/poly (butylene adipate-co-terephthalate) films for active food packaging. *Silicon* [Internet], *11*, 2233–2239. https://www.scopus.com/inward/record.uri?eid=2-s2.0-84964326638&doi=10.1007%2Fs12633-015-9402-8&partnerID=40&md5=320d4a64c7d77dfebba59bc8ea9d2da0.

Vidal, C. P., Luzi, F., Puglia, D., López-Carballo, G., Rojas, A., Galotto, M. J., et al. (2023). Development of a sustainable and antibacterial food packaging material based in a biopolymeric multilayer system composed by polylactic acid, chitosan, cellulose nanocrystals and ethyl lauroyl arginate. *Food Packaging and Shelf Life, 36*, 101050.

Videira-Quintela, D., Martin, O., & Montalvo, G. (2021). Recent advances in polymer-metallic composites for food packaging applications. *Trends in Food Science and Technology, 109*, 230–244.
Villaneuva, C. T., James, L. A., & Fuentes, N. C. (2019). Edible bio-active films based on chitosan or a mixture of quinoa protein-chitosan; sheets having chitosan-tripolyphosphate-thymol nanoparticles; production method; bio-packaging comprising same; and use thereof in fresh fruit with a low ph. *Google Patents*.
Wang, L., Fogliano, V., Heising, J., Meulenbroeks, E., & Dekker, M. (2020). Volatile antimicrobial absorption in food gel depends on the food matrix characteristics. *Food Hydrocolloids, 107*, 105933.
Wang, L., Porto, C. L. O., Palumbo, F., Modic, M., Cvelbar, U., Ghobeira, R., et al. (2021). Synthesis of antibacterial composite coating containing nanocapsules in an atmospheric pressure plasma. *Materials Science and Engineering: C, 119*, 111496.
Widsten, P., Mesic, B. B., Cruz, C. D., Fletcher, G. C., & Chycka, M. A. (2017). Inhibition of foodborne bacteria by antibacterial coatings printed onto food packaging films. *Journal of Food Science and Technology, 54*, 2379–2386.
Wieczyska, J., & Cavoski, I. (2018). Antimicrobial, antioxidant and sensory features of eugenol, carvacrol and trans-anethole in active packaging for organic ready-to-eat iceberg lettuce. *Food Chemistry, 259*, 251–260.
Wieczyska, J., Luca, A., Kidmose, U., Cavoski, I., & Edelenbos, M. (2016). The use of antimicrobial sachets in the packaging of organic wild rocket: Impact on microorganisms and sensory quality. *Postharvest Biology and Technology, 121*, 126–134.
Wyrwa, J., & Barska, A. (2017, October 1). Innovations in the food packaging market: active packaging. European Food Research and Technology [Internet], *243*(10), 1681–1692. https://link.springer.com/article/10.1007/s00217-017-2878-2. Cited 9 Jan 2024.
Yadav, P., Pandey, P., & Parashar, S. (2017). Pectin as natural polymer: An overview. *Research Journal of Pharmacy and Technology, 10*(4), 1225–1229.
Yam, K. L., & Lee, D. S. (2012, January 1). Emerging food packaging technologies: An overview. In *Emerging food packaging technologies* (pp. 1–9). Woodhead Publishing.
Yan, Y., Zhang, J., Ren, L., & Tang, C. (2016). Metal-containing and related polymers for biomedical applications. *Chemical Society Reviews, 45*(19), 5232–5263.
Yan, D., Li, Y., Liu, Y., Li, N., Zhang, X., & Yan, C. (2021). Antimicrobial properties of chitosan and chitosan derivatives in the treatment of enteric infections. *Molecules, 26*(23), 7136.
Yang, W., Fortunati, E., Dominici, F., Giovanale, G., Mazzaglia, A., Balestra, G. M., et al. (2016). Synergic effect of cellulose and lignin nanostructures in PLA based systems for food antibacterial packaging. *European Polymer Journal, 79*, 1–12.
Yildirim, S., Röcker, B., Pettersen, M. K., Nilsen-Nygaard, J., Ayhan, Z., Rutkaite, R., et al. (2018, January 1). Active packaging applications for food. Comprehensive Reviews in Food Science and Food Safety [Internet], *17*(1):165–199. https://onlinelibrary.wiley.com/doi/full/10.1111/1541-4337.12322. Cited 9 Jan 2024.
Yin, M., Wang, Y., Zhang, Y., Ren, X., Qiu, Y., & Huang, T. S. (2020). Novel quaternarized N-halamine chitosan and polyvinyl alcohol nanofibrous membranes as hemostatic materials with excellent antibacterial properties. *Carbohydrate Polymers, 232*, 115823.
Yunilawati, R., Handayani, W., Oktarina, E., Imawan, C., Rahmi, D., Umar, A., et al. (2023). Antibacterial sachet from the cyclodextrin/lemongrass oil inclusion complex for shrimp freshness. *International Journal of Food Science and Technology*.
Zabihzadeh Khajavi, M., Ebrahimi, A., Yousefi, M., Ahmadi, S., Farhoodi, M., Alizadeh, A. M., et al. (2020). Strategies for producing improved oxygen barrier materials appropriate for the food packaging sector. *Food Engineering Reviews, 12*(3), 346–363.
Zamindar, N., Anari, E. S., Bathaei, S. S., Shirani, N., Tabatabaei, L., Mahdavi-Asl, N., et al. (2020). Application of copper nano particles in antimicrobial packaging: A mini review. *Acta Scientific Nutritional Health, 4*(5), 14–18.

Zanganeh, H., Mortazavi, S. A., Shahidi, F., & Alizadeh Behbahani, B. (2021). Evaluation of the chemical and antibacterial properties of Citrus paradise essential oil and its application in Lallemantia iberica seed mucilage edible coating to improve the physicochemical, microbiological and sensory properties of lamb during refriger. *Journal of Food Measurement and Characterization, 15*(6), 5556–5571.

Zhang, M., Biesold, G. M., Choi, W., Yu, J., Deng, Y., Silvestre, C., et al. (2022). Recent advances in polymers and polymer composites for food packaging. *Materials Today, 53*, 134–161.

Zhang, X., Xiaoping, M. A., Maoyang, L. I., Peiyu, J. I., Huang, T., Zhuge, L., et al. (2023). Enhanced antibacterial activity of cotton via silver nanocapsules deposited by atmospheric pressure plasma jet. *Plasma Science and Technology, 25*(3), 35503.

Zhao, Y., Li, B., Zhang, W., Zhang, L., Zhao, H., Wang, S., et al. (2023). Recent advances in sustainable antimicrobial food packaging: Insights into release mechanisms, design strategies, and applications in the food industry. *Journal of Agricultural and Food Chemistry, 71*(31), 11806–11833.

Zhou, S., Hu, C., Zhao, G., Liu, L., Sheen, S., & Yam, K. L. (2018). A novel gaseous chlorine dioxide generating method utilizing carbon dioxide and moisture respired from tomato for Salmonella inactivation. *Food Control, 89*, 54–61.

Zhu, F. (2021). Polysaccharide based films and coatings for food packaging: Effect of added polyphenols. *Food Chemistry, 359*, 129871.

Part IV
Key to Success and Future Perspectives

Influence of Experimental Design in Antimicrobial Efficacy Estimation

Zejia Lin and Dan Li

1 Introduction

Researchers have developed many in vitro approaches to assess the effectiveness of antimicrobial strategies in published studies. These methodologies are employed to study the effectiveness of particular antimicrobial agents in both food models and culture media. However, numerous studies have failed to take into account these essential precautions throughout the experiment. Therefore, in certain situations, the experimental design is invalid due to the absence of standard criteria and methodologies. These designs unquestionably led to obstacles when analysing the results, which highlights the need to design appropriate in vitro evaluation experiments for antimicrobial strategies tailored towards specific foods or target microorganisms. While certain industrial standard guidelines, including EN 13697:2015 + A1:2019, BS EN 1276: 2019, EN 1650: 2019, and ISO 16869, are accessible, they often fail to account for variations in food matrix composition or the specific microorganisms that a design targets. As a result, numerous modifications have been made to industrial standard guidelines in recent years, thereby expanding the plethora of methods described in the literature for assessing antimicrobial treatment. This chapter aims to provide a thorough overview of the process of developing in vitro experiments for antimicrobial methods, focusing on three important aspects. Recommendations are given for an extensive range of design techniques for various estimations of antibacterial strategies.

Z. Lin · D. Li (✉)
Department of Food Science & Technology, National University of Singapore, Singapore, Singapore
e-mail: dan.li@nus.edu.sg

D. Li (ed.), *Antimicrobial Strategies in the Food System: Updates, Opportunities, Challenges*, https://doi.org/10.1007/978-3-031-95056-8_14

2 Influence of the Food Matrix

Food is a complicated system, and its interactions with microorganisms may have a significant influence on their survival. Food can appear in many different physical forms, comprising liquids to gels to solids, which affects nutritional and microbial dispersion. Solid food, for example, might hinder the migration of germs and micronutrients across the medium, resulting in changes in microbial activity and growth patterns. Microniches occur in gels and solid foods when colonies use available nutrients. If these microorganisms are not motile, they may quickly exhaust limited resources, and their survival may be impeded by the buildup of byproducts (Acuff & Ponder, 2020). In addition, it has been demonstrated that food processing operations, including grinding, extruding, and mincing, modify the characteristics of intact food (e.g., meat and fruits), facilitating enhanced bacterial proliferation in comparison to that of the intact product. This is attributed to the increased accessibility of nutrients and the ability of mobile microbes to relocate (Carlez et al., 1994). In addition, the location of microbial habitats also has a significant impact on their survival. Daş et al. (2006) reported that *Salmonella* Enteritidis strains that are localized inside the stem scars of cherry tomatoes were protected from antimicrobial gases, leading to higher enumeration counts compared to *Salmonella* strains attached to the tomato surface. It may be challenging to monitor these variables in a confined laboratory setting; hence, one approach to designing an antimicrobial experiment is to use growth media and conventional growth conditions. However, conclusions drawn from these findings ought to be approached with care since interactions with food environments could be unrealistic. The food microenvironment often affects and dictates whether a microbial cell can survive, endure, or grow in an environment, all of which require particular metabolic processes. Water activity (a_w), pH, nutrient compounds, and background microflora are called intrinsic elements in the food matrix that impact the bacterial response, while extrinsic elements include temperature and atmospheric conditions. To perfect the antimicrobial experimental design, both internal and extrinsic elements must be carefully considered. The common intrinsic and extrinsic elements of foods, as well as their influence on the development and resistance of common foodborne pathogens, are outlined as follows.

2.1 Water Activity (a_w)

Before planning an antimicrobial experiment, it is critical to assess the impact of the food matrix on the targeted microorganisms. Water activity (a_w), or the ratio of free, unbound water to pure water in food, is an important indicator for predicting whether particular microbes may thrive on food.

Pure water has an a_w of 1.00. Most fresh produce has an a_w of 0.99, whereas low-moisture foods typically have an a_w below 0.60 (Mathlouthi, 2001; Syamaladevi et al., 2016). Shelf-stable foods usually have a_w values between 0.60

Table 1 Minimum water activity (a_w) thresholds for the growth of specific microbes and relevant food products

Microorganisms	a_w (min) for growth	Relevant food products
Lactobacillus	0.95	Beer, acid foods
Pseudomonas	0.95	Fish, dairy, eggs, fresh produce, refrigerated meat and poultry
Escherichia	0.95	Raw milk and cheeses, fresh produce, raw or undercooked ground meat
Shigella	0.95	Fruit, aquatic products, produce, poultry
Klebsiella	0.95	Raw meats
Clostridium perfringens	0.95	Undercooked meat
C. Botulinum E, G	0.95	Chilled foods, minimally heated
Salmonella	0.91	Raw dairy products, raw meat and poultry, eggs
Listeria monocytogenes	0.91	Raw dairy products, raw meat, poultry, seafood, produce and fruits
Vibrio	0.91	Seafood
C. Botulinum A, B	0.91	Low-acid foods
Bacillus cereus	0.91	Meats, bakery products, dairy, produce, fruits
Micrococcus	0.86	Meat products, aquatic products
Staphylococcus aureus	0.86	Seafood, eggs, raw dairy products, raw meat and poultry
Candida	0.86	Dairy products, seafood, fruits, meat
Penicillium	0.80	Meat, nuts, cereals
Zygosaccharomyces	0.60	Wine, high-sugar foods
Aspergillus	0.60	Low-moisture foods

Adapted from Troller (2012), Wareing (2010), and Sun (2011)

and 0.85 (Troller, 2012). Most spoilage bacteria do not grow at a_w levels below 0.91, and most cannot survive at a_w levels below 0.85. Common spoilage bacteria include *Lactobacillus* and *Pseudomonas*. Moreover, halophilic bacteria (e.g., *Vibrio* and *Micrococcus*) can still thrive and cause spoilage in high-salt foods with a_w levels below 0.75, such as salted aquatic products. Molds and yeasts may thrive in environments with a_w levels as low as 0.60, particularly xerophilic molds and osmophilic yeasts. They are the main microbes responsible for food spoilage when a_w levels drop below 0.85. *Zygosaccharomyces* yeasts as well as *Aspergillus* molds are also linked to food spoilage (Jay et al., 2008). Moreover, most foodborne pathogens cannot thrive when the a_w is below 0.86, yet gram-negative bacteria have a growth limit at an a_w of 0.95 (Maserati et al., 2017). *Staphylococcus aureus* and *Bacillus cereus*, along with other gram-positive pathogens, may grow at a_w levels of 0.86 and 0.91, respectively (MedVeďoVá et al., 2010). The minimum a_w values for microbial growth and associated food products are detailed in Table 1. However, even if the a_w is below the minimum growth condition for microbes, a low a_w may trigger the microorganisms to initiate a survival strategy or become dormant until they can revive, enabling them to survive in very low-a_w environments. Foodborne pathogens, including *Clostridium botulinum*, *S. aureus*, and *Bacillus cereus,* undergo sporulation when exposed to a low a_w environment (Bressuire-Isoard et al., 2018). *Cronobacter* spp. can survive in milk powder with an a_w of 0.20 by accumulating osmoprotectants and modifying membrane phospholipids to address osmotic and desiccation stress (Lin et al., 2023). Forghani et al. (2019) also reported that *Salmonella* and pathogenic *E. coli* remained alive in wheat flour (a_w 0.45) for up to 1 year of storage.

When incorporating heat treatment into an antimicrobial treatment, it is critical to consider that microorganisms may develop increased resistance to heat as the a_w decreases. Chen and Jiang (2017) discovered that when the a_w of a food system falls below 0.6, the thermal resistance (D-value) of *Salmonella* increases dramatically. Comparatively, the thermal resistance (D-value) of *C. botulinum* spores was greatest when the a_w was between 0.1 and 0.5, whereas it was least when the a_w was between 0.7 and 0.9.

2.2 *pH*

The pH, or acidity, of a food matrix is another crucial intrinsic element that affects the growth of microbes in foods. The ability of microorganisms targeted in antimicrobial tests to grow or live within the pH range of certain food models should be ensured. Food products are categorized as high-acid (pH below 4.6) or low-acid (pH above 4.6) foods. Differentiation is determined by the ability of the foodborne pathogen *C. botulinum* to grow above a pH of 4.6 (Andrés-Bello et al., 2013). High-acid foods include fermented food, some salad dressings, fruits, and fruit juices. Low-acid foods consist of cereal, milk, meat, and fresh vegetables.

The pH of a food matrix, such as water activity, may also impact the effectiveness of antimicrobial agents. When microorganisms are exposed to acid in food, they can trigger an acid-adaptation response that not only helps them resist low pH but also activates various physiological and genetic mechanisms that provide advantages in resisting other stresses, known as cross-protection responses. Previous studies have shown that acid adaptation of foodborne pathogens may enhance their ability to withstand various stressors, such as disinfectants, antibiotics, osmosis, cold, heat, and nonthermal treatments (Al-Nabulsi et al., 2011; Álvarez-Ordóñez et al., 2011; Chiang et al., 2014; Kang et al., 2015; Wang et al., 2019; Wemekamp-Kamphuis et al., 2004; Wu et al., 2022). Hence, it is imperative to consider that when planning an antimicrobial experiment, a method that proves effective in low-acid foods may become ineffective when applied to high-acid foods (Table 2).

Table 2 The pH range of certain microbes and relevant food products

Microorganisms	pH range	Relevant food products
Campylobacter jejuni	5.0–9.0	Raw meat and poultry, seafood, produce
Clostridium botulinum	4.6–8.5	Low-acid vegetables, canned food, meat products
Bacillus cereus	4.9–9.3	Dairy products, meat, starchy foods, ready-to-eat foods
Vibrio parahaemolyticus	4.8–11.0	Seafood
Shigella	4.8–9.3	Salads, milk, vegetables
Escherichia coli	4.4–8.7	Ground meat products, milk, sprouts, flour
Yersinia enterocolitica	4.4–4.6	Meat, seafood, milk
Listeria monocytogenes	4.1–9.6	Dairy products, meat, vegetables, fruits
Salmonella	3.8–9.5	Meat, eggs, fruits, vegetables
Staphylococcus aureus	4.0–9.8	Poultry, meat products
Lactobacillus	2.7–8.9	Dairy products, alcoholic drink
Acetobacterium	2.8–4.3	Fruits, alcoholic drink
Yeasts	1.5–8.5	Bakery, alcoholic drinks, fruits
Molds	1.5–11	Fruits, pickles, jams, salt meats,

Adapted from Rahman and Rahman (2020) and Demirci et al. (2020)

2.3 Nutrient Components and Microorganisms in Foods

Foods typically have four major nutritional components, protein, carbohydrates, and fats, along with some vitamins and minerals. The quantities and types of nutrients present in a food matrix often influence the efficiency of antimicrobial tests. On the one hand, some nutrients can improve the persistence of foodborne pathogens. Meat products, for example, have a high concentration of proteins and lipids, and Sugiyama (1951) revealed that the presence of fatty acids improved the heat resistance of *C. botulinum* spores, with longer chain fatty acids showing greater thermal resistance. Similarly, Lin et al. (2024) reported that the resistance of *Salmonella* to desiccation stress was positively correlated with the sugar content of low-moisture foods. Some components in foods, however, may restrict the growth of pathogens, such as starches and cellulose in vegetables, which cannot be directly utilized by pathogens and must be broken down further, or natural antimicrobials found in foods that can kill pathogens (Lucera et al., 2012; Roberts, 2009). Aside from nutritional compounds, the presence of natural microflora in food also impacts antimicrobial results. *S. Typhimurium* on vegetables was shown to be more vigorous when coinoculated with *Erwinia carotovor*a, most likely due to the increased availability of nutrients supplied by soft rot bacteria (Wells & Butterfield, 1997). Similarly, Simko et al. (2015) reported that *Bremia lactucae*, a plant microbe, promoted the persistence of *Salmonella* and *E. coli* O157:H7 on romaine lettuce. However, for every symbiotic connection, there is an antagonistic connection. Cooley et al. (2006) discovered that the presence of *Wausteria paucula* promoted the growth of *E. coli* O157:H7 on lettuce, but the presence of *Enterobacter asburiae* inhibited the growth of *E. coli* O157:H7, most likely owing to competition for carbohydrate and nitrogen sources.

2.4 Temperature

Extrinsic elements are those that are influenced by environmental conditions, with temperature being one of the most significant. Temperature may impact the survival of microorganisms in food systems. Thus, caution must be taken when determining the optimal temperature for antimicrobial treatment. Microorganisms are categorized according to their growth temperature ranges as follows: psychrophiles (−20–0 °C), psychrotrophs (0–20 °C), mesophiles (20–45 °C), and thermophiles (55–65 °C) (Farrell & Rose, 1967). Psychrotrophic microbes are cold-tolerant and can be found in refrigerated foods, whereas psychrophilic microbes thrive only in extremely cold situations (Berry & Foegeding, 1997). Compared with bacteria, yeasts and molds may grow within a wider temperature range of 10–35 °C (Marshall et al., 1998). The majority of foodborne pathogens thrive under mesophilic conditions, with optimal temperatures ranging from 20–45 °C (Adams & Moss, 2000). However, psychrotrophs, such as *Pseudomonas* spp., *Listeria monocytogenes,* and *St. aureus,* can thrive at cold temperatures (below 10 °C) and are responsible for spoiling refrigerated foods, including fresh produce and meat products, while thermophiles, such as *Clostridium* and *Bacillus*, can tolerate high temperatures (up to 50 °C) and are often found in canned or retorted goods (Jay et al., 2008). Finally, after determining the temperature throughout the experiment, it is vital to avoid temperature abuse. Even a slight temperature shift may cause considerable variation in the bacterial population. Andersson et al. (1995) discovered that a subtle temperature deviation from 2 to 8 °C during the transportation of chilled milk led to an increase in *B. cereus* bacteria to 3 Log CFU/ml. An effective method for monitoring temperature would be installing a thermometer in the incubator to continually record temperature data that can be accessed on a computer.

2.5 Atmospheric Conditions of Foods

The state of microorganisms in food matrices is highly dependent on atmospheric conditions, particularly oxygen availability. Foodborne pathogens are classified into two types based on their gaseous requirements: obligatory anaerobes (e.g., *Clostridium* spp.) and facultative anaerobes (*Bacillus* spp., *Staphylococcus, Shigella, Salmonella, Listeria*, and pathogenic *E. coli*) (Bintsis, 2017). Although facultative anaerobes are adaptable bacteria that can thrive with or without oxygen, investigations have revealed that oxygen availability may considerably affect their physiological condition. Abbas et al. (2014) discovered that *B. cereus* can produce spores under anaerobic circumstances, which are more resistant to high temperatures and low pH than are spores in aerobic environments. Similarly, *L. monocytogenes* grown anaerobically was more resistant to bile salt and acid, with improved adhesion and invasion (Sewell et al., 2015). Other foodborne bacteria, such as *S.* Typhimurium and enterohemorrhagic *E. coli* (EHEC), can initiate metabolic and physiological responses to adapt to anaerobicity (Wallace et al., 2016). Although facultative anaerobic pathogens exhibit distinct behaviors in anaerobic environments, the majority of antimicrobial studies targeting them have been conducted under aerobic conditions. Therefore, it is crucial, when designing an antimicrobial experiment, to consider the actual atmospheric conditions of foods, as they may have a significant impact on antimicrobial efficacy. Another critical consideration is that foods must remain in their original gaseous state throughout the experiment. Consider the obligate anaerobe *C. botulinum* as an example. When conducting an antimicrobial experiment, it is ensured that the food packaging is always airtight; otherwise, a false positive antimicrobial result may be obtained, while the truth is that it was killed by oxygen leakage.

3 Influence of the Inoculum

Antimicrobial research on food systems could be conducted on either native food microbes or selected pathogens. The former requires no microbial inoculation, while the latter involves the introduction of a targeted bacterium to the desired food. In general, the inoculum should be selected based on the food category and the epidemiological or ecological relevance of the microorganism. Other factors, such as inoculum level and temperature conditions, could also affect the effectiveness of the inoculation process. Moreover, the inoculation methods should include food packaging conditions (e.g., open packaging (aerobic), vacuum packaging (anaerobic), or modified atmosphere packaging), as well as the food matrix structure.

3.1 Inoculum Determination

The choice of microorganism is the first crucial factor when carrying out a food model antimicrobial test for nonnative microbes. The common pathogenic microorganisms associated with various food categories are listed in Tables 1 and 2. The specific microorganism of concern for a new experiment could always be designed by the researcher as long as practical significance or reasonable justification is provided. Although most established antimicrobial studies have focused on the inoculation of one particular strain, it is advisable to test more than three strains, possibly up to 10 strains, either individually or as a cocktail, to closely mimic real scenarios of food systems (National Advisory Committee On Microbiological Criteria For, 2010). To perform cocktail inoculation, specific bacteria are cultured individually in an appropriate broth. Following the necessary incubation period, an equivalent volume of each culture was added to a tube, homogenized, and subsequently utilized as a cocktail (Moradi et al., 2021). It is strongly advised to employ a collection of microbe strains already isolated from food comparable to that used in antimicrobial studies due to their similar growth conditions, which allows for better adaptation to the tested food model (National Advisory Committee On Microbiological Criteria For, 2010). To minimize microbiological contamination, some researchers may desire to employ standard strains of organisms from the American Type Culture Collection (ATCC). In such cases, it is also ideal to include a particular similar nonpathogenic strain known as a surrogate microbe as a substitute for the original foodborne pathogen. For instance, in antimicrobial research, *C. sporogenes* PA 3679, *S.* Typhimurium LT2 ATCC 700720, *E. coli* K-12, *E. coli* ATCC 25922, and *L. innocua* ATCC 33090 may serve as surrogates for *C. botulinum*, *S.* Typhimurium, *E. coli* O157:H7, and *L. monocytogenes*, respectively (El Jaam et al., 2016; Hu & Gurtler, 2017; Karagöz et al., 2014; Sheng et al., 2020). Upon selection of inoculated microbes, the inoculum level is usually determined by the aim of the food model study. Low inoculum levels (e.g., 2–3 Log CFU/ml) are often employed in growth/survival experiments (Fuster-Valls et al., 2008; Kusumaningrum et al., 2003; Noor et al., 2015), which should represent realistic microbial reactions, while high inoculum levels (e.g., 5–8 Log CFU/ml) are utilized in antimicrobial evaluations (Kim et al., 2016; Mith et al., 2014; Weerakkody et al., 2010). Finally, in certain cases, such as frozen or refrigerated foods, cold adaptation of the inoculum is necessary to reduce the lag period. For example, *L. monocytogenes* was cultured at 10 °C for 3 days before being tested for shelf life in refrigerated ready-to-eat foods (Álvarez-Ordóñez et al., 2015).

3.2 Inoculation Methods

When the atmospheric conditions of foods are not considered, several methods of inoculation are accessible for antimicrobial experiments. To inoculate by mixing, a certain amount of microbial culture is introduced to the food in a sterile tube or bag and thoroughly mixed using a vortex or stomacher to achieve the desired inoculation level (Joerger, 2007). During the inoculation process by dipping, the food is submerged either individually or in large batches into the suspension, and the duration of immersion in the bacterial suspension is determined by the form and texture of the food. For instance, 10 s was used for cheese slices, 1 min for beef, 10 min for shell eggs, and 15 min for raw chicken breasts (Arrioja-Bretón et al., 2020; Guitián et al., 2019; Lin et al., 2018; Zang et al., 2019). Surface inoculation, also known as spot inoculation, involves the even distribution of a certain amount of inoculum at a suitable level on the food surface (e.g., 10–15 μl on a 1 cm^2 area), which is suitable for meat, cheese, fruits, and vegetables (Göksen et al., 2020; Moradi et al., 2021; Razavi et al., 2020). The inoculation process needs to be effective while maintaining the intrinsic elements of the food. To avoid changes in the physicochemical characteristics of food, the volume of inoculum must not exceed 1% of the food's weight (or volume); otherwise, it might significantly impact the intrinsic elements of the product and therefore the growth characteristics of the inoculum (Spanu et al., 2014). All of the inoculation methods discussed above are appropriate for foods with high moisture content; however, for low-moisture foods, an effort should be made to limit the amount of the inoculation solution. For instance, Lin et al. (2024) concentrated a bacterial suspension tenfold before spot-inoculating it into flour and milk powder. Another viable approach involves spraying the inoculum on food surfaces (Michael et al., 2020). On the other hand, when vacuum or modified atmosphere packaging foods are involved, the original atmospheric conditions of foods need to be carefully maintained. Either the food is removed from its packaging, inoculated in the same way as open packaging foods, and then repacked under similar gas conditions as an unopened pack, or the food is inoculated through a septum that is immediately recovered by a second septum to protect the original gaseous environment (Álvarez-Ordóñez et al., 2015).

Due to the diverse structure of foods, numerous inoculation strategies may be implemented. In the case of foods that are homogeneous (such as infant formula or cheese) or are prepared by mixing multiple ingredients (such as dough or salad), inoculation should be performed at random depths throughout the food (Cao-Hoang et al., 2010; Snelling et al., 2020). Conversely, in the case of foods containing several layers or ingredients (e.g., hamburger), inoculation should be limited to a few that are related to pathogen contamination (Beuchat & Mann, 2015). Last, in the case of foods whose internal environments have been completely sterilized and sealed (e.g., canned food), inoculums should be applied to the surface, where contamination is only feasible (Osuntokun et al., 2021).

Following microbial inoculation, food samples should be stored at the appropriate temperature for 20–30 min to facilitate bacterial attachment or for up to 2–10 days to allow biofilm formation (Amaeze et al., 2020; Joerger, 2007). Last, it is crucial to conduct an enumeration of the pathogen population that was introduced into the food before conducting an antimicrobial experiment.

4 Influence of Experimental Design

When performing an antimicrobial experiment, the working concentration, working temperature, and contact duration must be determined based on the practical usage of the food. In particular, while the working concentrations could be as low as possible but still effective against microorganisms, they must not exceed the relevant regulatory limits or compromise food quality. Chlorine, for example, is a commonly used antimicrobial regent with a maximum concentration of 200 ppm for washing fruits and vegetables in the U.S. (Code of Federal Regulations Title 21 Part 178), and studies have shown that improper use of chlorine, even at low concentrations, can produce undesirable off-flavours and reduce the quality and shelf life of fresh produce (Chaidez et al., 2012; Delaquis et al., 2000; Gómez-López et al., 2013). Similarly, determining the working temperature and contact duration must assure food quality, even though these two factors are often not associated with regulatory compliance. It is recommended to choose working temperatures within the range of 4–40 °C to replicate the actual temperature conditions of food production, transit, or storage. The antimicrobial tests conducted at room temperature needed to fall within

Table 3 Neutralizers commonly used for antimicrobial experiments

Neutralizers
Dey-Engley neutralizing broth
Fresh egg yolk (5%, V/V)
Sodium thiosulphate (5 g/l)
L-cysteine (1.5 g/l)
Phosphate-buffered saline (0.25 Mol/l)
Polysorbate (30 g/l), lecithin (3 g/l)
Thiomalic acid (0.075%, V/V)
Tween 80 (0.1%, V/V)
Water (up to 1000 ml)
Buffer peptone water (1%, V/V)

Adapted from BS EN 13697:2015 + A1:2019 and Sheraba et al. (2012)

the range of 18–25 °C. Moreover, contact durations should be selected within a range of 1 to 5 min at 1-min intervals or a range of 5–60 min at 5-min intervals, considering that extended treatment times (>60 min) are uncommon in the food industry (Barsotii & Cheftel, 2001).

Following antimicrobial treatment, chemical treatment neutralization is required to avoid further inactivation processes. The common neutralizers used are summarized in Table 3. Typically, a neutralizer should be applied to completely cover or immerse the whole food with a period of 5 min ± 10, unless specified by the manufacturers or relevant studies (Barsotii & Cheftel, 2001; Dey & Engley Jr, 1994; Lin et al., 2023; Mehregan et al., 2006).

To quantify the efficacy of an antimicrobial treatment, as well as mitigate the impact of any extraneous, confounding variable, the inoculated food sample would undergo the same procedures under the same conditions as a formal antimicrobial test, except sterile deionized water should be used in place of the antimicrobial regent; this is referred to as the "water control" or "control" (BS EN 13697: 2015 + A1:2019). In addition, neutralizer control is also required to verify the absence of toxicity of the neutralizer. An inoculated food sample without antimicrobial treatment was subjected to neutralizer treatment for an equivalent duration and quantity (BS EN 13697:2015 + A1:2019). Finally, all experiments were independently repeated at least three times to ensure the consistency and reliability of the results.

The quantitative analysis of microorganisms in food samples is a final step in evaluating the effectiveness of antimicrobial treatment. Conventional culture-based methods rely heavily on microbial metabolism, growth, and colony formation abilities. (Holbrook, 2000). However, after treatment, some injured or stressed microorganisms might enter viable but nonculturable (VBNC) status and cannot directly grow on a routine medium (İzgördü et al., 2022). Although VBNC-state cells might be resuscitated by culture-based enrichment methods, a variety of other methods have also been developed for microbiological analysis of foods, including microscopic instrumentation, immunology, bioluminescence, and molecular biology, which will be discussed in detail as follows.

- Culture-based methods

Culture-based methods are the most commonly used approaches because of their simplicity and low cost. However, these methods may be time-consuming, laborious, and resource-intensive and may require specific expertise. Furthermore, their reliability might diminish when VBNC-state cells are present. Cultured-based enumeration is usually carried out via either the pour plate or spread plate technique (Belete et al., 2014). The pour plate technique is intended to count anaerobic or facultative anaerobic bacteria in food that can grow under certain circumstances (e.g., temperature, incubation period, and culture medium). The spread plate technique involves evenly spreading tiny aliquots (0.1–0.2 ml) or dilutions of the sample across the agar surface in a Petri dish using a sterile L-shaped cell spreader. When low counts are anticipated, a larger sample volume (0.3–1 ml) may be spread on several dishes at the same time or on a larger petri dish to promote sensitivity (Pepper & Gerba, 2015). The spiral plate method is a technique for counting microorganisms by dispensing a known amount of suspension onto a spinning agar plate. As the spiral moves closer to the edge of the plate, the volume of the sample decreases, diluting the material. To count the colonies within a certain area of the plate, a modified counting grid was used, which matches the area of the plate with the sample volume (Gilchrist et al., 1973). Finally, in the assessment of antimicrobial effectiveness, 2-log reductions may have practical implications (National Advisory Committee On Microbiological Criteria For, 2010). Notably, when samples were inoculated with a low-concentration population (10^4 CFU/ml), the 2-log reduction was greater than that in samples inoculated with a high concentration (10^8 CFU/ml) (Sjollema et al., 2018).

In some situations where the target microorganism coexists with other microbes in foods (e.g., unsterile food sample before inoculation), a selective or differential medium should be applied. Selective media facilitate the proliferation of certain types of organisms while impeding the development of others. Differential media are employed to differentiate various groups of organisms based on the reaction of specific dyes or chemicals in the medium. These substances prompt organisms to exhibit distinct changes or growth patterns that help in their identification or differentiation (George et al., 1979). Common selective and differential media and their targeted microorganisms are listed in Table 4.

An enrichment medium may be required when target organisms enter a VBNC state as a result of exposure to sublethal stressors such as low pH, desiccation, cold, and heat. Commonly used enrichment media in food microbiology include blood agar, chocolate agar, and tryptic soy agar (Baylis et al., 2000). An example of a selective enrichment approach is used to retrieve VBNC bacteria from complex food samples. In this approach, treated samples are inoculated on a selective medium that inhibits the growth of the background microflora and then poured over on an enrichment medium, stimulating the target cells to recover from the VBNC state. (Gracias & McKillip, 2004).

Table 4 List of common selective and differential media and the types of microorganisms isolated

Medium	Classification	Target
Violet red bile agar	Selective medium	Coliform bacteria
Chloramphenicol yeast Glucose agar	Selective medium	Yeast and mold
MacConkey agar	Selective and differential medium	Gram-negative enteric bacilli
Eosin methylene blue agar	Selective and differential medium	Gram-negative enteric bacilli
Tergitol-7 agar	Selective medium	*E. Coli*
Baird parker agar	Selective medium	*S. Aureus*
Mannitol- egg yolk polymyxin agar	Selective medium	*Bacillus cereus*
Brilliant green agar	Selective and differential medium	*Salmonella* spp
Xylose lysine Deoxycholate agar	Selective and differential medium	*Salmonella* spp
Deoxycholate citrate agar	Selective and differential medium	*Shigella* spp.
Thiosulphate citrate-bile salts-sucrose agar	Selective and differential medium	*Vibrio* spp

Adapted from Chauhan and Jindal (2020)

- Microscopic methods

Microscopes may be employed for qualitative analysis to detect or characterize microorganisms in food; on the other hand, they can also be used for quantitative analysis to enumerate bacteria in food products. Microscopic methods are valuable for examining VBNC bacteria because they can distinguish between living and dead microorganisms.

For the routine microbiological examination of foods, a number of microscopic methods utilizing bright field, dark field, or phase contrast microscopy have been implemented (Robinson, 2014). Bright-field microscopy, which can quickly determine basic morphology, view cell structures, and differentiate fundamental groups, is frequently employed for microbial examination in foods (Yousef & Carlstrom, 2003). In contrast, dark field microscopy is a method applied to visualize unstained materials, resulting in their illumination against an almost dark background (Harrigan, 1998). A quantitative study was adapted by Xu et al. (2018), who utilized a magnetic nanoparticle sensor and a dark field microscope to enumerate *E. coli* populations. Phase contrast microscopy is another valuable technique for viewing microorganisms in their natural, living condition without the need for staining. Specialized optics are used by phase contrast microscopy to improve specimen contrast. A common protocol involves placing a small quantity of liquid culture on a glass microscope slide and covering it with a thin layer of glass. The cells are then examined using a lens that provides the desired level of magnification (Association, 1976). Examples of phase contrast microscopy applications include viewing bacterial structures (Bulla et al., 1969), motility (Taute et al., 2015), and intracellular inclusions (Gaju et al., 1989).

Fluorescence microscopy examination of microorganisms uses a range of fluorescent dyes in conjunction with a fluorescence microscope, the most popular of which is the confocal laser scanning microscope (CLSM) (Winkelströter et al., 2014). Numerous studies have used CLSM as a primary method to investigate the internalization and location of microorganisms in fresh produce (Han et al., 2000). In addition, the identification of living pathogens in food products, the elimination of microorganisms from vegetables, the distribution of microbial populations in milk, the adhesion of pathogens to beef, and the comparison, growth, and identification of fungal hyphae are just a few examples of CLSM applications in food microbiology (Liu et al., 2022; Wang et al., 2022; Xiao et al., 1999). Furthermore, when specialized fluorescent dyes are used, bacterial viability can be visualized and enumerated. A live/dead BacLight viability kit employing SYTO9 and propidium iodide (PI) stains live cells green, whereas dead cells red. Green-fluorescing SYTO9 can enter all cells and is used for assessing total cell counts, whereas red-fluorescing PI enters only cells with damaged cytoplasmic membranes. The live/dead ratio detected by a kit has been commonly used to indicate changes in viable cell populations, including dormant VBNC-state cells, under stressful circumstances and/or during inactivation treatment (Cheng et al., 2023; Dolezalova & Lukes, 2015). However, this commercial kit may not be appropriate for evaluating the effectiveness of antimicrobial substances that may cause damage to the bacterial membrane, as red fluorescent cells are typically considered dead, although some studies have shown that membrane-damaged bacteria might still survive (Lin et al., 2022; Shi et al., 2007). In addition to viability enumeration, other quantitative analyses could be carried out using various fluorescent dyes. Minich et al. (2022) investigated the antimicrobial effects of two phenolic compounds (vanillin and syringic acid) on *Staphylococcus epidermidis* biofilms. To quantify the effects of vanillin and syringic acid on extracellular polymeric substances (EPSs) in biofilms, extracellular proteins and DNA were stained with SYPRO Ruby and DAPI, respectively, and subjected to CLSM imaging.

Flow cytometry (FCM) is a quick method for the flow-through fluorescence examination of individual cells to determine the survival of bacteria. (Comas-Riu & Rius, 2009). This method is perfectly appropriate for liquid food preparations, as bacteria suspended in liquid are sequentially illuminated as they pass through a stream in front of a laser beam (Gunasekera et al., 2003). The viability of bacteria can be assessed by measuring the fluorescence of particles in samples that have been stained with particular fluorescent dyes. The application of FCM in foods involves tracking the growth of microorganisms, identifying and counting pathogens, and evaluating the effects of antimicrobial strategies or common food processing methods on the physiology and viability of pathogenic microbes. (Boye & Løbner-Olesen, 1991; Kennedy & Wilkinson, 2017; Léonard et al., 2016).

Other advanced microscopy techniques, such as scanning electron microscopy (SEM), transmission electron microscopy (TEM), and atomic force microscopy (AFM), provide distinct benefits in comparison to light microscopy. An electron microscope, for instance, has a spatial resolution of 0.2 nm and can magnify microorganisms by approximately 100,000 times (Rosa Teixeira et al., 2013). AFM can provide high-quality imaging of distinct topographical characteristics of microbial cells and resolve structures at the nanoscale (Grzeszczuk et al., 2020). Although these microscopy methods are not ideal for quantitative investigation, they can be employed for high-resolution character studies (e.g., studies of microbe colonization, adhesion properties, and cellular structural alterations) (Liao et al., 2010; Pelling et al., 2004; Whitehead et al., 2010).

- Immunological methods

Immunological methods utilize the precise binding of antibodies and antigens to create an antigen-antibody complex. The antibody is engineered to selectively bind with the desired antigen, resulting in the formation of a detectable complex. This complex indicates the presence of the antigen or a particular target in a given matrix (Dowd et al., 2009). Enzyme-linked immunosorbent assays (ELISAs), the most common immunoassays, are now extensively used for the identification and quantification of pathogenic organisms from food sources. (Notermans & Heuvelman, 1985). ELISA uses a solid substrate, often a microtiter plate, to fix the antigens or antibodies. It utilizes antibody-enzyme conjugates to identify the existence of antigen-antibody complexes. ELISA encompasses several methods, such as competitive tests for quantitative antigen measurements and noncompetitive assays for straightforward antigen detection (Engvall, 1977; Lapeyre et al., 1988). Compared to conventional culture-based methods, ELISAs enable rapid high-throughput assays of food samples and are capable of detecting microbial toxins (e.g., botulinum toxins) in micrograms or picograms (Maslanka et al., 2011; Srinoulprasert et al., 2019). Nevertheless, their ability to accurately identify microorganisms at concentrations less than 10^4–10^5 CFU/ml is compromised because the antigens to be detected are present in trace amounts on the cells (Demirci et al., 2020). In addition, ELISA techniques necessitate specialized expertise and training to eliminate the potential for erroneous outcomes. False-negative outcomes occur when a specific antibody fails to form a binding interaction with the antigen. This phenomenon may arise when molecules or chemicals present in food samples obscure the antigen, impede or eradicate the development of the antigen-antibody complex, or incapacitate the working enzyme on the detecting antibody. Additionally, false-negative outcomes can occur when the target microorganism does not reach the minimum population threshold. False positive outcomes transpire when the antibody utilized in an experiment reacts with an antigen that is not the intended target. (Notermans, 1992).

- Bioluminescence methods

In food microbial analysis, bioluminescence is often employed to detect ATP, which may then be used to estimate the total viable bacteria in food (Costa et al., 2004). Within the microbial cells, the enzyme luciferase catalyzes the conversion of ATP to AMP in the presence of oxygen. Oxidized AMP is electronically excited and returns to its unexcited form by emitting a light photon, which can be detected by a bioluminometer and measured in relative light units (RLUs) (Nante et al., 2017). The reaction exhibits a rapid response, exceptional sensitivity (capable of detecting ATP at picogram levels), and remarkable specificity for ATP, with a linear correlation between light photons and ATP amount (Costa et al., 2006). Thus, the measurement could be converted to microbial density as a result of the consistent amounts of ATP (typically 1 femtogram) in microbial cells, and this method is effective in detecting microbial populations from 3–6 log CFU/ml (Hunter & Lim, 2010).

Nevertheless, the ATP bioluminescence method has certain constraints, particularly when used in the context of raw meats or fresh produce. This is because animal and plant cells, which typically contain higher levels of ATP than microorganisms, can greatly influence the accuracy of the assay. Thus, when ATP is measured, it may not be possible to determine the specific source of ATP, resulting in a lack of specificity (Møretrø et al., 2019; Ukuku et al., 2005). In addition, the presence of some ingredients (e.g., acid, alkaloids) or chemical residues from the sanitizing process might suppress enzyme activities and lead to variation, which could be addressed by the use of control assays and calibration curves (Calvert et al., 2000).

- Molecular Methods

Molecular methods involve the examination of microbial nucleic acids, such as RNA or DNA. Among these methods, real-time PCR, also referred to as quantitative PCR (qPCR), is the most frequently utilized technique for quantitative investigations in the field of food microbiology. While conventional PCR involves two processes, amplification and product detection, qPCR may incorporate those two steps in a single process, providing a real-time picture of the progression of the PCR products (Bustin & Mueller, 2005). This method necessitates the application of particular fluorescent dyes, which are incorporated into the PCR products during amplification. The fluorescent dyes can be detected by a luminometer, and the intensity of fluorescence is proportional to the quantity of PCR products (Postollec et al., 2011). Hence, the qPCR approach may be used for quantitative purposes, with a notable application being the enumeration of VBNC-state cells. Liu et al. (2018) demonstrated that a method combining real-time quantitative PCR with propidium monoazide was effective for quantifying VBNC-state *V. parahaemolyticus* in seafood models. Apart from the potential utilization of DNA, RNA may also be the object of study. In reverse transcriptase PCR (RT–PCR), RNA serves as the PCR target (Bleve et al., 2003). RT–PCR requires the reverse transcriptase enzyme to convert the target RNA to DNA, which generates a DNA-formatted copy of the RNA and is referred to as cDNA (Rodríguez-Lázaro & Hernández, 2006). After cDNA was generated, PCR was performed in the same manner as normal PCR. RT–PCR is often used in antimicrobial research to assess the effects of certain antimicrobial agents on the microbiological state, such as changes in virulence factors. Das et al. (2021) examined the impact of carvacrol, a prominent component of oregano essential oil, on the foodborne pathogen *Vibrio cholerae*. The researchers used RT–PCR to determine the effects of carvacrol on the expression of virulence factors (*tcpA*, *ctxB*, *hlyA*, and *toxT*), which showed significant downregulation in the expression of these factors, with a fold change ranging from 5 to 7 logs. However, despite optimization, these molecular approaches still possess some limitations. Due to their high sensitivity to minor changes in sample preparation, achieving optimal nucleic acid extraction and designing primers to prevent PCR biases require specialized skills and experience (Davis, 2014; Nagarajan & Loh, 2014). Furthermore, there is still a significant ongoing debate among scientists worldwide regarding which algorithms should be utilized to interpret qPCR results (Smith & Osborn, 2009).

5 Summary

In this chapter, we provide a review of the development of an experimental design for estimating antimicrobial effectiveness levels. Although numerous laboratory studies have investigated the antimicrobial effects of diverse substances, most of them use simplified growth medium matrices that may not necessarily accurately replicate foods. The interactions between the food matrix and foodborne pathogens are quite complicated. The internal and extrinsic elements in food items impact microbial growth and persistence. In addition to food characteristics, inoculum parameters have a considerable impact on experimental expectations. Although the objective of the experiment may influence inoculum selection, inoculation should always be performed with caution to preserve the original qualities of foods. Similarly, antimicrobial treatment settings should not compromise food quality or violate regulatory limits. Several methods have been developed and modified to evaluate antimicrobial designs, and after the assay, enumeration is required to compare the dead and living populations. While the most typical method is to culture and count the number of CFUs, several additional methods, such as microscopy, ELISA, ATP bioluminescence, qPCR, and RT–PCR, may be used to assess microbial alterations. However, various antimicrobial agents may cause diverse kinds of bacterial damage, leading to varying results in terms of quantifying bacterial mortality. Thus, it is recommended to assess bacterial death using culture-based approaches supplemented with at least one additional method.

References

Abbas, A. A., Planchon, S., Jobin, M., & Schmitt, P. (2014). Absence of oxygen affects the capacity to sporulate and the spore properties of Bacillus cereus. *Food Microbiology, 42*, 122–131.

Acuff, J., & Ponder, M. (2020). Interactions of foodborne pathogens with the food matrix. *Food Safety Engineering*, 129–156.

Adams, M. R., & Moss, M. O. (2000). *Food microbiology*. Royal Society of Chemistry.

Al-Nabulsi, A. A., Osaili, T. M., Elabedeen, N. A. Z., Jaradat, Z. W., Shaker, R. R., Kheirallah, K. A., & Holley, R. A. (2011). Impact of environmental stress desiccation, acidity, alkalinity, heat or cold on antibiotic susceptibility of Cronobacter sakazakii. *International Journal of Food Microbiology, 146*(2), 137–143.

Álvarez-Ordóñez, A., Fernandez, A., Bernardo, A., & Lopez, M. (2011). Efficacy of trisodium phosphate in killing acid-adapted salmonella typhimurium. *Journal of Food Safety, 31*(2), 250–256.

Álvarez-Ordóñez, A., Leong, D., Hickey, B., Beaufort, A., & Jordan, K. (2015). The challenge of challenge testing to monitor listeria monocytogenes growth on ready-to-eat foods in Europe by following the European Commission (2014) technical guidance document. *Food Research International, 75*, 233–243.

Amaeze, N., Akinbobola, A., Chukwuemeka, V., Abalkhaila, A., Ramage, G., Kean, R., & Mackay, W. (2020). Development of a high throughput and low cost model for the study of semi-dry biofilms. *Biofouling, 36*(4), 403–415.

Andersson, A., Ronner, U., & Granum, P. E. (1995). What problems does the food industry have with the spore-forming pathogens Bacillus cereus and Clostridium perfringens? *International Journal of Food Microbiology, 28*(2), 145–155.

Andrés-Bello, A., Barreto-Palacios, V., García-Segovia, P., Mir-Bel, J., & Martínez-Monzó, J. (2013). Effect of pH on color and texture of food products. *Food Engineering Reviews, 5*, 158–170.

Arrioja-Bretón, D., Mani-López, E., Palou, E., & López-Malo, A. (2020). Antimicrobial activity and storage stability of cell-free supernatants from lactic acid bacteria and their applications with fresh beef. *Food Control, 115*, 107286.

Association, A. P. H. (1976). *Compendium of methods for the microbiological examination of foods*. American Public Health Association.

Barsotii, L., & Cheftel, J. (2001). Chemicals desinfectants and antiseptics. Quantitative non porous surface test for the evaluation of bactericidal and or fungicidal activity of chemical desinfectants use in food, industrial, domestic and institutonial areas. Test method and requeriments.

Baylis, C., MacPhee, S., Martin, K., Humphrey, T., & Betts, R. (2000). Comparison of three enrichment media for the isolation of campylobacter spp. from foods. *Journal of Applied Microbiology, 89*(5), 884–891.

Belete, T., Crowley, E., Bird, P., Gensic, J., & Wallace, F. M. (2014). A comparison of the BAX system method to the US Food and Drug Administration's bacteriological analytical manual and International Organization for Standardization reference methods for the detection of salmonella in a variety of soy ingredients. *Journal of Food Protection, 77*(10), 1778–1783.

Berry, E. D., & Foegeding, P. M. (1997). Cold temperature adaptation and growth of microorganisms. *Journal of Food Protection, 60*(12), 1583–1594.

Beuchat, L. R., & Mann, D. A. (2015). Survival of salmonella in cookie and cracker sandwiches containing inoculated, low–water activity fillings. *Journal of Food Protection, 78*(10), 1828–1834.

Bintsis, T. (2017). Foodborne pathogens. *AIMS Microbiology, 3*(3), 529.

Bleve, G., Rizzotti, L., Dellaglio, F., & Torriani, S. (2003). Development of reverse transcription (RT)-PCR and real-time RT-PCR assays for rapid detection and quantification of viable yeasts and molds contaminating yogurts and pasteurized food products. *Applied and Environmental Microbiology, 69*(7), 4116–4122.

Boye, E., & Løbner-Olesen, A. (1991). Bacterial growth control studied by flow cytometry. *Research in Microbiology, 142*(2–3), 131–135.

Bressuire-Isoard, C., Broussolle, V., & Carlin, F. (2018). Sporulation environment influences spore properties in bacillus: Evidence and insights on underlying molecular and physiological mechanisms. *FEMS Microbiology Reviews, 42*(5), 614–626.

Bulla, L., Julian, G. S., Rhodes, R., & Hesseltine, C. (1969). Scanning electron and phase-contrast microscopy of bacterial spores. *Applied Microbiology, 18*(3), 490–495.

Bustin, S. A., & Mueller, R. (2005). Real-time reverse transcription PCR (qRT-PCR) and its potential use in clinical diagnosis. *Clinical Science, 109*(4), 365–379.

Calvert, R., Hopkins, H., Reilly, M., & Forsythe, S. (2000). Caged ATP – An internal calibration method for ATP bioluminescence assays. *Letters in Applied Microbiology, 30*(3), 223–227.

Cao-Hoang, L., Chaine, A., Grégoire, L., & Waché, Y. (2010). Potential of nisin-incorporated sodium caseinate films to control listeria in artificially contaminated cheese. *Food Microbiology, 27*(7), 940–944.

Carlez, A., Rosec, J.-P., Richard, N., & Cheftel, J.-C. (1994). Bacterial growth during chilled storage of pressure-treated minced meat. *LWT-Food Science Technology, 27*(1), 48–54.

Chaidez, C., Campo, N. C. D., Heredia, J. B., Contreras-Angulo, L., González-Aguilar, G., & Ayala-Zavala, J. F. (2012). Chlorine. In *Decontamination of fresh and minimally processed produce* (pp. 121–133).

Chauhan, A., & Jindal, T. (2020). *Microbiological methods for environment, food and pharmaceutical analysis*. Springer.

Chen, Z., & Jiang, X. (2017). Thermal resistance and gene expression of both desiccation-adapted and rehydrated salmonella enterica serovar Typhimurium cells in aged broiler litter. *Applied Environmental Microbiology, 83*(12), e00367–e00317.

Cheng, S., Su, R., Song, L., Bai, X., Yang, H., Li, Z., & Lü, X. (2023). Citral and trans-cinnamaldehyde, two plant-derived antimicrobial agents can induce Staphylococcus aureus into VBNC state with different characteristics. *Food Microbiology, 112*, 104241.

Chiang, M.-L., Chen, H.-C., Wu, C., & Chen, M.-J. (2014). Effect of acid adaptation on the environmental stress tolerance of three strains of Vibrio parahaemolyticus. *Foodborne Pathogens Disease, 11*(4), 287–294.

Comas-Riu, J., & Rius, N. (2009). Flow cytometry applications in the food industry. *Journal of Industrial Microbiology and Biotechnology, 36*(8), 999–1011.

Cooley, M. B., Chao, D., & Mandrell, R. E. (2006). Escherichia coli O157: H7 survival and growth on lettuce is altered by the presence of epiphytic bacteria. *Journal of Food Protection, 69*(10), 2329–2335.

Costa, P. D., Andrade, N. J. D., Passos, F. J. V., Brandão, S. C. C., & Rodrigues, C. G. F. (2004). ATP-bioluminescence as a technique to evaluate the microbiological quality of water in food industry. *Brazilian Archives of Biology and Technology, 47*, 399–405.

Costa, P. D., Andrade, N. J., Brandão, S. C. C., Passos, F. J. V., & Soares, N. D. F. F. (2006). ATP-bioluminescence assay as an alternative for hygiene-monitoring procedures of stainless steel milk contact surfaces. *Brazilian Journal of Microbiology, 37*, 345–349.

Daş, E., Gürakan, G. C., & Bayındırlı, A. (2006). Effect of controlled atmosphere storage, modified atmosphere packaging and gaseous ozone treatment on the survival of salmonella Enteritidis on cherry tomatoes. *Food Microbiology, 23*(5), 430–438.

Das, S., Chourashi, R., Mukherjee, P., Kundu, S., Koley, H., Dutta, M., & Chatterjee, N. (2021). Inhibition of growth and virulence of vibrio cholerae by carvacrol, an essential oil component of Origanum spp. *Journal of Applied Microbiology, 131*(3), 1147–1161.

Davis, C. (2014). Enumeration of probiotic strains: Review of culture-dependent and alternative techniques to quantify viable bacteria. *Journal of Microbiological Methods, 103*, 9–17.

Delaquis, P., Stewart, S., Cliff, M., Toivonen, P., & Moyls, A. (2000). Sensory quality of ready-to-eat lettuce washed in warm, chlorinated water 1. *Journal of Food Quality, 23*(6), 553–563.

Demirci, A., Feng, H., & Krishnamurthy, K. (2020). *Food safety engineering*. Springer.

Dey, B., & Engley, F., Jr. (1994). Neutralization of antimicrobial chemicals by recovery media. *Journal of Microbiological Methods, 19*(1), 51–58.

Dolezalova, E., & Lukes, P. (2015). Membrane damage and active but nonculturable state in liquid cultures of Escherichia coli treated with an atmospheric pressure plasma jet. *Bioelectrochemistry, 103*, 7–14.

Dowd, S. E., Halonen, M. J., & Maier, R. M. (2009). Immunological methods. In *Environmental microbiology* (pp. 225–241). Elsevier.

El Jaam, O., Fliss, I., & Aider, M. (2016). Application of electro-activated potassium acetate and potassium citrate solutions combined with moderate heat treatment on the inactivation of clostridium sporogenes PA 3679 spores. *Innovative Food Science & Emerging Technologies, 33*, 483–488.

Engvall, E. (1977). Quantitative enzyme immunoassay (ELISA) in microbiology. *Medical Biology, 55*(4), 193–200.

Farrell, J., & Rose, A. (1967). Temperature effects on microorganisms. *Annual Reviews in Microbiology, 21*(1), 101–120.

Forghani, F., den Bakker, M., Liao, J.-Y., Payton, A. S., Futral, A. N., & Diez-Gonzalez, F. (2019). Salmonella and enterohemorrhagic Escherichia coli serogroups O45, O121, O145 in wheat flour: Effects of long-term storage and thermal treatments. *Frontiers in Microbiology, 10*, 323.

Fuster-Valls, N., Hernández-Herrero, M., Marín-de-Mateo, M., & Rodríguez-Jerez, J. J. (2008). Effect of different environmental conditions on the bacteria survival on stainless steel surfaces. *Food Control, 19*(3), 308–314.

Gaju, N., Guerrero, R., & Pedrós-Alió, C. (1989). Measurement of cell volume of phototrophic bacteria in pure cultures and natural samples: Phase contrast, epifluorescence and particle sizing. *FEMS Microbiology Ecology, 5*(5), 295–302.

George, W., Sutter, V., Citron, D., & Finegold, S. (1979). Selective and differential medium for isolation of Clostridium difficile. *Journal of Clinical Microbiology, 9*(2), 214–219.

Gilchrist, J. E., Campbell, J., Donnelly, C., Peeler, J., & Delaney, J. (1973). Spiral plate method for bacterial determination. *Applied Microbiology, 25*(2), 244–252.

Göksen, G., Fabra, M. J., Ekiz, H. I., & López-Rubio, A. (2020). Phytochemical-loaded electrospun nanofibers as novel active edible films: Characterization and antibacterial efficiency in cheese slices. *Food Control, 112*, 107133.

Gómez-López, V. M., Marín, A., Medina-Martínez, M. S., Gil, M. I., & Allende, A. (2013). Generation of trihalomethanes with chlorine-based sanitizers and impact on microbial, nutritional and sensory quality of baby spinach. *Postharvest Biology and Technology, 85*, 210–217.

Gracias, K. S., & McKillip, J. L. (2004). A review of conventional detection and enumeration methods for pathogenic bacteria in food. *Canadian Journal of Microbiology, 50*(11), 883–890.

Grzeszczuk, Z., Rosillo, A., Owens, Ó., & Bhattacharjee, S. (2020). Atomic force microscopy (AFM) as a surface mapping tool in microorganisms resistant toward antimicrobials: A mini-review. *Frontiers in Pharmacology, 11*, 517165.

Guitián, M. V., Ibarguren, C., Soria, M. C., Hovanyecz, P., Banchio, C., & Audisio, M. C. (2019). Anti-listeria monocytogenes effect of bacteriocin-incorporated agar edible coatings applied on cheese. *International Dairy Journal, 97*, 92–98.

Gunasekera, T., Veal, D., & Attfield, P. (2003). Potential for broad applications of flow cytometry and fluorescence techniques in microbiological and somatic cell analyses of milk. *International Journal of Food Microbiology, 85*(3), 269–279.

Han, Y., Linton, R. H., Nielsen, S., & Nelson, P. (2000). Inactivation of Escherichia coli O157: H7 on surface-uninjured and-injured green pepper (Capsicum annuum L.) by chlorine dioxide gas as demonstrated by confocal laser scanning microscopy. *Food Microbiology, 17*(6), 643–655.

Harrigan, W. F. (1998). Laboratory methods in food microbiology: Gulf professional publishing.

Holbrook, R. (2000). Detection of microorganisms in foods—Principles of culture methods. In *Microbiological safety and quality of food* (Vol. 1, p. 2).

Hu, M., & Gurtler, J. B. (2017). Selection of surrogate bacteria for use in food safety challenge studies: A review. *Journal of Food Protection, 80*(9), 1506–1536.

Hunter, D. M., & Lim, D. V. (2010). Rapid detection and identification of bacterial pathogens by using an ATP bioluminescence immunoassay. *Journal of Food Protection, 73*(4), 739–746.

İzgördü, Ö. K., Darcan, C., & Kariptaş, E. (2022). Overview of VBNC, a survival strategy for microorganisms. *3Biotech, 12*(11), 307. https://doi.org/10.1007/s13205-022-03371-4

Jay, J. M., Loessner, M. J., & Golden, D. A. (2008). *Modern food microbiology*. Springer.

Joerger, R. D. (2007). Antimicrobial films for food applications: A quantitative analysis of their effectiveness. *Packaging Technology and Science: An International Journal, 20*(4), 231–273.

Kang, J., Wiedmann, M., Boor, K. J., & Bergholz, T. M. (2015). VirR-mediated resistance of listeria monocytogenes against food antimicrobials and cross-protection induced by exposure to organic acid salts. *Applied Environmental Microbiology, 81*(13), 4553–4562.

Karagöz, I., Moreira, R., & Castell-Perez, M. (2014). Radiation D10 values for salmonella Typhimurium LT2 and an Escherichia coli cocktail in pecan nuts (Kanza cultivar) exposed to different atmospheres. *Food Control, 39*, 146–153.

Kennedy, D., & Wilkinson, M. G. (2017). Application of flow cytometry to the detection of pathogenic bacteria. *Current Issues in Molecular Biology, 23*(1), 21–38.

Kim, D.-H., Jeong, D., Kim, H., Kang, I.-B., Chon, J.-W., Song, K.-Y., & Seo, K.-H. (2016). Antimicrobial activity of kefir against various food pathogens and spoilage bacteria. *Korean Journal for Food Science of Animal Resources, 36*(6), 787.

Kusumaningrum, H. D., Riboldi, G., Hazeleger, W., & Beumer, R. (2003). Survival of foodborne pathogens on stainless steel surfaces and cross-contamination to foods. *International Journal of Food Microbiology, 85*(3), 227–236.

Lapeyre, C., Janin, F., & Kaveri, S. (1988). Indirect double sandwich ELISA using monoclonal antibodies for detection of staphylococcal enterotoxins a, B, C1 and D in food samples. *Food Microbiology, 5*(1), 25–31.

Léonard, L., Bouarab Chibane, L., Ouled Bouhedda, B., Degraeve, P., & Oulahal, N. (2016). Recent advances on multi-parameter flow cytometry to characterize antimicrobial treatments. *Frontiers in Microbiology, 7*, 210347.

Liao, C. H., Cooke, P., & Niemira, B. (2010). Localization, growth, and inactivation of salmonella Saintpaul on jalapeno peppers. *Journal of Food Science, 75*(6), M377–M382.

Lin, L., Xue, L., Duraiarasan, S., & Haiying, C. (2018). Preparation of ε-polylysine/chitosan nanofibers for food packaging against salmonella on chicken. *Food Packaging and Shelf Life, 17*, 134–141.

Lin, Z., Wang, G., Li, S., Zhou, L., & Yang, H. (2022). Dual-species biofilms formed by Escherichia coli and salmonella enhance chlorine tolerance. *Applied and Environmental Microbiology, 88*(22), e01482–e01422.

Lin, Z., Wang, G., Zhang, K., Jiang, S., Li, S., & Yang, H. (2023). Metabolomics investigation of global responses of Cronobacter sakazakii against common sanitizing in infant formula processing environments. *Food Research International*, 113162.

Lin, Z., Jiang, S., Zwe, Y. H., Zhang, K., & Li, D. (2024). Glycogen plays a key role in survival of salmonella Typhimurium on dry surfaces and in low-moisture foods. *Food Research International, 175*, 113714. https://doi.org/10.1016/j.foodres.2023.113714

Liu, Y., Zhong, Q., Wang, J., & Lei, S. (2018). Enumeration of Vibrio parahaemolyticus in VBNC state by PMA-combined real-time quantitative PCR coupled with confirmation of respiratory activity. *Food Control, 91*, 85–91. https://doi.org/10.1016/j.foodcont.2018.03.037

Liu, Y., Yan, Y., Dong, P., Ni, L., Luo, X., Zhang, Y., & Zhu, L. (2022). Inhibitory effects of clove and oregano essential oils on biofilm formation of salmonella Derby isolated from beef processing plant. *LWT, 162*, 113486.

Lucera, A., Costa, C., Conte, A., & Del Nobile, M. A. (2012). Food applications of natural antimicrobial compounds. *Frontiers in Microbiology, 3*, 287.

Marshall, V., Poulson-Cook, S., & Moldenhauer, J. (1998). Comparative mold and yeast recovery analysis (the effect of differing incubation temperature ranges and growth media). *PDA Journal of Pharmaceutical Science Technology, 52*(4), 165–169.

Maserati, A., Fink, R. C., Lourenco, A., Julius, M. L., & Diez-Gonzalez, F. (2017). General response of salmonella enterica serovar Typhimurium to desiccation: A new role for the virulence factors sopD and sseD in survival. *PLoS One, 12*(11), e0187692.

Maslanka, S. E., Luquez, C., Raphael, B. H., Dykes, J. K., & Joseph, L. A. (2011). Utility of botulinum toxin ELISA a, B, E, F kits for clinical laboratory investigations of human botulism. *The Botulinum Journal, 2*(1), 72–92.

Mathlouthi, M. (2001). Water content, water activity, water structure and the stability of foodstuffs. *Food Control, 12*(7), 409–417.

MedVeďoVá, A., Valík, L., & StudenIčová, A. (2010). The effect of temperature and water activity on the growth of Staphylococcus aureus. *Czech Journal of Food Sciences, 27*(Special Issue 2), 28-S (II) 35.

Mehregan, H., Elmi, F., Fazeli, M. R., Shahverdi, A. R., & Samadi, N. (2006). *Evaluation of neutralizing efficacy and possible microbial cell toxicity of a universal neutralizer proposed by the CTPA*.

Michael, M., Acuff, J., Lopez, K., Vega, D., Phebus, R., Thippareddi, H., & Channaiah, L. H. (2020). Comparison of survival and heat resistance of Escherichia coli O121 and salmonella in muffins. *International Journal of Food Microbiology, 317*, 108422.

Minich, A., Levarski, Z., Mikulášová, M., Straka, M., Liptáková, A., & Stuchlík, S. (2022). Complex analysis of vanillin and syringic acid as natural antimicrobial agents against Staphylococcus epidermidis biofilms. *International Journal of Molecular Sciences, 23*(3), 1816.

Mith, H., Duré, R., Delcenserie, V., Zhiri, A., Daube, G., & Clinquart, A. (2014). Antimicrobial activities of commercial essential oils and their components against food-borne pathogens and food spoilage bacteria. *Food Science & Nutrition, 2*(4), 403–416. https://doi.org/10.1002/fsn3.116

Moradi, M., Kousheh, S. A., Razavi, R., Rasouli, Y., Ghorbani, M., Divsalar, E., et al. (2021). Review of microbiological methods for testing protein and carbohydrate-based antimicrobial food packaging. *Trends in Food Science & Technology, 111*, 595–609.

Møretrø, T., Normann, M. A., Sæbø, H. R., & Langsrud, S. (2019). Evaluation of ATP bioluminescence-based methods for hygienic assessment in fish industry. *Journal of Applied Microbiology, 127*(1), 186–195.

Nagarajan, K., & Loh, K.-C. (2014). Molecular biology-based methods for quantification of bacteria in mixed culture: Perspectives and limitations. *Applied Microbiology and Biotechnology, 98*, 6907–6919.

Nante, N., Ceriale, E., Messina, G., Lenzi, D., & Manzi, P. (2017). Effectiveness of ATP bioluminescence to assess hospital cleaning: A review. *Journal of Preventive Medicine and Hygiene, 58*(2), E177.

National Advisory Committee On Microbiological Criteria For, F. (2010). Parameters for determining inoculated pack/challenge study protocols. *Journal of Food Protection, 73*(1), 140–203.

Noor, R., Malek, M., Rahman, S., Meghla, M., Acharjee, M., & Rahman, M. M. (2015). Assessment of survival of pathogenic bacteria in fresh vegetables through in vitro challenge test. *International Journal of Food Contamination, 2*(1), 15.

Notermans, S. (1992). *Quality control of immunological detection methods applied in food microbiology*.

Notermans, S., & Heuvelman, C. (1985). Immunological detection of moulds in food by using the enzyme-linked immunosorbent assay (ELISA); preparation of antigens. *International Journal of Food Microbiology, 2*(4), 247–258.

Osuntokun, O., Thonda, O., Akele, E., Adedokun, L., Adedayo, S., & Bello, O. (2021). Pathogenic bacteria found on surfaces of canned drinks and wines being sold in retail shops in Ondo state, Nigeria, health implications, food safety and quality assessment. *South Asian Journal of Parasitology, 5*(4), 68–94.

Pelling, A. E., Sehati, S., Gralla, E. B., Valentine, J. S., & Gimzewski, J. K. (2004). Local nanomechanical motion of the cell wall of Saccharomyces cerevisiae. *Science, 305*(5687), 1147–1150.

Pepper, I. L., & Gerba, C. P. (2015). Cultural methods. In *Environmental microbiology* (pp. 195–212). Elsevier.

Postollec, F., Falentin, H., Pavan, S., Combrisson, J., & Sohier, D. (2011). Recent advances in quantitative PCR (qPCR) applications in food microbiology. *Food Microbiology, 28*(5), 848–861.

Rahman, M. S., & Rahman, M. R. T. (2020). pH in food preservation. In *Handbook of food preservation* (pp. 323–332). CRC Press.

Razavi, R., Tajik, H., Moradi, M., Molaei, R., & Ezati, P. (2020). Antimicrobial, microscopic and spectroscopic properties of cellulose paper coated with chitosan sol-gel solution formulated by epsilon-poly-l-lysine and its application in active food packaging. *Carbohydrate Research, 489*, 107912.

Roberts, M. C. (2009). The evolution of antibiotic-resistant microbes in foods and host ecosystems. In *Food-borne microbes: Shaping the host ecosystem* (pp. 211–229).

Robinson, R. K. (2014). *Encyclopedia of food microbiology*. Academic.

Rodríguez-Lázaro, D., & Hernández, M. (2006). *Molecular methodology in Food Microbiology diagnostics: trends and current challenges*. Paper presented at the 13th World Congress of Food Science & Technology 2006.

Rosa Teixeira, K. I., Araújo, P. V., Almeida Neves, B. R., Bohorquez Mahecha, G. A., Sinisterra, R. D., & Cortés, M. E. (2013). Ultrastructural changes in bacterial membranes induced by nano-assemblies β-cyclodextrin chlorhexidine: SEM, AFM, and TEM evaluation. *Pharmaceutical Development and Technology, 18*(3), 600–608.

Sewell, D., Allen, S. C., & Phillips, C. A. (2015). Oxygen limitation induces acid tolerance and impacts simulated gastro-intestinal transit in listeria monocytogenes J0161. *Gut Pathogens, 7*, 1–5.

Sheng, L., Shen, X., & Zhu, M.-J. (2020). Screening of non-pathogenic surrogates of listeria monocytogenes applicable for chemical antimicrobial interventions of fresh apples. *Food Control, 110*, 106977.

Sheraba, N. S., Yassin, A. S., Fahmy, A., & Amin, M. A. (2012). Efficacy and toxicity of neutralizers against disinfectants and antiseptics used in vaccine production facility. *African Journal of Microbiology Research, 6*(36), 6565–6571.

Shi, L., Günther, S., Hübschmann, T., Wick, L. Y., Harms, H., & Müller, S. (2007). Limits of propidium iodide as a cell viability indicator for environmental bacteria. *Cytometry Part A: The Journal of the International Society for Analytical Cytology, 71*(8), 592–598.

Simko, I., Zhou, Y., & Brandl, M. T. (2015). Downy mildew disease promotes the colonization of romaine lettuce by Escherichia coli O157: H7 and salmonella enterica. *BMC Microbiology, 15*(1), 1–9.

Sjollema, J., Zaat, S. A., Fontaine, V., Ramstedt, M., Luginbuehl, R., Thevissen, K., et al. (2018). In vitro methods for the evaluation of antimicrobial surface designs. *Acta Biomaterialia, 70*, 12–24.

Smith, C. J., & Osborn, A. M. (2009). Advantages and limitations of quantitative PCR (Q-PCR)-based approaches in microbial ecology. *FEMS Microbiology Ecology, 67*(1), 6–20. https://doi.org/10.1111/j.1574-6941.2008.00629.x

Snelling, J., Malekmohammadi, S., Bergholz, T. M., Ohm, J., & Simsek, S. (2020). Effect of vacuum steam treatment of hard red spring wheat on flour quality and reduction of Escherichia coli O121 and salmonella Enteritidis PT 30. *Journal of Food Protection, 83*(5), 836–843.

Spanu, C., Scarano, C., Ibba, M., Pala, C., Spanu, V., & De Santis, E. P. L. (2014). Microbiological challenge testing for listeria monocytogenes in ready-to-eat food: A practical approach. *Italian Journal of Food Safety, 3*(4).

Srinoulprasert, Y., Kantaviro, W., Nokdhes, Y.-N., Patthamalai, P., Dowdon, L., Chawengkiattikul, R., & Wanitphakdeedecha, R. (2019). Development of inhibition ELISA to detect antibody-induced failure of botulinum toxin a therapy in cosmetic indications. *Journal of Immunological Methods, 473*, 112635.

Sugiyama, H. (1951). Studies on factors affecting the heat resistance of spores of clostridium botulinum. *Journal of Bacteriology, 62*(1), 81–96.

Sun, D.-W. (2011). *Handbook of food safety engineering*. Wiley.

Syamaladevi, R. M., Tang, J., Villa-Rojas, R., Sablani, S., Carter, B., & Campbell, G. (2016). Influence of water activity on thermal resistance of microorganisms in low-moisture foods: A review. *Comprehensive Reviews in Food Science and Food Safety Engineering, 15*(2), 353–370.

Taute, K., Gude, S., Tans, S., & Shimizu, T. (2015). High-throughput 3D tracking of bacteria on a standard phase contrast microscope. *Nature Communications, 6*(1), 1–9.

Troller, J. (2012). *Water activity and food*. Elsevier.

Ukuku, D. O., Sapers, G. M., & Fett, W. F. (2005). ATP bioluminescence assay for estimation of microbial populations of fresh-cut melon. *Journal of Food Protection, 68*(11), 2427–2432.

Wallace, N., Zani, A., Abrams, E., & Sun, Y. (2016). The impact of oxygen on bacterial enteric pathogens. *Advances in Applied Microbiology, 95*, 179–204.

Wang, Q., Buchanan, R. L., & Tikekar, R. V. (2019). Evaluation of adaptive response in E. Coli O157: H7 to UV light and gallic acid based antimicrobial treatments. *Food Control, 106*, 106723.

Wang, N., Jin, Y., He, G., & Yuan, L. (2022). Dynamic tracing of bacterial community distribution and biofilm control of dominant species in milk powder processing. *LWT, 154*, 112855.

Wareing, P. (2010). *Factors affecting the growth of micro-organisms in foods*.

Weerakkody, N. S., Caffin, N., Turner, M. S., & Dykes, G. A. (2010). In vitro antimicrobial activity of less-utilized spice and herb extracts against selected food-borne bacteria. *Food Control, 21*(10), 1408–1414. https://doi.org/10.1016/j.foodcont.2010.04.014

Wells, J., & Butterfield, J. (1997). Salmonella contamination associated with bacterial soft rot of fresh fruits and vegetables in the marketplace. *Plant Disease, 81*(8), 867–872.

Wemekamp-Kamphuis, H. H., Wouters, J. A., de Leeuw, P. P., Hain, T., Chakraborty, T., & Abee, T. (2004). Identification of sigma factor σB-controlled genes and their impact on acid stress, high hydrostatic pressure, and freeze survival in listeria monocytogenes EGD-e. *Applied Environmental Microbiology, 70*(6), 3457–3466.

Whitehead, K. A., Smith, L. A., & Verran, J. (2010). The detection and influence of food soils on microorganisms on stainless steel using scanning electron microscopy and epifluorescence microscopy. *International Journal of Food Microbiology, 141*, S125–S133.

Winkelströter, L. K., Teixeira, F. B. D. R., Silva, E. P., Alves, V. F., & De Martinis, E. C. P. (2014). Unraveling microbial biofilms of importance for food microbiology. *Microbial Ecology, 68*, 35–46.

Wu, R. A., Yuk, H.-G., Liu, D., & Ding, T. (2022). Recent advances in understanding the effect of acid-adaptation on the cross-protection to food-related stress of common foodborne pathogens. *Critical Reviews in Food Science Nutrition, 62*(26), 7336–7353.

Xiao, Y., Kreber, B., & Breuil, C. (1999). Localisation of fungal hyphae in wood using immunofluorescence labelling and confocal laser scanning microscopy. *International Biodeterioration & Biodegradation, 44*(4), 185–190.

Xu, H., Tang, F., Dai, J., Wang, C., & Zhou, X. (2018). Ultrasensitive and rapid count of Escherichia coli using magnetic nanoparticle probe under dark-field microscope. *BMC Microbiology, 18*, 1–6.

Yousef, A. E., & Carlstrom, C. (2003). *Food microbiology: A laboratory manual*. Wiley.

Zang, Y., Bing, S., Li, Y., Shu, D., Huang, A., Wu, H., & Wu, H. (2019). Efficacy of slightly acidic electrolyzed water on the microbial safety and shelf life of shelled eggs. *Poultry Science, 98*(11), 5932–5939.

The Challenges and Opportunities

Dan Li

1 The World Changes Rapidly

1.1 Microorganism Evolution

On our planet, microorganisms are extremely successful, inhibiting a diverse range of niches and prospering for billions of years. Facing numerous types of pressure, microorganisms have developed a remarkable ability to adapt by rapid mutation and evolution, allowing the most suitable microorganisms to survive. For instance, bacteria can transfer genetic information in various ways, such as through conjugation (which requires cell-to-cell contact between cells), transduction (which involves bacteriophage-facilitated transfer of genetic information), and transformation (which involves the uptake of free DNA from the environment). The transfer of mobile genetic elements encoding important proteins has been found in multiple important bacterial pathogens, including *Escherichia coli* O157:H7, *Salmonella*, *Staphylococcus aureus*, and *Listeria monocytogenes* (Kelly et al., 2009). For viruses which are obligate intracellular parasites, when replicating, virus genes undergo random "copying errors", which may be selected by host–virus interplay, including immune evasion, or by heterogeneity in the expression of virus receptors on host cells. In addition, recombination of different virus strains is frequently observed and is thought to be an important mechanism by which genetic diversity is generated.

In May 2011, an unusual foodborne outbreak involving a high number (782) of hemolytic uremic syndrome (HUS) cases was reported in Germany. Fenugreek sprouts were identified as a common source of this outbreak. The epidemiological investigation of this outbreak took much longer and involved more scientific forces

D. Li (✉)
Department of Food Science and Technology, National University of Singapore, Singapore, Singapore
e-mail: dan.li@nus.edu.sg

D. Li (ed.), *Antimicrobial Strategies in the Food System: Updates, Opportunities, Challenges*, https://doi.org/10.1007/978-3-031-95056-8_15

than usual, as this enterohaemorrhagic *E. coli* (EHEC) strain has characteristics of the less well-known diarrhoeagenic *E. coli*, enteroaggregative *E. coli* (EAEC). Unlike most known EHEC strains, this strain lacks both the locus for the enterocyte effacement pathogenicity island and the EHEC virulence plasmid. Instead, this strain possesses an aggregative adherence plasmid and demonstrates aggregative adherence to epithelial cells (Chattaway et al., 2011). Its ability to produce Shiga toxin was found to be due to the presence of a Shiga toxin 2a- (Stx2a) encoding prophage (Beutin et al., 2013). The appearance of this strain presented a new microbiological challenge. Numerous follow-up studies have been and are being conducted until very recent years (Ortiz et al., 2023).

Food microbiologists must closely monitor these changes and adjust detection and mitigation strategies accordingly. However, caution must be taken when evaluating the genuine influence of these mutation events on food safety. Not all hazards have equal risks. Priorities should be focused on the most relevant risks to achieve a more efficient food safety management system to protect public health.

The most well-known case is severe acute respiratory syndrome coronavirus 2 (SARS-CoV-2). Since the emergence of the virus, thousands of cumulative mutations have occurred, resulting in a tragic pandemic between 2019 and 2022. Numerous concerns have been raised over whether SARS-CoV-2 can be transmitted via food and/or food packaging materials (Yekta et al., 2021). The need for decontamination to reduce possible SARS-CoV-2 transmission via food has been debated. However, compared with those of foodborne viruses (e.g., human norovirus, hepatitis A virus, and hepatitis E virus), the chance of foodborne transmission of CoVs is considered low. This was because CoV infection has never been linked to food consumption according to epidemiological evidence. Although CoVs can remain infectious on food and/or food packaging materials long enough (from several days to several weeks) to potentially cause transmission (Jia et al., 2022), they are much less resistant to chemical and physical disinfections than foodborne viruses (Li et al., 2021). More importantly, CoVs mainly spread through the respiratory tract, and there is no clear evidence showing that CoVs can follow fecal-oral routes and cause infection in the human digestive tract. Therefore, in comparison with the mainstream precautions for SARS-CoV-2 spread, including masking, handwashing, and vaccination, decontamination of SARS-CoV-2 in food systems provides little benefit (Sobolik et al., 2022).

1.2 Climate Change

In 1994, climate change was defined by the United Nations as "a change of climate which is attributed directly or indirectly to human activity that alters the composition of the global atmosphere and which is in addition to natural climate variability, observed over comparable time periods" (Misiou & Koutsoumanis, 2022). The significant and urgent threat of climate change to food safety has been identified in countries worldwide in recent decades (Mirón et al., 2023; Tirado et al., 2010;

Uyttendaele et al., 2015; Duchenne-Moutien & Neetoo, 2021). In recent years, advanced technologies such as artificial intelligence (AI) have been employed to evaluate and monitor the influence of climate change on food systems (Karanth et al., 2023). Risk assessment and life cycle assessment tools were proposed to limit the environmental impact on food safety risks (Feliciano et al., 2022).

One of the most important phenomena of climate change is global warming. Largely due to carbon dioxide and other air pollutants being collected in the atmosphere and absorbing sunlight and solar radiation, the world is now warming faster than at any point in recorded history. Higher temperatures favour the proliferation of many microorganisms that can cause food-borne diseases. For example, summer peaks that coincide with higher temperatures are well documented in Europe due to the seasonal patterns of Campylobacteriosis and Salmonellosis, which are the two most frequent food-borne diseases (Mirón et al., 2023). Even in tropical countries such as Singapore, ecological time series analysis has confirmed the short-term influence of climatic conditions on *Salmonella* infections and suggested the potential impact of climate change on salmonellosis (Aik et al., 2018).

In addition to the direct effect of higher temperature on pathogen survival and circulation, global warming also leads to ocean elevation levels, changes in precipitation patterns, and changes in the contributions of rainfall to oceans and lakes, as well as other extreme meteorological phenomena, such as increases in the frequency of great storms. All the abovementioned changes may cause sewer system failure due to flooding and drainage overflow caused by floods, eventually leading to the uncontrollable spread of animal/human fecal materials to primary food production systems. Researchers have proven this point by both field studies and laboratory studies. For instance, Castro-Ibáñez et al. (2015) evaluated the effects of a flood event on the microbial contamination of leafy greens grown in floodplains and reported that the levels of coliforms and *E. coli* in irrigation water, soil and lettuce samples were greater than 5 and 3 log CFU/g or 100 mL, respectively, when the samples were collected 1 week after flooding. *E. coli* is a commonly used indicator of fecal contamination and foodborne pathogen presence. Indeed, in this study, a significant correlation ($P < 0.005$) was found between the presence of pathogens (*Salmonella*, Shiga toxin-producing *E. coli*) and the number of generic *E. coli* counts, highlighting a greater probability of detection of pathogens when high levels of *E. coli* are found.

Shiraz et al. (2020) investigated the influence of a precisely controlled laboratory setup on the safety of fresh produce. Flooding scenarios were simulated for the raised beds of strawberry plants. The plants were exposed to high volumes of floodwater, so strawberries were fully submerged, or to low volumes of floodwater where the berries did not come into direct contact with the floodwater. Generic *E. coli* was detected in the soil samples within 96 h after flooding. Coliforms were present in the strawberries and soil at 0, 48, 96, and 144 h in all treatment beds.

In addition, as reviewed by Misiou and Koutsoumanis (2022), for agriculture, high temperature and heavy precipitation will cause high relative humidity, affect crop production and increase fungal prevalence and thus mycotoxin production. During animal production, such changes will induce heat stress, leading to poor milk production and quality and increased microbial loads. The influence of higher temperature and higher relative humidity typically also affects the storage and distribution of food products, favouring bacterial and fungal growth and eventually posing a greater threat to food safety and quality.

Therefore, climate change can significantly impact food safety in various ways. Addressing these challenges requires comprehensive strategies, including improved food safety regulations, resilient agricultural practices, enhanced surveillance and monitoring systems, and investments in research and development for climate-resilient food systems.

1.3 New Food Systems

In the face of climate change, population growth, ecosystem degradation and increasing resource scarcity, most countries in the world have made great efforts to establish a more sustainable agri-fresh produce supply chain in recent years. However, food safety concerns are often overlooked in the debate on sustainable food. In fact, changes in dietary structure may expose consumers to certain hazards to a greater extent (e.g., more microbial food safety issues related to fresh produce along with the increased consumption of plant-based foods). Exploiting alternative protein sources, adopting other food production systems or increasing the reuse of water may introduce "new hazards" (e.g., exposure to enteric pathogens, including viruses or parasites, if organic manure or reclaimed wastewater is not sufficiently treated) or create "new routes of transmission" for existing hazards (e.g., aquaculture-related microbial pathogens introduced to plant-based foods via aquaponic agriculture) to enter the food supply chain.

Multitrophic farming systems refer to agroecosystems with combinations of different cultivation elements. The most common type of multitrophic system seen is still aquaponics, but moving forward, there might be other types, such as combining plant farming with insect or other animal farming. Apparently, the benefit would never come alone without bringing new challenges. Aquatic bacteria, including *Vibrio* spp. and *Aeromonas hydrophila*, as well as protozoans associated with aquatic animals, might develop new routes for entering the food chain by contaminating plants and eventually plant-based foods (Mori & Smith, 2019). However, the actual fate of pathogens in those systems remains largely unknown, and thus, no risk assessment can be performed accordingly.

Another example is cellular agriculture, which entails the production of animal products from cell culture rather than from livestock sources (Broucke et al., 2023). Although still in the experimental stage, many companies have demonstrated the potential to create cell-based meat, seafood, and milk products. Moreover, a great amount of research has focused on this topic (Pasitka et al., 2023; Messmer et al., 2022; Post et al., 2020), hopefully leading to more sustainable agriculture to address global environmental challenges and overuse of natural resources. Currently, in the agricultural sector, antibiotic overuse has become a major human health and food safety issue, inducing antibiotic-resistant bacteria and leading to untreatable superbug infections. A compelling feature of cellular agriculture is the potential for antibiotic-free meat and seafood production. However, in comparison with animal production, cell culture systems are even more susceptible to microbial contamination and require highly sterile environments. This is extremely challenging, especially as production levels shift from the laboratory scale to larger quantities. Therefore, the adoption of alternative antimicrobial solutions is of utmost importance for the development and application of cellular agriculture.

2 From Paper to Factory: Industrial Challenges

As a food microbiologist and a food technologist working in academia, I constantly sense the enormous gaps between scientific outcomes published as peer-reviewed manuscripts and technologies in use in the food industry in the real world.

For instance, the electrospinning of nanofibers with structural and functional advantages has drawn much attention in recent years due to their potential applications for active food packaging. To develop electrospinning techniques for active food packaging, electrospun nanofibers can be covalently or noncovalently functionalized for the loading of diverse bioactive compounds, including antimicrobial agents, antioxidant agents, oxygen scavengers, carbon dioxide emitters, and ethylene scavengers (Zhang et al., 2020). On April 2, 2024, the writer performed a literature search from the "Web of Science" database with the keywords "electrospinning" and "food packaging" and found 568 research articles and 88 review articles. However, to the best of our knowledge, this technology has not yet been used in the food industry.

Similarly, cold plasma, a nonthermal food processing method that inactivates contaminating bacteria on foods and packing materials by using charged, highly reactive gaseous molecules and species, has attracted considerable interest from many researchers in recent years. On April 2, 2024, the writer performed a literature search from the "Web of Science" database with the key words "cold plasma" and "food preservation" and found 234 research articles and 103 review articles. However, as reviewed by Birania et al. (2022), cold plasma processing has yet to be commercialized on a significant scale.

When a new food processing and/or food preservation method is adopted in the food industry, one of the most important considerations is the associated safety aspects. Indeed, static bactericidal and/or bacterial strategies can often largely enhance food microbial safety by mitigating the presence/proliferation of microbial pathogens in foods. However, one must rule out the possibility that antimicrobial strategies might introduce new chemical hazards to foods.

The antimicrobial components directly formulated in foods, which are considered food additives, are closely regulated by legislation. As reviewed by Batiha et al. (2021), regulators typically work closely with scientists in evaluating the safety of the proposed food additives to support their policies. In the United States, the Food and Drug Administration (FDA) has assessed the safety of unapproved food additives to determine whether they should be approved. The assessment involves the quantity of the substance that will usually be ingested, as well as the short- and long-term health consequences and other safety aspects. The FDA releases a regulation when a food additive is approved that can include the types of foods in which it may be used, the maximum permissible quantity and the correct labelling on food labels, and these regulations are issued in Title 21 of the Code of Federal Regulations. According to the FDA's Guidelines for Industry, the FDA has regulatory authority over food additives. In Europe, applicants need to submit requests with scientific safety details to the European Commission for authorization. The Commission would request that the European Food Safety Authority (EFSA) issue an opinion on the safety of substances for their intended uses upon approval of the application. In China, the Chinese government employs the China Food Additives Association (CFAA) as the only nationally registered food additive and ingredient industry organization to carry out assessments and help the government regulate the food and food additive sector. However, food additive policies vary significantly from country to country due to differences in regulations, cultural preferences, and public health priorities. These policies can have a substantial impact on international trade in food products. Some international organizations, such as the Codex Alimentarius Commission, work to harmonize food standards and regulations globally. Harmonization can facilitate trade by promoting uniformity in food additive policies across countries. However, achieving consensus among member states can be challenging due to differing priorities and regulatory frameworks.

Safety concerns not only originate from food additives directly formulated in foods but also from certain food processing procedures, of which most likely the most relevant example lies in ionizing food irradiation. The treatment of foods with ionizing energy is a well-explored antimicrobial strategy applicable to a wide range of fresh, dry, or frozen foods. The utility of ionizing food irradiation remains controversial because of the negative public and industry perceptions of "nuclear technology". From a scientific point of view, the most important safety issue concerning irradiated foods is the formation of a series of radiation-induced cyclic ketones, namely, 2-alkylcyclobutanones (2-ACBs), whose exact toxicity or carcinogenicity is still under investigation (Stefanova et al., 2010). As reviewed by Roberts (2016), approximately 60 countries currently permit irradiation of one or more food or food classes, whereas rationalization and greater consistency in labelling regulations are still needed for the future growth of applications of food irradiation.

Another important barrier to commercializing novel food antimicrobial strategies is cost. In the food industry, pricing is a critical factor influencing consumer purchasing decisions. To gain market share, companies may need to price their products competitively, which can further reduce profit margins. Novel antimicrobial strategies may face challenges in competing with existing antimicrobial methods or traditional food preservation techniques in terms of cost-effectiveness.

Most of the new technologies generate significant additional costs for food producers and eventually come to consumers. The cost may be directly from the production, including equipment and materials, as well as the operative cost, such as energy. For instance, Niemira (2012) calculated the price of power usage in kWh for cold plasma systems to be $0.05. This meant that the estimated electricity cost per 1000 h of operation would be $4500. Another example is the use of food waste and byproducts. Numerous studies in the scientific literature have demonstrated the marvellous potential of extracting bioactive components from food waste, including antimicrobial components, for use in food preservation and food packaging. However, the extraction of the components may require large volumes of chemical solvent, and/or involve laborious and time-consuming steps, incurring very high costs when combined. In addition to the cost of production, companies must still bear the costs of R&D, regulatory compliance, scale-up, marketing, and distribution. If the cost of production outweighs the expected returns, it may not be financially feasible to bring the product to market.

Last but not least, many of these technologies may induce unpleasant changes in the sensory properties of food products. The most renowned example would be heating. Although heat, which is delivered in numerous ways (e.g., through canning, pasteurization, microwave heating, and ohmic heating), is the most historical and commonly used method for reducing microbial loads in foods, heat can alter the texture of food, sometimes making the foods less palatable. For example, overcooking vegetables can make them mushy. Overheating or prolonged cooking can lead to the degradation or loss of certain nutrients in food, particularly water-soluble vitamins such as vitamin C and B. Therefore, great efforts have been made to develop methods for nonthermal processing of foods, intending to maintain the organoleptic properties of foods (especially fresh produce) to their maximum levels. However, changes in food composition are still inevitable. For example, as reviewed by Stefanova et al. (2010), ionizing irradiation may lead to the formation of radiolysis products by forming electrically charged species or neutral fragments with unpaired electrons. Radiation may also cleave the chemical bonds of food molecules and form unpredictable byproducts. These changes may cause the development of unpleasant off-flavours and aromas. The auto-oxidation of fat-containing food is initiated by oxygen radicals produced during ionizing radiation and leads to the formation of lipid oxides. Breaking of the sulfur containing amino acid residues of protein chains could be induced by hydroxyl radicals and electrons and thereby cause changes in the structure of proteins. Free radical scavengers such as vitamins C and E could also be depleted.

For chemical treatment/additives, many categories of antimicrobial compounds can cause unpleasant tastes/smells in foods, including but not limited to the astringency and bitterness of polyphenols, the acidity of organic acids, and the pungent smell of essential oils. However, it has been noted by the writer that in most of the research papers, extensive evaluation of the influence of food antimicrobial and food preservation technologies on food quality is lacking. We strongly recommend that researchers consider this important aspect in future studies and employ multiple methodologies in such evaluations, including both physiochemical analysis and sensory tests.

3 Challenges Are Opportunities

In the domain of human health, the "hygiene hypothesis" was proposed in 1989 and has become enshrined in popular culture: "We're too clean for our own good. People tend to believe that the drastic increase in chronic inflammatory diseases and allergic disorders is largely due to hygiene practices; thus, people lose the chance to be exposed to microbes in their early life to train their immune systems. Indeed, early exposure to a diverse range of "friendly" microbes-not infectious pathogens- is necessary to train the human immune system to react appropriately to stimuli. However, as noted by an increasing number of scientists (Scudellari, 2017), cutting back on personal hygiene will not have an impact on rates of chronic inflammatory and allergic disorders; it will, however, increase infections. The same is true for antimicrobial strategies in food systems.

The new food system is being driven not only by the primary production and food industry but also by consumers. For instance, in recent years, more people have been pursuing "healthy foods", which are "natural", and/or "clean-labelled", which are often devoid of food preservatives. However, while the clean label movement has gained traction due to perceived health benefits and environmental concerns, it also raises questions regarding food safety and quality. Clean-label foods may lack certain preservatives and antimicrobial agents that help inhibit bacterial growth and prolong shelf life. Without these additives, there is a risk of increased microbial contamination and foodborne illness if proper handling and storage practices are not followed. Without preservatives, foods may have shorter shelf lives and be more susceptible to spoilage, rancidity, or color changes. This limitation can impact product availability, distribution logistics, and consumer convenience.

Different microbial strategies have also been proposed for food processing. For example, chlorine is the most commonly used water disinfectant in general. It has remained popular due to its low cost, reliable availability, and good effectiveness against a diverse range of microorganisms. In the United States, chlorine is not only commonly used to sanitize food contact surfaces but also to sanitize raw fruits and vegetables during the washing or peeling process, as long as the concentration of sanitizer in the wash water does not exceed 2000 ppm hypochlorite (McGlynn, 2004). However, the use of chlorine may cause the generation of harmful disinfection byproducts (DBPs), such as trihalomethane, chlorate, chloramine, and chloroacetc acid. In Europe, the use of chlorine in fresh-cut produce washing is prohibited in countries such as Germany, Switzerland, the Netherlands, Denmark, and Belgium (Van Haute et al., 2013).

Although the concerns and possible side effects demonstrated above are valid, the writer still firmly believes in the vital importance of applying appropriate antimicrobial strategies in food systems. The real opportunities lie in the research efforts seeking more sustainable and safe alternatives and, more importantly, designing more reasonable procedures that consider the practical situations of the application scenarios.

For instance, in multitrophic farming systems, some entrepreneurs apply proactive mitigation strategies such as ultraviolet (UV) radiation. However, UV radiation may have inconsistent effects on the total microbial load and pathogen presence in certain parts of complicated farming systems. For instance, both Elumalai et al. (2017) and Moriarty et al. (2018) compared the coliform levels in aquaponic systems with and without UV radiation. Moriarty et al. (2018) reported that UV treatment reduced bacterial counts in water samples by approximately 1.5 and 3.0 log on 3-M petrifilm and m-Endo agar, respectively, while Elumalai et al. (2017) reported no significant reduction in aerobic plate counts or coliform counts when comparing UV-equipped systems to control system samples. This might be due to different experimental setups, as the UV sterilizer was installed at the outflow of the fish tank in the system of Moriarty et al. (2018), while for the system of Elumalai et al. (2017), the UV sterilizer was located at the inflow of the fish tank. In the future, extensive research is still needed to validate the influence of the technical details of the effectiveness of UV on the microbial quality of aquaponic systems.

It is also questionable whether long-term sublethal exposure to UV radiation induces mutations (Shibai et al., 2017) and/or selective stress on the microbiome, resulting in unpredictable consequences. Therefore, we believe that thorough cleaning and sanitization of indoor farming systems, at least between harvest batches, is necessary. Recently, our group explored sanitization solutions of nutrient film technique (NFT) hydroponic systems with both laboratory-scale and field-scale tests and found that sodium hypochlorite (NaOCl) exhibited superior efficacy in biofilm removal compared to hydrogen peroxide (H_2O_2) and sodium percarbonate (SPC). More importantly, we found that omitting the rinsing steps did not diminish the crop yield, but the harmful byproducts could remain at high levels in the hydroponic solutions and be internalized and accumulated in the edible parts of the crops grown in the systems, which are consequences that could be easily ignored by growers and consumers and could become a new food safety issue in the long run. It is therefore vitally important to set regulatory guidelines for the technical details of sanitizing hydroponic systems with NaOCl, especially to prevent overdose and to ensure the appropriate rinsing steps (Tham et al., 2024).

All in all, the world changes fast, and so do our foods. Food microbiologists must face a dynamic situation that is the result of the interplay between the rapid mutation of microorganisms, the global change in the total environment, and the response from humans reforming our food system. The antimicrobial strategies used to mitigate long-standing and emerging/reemerging microbial threats in food systems must be adjusted occasionally.

References

Aik, J., Heywood, A. E., Newall, A. T., Ng, L. C., Kirk, M. D., & Turner, R. (2018). Climate variability and salmonellosis in Singapore–a time series analysis. *Science of the Total Environment, 639*, 1261–1267.

Batiha, G. E. S., Hussein, D. E., Algammal, A. M., George, T. T., Jeandet, P., Al-Snafi, A. E., et al. (2021). Application of natural antimicrobials in food preservation: Recent views. *Food Control, 126*, 108066.

Beutin, L., Hammerl, J. A., Reetz, J., & Strauch, E. (2013). Shiga toxin-producing Escherichia coli strains from cattle as a source of the Stx2a bacteriophages present in enteroaggregative Escherichia coli O104: H4 strains. *International Journal of Medical Microbiology, 303*(8), 595–602.

Birania, S., Attkan, A. K., Kumar, S., Kumar, N., & Singh, V. K. (2022). Cold plasma in food processing and preservation: A review. *Journal of Food Process Engineering, 45*(9), e14110.

Broucke, K., Van Pamel, E., Van Coillie, E., Herman, L., & Van Royen, G. (2023). Cultured meat and challenges ahead: A review on nutritional, technofunctional and sensorial properties, safety and legislation. *Meat Science, 195*, 109006.

Castro-Ibáñez, I., Gil, M. I., Tudela, J. A., & Allende, A. (2015). Microbial safety considerations of flooding in primary production of leafy greens: a case study. *Food Research International, 68*, 62–69.

Chattaway, M. A., Dallman, T., Okeke, I. N., & Wain, J. (2011). Enteroaggregative E. coli O104 from an outbreak of HUS in Germany 2011, could it happen again? *Journal of Infection, 5*(06), 425–436.

Duchenne-Moutien, R. A., & Neetoo, H. (2021). Climate change and emerging food safety issues: a review. *Journal of Food Protection, 84*(11), 1884–1897.

Elumalai, S. D., Shaw, A. M., Pattillo, D. A., Currey, C. J., Rosentrater, K. A., & Xie, K. (2017). Influence of UV treatment on the food safety status of a model aquaponic system. *Water, 9*(1), 27.

Feliciano, R. J., Guzmán-Luna, P., Boué, G., Mauricio-Iglesias, M., Hospido, A., & Membré, J. M. (2022). Strategies to mitigate food safety risk while minimizing environmental impacts in the era of climate change. *Trends in Food Science and Technology, 126*, 180–191.

Jia, M., Taylor, T. M., Senger, S. M., Ovissipour, R., & Bertke, A. S. (2022). SARS-CoV-2 remains infectious on refrigerated deli food, meats, and fresh produce for up to 21 days. *Food Security, 11*(3), 286.

Karanth, S., Benefo, E. O., Patra, D., & Pradhan, A. K. (2023). Importance of artificial intelligence in evaluating climate change and food safety risk. *Journal of Agriculture and Food Research, 11*, 100485.

Kelly, B. G., Vespermann, A., & Bolton, D. J. (2009). The role of horizontal gene transfer in the evolution of selected foodborne bacterial pathogens. *Food and Chemical Toxicology, 47*(5), 951–968.

Li, D., Zhao, M. Y., & Tan, T. H. M. (2021). What makes a foodborne virus: comparing coronaviruses with human noroviruses. *Current Opinion in Food Science, 42*, 1–7.

McGlynn, W. (2004). *Guidelines for the use of chlorine bleach as a sanitizer in food processing operations*. Oklahoma Cooperative Extension Service.

Messmer, T., Klevernic, I., Furquim, C., Ovchinnikova, E., Dogan, A., Cruz, H., et al. (2022). A serum-free media formulation for cultured meat production supports bovine satellite cell differentiation in the absence of serum starvation. *Nature Food, 3*(1), 74–85.

Mirón, I. J., Linares, C., & Díaz, J. (2023). The influence of climate change on food production and food safety. *Environmental Research, 216*, 114674.

Misiou, O., & Koutsoumanis, K. (2022). Climate change and its implications for food safety and spoilage. *Trends in Food Science and Technology, 126*, 142–152.

Mori, J., & Smith, R. (2019). Transmission of waterborne fish and plant pathogens in aquaponics and their control with physical disinfection and filtration: A systematized review. *Aquaculture, 504*, 380–395.

Moriarty, M. J., Semmens, K., Bissonnette, G. K., & Jaczynski, J. (2018). Inactivation with UV-radiation and internalization assessment of coliforms and Escherichia coli in aquaponically grown lettuce. *LWT, 89*, 624–630.

Niemira, B. A. (2012). Cold plasma decontamination of foods. *Annual Review of Food Science and Technology, 3*, 125–142.

Ortiz, Y., Heredia, N., & García, S. (2023). Boundaries that prevent or may Lead animals to be reservoirs of Escherichia coli O104: H4. *Journal of Food Protection, 86*(3), 100053.

Pasitka, L., Cohen, M., Ehrlich, A., Gildor, B., Reuveni, E., Ayyash, M., et al. (2023). Spontaneous immortalization of chicken fibroblasts generates stable, high-yield cell lines for serum-free production of cultured meat. *Nature Food, 4*(1), 35–50.

Post, M. J., Levenberg, S., Kaplan, D. L., Genovese, N., Fu, J., Bryant, C. J., et al. (2020). Scientific, sustainability and regulatory challenges of cultured meat. *Nature Food, 1*(7), 403–415.

Roberts, P. B. (2016). Food irradiation: Standards, regulations and world-wide trade. *Radiation Physics and Chemistry, 129*, 30–34.

Scudellari, M. (2017). Cleaning up the hygiene hypothesis. *Proceedings of the National Academy of Sciences, 114*(7), 1433–1436.

Shibai, A., Takahashi, Y., Ishizawa, Y., Motooka, D., Nakamura, S., Ying, B. W., & Tsuru, S. (2017). Mutation accumulation under UV radiation in Escherichia coli. *Scientific Reports, 7*(1), 1–12.

Shiraz, S., Djebbi-Simmons, D., Alhejaili, M., Danos, K., Janes, M., Fontenot, K., & Xu, W. (2020). Evaluation of the microbial safety and quality of Louisiana strawberries after flooding. *Food Control, 110*, 106970.

Sobolik, J. S., Sajewski, E. T., Jaykus, L. A., Cooper, D. K., Lopman, B. A., Kraay, A. N., et al. (2022). Decontamination of SARS-CoV-2 from cold-chain food packaging provides no marginal benefit in risk reduction to food workers. *Food Control, 136*, 108845.

Stefanova, R., Vasilev, N. V., & Spassov, S. L. (2010). Irradiation of food, current legislation framework, and detection of irradiated foods. *Food Analytical Methods, 3*, 225–252.

Tham, C. A. T., Zwe, Y. H., Ten, M. M. Z., Ng, G. S. Y., Toh, J. Y. L., Poh, B. L., Zhou, W., & Li, D. (2024). Sanitization of hydroponic farming facilities in Singapore: what, why, and how. *Applied and Environmental Microbiology, 90*(7), e00672–24.

Tirado, M. C., Clarke, R., Jaykus, L. A., McQuatters-Gollop, A., & Frank, J. M. (2010). Climate change and food safety: A review. *Food Research International, 43*(7), 1745–1765.

Uyttendaele, M., Liu, C., & Hofstra, N. (2015). Special issue on the impacts of climate change on food safety. *Food Research International, 68*, 1–6.

Van Haute, S., Sampers, I., Holvoet, K., & Uyttendaele, M. (2013). Physicochemical quality and chemical safety of chlorine as a reconditioning agent and wash water disinfectant for fresh-cut lettuce washing. *Applied and Environmental Microbiology, 79*(9), 2850–2861.

Yekta, R., Vahid-Dastjerdi, L., Norouzbeigi, S., & Mortazavian, A. M. (2021). Food products as potential carriers of SARS-CoV-2. *Food Control, 123*, 107754.

Zhang, C., Li, Y., Wang, P., & Zhang, H. (2020). Electrospinning of nanofibers: Potentials and perspectives for active food packaging. *Comprehensive Reviews in Food Science and Food Safety, 19*(2), 479–502.

Index

D. Li (ed.), *Antimicrobial Strategies in the Food System: Updates, Opportunities, Challenges*, https://doi.org/10.1007/978-3-031-95056-8

The manufacturer's authorised representative in the EU is Springer Nature Customer Service Centre GmbH, Europaplatz 3, 69115 Heidelberg, Germany. If you have any concerns regarding our products, please contact ProductSafety@springernature.com

Printed and bound by CPI Group (UK) Ltd, Croydon, CR0 4YY
01/10/2026
02225915-0019